Birds of Washington

Status and Distribution

A gift from the estate of Ruby E. Egbert helped make publication of this book possible. The Oregon State University Press and the editors are grateful for this support.

Birds of Washington

Status and Distribution

Terence R. Wahl
Bill Tweit
Steven G. Mlodinow
Editors

Maps by Kelly M. Cassidy
Habitats in Washington by Christopher B. Chappell
History of avian conservation in Washington by Joseph B. Buchanan
Species description contributions from over 40 authors
Illustrations by Shawneen Finnegan and G. Scott Mills

Oregon State University Press
Corvallis, Oregon

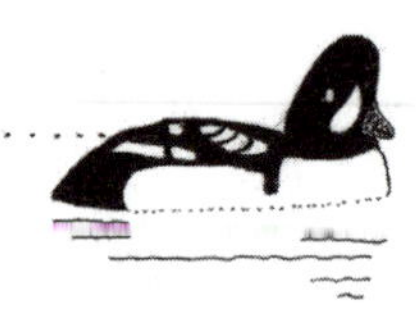

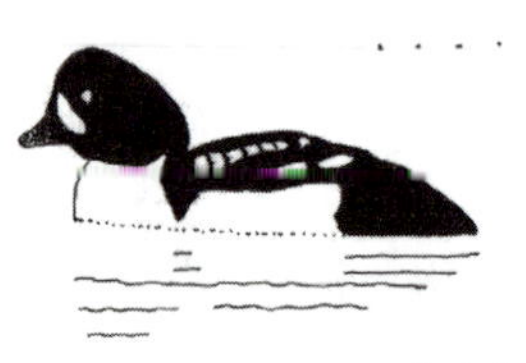

Illustration above: Barrow's Goldeneye (G. Scott Mills). Illustration on page i: Surfbird and Black Turnstone (G. Scott Mills).

The paper in this book meets the guidelines for permanence and durability of the Committee on Production Guidelines for Book Longevity of the Council on Library Resources and the minimum requirements of the American National Standard for Permanence of Paper for Printed Library Materials Z39.48-1984.

Library of Congress Cataloging-in-Publication Data
Birds of Washington : status and distribution / Terence R. Wahl, Bill Tweit, Steven G. Mlodinow, editors ; maps by Kelly M. Cassidy ; Habitats in Washington by Christopher B. Chappell ; History of avian conservation in Washington by Joseph B. Buchanan ; species description contributions from over 40 authors ; illustrations by Shawneen Finnegan and Scott Mills.— 1st ed.
p. cm.
Includes bibliographical references and index.
ISBN 0-87071-049-4 (hardcover : alk. paper)
1. Birds—Washington (State) 2. Birds—Washington (State)—Geographical distribution. I. Wahl, Terence R. II. Tweit, Bill. III. Mlodinow, Steven.
QL684.W2B57 2005
598'.09797—dc22
2004026511

© 2005 Terence Wahl
All rights reserved. First edition 2005
Printed in the United States of America

Oregon State University Press
102 Adams Hall
Corvallis OR 97331-2005
541-737-3166 • fax 541-737-3170
http://oregonstate.edu/dept/press

Table of Contents

Preface

Since Jewett et al. (*Birds of Washington State*, 1953) described the ornithological history and the status and distribution of birds in Washington there have been many changes in the environment, as well as in human populations and activities, along with increased knowledge of the populations and apparent diversity of birds in the state.

The purpose of this book is to describe the status and abundance, trends, and changes of species and populations occurring in Washington as of the year 2000, and to update and correct previous descriptions. We summarize information provided by systematic surveys, field observations from a vastly increased number of knowledgeable observers, scientific publications, wildlife agency reports, and other sources available today.

Acknowledgments

In addition to the pioneers and more recent scientists and observers we are indebted to a large number of persons. Over forty agency biologists, wildlife managers, and scientists contributed species accounts. Contributors also include many authors of papers and publications cited and listed in the Reference section. In addition hundreds, if not thousands, of observers not listed elsewhere performed field studies, participated in Christmas Bird Counts, Breeding Bird Surveys, Breeding Bird Atlas surveys, and other systematic field efforts such as the Marine Ecosystems Analysis project in the late 1970s. Their records submitted in the last quarter of the twentieth century and their reports published in seasonal journals supplied a great deal of valuable information.

Many of the contributing authors also reviewed species accounts for this book. Other reviewers contributing to regional status accounts included Jim Acton, Andy Stepniewski, Mike Denny, Dan Stephens, Dennis Paulson, Jack Bettesworth, Wilson Cady, and Bill and Nancy LaFramboise. Kelly Cassidy adapted and updated distribution maps, and Chris Chappell and Joe Buchanan contributed important text sections. Phil Mattocks took on the task of organizing and maintaining the growing number of state records and the Washington Bird Records Committee. Zella Schultz and, subsequently, Ed Miller served as organizers for the statewide Breeding Bird Surveys (BBS) begun in 1966. The Washington Department of Fish and Wildlife supported this project by providing data and encouraging staff to participate in the project. David Johnson and Thomas O'Neil, editors of *Wildlife Habitat Relationships of Oregon and Washington*, are due thanks for supplying the species:habitat table used as the basis for our table (see Appendix A).

The initial drafts of seasonal occurrence maps were created for Seattle Audubon Society's BirdWeb website (http://www.birdweb.org) and were made available for this book in exchange for map reviews by contributors to this book. The Seattle Audubon Society generously provided funds for the mapping project. In addition, people who supported Westport Seabirds offshore boat trips helped increase the knowledge of oceanic birds and enabled creation of one of the longest-running data sets of such species. Robin Wahl and Brendan Wahl assisted with production. Mary Braun and Jo Alexander at Oregon State University Press provided valuable advice and assistance with production of this book.

Introduction

Explanation of our Presentation

Species are discussed in most recent taxonomic order (AOU 2003, 2004).

Readers will note we have minimized or omitted descriptions and discussions of identification, biology, and behavior. We have done so because many of these aspects of ornithology are well covered in other recent references like the *Birds of North America* series, the four-volume *Birds of British Columbia* (Campbell et al. 1990a, 1990b, 1997, 2001), *Birds of Oregon* (Marshall et al. 2003), and many field identification guides. We include minimal citation of museum specimen records; many recovered specimens are now reported in field ornithology journals and Internet websites. Those specimens are almost all now cited from published sources.

We cite references in the standard style used in scientific papers, and the full details of each cited work can be found in the References. The references from seasonal field journals *Audubon Field Notes, American Birds, Field Notes, North American Birds,* and *Washington Ornithological Society Newsletter* (AFN, AB, FN, NAB, WOSN, respectively) are cited in Occurrence sections of species accounts. Note that any records not specifically referenced in Noteworthy Records sections are from those sources.

The primary author was the person responsible for each draft species account. Drafts were reviewed by regional reviewers, editors, and, in most cases, primary authors. The review process was intended to standardize style, format, and terminology and add updates and references, but was not intended to change meanings.

Organization of species accounts

Format of accounts

A. Accepted Species:

Species name, Taxonomic name

Brief statement of present status in Washington

Graphic showing known range of dates of occurrence, west and east, where occurrence apparently varies over the year

Subspecies occurrence in Washington

Maps showing seasonal distribution (for selected species)

Habitats: brief description, main habitats used for nesting, foraging, migration/winter if different

Occurrence: Range in western or eastern Washington. Seasonal status: breeding, non-breeding, winter visitor, migrant, year-round resident. Widespread or localized occurrence (e.g., nesting locations, roosts, limited foraging areas). Life zones/elevation, habitats utilized, variations, if any, from rest of species range. Associations with natural events like herring spawns, insect outbreaks and with human activities (farming, fishing, reservoirs, irrigation, garbage, sewage, impacts of human development, population, activities in Washington). Apparent changes in status and distribution over time (since 1950), explanations of differences from historical or published status (as per Jewett et al. 1953) and trends in numbers.

Remarks: Identification problems. Specific threats. Endangered, threatened designations: state/federal/other listings. Questions, conservation and management issues, habitat concerns. Needs for further studies. Taxonomic or common name changes since 1950.

Noteworthy records: Selected by authors, editors from systematic studies and from other reports.

B. Other Species

1. Escapes and unsuccessful introductions
2. "Hypotheticals": Species we believe correctly identified but not acted upon or accepted by WBRC, wild status and or origin questioned; requiring discussion.

Definitions of Terminology in Accounts

Abundance

This status assumes a competent observer, proper habitat, and good observation conditions. It does not allow for differences in observability, such as large, white swans in open areas being more visible than Song Sparrows; hence abundance cannot necessarily be compared across species. Our abundance codes may differ from those used in other references. We have tried to define these codes more closely than some other references.

Species are treated by regular and irregular occurrence and by the number of individuals seen/identified by sound per day per locality, averaged over many trips over several years in appropriate seasons and habitats. (Locality is defined as an area of a typical day's trip by an experienced observer, approximately countysized.)

Regular species (usually reported annually):

Abundant: seen/detected on every day afield; 1000 or more per day per locality

Common: seen/detected on >90% of days afield; 25-1,000 per day per locality

Fairly common: seen/detected on >50% of days; 1-25 per day

Uncommon: seen/detected on <50% of days; 0-10 per day

Rare: unlikely to be found by 1 observer every year; 1-10 per season seen on <10% of days statewide

Irregular species (not reported annually):

Very rare: 2 or more records for all time, of very infrequent occurrence

Casual: 1-2 but less than 5 records for all time; not expected to occur again

Seasonal Definitions

Dates are generalized ranges and, except where noted, do not include out-of-season occurrences or early or late outlier dates of migrants obviously ahead of or behind the main population movements: our purpose here is to characterize occurrence.

Status:

Resident: year-round/permanent in the area under reference

Summer: present only during summer, may breed

Winter: present only in winter

Migrant, spring and fall: migrates through area, does not winter or breed. In some cases, spring and fall are subdivided:

- early spring: March-April
- late spring: May
- summer: June
- early fall: July-August
- late fall: September-November
- winter: December-February

Visitor: irregularly present at certain times of the year

Vagrant: wanderer outside normal migration range of the species

Habitats

We generally follow simplified, common usage used by observers (e.g. Wahl and Paulson 1991; combined with Franklin and Dyrness 1973, Wahl et al. 1981, Wahl and Tweit 2002a). Our habitats are less precisely classified than those from Johnson and O'Neil (2001) and are discussed in more detail in the chapter on Bird Habitats of Washington, below. The detailed habitat descriptions of Johnson and O'Neil are displayed in the Appendix.

Abbrevations

AB	*American Birds*
ad.	adult
AFB	Air Force Base
AFN	*Audubon Field Notes*
AOU	American Ornithologists' Union
BBA	Breeding Bird Atlas
BBS	Breeding Bird Survey
B.C.	British Columbia
BL	Blue List - National Audubon Society
c./C.	central
CBC	Christmas Bird Count
CG	campground
Co.	County
CP	County Park
Cr.	Creek
CW	Count Week of Christmas Bird Count
e./E.	east or eastern
ENSO	El Niño Southern Oscillation
FC	Federal Candidate species
FE	Federal Endangered species
FEALE	Fitzner-Eberhardt Arid Lands Ecology Reserve
FN	*Field Notes*
FSC	Federal Species of Concern
FRG	Falcon Research Group
Ft.	Fort
FT	Federal Threatened species
GL	Green List - American Bird Conservancy
ha	hectare
HMU	Habitat Management Unit (U.S. Army Corps of Engineers)
hr	hour
HWI	Hawk Watch International
Hwy.	Highway
imm.	Immature
Is.	Islands or Isles
Jct.	Junction
juv.	juvenile
km	kilometer
L.	Lake
m	meters
MESA	Marine Ecosystems Analysis program (1978-79)
Mt.	Mountain or Mount
Mts.	Mountains
n./N.	north or northern
NAB	*North American Birds*
ne.	northeast or northeastern
NF	National Forest
nm	nautical mile
NP	National Park
NPS	National Park Service
nr.	near
nw.	northwest or northwestern
NWR	National Wildlife Refuge
OBRC	Oregon Bird Records Committee
p.c.	personal communication
Pen.	Peninsula
ph.	photographed
PHS	Priority Habitat Species - WDFW
Pk.	peak
ppt	parts per thousand
PSAMP	Puget Sound Ambient Monitoring Program
Pt.	Point (not Port)
R.	River
Res.	Reservoir
s./S.	south or southern
SC	State Candidate species
SE	State Endangered species
se.	southeast or southeastern
SP	State Park
SS	State Sensitive species
SST	sea-surface temperature
ST	State Threatened species
sw.	southwest or southwestern
USDA	U.S. Department of Agriculture
USDI	U.S. Department of Interior
USFS	U. S. Forest Service
USFWS	U.S. Fish and Wildlife Service
UPS	University of Puget Sound
UWBM	University of Washington Burke Museum
v.t.	video-taped
w./W.	west or western
WBRC	Washington Bird Records Committee
WDF	Washington Department of Fisheries (precursor of WDFW)
WDG	Washington Department of Game (precursor of WDFW)
WDFW	Washington Department of Fish and Wildlife
WDNR	Washington Department of Natural Resources
WDOE	Washington Department of Ecology
WDW	Washington Department of Wildlife (precursor of WDFW)
WL	Watch List (National Audubon Society)
WMA	Wildlife Management Area
WOSN	*Washington Ornithological Society Newsletter*
WWU	Western Washington University

Metric Conversions

1 meter	3.29 feet
1 kilometer	0.621 mile
1 hectare	2.5 acres

Political map of Washington, showing county boundaries, selected municipalities, and roads.

Bird Habitats of Washington

by Christopher B. Chappell

Habitat is an area with the resources (e.g. food, water) and environmental conditions (e.g. temperature, moisture, presence or absence of predators) needed by individuals of a given species to survive and reproduce (Morrison et al. 1992). For birds, this area may be more or less constant throughout the year or may consist of areas occupied only seasonally or even for very short periods during migration.

In terrestrial areas, bird habitat is typically strongly dominated or characterized by vegetation, the total of all plants living in an area. Strictly physical aspects of the environment such as climate and physical substrate may also be important to a species and are often correlated with the vegetation. Because vegetation integrates or reflects many dominant biological and physical aspects of a habitat, it is perhaps the single most useful aspect to consider when comparing terrestrial habitats. In the case of aquatic habitats, vegetation is much less of a prominent component and dominant biological communities (the aquatic equivalent of vegetation) are more difficult to readily observe. Therefore, physical attributes of aquatic bird habitats—such as water depth and movement, salinity, and substrate of the bottom or shoreline— are generally used to describe and compare them.

Bird habitats in Washington have been described in a variety of ways over the years. Simple classifications using widely understood (but often poorly defined) physical features and vegetation structural types have been used extensively since the beginning of Washington ornithology, whether explicitly defined (e.g., Wahl and Paulson 1991) or implicitly used (e.g., words like lake, pond, or evergreen forest). Historically, Merriam's life zones were prominently used as an organizing principle until the 1970s (Jewett et al. 1953); his loosely defined continent-wide zones corresponded with some important differences in vegetation and bird distribution when used within the state. The advent of widespread vegetation-classification work provided an improved framework for habitat description that was synthesized by Franklin and Dyrness (1973). Smith et al. (1997) used this framework as the basis for their description of bird distributions and habitat in the state, in conjunction with a classification of land cover (Cassidy 1997a). Most recently, Johnson and O'Neil (2001) used a wildlife-habitat type classification to characterize both the habitats and the use of them by birds.

The development of vegetation classifications in the form of vegetation zones and plant associations (Franklin and Dyrness 1973) allowed a new level of refinement in describing bird habitats, one that was tied to compositional and structural features of vegetation that were correlated with physical environmental features. The vegetation zones of Franklin and Dyrness (1973) are explicitly defined in terms of potential natural vegetation, i.e., the climate-driven climax (late-successional) vegetation that is projected to occur wherever the vegetation is in its natural condition and remains undisturbed long enough for species capable of replacing themselves to become dominant. The disadvantages of this approach have become increasingly apparent over the years and in relation to bird habitat are primarily twofold: (1) the importance of natural

Table I. Summary of methodological differences between vegetation zones and wildlife-habitat types.

	Vegetation Zones (Franklin and Dyrness 1973)	*Wildlife-Habitat Types (Johnson and O'Neil 2001)*
Habitats covered	Vegetated Barren Marine Aquatic	Vegetated
Break between Types	Different late-sucessional or environmental indicator plant species is potentially important or dominant	Significantly different group of wildlife species using habitat
Vegetation Classification	Potential natural vegetation (existing vegetation used where potential not understood)	Existing vegetation
VegetationStructure	Not classified	Classified

disturbance in shaping existing vegetation, which is what the birds actually experience, and (2) the importance of vegetation structure to birds, which is more or less ignored by the vegetation zones. The strength of vegetation zones is as an organizing principle for the physical environment that is related to the biological potential of an area, and one that can be observed easily by a person who is capable of identifying major vegetation species. In conjunction with a classification of land cover or vegetation structure, the vegetation zones remain a powerful tool for habitat description (Table 1).

A classification of wildlife-habitat types was developed using, as a starting point, groups of similar or related plant associations/communities (grouped by existing vegetation composition and structure, environmental characteristics, and processes) (Chappell et al. 2001, Johnson and O'Neil 2001). These groups of related associations were further grouped based on an analysis of their use by wildlife species: groups of vegetation with similar wildlife use were grouped together to form a single wildlife-habitat type (Johnson and O'Neil 2001). These broad wildlife-habitat types (30 total in Washington) were further split into structural conditions of importance to one or more groups of vertebrate wildlife species (O'Neil et al. 2001). A total of 26 forest structural conditions and 20 shrubland/grassland structural conditions were identified, considering dominant physiognomic vegetation (forb, grass, shrub, tree), height of dominant vegetation, relative density/openness of vegetation, vertical layering, and relative maturity. For the most part the names of the structural conditions thus defined are relatively intuitive and they will not be described in detail here (refer to O'Neil et al. 2001). Unfortunately for the ornithologist, the structural conditions do not distinguish between needle-leaf (conifer) versus broadleaf tree canopies, a difference that is of major significance to many bird species in habitat selection.

Most habitat descriptions used in this book are essentially based on ecoregions (Fig. 1) and general vegetation types (e.g. Wahl and Paulson 1991). Many bird records included here did not include details to accurately attribute them to one or more of the terrestrial wildlife habitats given by Johnson and O'Neil (2001). Consequently, species accounts in this book do not use a standard habitat classification. Each author may use general descriptive terms or may refer to specific vegetation zones or wildlife-habitat types.

The relationship between different classification schemes for bird habitats and vegetation types can be compared (Table 2).

The wildlife-habitat type and structural condition matrices from Johnson and O'Neil (2001) are useful for the reader interested in learning more about habitat relations of particular species (Appendix). The following summaries of the Johnson and O'Neil wildlife-habitat types as defined by Chappell et al. (2001) will help the reader use the habitat matrices in the appendix and better understand the relationship of the habitat types to other vegetation references (e.g., Franklin and Dyrness 1973). These wildlife-habitat types were defined based on their overall use by all wildlife species, not just birds. The reader should consider that within many of these wildlife-habitat types, as defined, dominant vegetation may be conifer (needle-leaf) or broadleaf or a mixture of the two. The distinction in the leaf structure of the canopy is important to many birds; different bird communities use the broadleaf versus the conifer variations of each wildlife-habitat type. This distinction in leaf structure could easily be considered another important division of habitat for birds. The species accounts note when a bird species has a particular association with one or the other leaf-structure variant.

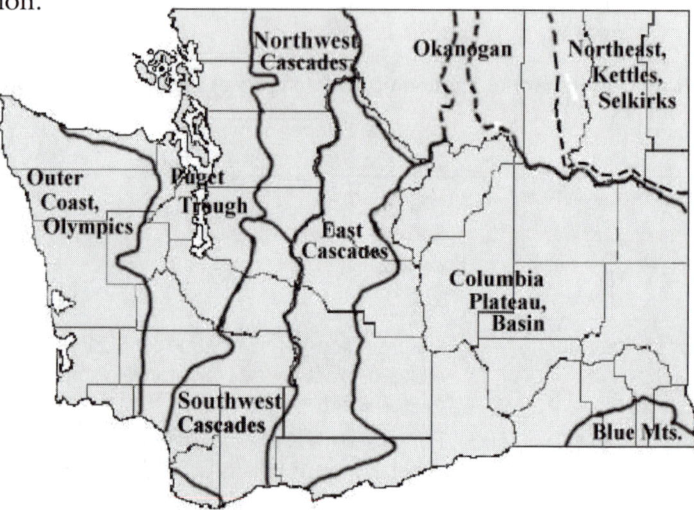

Figure 1: Washington State Ecoregions.

Westside Lowlands Conifer-Hardwood Forest

These lowland to low montane upland forests occurs over most of w. Washington, typically at elevations up to about 450-760 m (1500-2500 feet). These forests are dominated by one or more of the following conifer species: western hemlock, Douglas-fir, Sitka spruce, western redcedar, or lodgepole pine (the latter mainly just on coastal dunes), and/or the hardwood species red alder and bigleaf maple. This habitat does not, however, include dry-site Douglas-fir forests where western hemlock or western redcedar do not grow. It corresponds with the Franklin and Dyrness's (1973) western hemlock zone and Sitka spruce zone forests (and associated early-successional shrubby vegetation), exclusive of wetlands and riparian floodplains.

Westside Oak and Dry Douglas-fir Forest and Woodlands

These upland forests or woodlands (30-100% tree crown cover) are located mostly at low elevations in the Puget Lowland and the Portland-Vancouver Basin, though they also occur in small amounts on the e. and ne. Olympic Mts. and in the s. Cascades. They are dominated by Douglas-fir, Oregon white oak, Pacific madrone, and/or lodgepole pine. For Douglas-fir-dominated forests, this type only includes those that are located on dry sites where western hemlock and western redcedar are infrequent or absent, mainly in the San Juan Islands vicinity and around s. Puget Sound prairies. These habitats occupy a small percentage of the landscape in most areas and therefore are mostly referred to as special types within broader vegetation zones described by Franklin and Dyrness (1973).

Montane Mixed Conifer Forest

These upland forests are found at middle to high elevations (about 450-760 m up to 1400-1900 m in the mountains from the Cascade crest w. and at high elevations e. of the Cascade crest (mostly from about 1200 to 2000 m). They are dominated or co-dominated by one or more of the following indicator species: Pacific silver fir, mountain hemlock, subalpine fir, Engelmann spruce, Alaska yellow-cedar, or noble fir. Other common trees that may occur but are not diagnostic of this habitat include western hemlock, western redcedar, Douglas-fir, lodgepole pine, and western larch. These forests (and associated early-successional shrubby vegetation) are the major vegetation within the Pacific silver fir zone and the lower subzones of the mountain hemlock and subalpine fir zones of Franklin and Dyrness (1973).

Eastside Mixed Conifer Forest

These low- to mid-elevation (mostly about 550-1500 m) montane forests occur widely in e. Washington (e. Cascades, Blue Mts., Okanogan Highlands). They usually have a significant component of Douglas-fir, grand fir, western redcedar or western hemlock. Successional forests dominated by western larch or western white pine are part of this habitat, and ponderosa pine or lodgepole pine are sometimes co-dominant. This habitat includes much of the forests and their associated early-successional vegetation within the Douglas-fir, grand fir, and western hemock zones in e. Washington, except those strongly dominated by ponderosa pine (Franklin and Dyrness 1973).

Lodgepole Pine Forests and Woodlands

These forests or open woodlands occur on upland sites in e. Washington and are dominated by lodgepole pine. Lodgepole pine stands in w. Washington are considered part of other wildlife-habitat types. Several other tree species may co-occur. These are early- to mid-successional forests that have regenerated after fire or logging. Franklin and Dyrness (1973) mention these successional lodgepole pine stands as components of the Douglas-fir, grand fir, western hemlock, and subalpine fir zones.

Ponderosa Pine Forest and Woodlands (includes Eastside Oak)

These forests, woodlands, or savannas (10-100% tree cover) are dominated by ponderosa pine and/or Oregon white oak at lower to middle elevations in e. Washington. This habitat type can include areas adjacent to the steppe or shrub-steppe that are too dry for other conifers, as well as stands historically maintained by frequent fires. Douglas-fir, western larch, or grand fir are frequently prominent, though by definition subordinate. Franklin and Dyrness (1973) describe this habitat in their chapter on the ponderosa pine zone and note that fire-maintained stands occur also in the Douglas-fir and grand fir zones.

Table 2. Summary of corresponding units of habitat or vegetation from three references. Earlier references are related to the corresponding recent wildlife-habitat types classification.

Johnson and O'Neil 2001 *Chappell et al. 2001*	*Franklin and Dyrness 1973* *(Smith et al. 1997)*	*Wahl and Paulson 1991*
Westside lowlands conifer-hardwood forest	Sitka spruce zone forests Western hemlock zone forests (western WA only, including Puget Sound area) Willamette Valley conifer forests	Wet coniferous forest Broadleaf forest Shrubby thicket
Westside oak and dry Douglas-fir forest and woodlands	Puget Sound area (part of western hemlock zone) oak groves Puget Sound area pine stands Puget Sound area dry Douglas-fir forests Willamette Valley oak woodlands Willamette Valley conifer forests	Wet coniferous forest Broadleaf forest Shrubby thicket
Montane mixed conifer forest	Pacific silver fir zone forests Mountain hemlock zone forests (lower subzone only) Subalpine fir zone forests (lower subzone only)	Wet coniferous forest Shrubby thicket
Eastside mixed conifer forest	Douglas-fir zone forests Grand fir zone forests Western hemlock zone forests in eastern WA	Dry coniferous forest Wet coniferous forest Broadleaf forest Shrubby thicket
Lodgepole pine forest and woodlands	Lodgepole pine stands in eastern WA (occurs in Douglas-fir, grand fir, western hemlock & subalpine fir zones)	Wet coniferous forest Shrubby thicket
Ponderosa pine forest and woodlands	Ponderosa pine zone forest or woodland Ponderosa pine-dominated stands in Douglas-fir and grand fir zones Oak woodlands in eastern WA	Dry coniferous forest Shrubby thicket Broadleaf forest
Upland aspen forest	Not mentioned in WA (most aspen in WA is riparian or wetland)	Broadleaf forest
Subalpine parkland	Mountain hemlock zone parkland (upper subzone only) Subalpine fir zone parkland (upper subzone only) Subalpine meadow communities Rawmark and low herbaceous communities	Wet meadow Wet coniferous forest Shrubby thicket
Alpine grasslands and shrublands	Alpine communities Green fescue communities (often subalpine but included here) Grassy balds or parks in subalpine fir zone	Wet meadow
Westside grasslands	Puget Sound area prairies San Juan Islands grasslands	Dry grassland
Western juniper and mountain mahogany woodlands	Rare in WA (western juniper woodlands in Columbia Basin not described here)	Not addressed
Eastside canyon shrublands	Idaho fescue/common snowberry zone shrublands Shrub garlands on talus margins of Columbia Basin	Shrubby thicket

Johnson and O'Neil 2001 *Chappell et al. 2001*	*Franklin and Dyrness 1973* *(Smith et al. 1997)*	*Wahl and Paulson 1991*
Eastside grasslands	Grasslands in steppe and shrub-steppe zones of the Columbia Basin Grasslands in ponderosa pine, Douglas-fir, and grand fir zones	Dry grassland
Shrub-steppe	Big sagebrush associations Three-tip sagebrush/Idaho fescue association Bitterbrush associations Mountain big sagebrush communities	Sagebrush desert
Dwarf shrub-steppe	Associations on shallow soils in the Columbia Basin	Sagebrush desert Dry grassland
Desert playa and salt scrub shrublands	Saltgrass associations on saline-alkali soils Spiny hopsage/Sandberg's bluegrass association	Sagebrush desert Wet meadow
Agriculture, pasture and mixed environs	Not addressed, agricultural lands in any zone	Farmland
Urban and mixed environs	Not addressed, urban or suburban lands in any zone	Parks and gardens
Open water—lakes, rivers, streams	Not addressed	Freshwater, marsh and shore
Herbaceous wetlands	Freshwater wet meadows or marshes in all zones except timberline (upper subzones of subalpine fir and mountain hemlock) and alpine	Freshwater, marsh and shore Wet meadow
Westside riparian - wetlands	Forested swamps in Sitka spruce zone Sitka alder communities in Pacific silver fir zone Bogs and moors in mountain hemlock and silver fir zones Riparian communities of Willamette Valley Includes any shrubby or forested wetland or riparian floodplain in western hemlock or Sitka spruce zones	Riparian woodland Broadleaf forest Wet coniferous forest Wet meadow Shrubby thicket
Montane coniferous wetlands	Pacific silver fir/devils club association Includes any forested wetland dominated by conifers in the Pacific silver fir, mountain hemlock, subalpine fir, or eastern WA western hemlock zones	Wet coniferous forest
Eastside riparian - wetlands	Black hawthorn associations and related riparian types in the Columbia Basin Includes any shrubby or wooded wetland or riparian area in eastern WA	Riparian woodland Broadleaf forest Shrubby thicket
Coastal dunes and beaches	Sand dune and strand communities	Sandy shore, mudflats, salt marsh
Coastal headlands and islets	Herb- and shrub-dominated communities of ocean-front	Shrubby thicket
Bays and estuaries	Tideland communities Also includes open water and unvegetated tidal flats	Sandy shore, mudflats, salt marsh Open saltwater
Inland marine deeper water	Not addressed	Open saltwater
Marine nearshore	Not addressed	Open saltwater Rocky shore Sandy shore, mudflats, salt marsh
Marine shelf	Not addressed	Open saltwater
Oceanic	Not addressed	Open saltwater

Upland Aspen Forest

These forests or groves dominated by quaking aspen are not associated with streams, lakes, or wetlands, and are located in e. Washington. This habitat is uncommon to rare in Washington, where it is most obvious in the ne. Cascades. Most aspen forest in Washington is part of Eastside Riparian – Wetlands wildlife-habitat type. Franklin and Dyrness (1973) describe it only from se. Oregon, where it is more common.

Subalpine Parkland

This habitat is located in high mountains throughout the state and is that area below upper tree line but above continuous forest, with tree cover mostly 10-50% (savanna or woodland structure). Upper tree line is defined as the elevation above which trees are unable to grow in an upright form, though they may still be present as shrubs (krummholz). Tree line is usually found somewhere between 1500 and 2100 m, depending on the local climate. The subalpine parkland is often a mosaic of small patches of herbaceous- or dwarf-shrub-dominated vegetation (meadows) mixed with patches of open woodland, tree islands, or scattered trees. Major trees are subalpine fir, Engelmann spruce, mountain hemlock, Pacific silver fir, Alaska yellow-cedar, whitebark pine, and subalpine larch (the latter two diagnostic when prominent). Major shrubs and herbs include heathers, dwarf *Vaccinium* spp. (blueberries or huckleberries), sedges, green fescue, and many forbs. Included here also are subalpine and alpine wetlands (at comparable or less commonly higher elevations) characterized by plants such as black alpine sedge, marsh marigold, bog laurel, and willows. Franklin and Dyrness (1973) describe this habitat as the parkland in the mountain hemlock and subalpine fir zones (or the upper subzones of those zones), and they describe subalpine meadow communities in detail.

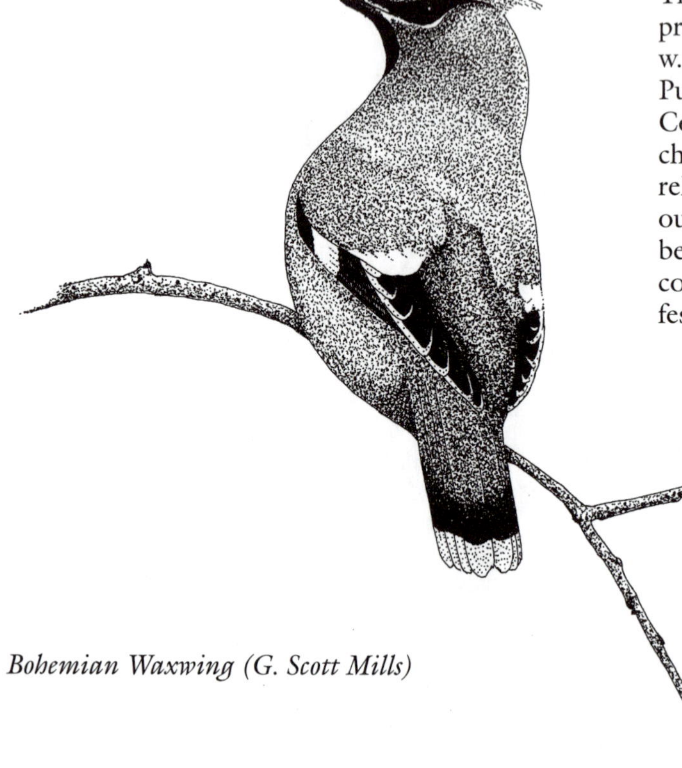

Bohemian Waxwing (G. Scott Mills)

Alpine Grasslands and Shrublands

This habitat includes all vegetated areas above upper tree line (ranging from about 1500 to 2100 m in the highest mountains throughout the region (the alpine zone). In addition to the alpine zone, this habitat also includes significant expanses of grassland (with less than 10% cover of trees) that occur below upper tree line within the subalpine fir zone. These subalpine grasslands are often dominated by green fescue, Idaho fescue, or Roemer's fescue (Franklin and Dyrness 1973). Alpine communities are dominated by various species of sedges, grasses, forbs, and/or dwarf-shrubs such as heather (Franklin and Dyrness 1973).

Westside Grasslands

These relatively uncommon grasslands are located primarily at low elevations, rarely middle elevations, w. of the Cascade crest. Most are located in the Puget Lowland, especially Thurston and Pierce Cos. and the San Juan Is. They can generally be characterized as either prairies that grow on relatively flat ground, especially deep gravelly glacial outwash, or as balds that grow on shallow soils with bedrock outcrops on sunny slopes. In their native condition, they are dominated mostly by Roemer's fescue, red fescue, and/or diverse native forbs. Exotic species now often dominate or co-dominate and can include up to about 25% cover of scotch broom, a shrub. The habitat also includes tree savanna with 10-25% cover of Douglas-fir, Oregon white oak, or rarely ponderosa pine. This habitat does not include grasslands located on spits or dunes, or along the outer coast on headlands. Franklin and Dyrness (1973) mention these as special types and unique habitats due to their relative rarity.

Western Juniper and Mountain Mahogany Woodlands

These are open woodlands dominated by western juniper or mountain mahogany with bunchgrass or sagebrush understory. The trees are mostly less than 12 m (40 feet) tall. Juniper woodlands are located mostly from the Columbia R. Gorge, Klickitat Co., southward into Oregon, with a major isolated stand of western juniper on dunes in Franklin Co. There are a few small stands of mountain mahogany woodland along the Snake R. in Asotin Co.

Eastside Canyon Shrublands

These are medium-tall to tall (0.5-2 m) shrublands dominated by black hawthorn, ninebark, oceanspray, chokecherry, rose and/or snowberry, and typically found in a mosaic with bunchgrass. This habitat is often associated with isolated ponderosa pine or quaking aspen. It is located mostly in canyons surrounding the Blue Mts. nr. or in the steppe zone, as well as small scattered stands around the edges of talus throughout the Columbia Plateau.

Eastside Grasslands

These grasslands are located throughout the open country of the Columbia Plateau and to a much lesser degree as balds or openings within the forested zones of e. Washington below the subalpine fir zone. They are dominated by bunchgrasses and/or annual grasses with varying quantities of forbs. Dominant bunchgrasses are bluebunch wheatgrass, Idaho fescue, dropseed, or threeawn. Annual bromes, primarily cheatgrass, and a variety of exotic forbs dominate or co-dominate with native or planted bunchgrasses, such as crested or intermediate wheatgrass. Shrubs taller than the grass layer provide less than 10% cover. This wildlife-habitat type includes both late-successional grassland associations (such as bluebunch wheatgrass-Idaho fescue association) described by Daubenmire (1970) and Franklin and Dyrness (1973), as well as existing grass-dominated vegetation resulting from fire or other disturbances to sagebrush or bitterbrush late-successional associations.

Shrub-steppe

This habitat type features open to closed shrublands (greater than 10% shrub cover) characterized by big sagebrush, antelope bitterbrush, or three-tip sagebrush. Shrubs are associated with bunchgrass or annual grass/forb understories. This type includes sand dune and sandy soils often without a shrub layer or with a shrub layer dominated by rabbitbrush. It is located entirely e. of the Cascades, primarily in non-forested zones of the Columbia Plateau. Daubenmire (1970) and Franklin and Dyrness (1973) describe several associations as part of the shrub-steppe, but this wildlife-habitat type does not include communities on very shallow soils (dwarf shrub-steppe) or those currently dominated by grasses (eastside grasslands).

Dwarf Shrub-steppe

These dwarf-shrublands or grasslands with scattered dwarf-shrubs (less than 0.5 m tall) are usually open in structure, with some even sparsely vegetated. In Washington, they are found entirely e. of the Cascades, primarily in non-forested zones of the Columbia Plateau, primarily on shallow soils (scablands) with stones or outcrops apparent at the surface. Diagnostic low shrub species include rigid sagebrush, any one of several buckwheat species, and low sagebrush (the latter very local in Washington). Sandberg's (curly) bluegrass is the dominant grass, though it also occurs commonly in other habitats. This type is equivalent to what Franklin and Dyrness (1973) describe as associations on shallow soils in the steppe and shrub-steppe zones.

Desert Playa and Salt Desert Scrub

These are primarily shrublands dominated by black greasewood or spiny hopsage, or grasslands dominated by basin wildrye or saltgrass. They are associated with hot salt desert and lakebed wetland environments. Alkaline and saline sedge and rush wetlands in salt desert are included here. This wildlife-habitat type occurs sporadically in the shrub-steppe zones of the Columbia Plateau of Washington, where part of it was described by Franklin and Dyrness (1973) as saltgrass associations on saline-alkali soils and the spiny hopsage/ Sandberg's bluegrass association.

Agriculture, Pastures, and Mixed Environs

This wildlife-habitat type is dominated by intensive agriculture, including planted pasture, row crops, orchards, and vineyards. It also includes associated scattered dwellings and intervening areas of weedy vegetation. In w. Washington lowlands, shrublands dominated by exotic species such as Himalayan blackberry and scotch broom are included here.

Urban and Mixed Environs

This wildlife-habitat type is dominated by residential, commercial, or industrial land uses. Residential development that is considered part of this habitat can vary in density down to about one house per 1.25 hectares. The habitat can include small patches of native vegetation.

Open Water – Lakes, Rivers, and Streams

These are areas of open freshwater: rivers, lakes, ponds, reservoirs, and streams. This wildlife-habitat type does not include areas of emergent or floating vegetation around the edges of any of these water bodies, which are considered part of the herbaceous wetlands habitat.

Herbaceous Wetlands

All freshwater wet meadows, marshes, fens, and aquatic beds are included here, except those that are found at high elevations and included within the subalpine parkland habitat. These are wetlands or portions of riverine floodplains that are dominated by herbaceous vegetation. Common dominants include cattails, sedges, grasses, bulrushes, rushes, or various forbs. Aquatic rooted plants that extend to the surface or floating aquatic plants are also included here as dominants.

Westside Riparian - Wetlands

This wildlife-habitat type includes all freshwater wetlands or riverine floodplains (not all floodplains are wetlands) that are dominated by trees or shrubs at low elevations (up to about 600 m on avg.) w. of the Cascades. Typical dominant species include Sitka spruce, western redcedar, western hemlock, red alder, black cottonwood, Oregon ash, willows, red-osier dogwood, and spiraea. This habitat type also includes sphagnum bogs (forested, shrub, and herb dominated), except those located in the subalpine parkland. Sitka alder and vine maple shrublands (often but not always wet) associated with avalanche chutes occur commonly at middle to high elevations in the Cascade Range and Olympic Mts. and are considered part of this wildlife-habitat type.

Montane Coniferous Wetlands

These forested wetlands occur at middle to high elevations in the mountains and are dominated by conifers. In addition to typical wetlands in depressions or adjacent to streams, this habitat can be found on steep slopes that are subirrigated and show little to no surface evidence of water (characteristic of devils club understory). Major diagnostic tree species are mountain hemlock, Pacific silver fir, and Alaska yellow-cedar w. of the Cascade Crest, and Engelmann spruce, western hemlock, and western redcedar e. of the Cascade Crest. Western hemlock and western redcedar on the westside, and quaking aspen, subalpine fir, or lodgepole pine on the eastside can be important but are not diagnostic.

Eastside Riparian - Wetlands

These are forests, woodlands, and shrublands influenced by streams, lakes, or wetlands e. of the Cascade crest, including the Columbia Plateau. They are mostly dominated by deciduous trees, especially black cottonwood, white alder, and quaking aspen, and shrubs, especially thinleaf alder, red-osier dogwood, and willow. Riparian Douglas-fir and ponderosa pine within the shrub-steppe zones of the Columbia Basin are included here, but conifer-dominated wetlands or riparian in the forested zones are not. This type occurs at all elevations below the subalpine parkland zone of Franklin and Dyrness (1973).

Coastal Dunes and Beaches

These are sand beaches above high-tide line, non-forested sand dunes, spits, and berms along the coast. This type is most common on the outer coast, but also occurs as small linear strips around the shores of Puget Sound and other inland marine waters. Vegetation varies from nonexistent to dense grasslands or shrublands with scattered lodgepole pine. European beachgrass, a non-native invader, now dominates most grasslands. The native species American dunegrass and red fescue are still dominant in spots around the inland waters. Shrublands are dominated primarily by salal and evergreen huckleberry, or the non-natives scotch broom or gorse. Wetlands within the dune landscape are not part of this type. Franklin and Dyrness (1973) describe the vegetated portions of this type as sand dune and strand communities.

Coastal Headlands and Islets

These are shrublands and grasslands that occur on exposed headlands and islands along the outer coast. This type also extends as small islands into the inland waters as far as Puget Sound. Dominant shrubs include salal, evergreen huckleberry, salmonberry, black twinberry, and California wax-myrtle. Native dominant grasses are red fescue and Nootka reedgrass.

Bays and Estuaries

This wildlife-habitat type includes areas with significant mixing of salt and freshwater, including lower reaches of rivers, intertidal sand and mud flats, salt and brackish marshes, and open water portions of associated bays. Only the most brackish portions of open water areas of Puget Sound adjacent to river mouths are included here. Essentially all of Grays Harbor and Willapa Bay are considered part of this type. Eelgrass is an important submerged or intertidal vegetation type associated with muddy substrates. Salt marshes (including brackish) are located in the intertidal zone of estuaries where differences in flooding frequency and duration, salinity, and substrate result in a variety of plant communities dominated by grasses (especially saltgrass and tufted hairgrass), forbs (perhaps most characteristically pickleweed), sedges, and rushes.

Inland Marine Deeper Water

This wildlife-habitat type includes the open waters deeper than 20 m in Puget Sound, Hood Canal, the Strait of Georgia, around the San Juan Is., and the Strait of Juan de Fuca e. of the Elwha River. Localized features attractive to birds include current boundaries ("tide rips" or "convergences"), which concentrate prey.

Marine Nearshore

This wildlife-habitat type includes marine water areas adjacent to shorelines that are less than 20 m deep and the intertidal zone up to high-tide line, but does not include those waters or shorelines significantly affected by freshwater (bays and estuaries). This includes shorelines along the inland marine waters such as Puget Sound as well as those along the outer coast. There is great variation in substrate character, from mud to rock, within this type, and this is important to the use of an area by different species.

Marine Shelf

This wildlife-habitat type consists of marine waters 20-200 m (65-656 feet) deep along the outer coast and in the Strait of Juan de Fuca w. of the Elwha R. Most of the marine shelf is a relatively shallow, flat area with a sandy bottom about 15-65 km wide. Here again, current boundaries, sea temperatures, and fishing activity mean the area is not uniform in productivity and this affects the distribution of marine birds. Birds here associate essentially only with surface and uppermost layers of the water body, and not with benthic prey. Fishing activities provide an important food source, from scavenging discards or "bycatch" to preying on fish brought to the surface.

Oceanic

This is the deep open ocean extending westward from the 200 m depth line. This line corresponds approximately with the edge of the continental shelf and the beginning of the steeper continental slope. This habitat type varies in bird usage with effects of water temperature and seasonal upwelling on prey species.

Seasonal Distribution Maps

by Kelly M. Cassidy

Range Limits

Most bird species in Washington have a breeding, wintering, and migration component to their distribution maps, but the sources and quality of information available for each map component vary. Breeding-range limits were derived primarily from the breeding-bird maps of Smith et al. (1997)—hereafter referred to as the Gap Analysis Project (GAP) breeding-bird maps—with some generalizations to reflect the lower resolution of maps in this book. GAP breeding-bird range limits were based on the Breeding Bird Atlas (BBA) for Washington researched by the Seattle Audubon Society and GAP between 1987 and 1996. Some breeding-range limits were modified from the GAP breeding-range limits because of recent observations, additional information from literature sources, or reviewer opinion. For more information about the BBA, see Smith et al. (1997). For more information about GAP, see Cassidy et al. (1997).

For winter ranges, the most valuable source of information was the Audubon Society's Christmas Bird Count. Only CBC records since 1990 were used, since the focus was on current distributions. The most serious deficiency of the CBC data for mapping statewide ranges is the biased location of count circles. More than a third of the Washington count circles are in the Puget Trough. Virtually all counts are centered on a city. Only one or two counts (depending on how "mountainous" is defined) occur well away from developed areas in mountainous terrain. Thus, CBC data provide little information about elevation distribution of birds in winter or the presence of species in poorly covered areas, such as the ne. corner of the state.

CBC data were supplemented by reviewer opinion, written accounts, bird checklists, and information gleaned from posts to the Internet birding lists "tweeters" and "inland-NW-birders." Specific written accounts included Wahl and Paulson (1991), Paulson (1993), Morse (1996), Contreras (1997), Patterson (1998), and Stepniewski (1999).

Bird checklists were obtained from many sources: Nisqually NWR, Turnbull NWR, Hells Gate SP (in Idaho, on the opposite side of the Snake R. from Washington), Ridgefield NWR, Columbia NWR, Conboy L. NWR, McNary NWR, Klickitat Co. (provided by Ken Knittle), ne. Washington and the Yakima R. Canyon (both from the Northern Prairie Wildlife Research Center, U.S. Geological Survey), Lincoln Co. (compiled by Jerry Hickman), and the Palouse region (compiled by the Palouse Audubon Society). Bird checklists vary widely in the amount of time over which they have been kept, the number of observers involved in compiling the list, and in how abundance categories are interpreted. Nonetheless, they provide valuable information for some areas of the state for which there are otherwise few sources.

For migration-range limits, there is no equivalent of the CBC or GAP breeding-bird maps. Refuge lists, written accounts, and expert review were the major sources of information. Sources were the same as for the winter-range limits, except for the sources specific to wintering birds.

Land Cover Base Map

As part of GAP, a land cover map—hereafter referred to as the GAP land cover map—was created from 1991 Landsat satellite thematic mapper imagery. The GAP land cover map had a nominal minimum mapping unit of 100 hectares, which meant that only land cover features of 100 hectares or larger were mapped. In practice, some smaller features, especially water bodies, were mapped, but many polygons were much larger than 100 hectares. For more details on the GAP land cover map see Cassidy (1997a).

The GAP land cover map was designed for conservation analysis. Display resolution was not a consideration in its creation. In contrast, clarity of display was the sole criterion for the maps of the Seattle Audubon Society BirdWeb site and this book. The BirdWeb images were displayed at a 600- by 400-pixel screen resolution, and the maps for this book were expected to be 2 to 3 inches wide. The GAP land cover map was a large file with more detail than could be seen at the planned display sizes. A more serious problem for the maps in this book was that narrow linear features, notably rivers, shorelines, and near-shore distributions, effectively vanished at the planned display sizes. The original GAP land cover map was consequently modified to create a more appropriate land cover map. Polygon boundaries were smoothed to reduce the file size. Rivers, shorelines, and near-shore distributions were exaggerated to be more visible at the display sizes. Some rivers missing from the GAP land cover map were included in the land cover map for this book.

The GAP land cover map included information about vegetation zones, or areas of similar climate and elevation within which the natural vegetation was reasonably similar (e.g., Ponderosa Pine Zone or Alpine/Parkland Zone). Zonal information was used in mapping bird ranges, since birds tend to respond predictably to climate and elevation variables, at least at the resolution of these maps. The GAP land cover map also included information about land cover in 1991. Information about transitory cover types—forest seral stages and logged areas—was not used for the present maps. Information about reasonably stable cover types—development and agriculture—was used for these maps, although at a lower level of resolution than for the GAP maps.

Brown Pelican (G. Scott Mills)

Range Map Creation

Breeding- and wintering-range maps were created by selecting appropriate habitat on the land cover map within the species' breeding- and wintering-range limits, respectively. Appropriate habitat sometimes occurred below the resolution of the land cover map. In those cases, maps were a compromise between over- and under-representing the species' range. Migration-range maps are less habitat specific, since migrating birds tend to be less choosy about habitat than breeding and wintering birds. If the species was known to stop over only in fairly specific locations, migration ranges were sometimes made more specific. For most species occurring in winter or summer and in migrations, the winter or summer seasons only were mapped, illustrating the seasons of consistent longer-term presence. Exceptions included migration when significantly large numbers consistently passed through areas unused in winter or summer (e.g. Sandhill Crane in e. Washington). Note that range maps can give a misleading impression of abundance: one species may be rare across a large range while another may be common within a smaller range.

All mapping work was done in ArcView, a Geographical Information System software package produced by ESRI (Redlands, CA).

All maps were reviewed at least once by this book's editors, contributors, and interested outside reviewers. The editors, Chris Chappell and Randy Hill, provided the majority of map reviews.

Avian Conservation in Washington

by Joseph B. Buchanan

When Jewett and colleagues published *The Birds of Washington* in 1953 it is doubtful they could have imagined the magnitude of change in the state that lay ahead. In the subsequent half-century Washington's human population increased from 2.4 million to 5.6 million (WDNR 1998). With this growth came a wave of urban and suburban expansion that quickly engulfed large areas of the state. Forests were cleared, wetlands filled, and shrub-steppe and grasslands plowed to accommodate an expanding human population. Riparian areas and other lowland habitats disappeared behind dams of all sizes. Paved roads, highways, parking lots, malls, housing developments, and power-line corridors eliminated or significantly altered habitats. Ironically, even agricultural areas—already altered by humans—were lost to conversion. Clear-cut harvesting became more prevalent and efficient.

The impacts of these changes represent a legacy of environmental modification that began with the Native Americans and accelerated with the arrival of European settlers. These changes have resulted in population declines of many bird species, the most prominent of which include the Marbled Murrelet, Spotted Owl, Sage Grouse, and Sharp-tailed Grouse. This chapter presents a brief overview of some of the factors that impact bird populations in Washington, as well as measures to protect birds or their habitats, emphasizing issues of significance over the last half-century.

Conservation Issues

Among the most prevalent ecosystem attributes of the region are the marine waters and shorelines of the outer coast and the Puget Sound, the coniferous forests, the shrub-steppe, and the multitude of great rivers and estuaries. Each of these ecosystems has been altered by humans. For an overview of the status and description of 30 habitat types in Washington, see Chappell et al. (2001).

Some of the most widespread and far-reaching human-related impacts have occurred in the oceans of the world. Although El Niño is a natural phenomenon, scientists believe that recent El Niño events have been greatly magnified by human factors such as elevated amounts of greenhouse gases (Melillo 1999). These periodic El Niño events are characterized by elevated oceanic temperatures and greatly altered regional weather patterns (see Buchanan et al. 2001 for review) and have been associated with decreased reproduction of some bird populations (Wilson 1991, Ainley et al. 1994).

The marine waters of Washington are also impacted by pollution. The two greatest sources of marine pollution may be the spilling of oil and plastic particles. Numerous oil spills have occurred in Puget Sound, the outer coast, and even on the Columbia R. (e.g., Speich and Thompson 1987, Larson and Richardson 1990). These spills foul or kill many birds. Billions of gallons of oil are transported through Puget Sound and outer coastal waters annually (WDOE 2000). As we have learned from the infamous *Exxon Valdez* spill in Alaska, a large spill in these areas could have catastrophic consequences for birds of our nearshore, coastal, and estuarine ecosystems (Wiens 1996). Plastic-particle pollution enters the marine environment from docks or from ships at sea. Plastic floats on the surface, and the small particles that do not wash ashore, including on the most isolated beaches in the region (Ribic et al. 1997), are available to be ingested by phalaropes, fulmars, and other marine birds. Ingestion of plastic particles causes lacerations and may lead to starvation of birds since the particles accumulate in the stomach and cannot be readily digested. Although plastic-particle pollution is known to impact individual birds (Day et al. 1985), it may also impact bird populations, although this has not been demonstrated.

The Pacific Northwest, and especially the w. slopes of the Cascade Mts., the Olympic Pen., and the lowlands of sw. Washington, support some of the most productive forests in the world (see Franklin and Spies 1991). At the midpoint of the last century, substantial areas of the Puget lowlands had been cleared for settlement, and timber harvest had occurred in many parts of the region. The use of clear-cut harvesting accelerated at about that time and significant areas of old-growth forest—for example, the coastal plain s. and w. of Forks—were logged. Estimates of the loss of old-growth since European settlement began range as high as 97% for the westside and 85% for the eastside (http://wa.audubon.org/new/audubon/). Concern about the Spotted Owl and other species associated with old-growth forests led to the implementation of a comprehensive federal forest-planning strategy designed to protect the habitats for these and other species and to maintain overall ecosystem function

(Forest Ecosystem Management Assessment Team 1993). Other protective measures have been established for state and private lands (see below). With modernization and technological advancement, timber harvesting has become more efficient. Millions of acres of forest in Washington are now managed on short rotations of about 50 years. This time span between harvests prevents trees from growing to the size that would eventually produce snags that are important for cavity-dependent species (Spies and Cline 1988). As a consequence, most managed industrial forests support bird communities that have smaller populations of cavity-dependent birds. Other species such as the Northern Goshawk have become rare in lowland forests (Watson et al. 1999).

The effects of fire suppression over the last 80 to 100 years have influenced habitat conditions in the interior dry forests of the state. Prior to European settlement, some interior forests, particularly those dominated by ponderosa pine, were characterized by frequent but low-intensity fires that burned through the understory and removed saplings and low shrubs. With increasing efforts to suppress fires, the frequency between fires increased substantially and this permitted an invasion of these forests by Douglas-firs and grand firs (and sometimes other species). The resulting forests contained more trees, had a well-developed understory, and, because of this greater stand-level complexity, if ignited would not burn as low-intensity ground fires, but instead as high-intensity crown fires (Agee 1993). Conversion to pre-European conditions is controversial because these forests now support Spotted Owls and other species of management concern. On the other hand, the changes in forest composition and structure may have resulted in degradation of open pine conditions used by species like the White-headed Woodpecker and Flammulated Owl. Some overstocked forests have been damaged by forest insects and disease (see Sallabanks et al. 2001), further complicating the debate about management of these forests for multiple resources.

In addition to the large expanses of forest that have been altered by short-rotation management, extensive forest areas have been cleared and converted to non-forest conditions. For example, 104,000 hectares of forest in Washington were converted to urban use between 1945 and 1970 (see Ferguson et al. 2001). Jewett and colleagues obviously saw some of this conversion in their day, but the sprawl has continued and has reached far out from the major urban centers. Landsat satellite imagery clearly shows the vast area of major conversion in the Puget lowlands.

Washington has thousands of miles of rivers and streams. The Columbia R. is one of the largest on the N. American continent. The Snake R. is famous for its deep canyon, and the Skagit and Nooksack Rs. support large concentrations of wintering Bald Eagles. A wide range of bird species are found on or along such rivers, from Bullock's Orioles to Caspian Terns. At least 1022 dams are present on Washington rivers and streams (WDNR 1998). Many of Washington's larger rivers have been altered by dams; at least 39 major dams are now present and several of these were built after 1950 (e.g. The Dalles, Mayfield). Since European settlement, 50% of streamside habitats have been lost (http://wa.audubon.org/new/audubon/).

Changes in water flow after dam construction likely influence bird use of riverine areas. For example, reservoirs may provide more foraging opportunities for species like the Osprey. On the other hand, shrub-steppe, riparian, or forest areas in valleys above dams that were flooded after dam creation are no longer available habitats.

In e. Washington, almost 3 million hectares of shrub-steppe and 0.4 million hectares of grassland were classified by Chappell et al. (2001). This represents a reduction of over 50% in the combined amount of these arid interior habitats, as 3.7 million hectares have been converted to agricultural and other non-native conditions since European settlement (Chappell et al. 2001, Vander Haegen et al. 2001). This conversion of large areas to agriculture has impacted certain species in the region, including the Ferruginous Hawk (Richardson 1996). In addition to the overall loss of shrub-steppe and grassland habitats, in many places the remaining patches of these habitats exist as relatively small fragments in landscapes dominated by agriculture (Vander Haegen et al. 2001). Some species such as the Sage Sparrow appear to be less common in areas fragmented by agricultural lands (Vander Haegen et al. 2000), and some species in such areas experience higher nest-predation rates (Vander Haegen et al. 2002).

Among the rarest and most vulnerable habitat types in Washington is the westside grassland (Chappell et al. 2001). This habitat, which may include Oregon white oak and other tree species in some areas, is found in the Puget Trough and the I-5 corridor—an area of very high human activity and population. A substantial number of species associated with grasslands have experienced population declines in recent decades (Altman et al. 2001), much of this due to habitat conversion (to agricultural or urban/suburban conditions) and alteration of disturbance regimes (e.g., fire, flooding) that maintained native plant communities (Altman et al. 2001).

Estuaries in Washington support high concentrations of birds, including waterbirds, waterfowl, and shorebirds (Buchanan et al. 2001). An estimated 70% of coastal wetlands have been lost to conversion (WDNR 1998, Buchanan 2000). Conversions in urban or industrial areas were essentially permanent whereas those in present-day agricultural areas could conceivably be reversed. Estuarine areas in locations such as Seattle and Tacoma have been permanently converted, whereas some estuarine habitats associated with many of the large estuaries (e.g. Port Susan Bay, Skagit Bay, Padilla Bay, Samish Bay, in Puget Sound; locations around Grays Harbor and Willapa Bay) have been converted to agriculture and are bordered by extensive series of dikes. Although many of these modifications may have occurred prior to 1950, some conversions have been more recent (e.g. Wahl 2002). Data relating to bird responses to these changes are generally lacking. Some estuarine areas are now recognized as Marine Protected Areas (see Buchanan et al. 2001).

A variety of bird species have been introduced or otherwise become established in Washington (see Witmer and Lewis 2001). Intentional introductions include a variety of game species (e.g., Mountain Quail in w. Washington, Ring-necked Pheasant, Chukar). Other species that were introduced elsewhere subsequently expanded their ranges to this state (e.g., European Starling, Eurasian Collared-Dove [*Streptopelia decaocto*]). A number of other species, including the Barred Owl and Black-throated Sparrow, have undergone natural range expansions into Washington. Some of these new species fill vacant niches whereas others use habitats important to native species. Concern about competition between European Starlings and native cavity users (e.g., Western Bluebirds) is well known. Similarly, Barred Owls likely compete with Spotted Owls for prey and habitat (Kelly et al. 2003). Even species that are thought to occupy vacant niches may influence other bird species in ways that have not been documented. Other introduced species that influence native bird populations include the red fox and domestic cats, the latter of which may kill millions of birds annually in Washington.

Other introductions or colonizations involve invertebrates and plants, and some of these pose potentially serious problems for bird populations in Washington. A large number of invertebrates have been accidentally introduced to Washington intertidal areas (see Cohen et al. 2001); some of these species have the potential to significantly modify the structure of intertidal communities (for summary, see Buchanan 2000). Exotic plants such as Japanese knotweed, purple loosestrife, cordgrass, and European beachgrass have become established in the region. Although the possible consequences of these exotics to birds are often unknown, some species (e.g., purple loosestrife, cordgrass, European beachgrass) are known to significantly alter the structure of habitats used by birds (for summary relating to shorebirds, see Buchanan 2000).

The Pacific Northwest offers many recreation opportunities, and our beaches, rivers, lakes, and forests have become crowded with humans. Many birds are sensitive to human disturbance and some human activity may facilitate undesirable interactions among birds (Raphael et al. 2002). A review of potential human disturbances to shorebirds in e. N. America indicates that a number of species alter their behavior in the presence of humans; the significance of these changes on fitness or survival are unknown (Buchanan 2000). Many other potential impacts may result from disturbance caused by human activities, including recreational pursuits and their technologies, which are often overlooked but are in sum far greater in magnitude than simple human population increases. Outdoor recreation is popular in Washington, with about 3.3 million residents engaged in activities annually. These potential impacts include, but are not limited to, boating (including kayaking, river rafting, personal watercraft), surfing, jogging and running, dog running, and off-road vehicles of many kinds. Mountain biking has become very popular, and Washington ranks 7th nationally (participation per capita) in dirt-road bicycling (www.outdoorindustry.org). There are many other sources of potential human disturbance including camping, hiking, rock climbing, and even bird-watching (Knight and Gutzwiller 1995, Sekercioğlu 2002). Over 260,000 boats were registered in Washington in 2001, and additional powered and unpowered small craft not requiring registration are used for sport fishing and other recreational activities. Recreation industry figures indicate Washington ranks 5th nationally in participation per capita in whitewater rafting, 5th in recreational kayaking, 6th in sea kayaking, and 7th in whitewater kayaking (www.outdoorindustry.org). These figures reflect substantial human activity by an increasing population in a diversity of habitat types in lowlands, mountains, and marine waters of the state. Many of these recreational activities did not exist in 1950, and those that did were enjoyed by many fewer people than they are today. The rate of increase in outdoor activities has been >10% per year since the 1950s (Walsh 1986) and may increase between 63 and 142% in the next 50 years (Flather and Cordell 1995).

A number of other factors have been identified that may influence bird populations in the region. Bird collisions with communications towers have been identified as a potentially significant source of mortality in other regions (Avery et al. 1980). Roads and powerline corridors change the nature of habitat blocks and may result in changes in bird populations (Forman and Alexander 1998). Chemical and other pollutants have been associated with population declines of species such as the Peregrine Falcon (e.g., Hayes and Buchanan 2002). Finally, West Nile Virus has killed birds elsewhere in N. America in nearly every avian order found in Washington. With the arrival of West Nile Virus predicted as imminent, it is likely that large numbers of corvids (in particular) and other species will succumb to this virus, which was first documented in N. America in 1999.

Habitat Protection

Many bird species in Washington find refuge in protected areas that are managed for a variety of purposes, including the protection of wildlife or wildlife habitats. Some areas are protected in a manner that also emphasizes human use through recreation or other activities (including resource extraction).

Federal Lands

Numerous important federal areas were protected prior to 1950, but many new areas have been designated since then. Mt. Rainier and Olympic NPs were established in 1899 and 1938, respectively. The "coastal strip" was added to Olympic NP in 1953 and N. Cascades NP, an area of over 200,000 hectares, was designated in 1968. Millions of acres of forests were owned and managed by the USFS in the mid-20th century, but large areas have been designated as wilderness areas since then. For example, in the Mt. Baker/Snoqualmie NF alone, eight wilderness areas (Alpine Lks., Boulder, Clearwater, Glacier Pk., Henry M. Jackson, Mt. Baker, Noisy Diobsud, Norse Pk.) have been established since 1964. Forest Service lands are currently managed under the Northwest Forest Plan (Forest Ecosystem Management Assessment Team 1993), a management strategy that protects certain areas as Late Successional Reserves for species associated with old-forest habitats.

National Wildlife Refuges have long been an important means to protect habitats for conservation of bird populations. A small number of refuges were present in 1950 (e.g. Turnbull NWR); this number has grown substantially, and now there are over 20 refuges that protect over 142,000 hectares of important habitat. Examples of additions since 1950 include Umatilla (1969; 10,258 hectares), Julia B. Hansen (1972, >2266 hectares) and Saddle Mt. (2000; 78,914 hectares) NWRs.

Several military installations in Washington (Bangor Naval Submarine Base, Fairchild AFB, Fort Lewis, McChord AFB, Whidbey Island Naval Air Station, Yakima Firing Center) provide important habitats to a broad array of species. Although the use of these lands is focused on military activities, areas of varying size function as (generally unappreciated) high-quality habitat. Several studies of birds have recently been conducted on these lands (e.g., Leu 1995, Rogers 2000, Pearson 2003).

State Lands

Three state agencies manage lands in Washington. The Department of Natural Resources manages over 2 million hectares, primarily forestlands but including non-forested terrestrial areas and much of the intertidal zone. More than half a million hectares of this land base is managed under a Habitat Conservation Plan developed for Spotted Owls and other terrestrial forest wildlife species (WDNR 1997). WDNR also manages 49 Natural Area Preserves (NAPs) and 28 Natural Resources Conservation Areas (NRCAs); these areas encompass about 46,500 hectares (see www.dnr.wa.gov). Protection of NAPs and NRCAs was allowed under legislation enacted in 1972 and 1987, respectively. NAPs are established to protect high-quality native ecosystems and rare plant and animal species while NRCAs protect areas for their scenic and ecological values, but allow low-impact public use (www.dnr.wa.gov).

Over 0.4 million hectares of land are managed by the Washington Department of Fish and Wildlife and the State Parks and Recreation Commission. WDFW manages about 338,900 hectares in wildlife areas across the state (Mark Quinn, p.c.), over 90% of this occurring e. of the Cascade Mt. crest. The Sinlahekin Wildlife Area, designated in 1939, was the first such area in Washington. At least 16 wildlife areas (in some cases representing complexes of areas) that total over 202,300 hectares have been designated since 1950, including Colockum (1953; 42,458 hectares), L.T. Murray (1966; 38,992 hectares) and Swanson Lks. (1990; 7689 hectares) wildlife areas. The number of state parks in Washington has grown from 79 in 1950 to 149 in 2004 (Janet Crader, p.c.). Similarly, the acreage in the State Parks system has increased from 53,029 hectares in 1953 to 105,208 hectares of developed parks in 2001 (Janet Crader, p.c.). A small number of the parks have not been developed. Although many state parks are managed for a variety of uses (e.g., human recreation) that may in some cases conflict with bird use of the area, others have emphasized management that should facilitate use by local species of management interest. For example, Bottle Beach SP has been proposed to function as a bird-watching area, as the location has long been known as an excellent area to observe shorebirds.

Other Lands

Lands of many other designations provide important habitats for birds in ways that have changed since the mid-20th century. Numerous county and city parks have been established since 1950 that are locally important areas or that are frequented by bird-watchers. Examples of such locations include: Lighthouse CP at Pt. Roberts, Discovery Park in Seattle, Vancouver L. CP, Kamiak Butte CP, and several riverfront parks at the Tri-Cities (see Opperman 2003). Environmental groups and other non-governmental organizations have purchased small areas that are now protected (e.g., Port Susan Bay Preserve). Although nearly all private industrial forests provide different habitat conditions than formerly existed, these areas continue to provide habitats for many species that no longer exist in areas converted from forestry.

Regulations

When Jewett et al.'s *The Birds of Washington* was published in 1953, with notable exceptions involving game birds, some wading birds, and shorebirds, relatively little attention was afforded to monitoring bird populations and there were few regulations to protect bird populations. A lot has changed in the last 50 years. A number of regulations now protect birds or their habitats, either directly or indirectly. Examples of regulations include the federal Migratory Bird Treaty Act (multiple updates since being enacted in 1918) and Endangered Species Act (1973), and state acts including the Shoreline Management Act (1971), the State Environmental Policy Act (1971) and associated Forest Practices Rules (1974; a Forest Practices Act requiring reforestation was developed in 1946), and the Growth Management Act (1990). In addition to these changes, the Washington State Fish and Wildlife Commission regulates hunting seasons and bag limits of numerous game species including waterfowl, upland game birds, and Wilson's Snipe, and makes decisions about state-level listings (or down-listings) of species.

Initiatives

Among the greatest positive changes in the last 50 years are four types of initiatives that have resulted in on-the-ground improvements to protect birds or their essential habitats. These initiatives reflect societal concern about the well-being of wildlife, and represent an extension of the legacy, although imperfect, of interest in, and care for, the natural heritage of this country (see Farber 2000, Moring 2002). These initiatives do not necessarily improve conditions over those in place 50 years ago; in some cases the initiatives simply disclose impacts or attempt to offset them to some degree.

Investigation. Some would argue that we already know enough about the natural world to adequately protect bird populations, and that the impediment to protection is the uncertain tension between development/resource use and protection. This long-standing (and persisting) tension acts as a constraint, however, and it seems fair to say that this constraint imposes a greater need for knowledge of most species. For this reason, research on the natural history of birds and their response (at the level of individuals, populations, or communities) to different stressors has been, and will continue to be, essential. Nearly anyone currently working with birds in the resource-management arena will agree that research is often needed to better understand species' responses to potential human impacts. Jewett and colleagues probably understood this at some level, but the magnitude of this issue has clearly changed, both in scope and complexity, over the last 50 years. For example, although resource agencies and universities obviously funded research in 1950, the variety and complexity of avian research has clearly expanded (e.g., to provide greater focus on topics such as evolutionary ecology, genetics, behavioral ecology, and conservation biology). Moreover, research is now conducted by consulting firms that offer their services to a broad clientele, and many development or resource-use companies conduct their own research activities. Finally, avian research in the last 50 years has evolved from the rather simple and descriptive to highly quantitative studies. Present-day research activities often involve predictive and spatially explicit models, simulations, evaluation of community dynamics, and multi-scale assessments. A valuable long-term monitoring tool used in many recent conservation planning efforts, the Breeding Bird Survey, was implemented in 1965 (Robbins et al. 1986).

Mitigation. Whether the result of litigation or regulation, mitigation of impacts has become an important process for addressing changes to the natural world. Mitigation generally is used to address either planned activities that are expected to impact species of concern or sensitive habitats, or in response to accidents that cause environmental damage.

Planned mitigation has become commonplace since 1950 and is often implemented in response to activities such as wetland conversion, placement of powerline corridors, and development of hydroelectric facilities. Mitigation often occurs after the extent of the potential impact has been evaluated in an Environmental Impact Statement (EIS), a process that was developed about 35 years ago.

One of the more prominent forms of mitigation in recent years was developed as part of a clause in the Endangered Species Act. This clause allows for the lawful "incidental take" of a listed species (i.e., through impacts to its habitat) if the impact occurs in accordance with a formal management plan (i.e., a Habitat Conservation Plan, or HCP) that describes the impact and the measures used, in some way, to address those impacts. Several HCPs, with a combined total of over 0.8 million hectares, have been developed on non-federal lands in Washington for Spotted Owls and other forest-dwelling species. Several other HCPs address salmon species and the protective measures and mitigation likely benefit a variety of bird species, although much of this benefit has not been evaluated. Implementation of HCPs, although controversial due to concerns about the extent of allowable impacts and the adequacy of mitigation, has occurred only since the early 1990s.

As mentioned earlier in this chapter, huge amounts of oil are transported through marine waters and freshwater systems like the Columbia River. When oil spills occur, damage assessments are made and mitigation is required to compensate for impacts. Like other forms of mitigation, oil-spill mitigation may occur off-site. Recent mitigation packages have collectively included funds for research, for improving spill response preparedness, and for purchase of property to protect habitats used by species potentially impacted by oil spills (Eric Cummins, p.c.).

Incentives. Many bird species and the habitats they use are found on property owned by private individuals or companies (Cassidy and Grue 2000). This can be something of a dilemma because the tension between development/resource use versus protection in these situations is often manifested in economic terms.

While there are many popular and effective incentive programs, two examples will be given here. The first is the Conservation Reserve Program wherein agricultural landowners are paid to retain a portion of their lands out of cultivation to allow for the short-term growth of plant species that provide important resources to native wildlife. This program was established in 1987 and is very popular, particularly in e. Washington. The enrollment in Washington as of 2002 included 9,559 contracts for 3,945 farms and a total area of almost half a million hectares (http://www.fsa.usda.gov/dafp/cepd/stats/FY2002.pdf). The second example is the federal Safe Harbor program developed in support of the Habitat Conservation Planning process (see Wilcove and Lee 2004). Safe harbor agreements, negotiated between the USFWS and private landowners, act as incentives to landowners by removing regulatory disincentives associated with the presence of a listed (and protected) species on private land. The agreement requires landowners to develop or maintain habitat conditions prior to occupancy by a federally listed species, in exchange for the right to engage in resource-extraction activities at a future date, even if the listed species has taken up residence. The safe harbor agreement therefore encourages habitat development or retention, where none existed or was likely to occur, without the requirement of long-term protection. A state-level version of the safe harbor concept exists for Spotted Owls and Marbled Murrelets (i.e., the Cooperative Habitat Enhancement Agreement; Washington Administrative Code 222-16-105).

Comprehensive Management Planning. The last decade has witnessed the development of unprecedented comprehensive management strategies for birds. A comprehensive strategy was developed for management of waterfowl populations in N. America in 1986 and updated in 1998 and 2002 (e.g., Canadian Wildlife Service et al. 1998). This plan, organized by waterfowl flyways—Washington is in the Pacific Flyway—is crucial for cooperative management of waterfowl populations in Canada, Mexico, and the United States.

Conservation plans for taxa other than waterfowl did not exist until the emergence of Partners in Flight (PIF) in the 1990s. Landbird conservation plans developed by PIF for Washington cover westside coniferous forests (Altman 1999), coniferous forests of the e. Cascade Mts. (Altman 2000a), westside lowlands and valleys (grasslands and oak woodlands; Altman 2000b), the "Rocky Mt." areas of se. and ne. Washington (Altman 2000c), and the Columbia Plateau (shrub-steppe and riparian; Altman and Holmes 2000). These plans describe species-habitat associations, identify species that are potentially at risk (see Carter et al. 2000), specify conservation issues of importance, introduce biological objectives and potential conservation strategies, and outline general processes for implementing landbird conservation. The local plans are linked to a national plan (Rich et al. 2004). The conservation documents are non-regulatory, and the emphasis is to implement conservation strategies that work at various spatial scales (B. Altman, p.c.). An element of PIF conservation planning was the recognition of Important Bird Areas (IBA). An initial list of 64 IBAs was presented by Cullinan (2001).

Following the lead of PIF, shorebird and waterbird (including seabirds) plans were developed both nationally (Brown et al. 2001, Kushlan et al. 2002) and locally (Drut and Buchanan 2000, McNaughton et al. 2004, Ivey and Herziger 2005). As these groups were being organized it became apparent that an effort was required to develop a unifying theme that would allow the various groups to work together. The N. American Bird Conservation Initiative (NABCI) evolved to serve this purpose. One of the first orders of business was to standardize the boundaries of ecoregions for planning purposes. The result was the development, by NABCI, of standard Bird Conservation Regions (BCRs) over the N. American continent. Portions of three BCRs occur in Washington: w. Washington is within BCR 5, ne. and se. Washington are in BCR 10, and the e. slope of the Cascade Mts. and the Columbia Plateau are in BCR 9. Participants in these bird conservation initiatives have recently begun to organize in regional "all-bird" conservation groups, another unifying theme to maximize efficiency. The bird conservation initiatives (including "all-bird" efforts) often seek the most parsimonious means to secure funds to implement bird conservation in the different BCRs. This is often accomplished by working with the Joint Ventures (another initiative since 1953) to package multi-species or ecosystem-type strategies to present to conservation funding entities.

With funding for purchase or protection of important habitats becoming scarce, the ability to prioritize different areas according to their conservation value is imperative. The Nature Conservancy and other partners (e.g., WDFW) have developed a procedure for prioritizing such areas. Ecoregional conservation assessments (e.g., Floberg et al. 2004) are expected for the entire state (comprising 9 regions) in 2005.

The U.S. Congress recognized that resource management agencies face immense pressures to adequately protect wildlife resources. Funding will be allocated to states in the near future, but to be eligible, each state is required to draft a Comprehensive Wildlife Conservation Strategy (CWCS). The Washington document will include information synthesized from all possible sources, including the various bird conservation initiatives, regional management plans, and species recovery plans. The CWCS report will include information on species occurrence and status; the distribution, quality, and status of habitats; limiting factors; needed conservation actions; and strategies for monitoring (Joe La Tourette, p.c.). The CWCS report must be submitted to the USFWS by October 2005. Approval of the report will make Washington eligible to receive substantial federal funds for habitat acquisition and a variety of activities that relate wildlife conservation with education and recreation.

Conclusion

What does the future hold in store for Washington's birds? The answer to this question is not clear, and some of the news is not very good. Although the initiatives described above (and others) should be helpful to address some of the present-day conservation issues, many other issues will likely pose far greater challenges. The potential effects of global warming on our bird populations and ecosystems are unknown but could be dramatic. The continued increase in the human population, both in Washington and elsewhere, will place more stress on the earth's resources and new resource use/ protection conflicts are inevitable. Our problem is one of globally sustainable resource use. Ecologists have been telling this story for decades. In short, population growth and resource-use patterns must change or we will witness more impacts to bird populations. If humans can find a way to coexist in a sustainable manner in the ecosystems in which they are a part, the future for Washington's birds may be bright.

Evolution of Field Ornithology in Washington

This book is by necessity a blend of systematically acquired data and non-systematic reports and records used to describe as completely as possible our knowledge of the current status and distribution of Washington's avifauna. We have gleaned reports, many gathered opportunistically by bird-watchers, to augment the incomplete coverage of systematic surveys. The resulting contribution by Washington bird-watchers has been enormous, far in excess of what has been available in earlier times. In the 50 years since Jewett and his co-authors summarized the field efforts that formed the basis of their description of the status and distribution of Washington's avifauna, there has been a marked increase in the scope of fieldwork in Washington. There have been equally pronounced changes in our knowledge of the status and distribution of much of the state's avifauna, resulting in difficulties in determining whether the knowledge changes are a result of distributional changes or observer awareness. A determination of whether these changes are real or an artifact of the increase in the number and sophistication of field observers is a necessity. Without an understanding of the increased observer effort and skill, changes in our knowledge of the range and status of populations can be misinterpreted.

The increase in the numbers and range of skilled field observers, resulting from improved field guides, optics, and transportation, lifestyle, and affluence provides a much greater database on status and distribution than was available to Jewett et al. An obvious example of that change in observer coverage has been in the pelagic waters of Washington. Jewett et al. relied upon data from about 10 days of work on the ocean covering a half-century and, as a result, they reported only single records of Flesh-footed and Buller's Shearwaters, species that are now known annually due in large part to records from approximately 425 days of pelagic surveys used for this book. While not quite as striking, the changes in observer coverage of terrestrial habitats has been as profound. The continent-wide increase in knowledge of Boreal Owl range (Hayman and Verner 1994), almost certainly due to improvements in fieldwork for this species, was mirrored in Washington. Jewett et al treat this species as hypothetical; recent studies find it a widespread, uncommon breeder in appropriate habitat.

Early Fieldwork

The first era of ornithological investigation and description in Washington was the period prior to about 1950, well described by Jewett et al. This included explorers beginning with Lewis and Clark, many early scientists pioneering the region, and, later, federal and state wildlife biologists, academic investigators, and amateur observers. These are described in depth and credited by Jewett et al. for important contributions to knowledge of the avifauna of Washington (archived at Slater Museum of Natural History, University of Puget Sound; D. R. Paulson, p.c.). They describe the basis of their information as the "mass of original information from activities" of the U.S. Biological Survey/ USFWS from 1889 to 1950 and cooperative work of the U.S. Biological Survey and State College of Washington from 1917 to 1921. The bibliography provided by Jewett et al. indicates the depth of efforts and how fortunate Washington is to have the amount of historical knowledge accumulated through this period.

Fieldwork During the 1950s and 1960s

A great expansion of fieldwork by amateurs helped refine knowledge of occurrence and distribution during the period from about 1950 to 1970, greatly increased the importance of field ornithology, and ultimately led to our heavy emphasis on field observations in this book. The Pacific Northwest Bird and Mammal Society published the *Murrelet* and was the focal point for active observers during these two decades. In the early 1960s, only a few observers submitted field observations to the seasonal reports in *Audubon Field Notes* and these came mainly from e. Washington, the Seattle area, and Whatcom County. Primary fieldworkers in e. Washington during this timespan included Jim Acton, Warren Hall, Wayne Doane, and Robert E. Woodley. Thomas Rogers also served as *Audubon Field Notes* regional editor. Garrett Eddy was active in the field in w. Washington, as were Bob and Elsie Boggs, who also served as *Audubon Field Notes* regional editors. Zella (McMannama) Schultz was active in field observations and writing about birds. A well-known bird artist, fieldworker, and bird bander and an early bird rehabilitator, she inspired a large group of people who both supported her work and learned from her, helping band thousands of gulls and swallows, and acquiring interest in avian biology and conservation.

Frank Richardson contributed field observations including many from the San Juan Is. Gordon Alcorn built museum collections for University of Puget Sound and John Slipp was active in the field and collected for Pacific Lutheran College, primarily in the Pierce Co. area. Statewide, primary observers included Lynn LaFave, Earl Larrison, Bart Whelton, and Terence Wahl, who also banded birds and organized pelagic seabird trips. Steve Speich, though he resided in Washington only a few years, helped bring scientific discipline to fieldwork done by volunteers, encouraged benchmark work on bird populations and inspired other workers.

State and federal agency work was mostly limited to studies of waterfowl and game birds. Museum collection activities were less extensive than in previous decades, as the transition from amateur collectors to amateur observers and banders had begun. Birding as a hobby had taken hold by the late 1960s, although only a handful of observers made observations on a statewide basis. Several statewide updates were published in the 1960s (Alcorn 1962, Larrison and Francq 1962, Larrison and Sonnenberg 1968), although these contained significant errors arising from incorrect assumptions about distribution in areas that had received little coverage and from mistaken visual identifications.

The impressive increase in field observers seen in recent years was not evident early in this period, but the seeds of this increase were sown in the late 1960s, with an influx of talented, serious field observers from other parts of America and abroad. The involvement of Dennis Paulson, Phil Mattocks, David L. Pearson, Ted Stiles, Susan Smith, John Wingfield, Michael Perrone, Edwin O. Willis, Robert Furrer, Eugene Hunn, Jim Duemmel, and others with talent and enthusiasm led to many more field trips and identification classes, and much greater knowledge of birds in Washington. This huge expansion of fieldwork led to a similar increase in reports published in Audubon Society newsletters and national publications like *Audubon Field Notes.* Among regional editors were Thomas H. Rogers, Bob and Elsie Boggs, Werner and Hilde Hesse, John Crowell, and Harry Nehls.

Fieldwork from the 1970s through Present

Beginning in the 1970s, the rise in numbers of both birders and conservationists had a profound effect on our knowledge of Washington's avifauna. Mirroring nationwide trends (Cordell and Herbert 2002), the number of birders in Washington has steadily increased since 1970. As bird-watching has increased in popularity, so have observer proficiency and the quality of optics. Many species formerly considered unidentifiable or extremely difficult to identify in the field are now fairly routinely identified with considerable accuracy by most observers. Improvements in optics have greatly expanded the distance at which species can be recognized. And recordings and identification by sound have helped in the identification of many birds. These factors increased the scientific value of birders' observations and changed the nature of bird-watching. Participation in systematic survey efforts designed to track changes in population status have increased, as has the emphasis on finding and reporting uncommon and rare species. The pursuit of "lists" motivated many bird-watchers to travel more extensively through the state and to attempt to increase their species totals. The scope and accuracy of the several statewide works produced during this time period (Table 3) was considerably improved as a result of these changes.

The editors of this work have drawn heavily upon data and analyses from numerous systematic efforts by bird-watchers. Virtually all of these were initiated or increased greatly from 1970 on: the N. American Breeding Bird Survey (BBS), Christmas Bird Count (CBC), the Marine Ecosystem Analysis Project (MESA), Westport Seabirds pelagic trips, Breeding Bird Atlas (BBA), winter raptor surveys, and spring and fall migratory raptor counts. Other systematic work conducted largely by volunteers in Washington includes bird banding and the Monitoring Avian Productivity and Survivorship Program (http://www.birdpop.org/maps.htm), though these sources were not relied upon as extensively. The BBS effort began in 1966, and showed almost a threefold increase in survey effort in Washington state from 1966-79 to the 1980-2003 period (Sauer et al 2003). While a few CBCs were conducted in Washington over a half-century ago, the number of CBCs and the number of observers increased eight- to tenfold from the 1960s to the 1990s. The result has been a steady, linear increase in field effort continuing into the present. The MESA project relied on extensive volunteer participation to survey marine bird populations in 1978 and 1979 (Wahl et al 1981). Westport Seabirds pelagic trips began censuses of seabirds encountered during organized pelagic trips off Westport in 1971 (Wahl and Tweit 2000a).

Table 3. Post 1970 publications describing statewide distribution and population status.

Year	Title	Authors
1971, revisions through 1991	*A Guide to Bird Finding in Washington*	T. R. Wahl and D. R. Paulson
1976	*A Checklist of the Birds of Washington State*	P. W. Mattocks, Jr., E. S. Hunn and T. R. Wahl
1997	*Breeding Birds of Washington State: Location Data and Predicted Distributions*	M. R. Smith, P. W. Mattocks, Jr. and K. M. Cassidy
2003	*A Birders Guide to Washington*	H. Opperman

Other surveys from University of Washington and National Oceanic and Atmospheric Administration research vessels provided further opportunities for sampling marine bird populations. The BBA incorporated data from fieldwork from 1987 to 1996 (Smith et al. 1997). Surveys of wintering raptor populations have been conducted by the Falcon Research Group in w. Skagit and Snohomish counties annually for over a decade (http://www.frg.org/frg/newsletter.html). Fall raptor migrations have been studied at Chelan Ridge on the e. slope of the Cascades from 1997 to 2003 (Hawk Watch International, available at http://www.hawkwatch.org/) and spring migrations at Cape Flattery on the outer coast were studied from 1990 to 1997 (Clark et al. 1998). Other continuing studies include beach surveys for dead birds begun in the late 1970s. Information from many studies forms a basic part of the data used by authors covering associated species in this book.

Numerous publications describing distribution and status for regions of Washington (Table 4) have appeared in recent years. These publications have been indispensable for our work.

Increased attention by bird-watchers to the status of uncommon and rare species in Washington has produced a dramatic change in our understanding of species in this category. Jewett et al. (1953) documented the presence of 338 species in Washington. Writing two decades later, Mattocks et al (1976) added 39 species to the total, about two additions per year. Another quarter-century later, our work covers 483 species, a rate of almost four new species per year since then. The materials documenting these additions have been evaluated and archived by the Washington Bird Records Committee (WBRC). This body, a standing committee of the Washington Ornithological Society (WOS), began work in 1989 and periodically publishes its results in *Washington Birds.* We have relied on their work as the basis for most of our treatment of the status of casuals and accidentals. The level of bird-watcher reporting to periodicals that compile and summarize regional observations has also increased, providing us with an excellent source of reports that we have drawn heavily upon for this work.

The Washington Ornithological Society was formed in 1988, "to increase our knowledge of the birds of Washington and to enhance communication among all persons interested in those birds," and had about 500 members by 2002. As a communication forum for bird-watchers, it replaced the Pacific Northwest Bird and Mammal Society, which for decades had included most of the amateur and professional ornithologists in the region, and had provided Jewett et al. with many observations. WOS publishes the peer-reviewed journal *Washington Birds* and a newsletter, *WOSNews*, which includes seasonal sightings columns, and hosts a website (http://www.wos.org/). Another recent communication forum is the electronic listserve Tweeters (http://www.wos.org/Tweeters.htm), but as it lacks any editorial or peer review, we drew very little material from this source.

State and federal agencies that had traditionally directed their management and research at hunted species greatly broadened their scope of activities during this period at the urging of conservationists, resulting in the creation of legislation and programs

Table 4. Publications describing distribution and population status for regions of Washington

Year	Title	Authors
1981	*Marine Bird Populations of the Strait of Juan de Fuca, Strait of Georgia and Adjacent Waters, 1978 and 1979*	T. R. Wahl, S. M. Speich, D. A. Manuwal, et al.
1982	*Birding in Seattle and King Co.*	E. S. Hunn
1987	*Birding in the San Juan Islands*	M. G. Lewis and F. A. Sharpe
1989	*Catalog of Washington Seabird Colonies*	S. M. Speich and T. R. Wahl
1991	*Birds of the Tri-Cities and Vicinity*	H. Ennor
1995	*Birds of Whatcom County*	T. R. Wahl
1999	*The Birds of Yakima County*	A. Stepniewski
2001	*Birder's Guide to Coastal Washington*	R. W. Morse

Table 5. Periodicals providing regular information on Washington birds.

Time span	*Periodical Name*	*Coverage*
1944 - present	*Audubon Field Notes, American Birds, North American Birds*	Statewide, four seasonal columns per year
1988 - present	*WOSNews*	Newsletter, four seasonal columns per year
1989 - present	*Washington Birds*	Peer-reviewed journal
1993 - present	*Washington Birder*	Self-published newsletter, primarily bird finding and lists

that focused on non-game species. In Washington, Federal Endangered Species Act (ESA) listings of Brown Pelican, Bald Eagle, Peregrine Falcon, Snowy Plover, Marbled Murrelet, and Spotted Owl have resulted in detailed knowledge of their status through the work of agency biologists and others.

Museum activities, particularly collecting, which had formed the basis of fieldwork during the first half of the 20th century, decreased in importance in recent decades relative to many aspects of status and distribution. Collection activities focused on particular species groups, such as the studies of Hermit and Townsend's Warbler overlap conducted by the Burke Museum staff at the University of Washington, or of shorebird migration conducted by Slater Museum staff at the University of Puget Sound. Field studies by professionals increased, with fieldwork covering threatened species and habitats and results of events such as the Mt. St. Helens volcanic eruption in 1980. Active investigators included David A. Manuwal, Steve Herman, Gordon Orians, Richard Johnson, and Julia Parrish.

The variety of data sources results in an unfortunate unevenness in coverage between species. We are very confident that the status and distribution of some species are very well known, for varying reasons. Any species listed by the ESA, such as Peregrine Falcon, has received a great deal of attention. Some species like the Snowy Owl are avidly sought by bird-watchers, or well studied by academics (e.g., Common Murre). However, somewhat paradoxically, there is less confidence about the status of many common species; they receive relatively little attention from government agencies and relatively little mention by bird-watchers. The regional summaries published in *North American Birds* or *WOSNews* (Table 5) generally contain more information about rarities and unusual sightings than about common species. Bird-watchers' efforts at systematic data collection through BBS, atlas efforts, or CBCs counteract this tendency to some extent. Still, the primary contribution of non-systematic bird-watching remains evidence of unusual and conspicuous changes: range expansions, early arrival dates, irruptions, changes to migration patterns, and the like. Evidence of subtle changes, such as long-term declines or increases in the population of common species, must come from systematically collected data, which represent the minority of bird-watching effort.

Sources of Information

Sources include systematically acquired statewide, regional, or local quantitative data both published and unpublished, a greatly expanded number of agency reports, and peer-reviewed papers in scientific journals and books. Journal sources include the *Auk*, the *Condor*, the regional *Murrelet* (now *Northwestern Naturalist*), *Western Birds*, and *Washington Birds.*

Where possible we have used or referred to analyses of quantitative data from the longest-term available (1966-2000) rather than shorter spans of years. Breeding Bird Surveys (BBS) began in 1966, with 48 routes eventually established, and most were covered annually over the years. Between 1960 and 1999, 756 Christmas Bird Counts (CBCs) were done at least once in 51 different locations/circles in Washington. Long-term data also included results from 382 one-day survey trips off Grays Harbor to ca. 100 km offshore from 1971 to 2003, with about 7500 censuses, 65,000 records of more than 3.3 million birds (see Wahl and Tweit 2000a). Waterfowl surveys by wildlife agencies also provided data, as did shorebird censuses (e.g. Buchanan 2000) and raptor surveys performed in strategic locations by raptor researchers.

Efforts covering 10-15 years included marine bird survey flights (Puget Sound Ambient Monitoring Project; e.g., Nysewander et al. 2001). Data from studies covering two to five years included several marine bird survey projects (e.g., Marine Ecosystems Analysis program [MESA]: Wahl et al. 1981; Wahl and Speich 1983, 1984; and also by Briggs et al. 1992), specialized surveys of forest birds (e.g., Manwual et al. 1987) and shrub-steppe birds (e.g. Knick et al. 2003) and BBA surveys (Smith et al. 1997). Many shorter-term efforts provided benchmarks: the MESA surveys in 1978-79 produced over 6000 censuses of 3,900,000 birds, using shore counts, ferries, small boats, and aircraft (Wahl et al. 1981).

Many non-systematically acquired records such as seasonal reports in *Audubon Field Notes, American Birds, Field Notes, North American Birds*, and WOS newsletters were valuable. On-line reports, popular birding magazines, and selected unpublished reports were also utilized. In addition to reviews by local and regional journal and newsletter editors and the WBRC, records included in accounts were also reviewed by primary authors of draft species accounts and the editors.

Analysis

We used the only long-term indicators of what is happening to regularly occurring, common species that are inconsistently addressed in current field journals due to focus largely on rare and uncommon species and occurrences. We have used, with knowledge of shortcomings, what is available on a wide geographic scale. Our long-term BBS, CBC, and offshore data sets are less than ideal in some respects but they are what is available and certainly greater in amount and value than what was available 50 years ago.

BBS analyses by a number of researchers looking at data covering the N. American continent and regionally are cited in species accounts (e.g. Sauer et al. 2000).

Data from 23 CBC locations that had at least five counts per decade (Table 6) were used to look at possible trends in populations. We assumed that effort and coverage essentially reached standardization relatively early, in the 1970s in most cases, and data can be meaningfully compared. We used avg. count per year for conspicuous species in open habitats and numbers per 100 party hours for species in other habitats, with the former usually representing censuses, the latter samples. Tables of 10-year avgs. in species accounts illustrate changes by CBC circle, between circles and variations between westside and eastside circles.

We also used CBC data from 12 locations (seven westside, five eastside) that had a minimum of five samples each over at least three decades (Table 6). We attempted to minimize variations in effects of weather and observer effort over the long term by using linear regression to show long-term statistical trends. Patterns and statistical trends were calculated for all common species but were presented only when changes were apparent and these usually corresponded with non-systematic reports and intuition of experienced observers. And descriptions of these in the text often employ words like "suggested" and "indicated" where statistical test was not significant (significance noted as $P<0.01$ or $P<0.05$).

Interpretation

There are noticeable differences in depth of details and lengths of species accounts. These result from the amount of attention given by researchers, management agencies, and field observers; endangered, threatened, or other special status; change in status (especially since 1950); taxonomic changes; published findings in scientific literature and other references; varying availability of data between species; and changes of emphasis and resultant lack of availability of data on common species over time in seasonal publications. There has also been a disproportionate focus on rarities, raptors, shorebirds, marine waterbirds, and "glamour" species whose status changed dramatically during recent decades. Perhaps inevitably, the excitement of new birds found by hundreds of observers has resulted in this change of emphasis and a resulting relative lack of availability of information on common species over time in popular publications. When less was known about status, common species were covered. In *Audubon Field Notes* in the 1960s, for example, Thomas H. Rogers regularly sought data from wildlife managers in e. Washington. On the whole, however, the increase in effort by field observers has led to better information on seasonal dates and locations of abundant occurrence of common species. We have tried to be conservative in judging real vs. apparent changes, especially with rare species.

There are differences in interpretation between data from short-term and long-term studies and informal, opportunistic observations/reports, and the value of the data may also be affected by the sampling frequency employed. There are also possible problems with not knowing what stage of a possible cycle is represented. Effects of major climatic events and changes have been suggested for a number of groups of birds, including seabirds (e.g. Wahl and Tweit 2000a) and recently for breeding landbirds (Ballard et al. 2003).

BBS sampling methods are quite systematic though the routes cover roadside habitats and may not represent what is happening to adjacent habitats due, for example, to logging. The BBS coverage varies in suitability between species and may not be useful for species such as nocturnal owls. Results may in some cases be of value only as presence or absence indicators. Though BBS censuses are relatively rigorous in attempts to control variables due to observer, weather, time of day, and methods, they are conducted just once a year. Though results may not always agree with impressions of other observers, they often confirm trends in the state, region, or continent. We present results or authors' qualifications in species accounts.

Table 6. CBCs and party hours in 1960-99 used to calculate trends (minimum 5 counts per decade *) and statistical significance (linear regression).

	Number of counts				*Party hours*			
	1960s	*1970s*	*1980s*	*1990s*	*1960s*	*1970s*	*1980s*	*1990s*
Bellingham*	3	10	10	10	239	1657	1844	1899
Cowlitz-Columbia	0	0	7	10	0	0	307	460
E. Lk. Washington	0	0	9	8	0	0	438	508
Edmonds	0	0	6	10	0	0	251	562
Ellensburg	0	2	10	10	0	115	369	505
Grays Harbor*	0	8	8	10	0	971	987	1593
Kent-Auburn	0	0	10	10	0	0	1056	1207
Kitsap County*	0	6	10	8	0	534	960	658
Leadbetter Pt.*	0	7	9	8	0	213	522	342
Olympia	2	2	9	10	64	330	1712	1997
Padilla Bay	0	3	9	10	0	247	953	1606
Port Townsend	0	3	10	9	0	164	581	675
Sequim-Dungeness*	1	4	10	9	9	241	991	1355
Seattle*	10	10	10	10	876	1445	1505	1684
San Juans ferry*	0	9	5	7	0	66	38	51
San Juan Islands	0	0	8	10	0	0	844	969
Spokane*	10	10	10	10	1052	1212	1245	1322
Tacoma*	1	10	10	10	56	908	1265	1539
Tri-Cities*	3	10	10	10	139	529	638	815
Toppenish	0	0	7	10	0	0	305	594
Wenatchee*	7	10	10	10	181	624	885	681
Walla Walla*	0	7	10	10	0	397	537	481
Yakima Valley	0	10	10	10	0	434	558	758

Red-throated Loon (G. Scott Mills)

CBC data and analyses are subject to many variables: few counts are rescheduled due to weather conditions, for example. Other variables include habitat differences between circles, changes in habitats not covered during CBC counts, observer skills, and identification problems, observer effort (Table 6), and data/results typically are not equally suitable for analysis for all species. Additionally, over time there are increases and evolution of skills, effort by observers and parties, numbers of bird feeders, and habitat changes and degradation. Trends in abundances may in some cases reflect changes in habitats within circle(s) and consequently real changes in abundances there, though not necessarily changes in overall regional populations (see Sauer and Link 2002, Thompson 2002, Bart et al. 2004, Sauer et al. 2003b, for discussion of problems of interpretation of various types of censuses/surveys). It should be noted that long-term CBCs used for analyses are urban centered and, while probably accurate in reflecting changes there, may not illustrate changes in other habitats. Thus, CBC counts representing samples (birds/100 party hours) may have consistently covered habitats and routes over time but, due to factors like urbanization, may not represent bird populations within the count circle or area as a whole. Overall, relatively greater concentration on the number of species, rarities, and unusual occurrences rather than "censusing" may result in lower numbers of other species than might be possible. Lower numbers, however, may very likely represent real trends more than higher numbers do (i.e., numbers could increase with greater effort). Beyond what we present here, values of birds/party hour or birds/km^2 need to be applied to area covered by samples to determine birds/km^2 and the results then applied to changes in habitats over time.

Apparent increase in or more frequent occurrence of rare species may be due to the much higher level of field effort (i.e., rarities may have been/almost certainly were more likely to have been missed in the past). The number of rarities reported is likely higher where observer numbers are higher, and observer effort is likely biased toward the westside. Year-to-year comparisons of the number of some reports (e.g., Prairie Falcons in winter in w. lowlands) must be cautiously interpreted: there is no knowledge of the number of observers afield, how many hours or miles they spent looking for birds, or how many reports resulted from observers communicating sightings which resulted in additional reports.

We have looked for region-wide changes in locations where numbers are regular and sizeable enough to measure reasonably. We have tried to be conservative in judging real vs. apparent changes, especially with "rare" species. Our confidence in our analysis results from the general agreement of findings with non-quantitative observations and specialized or localized studies. And, though we present calculated values in our tables, we stress attention to general trends.

Changes in Status and Distribution since 1950

A number of publications have summarized general trends in species occurrence and abundance over time, which types of species have apparently increased, and which have decreased, and why (e.g. Paulson 1992b). Accounts in this book indicate or suggest that over one-half of the species currently known in Washington have changed in status or populations over the past 50 years. Apparent changes are noted in species accounts, along with sources of supporting data and studies. It should be noted that in some cases knowledge of certain change is probably related to a species' "glamour" or special attraction to human eyes or emotions, large size, conspicuousness, or special legal status, which may well reflect in the amount of data available. Though many changes are suggested on the bases of limited studies or comparisons with non-quantified statements in historical references such as Jewett et al. (1953), there are a number of species, 50 or more, that have changed dramatically, unquestionably, over time. While trends of some species are obvious, others indicate seasonally opposite trends. For example, a number of waterbirds, both summer and winter residents, benefitted from creation of reservoirs and irrigation from dams built in the mid-20th century, while populations of the same species suffered from human-caused changes in another season or region. The Common Loon is just one example of this.

There are a number of species with very evident, data-supported changes. Species that have increased are those adaptable to human actions, though many if not all such increases were unintentional, as in the cases of provision of hatchery fish and fisheries discards. Species increasing include the Double-crested Cormorant, which increased due to legal protection and more available habitat. Canada Goose increases resulted from introduction of breeding populations, and availability of habitats in areas protected from hunting. Trumpeter Swan increased after protection from hunting. Wood Duck and Hooded Merganser benefitted from provision of nest sites and protection. Osprey, Bald Eagle, Red-tailed Hawk, and Peregrine increased following protection from killing and/or elimination of toxic chemicals affecting reproduction. Brown Pelican increased following reduction of toxics on breeding grounds and likely returned in part due to oceanic productivity changes. White Pelican, Black-necked Stilt, American Avocet, and Ring-billed and California gulls increased with provision of habitats resulting from dams and irrigation. Glaucous-winged Gull populations increased with protection and garbage and other human-provided food sources. Caspian Terns increased greatly through range expansion and taking advantage of nesting and feeding opportunities. Barred Owl, Anna's Hummingbird, European Starling, and House Finch expanded their ranges by taking advantage of human-provided habitat changes. Steller's Jay, Western Scrub-Jay, Common Raven, and *Zonotrichia* sparrows increased due to habitat change and factors such as legal protection. Apparently natural range expansions, perhaps related to climate change, are shown for a few species such as Black-headed Grosbeak, Lazuli Bunting, and Tri-colored Blackbird. Other apparent increases may result simply from newly discovered populations or seasonal occurrence by a great increase in field effort by competent observers.

Blackfooted Albatross (G. Scott Mills)

Reasons for species' decreases were also varied. Some were due to apparent changes in marine environments. These included scoters, alcids such as Common Murre, Cassin's Auklet, and Tufted Puffin. Habitat loss affected many species, particularly American Bittern, Greater Sage-Grouse, Sharp-tailed Grouse, Upland Sandpiper, Long-billed Curlew, Marbled Murrelet, Spotted and Short-eared owls, Common Nighthawk, Lewis' and White-headed woodpeckers, Loggerhead Shrike, Brewer's and Sage sparrows, Western Meadowlark, and Olive-sided Flycatcher. Yellow-billed Cuckoo essentially disappeared perhaps due to habitat loss and range contraction. Snowy Plover apparently decreased primarily due to human activities in their breeding habitat. Purple Finch decreased, perhaps locally due to competition with House Finch but likely also from other causes. A number of species decreased due likely to climatic changes over their ranges: these include Rock Sandpiper, Horned Lark, Bohemian Waxwing, Lapland Longspur, Snow Bunting, Pine Siskin, and Evening Grosbeak. And intentional introduction of foxes on San Juan I. eliminated Sky Larks which had immigrated from introduced populations on Vancouver I.

Lack of data for some species very likely affects status known today while, on the other hand, some apparent changes are likely due simply to a great increase in studies and fieldwork. In the species accounts that follow there are many more descriptions of apparent—and likely—changes in status over decades. We hope these encourage further studies.

Species Accounts: Accepted Species

Fulvous Whistling-Duck *Dendrocygna bicolor*

Casual vagrant.

There is one Washington record, a specimen from the s. shore of Grays Harbor on 3 Oct 1905 (Jewett et al. 1953). Though this species is broadly distributed across the Americas, Africa, and India, the sw. U.S. population has declined in recent decades (Kaufman 1996), decreasing the likelihood of another Washington record.

Bill Tweit

Bean Goose *Anser fabalis*

Casual vagrant.

A Bean Goose of the race *middendorffi* was seen and photographed in Hoquiam from 7 to 17 Dec 2002 (Mlodinow 2004). It was most often with a small flock of Canada Geese that were probably of the race *fulva*. There are five previous accepted N. American records away from Alaska, three of which were *middendorffi* (Mlodinow 2004). A bird probably of this species was seen at the same location on 26 Apr 1993, but it was rejected by the WBRC due to inadequate description and because it was a single observer report (Tweit and Skriletz 1996).

Steven G. Mlodinow

Greater White-fronted Goose *Anser albifrons*

Fairly common spring and fall migrant in w. Washington; uncommon migrant e. Uncommon to rare in winter; very rare in summer.

Subspecies: *A. a. frontalis,* Greater White-fronted Goose, and *A. a. gambelli,* Tule White-fronted Goose, occur in Washington.

Habitat: Coastal estuaries, lakes, wetlands, and flooded agricultural fields.

Occurrence: Regular, fairly common w. Washington fall and spring migrant. Coastal areas such as Grays Harbor and Nisqually, Snohomish, Hamma Hamma, and Skokomish R. deltas inland host migrants in fairly large numbers briefly during spring and fall migrations. Jewett et al. (1953) considered the species a coastal migrant, and a rare migrant in e. Washington. Many white-fronts now migrate on an inland route, though birds are more common along the outer coast. Inland, McNary NWR has consistently reported birds through the past few decades, with up to 1200 seen in fall 1998 and 450 in spring of the same year.

Population trend in the Pacific Flyway was a decline of 85%, from reported historical peak estimates of 480,000 in 1966-68 to a low of 73,100 in 1979. Due to cooperative efforts to reduce subsistence and sport harvest on the flyway, numbers on the fall 1998 population survey increased to 413,050 (PFC 1987). White-fronted Geese are rarely taken by Washington goose hunters. Status of *gambelli* is not well documented but is currently estimated at approximately 7200 birds (PFC 2003). This race is thought to migrate earlier than the more numerous *frontalis.*

CBC data in recent decades show small numbers occurring irregularly but widely across w. Washington, most frequently at Grays Harbor. Small numbers are recorded occasionally e. of the Cascades. Occurrences and numbers statewide suggest increasing numbers wintering in the 1990s.

Though records of the Tule White-fronted Goose in B.C. are from coastal areas (Campbell et al. 1990a), the subspecies has been reported inland in w. Washington at the Ridgefield NWR on the Columbia R.

Remarks: Status of the larger, darker Tule White-fronted Goose in Washington has not been well

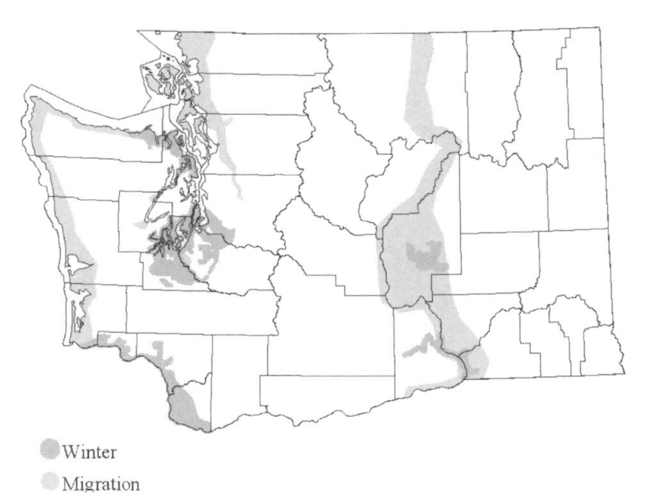

documented. Most observers are not aware of field identification criteria, though Orthmeyer et al. (1995) have developed techniques to differentiate the two subspecies. Ely and Dzubin (1994) describe current taxonomic status. Classed PHS.

Noteworthy Records: West, *Spring*: 9690 on 26 Apr 1996 at Cape Flattery; 1800 on 30 Apr 1997 at the Humptulips R. mouth. *Summer*: 1 on 11 and 24 Jun at Ocean Shores; 1 on 5 Jul 1988 at Dungeness; 1 in 1996 at Keystone (SGM). *Fall*: 1000 on 24 Sep 1978 at Ocean Shores. *Winter*: 21 on 26 Dec 1998 at Steigerwald L. NWR. *CBC, high counts*: 25 in 1993, 21 in 1976 at Grays Harbor. **East**, *Spring*: 20 on 23 Mar 1996 at Atkins L.; 70 on 12 Mar 1997 at McNary NWR. *High count*: 450 on 3 Apr 1998 at McNary NWR. *Fall, high count*: 1200 on 6 Sep 1998 at McNary NWR. *Winter*: 2 in early Dec 1969 at Richland; 1 on 18 Feb 1984 at Priest Rapids; 1 on 17 Jan 1988 at Toppenish NWR; 7 on 24 Dec 1988 at Lyle. Tule Goose: 75 on 17 Sep1996 at Ridgefield NWR; 1 on 18 Sep 1998 at Steigerwald L. NWR; 1 banded bird on 20 Dec 1997 at Steigerwald L. NWR (SGM).

Diann MacRae and Don Kraege

Emperor Goose *Chen canagica*

Rare migrant and winter resident in w. Washington.

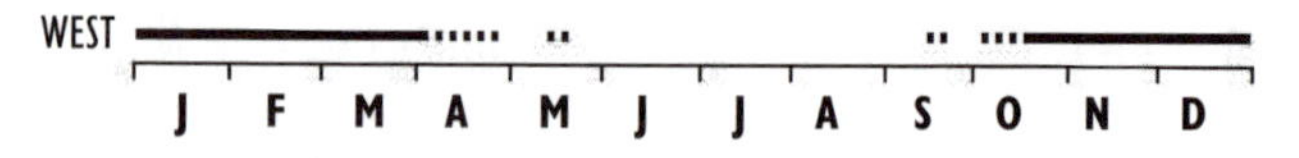

Occurrence: The first Emperor Goose recorded in Washington was one shot in early Jan 1922, and only two more had been found by the time of Jewett et al. (1953). The species is now annual in Washington, with 29 records between 1982 and 1999. Most are from the outer coast, Puget Sound lowlands, Olympic Pen., and the lower Columbia R. County records are Clark (3), Pacific (7), Grays Harbor (2), Clallam (7), Thurston (1), King (2), Snohomish (4), and Whatcom (1). Two additional published records give location as Puget Sound.

Many birds have been in the company of Cackling Geese, another coastal Alaskan breeder. Though escaped birds are possible, most Washington records appear to have involved wild birds (Mlodinow 1995). Classed GL.

Noteworthy Records: *Out of season*: Summer 1987 at Olympia; to 29 May 1994 at Everett (Mlodinow 1995); 5-13 May 1973 at Tokeland; 13 Sep 1998-16 Jan 1999 at Sequim. *High counts*: 3 on 9-16 Mar 1985 at Neah Bay; 3 in early Oct 1990 at Oceanside.

Steven G. Mlodinow

Snow Goose *Chen caerulescens*

Locally abundant migrant and winter resident in w. Washington. Rare on eastside. Very rare in summer.

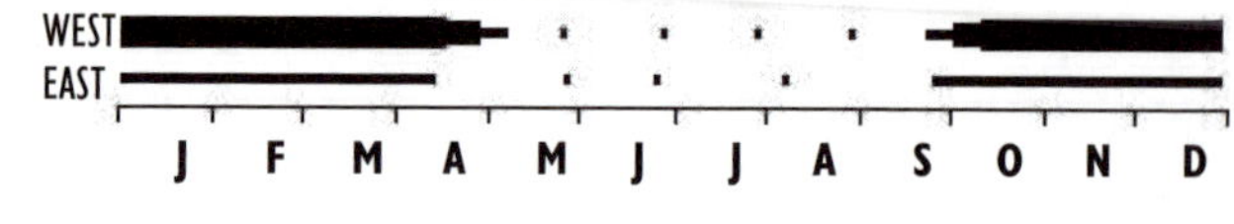

Subspecies: Lesser Snow Goose, *C. c. caerulescens* (see Mowbray et al. 2000).

Habitat: During winter, estuarine intertidal emergent vegetation, primarily three-square bulrush rhizomes. As winter progresses, more upland farmland plants, especially nutrient-rich winter-wheat fields, also pasture and small-grain (e.g. barley) fields nr. saltwater. Large flocks typically feed in upland fields during daylight and roost on saltwater.

Occurrence: Washington's birds originate almost entirely at Wrangel I., at 71° N, in the Chukchi Sea off n. Siberia. Birds from there also winter in B.C.'s Fraser R. delta, Oregon's Willamette Valley, and in California. In Washington, most are in two massive flocks at the Stillaguamish and Skagit R. deltas. A smaller number winter at Ridgefield NWR and adjacent Sauvie I., in Oregon. Birds appear at these locations during late Sep, and numbers continue to build into late Nov. Spring departure mostly during Apr. Away from main wintering areas, Snow Geese are uncommon migrants in w. Washington, during Oct-Nov and in Apr. In e. Washington, they are rare to uncommon migrants, and rare to very rare in winter. Very rare anywhere in the state in summer.

Although the Wrangel I. population reportedly declined since the 1960s (PFSC 1992), the wintering Skagit-Fraser population has increased (see Mowbray et al. 2000) and is much larger than reported by Jewett et al. (1953). The population averaged 29,171 for the period 1948-98. Although the flock declined to a low of 12,346 during 1974-75, it has averaged 41,877 during 1981-90 and

40,298 during 1991-98, with a high of 55,350 in Nov 1987 (PFSC 1992, WDFW 1998). The flock interchanges frequently between the Skagit and Fraser deltas during winter and, infrequently during extremely cold winters, part of the flock has dispersed into sw. Washington. Unlike birds from other wintering areas, Skagit-Fraser geese show red-orange facial feathers stained during feeding in iron-rich sediments. The flock along the Columbia R. in the Ridgefield NWR-Sauvie I. region typically numbers up to 2000. Numbers taken by hunters in Island, Skagit, and Snohomish Cos. ranged from about 500 to 2000 from 1993 to 2000, with 1995 taken in 2000 (WDFW 2001). Noteworthy CBC numbers occur only at Skagit Bay, with averages of about 24,000 there from 1986 to 1999. Birds at Ridgefield NWR are included in a CBC at Sauvie I., Oregon.

The blue phase, a.k.a. Blue Goose, is very rare throughout the state. Of approximately 18 records, 10 have been from e. Washington and most from Oct to Jan. Most were single birds, but 20 were noted with white morph birds over Bellingham on 3 Oct 1997. Westside occurrence appears to have increased in recent years, with seven of eight records since 1995. On two occasions birds appearing intermediate between white and blue morphs were seen. Blue Geese are found only in Canadian populations of Lesser Snow Goose, and some Lessers from the w. Canadian arctic regularly migrate through e. Oregon (Bellrose 1976). It is not unlikely that occasional birds of both morphs occur in Washington, particularly on the eastside.

Remarks: Acreage of secure upland feeding Skagit/Stillaguamish areas has increased in recent years due to establishment of inland reserves on Fir I. The invasion of cordgrass (*Spartina* spp.), however, threatens to reduce the carrying capacity of Skagit/Stillaguamish estuaries due to displacement of native three-square bulrush. Maintenance of the Wrangel I. reserve is also essential to the population. Classed PHS.

Noteworthy Records: *High counts*, **West**: 40,000 on 27 Feb 2001 at Stillaguamish R. delta (SGM, T. Aversa). **East**: 250 in early Oct 1968 at McNary NWR; 150, St. Andrews, 17 Apr 1986. *Spring, late*: 2 on 30-31 May 2001 at Sprague L.; 1-2 in late May 1971 n. of Reardan; 60 on 19 May 1982 nr. Seattle. *Summer*: 31 Jul 1966 at Turnbull NWR; 26 Jun 1987 at Columbia NWR. *Fall, early*: 30 Aug 1999 at Fir I. (SGM). *Blue Goose* - **West**: 1 on 10-15 Nov 1985 at Dungeness; 1 on 19 Oct 1996 at Stanwood; 1 on 23 Nov 1996 at Everett; 1 on 7 Jan 1996 at Post Office L. (SGM); 3 in Nov 2002-16 Mar 20023 at Fir I.; 2 on 27 Mar 2003 at Stanwood; intergrade BlueXSnow through summer 1997 at Everett; 20 with 15 white morph on 3 Oct 1997 going s. at Bellingham (TRW). **East**: 5 in Jan 1968 at McNary NWR; 2 in mid-Jan 1969 at Hanford Is.; 2 in late Oct 1969 at McNary NWR; 2 in Dec. 1969-Jan 1970 at Richland; 2 in late Oct 1970 at Turnbull NWR; 3 on 3 Oct 1971 at Stratford; 2 in fall 1975 at McNary NWR; 1 intergrade on 8 Feb 1976 at Coulee City; 1 in fall 1981 at Richland for the 14th year; 7 on 10 Jan 1983 at Turnbull NWR.

Don Kraege

Ross' Goose *Chen rossii*

Rare spring and very rare fall migrant and winter visitor, increasing in occurrence.

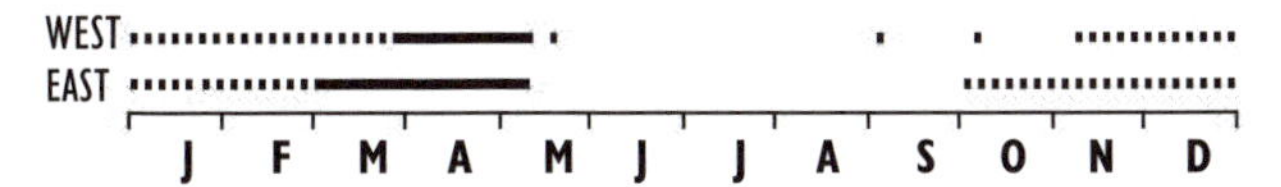

Habitat: Estuarine habitats, upland fields, pastures.

Occurrence: First recorded in Washington in 1950, a spring migrant in the Palouse (Weber and Larrison 1977). A decade elapsed before the second record, one at McNary NWR on 2 Dec 1962. The first westside record was one at Leadbetter Pt. on 18 May 1971. Through 1989, there were only 31 records statewide, less than one per year. From 1990 to 1999, there were an additional 84 records, a dramatic increase. This Washington pattern parallels the strong increase in the world population, which began in the 1950s and was due at least partially to reductions in hunting mortality (Ryder and Alisauskas 1995).

The majority (75%) of the 115 records through 1999 are from spring migration, with the remainder evenly split between the fall and winter. Eastside records comprise 58% of the total, but the number of westside records increased strongly during the 1990-99 period. About 90% of the westside records are from the most recent decade, compared to about 58% of the eastside records.

Remarks: All Washington records to date have been of white-phase birds. A bird at Ridgefield on 28 Mar 2003 "was likely" a blue-morph Snow x Ross' Goose (NAB 57:395).

Noteworthy Records: *High counts*: 60 on 8 Apr 2000 at Rock L., Whitman Co.; 17 on 31 Mar 2001 at McNary NWR.

Bill Tweit

Brant *Branta bernicla*

Fairly common to locally abundant migrant and winter resident in w. Washington marine habitats; locally uncommon in summer. Very rare e.

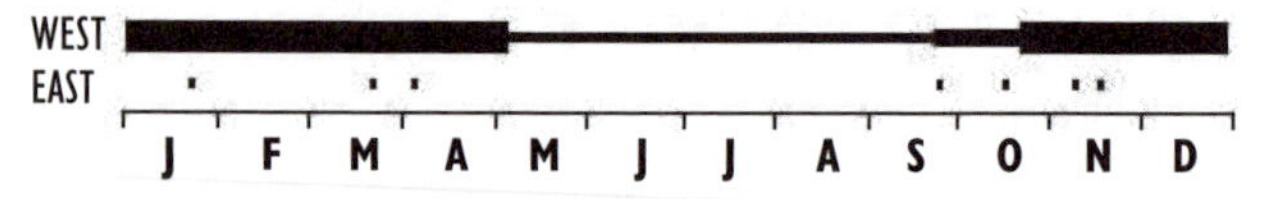

Subspecies: *B. b. nigricans* and "Western High-Arctic Brant" (see Reed et al. 1998). *B. b. nigricans* winters from Alaska to Baja California, including inland marine waters, coastal estuaries, Western High-Arctic Brant locally in Padilla, Samish, and Fidalgo bays.

Habitat: Protected intertidal estuaries with abundant eelgrass and sea lettuce and undisturbed grit-access sites exposed during daylight low tides, raft on open water during other tide stages and evening hours.

Occurrence: Numbers relate to the size of eelgrass beds, now locally reduced. During migration and winter, use the same sites annually, rarely venturing far from saltwater, although vagrants have been sighted nr. Ridgefield and in e. Washington. In Padilla/Samish/Fidalgo bays, primary grit-access areas include Swinomish Spit, Bayview, and Samish Spit. Areas in Willapa Bay including Nahcotta and Leadbetter Pt., at Dungeness Spit, beaches nr. Graysmarsh, Nisqually Reach, and s. Hood Canal. Elimination of grit areas can reduce local usage: after conversion of a gravel spit to a marina at Drayton Harbor, spring numbers declined from 3000-3900 in 1978-79 to just over 200 in the early 1980s (Wahl 1995). Disturbance can also cause abandonment of wintering sites (Einarsen 1965).

Significant numbers arrive in major wintering areas during late Oct-early Nov. In contrast to spring migration, fall migration is more rapid and brant tend to fly more directly to seasonal destinations. Observations in the Padilla Bay area, however, indicate that a number of *nigricans* from various breeding locations migrate through the area in late Nov-early Dec, and may winter there during some years. Highest use sites during winter and migration are n. Puget Trough bays (ca.10,000 birds), primarily Padilla/Samish/Fidalgo bays, also isolated bays n. to the Canadian border. Use areas appear to have expanded in n. Puget Sound since Jewett et al. (1953) described Smith I. and Dungeness as primary wintering sites. Other areas include Willapa Bay/Grays Harbor (ca. 1500 birds), Dungeness (1000), Hood Canal (500), and isolated areas in Puget Sound s. to Olympia.

From color tarsus banding and belly color analysis, it has been shown that from late Dec through Feb brant wintering in Padilla/Samish/Fidalgo bays in most years are almost all Western High-Arctic Brant (Reed et al. 1989). Other Washington wintering flocks are composed of brant from the Canadian low arctic (Banks and Victoria Is. and Mackenzie Delta) and Alaska.

Largest numbers occur during spring migration, as early as Feb and peaking in Apr. High numbers occur throughout traditional wintering locations, but also use many dispersed sites throughout Puget Sound. Most brant leave by the end of May. Small groups reported during summer are undoubtedly subads. (<3 yrs. old) or other non-breeders not returning to arctic molting locations.

Long-term Pacific Flyway population trend is measured annually in early Jan through a coordinated rangewide survey, which includes major wintering areas in Mexico. The long-term index generally declined since the 1960s (see Reed et al. 1998), particularly in coastal estuaries of Oregon and California. The population has been relatively stable recently, at approximately 130,000 (Trost 1998). During Apr of 1975, over 65,000 brant utilized Washington habitats. The last coordinated spring count projected approximately 37,000 brant in late Apr 1993. State populations have declined from the 1936-60 statewide average of 24,000. Since then there have been large annual fluctuations since 1970 with a low of 3000 in 1983 and a high of 20,000 in 1995, averaging 13,000 during the 1990s. Nysewander et al. (2001) indicated a decline from 1978-79 to 1994-95. Winter numbers may in part reflect severe weather in arctic breeding areas.

Remarks: *B. b. hrota* specimens with light belly feathers described by Jewett et al. (1953) and reported occasionally by field observers in n. Puget Sound are likely Western High-Arctic Brant instead of *hrota* (see Reed et al. 1998), which has occurred in California (Small 1994): taxonomic status of Western High-Arctic Brant needs clarification. Interior records are almost entirely from fall and of unknown geographic origin. The quantity of eelgrass in an area is known to influence brant use though trends in wintering habitat, particularly eelgrass beds, are not well documented. Washington hunting has been progressively restricted since 1976, with the season reduced from 49 days in 1980 to 5 days in 1998, and only about 300 birds taken annually from 2000 to 2002. Classed PHS.

Noteworthy Records: West, *high counts, Spring*: 56,602 in late Apr 1975 at Padilla/Samish/Fidalgo bays. *Summer*: 10 on 9 Jul 1972 at Ocean Shores; 17 on 24 Aug 1974 at Ocean Shores; 1 on 22 Jul 1974 at Samish I.; 15 on 28 Jun 1964 at Leadbetter Pt.; 10 in summer 1991 at Dungeness; 4 on 22 Jun 1997 at Sequim; 1 on 3 Jul at Everett; 5 at Ocean Shores, 6 at Westport and 1 at Tokeland on 18 Jul

1998; 3 in 1999 at Ocean Shores; 21 on 12-13 Jul 2000 in Clallam Co. *Inland*: Ridgefield/Woodland: 4 Apr 1996; 2 on 12 Feb 1997; 26-27 Oct 1999; Dec 1999-10 Apr 2000; Dec 2000-11 Apr 2001; 16 Mar 2001; 26 Dec 2001; 24-28 Feb 2002. **East**, *Fall*: 1 on 6 Nov 1977 at Walla Walla; 1 on 15 Nov 1987 at Richland; 3 on 26 Sep 1995 at Stratford L. Winter: 1 on 26 Jan 1994 at Asotin.

Don Kraege

Canada Goose *Branta canadensis*
Cackling Goose *Branta hutchinsonii*

Widespread, abundant year-round resident, migrant and winter visitor throughout Washington.

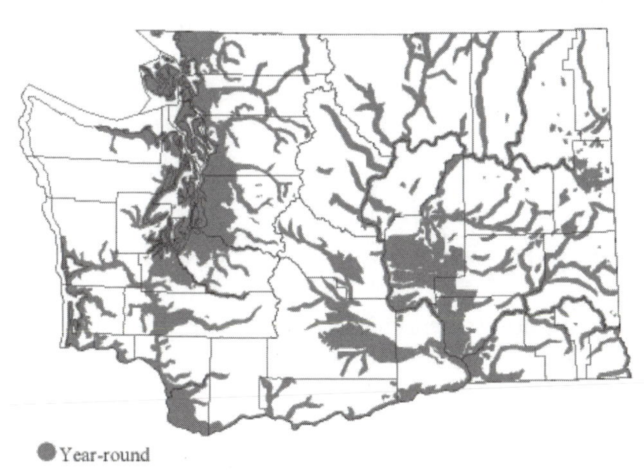

Subspecies: Western Canada Goose *B.c. moffitti*, is a migrant and year-round resident. Taverner's, *taverneri*, Lesser, *parvipes*, Cackling, *hutchinsonii* (*minima*), Dusky, *occidentalis*, Vancouver, *fulva*, and Aleutian Canada Goose, *leucopareia*, are migrants and winter visitors.

Habitat: For breeding, prefers riverine areas and islands of the Columbia R. and tributaries. Also nests in a wide variety of other wetlands, from alpine lakes to suburban/urban parks to lowland livestock ponds. Migration and winter foraging areas typically farmlands and natural wetlands, with preferred crops including small grains, winter wheat, and pasture. Some local populations forage on eelgrass.

Occurrence: Status complicated by the number of subspecies involved, but overall distribution has expanded and populations of breeders and visitors increased greatly statewide. From late spring into early fall, almost all Canada Geese are resident *moffitti*, though small numbers of other subspecies occasionally summer. *Occidentalis* and *maxima* were introduced into sw. Washington in the 1970s, but these have likely interbred with local *moffitti*. In e. Washington, breeding distribution expanded significantly since the early 1900s (Ball et al. 1981) and is much larger than described in Jewett et al. (1953).

Geese nest throughout the Columbia R. system in e. Washington, the Columbia Basin, and channeled scablands, and have expanded into Cascades drainages. Dams constructed in the 1950s and 1960s led to a reduction in goose-nesting islands (Ball et al. 1981), but survey counts increased from about 700 nests in the 1970s to approximately 2500 in the 1990s.

In w. Washington, relatively unknown as breeders (Jewett et al. 1953) until the 1960s. They are now common, with increased exploitation of urban and suburban habitats and natural range expansion downstream along the Fraser and Columbia Rs. Along the Columbia, for example, the number of nests from Longview downstream increased from less than 50 in 1985 to over 400 in 1995. Birds now breed opportunistically in a wide range of situations from urban parks to relatively high elevations (Smith et al. 1997).

Migrants may begin to arrive in late Aug, but the movement usually begins in late Sep and peaks from late Oct to late Nov. Spring migration begins in late Feb, peaks in Mar-Apr, and a few northbound birds can be found into late May.

Eastside migrants and winter visitors include *taverneri*, *parvipes*, and *moffitti*, with *hutchinsonii* uncommon and *leucopareia* a rare vagrant (Johnson et al. 1979). Primary areas include the Columbia Basin and dryland agricultural areas. Special late-Oct aerial surveys during 1983-91 found an average of approximately 115,000 birds, primarily at small lakes of Douglas, Grant, and Lincoln Cos. Typically 30,000-40,000 are also recorded during Nov aerial surveys in the Columbia Basin and Yakima Valley. Later, many migrants using n. areas eventually move to Columbia R. wintering areas from Ringold to Paterson, and along the lower Snake and Yakima Rs., depending on weather and food availability. Winter counts increased significantly following development of the Columbia Basin Irrigation Project in the late 1950s and early 1960s, and have averaged approximately 50,000 since then.

In w. Washington, all seven subspecies occur during migration and winter. Migrants are mostly found along the outer coast. Resident *moffitti* are common throughout during winter (some have been recovered in Oregon, California, Idaho, and other locations). Non-breeding subspecies winter mostly along the Columbia R. from Vancouver to Ilwaco, and to a lesser extent Willapa Bay/Gray's Harbor. Sw. winter counts increased from approximately 2000 in the early 1960s to approximately 25,000 in the 1990s. These are mainly *hutchinsonii, taverneri, parvipes*, and *moffitti*. A small flock of *fulva*

Table 7. Canada Goose: CBC average counts by decade.

	1960s	1970s	1980s	1990s
West				
Bellingham		37.1	291.7	1182.4
Cowlitz-Columbia			1352.6	1584.6
Edmonds			166.3	355.2
E. Lk. Washington			238.0	1051.2
Grays Harbor		44.1	156.4	942.9
Kent-Auburn			261.4	1039.5
Kitsap Co.		0.7	45.3	242.6
Leadbetter Pt		51.4	158.4	228.4
Olympia			296.3	1112.5
Padilla Bay			78.6	56.5
San Juan Is.			15.9	375.8
Seattle	21.2	219.4	778.3	1216.1
Sequim-Dungeness			196.7	450.9
Tacoma		96.8	391.4	1459.4
East				
Ellensburg			131.5	330.0
Spokane	15.6	91.3	267.7	748.7
Toppenish			191.6	1219.8
Tri-Cities		3547.3	3604.1	6867.9
Walla Walla		25.7	171.2	285.2
Wenatchee	67.2	199.4	505.1	1213.5
Yakima Valley		27.5	192.4	731.7

wintered at Port Susan Bay in recent years (SGM). *Parvipes* are most often seen during migration in Oct and Nov on the Long Beach Pen. and s. Willapa Bay, and almost all continue s. as far as the Central Valley of California. Marking studies of *occidentalis* and *hutchinsonii* have shown occasional use of Puget Sound areas during fall migration (e.g. Dungeness, Skagit Flats), but the majority are thought to follow a coastal route. Small groups of *occidentalis* are known to winter as far n. as the Elma area and s. Lewis Co. *Occidentalis* was the most common subspecies encountered in se. Washington, but has declined in numbers and in relative proportion of the wintering population since 1970. This was due to increases in *moffitti*, and beginning in the 1980s, lingering *hutchinsonii* which had previously wintered in California: *occidentalis* no longer predominates in the area. Much of the increase in numbers is due to additional available habitat (e.g., the Columbia Basin project and associated agriculture), shifts of geese from more southerly wintering areas, and effects of human actions of many kinds.

Hunting harvests and seasons increased with expanding populations, except that of *occidentalis* which is very restricted while numbers in Washington and Oregon increased to about 17,000, and *leucopareia* which is classed as threatened and protected in both states.

Statewide counts recorded each year during the U.S.-wide multi-agency cooperative winter waterfowl survey increased overall from about 40,000 in the 1950s to approximately 80,000 in the 1990s. Winter numbers represented by CBCs with consistent coverage indicate large increases, particularly in urban areas. Statewide, counts had significant increases (P <0.01). Averages by decade also indicated huge increases statewide (Table 7). Identification of subspecies is generally not possible during aerial surveys, and seldom recorded on CBCs. Summer population increases statewide are shown by BBS data—a significant increase of 9.7% per year from 1982 to 1991 (Peterjohn 1991).

Remarks: The Cackling Goose *B. hutchinsonii* (noted previously as *minimia*) is now designated a distinct species (AOU 2004). Determination of subspecies status and distribution is complicated by identification issues. Increasing numbers in many urban and agricultural areas are now regarded as a nuisance. Several control efforts were initiated to reduce numbers of resident and migrant geese during the 1990s (PFC 1998). Classed PHS; with the Aleutian Canada Goose classed ST, FSC.

Don Kraege

Trumpeter Swan *Cygnus buccinator*

Uncommon to rare migrant and winter resident in w. Washington, but locally common in lower river valleys from Snohomish Co. n. Locally uncommon to rare migrant and rare summer resident in e. Washington.

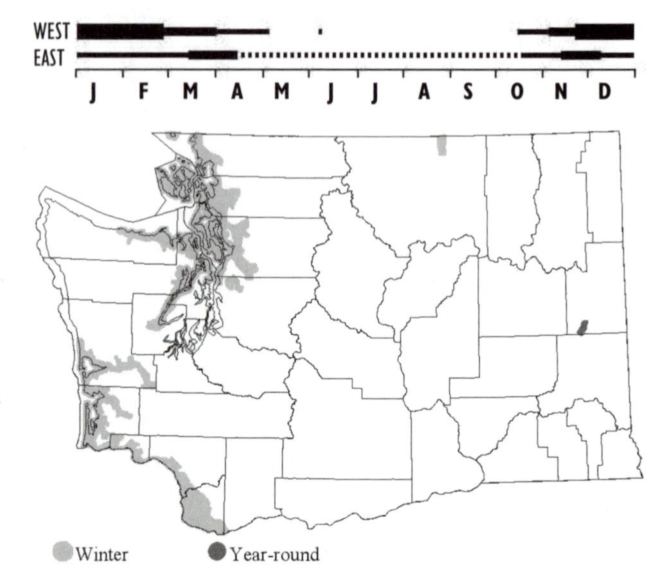

Habitat: Wetlands, fields in lowland river bottoms in winter. Areas temporarily flooded by snowmelt and spring rains attractive in spring. Permanent marshes with desirable vegetation like pondweeds, roots, and tubers selected in spring or fall. Move readily from freshwater to salt-water estuaries during severe freezes.

Occurrence: This species is much more widely distributed than in the recent past; the Trumpeter Swan's story includes a decline almost to extinction and a dramatic recovery over the last half of the 20th century. Jewett et al (1953) described historical abundance, but stated "no records in recent years." Discovery of a winter population nr. Clear L., in Skagit Co., in the late 1960s indicated the start of recovery in Washington.

The Pacific population breeds in central Alaska, winters primarily in coastal Washington and B.C., and has been increasing steadily over the last 20 years. A large segment of the Alaskan breeding population winters in the valleys and estuaries of n. Puget Sound, mainly between Lynden and Snohomish, but birds can be found on freshwater lakes on the Olympic Pen., at Ocean Shores, and Willapa Bay and in the Chehalis R.Valley. Although apparently increasing in the 1990s, they are relatively rare in the lower Columbia areas where they apparently abounded a century previous (Jewett et al. 1953). Birds arrive in n. Puget Sound estuaries and the coast by the end of Oct into mid-Nov. They begin leaving in Mar and are essentially gone by Apr.

Much less common in e. Washington. Small numbers migrate through e. Washington, and are periodically observed on Alkali and Lenore Lks. in the Columbia Basin and at Umatilla and McNary NWRs in spring. The Yakima Valley occasionally supports a few migrating Trumpeters and birds are periodically recorded locally along the Columbia R.

It is unclear if Trumpeter Swans bred in Washington historically. Past attempts to establish a breeding population on Turnbull NWR in the 1960s-70s failed; one bird remains through the winter there. Attempts to broaden the breeding and wintering range of the Red Rks. L. population by capturing and moving birds have resulted in repeated sightings of marked individuals during summer in the Columbia Basin.

Annual midwinter surveys record an average of 1245 Trumpeter Swans in the state. Comprehensive WDFW five-year swan surveys in w. Washington recorded 1000-3000 Trumpeter Swans. CBC totals from 1996 to 1999 average about 1600 birds per year statewide and do not include large numbers wintering outside CBC circles (e.g., in Whatcom Co.) or any of hundreds of unidentified swans counted within circles, suggesting swan survey numbers are conservative. Long-term CBC averages at major wintering locations in nw. Washington (Table 8) illustrate increases by decade.

Table 8. Trumpeter Swan: CBC average counts by decade. (Numbers at Bellingham and Skagit Bay are likely understated due to occurrence of large numbers of unidentified swans there.)

	1970s	*1980s*	*1990s*
Bellingham	0.9	81.7	257.4
Padilla Bay		7.4	181.4
Skagit Bay*		100.0	461.9
San Juan Is.		38.6	41.4

** Skagit Bay had only four CBCs in the 1980s.*

Trumpeter Swan and Tundra Swan (G. Scott Mills)

Remarks: Distinguishing Trumpeters from Tundra Swans requires care and some records may include both. Swans have strong site fidelity to migration stop-over areas and wintering areas. This emphasizes the importance of factors like the decrease in area of agricultural lands and changes from dairy and small grains to hybrid poplar orchards in the n. Puget Sound which reduce winter habitat. At least 868 swans, mostly Trumpeters, died from ingesting lead shot in Whatcom Co. and Sumas Prairie, B.C., during winters of 1999-2003. State and federal agencies, as well as non-government organizations and individuals, have been searching for the source of this problem through large-scale telemetry, color marking, and survey efforts. Classed PHS.

Noteworthy Records: *Summer*: 2 on 11 Jun-12 Jul 1977 nr. Elma; 1 on 17 Jun 1991 at Calispell L., Pend Oreille Co. *CBC high counts*: 1570 in 1996, 976 in 1993, 649 in 1999 at Skagit Bay; 965 in 1998 at Padilla Bay; 588 in 1998, 522 in 2000, 512 in 1995 at Bellingham.

Matthew J. Monda

Tundra Swan *Cygnus columbianus*

Locally common migrant in n. Puget Sound, Columbia Basin, and Columbia R. estuary. Locally uncommon to common winter resident. Very rare in summer.

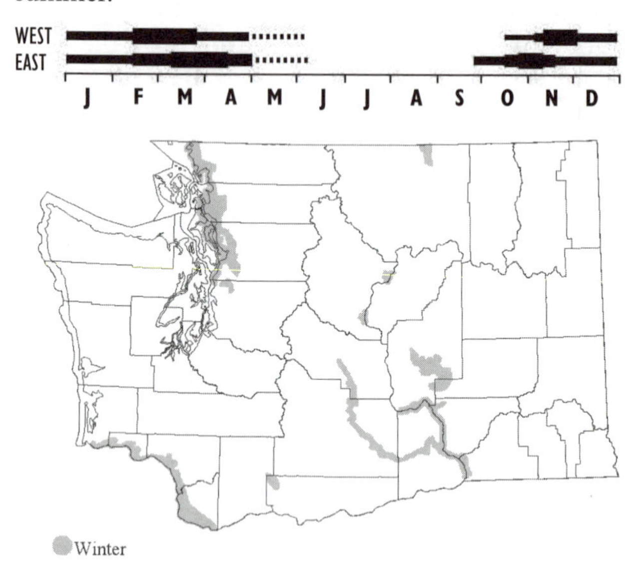

Subspecies: *C. c. columbianus.*

Habitat: Forages on aquatic vegetation, roots, and tubers from pond bottoms, in agricultural feeds, most commonly when sheet water floods part of the fields. Migrants regularly stop over at suitable permanent marshes and fields temporarily flooded by snowmelt and spring rains. Water bodies with pondweeds are selected.

Occurrence: The w. population of Tundra Swans nests in the lowlands of w. Alaska and winters primarily in California, Utah, and the Pacific Northwest. This population is currently averages over 100,000 birds and has been increasing at about 6% per year since 1990 (USFWS 2000c, Waterfowl Population Status Rep.).

A common wintering species in ne. Puget Sound and Columbia R. estuaries. The annual Jan midwinter USFWS waterfowl survey recorded an average of 57,307 in the 11 Pacific Flyway states between 1989 and 1999. Populations have been increasing, peaking at 122,521 in 1997. Depending on weather conditions, the number statewide varied from 900 to 3400, with most of these in n. Puget Sound estuaries. Comprehensive WDFW swan surveys in w. Washington every five years recorded totals of 2000-4000 Tundras. Some 500-1000 winter at Ridgefield NWR (SGM). The Skagit Bay CBC regularly records large numbers with a maximum of 1343 in 1991. Other CBCs that regularly record the species are Bellingham, Cowlitz-Columbia, Everett, and Padilla Bay (see Table 9). Birds are uncommon in winter in e. Washington. Regularly found at Turnbull, Columbia, and McNary NWRs, at Lenice L., Grimes L., and the Pend Oreille R. during mild winters, and other locations with adequate habitat that remain ice free.

Table 9. Tundra Swan: CBC average counts by decade.

	1970s	*1980s*	*1990s*
Bellingham	29.8	105.3	61.1
Cowlitz-Columbia		111.4	107.8
Padilla Bay		10.2	21.0
Skagit Bay*		292.5	578.9

** Skagit Bay had only four CBCs in the 1980s.*

During migration large numbers occur in intermittent wetlands associated with agricultural fields and large marshes across Washington. Large numbers occur nearly anywhere with adequate habitat w. of the Cascades, and are regular in the estuaries in n. Puget Sound (e.g., Skagit Flats, Nooksack delta, Snohomish estuary) and coastal estuaries (e.g. Willapa Bay, Grays Harbor, and the lower Columbia R.). More Tundra Swans migrate through e. of the Cascades in spring than fall, peaking from late Feb through Apr in n.c. Washington and the Columbia Basin. Calispell L. nr. Usk and flood waters along Crab Cr. between Wilson Cr. and Marlin attract large numbers (>1000) in the spring. Single non-breeders are occasionally found over-summering statewide.

Remarks: Formerly the Whistling Swan. Classed PHS by Washington State. Mortality due to lead shot ingestion noted in the late 1990s in Whatcom Co. and adjacent B.C. Bewick's Swan (*C. bewickii*), considered by various authorities to be a subspecies or a separate species, has been reported at least three times in Washington. Differentiation between the two forms based on bill coloration requires caution.

Noteworthy Records: Bewick's Swan: 9-21 Dec. 1984 at Lummi Flats; 30 Mar 1986 at Jameson L., 1 Apr 2002 at Dodson Rd.

Matthew J. Monda

Wood Duck *Aix sponsa*

Fairly common to common summer resident in w. lowlands, uncommon in e. Uncommon in winter w. Uncommon in migration, locally rare to fairly common in e. in winter.

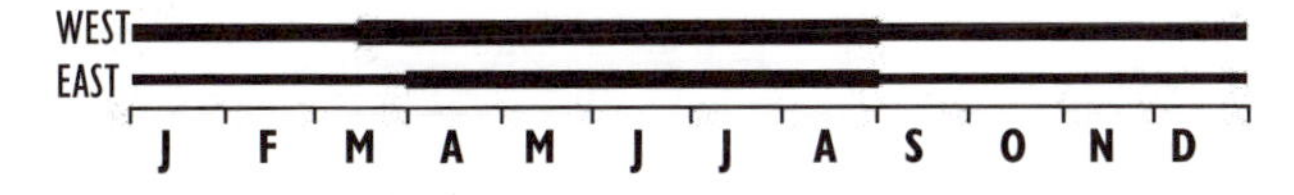

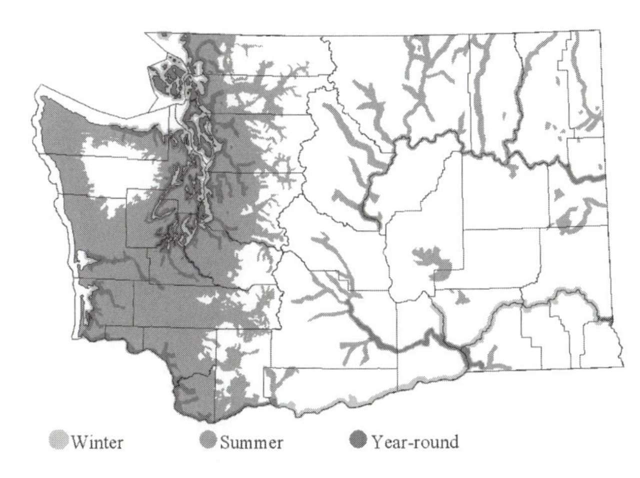

Habitat: Ponds, sloughs, swamps, and sewage lagoons, generally with overhanging brush. Loss of dead trees and snags with nest cavities compensated for increasingly by nest boxes.

Occurrence: Change in status over time appears entirely due to human influences. First threatened with extinction, completely protected from hunting from 1918 to 1941 with passage of Migratory Bird Treaty Act, and hunting limited since then (Jewett et al. 1953). Numbers increased due to habitat preservation and nest-box programs by agencies, conservation groups, and private landowners in recent decades. Through 1970s observers judged individual nests, broods, or small flocks noteworthy; reports in the 1980s began describing seasonal concentrations. Now found locally through lowlands statewide, locally in Columbia Basin and se. corner, with status at moderate elevations imperfectly known (Smith et al. 1997).

Nest boxes used in the Columbia Basin, including Columbia NWR, Desert WMA, and parks in Clarkston (Smith et al. 1997), where habitat was essentially nonexistent historically. In fall 1988 a count of 75 birds along the Snake R. nr. the Tri-Cities was attributed to good response by birds to a nest-box project and reports of summer and fall numbers increased, noticeably in the e. In B.C., breeding populations also increased dramatically following program to provide additional nest sites (Campbell et al. 1990a).

Table 10. Wood Duck: CBC average counts per 100 party hours by decade.

	1960s	*1970s*	*1980s*	*1990s*
West				
Bellingham		0.24	1.36	9.06
Edmonds			1.20	7.48
Kent-Auburn			1.99	21.21
Kitsap Co.		1.87	1.46	4.41
Olympia			1.29	9.17
Seattle	0.57	1.38	7.64	14.67
Sequim-Dungeness			5.25	31.88
Tacoma		0.88	3.95	18.78
East				
Spokane	3.23	2.64	9.72	8.92
Toppenish			4.59	13.13
Tri-Cities		0.00	20.69	153.25
Yakima Valley		0.00	3.23	101.25

Present all year, but abundance varies seasonally, regionally, and locally. Jewett et al. (1953) stated that scattered records indicate a few winter e. and w., and that the bulk of the population evidently left for winter. Now recorded all year, more localized in winter (e.g. numbers in the San Juan Is. much reduced in winter [Lewis and Sharpe 1987]) while in King Co. peak numbers are from Oct to Apr (Hunn 1982), concentrating in suitable urbanized areas and parks.

Long-term CBC data (Table 10) indicate winter numbers increased starting in the mid-1980s ($P < 0.01$), with relatively consistent trends noted especially in urban areas. In the w., winter concentrations of 20-70 birds were found in the 1990s at each of five different ponds in Whatcom Co., which has the coldest lowland winter climate w. of the Cascades (Wahl 1995). As in sw. B.C. (Campbell et al. 1990a), numbers of overwintering birds have increased by taking advantage of artificial feeding stations.

Remarks: Maintenance of nest boxes essential for continued reproductive success (Campbell et al. 1990a). Classed PHS.

Noteworthy Records: West, *Fall*: 85 on 26 Jul 1997 at Washougal; 90 on 3 Aug 1999 at Ridgefield NWR. *CBC high counts*: 95 in 1994 at Sequim-Dungeness, 70 in 1991 at Tacoma, 68 in 1991 at Seattle, 61 in 1992 at Kent-Auburn, 58 in 1996 at Olympia. **East**, *Summer*: 50 on 27 Jun at McNary NWR and 50 on 25 Jul 1998. *Winter*: 50 on 1 Mar 1999 at Sacajawea SP. *CBC high counts*: 236 in 1999, 232 in 1995, 179 in 1997, and 117 in 1998 at Tri-Cities; 147 in 1999, 130 in 1997 and 120 in 1998 at Yakima Valley.

Terence R. Wahl

Gadwall *Anas strepera*

Widespread, common to locally abundant breeder in e. Washington, uncommon breeder in w. Common to abundant winter resident and migrant statewide. Increasing.

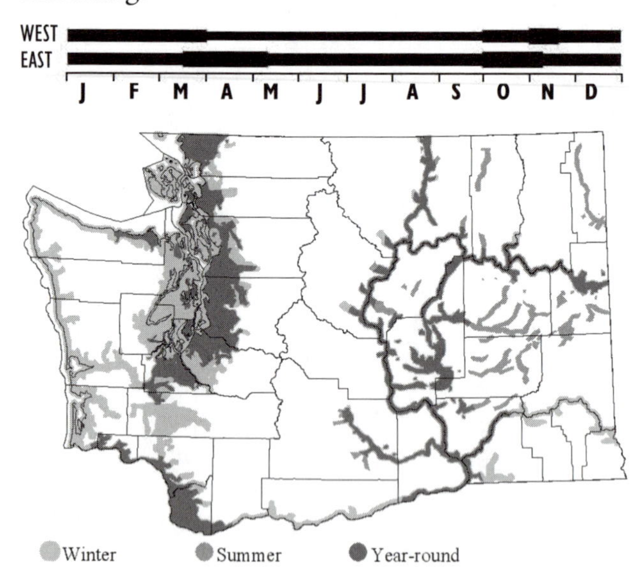

Habitat: Freshwater, wetlands with dense stands of aquatic vegetation or filamentous algae, less frequently on saltwater or in estuaries. Unlike other dabbling ducks, does not frequent agricultural fields.

Occurrence: Much more widespread both in summer and in winter than in the 1950s, when it was a summer resident on lakes in the Big Bend region and an uncommon migrant and casual winter visitor w. of the Cascades (Jewett et al. 1953; see Leschack et al. 1997). Now a common breeder in e. Washington in lowlands of the Columbia Basin, with smaller numbers in river valley habitats in ne. counties. Summer breeding numbers w. of the Cascades noticeably increased in the mid-1960s, especially concentrating in the urbanized s. Puget Trough, with first nesting noted also at Vancouver, B.C., and in Clark Co. (Canning and Herman 1983). Habitat changes, including the increase in a favored food, Eurasian milfoil, were advanced as reasons for major population increases (Smith et al. 1997).

Breeding duck surveys record an average of 11,300 Gadwalls in e. Washington and about 1000 on the westside. Highest densities of breeders occur in s.c. Okanogan, Grant, n. Douglas, n.c. Lincoln, nw. Whitman and sw. Spokane Cos. (WDFW 2000a). In Snohomish Co., Mlodinow reported 100 birds per day in summer, with a breeding population almost as large as that of the Mallard.

Winter numbers have also increased statewide. Fall and winter concentrations of up to 20,000 occur in the n. Columbia Basin (USFWS 2000b). Jan waterfowl surveys average about 5000 Gadwall in Washington with 19% of these w. of the Cascades (USFWS 2000b). Large numbers can be found during winter in the Columbia Basin in localized areas (e.g. Tri-Cities) that remain ice free and have adequate foods. Trends on counts in s. c. Puget Sound are particularly apparent, and birds occur widely in lowland river valleys and on lakes. Long-term CBC data (Table 11) indicate increases. Continent-wide, the N. American population has been building since the mid 1980s and is at 3,158,000, the highest level ever recorded since N. American breeding-duck surveys were initiated in 1955 (USFWS 2000c).

Table 11. Gadwall: CBC average counts by decade.

	1960s	*1970s*	*1980s*	*1990s*
West				
Bellingham		3.6	12.8	51.3
Cowlitz-Columbia			0.9	24.2
Edmonds			87.3	136.6
E. Lk. Washington			36.2	61.2
Grays Harbor		13.9	61.4	63.9
Kent-Auburn			90.5	122.6
Olympia			88.9	132.1
Padilla Bay			18.8	29.0
Port Townsend			3.9	42.2
San Juan Is.			15.4	72.7
Seattle	25.1	265.9	997.3	1339.7
Sequim-Dungeness			9.7	26.0
Tacoma		19.1	59.4	99.9
East				
Spokane	10.1	6.1	22.6	24.2
Toppenish			1.3	23.8
Tri-Cities		113.6	424.4	257.0
Wenatchee	1.0	0.0	26.4	61.2

Noteworthy Records: *High counts,* **West**, *Summer*: 68 on 3 Jun 2000 at Dungeness. **East**: 400 in fall 1970 at Turnbull NWR; 1462 in spring 1984 at Turnbull NWR and 350 in spring 1984 at Columbia NWR; 400 on 4 Jan 1987 at Moses L.; 470 in summer 1987 at Turnbull NWR. CBC high counts: 3089 in 1994 at Seattle; 1397 in 1989 at Moses L.; 905 in 1983 at Tri-Cities; 663 in 1992 at Edmonds; 402 in 1999 at Everett.

Matthew J. Monda

Falcated Duck *Anas falcata*

Casual vagrant.

Breeding in ne. Asia, the Falcated Duck has been documented three times in Washington. The first record, a specimen from the Naselle R., Pacific Co., on 3 Jan 1979 (Tweit and Paulson 1994) was followed 14 years later by sight record of a male at Dungeness on 3 Jul 1993 (Aanerud and Mattocks 2000) and then by a male in the Samish flats 21 Feb-26 Mar 2002 (NAB: 349). A report of a male from Willapa Bay, nr. Nahcotta, Pacific Co., on 27 Oct 1992 (AB 47:140) was rejected by the WBRC (Tweit and Paulson 1994) for lack of sufficient detail.

Bill Tweit

Eurasian Wigeon *Anas penelope*

Locally fairly common winter resident in w. Washington. Rare winter resident in e.

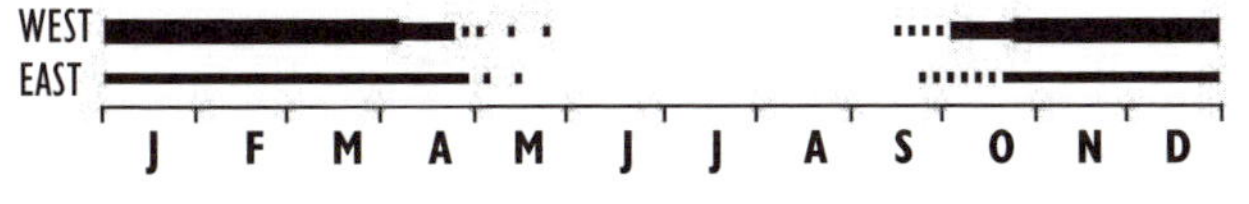

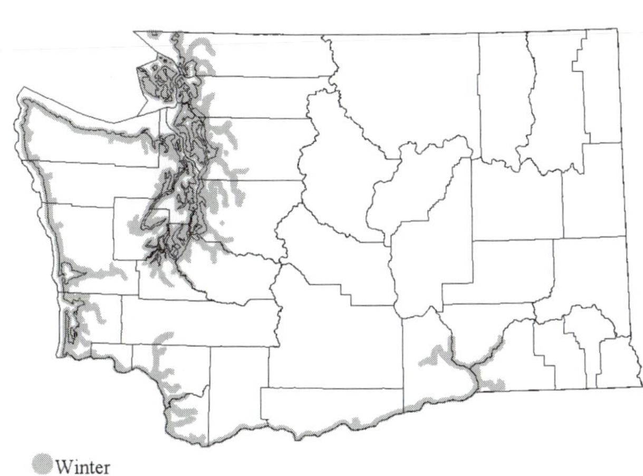

Habitat: Shallow eelgrass embayments, estuaries, lakes, and open, wet agricultural fields.

Occurrence: Though Eurasian Wigeon have not been found nesting in N. America, they are common locally in winter in Washington. This has not always been so—Jewett et al (1953) list Eurasian Wigeon as a "casual autumn and winter visitant in the Puget Sound country." Relatively low numbers were noted throughout w. N. America until 1965, when the population increased sharply and continued to do so at least into the early 1980s (Edgell 1984). The reason for this increase is unclear, but it certainly has been apparent in Washington.

On the westside, most numerous from early Feb to early Mar, with largest numbers in the Puget Trough. The most consistently high numbers have been noted on Padilla Bay and the Samish Flats, with 59 birds reported on the CBC in 1996. Other notable CBC totals have been at Kitsap Co. and Sequim-Dungeness. E. of the Cascades, numbers peak from late Dec to mid-Apr, and most records come from Spokane Co. and along the Columbia R. between Klickitat and Walla Walla Cos. Typically, five to 15 Eurasian Wigeon are seen annually in e. Washington.

Eurasian Wigeon are almost always with American Wigeon. Based on local field counts, the ratio of American to Eurasian wigeon in w. Washington is about 400:1 on average. Hybrids between Eurasian and American wigeon are not infrequent, with a pure Eurasian to hybrid ratio of about 20:1.

Though an overall population increase is apparent, CBC numbers on long-term counts show inconsistent trends. Sequim-Dungeness and Seattle (both $P < 0.01$) and Tacoma ($P < 0.05$) showed significant increases, and Bellingham, Kitsap Co., and Padilla Bay apparently increased. E. Lk. Washington numbers declined ($P < 0.05$), and declines were apparent at Edmonds, Grays Harbor, and Kent-Auburn.

Noteworthy Records: West, *Winter, high counts*: Samish Flats: 160 on 1 Mar 2002; 78 on 7 Feb 1992; 60 on 17 Jan 1994; 60 on 15 Mar 1999. *Fall, high count*: 22 on 31 Oct 1999 at Dungeness Bay. *Early*: 13 Sep 1995 at Longview; 16 Sep 1996 at Crockett L.; 17 Sep 1997 at Westport. *Spring, late*: 11 May 1974 at Ocean Shores; 11 May 1994 at Washougal; 27 May 1989 at Auburn. **East**, *high counts*: 5 on 28 Mar 2001 at Columbia NWR; 4 on 15 Mar 1986 nr. Reardan; 4 on 7 Mar 1995 at Spokane; 4 on 16 Mar 1996 at Spokane; 4 on 27 Feb 1999 nr. Wallula. *Fall, early*: 24 Sep 1995 at Wells Dam; 25 Sep 1994 at Richland. *Spring, late*: 11 May 1991 in se. Douglas Co.; 20 May 1997 at Reardan; 22 May 2002 at Richland.

Steven G. Mlodinow

American Wigeon *Anas americana*

Abundant winter resident and migrant in Puget Trough, outer coast, and along the Columbia R., common in e. Washington. Widespread, uncommon to locally common breeder in e. Washington, rare breeder in w.

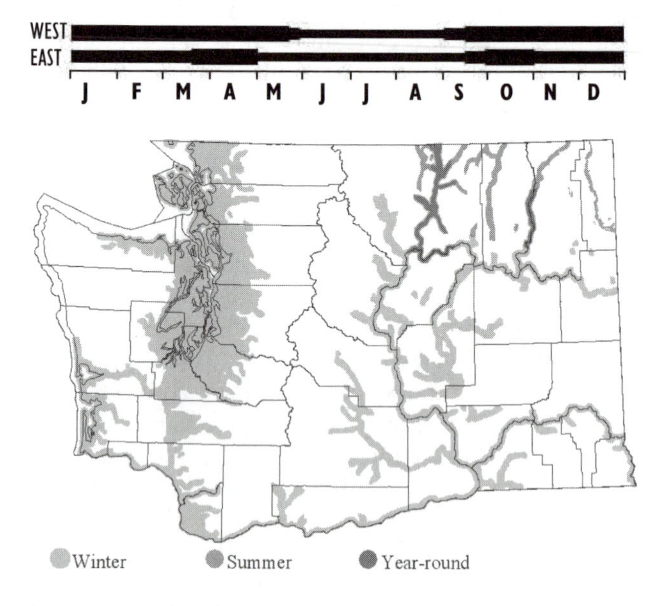

Habitat: Wetlands with dense stands of aquatic vegetation, pastures, and park lawns, lakes; also marine estuaries with eelgrass, flooded pastures, and fields in winter.

Occurrence: Breeding in N. America from arctic tundra to c. U.S., wigeon are abundant in winter in Washington. Large numbers winter on Puget Sound, coastal estuaries, and lowland river valleys. Most of the diet is vegetation, with eelgrass beds important, and large flocks are often noted grazing also in parks and pastures. Local distribution in winter may depend on inland freshwater freezes, and hunting seasons and hours. Numbers reportedly increase in the San Juan Is. when hunting starts at mainland areas (Lewis and Sharpe 1987).

Aerial surveys in n. Puget Sound can record 50,000 wigeon during fall and winter. Annual Jan waterfowl surveys average 112,000 in Washington, with 84% w. of the Cascades (USFWS 2000b). Birds winter in e. Washington where waters remain open and aquatic vegetation or lawns are available, with 40,000 recorded in the Columbia Basin. Monthly fall and winter aerial surveys in the n. Columbia Basin record 10,000 to 21,000 wigeon. Most of these birds are intermixed with coots along the Columbia R. and associated with beds of aquatic vegetation.

Breeds primarily in ne. counties, though not in the Methow Valley. A rare breeder in the Columbia Basin, rare to absent in the Palouse and se. Eastside surveys record an average of 6300 wigeon, with highest densities in n. Douglas, s.c. Okanogan, and n.c. Lincoln Cos. (WDFW 2000a). Very rare and local in w. Washington—not often recorded on breeding-duck surveys. Broods have been recorded on the Everett sewage ponds, in King Co., and birds possibly nest elsewhere in the Puget Trough and Clallam Co. (Smith et al. 1997).

The N. American population has recovered slowly after drought conditions of the 1980s. Numbers increased, but at a much smaller rate than most other dabbling ducks. The population is very nr. the long-term average of 2,647,000 (USFWS 2000a), though a report of a w. Washington flock of 500,000, for example, reported by Dawson in 1908 (in Jewett et al. 1953), suggests historic numbers in Washington may have been higher. Apparent increases in average CBC counts may in part reflect mild weather in the 1990s, particularly in e. Washington. About 68,000 birds (14% of all ducks) taken by hunters in 1999-2000 were wigeon (WDFW 2000b).

Remarks: Local numbers declined when dogs and recreationists took over traditional wintering sites, as at Bellingham and Sidney, B.C. (TRW). Birds banded nesting in Alberta were recovered in w. Washington, but not in the e. (Jewett et al. 1953). Classed PHS.

Noteworthy Records: *CBC high counts*: 9748 in 1981 at Bellingham, 6552 in 1996 at E. Lk. Washington, 19,375 at Edmonds, 6512 in 1981 at Everett, 7230 in 1985 at Grays Harbor, 6283 in 1980 at Kent-Auburn, 6229 in 1987 at Kitsap Co., 6176 in 1986 at Leadbetter Pt., 10,083 in 1987 at Olympia, 33,905 in 1998 at Padilla Bay, 10,752 in 1996 at Skagit Bay, 13,039 in 1994 at Sequim-Dungeness, 11,414 in 1978 at Tacoma, 13,560 in 1999 at Two Rivers.

Matthew J. Monda

Mallard *Anas platyrhynchos*

Widespread, common to locally abundant breeder, abundant winter resident and migrant, in most wetland and inshore marine habitats.

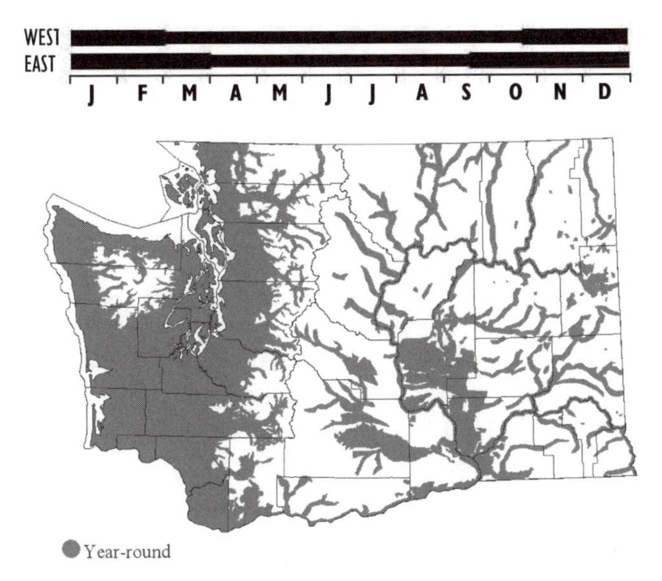

Subspecies: *A. p. platyrhynchos.*

Habitat: Lowland and mid-elevation wetlands, lake shores, nearshore marine waters. Nests in many upland habitats including irrigated agricultural areas, urban saltwater industrial waterfronts, shorelines with emergent vegetation, and in pristine marshes. Winters on freshwater lakes, estuarine areas, moving between foraging fields and offshore resting areas during hunting season.

Occurrence: The most common breeding duck in Washington and, like Canada Geese and gulls, has adapted better than almost any other native species to human populations, growth, and development. The origins of large winter populations are uncertain but presumably from interior N. America n. of Washington. The Columbia Basin supports large numbers. Westside birds shift to saltwater during severe inland freezes, but large numbers are present on marine waters all winter, with highest numbers present on Drayton Harbor, for example, from Aug to Feb (Wahl 1995). Large resting flocks occur on many embayments especially during hunting hours in season, moving inland to feed at night. Birds feed on a diversity of aquatic and terrestrial foods, including aquatic invertebrates, waste grain and corn stubble in agricultural fields, also potatoes and truck crops. Locally urban opportunists like Canada Geese, small numbers of Mallards nest on saltwater, often in urban waterfront situations.

The Columbia Basin is the most important wintering area for Mallards in the Pacific Flyway. Fall and winter concentrations of 100,000-500,000 occur in Columbia Basin, Yakima Valley, and n. Puget Sound. Annual Jan waterfowl surveys average 535,000 Mallards in Washington, with 28% w. of the Cascades (USFWS. 2000b).

Breeding-duck surveys record an average of 55,600 Mallards in e. Washington, with highest densities in Grant, n. Douglas, w. Okanogan, n.c. Lincoln, nw. Whitman, and sw. Spokane Cos. About 20,000 Mallards are recorded in w. Washington (WDFW 2000a). Monthly fall and winter aerial surveys in the n. Columbia Basin record 50,000 to 300,000 Mallards and surveys in n. Puget Sound can record 200,000 birds during fall migration and winter. An example of local concentrations: 40,000 at Padilla Bay on 9 Jan 1996 (SGM). About 270,000 Mallards were taken by hunters in Washington in 1999-2000, about 56% of all ducks harvested (WDFW 2000b).

The N. American Mallard population has been building since the drought of the 1980s and is at 9,470,000, nearly a record high (USFWS 2000c). Numbers recorded on CBCs are affected by weather, observation conditions, local distribution shifts due to hunting, and large numbers of unidentified ducks at many locations, and overall trends are uncertain. Classed PHS.

Matthew J. Monda

Blue-winged Teal *Anas discors*

Locally fairly common breeder and summer resident in e. Washington. Uncommon breeder w. Early migrant, rare in fall. Very rare in winter.

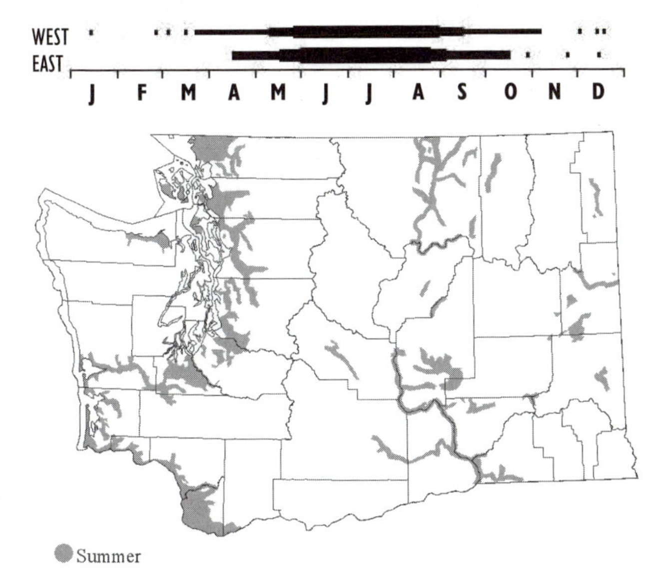

Subspecies: *A. d. discors.*

Habitat: Attracted to small shallow wetlands. Forages on seeds of wetland plants and invertebrates. Ponds with dense aquatic vegetation and algal mats nr. the surface are heavily used.

Occurrence: A neotropical migrant described as accidental in Washington in 1900 (Smith et al. 1997). Jewett et al. (1953) listed no winter occurrence and only a few reports w. of the Cascades where year-round reports were noteworthy by 1972. Now a fairly common breeder on freshwater wetlands and ponds statewide, occurring in most steppe zones along river valleys in the e., limited to the Puget Trough and coastal locations in the w. (Smith et al. 1997).

Breeding-duck surveys in e. Washington record an average of 7,900, with highest densities in w. s.c. Okanogan, Grant, n. Douglas, n.c. Lincoln, nw. Whitman, and sw. Spokane Cos. Surveys in w. Washington record about 500 (WDFW 2000a). Many birds w. of the Cascades in summer are non-breeders (Smith et al. 1997). Though this species is now more common as a breeder than the Green-winged Teal, BBA surveys did not find breeders in a number of counties e. or w. of the Cascades (see Smith et al. 1997).

Late migration and arrival and early departure from breeding grounds may confuse local status reports in the spring and, after Jul, identification problems confuse it with the Cinnamon Teal. Birds are rarely recorded in the annual Jan waterfowl surveys (USFWS 2000b). Though CBC data for this species are questionable, birds were reported 38 times from 1961 to 1999: 27 of these in the w. and 11 e. As in migration, birds may be seen on saltwater in winter (Wahl 1995).

Washington populations increased from the late 1940s to early 1960s, roughly coinciding with construction of dams and expansion of irrigation and agriculture in the Columbia Basin. Following resulting habitat changes in emergent vegetation Blue-wings became more numerous and outnumber Cinnamon Teal by 2:1 in some parts of e. Washington (see Connelly 1978, Smith et al. 1997).

Continent-wide, the population has been building since the early 1990s and is currently at 7,431,000, the highest level ever recorded since N. American breeding-duck surveys were initiated in 1955 (USFWS 2000c).

Noteworthy Records: West, *Summer*: 46 on 4 Jun 1998 at Stanwood; 45 on 27 Jun 2002 at Everett STP (SGM). **East**, *Winter*: 2 on Tri-Cities CBC in 1970.

Matthew J. Monda

Cinnamon Teal *Anas cyanoptera*

Locally common breeder and summer resident in e. Washington; local, uncommon breeder in w. Early migrant, rare in fall, very rare in winter.

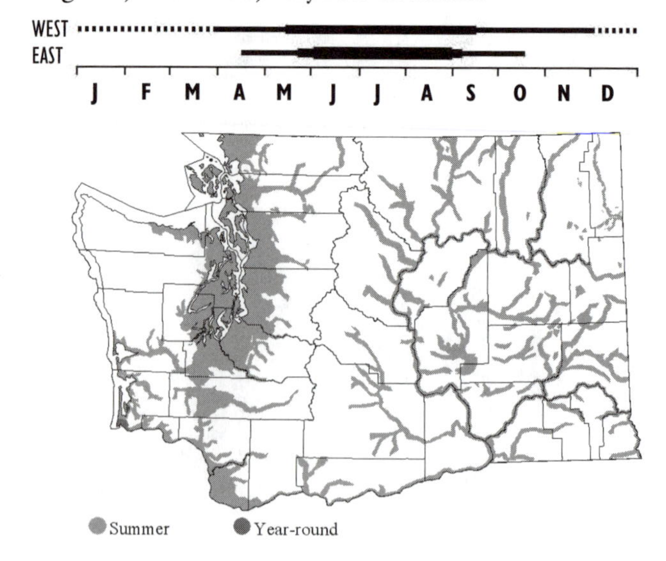

Subspecies: *A. c. septentrionalium.*

Habitat: Freshwater ponds, lakes; rarely on saltwater. Nests on ponds with dense aquatic vegetation and algal mats nr. the surface are heavily used. Forages on seeds of wetland plants and invertebrates.

Occurrence: A neotropical migrant, wintering s. to s. Mexico and n. S. America, the most common breeding teal in the state is a notoriously late arrival on breeding grounds and one of the first to leave. Ecologically similar to Blue-winged Teal, the species uses similar habitats up into the ponderosa pine, oak and sometimes Douglas-fir zones (Smith et al. 1997).

Breeding range in e. Washington includes the Columbia Basin e. to channeled scablands, the Palouse, n. into major river valleys, w. to Trout L. in Klickitat Co. (Smith et al. 1997). Highest breeding densities are in s.c. Okanogan, Grant, n. Douglas, n.c. Lincoln, nw. Whitman, and sw. Spokane Cos. (WDFW 2000a). In the w., local nesting recorded in suitable habitats in the Puget Trough, w. to Grays Harbor and along the lower Columbia R. (Smith et al. 1997). Cinnamon Teal and Blue-winged Teal numbers are combined in most waterfowl surveys: breeding season surveys record an average of 14,200 in e. Washington and about 2500 w. of the Cascades.

Not noted as occurring in Washington in winter by Jewett et al. (1953), now rare (SGM). Occurs mainly on lowland lakes and major river valleys. On CBCs from 1972 to 1999 birds have been reported 35 times w. of the Cascades, from Bellingham s. to

Wahkiakum, and three times in the e. Except for ad. males, some identifications may have been questionable.

Washington data indicate a long-term decline in numbers. This is unlike other Pacific Flyway states, where numbers appear to be fairly stable (USFWS 2000c). Causes are uncertain. Changes in e. Washington suggested due in part to habitat changes resulting from construction of large dams that inundated small wetlands and created large reservoirs, and irrigation and agricultural practices that are believed to have reduced populations while favoring the Blue-winged Teal (Connelly 1978, and see Smith et al. 1997). Subsequent maturation of extensive, newly created wetlands, however, favored the Cinnamon Teal (Connelly 1978).

Remarks: When not in breeding plumage males are difficult to distinguish from Blue-winged Teal; hens are always difficult to separate.

Noteworthy Records: *High counts*: 50 on 21 Mar 1988 at Ridgefield NWR; 22 on 1 May 1999 at Dodson Rd.; 34 on 28 Aug 2000 at Steigerwald L. NWR; 40 on 17 Jun 2001 at Ridgefield NWR.

Matthew J. Monda

Northern Shoveler *Anas clypeata*

Locally uncommon breeder and summer resident in e. Washington, rare breeder in w. Fairly common to locally abundant migrant and winter resident.

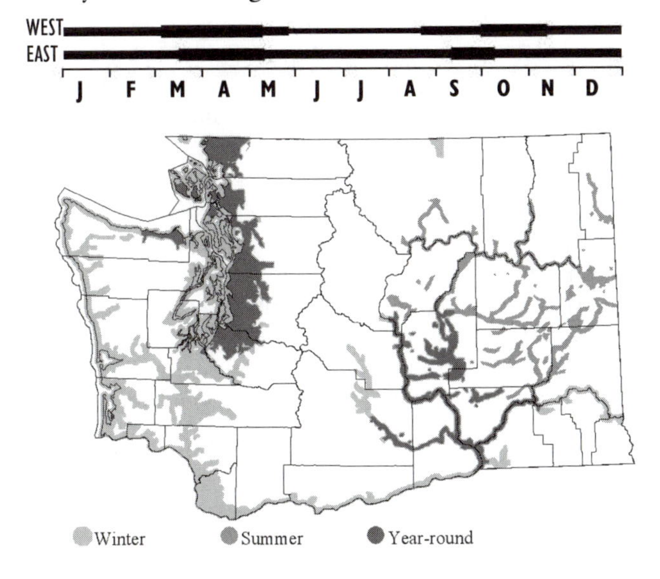

Habitat: Shallow lakes, freshwater marshes, sloughs, and potholes. Forages often in shallow wetlands with dense populations of small invertebrates and large numbers can be found foraging on sewage lagoons.

Table 12. Northern Shoveler: CBC average counts by decade.

	1960s	1970s	1980s	1990s
West				
Bellingham		18.4	33.6	156.2
Cowlitz-Columbia			49.4	159.9
Edmonds			17.0	144.1
E. Lk. Washington			25.1	33.9
Grays Harbor		73.4	137.0	211.1
Kent-Auburn			200.4	208.5
Olympia			36.0	118.6
Padilla Bay			36.1	17.3
Port Townsend			26.7	15.4
San Juan Is.			20.9	29.9
Seattle	189.3	104.0	269.3	154.0
Sequim-Dungeness			71.9	148.6
Tacoma		468.3	400.1	477.9
East				
Ellensburg			5.1	49.8
Tri-Cities		36.6	64.5	102.4

Occurrence: Shovelers use habitats similar to those of Blue-winged and Cinnamon teal during most of the year. Breeding populations in Washington are highest on the eastside. Breeding-season surveys in e. Washington record an average of 6,700 shovelers, with highest densities in the c. Columbia Basin and channeled scablands, s.c. Okanogan, Grant, n. Douglas, n.c. Lincoln, nw. Whitman, and sw. Spokane Cos. (WDFW 2000a). Less common at higher elevations and in n. areas. Surveys for w. Washington record about 800 birds. Shovelers breed locally in westside lowlands in the Puget Trough and e. Clallam Co. (Smith et al. 1997).

Migration of birds obviously not either summer or winter residents is notable over the continental shelf, and is about as frequent as more-abundant dabbling ducks. Groups of up to 25 birds were noted migrating about 20 times in Apr-May and mid Aug-early Oct 1979-97, up to 90 km offshore (TRW).

Annual Jan waterfowl surveys average 1700 shovelers in Washington, with 72% w. of the Cascades (USFWS 2000b). Birds are abundant in winter on Soap and Lenore Lks. in c. Washington. The N. American population has been building since the early 1990s and is at 3,521,000, the highest level ever recorded since breeding-duck surveys were initiated in 1955 (USFWS 2000c). Local winter increases are noted (e.g. Wahl 1995) and a general increase statewide is indicated by average counts on long-term CBCs (Table 12).

Noteworthy Records: *High counts*: 5000 on 5 Oct 1997 and 24 Oct-6 Nov 1993 at Everett STP, and 5000 on 6 Nov 1998 at Stanwood (SGM). CBC high counts: 505 in 1995 at Bellingham, 339 in 1997 at Columbia-Cowlitz, 327 in 1992 at Edmonds, 505 in 1995 at Everett, 437 in

1997 at Grays Harbor, 930 in 1975 at Padilla Bay, 8000 in 1991 at Skagit Bay, 1114 in 1982 at Seattle, 1267 in 1986 at Tacoma, 523 in 1999 at Kent-Auburn.

Matthew J. Monda

Northern Pintail *Anas acuta*

Locally abundant winter resident in n. Puget Sound, common winter elsewhere in w. Uncommon winter resident in e. Washington. Abundant migrant across Washington. Uncommon eastside breeder, uncommon summer visitor and very rare breeder in w.

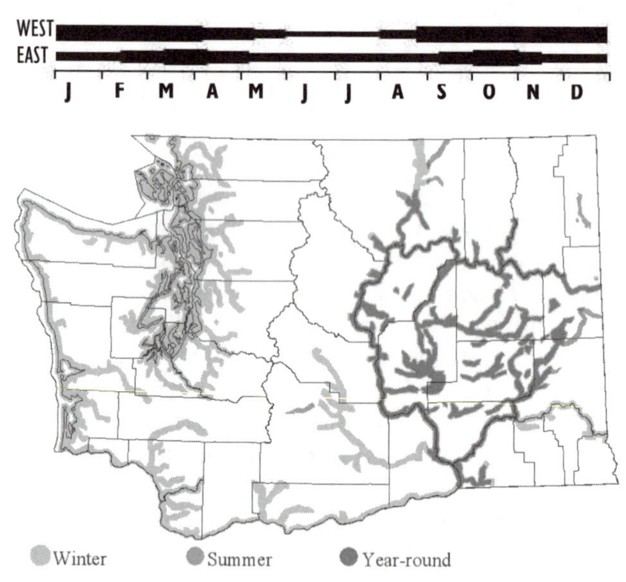

Subspecies: *A. a. acuta.*

Habitat: Most in marine estuaries, also found in most wetland and inshore marine habitats.

Attracted to shallow flooded areas during migration and winter. Feeds on seeds from moist-soil plants, waste grain in agricultural fields. Nests in all steppe zones (Smith et al. 1997).

Occurrence: The Northern Pintail, with the largest range of all waterfowl species, is the most common breeding dabbling duck in the Arctic and widespread in migration and winter when it occurs as far s. as Colombia and the Hawaiian Is.

Primarily a winter bird in Washington, with estuaries in n. Puget Sound supporting most migrant and wintering pintails—aerial surveys record up to 100,000 pintails in w. Washington during fall and winter. Like other dabblers, flocks move from forage fields to estuaries during hunting hours. Eastside winter monthly aerial surveys in the n. Columbia Basin record 100 to 5000 birds. The species is an early migrant in e. Washington with most moving out as winter proceeds. Annual Jan waterfowl surveys average 62,000 in Washington, with 93% occurring w. of the Cascades (USFWS 2000b).

An upland nester, found in the Columbia Basin, and at scattered locations in Okanogan and Pend Oreille Cos., locally nr. the ponderosa pine zone (Smith et al. 1997). Breeding-duck surveys in e. Washington record an average of 2100 pintails. Highest breeding densities occur in n. Douglas, n.c. Lincoln, nw. Whitman, and sw. Spokane Cos. (WDFW 2000a). Pintails nest very rarely w. of the Cascades, with records at Nisqually NWR, the Kent sewage-treatment ponds, and Ocean Shores, and old records from the Puget Sound region (Smith et al. 1997). Surveys record >100 summering pintails w. of the Cascades.

The N. American population has not recovered after the drought of the 1980s. Numbers are well below the long-term average of 4,320,000 (USFWS 2000c). Long-term CBC data trends are uncertain due to large numbers of unidentified ducks often recorded. About 6% of all ducks taken by hunters in 1999-2000, 27,769 birds, were pintails (WDFW 2000b). Classed PHS.

Noteworthy Records: *High count*: 40,000 on 9 Jan 1996 at Padilla Bay (SGM).

Matthew J. Monda

Garganey *Anas querquedula*

Casual vagrant.

This highly migratory Eurasian teal has been recorded four times in Washington: an ad. male seen and collected nr. Mt. Vernon, 27-30 Apr 1961 (Spear et al. 1988); an ad. male at Elma, on 12 Apr-15 May 1991; an ad. female at Richland on 15 Dec 1994-1 Jan 1995; and one at the Walla Walla R. delta, 26 Sep 1997.

Peak fall occurrence in the contiguous U.S. is from mid-Sep to early Nov (Mlodinow and O'Brien 1996). Most Garganeys in N. America are found with Cinnamon and Blue-winged teals. Identification is easy only when males are in breeding plumage from early Mar into midsummer.

Steven G. Mlodinow

Green-winged Teal *Anas crecca*

Widespread, abundant winter resident and migrant, especially in the Columbia Basin, Yakima Valley, and n. Puget Sound. Locally uncommon to rare breeder in e. Washington, rare breeder w.

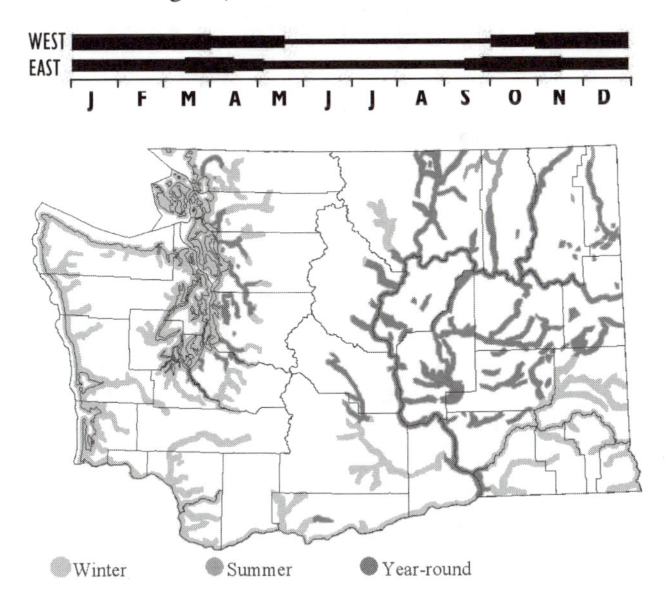

Subspecies: *A. c. carolinensis*; Eurasian, *A. c. crecca* or *nimia*, in small numbers.

Habitat: Wetland and inshore marine habitats, foraging in shallow wetlands, mud flats, and marshes, feeding on small seeds of moist-soil plants such as sedges, water grass, and smartweeds. Occasionally feeds in agricultural fields. Nests in upland habitats.

Occurrence: Less common as a breeder than other teals, the Green-winged Teal is one of the four most numerous dabbling ducks wintering in Washington. Flocks are widespread, found often in intertidal mud habitats such as Dungeness and Discovery bays and n. inland embayments. Apparent fall migrants arrive in mid-Jul, increase in early Aug, and peak by Oct on the westside. Numbers decrease in Apr, with birds common through mid-May. Birds are uncommon summer residents and rare breeders.

Annual Jan waterfowl surveys average 11,000 birds in Washington, with 59% w. of the Cascades (USFWS 2000b). Aerial surveys in n. Puget Sound can record 10,000 Green-winged Teal during fall migration and winter, and monthly fall and winter aerial surveys in the n. Columbia Basin record 500-7000. CBCs have recorded maxima of 5900 birds w. of and 1800 e. of the Cascades. Migration off Grays Harbor noted in May and Aug-Oct, with individuals and small groups seen to about 100 km offshore (TRW).

Summer duck surveys record an average of 3200 birds e. of the Cascades and about 100 w. Important eastside areas include Potholes Res., L. Lenore and Okanogan/Columbia confluence, with other widely spread locations except in Yakima Co. and se. corner of the state (Smith et al. 1997). In the w., nesting is variable in numbers, and irregular but widespread in locations from the lowlands including the San Juan Is. and ne. Olympic Pen. s. to Ridgefield.

Long-term CBCs in Washington have shown variability, likely due in part to numbers of unidentified ducks recorded, and no trends in winter numbers. The N. American breeding population has been increasing since the early 1990s and is estimated at 3,194,000, the highest level ever recorded since breeding duck surveys were initiated in 1955 (USFWS 2000c). About 52,000 Green-winged Teal (11% of all ducks) were taken by hunters in 2000 (WDFW 2000b).

First identified in 1968, the Eurasian Green-winged or Common Teal winters in small numbers, averaging about three to four reports annually. Most records are from the Puget lowlands, with reports also from Klickitat Co., Vancouver, and Walla Walla R. delta. Recorded from late Nov through early May, with molt likely obscuring identification of earlier males. Frequency of reports has apparently increased with the number and skills of observers, with three to four reports annually. Hybrids of the two races are reported on occasion, in a ratio of about 1:2-3 to Common Teal (SGM). High count of Common Teal was four birds at Samish Flats on 22 Feb 2001.

Noteworthy Records: East, *Winter*: 600 on 4 Jan 1987 at Potholes Res.; 800 on 9 Mar 1996 at Walla Walla R. delta; 235 on 4 Feb 1998 at Dallesport; 850 on 10 Dec 1999 at Iowa Beef, Walla Walla. *CBC high counts*: 4083 at Tri-Cities in 1978; 1814 at Moses L. in 1998. **West,** *CBC high counts*: 5947 at Bellingham in 1988; 4165 at Sequim-Dungeness in 1988; 3615 at Padilla Bay in 1999; 2850 at Skagit Bay in 1988; 2188 at Grays Harbor in 1978. Eurasian Green-winged Teal - Early: 23 October 1995 at Seattle. Late: 25 May 2001 at Othello.

Matthew J. Monda

Canvasback *Aythya valisineria*

Locally fairly common migrant and winter visitor on protected marine waters; rare in summer in w. Washington. Common migrant, locally common winter and rare breeder in e.

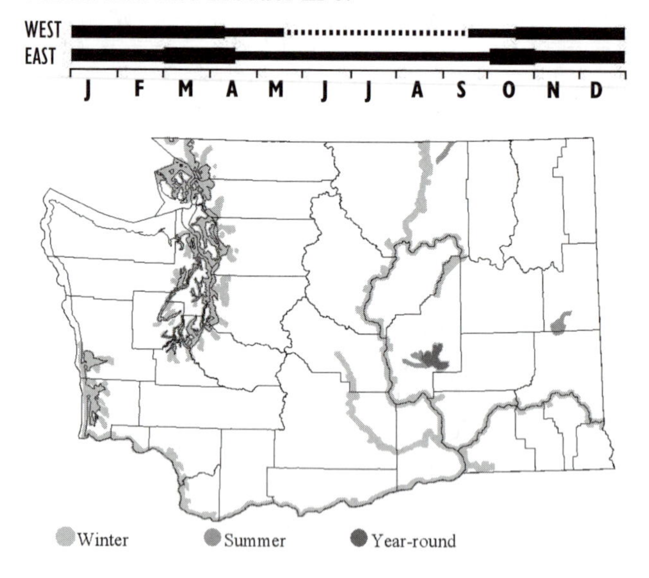

Habitat: Breeds in shallow prairie marshes with cattails, bulrushes, and similar emergent vegetation (Johnsgard 1975). Winters primarily in brackish marine water at river mouths or in relatively shallow bays or inlets with sand substrates and eelgrass (Hirsch 1980).

Occurrence: A rare and local breeder in e. Washington (Jewett 1953, Smith et al. 1997), with non-breeders rare in summer in w. Washington; essentially a winter bird. It is apparently much reduced in numbers, at least locally, in Washington. Kobbe (1900) reported immense flocks on the lower Columbia R. nr. Cape Disappointment in 1897 and McAtee noted flocks of 700-800 in Oyster Bay nr. Olympia in Dec 1914 (Jewett 1953): such numbers are not seen now. The most seen in Oyster Bay recently were 170 in winter 1992-93. Fewer than 2000 Canvasbacks now winter in marine waters and larger lakes in w. Washington (PSAMP, WDFW midwinter survey). Canvasbacks commonly winter in e. Washington, with up to 5300 reported on WDFW midwinter inventories.

Largest numbers (150-1000) on marine waters at Drayton Harbor, Lummi Bay, Bellingham Bay, Samish Bay, Skagit Bay, Snohomish R. delta, Everett sewage treatment ponds, Skokomish R. delta, Dyes Inlet, Budd Inlet, Grays Harbor, Willapa Bay, and lower Columbia R. with moderate numbers (25-100) occurring regularly at Liberty Bay, Totten Inlet, Eld Inlet, Penn Cove/Oak Harbor, Padilla Bay, and L. Stevens (PSAMP, WDFW midwinter survey data, SGM). Rafts numbering between 1500 and 2800 have been seen in winter on certain larger lakes and rivers of e. Washington, but numbers are usually lower at most sites there (WDFW midwinter survey data). Hunting harvest numbers not available.

N. American population estimated at 619,700 and increasing (USFWS/CWS 1996). Washington CBC counts are extremely variable between locations and show mixed long-term trends. Classed PHS.

Noteworthy Records: *Winter, high counts*: 1500 on 7 Feb 1963 36 km up Snake R. from McNary NWR; 200-600 in winter 1966-67 at McNary NWR; 2800 in winter 1966-67 at Turnbull NWR; 1000 in Dec 1974 off Samish I.; 700 on 1 Mar 1996 at McNary NWR. *PSAMP winter aerial high counts, 1992-99*: 425 in 1997 at Dyes Inlet; 350 in 1997 in Lummi Bay; 325 in 1993 in Skagit Bay; 207 in 1995 in Drayton Harbor; 206 in 1994 at Skokomish R. delta; 170 in 1992-93 at Totten Inlet; 169 in 1999-2000 at Everett/Snohomish R. delta; 45 in 1994 at Budd Inlet; 39 in 1998 at Eld Inlet; 35 in 1993 at Padilla Bay; 33 in 1999 at Liberty Bay; 20 in 1993 and 1996 at Penn Cove/Oak Harbor; 25 in 1994 w. of Nisqually R. delta; 15 in 1997 at Holmes Harbor.

David R. Nysewander

Redhead *Aythya americana*

Locally common migrant and breeder and uncommon winter in e. Washington. Uncommon winter visitor and very rare breeder in w. Washington, rare on the outer coast.

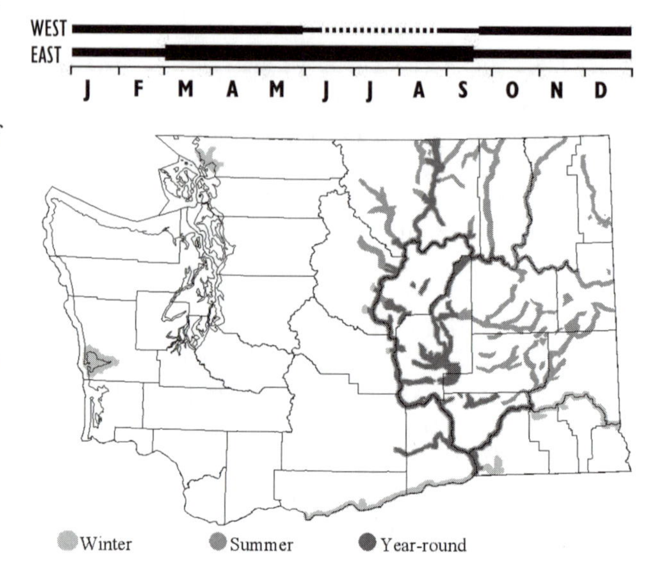

Habitat: Breeds on lakes, ponds, wetlands at low elevations with water depth sufficient for permanent, fairly dense emergent vegetation for nesting cover. Winters on lakes and larger rivers and westside sewage treatment ponds. Small numbers in winter on saltwater when freshwater freezes.

Occurrence: Redheads are the most numerous diving ducks breeding in e. Washington and exceeded in number of duck broods only by Mallards and Cinnamon/Blue-winged teals. Duck brood indices between 1993 and 2000 (Monda 2000) in e. Washington ranged between 143-726 broods per year, with no discernible trend over the last decade. Apr-May population indices (WDFW data) averaged 16,588 per year during 1979-99. Numbers may have declined somewhat during the 1993-2000 period with an average of 13,660 recorded. N. American population recently estimated at 691,400 and increasing (USFWS/CWS 1996). Most apparent from early-late May in w. Washington (SGM), the species has been noted breeding there at only a few locations (Smith et al. 1997).

Most Redheads on the Pacific coast winter s. of Washington. Fewer than 100 are usually recorded on midwinter inventories in w. Washington between Dec-Mar, somewhat fewer than the 300 reported to winter in Puget Sound (Bellrose 1976) in the past. Birds winter in relatively low numbers in e. Washington compared to parts of their winter range farther s., with an average of 4887 reported in 1990-95 and an average of 2927 in 1996-2000 on midwinter inventories. Midwinter inventories 1983-89 indicated four years when numbers varied between 5336 and 7463 (1983-86) and three years when numbers fell between 1354 and 2458 (1987-89). This suggests a cyclic variation that possibly relates to drought, water levels, and presence of dense stands of emergent vegetation in marshes on nesting grounds. Few CBC locations recorded appreciable numbers but long-term data suggest declines. Numbers taken by hunters not available. Classed PHS.

Noteworthy Records: *Winter*, **West**: 20 on 30 Jan 1966, 25 on 9 Mar 1969, 24 on 30 Jan 1988 at Beaver L., Skagit Co.; 12 in 1985-86 at Ocean Shores; 10 on 5 Feb 2000 at Ridgefield NWR. **East**: 300 on 9 Mar 1996 at McNary NWR. *Summer*, **West**: brood on 17 Jul 1988 at Kent ponds; 1 on 9 Jun 1998, 7 on 6 Jun and 4 on 12 Jun 2000 at Ridgefield NWR; 2 on 4-10 Jun 2000 at Everett STP. *CBC high counts*, **West**: 16 in 1984 at Bellingham; 10 in 1980 at Padilla Bay; 57 in 1989 at Skagit Bay; 14 in 1972 at Seattle. **East**: 353 in 1994 at Chelan; 298 in 1997 at Grand Coulee; 150 in 1997 at Moses L.; 40 in 1964 at Spokane; 558 in 1978 at Tri-Cities; 17 in 1998 at Toppenish NWR; 152 in 1993 at Two Rivers; 22 in 1980 in Wenatchee.

David R. Nysewander

Ring-necked Duck *Aythya collaris*

Common migrant and winter visitor, rare breeder in w. Washington; locally common migrant and winter visitor and fairly common breeder in e.

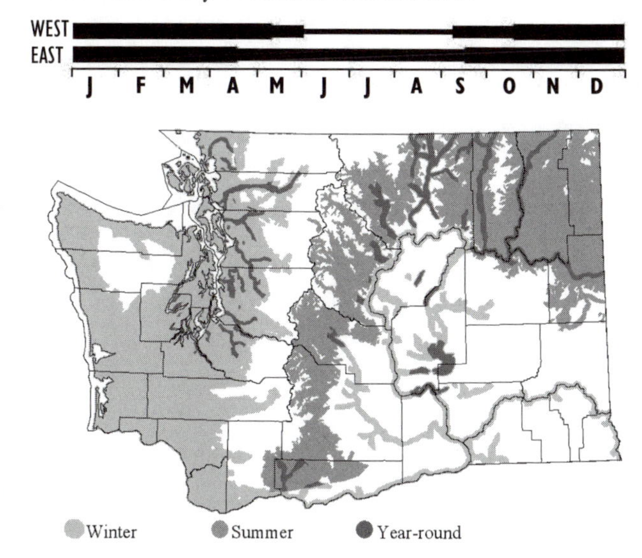

Habitat: Breeds on small lakes, bogs, swamps, or marshes nr. forested habitats. Feeds in shallower water than other diving ducks, usually less than 2 m deep (Bellrose 1976). Winters primarily on freshwater lakes, ponds, and rivers in w. Washington (seldom found in salinities >5 ppt: Hohman and Eberhardt 1998), though birds may move to saltwater during severe freezes.

Occurrence: Most of the Pacific breeding population winters from Washington s. into the Central Valley of California. A fairly common winter resident on freshwater lakes and rivers in w. Washington, with an average of 1528 ducks recorded on westside midwinter inventories between 1993 and 2000, with a total of over 8000 likely when numbers at sewage and industrial treatment ponds are included (SGM). Winter survey numbers slightly higher in e. Washington, averaging 2916 reported 1993-2000 on midwinter inventories. Some locations such as Columbia NWR, sites along the Columbia R., and larger lakes contain wintering numbers of this species that are second in abundance only to Ruddy Ducks or Lesser Scaup.

The species has expanded its breeding range the last century in N. America (Hohman and Eberhardt 1998) and may be increasing in Washington as well. Jewett et al. (1953) listed only one nesting location, in Pend Oreille Co. Now a widespread breeder in e. Washington, where duck brood indices from 1993 to 2000 (Monda 2000) averaged 59 (range of 0-79), mostly in the ne. counties, with no discernible trend. Apr-May population indices (WDFW data) averaged 2772 per year during the 1979-99 period.

Table 13. Ring-necked Duck: CBC average counts by decade.

	1960s	*1970s*	*1980s*	*1990s*
West				
Bellingham		49.1	106.6	200.1
Edmonds			63.2	110.8
E. Lk. Washington			24.2	72.3
Grays Harbor		55.1	71.5	59.1
Kent-Auburn			81.7	189.6
Kitsap Co.		1.7	19.6	39.3
Olympia			182.0	166.6
Padilla Bay			68.9	70.4
Port Townsend			4.7	33.6
San Juan Is.			115.0	194.8
Seattle	18.4	60.5	123.4	260.6
Sequim-Dungeness			26.2	67.4
Tacoma		53.3	100.1	197.3
East				
Ellensburg			3.5	20.7
Spokane	131.3	74.8	36.2	12.7
Toppenish			91.0	182.2
Tri-Cities		59.8	105.4	149.3
Wenatchee	27.7	98.6	296.6	250.2
Yakima Valley		8.7	8.2	103.7

Listed as a local, rare breeder in w. Washington by Smith et al (1997). N. American population estimated at 772,700 and stable (Hyslop and Kennedy 1996).

Winter surveys have not always covered the species' habitat effectively but range-wide populations are believed to be stable or slightly increasing, though potentially vulnerable to overharvest (Hohman and Eberhardt 1998). Hunting take numbers not available for Washington. CBC data on long-term counts (Table 13) are subject to weather conditions, freezes, and hunting, but local increases are apparent for 15 of 20 counts.

Noteworthy Records: *Winter*, **West**: 750 on 5 Nov 1994 and 650 on 24 Nov 2001 at Vancouver (SGM); 400 on 20 Nov 1999 at Columbia NWR; 425 on 3 Jan 1998 at Snohomish; 543 on 10 Nov 1973 nr. Olympia; 460 on 31 Dec 1995 nr. Everett. **East**: 255 on 9 Mar 1996 at McNary NWR. *Summer*, **West**: 4 males and 1 female with brood on 14 Jun 1978 on San Juan I.; female with brood of 5 on 2 Jun 1980 on Riffe Res., Lewis Co.; 1 female and 9 young on 2 Jul 1983 at Ft. Lewis; pairs on 5 and 19 Jun 1986 on Killibrew L., Orcas I.; brood in summer 1986 on Chambers L.; 2 broods on 30 May-1 Jun 1988 on Ft. Lewis.

David R. Nysewander

Tufted Duck *Aythya fuligula*

Rare migrant and winter visitor on westside. Very rare migrant and winter visitor in e. Washington.

Habitat: Lakes, ponds, sewage lagoons, and sheltered saltwater habitats.

Occurrence: First recorded in the contiguous U.S. in Alameda Co., California, in 1948 (Orr 1962). Two males ph. at Seattle, 31 Dec 1967 (Schultz 1971), were the first state record. Through 2001, about 44 more were recorded, with 35 of these since 1982. The increase in records may be attributed at least in part to greater observer awareness and effort.

Almost always found in the company of other *Aythya* ducks, especially scaup. Most records have been w. of the Cascades, with most from the greater Puget Sound region, Grays Harbor, Willapa Bay, and the lower Columbia R. in the interior; six of eight records are from Columbia R. reservoirs and nearby lakes. Similar to the pattern elsewhere in the contiguous U.S., records on freshwater outnumber those from saltwater by about 3:1 (Mlodinow and O'Brien 1996). Records by county: San Juan (1), Whatcom (1), Snohomish (5), Skagit (1), Island (1), Jefferson (1), King (3), Pierce (3), Grays Harbor (9), Lewis (1), Pacific (3), Wahkiakum (4), Clark (3), Skamania (1), Okanagan (2), Chelan (1), Douglas (1), Klickitat (2), Grant (1), Adams (1), and Lincoln (1).

Remarks: Identification problems can complicate occurrence records. Tufted Ducks sometimes hybridize with scaup (see Mlodinow and O'Brien [1996], Harris et al. [1989]). One of the first hybrids identified in N. America spent the winter of 1982-83 at Green L., Seattle, and returned for at least the following eight winters. Another hybrid male was seen at Everett during late winter/early spring 1979 and 1980, and hybrid females have been seen nr. Silvana, on 5 Apr 1997, and in Everett from 3 Jan into mid-Mar 1998 (SM).

Noteworthy Records: West, *Fall, early*: 10 Oct 1994 at Columbia NWR. *Spring, late*: 1 May 1996 at Chehalis; 1 May 1999 at Oak Harbor; 2 May 1993 at Snohomish; 15 Apr-14 May 2000 at Ridgefield NWR. **East**: 17-19 Jan 1986 at Wenatchee; 14-20 Mar 1988 nr. Pateros; 30 Mar 1988 at Columbia NWR; 29 Nov-2 Dec 1992 at Turtle Rk., Douglas Co.; 13-23 Apr 1994 at Reardan; 10-17 1994 at Columbia NWR; 2 Apr 1995 at Omak; 8-24 Nov 1996 at Bingen, returning 31 Jan-10 Mar 1998, and 9 Nov 1998-Mar 1999, and a male again on 9 Nov 2000-28 Feb

2001; with a female there 20 Feb 2001. CBC: 2 in 1967, 1 each in 1968, 1969, 1970, 1 in 1982 at Seattle; 1 each year 1983-85 at Leadbetter Pt. Some were likely the same bird returning.

Steven G. Mlodinow

Greater Scaup *Aythya marila*

Fairly common to common migrant and winter visitor, mostly in marine waters, rare in summer in w. Washington; fairly common migrant and winter visitor and very rare in summer on larger eastside lakes and rivers.

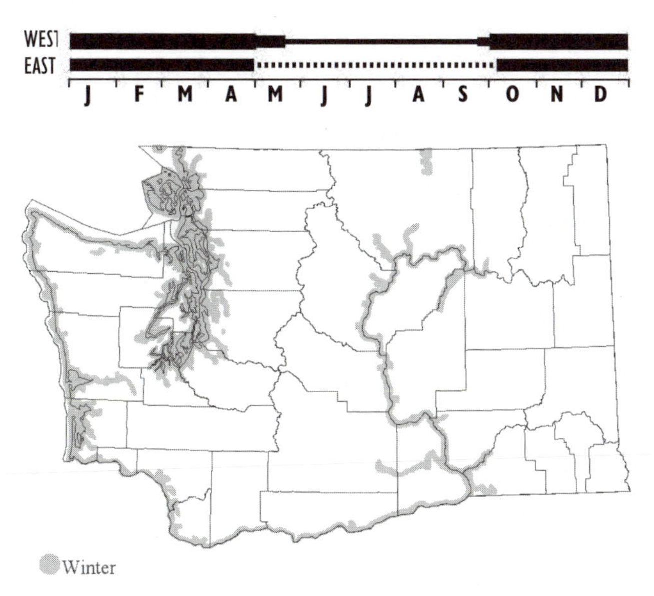

Subspecies: *A. m. nearctica.*

Habitat: Winters in shallow nearshore waters, particularly with soft substrate and eelgrass, in open to protected embayments (Nilsson 1969, Vermeer and Levings 1977, Hirsch 1980).

Occurrence: Largest flocks in Washington winter in n. inland waters at Drayton Harbor, Boundary Bay, Lummi Bay, portions of Bellingham, Samish, and Fidalgo/Padilla bays. Largest concentrations in c. and s. Puget Sound were in bays and estuaries, including Totten and Little Skookum Inlets, Lynch Cove, Annas Bay in Hood Canal, Quilcene Bay, Dabob Bay, Squamish Harbor, Dyes Inlet, Port Susan, Penn Cove/Oak Harbor, Skagit Bay, Port Townsend vicinity, and Dungeness/Jamestown nearshore waters. Grays Harbor, Willapa Bay, and lower portions of the Columbia R. also had important numbers. Mean overall densities in nearshore waters for Greater Scaup varied annually from 13.6 to 29.1 birds per km² during the 1992-99 period (PSAMP), with highest densities between 250 and 2000 scaup per km².

Scaup species cannot be separated in midwinter waterfowl inventories and aerial surveys. Based on hunters' bag counts, Bellrose (1976) estimated Greater Scaup numbers wintering in Washington at 15,000. Hirsch (1980) estimated scaup comprised 23.7% of diving ducks and projected a total of 35,194 scaup in the Strait of Juan de Fuca and n. inland marine waters. Aerial surveys throughout the inland marine waters during 1992-99 winters (PSAMP; Nysewander et al. 2001) indicated scaup comprised 8% of diving ducks. The highest numbers recorded on transect were 9969 in 1994-95, with a decline to 3500 by 2001.

An average of 15,186 scaup species combined was recorded on midwinter inventories in w. Washington between 1993 and 2000. Numbers are slightly higher in e. Washington, with an average of 17,270 reported on 1993-2000 midwinter inventories. Formerly, Greater Scaup were reportedly less common than Lessers in migration and winter in e. Washington, but they have become fairly common in recent years along the larger rivers and lakes (e.g. Stepniewski 1999), with larger flocks ranging from 1200 to 3500 recorded locally along the Columbia R. between Biggs, Oregon, and the Tri-Cities areas n. to the Wenatchee/Chelan vicinities.

Comparisons of density indices on recent surveys (PSAMP) with similar surveys in 1978-79 (MESA) in the n. portions of Puget Sound (Nysewander et al. 2001) suggest that scaup declined between 1978-79 and 1999. Nysewander and Evenson (1998) documented a similar decline in other portions of the Pacific Flyway such as San Francisco Bay. CBC data totals of both scaup species at westside long-term locations (where 57-94% of scaup were identified as this species) indicate general declines in most cases and an overall statewide decline ($P < 0.01$). Small numbers of summering non-breeders are sometimes reported, primarily from areas with large wintering concentrations. Classed PHS.

Noteworthy Records: East, *Winter*: up to 200 in Feb-Mar 1967 at Spokane; 2000 on 8 Feb 1995 at L. Chelan; 3500 on 27 Feb 1999 at Walla Walla R. delta. **West**, *Summer*: 25 on 10 Aug 1971 at Lummi I.; 8 on Jun 12-18 1988 at Dungeness; 30 at Livingston Bay and 1 at Crockett L. on 4 Jun 1988; 36 on 11-19 Jun 1998 at Hoquiam. *CBC high counts*: 2387 in 1989 at Bellingham; 230 in 1991 at Cowlitz/Columbia; 769 in 1998 at Columbia Hills/Klickitat; 1120 in 1994 at Chelan; 252 in 1995 at Everett; 395 in 1977 in Kitsap Co.; 348 in 1991 at Oak Harbor; 252 in 1997 at Port Gamble; 341 in 1986 at Skagit; 235 in 1984 in San Juans; 217 in 1993 in Wenatchee; 3169 in 1998 in Wahkiakum.

David R. Nysewander

Lesser Scaup *Aythya affinis*

Common migrant and winter and fairly common breeder locally. Common migrant and winter resident; irregular, fairly common summer visitor, rare breeder in w. Washington.

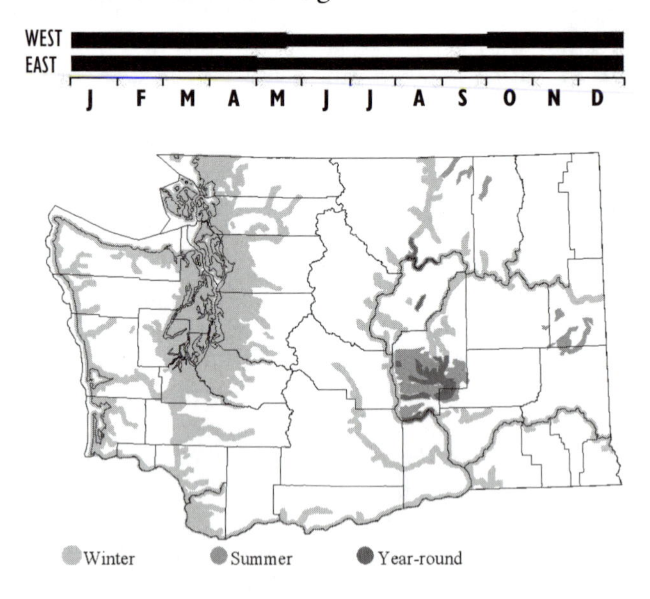

Habitat: Lakes, ponds, with islands, and moist meadows. Nesting pairs favor fresh to moderately brackish wetlands and lakes with emergent vegetation (Monda and Ratti 1988, Austin et al. 1998). In winter most often along shorelines in areas with fresh or low-salinity brackish waters, and on sewage ponds.

Occurrence: Common winter resident on freshwater lakes and rivers in w. Washington, with smaller numbers on sheltered saltwater habitats where it is generally outnumbered by Greater Scaup. Sewage-treatment ponds seem especially favored. Some locations in e. Washington such as Columbia NWR, sites along the Columbia R., and larger lakes contain larger wintering numbers of this species, with some CBCs such as Chelan and Columbia Hills/Klickitat regions recording 1013 in 1995 and 1211 in 1998 respectively. Eastside CBC counts indicate a significant long-term decline ($P < 0.01$).

Though Jewett et al (1953) did not list this as a breeding species, moderate numbers nest in ne. Washington. Duck brood indices between 1993 and 2000 (Monda 2000) averaged 62 broods (range of 0-228). Apr-May population indices (WDFW data) averaged 9189 per year during the 1979-99 period. Nesting first noted w. of the Cascades in 1978. Breeding is rare and local, noted most often around sewage-treatment ponds, recently especially at Deer Lagoon and Everett SP (SGM).

Bellrose (1976) listed Washington as containing 22,000 wintering Lesser Scaup: 59% of scaup found in hunters' bags. An average of 15,186 scaup species has been recorded on midwinter inventories in w. Washington between 1993 and 2000. Scaup winter in slightly higher numbers in e. Washington, with an average of 17,270 reported in 1993-2000 on midwinter inventories. Hunting harvest numbers not available.

Population estimates throughout N. America for both scaup species combined have declined dramatically since early 1980s to record lows by 1998 (Austin et al. 1998); declines in breeding populations are most pronounced in the w. Canadian boreal forest, where Lesser Scaup may be declining as much as or more than Greater Scaup. Since broods of scaup are uncommon in e. Washington, duck brood indices provide little trend information, but scaup numbers during the Apr-May survey period in 2000 are six times higher than they were in the early 1980s (Monda 2000). Classed PHS.

Noteworthy Records: *Winter, high count,* **East**: 1100 on 27 Feb 1999 at Walla Walla R. delta. *Summer*, **West**: brood on 3 Jul 1978 at Everett STP; brood in 1981 at Kent STP; broods in Sep 1990, Aug 1993, summer of 1994, and Jul 1995 at Everett STP; broods in 1994, 1995 and 1997 at Deer Lagoon, Whidbey I.; brood of 9 on 21 Aug 1993 at Steigerwald NWR; female with 5 young in Jul 1998 at Grays Harbor STP. *CBC high counts*, **West**: 958 in 1993 at Bellingham; 487 in 1984 at Edmonds; 330 in 1985 at Grays Harbor; 524 in 1977 in Kitsap Co.; 413 in 1990 at Oak Harbor; 643 in 1989 at Padilla Bay; 581 in 1997 at Skagit Bay; 900 in 1966 at Seattle; 392 in 1981 at Tacoma. **East**: 278 in 1998 at Bridgeport; 1211 in 1998 at Columbia Hills/Klickitat; 1013 in 1995 at Chelan; 392 in 1997 in Lyle; 1092 in 1994 at Moses L.; 2037 in 1961 at Spokane; 311 in 1987 at Tri-Cities; 189 in 1993 at Two Rivers; 402 in 1981 in Wenatchee.

David R. Nysewander

Steller's Eider *Polysticta stelleri*

Casual vagrant.

Two Washington records: a male ph. at Ft. Worden SP from 18 Oct 1986 to 8 Feb 1987 (Tweit and Paulson 1994), and an eclipse-plumaged male photographed while present from 9 to 13 Sep 1995 at the Walla Walla R. delta. The latter bird represents the only inland record from N. America s. of Alaska.

There are few other N. American records s. of Alaska. These include three from B.C., one from Oregon, and three from California, and have occurred from Oct to May. Classed GL.

Steven G. Mlodinow

King Eider *Somateria spectabilis*

Very rare fall, winter, and spring visitor on the coast and n. Puget Sound.

WEST

J F M A M J J A S O N D

The first state record was an imm. male collected at Seattle 23-30 Oct 1948 (Jewett et al. 1953). The next were almost two decades later: a female at Bellingham on 28 Oct 1965 (not Blaine, as per Roberson 1980), and two females at Orcas I. on 22 Feb 1967. Occurrence increased in the 1970s, with a total of five, including a first-year male and female nr. Bellingham from 29 Dec 1973 to 19 Jan 1974, a subad. male at Port Angeles on 6 Apr 1977, a female at Pt. Roberts from 23 Oct to 4 Nov 1977 and a young male there 1-15 Jan 1979. An additional seven reports came from the 1980s: a subad. male at Westport on 11 May 1980, a subad. male off Restoration Pt., Kitsap Co., on 3-4 Jan 1981, a female off Rosario Beach nr. Anacortes 3 Feb 1981 (Mlodinow 1999), a male at Port Williams, Clallam Co., on 13 Feb 1983, a female at Lopez I. on 29 Oct 1986, a subad. male at Birch Bay on 4-6 May 1987, and a male at Dungeness on 26 Dec 1988. Only two were reported during the 1990s: an ad. male at LaPush on 28 May 28 to 4 Jun 1990 and an imm. male n. of Pt. Partridge on 24 Dec 1995 to 6 Apr 1996 (Aanerud and Mattocks 2000).

Mlodinow (1999) suggested a decline in w. coast vagrancy of King Eiders since the mid-1980s was potentially explained by a 56% decline in numbers of the Beaufort Sea population from 1976 to 1996 (Suydam et al. 2000).

Bill Tweit

Common Eider *Somateria mollissima*

Casual vagrant

Washington's only documented record is of a well-photographed, eclipse-plumage male of the race *v-nigra* at Port Angeles 3 to 13 August 2004. This bird appeared shortly after California's first, which was present during much of July in Del Norte Co. and may have been the samc individual.

Bowles (1906a) and Dawson (1906) reported birds from the Nisqually area during Jan and Feb 1906. Apparently at least 10 birds were involved, including both males and females. However, no specimens were secured, and no written details exist, but these birds may well have been correctly identified. There are four records from B.C., including one from the Vancouver area, and *v-nigra* has strayed as far as inland as N. Dakota, Minnesota, Iowa, and Kansas and as far e. as Newfoundland and Greenland (A.O.U. 1998, Mlodinow 1999).

Steven G. Mlodinow

Harlequin Duck *Histrionicus histrionicus*

Fairly common breeder along turbulent mountain streams in Olympics, Cascades, and ne. mountains. Locally uncommon to common in winter in n. inland marine waters and the outer coast. Very rare in winter in e. Washington.

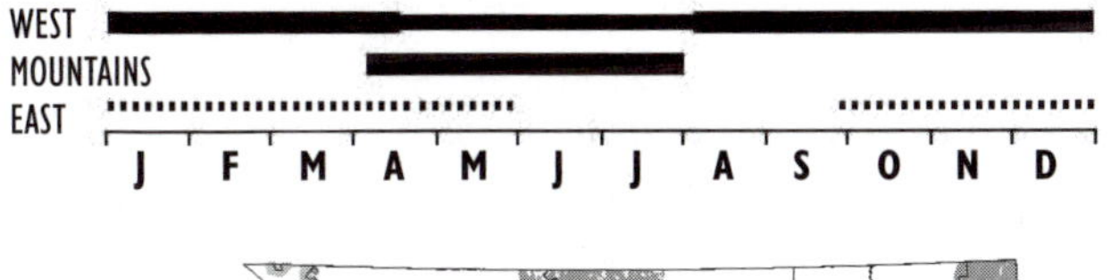

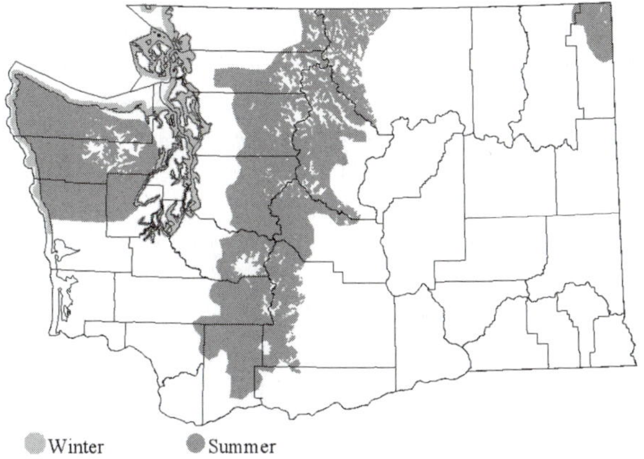

Habitat: Nest along fast-moving streams in forested riparian habitats. Winter along rocky shores and islands with kelp beds, small-size gravel shorelines with crabs and other winter food sources (Gaines and Fitzner 1987).

Occurrence: Many Harlequins noted in Washington cover long distances during migrations. The Alaska part of the Pacific breeding population has not been located wintering in the continental U.S. but, based on band returns and re-observations, the interior N. American population occurs in the Strait of Georgia and Puget Sound where the wintering population includes birds banded in Banff, Glacier, and Grand Teton NPs. Wintering birds occur locally in suitable habitat including rock jetties at the mouth of the Columbia R. and Grays Harbor. Winter numbers along the n. Olympic coast, from the Hoh R. n., although this is often thought of as classic harlequin habitat, appear much smaller than those in the n. inland waters.

Analysis of relative seasonal abundance is complicated by local distributions, early return to winter sites by post-breeding males in Apr-Jun, and unknown numbers of non-breeders using winter sites during summer molts. These molting sites are localized to specific points and shorelines, with individuals showing high site fidelity. Summer birds there are not necessarily non-breeders, but include failed female breeders, subad. males, and some subad. females, as well as post-breeding males and, later, females and young. Known molting sites include Protection I., Smith I., Penn Cove, Oak Harbor, and Mandarte I. (B.C.) in inland waters and at Cape Alava and Destruction I. on the outer coast. WDFW flights, however, did not detect molting birds in some outer-coast or other areas where birds are typically found a week or two before molt.

Jewett et al. (1953) described this species as "much more common than would be thought even by individuals that spend a great deal of time in the out of doors" and the harlequin is found on over 137 streams in Washington (Cassirer 1993), far more than the 60 recorded historically. Old stream sites still have harlequin pairs today, often in the same riffle or stream confluence as originally observed. Today, 32 locations are in the Olympics, 69 in the w. Cascades, 31 in the e. Cascades, four in the Selkirks, and one in the Blue Mts.

Large numbers historically congregated at herring spawn locations in spring (e.g. 205 with 25,000 scoters at Pt. Whitehorn, Whatcom Co., in Apr 1979 [MESA]). Failure of Washington herring runs has not yet been corroborated with impacts on Harlequin Ducks.

The early wintering population has been estimated using a mark recapture technique at about 3000 (Schirato and Hardin 1998), indicative of the importance of Washington waters to the w. N. American population. Wintering numbers increased throughout the 1990s (Nyeswander et al. 2001), and CBC numbers on long-term counts indicate interdecadal increases (P <0.01), particularly at Sequim-Dungeness.

Remarks: The w. N. America population was estimated at 100,000-200,000 by Chadwick and Littlehales (1993). Classed PHS, GL.

Noteworthy Records: *CBC high counts*: 401 in 1985 at Sequim-Dungeness; 172 in 1987 at San Juan Is.; 153 in 1989 at Port Townsend; 114 in 1991 Oak Harbor.

Gregory Schirato

Surf Scoter *Melanitta perspicillata*

Widespread common to abundant migrant and winter visitor in nearshore inland marine waters and coastal estuaries, locally common summer concentrations in w. Washington; uncommon fall migrant and rare winter visitor e.

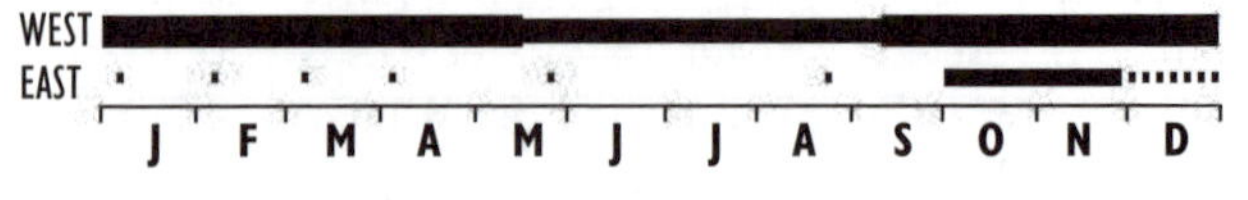

Habitat: Shallow marine coastal waters (<20 m deep) over varied substrates: pebbles, sand, mud, cobbles, and rock, and eelgrass habitats especially during herring-spawn events; rivers, lakes in migration.

Occurrence: The Surf Scoter is the most abundant diving duck wintering in marine waters of Washington, with winter aerial surveys in inland marine waters in 1992-99 (PSAMP: WDFW) indicating it averaged 68% of scoters identified to species.

Scattered groups or large numbers concentrate along shorelines, in estuaries, and nr. offshore shoals and banks. N. inland marine waters had concentrations in the 1990s at Boundary Bay, Lummi Bay, portions of Bellingham Bay, Padilla-Samish Bays, and n. portion of E. Sound and e. side of Cypress I. in the San Juans, but higher densities and larger total numbers of scoters were found on aerial surveys to the s. at sites such as Penn Cove/ Oak Harbor/Holmes Harbor portions of Whidbey I., a number of bays and inlets in s. Puget Sound, and s. Hood Canal. Mean overall densities in nearshore waters varied annually from 55 to 70 birds/km² during the 1992-99 period (PSAMP, WDFW), with highest densities between 250 and 1000/km². On the outer coast, numbers concentrate at locations such as the entrance to Grays Harbor in migration and winter. Numbers normally increase in early Sep and decrease in May when birds move inland and north to breed.

Summer surveys in 1978-79 and the 1990s recorded sizable non-breeding flocks at primary wintering locations such as Boundary Bay and Penn Cove, and small, scattered numbers elsewhere. There are few records from e. Washington, mostly in fall when season totals ranged as high as 20 in the 1990s. Winter and spring records are much less numerous.

Comparisons of 1990s surveys (PSAMP) with surveys in 1978-79 (MESA) in the n. greater Puget Sound (Nysewander et al. 2001) suggest that scoters have declined significantly (at least 57%) between 1978-79 and 1999. Nysewander and Evenson (1998) documented an evident decline in other portions of the Pacific Flyway such as San Francisco Bay. CBC data for Washington show no overall trend, likely due to large numbers of scoters unidentified to species and possibly to shifts in wintering areas (e.g. at Bellingham Bay [Wahl 2002]).

Ads. routinely feed on mollusks in marine wintering areas, and on eggs at herring-spawning events (e.g. Campbell et al. 1990a, Wahl 1995). Declines of key herring-spawn events are suggested as important factor in scoter declines. Wahl et al. (1981) recorded flocks of 20,000-22,000 scoters feeding at Pt. Whitehorn in Georgia Strait in 1978-79. A collapse of that herring stock followed, and fewer than 5000 scoters were seen in there in 1999-2000. These spring events may play crucial roles in building fat reserves critical for migration and breeding. Whether representative of overall scoter numbers is uncertain, but there were increased numbers (e.g. 12,000) in n. Bellingham Bay in winter-spring of 1999-2000 and 2000-01 (TRW), perhaps suggesting relocation of pre-migration flocks.

Remarks: Association with herring spawn was not mentioned by Jewett et al. (1953), when numbers were smaller than subsequent surveys found. The species' annual range was indicated by two marked birds recorded in both n. California and n. Canada and noted in Whatcom Co. in fall 2003 (D. Given-Seymour p.c.). N. American population was estimated at 765,000 and stable (Ellis-Joseph et al. 1992). Classed PHS.

Noteworthy Records: West, *CBC high counts*: 2410 in 1996 at Bellingham; 2595 in 1995 at Kitsap Co.; 4774 in 1993 at Oak Harbor; 2950 in 1979 at Olympia; 2010 in 1993 at Sequim/Dungeness; 1149 in 1980 via Anacortes/ Sidney ferry. *Summer*: 2180 on 13 Jul 2002 at Padilla Bay. **East**, *Winter-Spring*: 1 on 8 Feb 1970 nr. Richland; 1 on 5 Jan 1980 nr. Wenatchee; 2 on 2 Apr 1986 on Millcreek L., Walla Walla Co.; 1 on 12 Mar 2001 at Vantage. *Fall, early*: 1 on 1 Aug 1995 at Wenas L.; 1 on 12 Aug 1976 on Naches R.; 1 on 26 Aug 2001 at Orondo. *Fall, high counts*: 5 in late Oct 1975 on Medical L.; 17 in fall 1985 in Walla Walla Co.; 7 on 19 Aug 1986 at L. Lenore; 13 on 14 Oct 2000 at Rock L., Whitman Co.

David R. Nysewander

White-winged Scoter *Melanitta fusca*

Fairly common to locally abundant migrant and winter visitor in nearshore inland marine waters and coastal estuaries; fairly common locally in summer in w. Washington. Uncommon fall migrant and rare winter-spring on eastside.

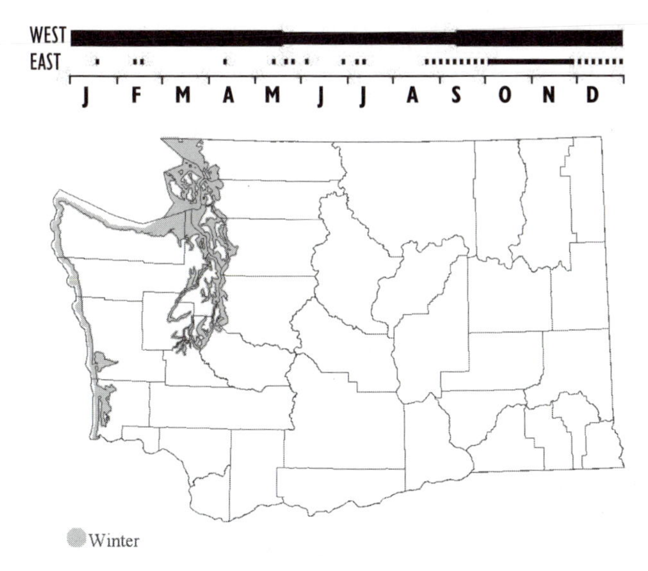

Subspecies: *M. f. deglandi* most widely distributed in N. America.

Habitat: Nearshore marine waters (<20 m) most often with gravel, sand, and mud substrates.

Occurrence: The White-winged Scoter is the second-most abundant of the three scoter species wintering in the inland marine waters. Like the Surf Scoter, winters locally along the outer coast and noted migrating offshore in fall to 125° W at the outer limit of one-day surveys (TRW). It is also the scoter most often seen, irregularly and in low numbers, in e. Washington.

Aerial surveys throughout the inner marine waters of Washington during 1992-99 winters (PSAMP: WDFW) indicated this species averaged 26% of scoters identified to species. Birds often associate with Surf Scoters, although they show a tendency to select larger prey than the other scoters, and feeding tends to be more restricted to sand/mud/gravel combinations. Wintering birds feed routinely on mollusks, especially bivalves, and herring spawn in spring. Herring spawn may well be important in building and storing fat reserves prior to migration and to reproductive success in n. breeding grounds (Brown 1981, Brown and Fredrickson 1997).

In winter, n. inland waters with large concentrations include portions of bays from Drayton Harbor s. to Samish Bay, and e. side of Cypress I. S. and c. portions of greater Puget Sound also contain significant winter densities with highest concentrations at Penn Cove/Oak Harbor/Holmes Harbor/Port Susan e. of Whidbey I., Puget Sound from westside inlets s. to Carr and Case Inlets and Eld, Totten. and Little Skookum Inlets and s. Hood Canal. Highest densities from PSAMP aerial surveys were between 100 and 368/km^2 in these concentrations. More often than other scoters, White-winged Scoters are often noted in Apr-early May "getting in shape" with short flights before migration (Wahl 1995).

Summer concentrations occur primarily along the shorelines of the s. and e. Strait of Juan de Fuca, at Boundary Bay, Drayton Harbor, Penn Cove/Oak Harbor, and in habitats at Indian and Marrowstone Is. in Admiralty Inlet. Migrants are noted inland, as at Ross L. (Wahl 1995). The species is a rare spring and fall migrant in e. Washington (e.g., Stepniewski 1999).

The overall declines noted for scoters, at least 57% between 1978-79 and 1992-99, apparently also include this species as there has been no noticeable difference in relative percentage of scoter species seen over this same period (Hirsch 1980, Nysewander et al. 2001). CBC data also indicate declines in most locations, with the San Juans Ferry count, for example, reflecting the disappearance of a large wintering flock in Harney Channel noted on MESA censuses in 1978-79.

The only scoter species reported in the interior by Jewett et al (1953), and still the most frequently noted, largely in the fall. Most fall counts provide totals of 10-20, with a peak of over 40 in 1984. Much smaller numbers remain through the winter or are noted as spring migrants. Consistent with their more southerly breeding distribution, s. to the Thompson/Okanagan plateau in B.C. (Campbell et al. 1990a), there have been at least 16 summering records in the interior.

Remarks: Predation by scoters at shellfish farms, as at Penn Cove, noted but not quantified (e.g. USDA 2002). Prior to about 2000, USFWS issued permits for control kills, but this was stopped as populations of sea ducks declined. Numbers of birds killed under control programs are unknown.

Noteworthy Records: *CBC high counts*: 916 in 1980 at Bellingham; 619 in 1977 at Grays Harbor; 1823 in 1995 in Kitsap Co.; 600 in 1973 at Leadbetter Pt.; 2446 in 1995 at Oak Harbor; 1248 in 1982 at Olympia; 628 in 1997 in Port Gamble; 764 in 1986 at Port Townsend; 1310 in 1987 at Skagit Bay; 2400 in 1963 at Sequim/Dungeness; 638 in 1980 via Anacortes/Sidney ferry. **East**, *Spring*: 1 on 11 Apr 1970 and 4 on 24 May 1974 at Turnbull NWR; 3 on 25 May 1981 at Colville; 3 on 20 May 1987 at McNary NWR; 8 on 13 May 1995 at Priest Rapids Res. (Stepniewski 1999). *Summer*: 1 on 11 Jul 1968 at McNary NWR; 1 on 5 Jun 1969 at Medical L.; 3 on 30 Jun 1974 at Keller; 1 on 8 Jul 1981 on Sidley L., Okanogan Co.; 10 on 5 Jul 1982 at Sullivan L. *Fall-Winter high counts*: 9 on 18 Oct 1976 at Medical L.; 16 on 19 Oct 1979 nr. Brewster; 15+ in fall 1984 on Moses L. and 20+ on Alkali L.; 13 in fall 1987 at Columbia NWR; 15 in Dec 1989 on Fish L., Chelan Co.; 13 on 14 Oct 1990 at L. Lenore; 20 on 28 Oct 1997 at Yakima R. mouth. CBC: 1 in 1980 and 1986 in Tri-Cities; 2 in 1986 at Wenatchee.

David R. Nysewander

Black Scoter

Melanitta nigra

Locally fairly common migrant and winter visitor in marine nearshore waters, locally rare in summer in w.

Washington. Very rare migrant e.

Subspecies: *M. n. americana.*

Habitat: Sandy substrates (Hirsch 1980), rock/ cobble habitats <10 m depth.

Occurrence: The w. Black Scoter population winters primarily along coastal Alaska, B.C., and Washington, with much higher proportions of the population wintering farther n. than either of the other two scoter species. It is the least abundant of the three scoters in Washington, with surveys throughout inland marine waters during 1992-99 winters (PSAMP, WDFW) indicating this species comprised 5-6% of scoters identified to species, with flocks usually in the low hundreds or less. Birds were distributed throughout marine nearshore waters and tended to segregate themselves from the other two species (Bordage and Savard 1995). PSAMP surveys and non-systematic observations (Wahl 1996, SGM) showed widespread, localized groups in s. Puget Sound, s. Hood Canal, Oak Harbor, locations in Whatcom Co. including Lummi Bay and nearby Portage, Pt. Roberts, the n. San Juan Is., and nearshore along the Strait of Juan de Fuca.

There are only six records from e. Washington, all from the Columbia Basin or lower Yakima Valley area, and most (four) from fall—this is much less common than the other two scoter species in the interior.

Trend for the Pacific portion of this species' population categorized as declining from 1984 to 1994 (Bordage and Savard 1995). Overall declines estimated at up to 57% between 1978-79 and 1992-99 of scoters in Washington apparently included this species as there was no noticeable difference in relative percentage of scoter species seen over this same period (Hirsch 1980, Nysewander et al. 2001). Longer-term CBC data indicate variability by location and time, but little change overall.

Remarks: Classed PHS, GL. Formerly Common Scoter.

Noteworthy Records: West, *Spring*: 500 at Pt. Whitehorn in 1979 with 25,000 other scoters foraging on herring spawn (MESA). *Summer/early Fall* : 8 on 15 Aug 1965 at Pt. Roberts; 5 on 20 Jun 1998 at Edmonds; 4 on 3 Jun 1972, 5 on 20 Jun 1976, 4 on 26 Jun 1974 and 1-2 on 10, 18, and 29 Jul 1998 at Ocean Shores; 1 on 11 Aug 1968 at n. jetty of Columbia R.. **East**: 1 collected on 20 Oct 1977 on Columbia R. nr. Brewster; 2 in spring 1983 at Yakima R. delta; 1 on 28 Oct 1997 at Yakima R. delta; 1 on 4 Sep 1998 at Richland; 1 on 7 Nov 1985 at Wapato 1985 (Stepniewski 1999). *CBC high counts*: 254 in 1983 at Bellingham; 207 in 1995 in Kitsap Co.; 752 in 1998 at Oak Harbor; 882 in 1995 at Olympia; 269 in 1991 at Seattle. *PSAMP survey counts in high year 1994-95*: 50 in Everett/ Snohomish R. delta area; 98 in Great Bend/Lynch Cove part of Hood Canal; 108 in Seabeck/Dabob Bay and central portions of Hood Canal; 117 in Des Moines/ Poverty Bay/Dumas Bay vicinity; 299 in Carr Inlet; 100 in Case Inlet; 90 in Totten and Little Skookum Inlets; 33 in Peale Passage; 161 in greater Port Orchard/Rich Passage/ Bainbridge I. marine waters; 121 in Quartermaster Harbor vicinity.

David R. Nysewander

Long-tailed Duck

Clangula hyemalis

Fairly common migrant and winter resident

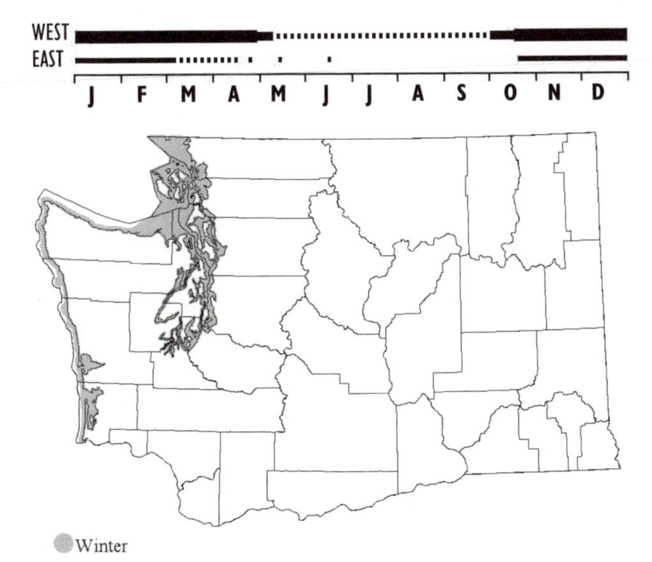

primarily in more open marine waters in w. Washington; very rare in summer. Rare in fall and winter, very rare in spring in e.

Habitat: Marine waters, often offshore waters, of embayments and channels; found nearshore locally, and especially at seasonal foraging opportunities. Occasional on lakes, sewage ponds.

Occurrence: A species noted to frequent more open waters and forage deeper than other diving ducks (e.g., Cottam 1939, Ellarson 1956, Gabrielson and Lincoln 1959, Bellrose 1976, Cramp et al. 1977).

MESA surveys in 1978-79, for example, found Long-tailed Ducks in deeper waters (>70 m) of Georgia Strait than other species and many reports confirm deep-water occurrence in other locations. As with other species, local situations affect distribution Along the shoreline of the e. Strait of Juan de Fuca, heavily used by the species, Hirsch (1980) found more Surf and White-winged scoters than Long-tailed Ducks in relatively "deeper" waters (mean 9-10 m), though the latter was farthest from shore (mean 490 m). During feeding events like herring spawns, this species forages in large numbers near shore in shallow water (e.g., Pt. Roberts, Cherry Pt., Drayton Harbor: MESA).

Hirsch (1980) estimated 6386 Long-tailed Ducks (4.3% of diving ducks in marine waters) in the Strait of Juan de Fuca and n. half of inner marine waters. Aerial surveys covering all inner marine waters during 1992-99 winters (PSAMP) showed Long-tailed Ducks to average 1-2% of diving ducks. These surveys included c. and s. Puget Sound waters, which attract fewer Long-tailed Ducks than those areas covered by Hirsch and the MESA studies 1978-79. Birds are very rare or irregular in summer in w. Washington and rare during migration and winter in e. Washington.

N. areas, notably Boundary Bay, Lummi Bay, Samish and Padilla bays, E. Sound in the San Juan Is., Smith I., and waters between Dungeness Spit and Protection I. contained largest numbers. S. and c. Puget Sound contained many fewer birds, with only a few sites such as Agate Pass, Kilisut Harbor, and Pitt Passage showing regular annual occurrence. Mean densities were often low (< 5 birds/km^2), even in the n. waters during the 1992-99 period (PSAMP, WDFW), but highest densities reached between 25-112/km^2 in Boundary Bay and the waters between Protection I. and Dungeness Spit.

Comparisons of surveys in the 1990s (PSAMP) with similar surveys conducted in 1978-79 (MESA) in the n. portions of greater Puget Sound (Nysewander et al. 2001) suggest that numbers there declined over the last 20 years, and they reportedly declined in other parts of the Pacific wintering range. Long-term CBC data are variable, but indicate noticeable increases at Sequim-Dungeness and from the San Juans ferry.

Eastside records are primarily of fall migrants, with smaller numbers remaining into winter, and six spring records. Fall numbers are variable, averaging >10 annually with a maximum of 16. Most fall records come from the Columbia R. and Soap, Lenore, and Banks Lks.

Remarks: Classed PHS. Formerly the Oldsquaw.

Noteworthy Records: West, *Winter, high counts*: 1200 on 20 Feb 1987 on Bellingham Bay (Wahl 1995) and 400 on 20 Feb 1987 on Manzanita Bay, Kitsap Co. CBC (>100): 344 in 1993 at Bellingham; 1143 in 1997 at Sequim/Dungeness; 309 in 1998 from Anacortes-Sidney ferry. *PSAMP High counts 1994-95 and 1996-97*: 271 between Dungeness Spit and Protection I.; 427 in ne. East Sound, Orcas I.; 171 in nw. Bellingham Bay and Lummi Bay nr. Portage I. and Hale Passage; 318 in Samish and Padilla Bays; 488 over Alden Bank w. of Sandy Pt. and Lummi Bay. **East**, *Fall-Winter, high counts*: 6 in Nov-Mar 1970-71on Spokane R.; 3 and 9 in fall 1965 at McNary NWR and Banks L.; 8 on 5-7 Dec 1992 at Soap L., Grant Co.; 9 on 22-23 Nov 1996 at Coulee L.; 6 on 28 Oct 1997 at Yakima R. mouth; 6 from 18 Dec 1997 to 7 Jan 1998 at Walla Walla R. delta; 8 on 18 Feb 2002 at Vantage. *Spring*: 1 in spring 1988 at Yakima R. mouth; 1 on 8-11 May 1988 at Reardan; 4 on 28 Mar to 23 Apr 1995 at Medical L.; 2 on 7 Mar 2002 nr. Asotin; 1 on 24 Mar 2002 at Clarkston; 1 on 11 May 2002 at Nile L., Stevens Co. *Summer*: 1 in mid-Jun 1963 at Columbia NWR; 1 from 20 May to 21 Aug 2001 at Bridgeport.

David R. Nysewander

Long-tailed Duck
(Shawneen Finnegan)

Bufflehead *Bucephala albeola*

Widespread, common to abundant migrant and winter resident in nearshore marine waters and coastal estuaries; uncommon in summer in w. Washington. Common migrant, fairly common winter resident, very rare to rare breeder in e.

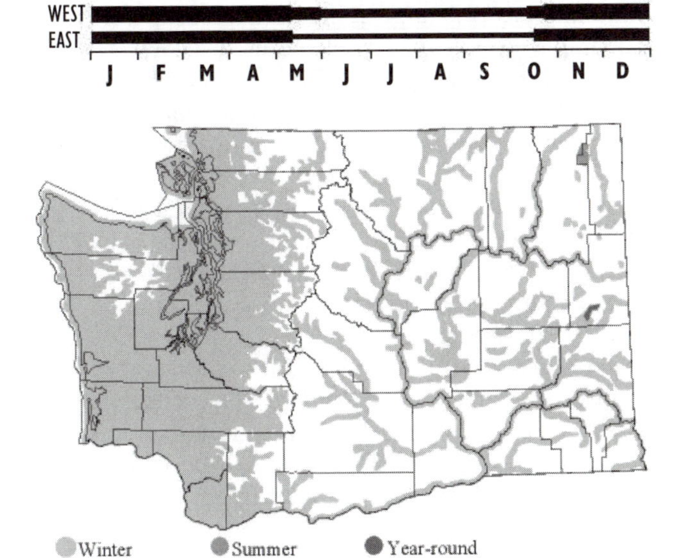

Habitat: Nearshore marine waters in estuaries and along open shorelines, lakes, and rivers. Hirsch (1980) found Bufflehead to prefer shorelines with gradual slopes and either cobble-rock or sand and eelgrass substrates.

Occurrence: The Bufflehead is the second-most numerous overall of the diving ducks wintering in nearshore marine waters, predominating locally in areas like Padilla Bay, protected bays in the San Juan Is., and the s. shores of the Strait of Juan de Fuca. Birds also occur on inland lakes below 300 m elevation. Aerial surveys throughout inner marine waters during winters in 1992-99 (PSAMP) found this species comprised, on average, 23% of diving ducks. Bellrose (1976) stated that half of the 30,000 wintering Buffleheads found s. of B.C. were in Washington, with three-quarters of these in Puget Sound. PSAMP survey transects in 1992-99 recorded numbers up to 22,332 in 1994-95. Hirsch (1980) projected 49,086 in the Strait of Juan de Fuca and n. inland marine waters. Hyslop and Kennedy (1996) estimated a stable N. American population of 8,864,000.

Comparisons of recent PSAMP surveys with similar surveys conducted in 1978-79 (MESA) in n. inland waters (Nysewander et al. 2001) suggest that Bufflehead numbers have not changed significantly over the last 20 years, unlike many other marine diving-duck species. Bufflehead have different feeding preferences than those of declining species like scoters, preying largely on crustaceans and gastropods during winter (Hirsch 1980).

The n. inland marine waters contained notable concentrations nr. Pt. Roberts, Semiahmoo Bay and Drayton Harbor, Samish and Padilla bays, heads of bays in the San Juan Is., and numerous places along the Strait of Juan de Fuca. S. and c. Puget Sound supported numbers at the heads of bays or estuaries such as Lynch Cove, Totten Inlet, and many others. Mean overall densities in nearshore waters ranged from 34 to 64 birds per km^2 during the 1992-99 period (PSAMP), with densities of between 250 and 972 Buffleheads per km^2 for the higher concentrations. These data (Nysewander et al. 2001) and long-term CBC data show increases statewide ($P < 0.01$) and indicate widespread distribution.

Usually fewer than 2500 Buffleheads have been recorded in e. Washington on the national midwinter waterfowl inventories during the last decade (D. Kraege p.c.). Unrecorded as a breeder at the time of Jewett et al (1953), now known as rare breeders locally in ne. Washington. They are fairly common in low numbers during migration along the bigger rivers and lakes in e. Washington. Classed PHS.

Noteworthy Records: *CBC high counts*, **West**: 1518 in 1979 at Bellingham; 2126 in 1996 at Olympia; 2423 in 1987 at Padilla Bay; 3362 in 1995 at Sequim/Dungeness; 3389 in 1988 in San Juans. **East**: 117 in 1998 at Bridgeport; 336 in 1994 at Chelan; 419 in 1996 at Tri-Cities; 185 in 1994 at Wenatchee. *Winter*: 607 in 1979 at Griffin Bay, San Juan I. (Wahl 1996); 2500 on 2 Jan 1999 at Birch Bay. *Summer, nesting reports*: 19 Jun 1971 at Turnbull NWR; broods on 10 Jun 1990 and 14 Jul 1991at Turnbull NWR; broods on 28 Jun 1990 and 26 Jun 1993 and in 1995 at Big Meadow L., Pend Oreille Co.; female with 7 ducklings on 7 Jun 1998 at Chewelah; female with 5 ducklings on 23 Jun 1998 at Bonaparte L.; 2 on 30 May 1999 at Conconully; female with 7 ducklings on 9 Jun 1999 at Republic; brood in 2000 nr. Chesaw.

David R. Nysewander

Common Goldeneye *Bucephala clangula*

Widespread common migrant and winter resident in westside nearshore marine waters and coastal estuaries, rare in summer. Very rare to rare breeder, common migrant, fairly common in winter in e. Washington.

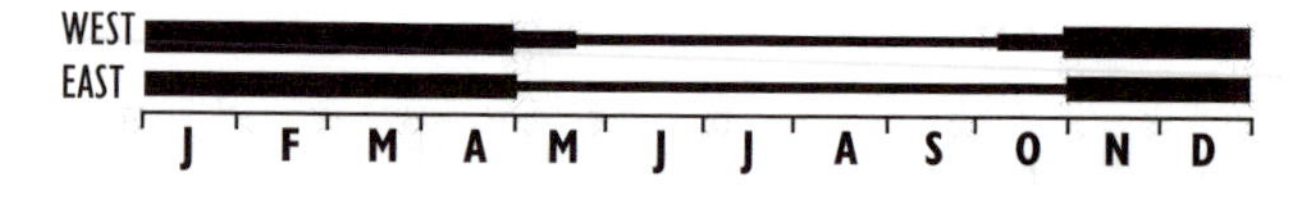

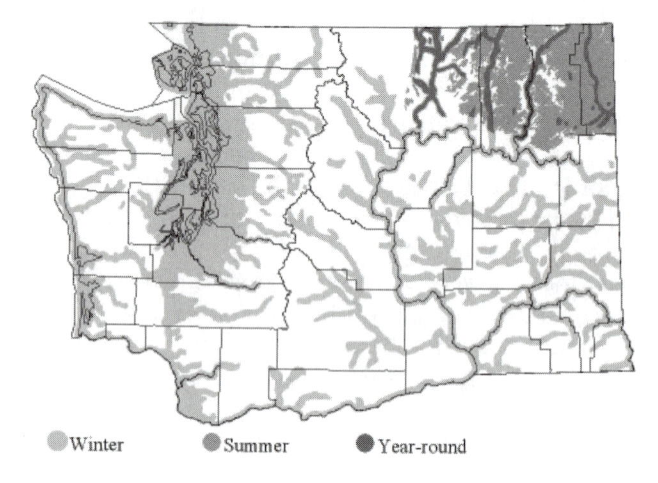

Subspecies: *B. c. americana.*

Habitat: Breeds at lakes and ponds in mature forests providing tree cavities, as well as wetlands where abundant invertebrate prey are available. Winters in intertidal areas, rivers, and ice-free lakes. Often found in more open, rougher water than Barrow's Goldeneye (Wahl 1995).

Occurrence: One of the more widely occurring ducks in winter in w. Washington, with birds found along many shorelines. Analysis of aerial survey data is complicated by the difficulty of separating Common from Barrow's goldeneyes and assumptions about unidentified goldeneyes may vary to unknown degrees between different survey programs.

Common Goldeneyes are apparently the third or fourth most-numerous overall of the diving duck species wintering in the nearshore marine waters. Aerial surveys in Washington inland marine waters during 1992-99 winters (PSAMP, WDFW) documented both goldeneye species to comprise, on average, 17% of diving ducks. CBC data from 1985 to 2000 for different sites found Common Goldeneyes comprised from 55% of goldeneyes identified to species at Oak Harbor and Olympia to 93% at Tri-Cities and Wenatchee and 98% at Dungeness/Jamestown area.

Bellrose (1976) estimated 12,000 wintering in Puget Sound, 3000 in outer coastal bays, and 3000 on the Columbia and Snake Rs. Hirsch (1980) projected 7390 Common Goldeneyes for the Strait of Juan de Fuca and n. inland marine waters of Washington (1980). Aerial surveys 1992-99 recorded both goldeneye species in numbers ranging up to 17,336 in 1994-95.

No definite Washington breeding records were known in the early 1950s (Jewett et al. 1953). Recent records indicate that Common Goldeneyes are rare breeders in a few isolated areas from Okanogan to Pend Oreille Cos. (Smith et al. 1997, WDFW PHS, Lewis and Kraege, SGM), but though recorded breeding at Soap L. (Smith et al. 1997), other reports from tree-less locations appear to be a result of confusion with Barrow's Goldeneye. Westside summer records are almost all from sewage-treatment ponds (SGM).

Long-term CBC counts suggest westside increases, and an overall eastside decline reflects a large decline at Spokane (P <0.01). An average of 4281 unidentified goldeneyes (from 1407 in 1998-99 to 7232 in 1995-96) was recorded in e. Washington on national midwinter waterfowl inventories during the 1990s (Kraege). Since eastside wintering Barrow's Goldeneyes are localized, a high percentage of these are likely Common, indicating a minimum population estimate of 3500-4000. N. American population estimated at 700,000-800,000 (Hyslop and Kennedy 1996). Classed PHS.

Noteworthy Records: *CBC high counts*, **West**: 571 in 1979 at Bellingham; 1813 in 1994 in Kitsap Co.; 645 in 1990 at Oak Harbor; 1013 in 1991 at Olympia; 504 in 1985 at Padilla Bay; 1008 in 1995 at Sequim/Dungeness; 892 in 1988 in San Juan Is.; 859 in 1988 at Tacoma. **East**: 253 in 1996 at Col. Hills/Klickitat Valley; 273 in 1991 at Ellensburg; 611 in 1996 at Grand Coulee; 910 in 1963 at Spokane; 615 in 1997 at Tri-Cities. *Winter and migration*: 600 on 2 Jan 1971 at Pt. Roberts; 1200 on 25 Oct 1972 at Olympia; 5700 in mid-Mar 1973 at Turnbull NWR. *Summer*: female, young in 1976 and 1986 on Pend Oreille R.; pair nested in 1976 at L. Wenatchee; brood with ad. on 13 May 1984 at sewage pond at Colville; broods in Jun-Aug 1992 at Winthrop and Sinlahekin WMA; broods in Jun 1993 and 1995 on Big Meadow L., Pend Oreille R., and Usk, Pend Oreille Co.; 2 broods on 6 Jun 1995 at Oroville; uncommon breeders in Jun-Jul 1997: 7 at Bonaparte L. and 1 at Winthrop; 4-5 on 1 Jun-31 Jul 1998 at Stanwood; pr, young on 23 Jun 1998 at Oroville/Toroda Cr. Rd., Okanogan Co.; brood on 17 Jul 1997 at Winthrop; 6 pairs on 7 Jul 2000 nesting at Republic; female with brood on 25 Jun 2001 at L. Lenore.

David R. Nysewander

Barrow's Goldeneye *Bucephala islandica*

Locally common migrant and winter resident in nearshore inland marine waters, locally on freshwater lakes; very rare in summer in w. Washington. Uncommon breeder in w. Cascades; uncommon breeder, common migrant, and fairly common in winter on larger rivers in e.

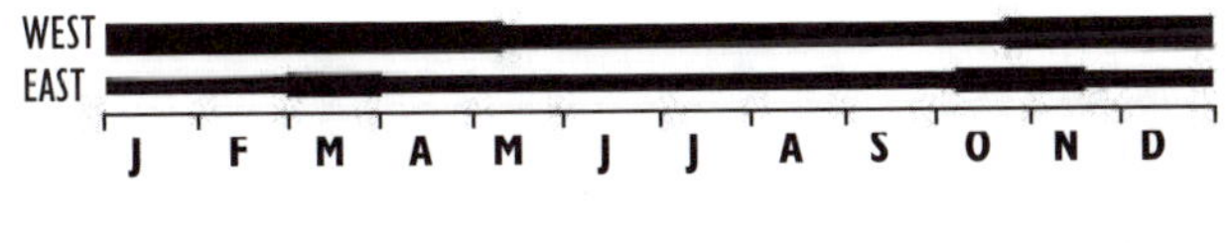

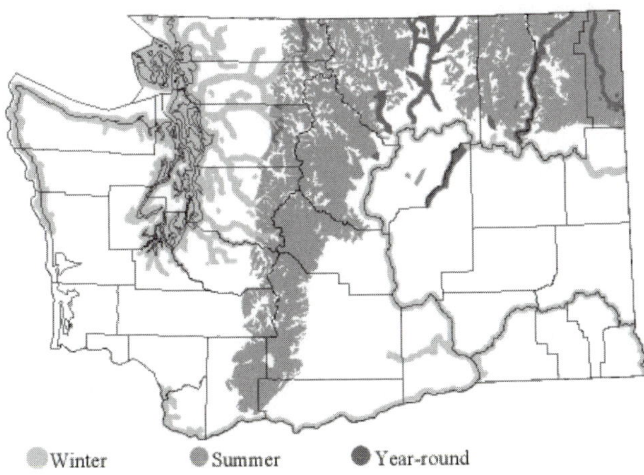

Habitat: Winters on semi-protected marine waters with rocky shorelines, especially with mussel beds, foraging also on sea life attached to pilings in winter and locally on larger rivers and lakes. Breeds usually in late-successional forests and riparian areas adjacent to low-gradient rivers, sloughs, lakes, and beaver ponds, nesting in tree cavities, locally in cliffs and talus slopes adjacent to alkali lakes.

Occurrence: Less widespread than the Common Goldeneye in westside marine waters, though locally predominant in areas such as Penn Cove/Oak Harbor. Often found in more protected water inland marine waters than the Common Goldeneye in winter (Wahl 1995), with birds congregating at creek mouths and other freshwater entering saltwater and attracted to herring-spawn events in the spring (Campbell et al. 1990a).

Barrow's Goldeneyes are fairly common breeders nr. large lakes, ponds, and reservoirs at moderate and higher elevations in Cascade Mts., Okanogan and Methow valleys, and ne. mountains and river valleys. Disjunct populations nest in cavities within talus slopes and basalt cliffs at L. Lenore, Jameson and Alkali Lks. at low elevations in the shrub-steppe zone of c. Washington (Smith et al. 1997, WDFW PHS, Lewis and Kraege 2000). Not recorded breeding in the Olympics, Blue Mts., or at Turnbull NWR. Males leave for molting grounds as females are incubating (Smith et al. 1997), though molting areas are not well known.

Low numbers migrate and winter locally along the larger rivers and lakes in e. Washington. Fewer than 1000 estimated to have wintered in e. Washington during the national midwinter waterfowl inventories during the last decade. Birds banded in B.C. recovered in both w. and e. Washington (Campbell et al. 1990a).

Hyslop and Kennedy (1996) estimated a w. N. American population of about 150,000. Bellrose (1976) estimated 2825 Barrow's wintering in Washington. Surveys during the 1992-99 period (PSAMP) suggested identified and unidentified goldeneye combined comprised 17% of all wintering diving ducks.

Long-term CBC data suggest an overall westside increase, with local decreases at Olympia, Edmonds, and Bellingham. Classed PHS.

Noteworthy Records: *CBC, high counts,* **West**: 365 in 1977 at Bellingham; 291 in 1997 at Edmonds; 261 in 1998 at Everett; 347 in 1996 in Kitsap Co.; 25 in 1988 at N. Cascades; 480 in 1992 at Oak Harbor; 1183 in 1978 at Olympia; 377 in 1997 at Seattle; 601 in 1995 at Tacoma. **East**: 187 in 1998 at Col. Hills/Klickitat Valley; 140 in 1969 at Spokane. *Winter/migration*: 600 on 8-14 Nov 1970 and 663 on 3 Mar 1978 at Turnbull NWR; about 100 in Nov 1999 and 2000 nr. Maryhill/John Day Dam (SGM); up to 1500 on 1 Nov 1974 at Olympia, with 200 there by mid-month. *Summer, high count*: 77 on 2 Jul 2000 at L. Lenore.

David R. Nysewander

Smew *Mergellus albellus*

Casual vagrant.

Two male Smews were reported at Friday Harbor on 22 Feb 1981 (AB 35:329, Lewis and Sharpe 1987). There are two subsequent Washington records, both of ad. males. One was at Willard, Skamania Co., on 28 Dec 1989 and probably the same bird was at nearby Stevenson 26 Jan-13 Feb 1991 (Tweit and Paulson 1994). This individual was also seen along the Oregon side of the Columbia R. during the winters of 1991 and 1992 (Gilligan et al 1994). One was photographed at McKenna, Pierce Co., 14-20 Mar 1993 (Tweit and Skriletz 1996).

There are about three records from B.C., two from California, and two from Oregon, all during winter. Smews are kept in some 40 waterfowl collections in the U.S. and Canada (G. Toffic p.c.) and escapes are known from B.C. (R. Toochin p.c.). All records should be treated with suitable caution.

Bill Tweit

Hooded Merganser *Lophodytes cucullatus*

Widespread fairly common migrant and winter resident in nearshore marine waters and freshwater ponds and rivers, uncommon breeder on westside; uncommon breeder, fairly common migrant and winter resident in e.

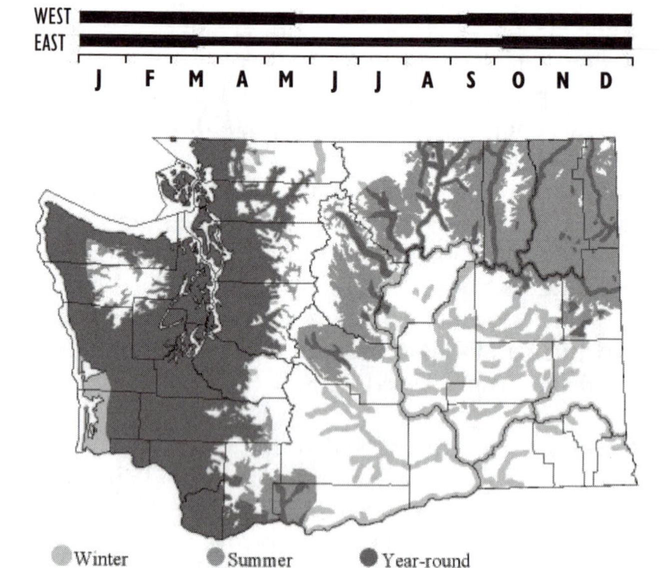

Habitat: Breeds in forested wetland systems, ponds, and riparian corridors, with brood habitat favoring shallow freshwaters that have rocks, logs, or unvegetated bars for loafing sites. Shallow, unfrozen freshwaters, brackish bays, estuaries, and tidal creeks and ponds in winter (Dugger et al. 1994).

Occurrence: Hooded Mergansers are the least abundant of the three merganser species wintering in w. Washington. Birds were widespread in low numbers along the shallow marine shorelines, with local densities varying from about <1-10 birds per km² during the 1992-99 period (PSAMP, WDFW), though surveys did not cover birds wintering in lowland freshwater locations, including valleys in sw. Washington. Westside numbers peak from about early Nov to Apr (SGM). Birds on freshwater move to saltwater following severe freezes. Bellrose (1976) estimated 600 wintering birds in w. Washington and 150 in e. Washington. PSAMP data indicate areas of higher numbers in marine areas include the protected waters of c. San Juan Is., around Whidbey and Camano Is., and some of the inlets of sw. and c. Puget Sound. The origin of many wintering birds may be from n. of Washington but pertinent information is apparently unavailable (e.g., Campbell et al. 1990a).

Jewett et al. (1953) felt that numbers had decreased between 1860 and 1953, possibly due to loss of nesting trees through logging and development. About 2500 were estimated to breed statewide in wooded areas in the 1960s (Jeffrey and Bowhay 1972). Birds now nest locally in the Puget trough, the San Juan Is., lower elevations w. of the Cascade crest, along river valleys in ne. Washington and locally elsewhere in e. Washington (Smith et al. 1997). Though breeding surveys are inadequate to determine trends, recent winter surveys suggest numbers have increased from 50 years ago. This is likely due to reforestation of lowland areas, protection from hunting, and appropriation of Wood Duck nest boxes (e.g. Wahl 1995, Smith et al. 1997). CBC counts show statewide significant increases (P <0.01, and see decadal averages, Table 14). Classed PHS.

Table 14. Hooded Merganser: CBC average counts by decade.

	1960s	*1970s*	*1980s*	*1990s*
West				
Bellingham		28.3	55.0	84.7
Cowlitz-Columbia			6.9	18.4
Edmonds			15.7	40.9
E. Lk. Washington			37.7	53.1
Grays Harbor		28.4	23.9	40.0
Kent-Auburn			33.4	95.5
Kitsap Co.		31.5	68.8	59.6
Leadbetter Pt		5.6	14.1	20.6
Olympia			59.2	119.0
Padilla Bay			62.8	91.6
Port Townsend			30.4	60.7
San Juan Is.			207.4	142.1
Seattle	2.3	27.2	59.4	57.4
Sequim-Dungeness			28.6	59.8
Tacoma		26.0	72.4	153.9
East				
Ellensburg			4.1	18.6
Spokane	5.4	7.4	25.5	38.4
Tri-Cities		2.0	15.4	36.5
Wenatchee	3.7	4.3	15.4	38.8
Yakima Valley		0.0	3.3	25.4

Noteworthy Records: *CBC high counts,* **West**: 138 in 1987 at Bellingham; 102 in 1998 at Everett; 146 in 1998 at Kent/Auburn; 116 in 1983 in Kitsap Co.; 128 in 1994 at Oak Harbor; 168 in 1998 at Olympia; 213 in 1994 at Padilla Bay; 108 in 1981 at Seattle; 353 in 1988 in San Juans; 202 in 1996 at Tacoma. **East**: 65 in 1998 at Bridgeport; 76 in 1994 at Chelan; 67 in 1997 at Ellensburg; 181 in 1993 at Grand Coulee; 64 in 1993 at Tri-Cities; 79 in 1989 at Wenatchee; 57 in 1991 at Yakima Valley. *Winter/migration high counts*: 80 on 3 Nov 1973 at Bellingham; 60 on 27 Sep 1974 at Willapa NWR; up to 195 on 31 Oct 1986 at Turnbull NWR; 132 on 7 Nov 1998 at Ebey's Landing; 52 on 13 Feb 1999 at McNary Pool; 60 on 23 Nov 2001 s. of N. Beach CP, Jefferson Co.

David R. Nysewander

Common Merganser *Mergus merganser*

Fairly common migrant and winter resident in nearshore marine waters, rivers, and larger lakes and uncommon breeder on westside; uncommon breeder, fairly common migrant and winter resident on large rivers and lakes on eastside.

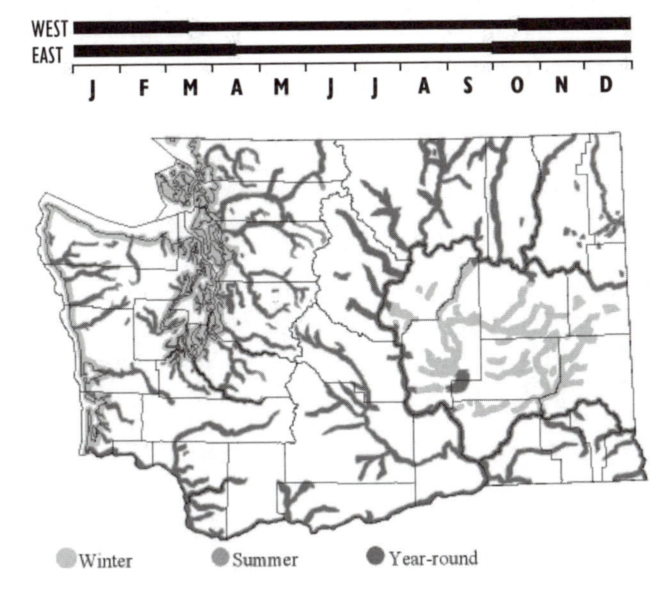

Subspecies: *M. m. americanus.*

Habitat: Nests in tree or rock cavities, in brush on the ground or among tree roots, on cliff ledges, or in holes in banks, associated with ponds nr. upper portions of rivers in forested regions or clear freshwater lakes with forested shorelines (Bellrose 1976, Mallory and Metz 1999). Winters in brackish water at river mouths, nearby marine shorelines, lowland lakes (Johnsgard 1975).

Occurrence: Common Mergansers were the second-most abundant of the three merganser species in winter 1992-99. They were found along many westside marine shorelines, with highest concentrations (50-263 birds per km^2) nr. river mouths, nearby estuaries, and passages such as Tacoma Narrows and Hammersley Inlet. Mean local densities varied from 1-25 per km^2 during the 1992-99 period (PSAMP, WDFW). Local movements may affect numbers observed on saltwater. Birds are locally numerous at lowland inland lakes that remain unfrozen (e.g. Wahl 1995), and a sizable portion of CBC totals may be from lakes then. In early spring, flocks may concentrate at herring-spawn events (Lewis and Sharpe 1987) and smelt spawning in the lower Columbia.

Common Mergansers are uncommon breeders in Washington, seen mostly nr. large lakes and rivers, especially noticed as they float down river with their broods. There are few breeding records in dry parts of e. Washington (Smith et al. 1997), but every major river in forested regions will likely have some broods, noted in midsummer where rivers enter into marine waters or large lakes. Bellrose (1976) estimated that Washington contained 300 breeding and 2300 wintering birds. Sex ratios observed locally vary seasonally relative to differences in migration and molting areas. Males leave nesting grounds when incubation begins and concentrate at poorly known molting areas (Campbell et al. 1990a).

Recent winter surveys led to estimates of about 1200 to 3800 birds wintering in e. Washington and about 1500 to 2500 birds w. of the Cascades (WDFW midwinter survey data), suggesting increasing numbers. Additional indications include flocks in recent decades (e.g. MESA, PSAMP) larger than concentrations reported by Jewett et al. (1953), along with the creation of large reservoirs that helped increase winter numbers throughout N. America (Bellrose, 1976). Long-term CBC data indicate a statewide increase ($P < 0.01$).

Remarks: This species is one of fish-eating species subject to control programs at fish hatcheries and at salmonid runs on the Columbia R. An average of 268 Common and Hooded mergansers per year were eliminated from 1997 to 2001 (USDA 2003).

Noteworthy Records: *CBC high counts,* **West**: 799 in 1992 at Bellingham; 1379 in 1988 at Olympia; 482 in 1988 at Seattle; 412 in 1985 via Anacortes-Sidney ferry; 1070 in 1987 in San Juan Is.; 1,486 in 1995 at Tacoma. **East**: 142 in 1998 at Bridgeport; 548 in 1987 in Cowlitz/Columbia R.; 383 in 1995 at Chelan; 782 in 1993 at Grand Coulee; 293 in 1997 at Lyle; 377 in 1994 at Moses L.; 400 in 1992 at Spokane; 203 in 1992 at Tri-Cities; 330 in 1996 at Toppenish NWR; 321 in 1998 in Yakima Valley. *High counts, Winter/migration,* **West**: 1600 on 27 Nov 1977 at Tacoma; 600 on 15 Sep 1999 at Possession Bar, Island Co.; 450 on 10 Sep 2000 at Skagit WMA. **East**: 250 on 8 Mar 1964 on lower Columbia R. nr. Stella; 500 in winter 1963 at Spokane and McNary NWR; 2000 on 26 Oct 1980 nr. Saddle Mt. NWR; 1500 on 4 Jan 1987 at Banks L.; 1,350 on 10 Dec 1998 and 3790 on 10 Jan 2003 at Moses L. *Summer*: 140 on 13 Jul 2002 at Dugualla Bay.

David R. Nysewander

Red-breasted Merganser *Mergus serrator*

Widespread, common migrant and winter resident and rare in summer in nearshore marine waters; locally uncommon migrant and rare winter in e.

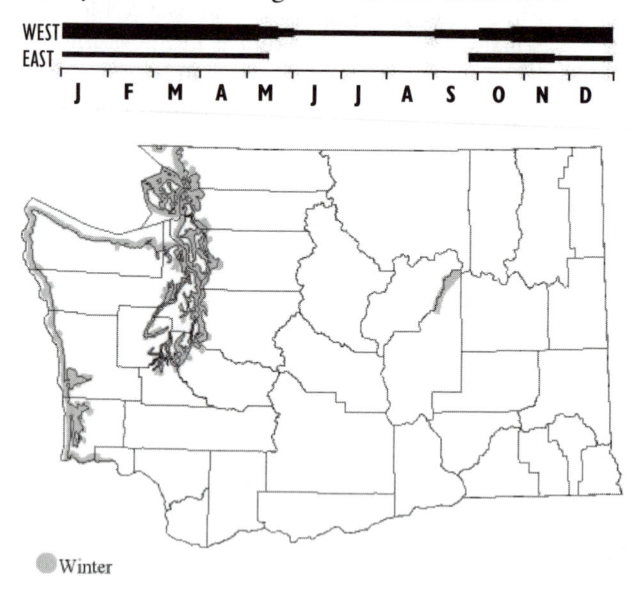

Subspecies: *M. s. serrator.*

Habitat: Nearshore marine areas, particularly protected bays and estuaries (Titman 1999) and large freshwater bodies.

Occurrence: Red-breasted Mergansers are widespread in winter in nearshore inland marine waters of Washington, with mean densities in 1992-99 that varied from 1 to 25 birds per km² (PSAMP), and are usually the most abundant of the three merganser species. Birds concentrate in relatively protected areas like the San Juan Is., Hood Canal estuaries, embayments such as Kilisut Harbor and Sequim Bay, and s. Puget Sound sites nr. estuaries and locations such as Hale Pass, Hammersley Inlet, and Peale Passage. MESA shore surveys in 1978-79 found large numbers in locations such as Fidalgo Bay (655 birds in Feb 1978) and Padilla Bay, where PSAMP aerial surveys found much lower numbers.

Flocks occur where fish are spawning (Campbell et al. 1990a) and overall distribution may vary noticeably during such events, particularly in spring when numbers concentrate at herring spawns (e.g. 239 in Hale Pass, Whatcom Co., in May 1979 and 169 at Lummi Bay in April 1978; Wahl 1995). Small numbers occur on lowland freshwater and numbers are notable on the lower Columbia R. when smelt runs occur upstream to the Cowlitz R. in late winter.

Red-breasted Mergansers are locally uncommon in migration and winter on large rivers and lakes in e. Washington and, based on midwinter waterfowl inventories over the last decade, probably few winter there (SGM). Largest seasonal concentrations are on Banks L., sometimes exceeding 100 birds. The mean occurrence (<0.001% for three sites) on eastside CBCs is quite low.

Bellrose (1976) estimated 1700 wintering Red-breasted Mergansers in Washington while estimates derived from 1992-99 PSAMP winter aerial surveys ranged from 1659 to 2805 birds in w. Washington (Kraege). Comparisons of recent PSAMP surveys with MESA surveys in 1978-79 in n. inland waters (Nysewander et al. 2001) suggest that numbers of all merganser species, including this one, increased by 55%. Long-term CBC data indicate increases in the 1990s at Sequim-Dungeness, Seattle, and Tacoma but decreases elsewhere in w. Washington.

Remarks: There are no definite breeding records for Washington: infrequent reports of breeders appear to all result from misidentification of Common Mergansers.

Noteworthy Records: *CBC high counts*, **West**: 430 in 1980 at Bellingham; 243 in 1996 at Edmonds; 296 in 1977 in Kitsap Co.; 873 in 1984 at Olympia; 542 in 1986 at Padilla Bay; 216 in1997 at Port Gamble; 494 in 1986 at Port Townsend; 1,846 in 1997 at Sequim/Dungeness; 395 in 1979 at Seattle; 300 in 1978 via Anacortes-Sidney ferry; 748 in 1985 in San Juan Is.; 378 in 1980 at Tacoma. **East**: 9 in 1993-94 at Grand Coulee; 2 in 1997 at Lyle; 4 in 1995 at Moses L.; 1 in 1972 at Spokane; 1 in 1975, 1980, and 1987 at Tri-Cities; 2 in 1997 at Two Rivers; 2 in 1974 and 1978 at Wenatchee. *Winter/migration:* 150 on 8 Mar 1964 on lower Columbia R. nr. Stella; 75 nr. Longview at smelt run 1966; MESA: 655 on 16 Feb 1978 in Fidalgo Bay; 169 on 30 Apr 1978 at Lummi Bay and 239 on 3 May 1979 nearby in Hale Passage. 165 on 15 Nov 1992 at Banks L. *Summer*: one on 23 Jun 1998 at Umatilla NWR.

David R. Nysewander

Ruddy Duck *Oxyura jamaicensis*

Locally common migrant and winter visitor; locally uncommon breeder on lakes and sewage-treatment ponds, rare elsewhere in summer in w. Common migrant and breeder in marsh habitats and fairly common in winter on large lakes and rivers in e. Washington.

Subspecies: *O. j. rubida.*

Habitat: Breeds in marshes with relatively stable water levels, utilizing emergent vegetation (Johnsgard 1975); winters primarily in brackish water in relatively shallow bays or inlets and on freshwater ponds and lakes and sewage-treatment ponds.

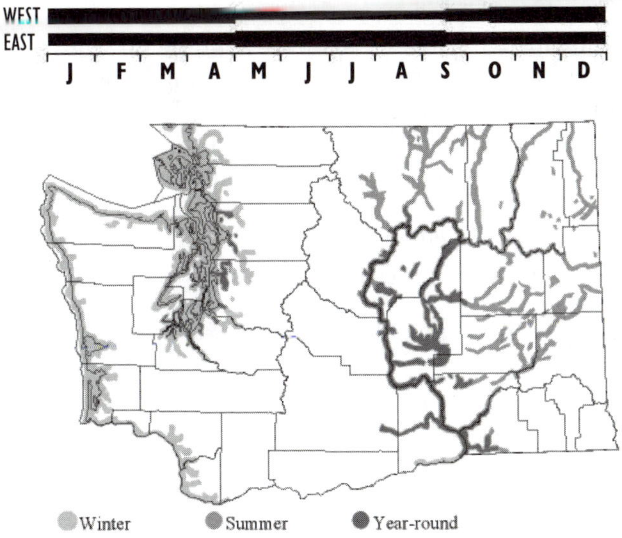

Occurrence: Most Ruddy Ducks on the Pacific coast winter s. of Washington, with largest concentrations in California and Mexico. Usually fewer than 3000 birds were estimated to occur in small flocks in marine waters and larger lakes in w. Washington between Dec and Mar, and Bellrose (1976) reported 3700 in the past for this region. Aerial surveys throughout inland marine waters during 1992-99 (PSAMP) indicated Ruddies comprise, on average, 2.5% of wintering diving ducks. Low numbers winter in e. Washington, with a maximum of 1244 reported on midwinter inventories. Large migratory concentrations are found in fall on eastside alkaline lakes, particularly Soap L. and L. Lenore, totaling up to 5000 birds.

Estimated westside winter numbers may be conservative. Up to 1000 winter on the Everett sewage treatment ponds and similar numbers use habitats on Whidbey I. (SGM).

Other large westside concentrations (up to 2000+ per flock) have occurred in the past, primarily at Drayton Harbor, Birch Bay, Padilla Bay, Skagit/Similk bays, Snohomish R. area, Port Townsend and Marrowstone I. vicinity, Lynch Cove in Hood Canal, Budd Inlet, Eld Inlet, Totten Inlet, Peale Passage nr. Squaxin I., and Grays Harbor. Smaller numbers (usually <100/flock) occur locally in bays and inlets throughout Puget Sound and Hood Canal and in the San Juan Is. Birds occur on lakes of varying size, from Green L. to L. Washington. Eastside numbers are lower, though rafts of 1150-5000 are noted locally on larger lakes and rivers. The origin of many wintering birds may be from n. of Washington (see Campbell et al. 1990a).

A common breeder in e. Washington: the second-most numerous diving duck there. Common in the c. Columbia Basin, the Palouse, and river valleys in n. counties, while essentially absent from parts of the w. Columbia Basin (Smith et al. 1997). Duck brood indices between 1993 and 2000 (Monda 2000) list between 102 and 530 broods per year, with no discernible trend over the last decade. Apr-May population indices (WDFW data) averaged 11,102 per year during the 1979-99 period. Westside birds are locally fairly common summer residents on larger freshwater marshes and sewage-treatment ponds (SGM).

Long-term, CBC numbers at w. Washington increased significantly ($P < 0.01$) and decreased e. of the Cascades ($P < 0.01$).

Remarks: The N. American/Caribbean population was estimated at about 680,300 (Hyslop and Kennedy 1996). Classed PHS.

Noteworthy Records: *Winter*, **West**: 350 on 5 Oct 1965 at Green L.; 1200 on 5 Oct-14 Nov 1998 at Everett STP; 840 0n 28 Nov 1998 at Penn Cove. East: 1150 in Nov 1986 at McNary NWR; 3000-5000 in Oct-Nov 1986 at Soap L. and L. Lenore. *Spring*, **East**: 1000 peak in late Apr 1974 at Turnbull NWR. *CBC high counts*, **West**: 1162 in 1990 at Olympia; 1201 in 1985 at Padilla Bay; 1136 in 1989 at Port Townsend; 1039 in 1978 at Tacoma. **East**: 55 in 1998 at Bridgeport; 52 in 1998 at Columbia Hills/Klickitat; 52 in 1998 at Chelan; 169 in 1989 at Moses L.; 104 in 1995 in Wenatchee.

David R. Nysewander

Chukar *Alectoris chukar*

Common along most major rivers and steep canyons in e. Washington, rare and local w.

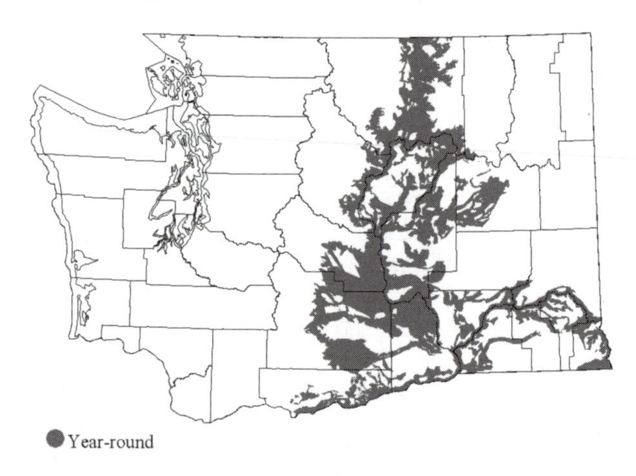

Subspecies: Unknown which of 14 subspecies or intergrades (Christensen 1996) occur.

Habitat: Characteristically talus areas containing annual grasses, bunch grasses, and sagebrush. Riparian areas, springs, and mesic slopes containing shrubs are also important. Cheatgrass is the main food source for Chukars throughout their range.

Occurrence: First introduced in Washington in 1931 in Grant and Garfield Cos., Chukars were released in 20 counties in the late 1930s and early 1940s. Chukar introductions were successful in e. Washington and by the early 1960s self-sustaining populations were well established along most major river drainages. They adapted well to Washington's semi-arid rocky canyons especially in areas of cheatgrass invasion due to intensive livestock grazing by early settlers. Chukars are most abundant in Washington along the middle sections of the Yakima R., the Columbia R., and the e. half of the Snake R. Healthy populations also occur along Oak Cr. in Yakima Co., Rock Cr. in Klickitat Co., Glade Cr. in Benton Co., Douglas Cr., Moses Coulee, and Banks L. in Douglas and Grant Cos., and Asotin Cr. and the Grand Ronde R. in Asotin Co. (WDFW, see also Weber and Larrison 1977, Smith et al. 1997).

Introductions in w. Washington were considered failures (as in the San Juan Is., see Lewis and Sharpe [1987]) and it has been recognized that suitable Chukar habitat does not occur w. of the Cascades. Privately released birds are occasional.

Population densities peaked in the late 1970s and have since declined in most areas. Several areas where Chukars were apparently successfully established in the 1950s and 1960s such as in Okanogan and Adams Cos. no longer support significant densities. Populations appear to be stable overall but they continue to demonstrate fairly dramatic annual fluctuations depending on winter severity and rainfall. Surveys of hunting harvest showed 26,913 taken in 2000-01 (WDFW 2001). BBS data show a significant ($P < 0.05$) long-term decrease (Sauer et al. 2001). CBC data indicate scattered winter occurrence with birds regular only at Wenatchee, where numbers appeared to decline in the 1990s. Field reports and non-systematic observations were not included in published seasonal reports after winter 1974-75.

Remarks: Called Rock Partridge by Jewett et al. (1953). The third-most harvested gallinaceous species in the state (WDFW 2001). Classed PHS.

David Ware

Gray Partridge *Perdix perdix*

Fairly common resident in agricultural areas of e. Washington.

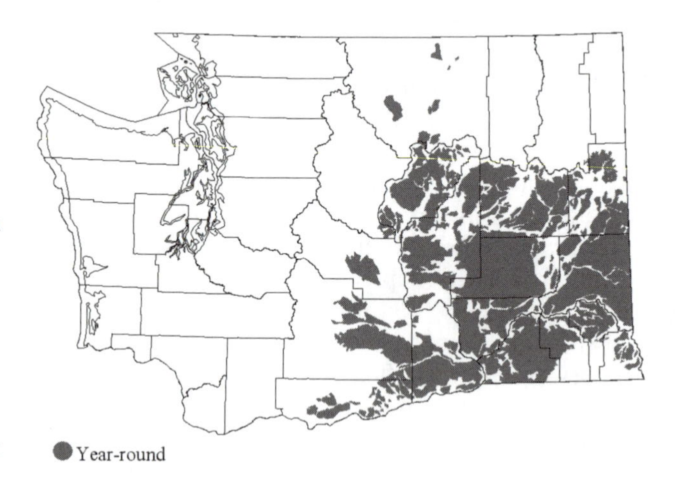

Subspecies: *P. p. perdix.*

Habitat: Shrub-steppe, dryland wheat, and other farming areas. Ponderosa habitats in Spokane and Lincoln Cos. (Smith et al. 1997). Most successful at elevations >335m, occurs to 1040 m on w. side of Blue Mts. Found in lightly forested grasslands in summer and fall, foraging on insects and seeds.

Occurrence: The status of the "Hungarian Partridge" is confused by widespread, long-term introductions in many areas of Washington (Smith et al. 1997). Believed first introduced in 1897 and 1906, this species was described as the "most successful" game bird in e. Washington and widespread on the westside (Jewett et al. 1953). It was first hunted in 1915 following releases in many locations where it is now extirpated, including the San Juan Is. (Lewis and Sharpe 1987), Skagit Co. (Jewett et al. 1953), and Whatcom Co. (Wahl 1995). Birds are fairly common in open country in e. Yakima Co. and the Yakima Training Center (Stepniewski 1999) and common in agricultural fields in the Palouse country (Weber and Larrison 1977), but rare in very hot areas and the lower Columbia Basin (Smith et al. 1997). Overall, BBA surveys indicated distribution from s. Okanogan e. to Spokane Cos., in Whitman, Adams, and Kittitas, Yakima, Klickitat, and Benton Cos. e. to n. Asotin Cos. (Smith et al. 1997). Campbell et al. (1990b) attributed populations in B.C., now restricted to the Okanagan Valley, to birds spreading from Washington.

Field observers reported declines in numbers due to deep snow cover in severe winters, as in 1968-69, and extended cold spring rains reduce chick survival. Hunted over its range in se. counties, cyclic in populations, seldom found in stable numbers anywhere. Hunting harvest surveys in 2000-01 showed 18,119 taken, in e. counties w. to Klickitat (WDFW 2001). The species was recorded fairly consistently, in highly variable numbers, between 1971 and 1999 on CBCs at Ellensburg, Spokane, Wenatchee and Walla Walla, and seven times at Yakima Valley in the 1980s and 1990s.

Terence R. Wahl

Ring-necked Pheasant *Phasianus colchicus*

Abundant resident in agricultural areas of e. Washington and uncommon with sporadic distribution in the Puget Sound lowlands w.

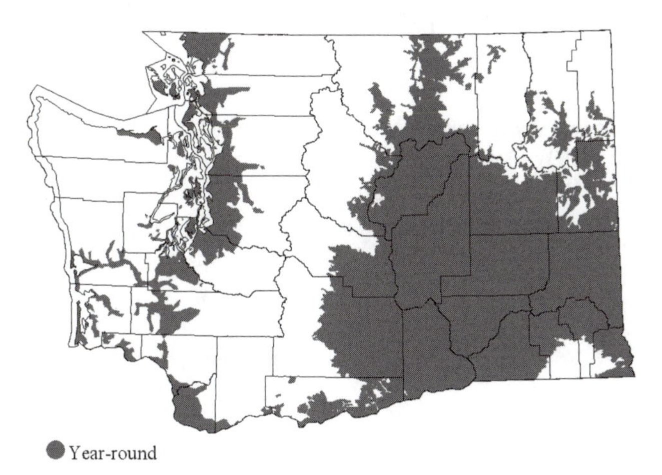

Habitat: Most numerous in agricultural areas where grain is the predominant crop and cropland is interspersed with permanent cover and unfarmed areas. Riparian zones and wetlands provide several important features of high-quality pheasant habitat. Brood-rearing habitat that includes broad-leaved plants and abundant insects critical for production of young and maintenance of self-sustaining populations.

Occurrence: Pheasants have been widely and repeatedly introduced throughout Washington since the late 1800s (see Jewett et al. 1953). At their peak in the early 1960s, 10 state-operated game farms were raising and releasing over 100,000 pheasants per year. WDFW records indicate that pheasants have been released in every county and that pheasant numbers peaked in the 1960s and 1970s.

Hunters reported harvesting over 500,000 pheasants annually during these peak years. Since then, harvest has steadily declined to just over 100,000 per year in the late 1990s and several counties are no longer represented in harvest statistics. The practice of releasing pheasants continues with about 35,000 pheasants released by WDFW each year in w. Washington and 20,000 in e. Washington. In addition, it is estimated that hunters, farmers, and other enthusiasts release another 20-30,000 pheasants each year.

Pheasant numbers and distribution have declined significantly in w. Washington with human population growth and declining grain production. Pheasant production is naturally limited in w. Washington due to high rainfall during peak hatching periods. Currently, self-sustaining pheasant populations persist in very limited areas of w. Washington. Without continual supplementation, pheasants may not persist w. of the Cascade Mts.

Many areas in e. Washington have also seen dramatic declines in pheasant numbers and distribution since the peaks of the 1960s and 1970s. Initially, as the landscape of e. Washington was changed to crop production, introduced birds such as pheasants fared very well. As farming technology, cultural practices, and cropping patterns changed, pheasant densities declined. Pheasants flourished with less-sophisticated farming, greater proportions of non-crop areas, and where irrigation runoff created small wetland habitats. Long-term CBC data indicate declines statewide, particularly in areas being urbanized. Pheasant populations in e. Washington appear to be stable over the past 10 years and are likely to persist over the long term, especially in fringe areas surrounding irrigated cropland in the Yakima and Columbia R. basins and in the dryland wheat farming areas of the Snake R. basin. Stepniewski (1999) pointed out loss of edge habitats as affecting populations in Yakima Co.

Remarks: Classed PHS. Subspecies occurring not described. The Japanese Green Pheasant (*P.c. versicolor*), a subspecies, has been introduced in some areas.

David Ware

Ruffed Grouse *Bonasa umbellus*

Widespread, fairly common resident in deciduous and mixed-conifer shrublands and forests throughout most of Washington.

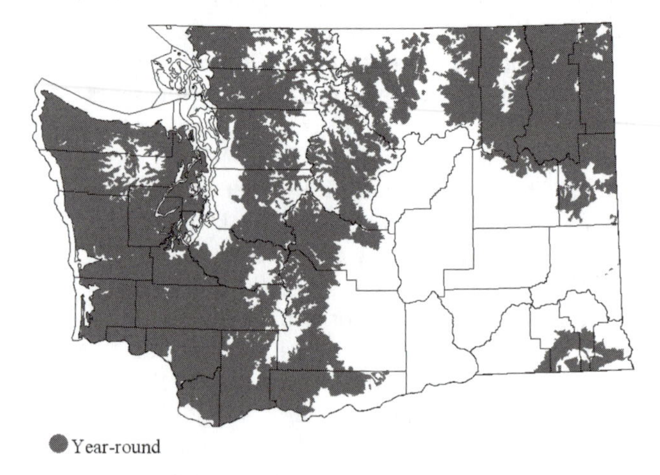

Subspecies: *B. u. brunnescens* in Puget Trough lowlands, *castanea* on Olympic Pen. s. to w. Oregon, *sabini* on w. side of Cascades from B.C. to Oregon, *affinis* on e. side of Cascades, n. Washington, and Blue Mts., and *phaia* in ne. corner.

Habitat: Deciduous and mixed deciduous/conifer forest containing birch, alder, and/or poplar. Fairly common in fire seres, brushy stream courses, alder thickets, and brushy forest edges. Birds shift habitats seasonally within home ranges; in winter they tend to use brushy areas and in summer areas that are more open.

Occurrence: Indigenous to all state regions except expansive areas of shrub-steppe. Except in developed w. lowlands, widespread in all forested zones below the silver fir zone; from low elevations into grand fir zone on Cascades e. slope, across n. counties to Spokane Co., and in Blue Mts. Common in hardwood and mixed forests at lower elevations and in riparian corridors at higher elevations, mostly absent from riparian corridors below the ponderosa pine zone in the Columbia Basin (Smith et al. 1997). Unrecorded in the past 50 years in the San Juan Is. (Lewis and Sharpe 1987). Though most common in lowlands, occurs in suitable habitats to mid-elevations (e.g. to 1300 m at Tunk Mt., 1600 m in the Blue Mts.; Jewett et al. 1953) and to about 1250 m in Yakima Co. (Stepniewski 1999). Overlaps with Blue Grouse up to elevations where hardwoods decline (Smith et al. 1997). Core areas include Indian Dan Canyon, Pogue Mt., Scotch Cr., Sinlahekin, and Chesaw WMAs in Okanogan Co.; Badger Mt. in Douglas Co.; along the Hoh R. in Jefferson Co. (MAS). Birds are sedentary, occupying the same territory year round, though juvs. can disperse as far as 19 km and birds occasionally occur in formerly occupied portions of their range such as urban and agricultural habitats. Down-slope movements in winter unknown.

Numbers have declined over w. N. America (Rusch et al. 2000). Population trends in Washington are uncertain: inter-annual variability is considerable. Smith et al. (1997) suggest increases in w. following logging and conversion of mature conifer forests to second-growth hardwood and mixed tracts. Statewide, declines are suggested due to decreases in brush fires (Jewett et al. 1953), habitat loss to urbanization and other developments (Smith et al. 1997), removal of riparian habitat and over-hunting in Yakima Co. (Stepniewski 1999), and likely predation on young by cats and dogs in urbanized areas (Wahl 1995).

Long-term data apparently minimal and restricted to relatively developed areas. Numbers declined on all long-term westside CBCs: these were in or nr. populated areas. CBC eastside numbers also lower at Spokane and Walla Walla. No birds were found at Wenatchee after 1985.

Remarks: Although populations fluctuate dramatically annually, they appear to be relatively unaffected by most management practices. The gray color phase is most common in e. Washington and the reddish-brown phase most common in coastal w. Population data and hunting harvest numbers are uncertain for this species, as well as Blue and Spruce grouse, which are all classed as and data lumped under Forest Grouse (WDFW 2000b). Harvest surveys for 2000-01 indicated 148,193 Forest Grouse taken.

Noteworthy Records: *Peripheral areas*: Turnbull NWR in 1968-1972 ; n. portion of Pend Oreille Co. in 1980-1986 (MAS); 1 on 26 Aug 1972 in the low dunes at Leadbetter Pt.; 1 record at Kamiak Butte (Weber and Larrison 1977).

Michael A. Schroeder

Greater Sage-Grouse *Centrocercus urophasianus*

Very local, uncommon resident in good-quality shrub-steppe habitats of c. Washington.

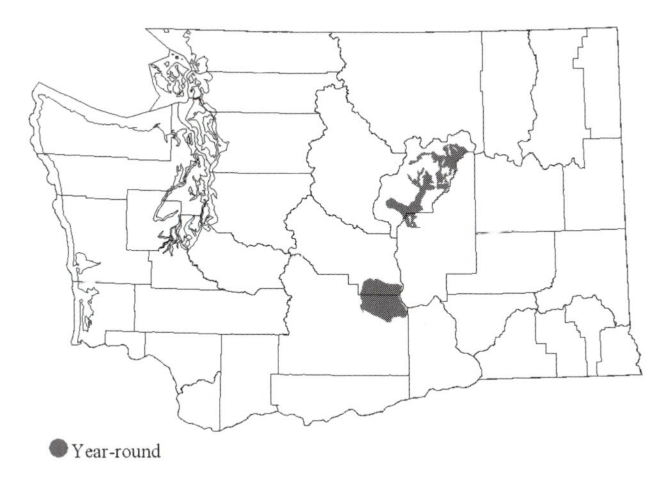

Subspecies: Western Sage-Grouse, *C. u. phaios.*

Habitat: Shrub-steppe and meadow steppe habitat dominated by sagebrush and bluebunch wheatgrass. Fairly common along edges of wheat fields associated with native shrub/grass habitat. Also croplands planted in grass as part of agricultural set-aside programs such as the Conservation Reserve Program.

Occurrence: One of the most habitat-specific birds in Washington and illustrative of the effects of human-caused habitat changes on natural ecosystems. Jewett et al. (1953) noted extirpation from parts of original range and decrease in numbers from descriptions in the mid-1800s and early 1900s, and Smith et al. (1997) and Stepniewski (1999) described similar changes to the present. Historically distributed in sagebrush-dominated habitats of w. N. America including portions of Montana s. to Colorado and California and w. to Oregon and Washington (see Schroeder et al. 2004). Current range in Washington essentially limited to shrub-steppe/wheat areas of one area in Douglas and Beezley Hills area of Grant Co., and on the Yakima Training Center in Kittitas and Yakima Cos. (MAS). Peripherally, birds are occasionally observed in areas adjacent to known populations, including Bridgeport SP, Dyre Hill, Badger Mt., and e. side of Banks L. (MAS). Additionally, recent observations in areas with extirpated populations include the Methow and Scotch Cr. WMAs and Colville Indian Agency area in Okanogan Co., Rattlesnake Hills, Quilomene WMA and Umtanum Ridge, Saddle Mts., and Swanson Lks. WMA, Lincoln Co. (MAS).

Gregarious males form flocks associated with specific leks (communal display sites) and defend territories on leks: ≥100 males may appear on a single lek. Females visit leks in small flocks. During winter, flocks of 4-50 birds are common, occasionally as many as 200 birds. Strutting activity is most intense during early-morning hours and period of peak female lek attendance. Males display fidelity to lek sites that are used repeatedly from year to year. Sagebrush is eaten throughout the year, and in winter it comprises 100% of the diet. Flowers, leaves, and invertebrates are also consumed during summer. Birds may migrate between winter and summer ranges in late Apr-early May, and return to winter areas in late Aug-early Oct, and to breeding areas in mid-Feb and early Mar. Movements are as far as 75 km.

Populations have declined dramatically due to conversion of native habitat to cropland and degradation of remaining uncultivated habitats by overgrazing and shrub removal (Schroeder et al. 1999, 2004; see Knick et al. 2003). Smith et al. (1997) estimated that just 10-30% of historic Washington range is now occupied. Estimated 1998 breeding population in Washington is about 1000 birds, a decline of nearly 80% since the 1960s.

Remarks: Classed ST, FC, PHS. A Washington State Sage Grouse management plan was developed in 1995 (WDFW 1995a). A statewide harvest closure was implemented in 1988.

Noteworthy Records: 1-2 birds reported in 1968 nr. Clarkston (Weber and Larrison 1977).

Michael A. Schroeder

Spruce Grouse *Falcipennis canadensis*

Locally fairly common resident in moist lodgepole pine forests along e. slope of Cascades and n. Washington.

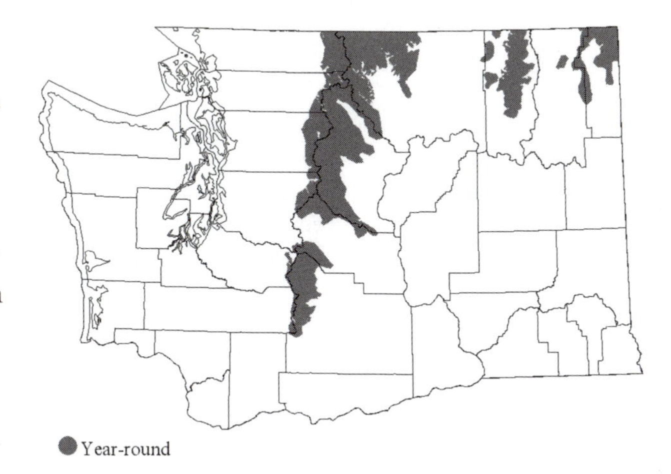

Subspecies: Franklin's Spruce Grouse, *F. c. franklinii.*

Habitat: Conifer forests, especially fire-adapted lodgepole pine forests. Greatest densities appear to be in young successional stands of dense lodgepole pine, 7-14 m height, with a well-developed middle/ under story of spruce, fir, and/or deciduous shrubs.

Occurrence: Range includes the upper e. slope of Cascades from the U.S.-Canada border s. to Yakima Co. and at higher elevations in Okanogan, Ferry, Stevens, and Pend Oreille Cos. Occurrence s. of Yakima Co. doubtful (Smith et al. 1997, Stepniewski 1999). Though at least one "very good description" exists of a bird in the Olympics (Smith et al. 1997), occurrence there remains questionable. And, though the species occurs in the Wallowa Mts. in ne. Oregon (Gilligan et al. 1994), it is not recorded in the Blue Mts.

Though resident with some birds remaining on territory year round, there are also some minor seasonal shifts in habitat and altitude. Some migrate up to 11 km between breeding and winter areas. Females tend to migrate more frequently and over greater distances than males. Timing of migration varies among populations but in general birds depart winter range mid-Feb to mid-May, depart breeding range mid-Aug to late Dec. Territorial behavior peaks during the breeding season when birds forage on forbs and invertebrates. Arboreal in winter, foraging primarily on lodgepole pine needles.

Although populations in Washington have not been adequately monitored, available transect and harvest information indicates that overall trends are downward (e.g. see DeSante and George 1994) because of forest management issues such as succession, timber harvest, fire suppression, and overgrazing.

Remarks: Reports of birds along Elwha R. on 4 May 1975, observations on Hurricane Ridge, at Iron Springs on 9 Sep 1979 were outside the known distribution and unverified. Numbers harvested by hunters are included under Forest Grouse (WDFW 2001) and are uncertain.

Noteworthy Records: *Peripheral area*: fall 1988 on Chelan Mt. and nr. Mission Pk., Chelan Co. (F. C. Zwickel p.c.); fall 1988 on Red Top Mt. Kittitas Co.; 2 on 12 May 1990 on Pine Grass Ridge, 1 nr. Bear Creek Mt. (F. Martinsen p.c.), and others nr. Bumping L. (W. Oliver p.c.), Yakima Co.

Michael A. Schroeder

White-tailed Ptarmigan *Lagopus leucurus*

Locally fairly common resident in alpine tundra in the Cascades.

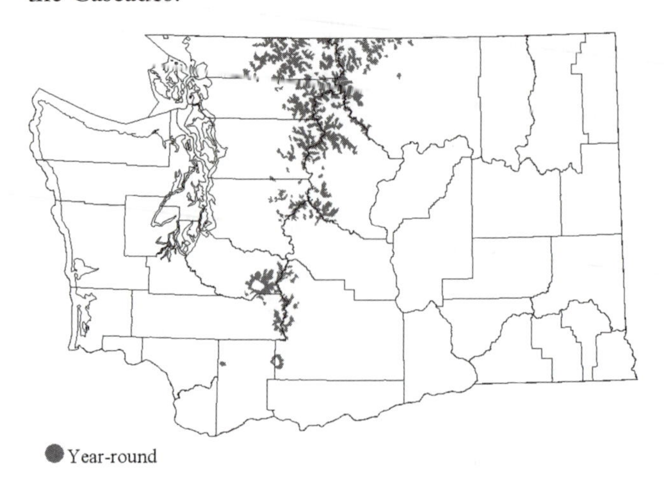

Subspecies: N. White-tailed Ptarmigan, *L. l. leucurus*, in n. Cascades and Mt. Rainier White-tailed Ptarmigan, *rainierensis*, in the s. Cascades.

Habitat: Alpine tundra consisting of moist vegetation nr. snow and/or boulder fields, willow-dominated plant communities, and rock meadows.

Occurrence: A resident of alpine tundra in w. N. America from Alaska s. to Washington, e. to Yukon and to Colorado, the White-tailed Ptarmigan is widespread and locally fairly common in the Cascades of Washington, where it is recorded to 2400 m. Reports of birds in the Olympic Mts. are not confirmed.

Birds arrive on breeding areas in early Apr-early Jun and depart for winter sites in late Sep to mid-Nov. Local populations may migrate between low-elevation winter areas and high-elevation summer areas, with willow buds, leaves, and twigs common diet items. Due to the species' high-elevation distribution, local occurrence and migration may be incompletely known. Birds may move considerable distances: a bird was photographed 30 km from nearest breeding habitat in Yakima Co. (see Stepniewski 1999). Flocks often segregate by sex in winter, with females tending to move farther than males. Males arrive on breeding range before females and defend territories by ground and aerial displays. Loosely organized flocks of broods, unsuccessful females, and males form in late summer.

Reports from non-ptarmigan habitat have usually been due to mistaken identification, but birds are occasionally observed at relatively low elevations in clear-cuts during winter and rarely at locations at low elevations (e.g., Steveston, B.C., at the mouth of the Fraser R., in 1990).

Remarks: There is little evidence of population fluctuations in Washington due to human-related activity, though overgrazing by domestic sheep may be a problem in some areas. Not hunted in Washington. There is a record of a Rock Ptarmigan, *L. mutus,* in Whiterock, B. C., just n. of Blaine, on 9-18 Jan 1976 (Campbell et al. 1990b).

Noteworthy Records: *Core areas*: Whatcom Co., Twin Sisters Mt.; 2 on 16 Oct 1977, 1 on 10 May 1978; Table Mt.: 1 on 13 Dec 1981; Mt. Terror: 1 on 21 Jul 1978, ≥ 1 on 25 Jul 1989; Crater Mt: 1 on 15 Jul 1985;. N. Pickets: 2 on 9 Aug 1971; Copper Mt.: female, chicks on 21 Aug 1991; Mt. Baker: 2 on 19 Sep 1971, 7 on 10 Sep 1973, 1 on 1 Nov 1987. Skagit Co., Cascade Pass: 1 on 13 Aug 1974; Forbidden Pk.: ≥1 on 28 Jul 1979; Overdrive Tower: 1 on 23 Jul 1987; Diobsud Butte: ≥ 1 on 8 Jun; Red Rks.: ≥ 1 on 7 Jan 1990; Monogram L.: female, chick on 18 Aug 1994. Snohomish Co., Glacier Pk.: 1 on 26 Jun 1983; Merchant Pk.: 1 on 6 May 1989. King Co., Snohomish Pk.: ≥ 1 on 21 Jun 1980; Big Snow Mt.: ≥ 1 on 21 Sep 1994; Granite Mt.: female, 5 chicks on 25 Jul 1986. Pierce Co., Panarama Pt./Nisqually Moraine: 1 on 13 Jul 1969, 1 on 22 Jul 1970, female, 6 chicks on 21 Jul 1978, female, 4 chicks on 7 Sep 1984, 5 on 12 Sep 1985, 4 on 10 Sep 1988, ≥ 1 on 10 Sep 1989, female, 3 chicks on 9 Sep 1990; Burroughs Mt.: female, 4 chicks on 22 Aug 1973, 4 on 7 Sep 1978, 9 on 9 Jul 1986. Okanogan Co, Slate Pk./Harts Pass: 13 on 2 Sep 1973, 1 on 11 Aug 1976, 2 on 5 Aug 1978, 1 in summer 1984, 1 on 24 Jul 1984, 1 on 1 Nov 1987, female, 5 chicks on 16 and 22 Jul 1998; Ptarmigan Pk.: ≥1 on 10 Aug 1991; Cathedral Pass: female, 8 chicks in summer 1983; Tiffany Mt.: 3 on 9 Aug 1988; Chopaka Mt.: 2 females, 6 chicks on 25 Jul 1991; Armstrong Pk.: many small flocks and broods in Aug and Sep 1997 and 1998; Horseshoe Mt.: flock of 7 on 5 Sep 1998. Chelan Co., Sahale Mt.: female, 4 chicks on 6 Aug 1974; Mt. Benzarino: 2 on 16 Sep 1984; Dumbell Mt.: 2 on 13 Aug 1989; Cardinal Pk.: 2 on 24 May 1987; High Pass: 1 on 29 Aug 1968, female, 5 chicks on 18 Aug 1969; Glacier Pk.: 1 in late Jun 1983; Alpine Lakes Wilderness: 1 in fall 1976. Kittitas Co., Bear's Breast Mt.: female, chicks on 4 Aug 1990. Skamania Co., Mt. St. Helens: 8 on 26 Jun 1966, 1 on 22 Jun 1986, and 1 on 4 Jan 1987 (MAS, Armstrong 1994, Hunn 1994, AFN, AB, FN, NAB, WOSN).

Michael A. Schroeder

Blue Grouse (G. Scott Mills)

Blue Grouse *Dendragapus obscurus*

Widespread, fairly common resident in shrub and forest habitats of Washington.

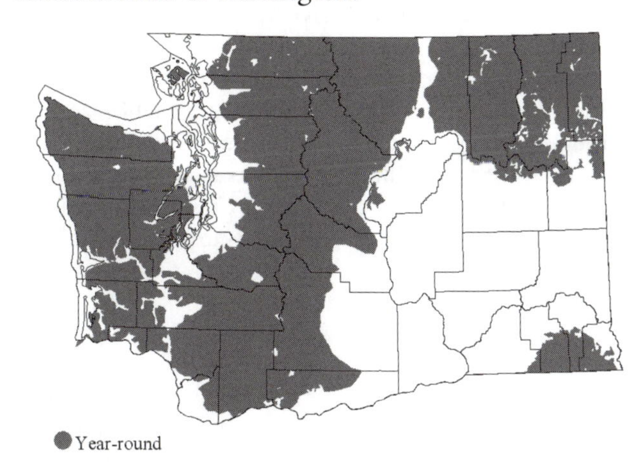

Subspecies: Sooty Blue Grouse, *D. o. fuliginosus,* in higher portions of Cascades in n. Washington and areas w. of Cascades, Sierra Blue Grouse, *sierrae,* in s. Cascades, and Oregon Blue Grouse, *pallidus,* in e. Washington including e. slopes of n. Cascades.

Habitat: *D. o. pallidus* found in shrub-steppe, meadow steppe, Douglas-fir, ponderosa pine, aspen, western larch, and pine-spruce forests; not in steppe habitats in winter. *D. o. fuliginosus* resident in forest communities from sea level to alpine; *sierrae* in alpine/subalpine forest communities. Sooty and Sierra Blue grouse tend to be more arboreal than Oregon Blue Grouse.

Occurrence: Breeds over much of montane region throughout w. N. America, generally between Ruffed Grouse at low elevations and Spruce Grouse in higher forests, though overlap and outright exceptions occur in certain habitats (Jewett et al. 1953, Smith et al. 1997). Fairly common in many regions of Washington except large expanses of relatively flat shrub-steppe habitat.

Mostly solitary but loose flocks of broods may form in late summer, and small flocks may form during winter. Birds mainly terrestrial in breeding season, more arboreal in winter. Varied diet in summer (leaves, flowers, invertebrates), winter diet primarily consists of conifer needles such as Douglas-fir. Although some are resident at upper elevations, most migrate between relatively low-elevation breeding ranges and high-elevation winter sites. Seasonal shift in habitat use particularly dramatic for Oregon Blue Grouse. Birds depart winter range late Mar to mid-Apr, depart breeding range mid-Jun to late Oct. Fifty percent of movements are greater than 8 km in n. c. Washington; longest known movement 50 km.

East of the Cascades, field surveys showed core areas of occurrence at Pinegrass and Jumpoff ridges in Yakima Co.; Manastash, Taneum, and Naneum ridges, Manastash, Robinson, and Reecer creeks, and Colockum Pass in Kittitas Co.; Chumstick Mt. and Chelan Butte in Chelan Co.; Methow and Chewack Rs., Harts, Freezeout, and Lone Frank Pass, Salmon Meadows, and Indian Dan Canyon, Chesaw, and Scotch Cr. Wildlife Areas in Okanogan Co.; Central Ferry Canyon Wildlife Area and Badger Mt. in Douglas Co. (MAS). BBA surveys found birds in forested w. lowlands, the Olympics, the Cascades, and e. to Pend Oreille and Spokane Cos., and Blue Mts. (Smith et al. 1997). In the San Juans they are found on Orcas and Stuart Is. (Lewis and Sharpe 1987) and nearby Lummi I. (Wahl 1995). Historical occurrence to nr. sea level suggested by current presence there on the Olympic Pen. (Smith et al. 1997) and records within recent decades nr. forested shorelines in Pacific and Whatcom Cos. (Wahl 1995).

Birds are occasionally observed in former portions of their range such as urban and agricultural habitats. Locally, birds may repopulate areas where logged-off forests are regenerating. Winter records include a few birds on CBCs at Leadbetter Pt. (18 in 1973), Grays Harbor, Sequim-Dungeness, Bellingham, N. Cascades, Chewelah, Spokane, Yakima Valley, and Kitsap Co.

Although populations in Washington have not been adequately monitored, available information indicates that overall trends are downward. Populations have been permanently eliminated in large areas of w. Washington due to residential development. Overgrazing by livestock may have a negative impact on Blue Grouse in e. Washington. BBS data showed a significant population decline in Washington ($P < 0.01$) from 1966 to 2000 (Sauer et al. 2001).

Remarks: Numbers taken by hunters are included in Forest Grouse in hunting harvest surveys, along with Ruffed and Spruce grouse (see WDFW 2000b, 2001). Classed PHS, GL.

Noteworthy Records: Birds hooting nr. mouth of Columbia R. in Pacific Co., 4 Jul 1973.

Michael A. Schroeder

Sharp-tailed Grouse *Tympanuchus phasianellus*

Very local, uncommon resident in good-quality grasslands in n.c. Washington.

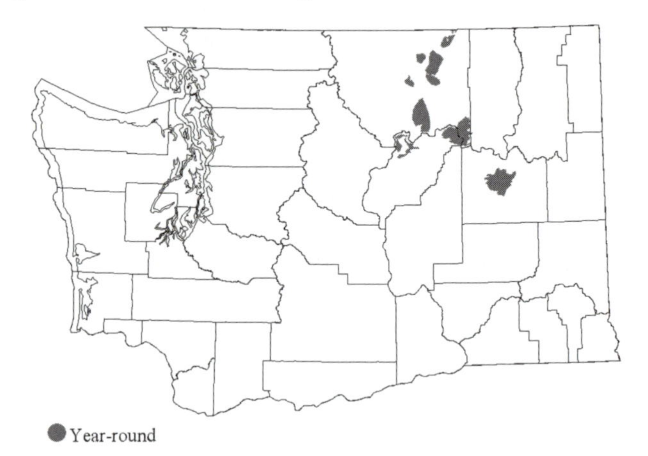

Subspecies: Columbian Sharp-tailed Grouse, *T. p. columbianus.*

Habitat: Shrub-steppe, meadow steppe, steppe, intermixed with deciduous mountain shrub, and riparian habitat. Important habitat components include deciduous shrubs and small trees, sagebrush, fescue, and wheatgrass. Frequently uses cropland adjacent to native habitat and croplands that have been planted with grass as part of agricultural set-aside programs such as Conservation Reserve Program.

Occurrence: Sharp-tailed Grouse were historically distributed in expansive areas of grass, shrub-steppe, mountain shrub, and early successional stages of forest throughout w. N. America from Alaska to Minnesota, Utah, and Oregon. Noted decreasing from the 1860s on (e.g., Jewett et al. 1953), with loss of bunch-grass habitat one cause, and later similarly described by Weber and Larrison (1977), Smith et al. (1997), Stepniewski (1999) and in B.C. by Campbell et al. (1990b). They are now greatly reduced in range with Washington birds limited to small remnant habitats in Okanogan Co. at Scotch Cr. and Chesaw Wildlife Areas and Nespelem, Tunk Valley, Horse Springs Coulee, Bonaparte Cr., and Siwash Cr. areas in n. Douglas Co. at Central Ferry Canyon and W. Foster Cr. Wildlife Areas and n. edges of the county, and at Swanson Lks. Wildlife

Area in Lincoln Co. (MAS). Birds are occasionally observed in adjacent areas including the Omak/Okanogan and Havillah areas, nr. Withrow, and nr. Odessa (MAS). Recent observations include remnant birds in areas with extirpated populations including the Twisp, Winthrop, and Oroville areas (MAS).

Loss and fragmentation of habitat affects movements. Males form flocks associated with leks during breeding season, females often visit leks in small flocks. Birds migrate up to 34 km between breeding and winter range, with females tending to move further than males. Juvs. in other regions may disperse as far as 150 km. There may be some local movement from lower elevations to higher, forested wintering areas with snow depths adequate for roosting (Campbell et al. 1990b).

Populations have declined dramatically due to conversion of native habitat to cropland and degradation of the remaining uncultivated habitat because of overgrazing, shrub removal, and flooding associated with dams (see Connelly et al. 1998). Estimated 1998 breeding population in Washington was about 1000 birds.

Remarks: A Washington State Sharp-tailed Grouse management plan was developed in 1995 (WDFW 1995b). A statewide hunting harvest closure was implemented in 1988. Classed PHS, ST, FSC.

Noteworthy Records: Observations in 1964 nr. Deer Park, Spokane Co., and in 1974 nr. Yakima appear to represent extinct populations.

Michael A. Schroeder

Wild Turkey *Meleagris gallopavo*

Locally common in many parts of e. Washington. Fairly common but uncertainly established locally in sw. Washington.

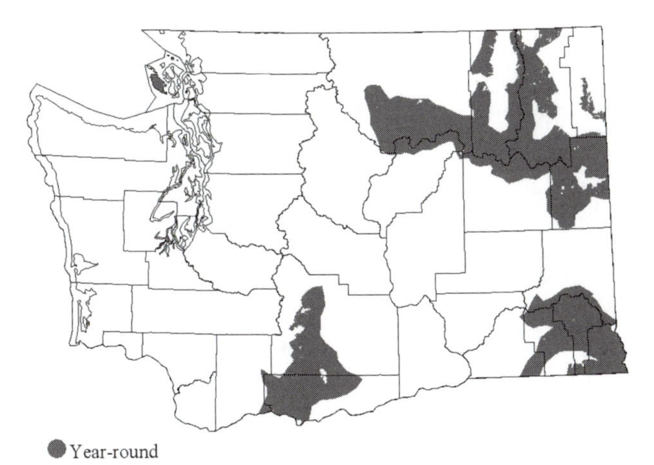

Subspecies: Rio Grande Turkeys, *M. g. intermedia,* released in Blue Mts. foothills, e. slope of the Cascades, Lincoln and w. Okanogan Cos., Merriam's, *merriami,* in ne. and Klickitat Co., Eastern, *silvestris,* in sw.

Habitat: Many habitat types with key components of forests and grasslands. Trees supply roosts and nuts and fruits and birds require grasslands for food and for brooding young or poults, and take advantage of grain crops and livestock pastures.

Occurrence: Introduced into various counties in Washington since at least 1913, though early releases were not successful. The first successful introduction was by the Department of Game in 1961 in Stevens and Klickitat Cos. Birds were subsequently trapped from Stevens Co. and moved to several other areas of the state with limited success. Populations were sustained in Klickitat and Stevens Cos. for many years, although they declined significantly through the 1970s. Beginning in 1984, the WDG started an aggressive effort to introduce turkeys: *intermedia* from Texas, *merriami* from S. Dakota and *silvestris* from Pennsylvania, Missouri, and Iowa. Additional turkeys are regularly trapped within the state and relocated to other areas.

Turkeys have been established in greatest densities in oak and ponderosa pine habitats in ne. Washington, especially in Stevens and n. Lincoln Cos. They are also doing very well in the foothills of the Blue Mts. and in Klickitat Co. (see WDFW 2000b). Introductions in other areas have been less successful, but reproduction is occurring and populations are likely to be sustained for a time, especially in areas of sw. Washington. The hunting harvest in 2000 was 1791, with 43 of these in w. counties (WDFW 2001).

Occurrences on CBCs have been sporadic except in the San Juan Is. where a well-known, almost-tame population was introduced in the 1970s (Lewis and Sharpe 1987), is protected and numbers on the CBC appear to have increased in the 1990s. Walla Walla was the only eastside CBC with relatively consistent occurrence of turkeys. Classed PHS.

Noteworthy Records: 77 on 18 Dec 1999 at Mill Cr., Walla Walla Co., 40 on 22 Jan 2000 at Joseph Cr., Asotin Co.

David Ware

Mountain Quail *Oreortyx pictus*

Uncommon and local on sw. Olympic Pen. and adjacent Puget Trough, most frequently observed in n. Mason and Kitsap Cos. Rare in sc. Washington.

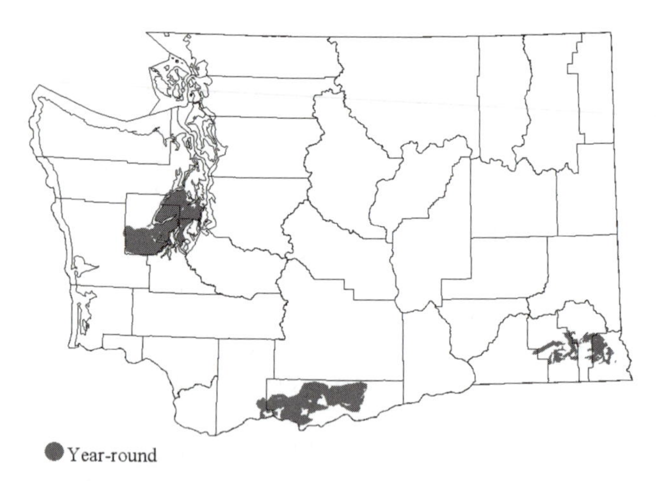

Subspecies: *O. p. palmeri* and *pictus* in Washington.

Habitat: Very dense brush cover, shrubby vegetation including clear-cuts, in forested areas and in riparian habitats. Noted especially in stands of Pacific madrone and scotch broom in some areas.

Occurrence: Resident in the Cascades of s. Washington through Oregon in the Coast and Sierra Nevada ranges s. through California and n. Baja California. Status n. of the Columbia R was long in question (e.g., Jewett et al. 1953), and the northernmost populations of Washington, Vancouver I., and Willamette Valley are now all believed likely to be from introductions beginning in the mid-1800s. Populations from the Great Basin extend into the Blue Mts. of se. Washington, where the abundance and range of the quail is declining (Vogel and Reese 1995). Very few birds remain in the se. except in riparian areas along the Snake R. and at Grand Ronde tributaries such as Deer Cr. Elsewhere: gone by the 1930s from Orcas and Waldron Is. in the San Juans (Lewis and Sharpe 1987), where birds were introduced in 1905. Apparently extirpated, for example, in Yakima Co. (Stepniewski 1999). Scattered sightings, as in Pierce Co. in 2000, suggest relict populations or birds recently released by private individuals. Their current status in the s. Cascades is unknown; there are no recent records. Thus, the core population is currently in Mason, Kitsap, and ne. Grays Harbor Cos. (Morse 2001).

Unlike populations to the s., which may make elevational migrations and utilize high mountain ranges (600-3000 m), elevational migrations are not apparent in w. Washington birds, which can be found throughout the year at or nr. sea level. Diet includes a variety of vegetative matter that does not suggest specialized foraging (Crawford and Pope 1999). Birds can be found with some ease in the right habitats of Kitsap and Mason Cos. where B. Angerman (WDFW) trapped birds along Puget Sound shorelines in madrona stands. Birds also are often found in brushy areas with scotch broom stands and have been surveyed and collected in the past few years in these locations. The consistency of occurrence in these habitats in part suggests that locally food could be one limiting factor: crops from birds harvested in fall show exclusive scotch broom seed in w. Washington samples. WDFW estimates harvests of 50-100 per year in Mason and Kitsap Cos.: these are likely Mountain Quail, the only quail consistently encountered there.

Remarks: Subspecies definitions are rather dubious and confounding (Guitierrez and Delehanty 1999). *O. p. eremophilus, russell,* and *confinus* occur elsewhere within the species range. Genetic studies of 26 loci did not provide any differentiation between the two subspecies found in Washington (K. Islam, Oregon State University). Species described as possibly indigenous in B.C. (Campbell et al. 1990b). Classed PHS.

Noteworthy Records: West: spring 1979 and spring 1980 at Nisqually; 1 heard in spring 1980 10 km w. of Mt. St. Helens; 2 ad. with 2 chicks on 20 Jul 2000 in Pierce Co., from a previously unreported population. **East**: ad. with 4 young 16 km e. of Lyle, Klickitat Co., in summer 1975; pair with 5-6 young on 6 Jul 1976 at Lyle.

Gregory Schirato

California Quail *Callipepla californica*

Introduced. Widespread, common to locally very common in e. Washington, locally common on westside.

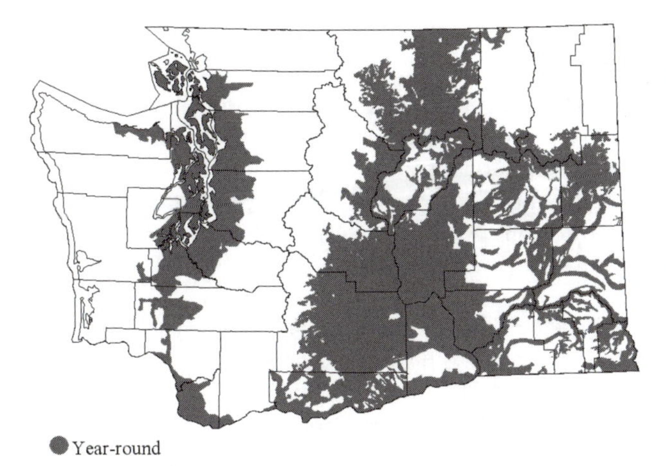

Subspecies: Possibly mixed *C. c. californica* and *orecta* (Jewett et al. 1953, see Smith et al. 1997).

Habitat: Open agricultural areas, irrigated areas, edges, shrubby thickets, riparian areas, rural and urban gardens, parks, powerline rights-of-way. Less frequent in sagebrush flats, ravines, deciduous bottomlands, open ponderosa pine forests.

Occurrence: Reported to have been widely introduced throughout Washington, as early as 1857 (Jewett et al. 1953), California Quail are widespread in e. Washington and locally in the humid transition zone in the Puget Trough, e. Clallam Co., the San Juan Is., sw. Skagit and Island Cos. (Jewett et al. 1953, Smith et al 1997, Stepniewski 1999). Adaptable to many human-created habitats, the species is better established than other gamebirds in urban areas (such as Seattle; Hunn 1982) and in the San Juan Is. (Lewis and Sharpe 1987).

Observers reported variations in numbers due to winter weather, as under extremely cold conditions in 1968-69 (Campbell et al. 1990b, and see Stepniewski 1999), and "huge" numbers in spring 1990 in Spokane following a mild winter. Long-term, habitat loss affects numbers, as at Pt. Roberts (Wahl 1995), if not eliminating populations. Like other species, releases of hand-raised birds occur widely, though survival and establishment of populations are doubtful.

Population trends reflected in surveys and harvest data are uncertain but suggest overall decline (WDFW 2000b). Hunting harvest surveys indicated 131,789 birds taken in 2000-01, virtually all from Klickitat Co. and e. Washington (WDFW 2001). BBS data indicated a significant population increase ($P < 0.05$) statewide from 1966 to 2000 (Sauer et al. 2001), differing from a calculated decrease from 1966 to 1994 (see Calkins et al. 1999). Largest CBC numbers from long-term sites occurred at Ellensburg, Spokane, Tri-cities, Wenatchee, Walla Walla, and Yakima Valley in e. Washington and at Sequim-Dungeness on the westside.

Remarks: Various authors (e.g., Jewett et al. 1953) suggest the species was possibly indigenous in sw. Washington.

Terence R. Wahl

Northern Bobwhite *Colinus virginianus*

Introduced, locally uncommon resident in southern Puget Trough. Variable numbers and distribution result from releases.

Subspecies: Three introduced: *C. v. virginianus, texanus, taylori.*

Habitat: Grassy and shrubby areas, agricultural habitats (Smith et al. 1997).

Occurrence: A widely introduced gamebird that has declined drastically in Washington in recent decades (e.g. Smith et al. 1997, Stepniewski 1999, and see Brennan 1999); the bobwhite population is self-sustaining only in Pierce and Thurston Cos. and very locally in Kitsap and perhaps King Cos. First introduced in 1871 and "only moderately successful" subsequently (Jewett et al. 1953); birds were released and no longer present in e. Washington, Whatcom Co., the San Juan Is. (Jewett et al. 1953, Lewis and Sharpe 1987, Wahl 1995). Distribution in most of the state now may be dependent on releases, including chicks given to customers at feed stores (Wahl 1995). Outside the Ft. Lewis area, field observations deemed worthy of note by editors included birds at Sequim and Spokane in 1976, Walla Walla in 1978, Hood Canal in 1983, and McNary NWR in 1987.

Terence R. Wahl

Red-throated Loon *Gavia stellata*

Widespread, fairly common to locally common winter and uncommon summer visitor on marine waters, rare migrant and winter visitor on westside freshwater. Very rare fall migrant and winter visitor in e. Washington.

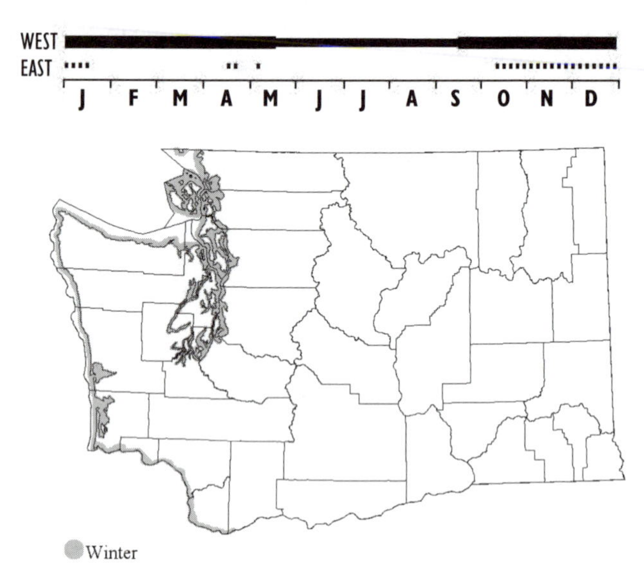

Habitat: Estuaries and shallow offshore marine waters. Concentrates locally in active tidal fronts. Irregular on freshwater in w. Washington, occurs regularly on the lower Columbia R. (Gilligan et al. 1994) and rarely on e. Washington reservoirs.

Occurrence: Widespread, on occasion the most numerous loon seen from shore in w. Washington. Birds appear in breeding plumage in Apr when their distinctive vocalizations are frequently heard before departure in spring. Most often seen singly or in small groups but hundreds may aggregate at prime feeding locations. High counts in the n. inland marine waters were in shallow estuaries, in nearshore waters during herring spawn events (MESA). Several hundred may be observed feeding locally in tidal currents (e.g. 2000 birds at Deception Pass on 26 Feb 2000). Sick or oiled birds occasional on sand banks and beaches.

On the outer coast large groups noted during peak migration periods. Migration censuses in central California recorded large numbers (e.g. 33,000 in Mar-May 1979 [AB 33:802]) and these probably passed Washington. Most migrate within a few km of the coast, with 2000 noted off Ocean Shores on 18 Oct 1986 (AB 41:132). A few occur well offshore, apparently on a nw.-se. route between Vancouver I. and sw. Washington, and have been recorded 60 or more km offshore (TRW). Main spring numbers peak in Mar and Apr (MESA, Wahl 1995).

Almost all eastside records are of single birds on the Columbia R. or nearby impoundments. Winter status is uncertain: there are only four records after mid-Jan. Occasional wintering is likely: there are inland winter records in B.C. (Campbell et al. 1990a). Long-term CBC data suggest no overall trends in numbers.

Remarks: Though Red-throated Loons have reportedly suffered low mortality in Washington from commercial fishing nets and oil spills to date (Speich and Thompson 1987, Ford et al. 1991), these remain potentially serious threats to seasonal populations. Classed PHS.

Noteworthy Records: West, *Outer coast, Spring*: 210 on 2 Apr 1994 off Grays Harbor; 200 on 18 Apr 1973 at Ocean Shores; 300 on 15 May 1976 in Grays Harbor. Fall: 2000+ on 18 Oct 1986 at Ocean Shores. Winter: 120 on 19 Jan 1963 in Willapa Bay; 151 on 19 Jan 1991 off Grays Harbor (TRW); 2000 on 26 Feb 2000 at Deception Pass. *Inland marine waters, high counts* (MESA): 651 on 24 Mar 1979 at Drayton Harbor; 326 on 27 Apr 1979 at Samish Bay; 170 on 22 Feb 1979 at Dungeness Bay. *CBC*: 551 on 29 Dec 1990 at Padilla Bay; 208 on 19 Dec 1992 at Oak Harbor; 170 on 22 Dec 1985 at Bellingham; 119 on 4 Jan 1997 at Kitsap Co. **East**, *Spring*: 1 in Apr 1975 nr. Wenatchee; 1 on 17 Apr and 7 May 1993 at Priest Rapids (Stepniewski 1999). *Fall-Winter*: 1 on 2 Jan 1965 at Wenatchee; 1 on 29 Dec 1969 at Chelan; 1 on 31 Dec 1977 at Wenatchee; 2 - 6 on 31 Dec-early Feb 1978 nr. Wenatchee; 1 on 13 Oct 1979, 2 on 20 Oct at Yakima R. mouth; 1 on 13 Feb 1983 at Richland; 1 on 10 Jan 1994 at Potholes SP; 1 each on 28 Oct 1984, in early Dec 1987, on 8 Nov 1988, and on 14 Dec 1991 at Yakima R. delta nr. Richland; 1 on 10 Jan 1994 at Potholes; 1 on 8 Oct 1994 at Palmer L.; 1 on 31 Dec 1994 at Priest Rapids Dam; 1 on 12 Nov 1995 at Saddle Mt., Grant Co.; 4 Nov 1998 at Two Rivers CP; 4-10 Jan 1999 at Bingen; 15 Apr 2000 at Desert Aire; 12 Nov 2000 at John Day Dam; 29 Nov 2001 at Wenatchee.

Terence R. Wahl

Arctic Loon *Gavia arctica*

Casual vagrant.

Following the separation of *G. pacifica* from this species (AOU 1985), all Arctic Loons previously reported in Washington were classed as Pacific Loons. The first certain identification was on the Columbia R. upstream from Priest Rapids Dam 16 Jan-15 Apr 2000. This represented the first interior N. American record. Subsequent records were at Point No Point on 4 Nov 2000, with 2 there 9-10 Dec 2000, and 1 on 19 Jan 2002. One was at

Edmonds 29 Jan-4 Feb 2002; 1 on 7 Feb 2002 at Port Angeles; and 1 at Sequim on 31 Mar 2002. It is possible that some of the later records may represent the same bird. Subspecies presumed to be *G. a. viridigularis* (Russell 2002).

Terence R. Wahl

Pacific Loon *Gavia pacifica*

Fairly common to locally abundant winter visitor and migrant on saltwater, uncommon in summer. Rare migrant on freshwater on westside. Uncommon fall migrant and rare winter visitor/spring migrant on large rivers and reservoirs, rare elsewhere in e. Washington.

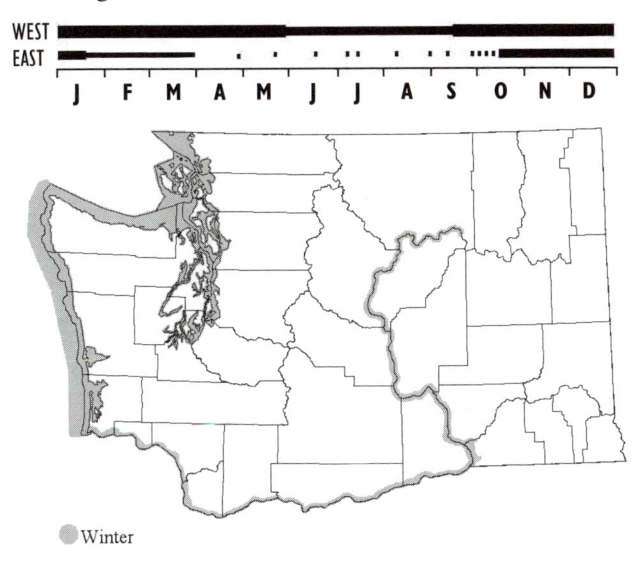

Habitat: Deeper marine waters (>20 m depth) in passages, surge channels. and embayments, secondarily in nearshore waters, shallower coves, lagoons, and inland lakes and reservoirs.

Occurrence: More often in large flocks than other loons and, with Brandt's Cormorant and Common Murre, one of the more abundant divers wintering in deeper inland marine waters. Flocks feed locally in the e. Strait of Juan de Fuca, Admiralty Inlet, s. San Juan Channel, Rosario Strait, Speiden Channel, and especially nr. spring herring spawns in the Strait of Georgia. Individuals or small numbers noted in many inland marine habitats. Records on freshwater in w. Washington are limited, except for fairly regular occurrence of small numbers of migrants and winter visitors on the lower Columbia R. and Vancouver L.

Migrates and forages farther offshore than other loons. Recorded to at least 185 km off Grays Harbor in Apr 1977 and as many as 35 birds together were recorded 50 km offshore in early May (TRW). Often forages with cormorants and shearwaters on the ocean. Largest numbers on the outer coast during spring migration, usually from mid-Apr to mid-May and, in some years, into late Jun. An estimated 24,000 passed the mouth of the Columbia R. on 1 Jun 1976 and 153,900 were seen off nw. Vancouver I. in May 1973 (Campbell et al. 1990a). Presumably similar numbers pass during the less-concentrated fall migration, peaking early Sep-early Oct. Small numbers are present irregularly in Jun and Jul (e.g. 25 birds noted at locations along Whidbey I. on 4 Jul 1997).

Jewett et al. (1953) cited only two e. Washington records: Oct 1901 and Nov 1919. Now apparently more frequent and widespread, at many reservoirs and lakes, with one or two birds per year in the 1960s and 1970s and 20-25 birds per year reported in the early 1990s (e.g . Stepniewski 1999). Most from mid-Oct and early Dec, with a few later in the winter. The creation of lakes and extensive reservoirs behind dams has likely been responsible for much of the increase, along with more coverage by skilled observers. Long-term CBC data in w. Washington indicate considerable variability but no certain trends.

Remarks: Past records of the Arctic Loon are now attributed to be *pacifica* (see AOU 1957, 1985). May suffer mortality from oil spills and other toxics and in gillnets, though known fatalities from these threats have been low to date (Speich and Thompson 1987, Ford et al. 1991). Classed PHS.

Noteworthy Records: West, *high counts, outer coast*: 800 on 10 May 1986 off Westport. *Inland marine waters, Spring*: 352 on 17 Mar 1978 at Whale Rks., San Juan Co.; 223 on 17 Mar 1978 in Mosquito Pass, San Juan Co.; 239 on 7 May 1979 in n. Rosario Strait; 1300-1800 May 18-24 1978 off Cherry Pt. and Sandy Pt. at herring spawns; 2000 on 6 May 1981 off Pt. Roberts. *Fall*: 200 on 6 Oct 1978 in President Channel, San Juan Co. *Winter*: 500 on 16 Dec 1978 in Rosario Strait (MESA). *CBC*: 2508 in Jan 1971 at Bellingham (2500 off Sandy Pt.); 1512 in 1992 at Bellingham; 1330 in 1995 at Port Townsend; 1016 in 1991 at Port Townsend; 917 in 1978 on San Juans Ferry; 884 in 1987 at Port Townsend; 860 in 1981 at Kitsap Co.; 656 in 1993 at San Juan Is.; 432 in 1996 at Padilla Bay; 339 in 1986 at Sequim-Dungeness. **East**, *Spring-Summer*: 2 on 18 May 2000 at Vantage; 20 May 2001 at Gingko SP; 22 May 2001 at Priest Rapids; 11 Jun 2002 at Wanapum Dam; 9 Jul 1979 at Suzy's Pond, Douglas Co.; 11 Jul 2002 at Orondo; 1 Aug 2000 at Yakima R. delta; 27 Aug 2000 at Alkali L. *High counts*: 8 on 18 Oct 1993 at Blue L., Grant Co.; 8 on 24 Nov 2000 at John Day Dam and 8 there on 23 Nov 2001; 7 on 18 Oct 1993 at Banks L.

Terence R. Wahl

Common Loon *Gavia immer*

Fairly common to common migrant and winter visitor and rare breeder in w. Washington. Uncommon to fairly common migrant and winter visitor and rare breeder in e.

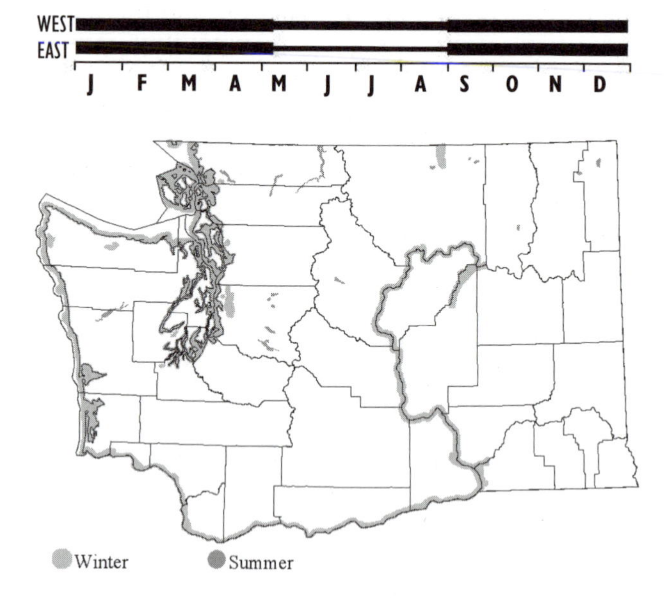

Habitat: Nearshore marine waters, including embayments and along open shorelines. Less numerous on lakes and reservoirs. Breeds on lakes and reservoirs with suitable shoreline vegetation and minimal development and disturbance.

Occurrence: In winter singles or small numbers occur in almost all nearshore marine and freshwater habitats in w. Washington, though up to 100 or more may congregate at prime foraging locations. In summer small numbers are scattered on both fresh- and saltwater. Numbers increase in late Aug-early Sep, more than double in early Oct with the appearance of family groups, then decline, with about half that number present from Nov until Mar or late Apr before peaking again (about 60% of fall peak) with migrants in mid-Apr to early May (see Wahl 1995). Birds are least numerous on marine waters in late Jun (MESA).

Migration usually small-scale over inland marine waters: most migrants appear, often feeding, where they had been absent before, or are heard calling in flight. Migration most noticeable at locations such as Admiralty Inlet or Rosario Strait. Along the outer coast and offshore, migration obvious, usually of individuals or small groups. Surveys off Grays Harbor from 1971 to 2000 recorded 1242 birds, 78% of which were over nearshore waters <20 m deep. During migration peaks in Apr-May and late Aug-Oct birds were seen from there to oceanic waters 100 km offshore (TRW).

Table 15. Common Loon: CBC average counts by decade

	1960s	1970s	1980s	1990s
West				
Bellingham		36.6	50.0	53.3
Grays Harbor		42.8	65.4	78.5
Kitsap County		22.8	18.0	37.0
Leadbetter Pt		18.1	20.9	21.9
Olympia			58.3	67.1
Padilla Bay			41.7	77.7
Port Townsend			92.3	67.3
San Juan Is.			80.5	70.2
San Juans Ferry		15.6	15.4	11.0
Seattle	2.4	8.3	7.8	12.6
Sequim-Dungeness			109.2	122.2
Tacoma		12.1	24.0	24.3
East				
Tri-Cities		7.1	11.5	11.7
Wenatchee	0.8	3.9	6.5	5.9

Jewett et al. (1953) gave a handful of fall reports and a single winter record for e. Washington, and Weber and Larrison (1977) considered the species uncommon and irregular in winter in the se. Fall and winter reports increased in the 1970s, with dozens to hundreds noted on occasion at Banks L. and other Columbia Basin impoundments.

Jewett et al. (1953) stated the species was a "fairly common" breeder in Washington, with few details. Kitchin (1934) believed the species nested statewide, more commonly in w. Washington. Kitchin (1949) presumed loons once bred at low-elevation lakes on the Olympic Pen., before logging and clearing, and suggested a few pairs might still breed in 1949, but had no recent records. Believed extirpated as a breeder in w. Washington in 1979, but found nesting in the late 1980s with 11 confirmed locations in w. Washington, five in e. Washington and probably eight in 1996. Nesting confirmed in Whatcom, King, Okanogan, Chelan and Ferry Cos., and probable in Pend Oreille and Grays Harbor Cos. (Smith et al. 1997).

Long-term CBC data (Table 15) suggest increases on the westside (P <0.01) and eastside (P <0.05) but Nysewander et al. (2001) reported a westside decrease of 64% from 1978-79 through 1994-95.

Remarks: No subspecies recognized (McIntyre and Barr 1997), though Jewett et al. (1953) listed *G. i. elasson* and *immer* for Washington. Human activities conflict directly with breeding loons. Breeding attempts, let alone success, are dependent upon availability of lakes with suitable food supply and undeveloped shorelines free from disturbance and significant wave action. These conditions exist only on protected reservoirs and lakes with little or no fishing activity, no fast motorized boats or personal watercraft. Such locations, now rare, exist mainly at municipal water-supply lakes (e.g. King Co.) and a few lakes managed for wildlife by government agencies (e.g. Lost L. in Okanogan Co.). Classed SS, PHS.

Noteworthy Records: West, *high counts*: 142 on 21 Oct 1979 at Drayton Harbor, 103 on 3 Oct 1978 at Dungeness Bay (MESA); 220 in Dec at Willapa NWR (Salo 1975). *CBCs by location*: 275 in 1984 at Port Townsend; 275 in 1995 at San Juan Is.; 171 in 1986 at Sequim-Dungeness; 159 in 1999 at Kitsap Co.; 140 in 1991 at Grays Harbor; 128 in 1997 at Padilla Bay. **East**, *high counts, Spring*: 27 in mid-May 1973 at Potholes Res.; 43 + on 19 Apr 1980 from c. Banks L. to s. end L. Lenore; 25 at Blue L. and 50 at Park L. on 24 Apr. 1997. *Fall*: 700+ on 4 Nov 1987 at Banks L.; 156 on 22 Oct 1994 at Coulee Lks.; 75-200 in Oct 1984 in Potholes area and Banks L.; 65 on 1 Oct 1995 at lower Coulee Lks.; 73 on 14 Oct 1995 at Blue L. *Winter, CBC high counts by location*: 24 in 1991 at Tri-Cities; 9 at Wenatchee in 1970.

Terence R. Wahl and Scott Richardson

Yellow-billed Loon *Gavia adamsii*

Very rare migrant, summer visitor. Rare winter visitor.

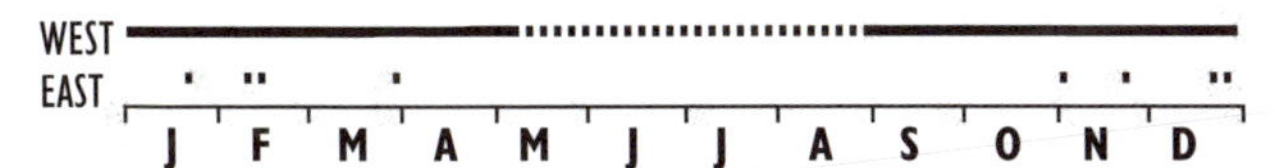

Habitat: Nearshore waters (<20 m [65 feet]depth) along open shorelines, in embayments and passages with moderate tidal flows; large lakes and reservoirs in e. Washington.

Occurrence: Classed as hypothetical by Jewett et al. (1953), has since occurred at least 135 times, starting in 1956 at L. Washington (Schultz 1970). As many as 90-100 individuals have occurred, some apparently returning to winter two or three times. Reports in N. America increased starting in the 1960s, likely due to better description of identification characters (e.g., Binford and Remsen 1974) and the huge increase in field observers. An increase in actual numbers is consequently uncertain.

Most records are from the Strait of Juan de Fuca and other n. interior marine waters. This pattern is likely biased, however, by observer effort (e.g. at least 35 of the 90 or so birds recorded on saltwater were observed at Pt. Roberts, Bellingham, and other locations in Whatcom Co.). Only about 21 have been recorded from the outer coast or statewide on freshwater. Of these, about 14 are from Grays Harbor and Willapa Bay and one 46 km offshore from Grays Harbor (TRW). Three more are from freshwater w. and nine are from eastside lakes or reservoirs.

Almost always seen individually. Two were together at Bellingham on 28 Dec 1980 (TRW) and two were noted also at Sidney, B.C., on 6 Dec 1979 (MESA). Ages of birds seldom reported but many apparently were pre-breeders. These include some birds seen in Apr and May. Birds in breeding plumage have been reported at least six times.

Reports peaked in winter 1979-80, during a period of intensive MESA field surveys, with five birds recorded from Admiralty Inlet n. to B.C. and w. to Cape Flattery, six were nr. Bellingham, and three more were elsewhere in Washington. Numbers above average also in 1983-84, 1984-85, and 1993-94. In spite of increased observer effort, reports in the 1990s were fewer than in the 1980s. Effects of an Alaska oil spill in 1989 on the species' population (McIntyre 1990, Piatt et al. 1990) were uncertain. Reports in Washington and Oregon during the 10 years preceding 1989 averaged about 8.5 per year while the average during the following nine years was about 4.5.

Remarks: As with other marine birds, this species is vulnerable to oil spills and entanglement in fishing nets. Classed GL.

Noteworthy Records: West, *Summer*: 14 Jul 1974 at Neah Bay; 22 Jun 1996 at Useless Bay, Whidbey I.; 8-13 Jul 1997 at Swantown, Whidbey I.; 15-19 Jul 1998 at Swantown; 14-31 Jul - Sep 2000 at Ocean Shores; 16 Jun 2001 at Diamond Pt., Clallam Co. *Winter, CBC high counts*: 5 in 1979 plus 1 nearby, at Bellingham; 4 in 1999 at San Juan Is. **East**: 14 Feb-17 Apr 1993, 12 Feb 2000, 30 Dec 2000 and 17-20 Jan 2001 at Priest Rapids Res.; 17-20 Nov 1993 at L. Chelan; 4 Nov-25 Dec 1995 at Blue L., Grant Co.; 4 Jan 1997 at Bridgeport; 29 May 1997 at Crescent Bar, Kittitas Co.

Terence R. Wahl

Pied-billed Grebe *Podilymbus podiceps*

Fairly common summer resident in e. Washington; local, uncommon w. Fairly common winter resident statewide.

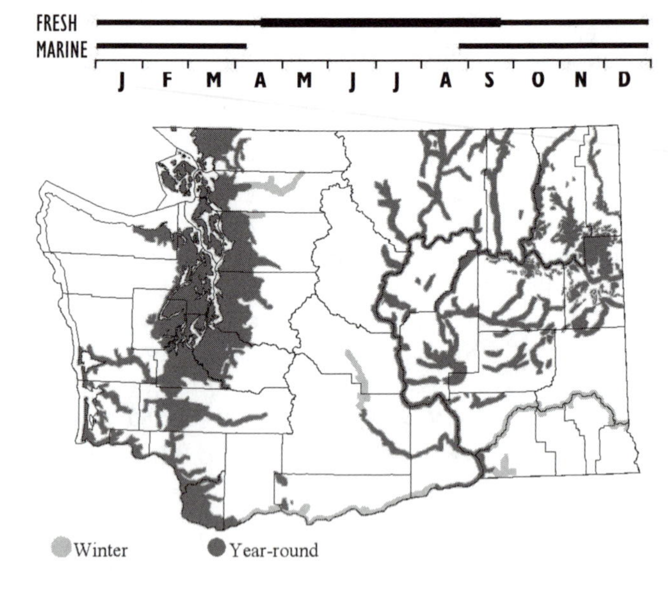

Subspecies: *P. p. podiceps.*

Habitat: Seasonal or permanent at ponds and lakes usually with emergent vegetation borders, including otherwise developed urban areas. Also shallow bays, protected marine waters, or slow-moving rivers in winter.

Occurrence: Locally common summer breeder in e. Washington. At Turnbull NWR breeding populations estimated at over 500 birds, while in Yakima Co. nesting is "surprisingly uncommon and very local" (A. Stepniewski). Status at higher elevations (e.g. Smith et al. 1997) is uncertain, though birds breed at Birch L., Yakima Co., at about 1400 m elevation (A. Stepniewski). Though widespread in summer, surveys in 1987-96 (Smith et al. 1997) did not locate birds in the se. counties (and see Weber and Larrison 1977).

A widespread, locally fairly common breeder w. of the Cascades, nesting opportunistically in situations from ponds at Ocean Shores, a small, marshy pond in a pasture nr. Lynden (Wahl 1995), to urban Seattle locations like Green L. (Muller 1995) and Union Bay (Ratoosh 1995). Has nested at Port Townsend (J. Hardin) but breeding status on the nw. Olympic Pen. uncertain (Smith et al. 1997).

The least numerous grebe on marine waters in w. Washington in winter. Birds arrive in late Aug and remain until Mar or early Apr, increasing in numbers when inland lakes freeze. Surveys in 1978-79 found them widespread from 4 Sep to 2 Apr, from Port Angeles n. and e. to Drayton Harbor, Padilla Bay, and the San Juan Is. Recorded on virtually every CBC count (30 of 32 counts in 1996-97). Rare to uncommon in most of e. Washington, occurring where there was open water and a food supply. CBCs in recent years show birds occurring on almost all eastside counts, including Chewelah, Walla Walla, Twisp, Chelan and Wenatchee, Ellensburg, and Spokane.

Historical and current populations are unknown. An estimated 20-40% reduction in Washington wetlands from pre-settlement to 1988 (Lane and Taylor 1997) has almost certainly reduced numbers. Weber and Larrison (1977) described this grebe as a common summer resident on scabland ponds in the Rock L. area, and Hurley (1921) reported widespread breeding in Yakima Co., but Smith et al. (1997) and A. Stepniewski do not support this today. Jewett et al. (1953) described summer distribution w. to Clallam Bay, though Smith et al. (1997) reported no certain records w. of Dungeness NWR on the Olympic Pen. Long-term CBC counts suggest increases in some westside urban situations, and in e. Washington, possibly reflecting milder winters.

Remarks: Breeding success affected by human activities including personal watercraft, water-skiing, canoe and kayak races, and fishing activities that threaten birds or create disturbance and waves sufficient to destroy nests or can lead to desertion of nests during or shortly after hatching (Wahl 1995, MJM). Introduced carp splashing at the surface during spawning destroy nests with eggs. Classed PHS.

Noteworthy Records: East, *Summer*: 500 on 1 Jun 1969 at Turnbull NWR; 272 on 30 Jun 1955 at Turnbull NWR, with 49 broods with 174 young. *Fall*: 150 on 19 Sep 1998 at Deer L., Whidbey I.; 40 on 19 Nov 1969 at McNary NWR; 55 on 5 Nov 1995 at Priest Rapids; 50 on 3 Nov 1974 at Judson L., Whatcom Co.; 53 on 27 Jul 2002 at Ridgefield NWR (SGM). *CBC, Winter, freshwater*: 64 on 7 Jan 1995 at Priest Rapids; 47 on 20 Dec 1987 at Richland. Winter, saltwater: 19 on 25 Oct 1979 at Padilla Bay (MESA).

Martin J. Muller

Horned Grebe *Podiceps auritus*

Widespread to common migrant and winter visitor in w., rare in summer. Widespread migrant and uncommon winter visitor e. Has bred very locally e.

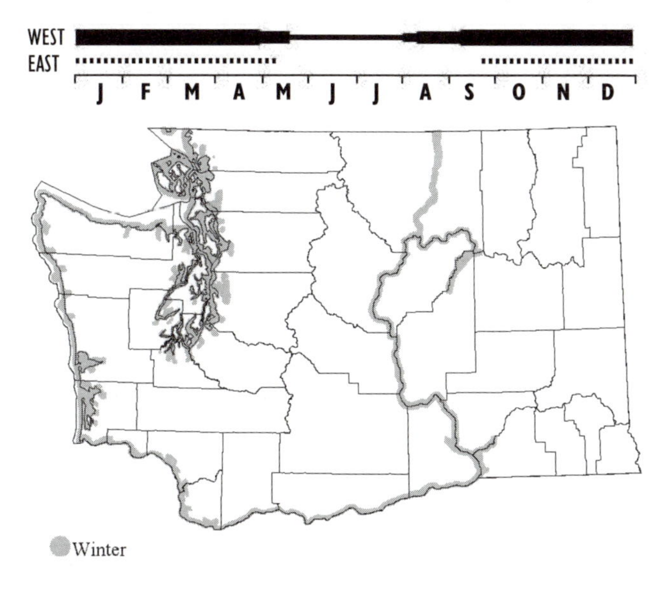

Subspecies: *P. a. cornutus.*

Habitat: Winters in bays and along shorelines in inland marine waters and coastal areas. Often forages in very shallow water. Winters on lakes and reservoirs in e. Breeds locally at marshy lakes.

Occurrence: CBC data indicate nw. Washington has the densest wintering population in N. America (Root 1988). Birds arrive as early as mid-Aug at main wintering westside locations and birds are widespread by Sep. Almost all depart by mid-May. Scattered birds noted in summer, with only five of 2440 records acquired in systematic censuses from Jun and Jul (MESA). Flocks less frequently than Eared and Western grebes but hundreds may concentrate at winter feeding areas. Surveys in 1978-79 (MESA) located aggregations at Drayton Harbor, Neah Bay, Dungeness, Sequim and Discovery bays, and at Mud and Griffin bays in the San Juan Is. Densities in limited protected outer coastal embayments are apparently similar. Though irregular, there are several records of small numbers in inland marine waters and along the outer coast in summer.

Occurrence and numbers of birds wintering e. of the Cascades appear to be increasing, likely due to habitat provided by reservoirs, but possibly biased by weather and increased field observation effort. There are scattered migrant and winter records of birds (e.g. at Asotin, Clarkston, Cheney, Yakima Valley, Kettle Falls, at Soap L. in Grant Co.).

Birds breed regularly in B.C. (Campbell et al. 1990a), but infrequently in Washington (e.g. 120 at Turnbull NWR in 1964 and see Jewett et al. 1953, Weber and Larrison 1977). Nesting not recorded during 1987-96 BBA surveys (Smith et al. 1997), but confirmed in 1998 at L. Chelan (R. Kuntz, NPS). Species less tolerant of salinity than the Eared Grebe at potential breeding lakes (Cannings et al. 1987), and this may limit breeding in Washington (Smith et al. 1997).

Surveys of overall inland marine waters numbers and some long-term CBCs indicate local declines in winter numbers (Nysewander et al. 2001, Wahl 2002) in n. inland waters though overall eastside and westside CBC data indicate increases (P <0.01).

Remarks: Monitoring of status and possible breeding is desirable; classed PHS, GL.

Noteworthy Records: West, *Fall-Spring, high counts:* 375 on 29 Oct 1995 on San Juan I.; 500 on 1 Apr 1978 at Port Townsend; 275 on 6 Apr 1997 along Whidbey I.; 300 on 18 Jan 1995 occurred at "San Juan"; 131 on 6 Oct 1998 at Pt. Roberts; 200 on 16 Nov 1998 at Morse Cr. *MESA surveys in 1978-79*: 382 on 16 Apr 1978 at Drayton Harbor; 329 on 10 Oct 1979 at Neah Bay; 320 on 23 Apr 1979 at Dungeness Bay. Summer: 1 on 7 Jun at Everett, 1 on 9 Jun at Ridgefield NWR, 1 on 18 Jun 1998 at Ocean Shores; 2 on 19 May 1999 at Penn Cove; 3 on 4 Jul 1997 at Ebey's Landing; 1 on 15 Jun at Ocean Shores, 1 on 31 Jul at Ediz Hook. **East**, *Summer*: 60 prs. on 1 Jun 1964 at Turnbull NWR; nested in 1998 at L. Chelan (R. Kuntz p.c.). Fall: 150 on 1 Oct 1995 on the lower Coulee Lks.; 35 on 6 Nov 1995 at Vantage.

Terence R. Wahl and Scott Richardson

Red-necked Grebe *Podiceps grisegena*

Fairly common to common migrant and winter visitor, rare in summer in marine waters, rare on freshwater w. Local, fairly common breeder, mostly in n. counties e. of the Cascades; widespread migrant, local and rare to uncommon in winter.

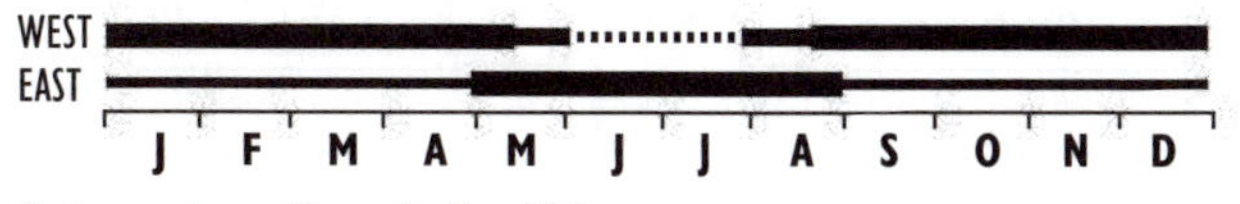

Subspecies: *P. g. holboellii.*

Habitat: Winters mainly on bays or other protected nearshore marine waters, often with kelp or eelgrass beds. Shallow lakes, ponds, reservoirs, still water with marsh vegetation in summer (Smith et al. 1997).

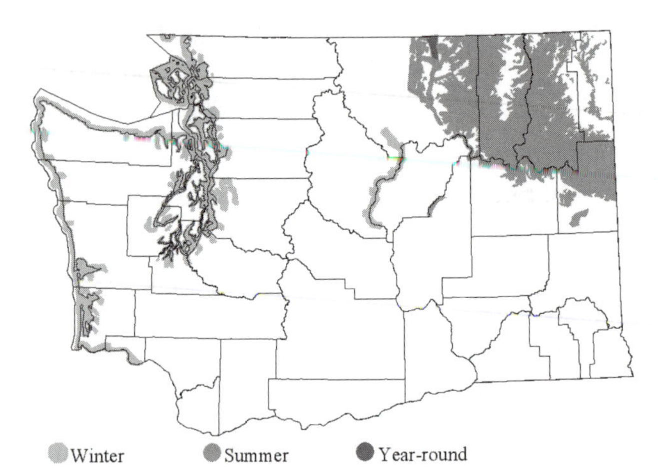

Table 16. Red-necked Grebe: CBC average counts by decade.

	1960s	*1970s*	*1980s*	*1990s*
Bellingham		73.9	66.5	25.6
Edmonds			62.0	39.2
Grays Harbor		8.6	11.6	10.5
Kent-Auburn			17.5	21.1
Kitsap Co.		56.5	88.9	74.0
Olympia			138.6	146.4
Padilla Bay			72.1	66.8
Port Townsend			116.7	60.4
San Juan Is.			197.1	74.9
San Juans Ferry		78.3	79.2	52.8
Seattle	32.6	107.5	151.2	130.2
Sequim-Dungeness			215.2	144.2
Tacoma		111.9	262.4	191.6

Occurrence: The inland marine waters probably support the largest wintering concentration in N. America (Root 1988), widespread along the Strait of Juan de Fuca and in inland waters to the s. end of Puget Sound. Local aggregations widespread in passages and bays in the San Juan Is. and in areas from Drayton Harbor s. to Samish and Padilla bays, but the center of abundance is along the Strait of Juan de Fuca shoreline, essentially between Port Angeles and Sequim Bay, with counts of up to 2500 recorded. Aggregations of birds stage in spring at a number of locations (e.g. the s. edge of the San Juans: Lewis and Sharpe 1987). Birds occur along the outer coast in relatively small numbers, in protected bays like Grays Harbor, and along exposed shorelines on occasion.

The species is a local, uncommon to common breeder in the Okanogan highlands e. to Pend Oreille Co. (see Smith et al. 1997), at Liberty and Newman Lks. as recently as 1978, at Turnbull NWR in 1996, possibly Douglas and n. Whitman Cos. (Weber and Larrison 1977), with non-breeders noted at other locations in summer. Breeders appear around mid-Mar and may leave as late as mid-Nov. Rare, scattered birds, usually individuals, found on larger lakes and rivers in e. Washington in winter. Non-breeding birds may be increasing in occurrence on reservoirs and lakes, though increases in observer field effort may affect numbers.

Surveys of winter numbers in inland marine waters indicate a decline from 1978 to 1999 regionally (Nysewander et al. 2001) and locally from 1970 to 2000 (Table 16; Wahl 2002).

Remarks: Oil spills, gillnets in areas of non-breeding concentrations, and disturbance at nesting areas are threats. After an oil spill at Port Angeles in Dec 1985, in the most important wintering area, over 65% of about 1600 oiled birds recovered alive were Red-necked Grebes, and the species suffered by far the greatest mortality of any species. Birds attempting to nest on some lakes may have limited success due to disturbance and nest predators (numbers at Liberty L. in Jun 1976 were reportedly low due to gull predation and that location suffered noticeable nesting failure several times in the 1960s and 1970s). Local nesting success has probably decreased due to human development and activity. Class PHS. Note: A breeding report for Ocean Shores on 28 Jun 1975 is assumed to have been in error.

Noteworthy Records: High counts, **West**: 2500 on 31 Aug 1988 between Dungeness and Port Angeles; 1000 on 19 Apr 1906 off Port Townsend (Dawson and Bowles 1909); 600 on 20 Oct 2001 at Marrowstone Pt.; 500-600 on 24 Aug 1976 at Green Pt., Clallam Co. CBC: 635 at Oak Harbor in 1993, 557 at Olympia in 1988, 525 at Everett in 1988, 468 at Sequim-Dungeness in 1987, 436 at Tacoma in 1981, 425 at Port Townsend in 1986, 418 at San Juan Is. in 1987, 300 at Sequim-Dungeness in 1985, 260 at Seattle in 1989. **East**: 26 on 10 Aug 1988 at Spectacle L.; 20 at Twin Lks., Ferry Co., on 1 Jun 1972. *CBC*: 19 in 1989 at Chelan.

Terence R. Wahl

Eared Grebe *Podiceps nigricollis*

Locally fairly common breeder, widespread migrant and rare local winter visitor in e. Washington. Locally uncommon winter visitor and migrant w. in marine waters, very rare in summer. Possibly decreasing.

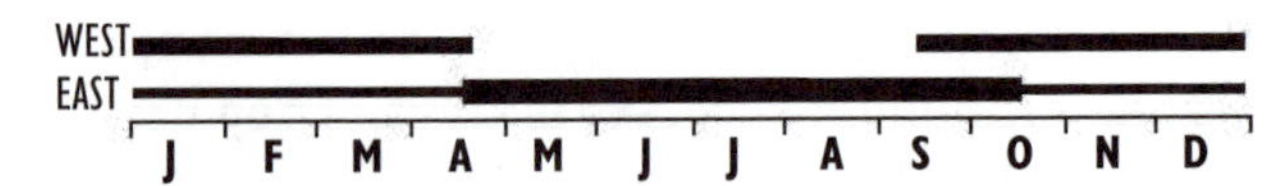

Subspecies: *P. n. californicus* breeds in Washington.

Habitat: Nests on small lakes, reservoirs, and ponds edged with dense vegetation. Winters on lakes, reservoirs, and saltwater bays.

Winter Summer

Occurrence: Fieldwork in recent decades showed Eared Grebes are very local summer residents e. of the Cascades, reported nesting in Adams, Grant, Lincoln and Spokane Cos. (Jewett et al. 1953), Okanogan, Douglas and Franklin Cos., mostly on NWR or other managed habitats (Smith et al. (1997). Local on ponds, lakes, some reservoirs in the Columbia Basin, the Okanogan Valley, nr. Turnbull NWR and the channelled scablands but not in the s.c. Columbia Basin (see Smith et al. 1997). Breeding numbers have been sporadically reported, with a maximum of about 200 pairs noted at Turnbull NWR in 1964, for example.

Soap L. had fall aggregations of up to 2000 migrants beginning in late Aug. Likely migrants from Canada (see Campbell et al. 1990a), as many as 500 birds remained there one winter until forced out by a Jan freeze-up. Small numbers reported on CBCs starting in 1948 show irregular occurrence in locations like Richland, Cheney, Spokane, Chelan, Wenatchee, and Walla Walla. Hudson and Yocom (1954) and Weber and Larrison (1977) indicated absence in winter in the Columbia Basin.

Though small numbers of migrants and widespread winter visitors occur w. of the Cascades, counts and reports starting in the 1970s (CBCs, MESA) showed small concentrations along the e. Strait of Juan de Fuca, especially in Port Angeles, Dungeness, and Sequim and Discovery bays, Port Townsend, and in Kitsap Co., Eld Inlet, Penn Cove, and Oak Harbor on Whidbey I., and very locally in the San Juan Is. Rare in Grays Harbor and Willapa Bay in winter. Noted in summer at Everett and Kent sewage lagoons.

Long-term CBC data show apparent declines ($P < 0.01$) at areas of concentrations. Non-systematic field counts reported also appeared to be substantially lower.

Remarks: Suffered mortality due to weather, disease, likely other causes in sw. states during the past century (Jehl 1996), mostly during migration and winter periods, with possible effects on Washington populations unknown. Designated PHS. Note: some CBC counts (e.g. Cape Flattery in 1966, Samish I. in 1968) are questionable and, based on status today, numbers relative to those of Horned Grebes on some other old CBCs appear unlikely (see Campbell et al. 1990a).

Noteworthy Records: East, *Spring, high counts*: 500-700 at Soap L. on 19 Apr 1945 (Jewett et al. 1953). *Summer*: first breeding record 1 Jun 1963 at Columbia NWR; 14 in 1988 at Royal Slough nr. Othello (reported nesting since 1963); 50 pairs, young on 28 Jun 1982 at Reardan; 400 in 1964 at Turnbull NWR; 110 on 25 Jun 2001 at Soap L.; 160 on 13 Jun 2001 at Kahlotus L.; 127 on 24 Jun 2001 at Cameron L. Rd, Okanogan Co.; 223 on 1 Jul 2001 at L. Lenore. Fall: 1200 on 27 Aug 2000 and 25 Aug 2001, 1000 on 11 Nov 1960 and 1500-2000 in Sep 1986 at Soap L. Winter: 400 on 27 Jan 1964, 500 on 15 Jan 1987 at Soap L. CBC: 12 in 1991 at Chelan; 11 in 1966 at Wenatchee. **West**, *Summer*: 1 on 26 Jul 1981, 2 on 3 and 26 June 1983 at Kent sewage lagoons; 7 on 17 Jun 1976 at Everett; 1 on 3 Jun at L. Kapowsin; 1 on 21 Jul 2001 at Washougal. Fall: 30 on 19 Oct at Everett sewage lagoons. *Winter*: 60 on 11 Dec 1976 at Penn Cove (TRW). *CBC*: 186 in 1977 at Port Townsend; 97 in 1977 at Sequim; 75 in 1985 at Kitsap; 17 in 1980 at Grays Harbor. *High counts, MESA*: 150 on 3 Oct 1978 at Dungeness Bay, 101 on 11 Mar 1979 at Penn Cove, 90 on 14 Oct 1978 at Port Angeles, 67 on 10 Nov 2001 at Utsalady Bay; 65 on 17 Mar 1978 at Westcott Bay.

Terence R. Wahl

Western Grebe *Aechmophorus occidentalis*

Common to locally abundant winter visitor on saltwater, uncommon to locally common on freshwater; uncommon non-breeding summer visitor w. Locally common summer breeder and migrant, locally rare to uncommon winter visitor e.

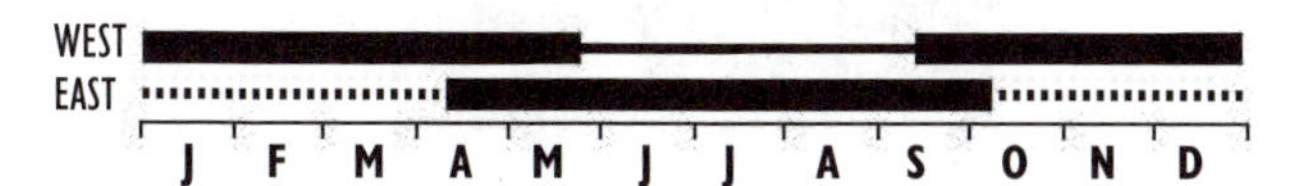

Subspecies: The nesting race is *A. o. occidentalis* (Smith et al. 1997).

Habitat: Marine waters with minimal tidal-current flow are primary habitat for largest winter numbers, though birds occur in many saltwater situations and on inland lakes. Large, shallow lakes and ponds with vegetated borders in summer breeding season.

Occurrence: Among the most abundant diving birds wintering in inland waters. Small numbers are widespread along shorelines, in bays, and in the

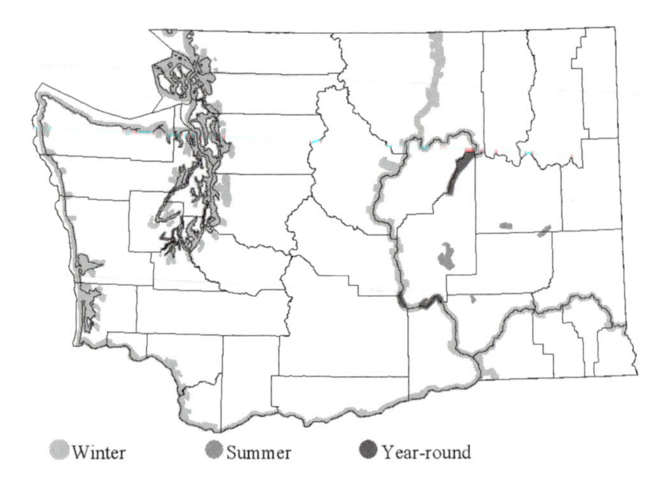

mouths of small streams, but flocks of hundreds or thousands winter from Boundary Bay s. through inland marine waters. Historical numbers particularly at Boundary Bay, Bellingham-Samish bays, Padilla and Skagit bays, Port Susan, Possession Sound, Port Townsend, Hood Canal, and Carr and Case inlets (Wahl et al. 1981, Wahl and Speich 1983).

More gregarious, foraging in deeper water than other grebes. Birds do not fly after Sep-Oct arrival, concentrating in bays lacking strong tidal flows that would displace roosting birds. In May, flocks make brief flights, apparently "getting in shape" before migrating inland. Birds uncommon on westside lakes and ponds; a bird on a lake nr. Mt. Baker at 1500-m elevation suggested a high-elevation attempt to cross the Cascades.

Flocks shift seasonally relative to prey availability, such as attraction to herring concentrations in spring. On Bellingham Bay flocks move considerable distances during winter flooding of the Nooksack R., moving as far as 10 km to Samish Bay where the surface layer is clear. Varying numbers, up to a few hundred, of non-breeders spend the summer in a few locations, such as Bellingham-Samish and Boundary bays. On the outer coast most birds occur in embayments, though some winter in nearshore waters along the exposed ocean shoreline, likely associating with outflows from bays and rivers. Several offshore records include a bird about 45 km off Grays Harbor, over a depth of about 160 m (TRW).

The species is a locally common eastside breeder, primarily in the Columbia Basin. Jewett et. al. (1953) noted breeding only at Brook L., with over 50 nests in 1905. Nesting at Moses L. first reported in 1980, and 139 ads. with 74 young were at Steamboat Rk. at Banks L. in 1988, 500 birds at Banks L. on 28 Aug 1977, and 300 at the Potholes on 29 Aug 1988. Jun records indicate possible nesting at L. Entiat and Rock L. Stepniewski (1999) reported breeding on Goose I., Priest Rapids L. Reported nesting at Tidewater, Frasier R. delta, B.C., in 1986 and 1988 (Weber and Ireland 1992).

Most breeding birds arrive in early Apr and leave in early Oct. Small numbers reported on CBCs at Richland, Yakima Valley, Two Rivers, Spokane, Wenatchee, and Chelan. Other winter records are from Clarkston, Marcus in Stevens Co., and Blue L. in Grant Co., and 25 birds wintered at Asotin in 1978, behind Lower Granite Dam.

Western Grebe
(G. Scott Mills)

Numbers from systematic censuses are much larger than records given by Jewett et al. (1953; see Wahl 1995), though this may reflect better coverage of distribution rather than population increase. Comparisons of counts in the late 1970s with the 1990s indicated large-scale regional and local declines (Hazlitt 2001, Nysewander et al. 2001, Wahl 2002). CBC counts indicated declines in n. inland waters.

Remarks: An indicator species subject to events like oil spills, gillnet mortality, predation, habitat loss, and, probably, climate change. Storer and Nuechterlein (1992) describe relationship with Clark's Grebe. Classed SC, FSC, PHS.

Noteworthy Records: *High counts,* **West**, *Fall-Spring*: 50,100 on 12 May 1957 in Admiralty Inlet; 40,000 on 23 Feb 1969 on Bellingham Bay. Summer: 2000 on 1 Jun 1967 on Bellingham Bay. *CBC*: 26,230 in 1991 and 21,137 in 1989 at Bellingham; 7276 in 1982 at Olympia; 5241 on 2 Jan 1983 at Tacoma. **East**, *Summer*: 2200 on 25 Aug 2001 at Potholes Res.; 1900 on 6 Sep 2000 at Potholes Res. Winter: 25 on Jan-29 May 1978 at Asotin.

Terence R. Wahl

Clark's Grebe *Aechmophorus clarkii*

Uncommon, locally fairly common breeder in e. Washington; rare to uncommon winter visitor in w.

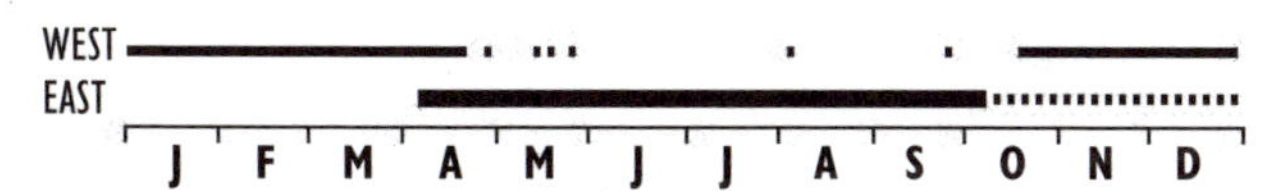

Subspecies: *A. c. transitionalis* breeds in Washington (Smith et al. 1997).

Habitat: Large, shallow lakes and ponds with vegetated borders in breeding season. Marine waters with minimal tidal-current flow primary winter habitat.

Occurrence: Following separation of the "light phase" of the Western Grebe as Clark's (Ratti 1981, Storer and Nuechterlein 1992), the first state record was on L. Washington on 18 Apr 1981. It has since been reported regularly. Recent apparently more frequent occurrence may be due to increased awareness of species status by observers.

First reported from e. Washington in Oct 1983 at Richland. Since 1984 confirmed nesting at Moses L. with 20-80 birds reported in Aug and Sep, and on N. Potholes Res. nr. Moses L., which had about 15 birds in May 1986. Nests also at Banks L. and Saddle Mt. NWR (Smith et al. 1997). Present in 1992 at Upper Hampton L. in the Columbia NWR. Other late spring-summer records from Scooteney Res., Franklin, and Sprague L. in 1995 and W. Medical L. Fall records include birds at Wallula and the Walla Walla R. delta, Priest Rapids, and Banks L. There are a number of reports, perhaps reflecting observer concentration, from the Tri-Cities area during the spring-fall season. Recorded once on CBCs at Chelan and Tri-Cities.

Rare in winter in w. Washington, with almost all records of single birds. Based on the distribution of fall and winter reports, most likely to occur along the s. outer coast, with a number of reports from Ocean Shores s. to Ilwaco. Winter reports and 10 CBC records show distribution from Pt. Roberts s. to the San Juan Is., Ediz Hook, and inland marine waters s. to Tacoma. There are spring records from the Columbia R. at Skamokawa and at L. Washington.

Remarks: Some reports may be questionable due to identification problems of Clarks's and Western grebes in non-breeding plumages. Intergrades are known (a bird at Tokeland 14 Jan-23 Mar 1990 was thought to be an intergrade/hybrid). Classed PHS, GL.

Noteworthy Records: West, *high counts*: 3 on 1 Nov 1994 at Ocean Shores; 6 on 31 Dec 1995 off Everett.

Terence R. Wahl

Shy/White-capped Albatross *Thalassarche cauta*

Casual vagrant.

A S. Hemisphere albatross, breeding off Tasmania and New Zealand, ranging widely across the s. oceans. One was collected about 60 km off Cape Alava on 1 Sep 1951 (Slipp 1952, Jewett et al. 1953) at a research vessel. That record, one in the Red Sea (Harrison 1983), and one report off Somalia in 1986 (Meeth and Meeth 1988) were apparently the only records of the Shy Albatross group for the N. Hemisphere until one was photographed at Heceta Banks, Oregon, in Oct 1996, followed by one off Pt. Arena, California, in Aug 1999 (Cole 2000). Washington's second record was on 22 Jan 2000 when a bird came to a boat about 54 km off Grays Harbor (see Cole 2000). Based on proposed taxonomic splits, available photographs, and descriptions, Cole (2000) suggests that the 1951 Washington record was *steadi*, the White-capped Albatross, while the 2000 bird, suggested as possibly the same bird as the one seen in Oregon and California, was *cauta*, the Shy Albatross (see Aanerud 2002).

Remarks: Formerly *Diomedea*, placed in genus *Thalassarche* (AOU 1997, 1998). Taxonomy of the *T. cauta* complex is not agreed upon (see Cole 2000).

Terence R. Wahl

Laysan Albatross *Phoebastria immutabilis*

Uncommon winter visitor in offshore oceanic waters; irregularly rare to uncommon in summer.

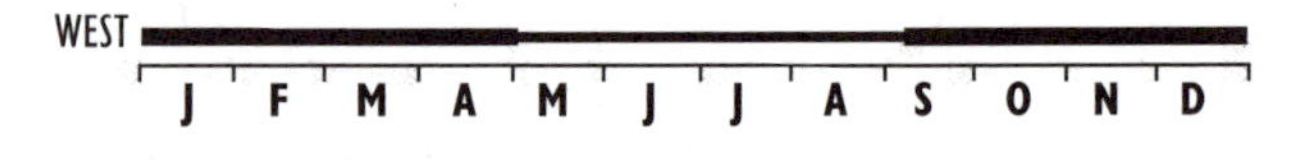

Habitat: N. Pacific oceanic waters e. to outer continental shelf waters off Washington.

Occurrence: The most abundant N. Pacific albatross, with an estimated world population of 2.5 million (McDermond and Morgan 1993). Outnumbers the Black-footed Albatross by six or seven to one in world population (McDermond and Morgan 1993), but much less frequent off the Pacific coast of N. America. Records, and observation effort, have increased in recent years.

Surveys off Grays Harbor between Sep 1971 and Oct 2000 had 2484 records of 23,844 Black-footed Albatrosses, just 68 records of about 88 Laysans (TRW). The latter's general distribution is more northerly, including the Bering Sea, and westerly than the Black-footed Albatross and relates to oceanographic conditions including salinity (Sanger 1974, Vermeer and Rankin 1984). Like other albatrosses, attracted to human-related activities: 55 of 68 records (TRW) occurred at chums or fishing vessels. One on 15 Apr 1977 followed a research vessel from offshore to within about 10 km (6 miles) of the Westport jetty (TRW), the most inshore record. One bird found on a barge arriving in Seattle in early Feb 1998 was followed within a few days by another found in a berry field at Gig Harbor. Neither could be rehabilitated (C. Sheridan p.c.). Off Grays Harbor, most records were from Sep into early Jun (TRW); Briggs et al.(1992) observed birds off Washington and Oregon in Nov, Jan, Mar, and May.

Starting in the 1970s breeding range expanded, with colonies established on the main Hawaiian Is. and in the w. Pacific Ocean at Torishima (McDermond and Morgan 1993). Small numbers nested in Mexico in 1983-84 on Guadalupe I., off Baja California (Pitman 1985, Pitman et al. 2002), and in 1988 on Isla San Benedicto and Isla Clarion in the Revillagigedos Is. (Gallo-Renoso and Figueroa-Carranza 1996, Pitman and Ballance 2002). The rapid increase of breeding birds on Isla de Guadalupe, from four pairs in 1984 to 131 birds in 1992 and 193 in 2000-2001, suggests recruitment from other colonies (Gallo-Renoso and Figueroa-Carranza 1996). This new population may at least in part explain more frequent occurrence off Baja California and California, but many other factors, including oceanographic events like ENSOs, may affect distribution there and farther n.

Remarks: Mortality at sea results from high-seas longline fisheries (see Whittow 1993) and ingestion of plastic debris is a threat (McDermond and Morgan 1993). Classed GL. Formerly in genus *Diomedea*.

Noteworthy Records, *Winter*: a number of dead birds were recovered along the s. Washington coast during a period of heavy seabird mortality in late 1995. 30 live birds were reported offshore between Ilwaco and Grays Harbor on 3 Dec 1995. *High counts*: 13 on 2 Feb 1997 at a fishing vessel off Grays Harbor were the most noted on surveys from 1971 to 2000 (TRW). 4 on 14 Dec 1998 from a research vessel about 60 km offshore between Cape Disappointment and Cape Shoalwater and 23 were noted between Grays Harbor and Queets R., Grays Harbor Co. (M. Force p.c.).

Terence R. Wahl

Black-footed Albatross *Phoebastria nigripes*

Seasonally uncommon to common year-round visitor in oceanic waters.

WEST
J F M A M J J A S O N D

Habitat: Outer continental shelf across the N. Pacific, concentrating locally at vessels discarding fisheries by-catch or garbage. Occasional nearer shore.

Occurrence: The most numerous Washington albatross, seldom seen except offshore. Feeds mostly on deepwater prey like squid in oceanic waters. Densities greater in outer continental shelf and slope waters, where both natural and human-provided food sources concentrate birds (see Wahl et al. 1989, Briggs et al. 1992). Of over 30,000 birds recorded from Sep 1971 to Oct 2000 off Grays Harbor, only 7% were over depths of <100 m (TRW). From aircraft surveys, Briggs et al. (1992) reported albatross concentrations off Washington-Oregon nr. Swiftsure Bank and off the mouth of the Columbia R.

Most commercial fishing activity is concentrated between 120 m depth and the shelf edge at about 200 m, and 66% of albatrosses were at fishing vessels or chums attracting birds (TRW, Wahl and Heinemann 1979, Wahl and Tweit 2000a). Concentrated at long-liners and trawlers providing larger bycatch items, less so than at shrimp trawlers where smaller, more adept shearwaters and gulls predominated (off Grays Harbor: average number of albatrosses was 44 per ground-fish trawler, 34 per long-liner, 14 per shrimp trawler). Like other *Procellariids*, attracted to food sources visually and by olfaction.

Variations in annual abundance may relate to location and magnitude of fishing effort but also are likely affected by oceanographic conditions. During ENSO events in 1982-84 and in the early 1990s, albatrosses were more abundant than during "normal" years, reflecting greater attraction to vessels during low ocean productivity. Long-term, from 1972 to 1998, overall abundance and concentration at fishing vessels increased (Wahl and Tweit 2000a).

Seasonal abundance varies. Numbers increase in late Apr-early May, peak from Jul to early Sep, decrease in late Sep-Oct as breeders withdraw to nest in the w. subtropical N. Pacific (Wahl 1975, Briggs et al. 1992, McDermond and Morgan 1993). In winter small numbers, apparently mostly subads., are present, though breeding ads. are known to forage then off the w. coast (Palmer 1962). Birds marked with leg bands are seen on occasion: these are presumably from colonies in the w. Hawaiian Is. chain.

Remarks: Albatrosses suffer mortality from fishing nets or long-lines, and are prone to eat plastic trash. Two gunshot, flightless birds were found nr. a long-liner off Westport (TRW). Confirmed nearshore Washington records are relatively rare, though there are several nearshore records in B.C. (Campbell et al. 1990a). Reported nesting at San Benedicto I., Mexico, in 2000 (Pitman and Ballance 2002). Note: a report from the Strait of Juan de Fuca on 20 Mar 1967 is questionable. Classed GL. Formerly in the genus *Diomedea*.

Noteworthy Records: *Off Grays Harbor*: 1 followed an inbound freighter to about 1000 m off the Westport jetty on 16 Apr 1977 (TRW). *Spring*: 268 off Grays Harbor on 25 Apr 1992 (TRW). Summer: 681 on 6 Jun 1996 (TRW). *Fall*: most of 907 off Grays Harbor to 125° W on 14 Aug 1993 were at fishing vessels (TRW). *Winter*: 30 offshore from Ilwaco n. to Grays Harbor on 3 Dec 1995.

Terence R. Wahl

Short-tailed Albatross *Phoebastria albatrus*

Very rare visitor.

Habitat: Oceanic waters and outer continental shelf. Historically also in nearshore and inland marine waters.

Occurrence: Jewett et al. (1953) cited seven Washington specimen records prior to 1898. Three were from oceanic areas. Inshore locations were Neah Bay, Shoalwater Bay, and Fort Canby. The seventh was an ad. collected on Cottonwood (Sinclair) I., Skagit Co., in 1896 (mounted specimen at WWU, Bellingham). Based on archaeological examination of kitchen middens, inshore occurrence was apparently frequent (McDermond and Morgan 1993).

Species believed extinct by the 1940s, due to feather hunters and periods of volcanic activity at the w. Pacific breeding colony on Torishima I. Non-breeders at sea during critical times reappeared to breed in 1950: the island was protected subsequently by the Japanese government and habitat-restoration efforts begun. By 1990 the breeding population was estimated at 123 pairs and a total population of 575 individuals (McDermond and Morgan 1993, and see Fisher et al. 1969, Hasegawa and DeGange 1982). By 2002 Hasegawa estimated a total population of about 1680 individuals.

Since the 1970s reported from waters off Alaska (where observers on fishing vessels congregate), B.C., Oregon, and California (see Roberson 1980), including one 80 km off n. Oregon on 9 Nov 1996. A bird reported off Washington on 3 May 1970 (Wahl 1970, Roberson 1980) was likely misidentified, though McKee and Pyle (2002) believed the initial identification was correct.

On 16 Jan 1993, 96 years after the last certain Washington record, a first-year bird came to a "chum" attracting birds about 60 km w. of Grays Harbor. On 27 Apr 1997 a subad. was off Edmonds (Aanerud 2002). On 27 Jan 2001 another first-year bird appeared at a chum 52 km w. of Grays Harbor. These followed a pattern of increasing, still infrequent, records in the ne. Pacific, reflecting a gradual rebuilding of the species' population. A high percentage of records from the e. Pacific are of imms. (McDermond and Morgan 1993).

Remarks: In addition to at-sea threats from long-line fisheries and ingestion of plastic debris, breeding birds are vulnerable to volcanic activity which began again in 2002 at Torishima, the main colony site (http://www.fakr.noaa.gov). Classed SC, FE, PHS, GL. Formerly in the genus *Diomedea*.

Terence R. Wahl

Northern Fulmar *Fulmarus glacialis*

Widespread, variably fairly common to abundant visitor over continental shelf and oceanic waters. Rare in outer coast nearshore waters; very rare in inland marine waters.

Subspecies: *F. g. rogersii.*

Habitat: Outer shelf, shelf edge, continental slope, oceanic waters.

Occurrence: Offshore surveys (Sanger 1970, Briggs et al. 1992, Wahl and Tweit 2000a) expanded status given by Jewett et al. (1953). Briggs et al. (1992) noted concentration year-round nr. the shelf break and shallow banks off n. Washington in summer and cool waters of the shelf break and edges of banks in winter. Off Grays Harbor from 1971 to 2000, 97% of fulmars recorded were over depths of 100-1000 m, the outer shelf and shelf-break area where commercial fishing generally occurs (TRW).

Abundance peaks noted particularly in 1973, 1985, 1986, 1990, 1993, 1995, and 1996-2000 (Wahl and Tweit 2000a, TRW). Abundance varies seasonally with Oct numbers generally greater than other months. Low numbers were obvious in El Niño years of 1983-84 (Wahl and Tweit 2000a) and 1991, 1992, and 1994. Numbers of non-breeders in summer also highly variable. Summer numbers in of 1972, 1985, 1986, and 1995 were high, but in almost all years there are at least low numbers present. Apparent variations in abundance are likely associated with regional and ocean-scale biological productivity.

Notorious fishing vessel scavengers (see Wahl and Heinemann 1979) with 78% of fulmars censused from 1971 to 2000 at vessels. Effects of this on regional distribution is uncertain. In Jul 1989 Wahl and Tweit (2000a) recorded very low numbers off Grays Harbor while aircraft surveys (Briggs et al. 1992) found large numbers off n. Washington in the area of a fishing fleet. Though foreign fish-processor fleets were prohibited off Washington in the 1980s, they continued working off B.C. with flocks of 5000-10,000 fulmars reported at vessels off sw. Vancouver I. (Vermeer et al. 1987). In 1990, when fulmars were abundant off Grays Harbor, Briggs et al. (1992) had lower numbers off n. Washington but numbers elsewhere off Washington and Oregon were larger than in 1989. Long-term, numbers increased significantly between 1972 and 1998, with birds increasingly concentrated at vessels (Wahl and Tweit 2000a).

Fulmars periodically suffer high mortality. In Washington, Harrington-Tweit (1979a) reported 63% of beached birds in a large-scale seabird die-off in early 1976 were fulmars. High mortality in 1995 die-off (J. Skriletz) and winter 2002-03 and apparently similar mortalities in 1889 and 1917 (Jewett et al. 1953) were likely due to poor food supply, possibly in combination with storms (which reduce foraging efficiency), disease, or oil spills. Birds noted in inland marine waters probably reflect oceanic food scarcity.

Remarks: Hatch (1993a) estimated that 80% of N. Pacific fulmars are dark-morph, with light-morph birds predominating in populations nesting n. of the Aleutians and in the ne. Sea of Okhotsk. Off Washington, about 91% were dark (TRW). Hatch and Gill (2003) suggested that there may be two separate subspecies of Pacific fulmars. Recorded mortality in Washington due to oil spills, ingestion of plastics, and other human-associated incidents has been low to date. Classed PHS.

Noteworthy Records: *Inland waters, Spring*: 1 on 18 Apr 1972 off Seattle. *Fall*: 1 on 26 Oct 1978 at Dungeness Spit (MESA); 1 (ph.) in Oct 1981 at Salmon Banks (Lewis and Sharpe 1987); 1 on 12 Oct 1985 at Bowman Bay, Skagit Co. (Lewis and Sharpe 1987); 1 on 31 Oct 1999 at Keystone; 1 on 31 Oct 2001 nr. Sequim. *Winter*: 1 in Haro Strait on 15 Dec 1973; 1 on 29 Feb at Montesano; 1 on 17 Apr 1976 dead at Elma (Harrington-Tweit 1979); 1 on 8 Jan 1985 in Port Angeles; 1 on 10 Dec 1986 at Semiahmoo Spit (Wahl 1995); 1 dead in early Dec 1987 in Samish Flats; Nov and Dec 1995 off Ediz Hook, Dungeness Spit, Sequim, Point No Point, Pt. Wilson; 1 on 1 Dec 1995 in Bellingham on fishing vessel arrived from the ocean (TRW); 1 on 1 Dec 1996 at Rosario Beach. CBC: 1 in 1995 at Sequim-Dungeness; 1 in 1973 from Port Angeles-Victoria ferry. *Offshore, high count*: 8295 on 12 Sep 1999 off Grays Harbor (TRW).

Terence R. Wahl

Solander's Petrel *Pterodroma solandri*

Casual vagrant

The Providence Petrel of Australia and New Zealand, this rare *Pterodroma* breeds at just two islands in the Tasman Sea. Non-breeding range includes the temperate N. Pacific Ocean, mainly w. of the international dateline but with certain records as far n. and e. as the n. Gulf of Alaska and off California (Wahl et al. 1989, Bartle et al. 1993, Wahl et al. 1993). Description and photograph of a bird noted on 11 Sep 1983, about 50 km w. of Grays Harbor (TRW) were not accepted or rejected by WBRC.

Terence R. Wahl

Murphy's Petrel *Pterodroma ultima*

Very rare visitor.

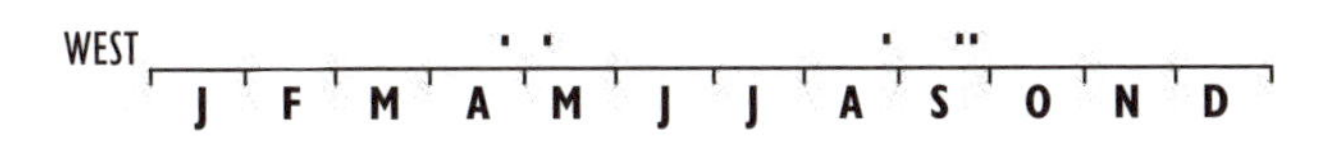

Breeding in French Polynesia and Pitcairn I. area, only in recent decades has world distribution been described. First recorded in Washington in Apr 1992, one of several species likely more frequent in Washington than the relatively low level of surveys of the outermost waters recorded. Non-breeding at-sea distribution is in the e. N. Pacific Ocean with records from Hawaii, Gulf of Alaska, and the w. coast from California to Washington (Bartle et al. 1993). Records have been annual off California in recent years, peaking from late Mar to late Jun (Mlodinow and O'Brien 1996).

Records: *Spring*: 24 on 25 Apr 1992 off Grays Harbor; 1 on 2 May 1998 off Grays Harbor (TRW) was listed as Murphy's/Solander's petrel by WBRC. *Summer*: 1 on 19 Jun and 1 on 21 Jun 2000 "off Cape Flattery" were in U.S. waters and listed as Murphy's/Solander (SGM). Fall: 2 on 31 Aug 1995 about 60-65 km off Grays Harbor (TRW); 1 on 21 Sep 1996 off Grays Harbor (TRW) listed as Murphy's/Solander's petrel by WBRC.

Terence R. Wahl

Mottled Petrel *Pterodroma inexpectata*

Rare migrant and winter visitor in oceanic waters. Accidental in continental shelf waters.

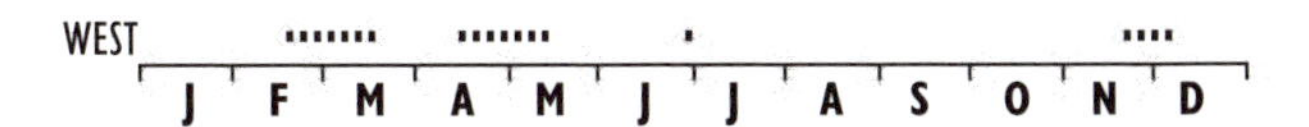

Habitat: Oceanic waters inshore to the shelf-break area.

Occurrence: Breeds nr. New Zealand, winters primarily in the n. and w. Pacific Ocean, deeper Bering Sea areas and the n. Gulf of Alaska (Bartle et al. 1993). Like a number of other marine species, only recently known to occur in Washington. Based on records from the ne. Pacific from Alaska to California (Ainley and Manolis 1979, Wahl et al. 1989, Campbell et al. 1990a), Mottled Petrels pass within Washington waters at least irregularly during migrations between breeding sites and wintering areas in the coldest, more northerly waters of the Pacific. Recent reports (e.g. Knue 2001) indicate rare occurrence in nearshore waters. Several records given in Wahl (1975) and Campbell et al. (1990a) and described as "off Washington" were actually on the Canadian side of the international fisheries boundary but these birds almost certainly travelled through U.S. waters. Mottled Petrels are more regular here than records so far indicate.

Noteworthy Records: *Spring*: 540 on 28 1972 ca. 112 km wnw. of Cape Flattery (Bourne and Dixon 1975) in Canadian waters; 2 Mar 1976 specimen at Westport (UPS 9952); 2 on 20 Apr 1977 about 140 km w. of Copalis Beach and 1 on 21 Apr 1977 about 20 km w. of Copalis Beach (TRW); singles dead at Ocean Shores, Leadbetter Pt., and n. jetty of the Columbia R. in Mar 1985; 1 on 25 Apr 1985 at Ocean Shores (Specimen 41985, UWBM); 1 on 11 May 1991 on Swiftsure Bank. *Summer*: 3 beached in late Jun-early Jul 1999 (C. Thompson p.c.). *Winter*: 1 on 23 Feb 1971 about 280 km w. of Olympic Pen. (Roberson 1980); 1 on 28 Feb 1976 at Ocean Shores (Harrington-Tweit 1979a); 1 on 2 Mar 1976 dead nr. Westport and 1 on 5 Mar 1976 dead nr. Copalis 1976 (Wahl 1975, Ainley and Manolis 1979); 4 on 2 Feb 1997 off Grays Harbor (TRW); 1 on 25 Feb 1991 at Westport (Mattocks 1999); 3 on 3 Dec 1995 offshore between Astoria and Grays Canyon; at least 10 on 26 Nov 2000 from shore at Ocean Shores (Knue 2001). *High count*: 34 on 19 Apr 1985 offshore between the Strait of Juan de Fuca and the Columbia R. (R.Pitman p.c.).

Terence R. Wahl

Cook's Petrel *Pterodroma cookii*

Casual vagrant.

Breeds nr. New Zealand, migrates into the N. Pacific, with winter records of birds off California and Baja California. Field identification requires awareness of subtle characteristics distinguishing this from similar *Pterodroma* species. A beached specimen was recovered on 15 Dec 1995 nr. Grayland following a period of severe storms. Two "Cook's/DeFilippi's (*P. defilippiana*) Petrels" on 1 Aug 2002 were seen about 140 km w. of Florence, Oregon. The species is annual off California, peaking from May to Oct (Small 1994).

Terence R. Wahl

Pink-footed Shearwater *Puffinus creatopus*

Fairly common to common migrant and summer resident in oceanic waters, rare in winter.

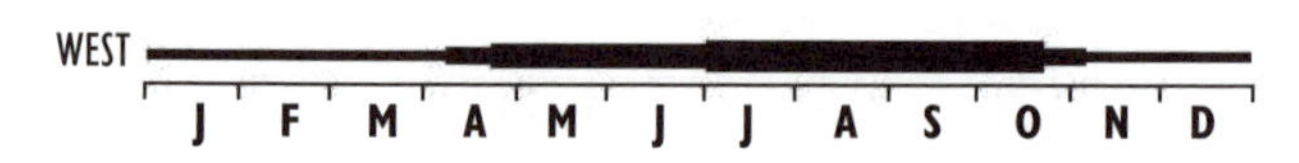

Habitat: Outer continental shelf and oceanic waters.

Occurrence: Noted in summer surveys across the subarctic Pacific Ocean and Bering Sea only along the N. American w. coast (Wahl et al. 1989), with small numbers and vagrants noted as far w. as Hawaii (Pitman 1986), Few noted n. of the Queen Charlotte Is. (Morgan et al. 1991). As elsewhere along the w. coast, the second-most abundant shearwater off Washington in season. Congregates with other species at feeding opportunities, and often roosts in mixed flocks. Unlike Sooty Shearwaters they are seen to dive after prey only infrequently. Briggs et al. (1992) found that Pink-footed and Sooty shearwaters occurred off Washington and Oregon in the same oceanographic conditions.

Surveys off Grays Harbor recorded birds from nearshore to 2000 m depth, with over 95% over 100-1000 m depth over the outer continental shelf and shelf-edge areas (TRW). Briggs et al. (1992) had similar distribution off n. Washington. Occasional birds are seen from shore. Trips off Grays Harbor indicated that the first sizable numbers of birds normally arrive in May, increase in Jun-Jul, and peak in Aug-Sep, suggesting that southbound migrants concentrated before departure in Oct when numbers dropped (TRW).

Birds forage regularly at fishing vessels (Wahl and Heinemann 1979), particularly shrimp trawlers, and are readily attracted to chums. One-day trips off Grays Harbor between late 1971 through Sep 2001 produced 2697 records of Pink-footed Shearwaters, totalling 50,011 individuals. Almost 76% of these birds were at fishing vessels or at chums attracting birds (TRW).

Data are limited, but numbers appeared to decline following the step change in oceanographic and biological factors in 1976 (Ebbesmeyer et al. 1991), and during the late 1980s to mid-1990s ENSO, with an increase beginning in 1998 (Wahl and Tweit 2000a, TRW).

Remarks: Classed PHS, GL.

Noteworthy Records: *Inland marine waters*: 1 on 22 Sep 1958 at Clover Pt., Vancouver I., B. C., after a storm (Campbell et al. 1990a). *Off Grays Harbor*: 2 on 3 Dec 1983, 1 on 21 Jan 1995, 1 on 27 Jan 2001, 25 on 7 Feb 2002 (TRW; noted in Jan 1990 from aerial surveys off n. Washington by Ford et al. 1991). *High counts, Spring*: 193 on 5 May 1981. Summer: 260 on 6 Jun 1996, 1138 on 19 Jul 2003 off Grays Harbor (TRW). *Fall*: 2286 on 7 Sep 1975 and 1560 on12 Sep 1982 off Grays Harbor (TRW). *Winter*: 25 on 8 Feb 2003 off Grays Harbor. *Nearshore*: 30 on 5 Sep 1965 at Leadbetter Pt.; 25 on 25 Aug 1997 at Ocean Shores.

Terence R. Wahl

Flesh-footed Shearwater *Puffinus carneipes*

Rare to fairly common migrant and summer visitor in oceanic waters, very rare in winter.

Habitat: Outer continental shelf and shelf edge habitats, oceanic waters.

Occurrence: A seldom-common S. Hemisphere breeder with a wide at-sea range into the Indian Ocean and the N. Pacific into the Bering Sea (Wahl et al. 1989, Everett and Pitman 1993). Jewett et al. (1953) reported one record for Washington. Surveys off Grays Harbor 1971-2000 had 172 records of 246 birds (TRW).

All but six birds were over depths of 200-1000 m, where fishing vessels also concentrated. Most often noted nr. vessels, primarily shrimp trawlers hauling nets, with flocks of other shearwaters and gulls (TRW). Over 76% of the birds were either at fishing vessels or at chums. Noted foraging solitarily on occasion and some attracted to chums appeared by themselves, apparently following scent tracks to the fish oil source. Apparently dives more frequently and more deeply than the Pink-footed Shearwater which was seldom noted diving off Grays Harbor (TRW).

Records suggest a northward migration peak in May and occurrence increasing again in early Jul, though survey effort was relatively low in Jun. Numbers observed, particularly during May, apparently declined following the 1976 step (Ebbesmeyer et al. 1991, Wahl and Tweit 2000a). Two birds offshore in Feb 2003 were exceptional. The only year during which none were observed was 1983, a major El Niño year. Annual numbers observed decreased in 1989 and remained low through 1994 during major changes oceanographic conditions and productivity, with noticeable increases in the early 2000s (TRW).

Remarks : Closely related to the Pink-footed Shearwater (considered by some authorities to be a subspecies of *carneipes*). Conservation issues include habitat destruction on Lord Howe I., introduced predators on nesting areas, and mortality in gill-nets (Everett and Pitman 1993). Threats to the small numbers occurring in Washington waters appear minimal. Classed PHS, GL. Formerly known as the Pale-footed Shearwater.

Noteworthy Records: *Winter-Spring*: 2 on 7 Feb 2003 off Grays Harbor was latest or earliest record (TRW). Fall: 1 on 19 Sep 1965 at Cape Disappointment is the only report from shore; 1 on 7 Nov 1981 off Grays Harbor (TRW). *High counts*: 15 on 6 May 1973 and 22 on 12 May 1973 off Grays Harbor (Wahl 1975), far above any other one-day totals observed.

Terence R. Wahl

Greater Shearwater *Puffinus gravis*

Casual vagrant.

One photographed on 24 Aug 2002 about 67 km off Grays Harbor (B. LaBar p.c., M. Donahue p.c., P. Anderson p.c.). The bird was with Black-footed Albatrosses, fulmars and about 20 Pink-footed and Sooty shearwaters. This was about the eighth record of the species for the e. Pacific Ocean (see Pearce 2002). Classed GL.

Terence R. Wahl

Wedge-tailed Shearwater *Puffinus pacificus*

Casual vagrant.

A dark-phase bird found dead nr. Ocean City on 10 Sep 1999 is the only state record of this tropical shearwater (C. Wood, T. Hass; UWBM 63735). Recorded twice in Oregon in 1999, both dark-phase birds: first found dead nr. Yaquina Bay, 26 March (Leal 1999), the second ca. 55 km w. of Newport, in October. The two nearest breeding areas are in Hawaii and the Islas Revillagigedo, Mexico. Over 95% of Hawaiian birds are light phase (Hawaii Audubon Society 1993) while 70-90% of the Revillagigedo birds (Howell and Webb 1995) and all S. Hemisphere breeders are dark phase (Pratt et al. 1987).

Steven G. Mlodinow

Buller's Shearwater *Puffinus bulleri*

Fairly common to common, occasionally abundant fall migrant in oceanic waters. Very rare in winter.

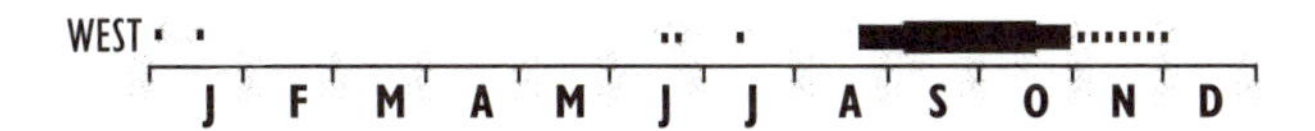

Habitat: Continental shelf and shelf-edge waters, concentrating over depths between 100 and 1000 m.

Occurrence: First recorded in Washington in 1932 (Jewett et al. 1953) but, likely due to limited offshore surveys, not again until 1968 (Wahl 1975). Surveys off Grays Harbor from late 1971 to 2000 had 948 records of 13,858 birds (TRW). Numbers peaked in the mid 1970s, and then remained relatively low, with apparent increases in 1988-89 and 1995 (Wahl and Tweit 2000a).

An increased breeding population likely explains at least part of the increase in numbers observed off the w. coast in the 1970s. Following elimination of pigs from nesting colonies in n. New Zealand, numbers of breeders increased from about 100 pairs in 1938 to about 200,000 pairs in 1981 and to a recently estimated total population of 2.5 million (see Everett and Pitman 1993). A decline in numbers in Washington noted after 1977 (Wahl and Tweit 2000a) likely reflected major, large-scale changes in ocean productivity (Ebbesmeyer et al. 1991).

Birds regularly appear in low numbers in mid-Aug and increase into Oct, with small numbers as early as Jun. Average number per day in late Aug was 14 birds, in Sep 63, and 172 in Oct (TRW). Off B.C.

there are records as early as 24 May (Morgan et al. 1991). Briggs et al. (1992) reported birds off Oregon and Washington in Mar through Nov, with highest populations from Jul through Nov.

Rare in the Grays Harbor channel; birds concentrate offshore, with 97% recorded at depths over 50-1000 m. Though numbers then decrease abruptly with increasing depth, migration is spread across much of the Pacific Ocean. Birds are often seen with other shearwaters but in years of abundance pure flocks may be seen (Wahl 1985). Do not plunge-dive but appear to take prey from the surface (Wahl 1986). They are probably at a disadvantage competing with gulls, fulmars, and other shearwaters for items available at vessels and they do not forage on discards (Wahl and Heinemann 1979, TRW), though birds often occur nr. fishing vessels.

Remarks: Classed PHS, GL. Formerly New Zealand, Gray-backed, and Gray shearwater.

Noteworthy Records: *Summer*: singles on 18, 19, 21 Jun 2000 off the Olympic Pen. *Fall*: singles on 14 and 15 Jul 2000 were earliest fall records off s. Washington (TRW); 5 on 7 Nov 1981 off Grays Harbor were latest. *From shore*: 2 on 18 Sep 1965 from Cape Disappointment. *Winter*: 1 dead on 16 Jan 1983 at Ocean Shores. Noted during aerial surveys off n. Washington in Jan 1990 by Ford et al. (1991). *High counts*: 860 on 7 Oct 1973, 935 on 10 Sep 1977, and 1232 on 3 Oct 1976 off Grays Harbor (TRW). *Inland marine waters*: singles on 24 Sep 1990 and 15 Jul 1997 in the Strait of Juan de Fuca.

Terence R. Wahl

Sooty Shearwater *Puffinus griseus*

Common to abundant migrant and summer visitor in oceanic and outer coastal nearshore marine waters; uncommon to common in inland marine waters. Rare to uncommon winter visitor.

Habitat: Oceanic, most over the continental shelf and shelf-break areas, seasonally inshore waters, major coastal estuaries, deeper Strait of Juan de Fuca and adjacent waters.

Occurrence: Except in winter, the most abundant seabird occurring over the continental shelf off Washington. Breeding mainly nr. New Zealand, Australia, Chile and at Cape Horn (Everett and Pitman 1993), an estimated 30 million visit the N. Pacific (Morgan et al. 1991), occurring across the N. Pacific in summer (Wahl et al. 1989).

Surveys off Grays Harbor from 1971 to 2001 provided 6365 records of 1,618,834 birds (TRW). Wahl (1984) estimated about one million shearwaters, very largely *griseus*, present off Washington in Sep. Widespread flocks feed on schooling prey and at fishing vessels, mainly over the continental shelf, though individuals occurred to the outer limit of surveys 183 km offshore (TRW, Briggs et al. 1992).

In Apr-May low numbers concentrate offshore over depths of 100-200 m (TRW). Flocks may be noted inshore in spring, however, with large numbers in Willapa Bay in late May (Jewett et al. 1953), 10,000 inside Grays Harbor in May 1976, and high densities off Willapa Bay in May 1989 and the Columbia R. in 1990 (Briggs et al. 1992). Studies indicate that spring migration along the w. coast is in "waves" of different age classes (Guzman and Myres 1983, Campbell et al. 1990a).

Numbers increase in Jul. By late Aug many occur in nearshore waters, with flocks feeding on anchovies in coastal harbors and the Columbia R. entrance. Large numbers congregate seasonally over La Parouse and Swiftsure Banks off the Strait of Juan de Fuca (e.g., Briggs et al. 1992). In Oct overall numbers decrease, with most offshore (TRW). In late fall and early winter small numbers occur in the e. Strait of Juan de Fuca and from Pt. Roberts s. as far as Steilacoom. Individuals occur out of normal habitat (e.g., at the s. end of Padilla Bay and flying over a river valley on the Olympic Pen. [J. Skriletz p.c.]). Low numbers occur in winter.

Abundance varies in association with oceanographic events, with a noticeable decline in numbers during the 1983-84 El Niño, an increase from 1985 to 1990, and a decline starting in 1991 to the lowest level of the 1971-91 period (Wahl and Tweit 2000a). Between 1987 and 1994 numbers off Washington and c. and s. California were estimated to have declined 90% (Veit et al. 1997). Though inshore flocks were reported in the 1990s (e.g. 10,000 nr. Tokeland on 21 Sep 1993) these were well below numbers in earlier decades, when an estimated one million birds were reported at the entrance to Willapa Bay in Sep 1968. Numbers increased in 1998, suggesting more normal ocean conditions: 200,000 at the Willapa Bay entrance on 13 Jul 2001 and 60,000+ off the Olympic Pen. on 18 Jun 2002. Briggs et al. (1992) described variations along the w. coast, and reports of relatively large numbers in inland waters in 1992 indicated displacement likely due to food shortage. Effects of oceanographic events on the world population are unknown.

Remarks: Birds enmeshed in gill-nets have been recovered on the outer coast (TRW, see Everett and Pitman 1993). Oil spills during seasons of abundance could cause major mortality. Overall mortality from an oil spill in Dec 1988 off Grays Harbor indicated low numbers of shearwaters present at that season: 13 shearwaters represented 0.01% of birds recovered (Ford et al. 1991). Classed PHS.

Noteworthy Records: *High counts, surveys off Grays Harbor, Spring*: 16,710 on 15 Apr 1972. *Fall*: 52,080 on 8 Sep 1973 and 103,600 on 11 Sep 1978. *Winter*: 84 on 7 Feb 2003 (TRW). *Nearshore, Fall*: 200 per minute on 7 Sep 1964 passed Long Beach for an hour; 500,000 on 5 and 19 Sep 1965 at Willapa Bay entrance; 100,000 on 10 Sep 1966 at Columbia R. mouth; 1 million on 8 Sep 1968 at Willapa Bay; 4000 per minute "with no end in sight" passed Grays Harbor for 25 minutes on 11 Sep 1978; 200,000 on 21 Aug 1989 at Grays Harbor; ca. 45,000 on 7 Sep 1996 at Tokeland. *Inland marine waters, Fall*: 80 on 27 Sep 1992 at Dungeness Spit; 200 on 27 Sep 1992 at Salmon Banks; 300 on 29 Sep 1992 off Seattle; over 400 on 2 Oct 1992 at Point No Point. *Winter*: 1 on 4 Dec 1977 nr. Whidbey I.; 1 on 10 Dec 1981 off Seattle (Hunn 1982); 1 on 5 Feb 1984 at n. end of Swinomish Channel.

Terence R. Wahl

Short-tailed Shearwater *Puffinus tenuirostris*

Uncommon, occasionally abundant, migrant and uncommon winter visitor in oceanic waters. Very rare, local winter visitor in inland waters.

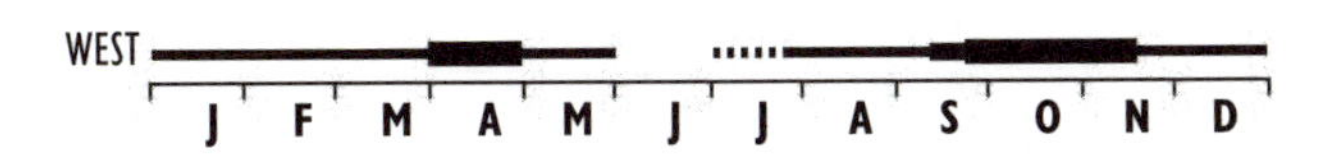

Habitat: Continental shelf and oceanic waters. Deeper passages and tidal convergences in inland marine waters.

Occurrence: One of the most abundant Pacific Ocean birds. After breeding on islands off Tasmania and s. Australia, birds cross the tropics to winter in the N. Pacific and Bering Sea as far as Pt. Barrow. Though their distribution overlaps that of Sooty Shearwaters and they are similar in appearance, Short-tailed Shearwaters (sometimes known as "Tasmanian Muttonbirds") are generally found in cooler waters (Wahl et al. 1989) and are at times abundant off Washington.

Numbers usually occur after most Sooty Shearwaters have passed southward: big flocks of shearwaters in Nov may be *tenuirostris*—winter records of this species in inland marine waters outnumber those of *griseus*. Small numbers occur in fall through early spring (Campbell et al. 1990a, Morgan et al. 1991).

Large numbers appear associated with changes in ocean productivity. A record of over 3000 in Oct 1977 off Grays Harbor (TRW), twice as many as *griseus*, was followed by widespread records in inland marine waters that fall and winter. This was during the 1976 oceanographic step/regime shift in the N. Pacific (Ebbesmeyer 1991), perhaps reflecting changes in feeding conditions in the Gulf of Alaska. In the 1982 El Niño year, 40 birds seen on 28 Aug off Grays Harbor were followed by lower numbers into Nov. Though subsequent numbers were smaller, lasting ecological changes, along with improved observer effort, resulted in many records in inland marine waters. A long-lasting period of declining productivity during the 1990s (Wahl and Tweit 2000a) also featured an increased number of records on the ocean and, particularly in winter, in inland marine waters. Large numbers noted off B.C. (e.g. in 1945, 1971, and 1977; Campbell et al. 1990a) likely also related to oceanography. Recent records off Washington appeared to show above-average occurrence in fall 1999, and in Jan-May and late Sep 2001 (TRW).

Timing of occurrences may indicate age classes involved. Northward migration of ads. is believed to spread across the Pacific, with first-year birds moving n. off Asia (Campbell et al. 1990a). Southbound breeders apparently move directly toward Australia, arriving as early as late Sep. Though some Aug birds may be breeding ads. dislocated by oceanographic changes, birds off Washington later—especially those in winter—are certainly non-breeders.

Remarks: Specimen records given by Jewett et al. (1953) are within known seasonal dates, though their sight records fall within the Sooty Shearwater season. Both species can occur together and differentiation can be difficult (Wahl 1982, Morgan et al. 1991). Beached-bird surveys have recovered specimens essentially year-round (C. Thompson p.c.). Though thousands of shearwaters have been killed annually in high-seas gillnets (see Everett and Pitman 1993), this is unlikely in Washington due to the species' low numbers in inland waters. The species is vulnerable to oil spills: 28 dead oiled birds were recovered following a spill off Grays Harbor on 22 Dec 1988. Formerly Slender-billed Shearwater. Classed PHS.

Noteworthy Records: *Off Grays Harbor, Spring*: 14 on 2 Apr 1994; 36 on 9 Apr 1989. *Summer*: 1 each on 4 Jul 1978, 22 Jul 1989, 20 Jul 1996. *Fall, high counts*: 3352 on 9 Oct 1977; 551 on 28 Aug 1982. Winter: 17 on 20 Jan 1990; 7

and 14 on 15 and 16 Jan 1994. *Inland marine waters, northern*: 1 on 11 Nov 1983 at Pt. Roberts (Wahl 1995); 1 on 4 Dec 1977 nr. Bird Rks.; 1 on 4 Dec 1977 off Whidbey I.; 1 on 10 Dec 1985 between Port Angeles and Victoria, B.C.; 2 on 6 Dec 1995 at Dungeness; 1 on 1 Nov 1998 at Ediz Hook; 22 reported on 16 Dec 2002 off Dungeness. *Central*: 1-3 on 4-5 Dec 1993, 25 on 8 Oct, 1-2 on 18 Nov-15 Dec 1994, 2 on 17 Dec 1997, 2 on 14 Dec 1998 at Point No Point; 1 each on 15 Jan 1995, 15 Jan 1996, 16 Dec 1997, and 6 Dec 1998 at Edmonds; 1 on 18 Jan 1942 off Vashon I. (Slipp 1942a); 1 on 20 Nov 1976 off Seattle, and 1 on 17 Nov and 4 on 18 Nov 1977 off Vashon I. (Hunn 1982); 1 on 1 Jan 1984 at Seattle; 1-2 on 30 Oct-6 Nov 1993 off Seattle. *Southern*: specimen 20 Nov 1941 from Tacoma Narrows (Jewett et al. 1953); 1 on 17 Dec 1979 off DuPont; 1 on 14 Dec 1983 between Nisqually Reach and Drayton Passage (BT, S. Speich); 1 in 1987 CBC at Tacoma; 1 on 7 Nov 1993 nr. Tacoma; 2 on 2 Dec 2000 at Pt. Defiance; 2 on 17 Dec 2000 at Tacoma; 1 on 1993 and 1998 CBCs at Olympia; 1 on 23 Dec 2000 at Capitol L., Olympia.

Terence R. Wahl

Manx Shearwater *Puffinus puffinus*

Rare visitor to oceanic, nearshore coastal, and inland marine waters.

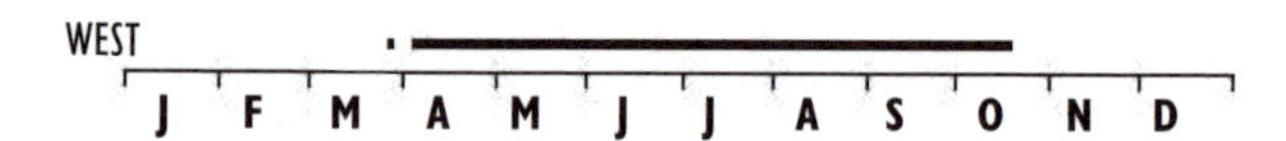

Habitat: Outer coastal waters from harbor entrances to outer continental shelf waters.

Occurrence: Breeds primarily in the e. Atlantic, with small numbers off Massachusetts and Newfoundland. Non-breeding range extended in recent decades from the S. Atlantic to s. S. America, off the w. coast of Chile (Roberson 1996) and to the n. Gulf of Alaska. Small numbers recorded annually off Washington in the late 1990s, mostly in Jun and Jul. Through Jul 2002, there were 31 records of 36 birds between 24 Mar and 10 Oct. Noted usually with Sooty Shearwaters, in some cases at fishing vessels offshore, about seven times by observers at the entrance to Grays Harbor and three times in inland waters from Admiralty Inlet, Port Townsend, and Port Angeles. Increasing effort by more observers along the w. coast may at least partially account for the increased number of reports, but whether this increase might indicate possible breeding range extension into the Pacific Ocean is unknown.

Remarks: Several N. Pacific species of shearwater (Black-vented *P. opisthomelas*, Townsend's, *P. auricularis*, and Newell's, *P. newelli*) formerly considered subspecies. Sightings of small, "white-bellied" shearwaters were not infrequent but identification remained uncertain and controversial (see Roberson 1980, Howell et al. 1994, Roberson 1996). Some earlier reports remain uncertain but many recent records, particularly following photographic documentation off Washington in 1992 (Tweit and Paulson 1994) and subsequently off California, established the certain occurrence of the Manx Shearwater in the N. Pacific. Classed PHS.

Records: *Outer coast*: 1 on 7 Sep 1973 from Westport jetty; 1 on 14-15 Sep 1990 in Grays Harbor channel; 1 on 6 Oct 1990 (Tweit and Paulson 1994); 1 on 17 Sep 1992 at Ocean Shores (Tweit and Paulson 1994); 1 ph. on 10 Oct 1992 about 20 km off Grays Harbor (Tweit and Paulson 1994); 4 ph. on 2 Apr 1994 nr. fishing vessels about 50 km off Grays Harbor; 1 on 20 and 27 Aug 1994 off Grays Harbor entrance; 1 on 20 Jul 1996 off Grays Harbor; 1 on 8 Jul 1997 from Ocean Shores; 1 on 30 Jun 1998 off Toleak Pt (Mattocks 1999); 1 on 25 Jul 1998 off Westport; 1 on 22 Aug 1998 off Westport; 3 on 10 Jul 1999 off Westport; 1 Manx/Black-vented Shearwater on 1 Aug 1999 at Ocean Shores; 1 on 9 Apr at Hoquiam and 1 on 14 May 2000 at Ocean Shores; 1 on 21 Jun 2000 ca. 18 km off the Hoh R.; 1 on 23 Jun 2000 55 km off Westport; 1 ca. 5 km off Kalaloch and 1 ca. 18 km. off Raft R., Grays Harbor Co., on 22 Jun 2000; 1 on 9 Jul and 23 Jul 2000 at Ocean Shores; 1 on 24 Mar 2001, 2 on 6 May 2001 off Westport; 2 on 18 May 2002 off Westport; 2 on 6 May 2001 off Westport; 1 on 21 Jul 2001 at Ocean Shores; 1 on 4 Aug 2001 off Westport; 1 on 10-16 Aug 2001 at Ocean Shores; 2 on 18 May 2002 off Westport; 2 on 12 Jun 2002 off the Olympic Pen.; 1 off LaPush on 10 May 2002; 2 at Ocean Shores on 12 May 2002. Inland waters: 1 in 9 Jul 1997 at Port Townsend; 1 on 3 Jun 1999 in Admiralty Inlet; 1 on 22 Jul 2000 off Port Angeles.

Terence R. Wahl

Wilson's Storm-Petrel *Oceanites oceanicus*

Very rare visitor in oceanic waters.

Breeds in the S. Hemisphere, on subantarctic islands and coastline of Antarctica; small numbers have been noted regularly in recent decades in small numbers as far n. as c. California and presumably are likely to move farther n. during periods when biological productivity is lower than normal.

Three state records: 1 about 48 km wnw. of Grays Harbor on 23 Jul 1984, 1 off Westport on 6 Sep 2001, 1 off Cape Shoalwater on 12 Jul 2003. There are one or two unaccepted records. Most California records are from Jul to Nov, peaking in Sep-Oct (Small 1994). Oregon has two records: on in May 1976 and another in Jul 1996 (Marshall et al. 2003).

Terence R. Wahl

Fork-tailed Storm-Petrel *Oceanodroma furcata*

Locally common breeder on islands off n. outer coast. Fairly common to common summer resident and rare to uncommon winter visitor in oceanic waters. Very rare to uncommon in inland waters.

OUTER COAST

INLAND MARINE WATERS

J F M A M J J A S O N D

Subspecies: *O. f. plumbea* is resident. N. race *furcata* recorded, occurs in winter (see Paulson 1992a).

Habitat: Breeds in holes and crevices on offshore rocks. Forages over outer continental shelf and offshore deeper oceanic waters. Deep passages, channels, and tidal convergences in inland marine waters.

Occurrence: Occurs across the subarctic Pacific and, in summer, the Bering Sea (Wahl et al. 1989), generally over cooler SSTs than Leach's Storm-Petrel. Confirmed breeding in Washington first in 1959 (Richardson 1960). Most abundant off Washington from spring through fall (Wahl 1975, 1984, Briggs et al. 1992). Briggs et al. (1992) found very few off Washington in Nov and none in Jan. About 98% of birds noted off Washington were over depths greater than 100 m (TRW).

Of a total population estimated at 5-10 million (Boersma and Groom 1993), about 3900 birds were estimated to breed at five sites in Washington, with an estimated 1600 on Carroll I., 1900 on the Bodelteh Is., and 200 each on Tatoosh and Alexander Is. (Speich and Wahl 1989).

Birds often forage singly or in small groups, but flocks of several hundred occur. Readily forage behind vessels in areas such as the Bering Sea (Gould et al. 1982, TRW). Wahl and Heinemann (1979) found no significant attraction to vessels: just 8% of some 27,000 total birds were observed at fishing vessels during 1972-2001 surveys off Washington (TRW). Birds are, however, attracted to fish-oil "chums."

Numbers varied off Grays Harbor, with large concentrations in 1987, 1988, and 1990 (Wahl and Tweit 2000a). In Apr and May 1985 birds foraged in Grays Harbor (TRW) feeding on crab larvae (D. Samuelson p.c.) inside the Westport marina. Others were in the Strait of Juan de Fuca nr. Neah Bay and off Edmonds. In fall 1997, following several years of widespread below-average productivity in the N. Pacific, unprecedented numbers appeared in inland waters. On 11 Oct hundreds passed Point No Point and birds were found s. to the Tacoma Narrows, with an estimated 500-1000 off Seattle. By 18 Oct they had disappeared.

Fork-tailed Storm-Petrel (Shawneen Finnegan)

Remarks: Ford et al. (1991) reported very low numbers of dead storm-petrels recovered following a large oil spill off Washington in Dec 1988, a season when few are normally present (and see Boersma and Groom 1993). Formerly the Fork-tailed Petrel. Classed PHS.

Noteworthy Records: *Off Grays Harbor, high counts, Spring*: 596 on 12 May 1985; 412 on 26 Apr 1997. *Summer*: 537 on 6 Jun 1996. Fall: 602 on 17 Jul 1993; 1239 (flock of 1200) on 18 Aug 1990; 1529 (flock of 1500) on 12 Sep and 1807 on 13 Sep 1987 (all TRW). *Nearshore, Spring*: numbers in Mar-Apr 1985 inside Grays Harbor (TRW). *Fall*: 1 on 10 Nov 1920 dead nr. Aberdeen (Jewett et al. 1953). *Winter*: 1 on 27 Jan 1990 in Willapa Bay (TRW), 2 on 8 Feb 2002 off Grays Harbor. *Inland waters, Spring*: 3 on 22 Apr 1967 in the Strait of Juan de Fuca; 1 on 24 Mar 1985 nr. Bird Rks. (Lewis and Sharpe 1987); 1 on 6 Apr off Edmonds, 1 on 14 Apr and 25 Apr 1985 in Bellingham Channel. *Summer*: 5 on 29 Jun and 2 on 7 Jul 1978 (MESA), 1 on 8 Jul 1994, and 11 on 7 Jul 1994 between Port Angeles and Victoria. *Fall*: 1 on 15 and 16 Aug 1981 in n. Rosario Strait (Lewis and Sharpe 1987); 1 each on 13 Nov 1982 in Upright and San Juan channels (Lewis and Sharpe 1987). *Winter*: 1 on 7 Dec 1953 beached on Eliza I. (Wick 1958); 1 on 16 Feb 1976 in mid-Georgia Strait (Wahl 1995); 1 on 1 Jan 1990 nr. Kingston; 1 on 14 Jan 1990 off Seattle; 1 on 10 Feb 1990 at Port Angeles; several in mid-Dec 1995 off Dungeness and Point No Point (R. Rowlett p.c.); 1 on 16 Dec 1995 at Ediz Hook; 2 on 1 Dec 1995 at Rosario Beach. High counts: 100-300 in Mar-Apr 1985 nr. Neah Bay in the Strait of Juan de Fuca.

Terence R. Wahl

Leach's Storm-Petrel *Oceanodroma leucorhoa*

Locally common to abundant breeder on outer coast, uncommon to common offshore; very rare in winter. Very rare in inland marine waters.

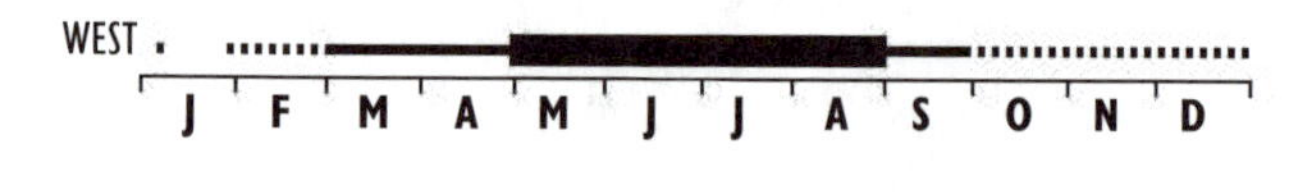

Subspecies: *O. l. leucorhoa* breeds in Washington.

Habitat: Breeds in burrows, less frequently in crevices on offshore rocks. Most forage over the outer continental shelf and oceanic waters.

Occurrence: Abundant and widespread across the subarctic Pacific Ocean and Bering Sea (Wahl et al. 1989), wintering to 10° S in the e. tropical Pacific. Non-breeders occur across the Pacific during the nesting season (Briggs et al. 1992, Boersma and Groom 1993).

Off Grays Harbor, recorded only about 10% as frequently and in only about 5% of the numbers of Fork-tailed Storm-Petrels (TRW), likely because Leach's forage more nocturnally, farther offshore, and associate with higher sea temperatures (see Vermeer et al. 1987). Migrants in Apr over waters colder (9.3-10° C) than those used by summer residents (TRW) were likely going to colonies n. of Washington. Locally nesting birds are at colonies in early Mar (Jewett et al. 1953) and likely accounted for concentrations off the Columbia R. and Juan de Fuca Canyon in Apr 1989 when SSTs were higher than normal (Briggs et al. (1992).

Up to 50,000 nest at 11, perhaps 25, colonies off n. Washington's outer coast (Speich and Wahl 1989; see Dawson 1908). Large colonies known at Jagged, Carroll, Alexander and Tatoosh Is., Kohchaa(uh), Petrel Rk., and possibly Cake and Rounded Is. Nesting success variable, with fledging from Aug to Nov, likely dependent on food availability (Jewett et al. 1953, Speich and Wahl 1989). Washington numbers are modest compared to 3.5 million estimated in Alaska, 1.1 million in B.C. and 350,000 in Oregon (Boersma and Groom 1993).

Sometimes called "tuna bird": numbers often noted along water-mass boundaries where albacore occur and a storm-petrel transition may be noted, with *furcata* in cooler, upwelled shelf water and *leucorhoa* offshore. In Jul-Sept 1971-2001 Leach's were recorded over SSTs of 13.0-18.5° C (TRW), above normal for the shelf. Briggs et al. (1992) found that densities increased with depth and steepness of the seafloor and that birds occurred nearer shore as SSTs increased there, though even during periods of highest SSTs inshore—Jul-Sep—91% were over depths >200 m (TRW).

Though birds may leave colonies as late as Oct or early Nov (Campbell et al. 1990a), Briggs et al. (1992) saw none offshore during surveys in Sep and Nov 1989, Jan and Sep 1990.

Nearshore and inland waters records more frequent in winter. During extended low ocean productivity in the 1990s, offshore numbers were low while reports from inland waters increased. Up to 10 birds were reported from Seattle to Olympia from 14 to 27 Dec, 1995 and one was on Padilla Bay in Jan 1996. Following fall 1981 storms birds noted in Willapa Bay and Quilcene.

Remarks: The subspecies *beali* (see Jewett et al. 1953) is considered merged into *leucorhoa* (Ainley 1980). Classed PHS.

Noteworthy Records: *High counts off Grays Harbor*: 50 on 29 Aug 1971; 66 on 17 Aug 1975; 51 on 22 Aug 1975; 51 on 22 Aug 1976; 56 on 15 Aug 1984; 54 on 26 Jul 1986; 210 on 20 Apr 1977; 182 far offshore on 17 Apr

1994 apparently migrating; 275 on 2 May 1998; 135 on 7 Aug 1999; 176 on 15 Jul and 192 on 4 Aug 2001 (TRW). *Winter*: 1 on 23 Feb 1971, 320 km w. of the Olympic Pen. *Outer coast, Spring*: 1 on 15 and 21 May 1972 at Ocean Shores (TRW); 1 on 19 Apr 1987 at Ocean Shores; 1 on 3 May 1975 at Westport. *Summer*: 100+, including dead, on 23 Jul 1963 at Destruction I. lighthouse; 2 dead on 2 and 6 Jun 1975 at Ocean City. Fall: 67 dead/dying, 15 live birds on 21 Sep 1965 at LaPush due to fog. *Winter*: 2 on 12 Dec 1995 at Naselle. *Inland waters, Summer*: 1 on 5 Jul 1979 in Strait of Juan de Fuca. *Fall*: 1 on 24 Oct 1981 in n. Rosario Strait (Lewis and Sharpe 1987); 1 on 2 Nov 1986 at Everett STP lagoons; 1 on 3 Aug 1993 in Strait of Juan de Fuca; 1 on 4 Nov 1993 off Seattle; 1 on 20 Nov 1997 at Edmonds. *Winter*: 1 on 14-25 Dec 1995 from Vashon-Southworth ferry (*Earthcare Northwest* Feb. 1996:10); 1 on 13 Dec 1995 nr. Deception Pass; 1 on 16 and 23 Dec 1995 at Boston Harbor; 1 on 16 Dec and 10 on 18 Dec 1995 at Tacoma; 1 on 20, 22 and 25 Dec 1995 at Seattle; 1 on 9 Jan 1996 at Padilla Bay; 1 on 31 Oct 2003 at Burlington.

Terence R. Wahl

Red-billed Tropicbird *Phaethon aethereus*

Casual vagrant.

The only state record of this tropical Pacific species was collected by a fisherman off Grays Harbor on 18 Jun 1941 (Flahaut 1947). It is a rare but regular visitor to s. California, mostly between mid-Jul and mid-Sep (Mlodinow and O'Brien 1996), with the next most northernmost record from 340 km w. of Bodega Bay, California, on 5 Oct 1979 (Roberson 1980). Subspecies presumably *P. a. mesonauta*.

Steven G. Mlodinow

Blue-footed Booby *Sula nebouxii*

Casual vagrant.

One Washington record—a specimen collected at Everett on 23 Sep 1935 (Jewett et al. 1953). Nesting locally in the tropical e. Pacific Ocean from Baja California to n. Peru, the Blue-footed Booby is rare in N. America, with most records in s. California at the Salton Sea from late Jul to late Oct (Mlodinow and O'Brien 1996). Subspecies presumably *S. n. nebouxii*.

Other than the Washington bird, the northernmost records are from Lincoln Co., Oregon, on 7-9 Sep 2002 (NAB 57:108) and Del Norte Co., California, on 16 Jan 1981.

Steven G. Mlodinow

Brown Booby *Sula leucogaster*

Very rare vagrant.

Three Washington records: an ad. at Protection I. on 18 Oct-9 Nov 1997, an ad. that landed on a boat in Puget Sound and rode into the port of Tacoma on 18 May 2002, and an ad. off Westport on 5 Oct 2002. The species is an almost-annual visitor to California, both coastally and inland. The coastal records are scattered throughout the year, apparently peaking from Aug through Oct (Mlodinow and O'Brien 1996). The only other Pacific coast record n. of California was ca. 28 km off Depoe Bay, Oregon, on 3 Oct 1998. Subspecies presumably *S. l. brewsteri*.

Steven G. Mlodinow

American White Pelican *Pelecanus erythrorhynchos*

Locally uncommon to common visitor and migrant, very local breeder in e. Washington. Rare visitor in w. Washington.

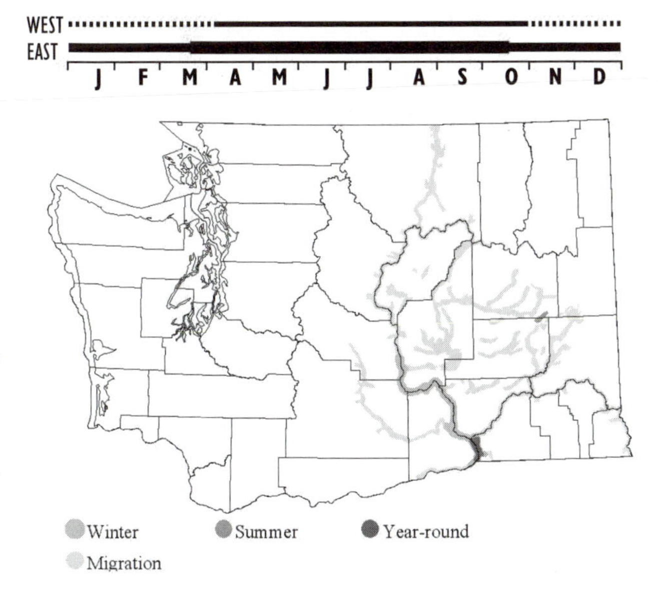

Habitat: Deltas and sandbars in slow-flowing rivers, nests on lakes and impoundments in e. Washington; shallow lakes, embayments, estuarine areas in w.

Occurrence: Jewett et al. (1953) noted few records, with a flock of 75 in 1921 the largest, and reports until the 1960s indicated the species was rare w. of the continental divide (see Evans and Knoff 1993). Up to 183 (at McNary NWR) were reported in fall in the 1960s. In the 1970s the maximum was 300 at the Potholes and in the 1980s as many as 500 fall migrants were reported. In summer 1990, 1000 were estimated in e. Washington, with over 700 on the Columbia R. Small numbers have occurred in many places, from Monse and Brewster to Sullivan L. and lakes in Spokane Co. Records of large numbers are from relatively few locations, including the Potholes and Columbia NWR, and the Columbia R. from Burbank and McNary NWR to the Walla Walla R. delta. Seasonal die-offs of whitefish in the Columbia NWR s. of Moses L. reportedly attracted hundreds and large numbers were also associated there with carp- and vegetation-control programs in the 1990s (Lowe 1997a).

Increasing human populations have impacted pelican populations. All five historically known nesting colonies in the state no longer exist or are threatened by high human use (Motschenbacher 1984, in WDFW 1997a). The early nesting colony at Moses L. (Brown 1926) was submerged behind a dam, and later there were uncertain references to nesting at several other locations. Breeding confirmed in 1994 at Crescent I., nr. Burbank (Ackerman 1994), with about 50 young present in late Aug. Birds also nested there in 1995 and 1996 (Ackerman 1997). That dredge spoil island had been covered with topsoil and planted for wildlife use in 1985 and offers relative freedom from disturbance and predation. In 1997 birds nested also on nearby Badger I., which was formed in L. Wallula in the 1950s; 75 imms. were present on 1 Aug 1997 (Ackerman 1997), and birds have nested there since.

Range contracts in winter, with birds found primarily from the Yakima R. delta along the Columbia R. to the Walla Walla R. delta, and CBC records of up to 24 birds at Tri-Cities, Two Rivers, and Toppenish NWR. Small numbers occur in w. Washington, with individuals sometimes lingering at a given location for months. Peak numbers from late Apr to late May in spring, and larger numbers from late Aug to mid-Oct. Most westside records from lowlands, Whatcom to Clark Cos., with only a few records from the Olympic Pen. and outer coast. Fewer than 10 of some 40 records w. of the Cascades were from freshwater. Numbers usually ranged from one to five, but 81 were at Lummi Bay and 102 nr. Ridgefield in fall 2001.

Marked birds from the breeding colony at Stum L., s.c. B.C., observed in both e. and w. Washington. Numbers here, particularly in summer, may reflect failed breeding attempts due to drought and poor water quality s. of Washington (see Ackerman 1994) and drought in the Great Basin may have caused a large influx of birds in spring-summer 2001.

Remarks: Populations affected positively through creation of habitat by dams and irrigation projects, and negatively by habitat destruction, utilization of wetlands and lakes for irrigation, hydroelectricity, disturbance of nesting colonies, decreases or fluctuations in food supply, shooting, predation by coyotes, pesticide contamination, powerline collisions, and fisheries management conflicts (Campbell et al. 1990a; Ackerman 1997; WDFW 1997a, 1999), e.g. shooting in Jul 2003 presumably because of perceived salmon predation (M. Denny p.c.). Classed SE, PHS.

Noteworthy Records: East, *high counts, Winter*: 40 or more in 1987-88 at Wallula; 22 in 1988-89 at Pasco; 28 on 20 Jan 1990 nr. Richland; 15 on 21 Jan 1995 nr. Richland; 38 on 15 Feb 1997 at Badger I.; 50+ on 8 Mar 1997 at Walla Walla R. delta; 28 on 20 Feb 1998 at Walla Walla R.; 22 on 5 Feb 1999 at McNary NWR; 45 on 5 Feb 2001 at Richland. **West**, *high counts*: 75-81 on 28-29 May 2001 at Lummi Bay; 30 on 21 Jun 2001 at Vancouver L.; 20-25 on 5 Sep 1912 at Nooksack R. delta (Jewett et al. 1953); 24 on 13 May 1976 at Dungeness; 8 on 2 May 1994 over Protection I. (J. Galusha, fide M. Denny p.c.).

Terence R. Wahl

Brown Pelican *Pelecanus occidentalis*

Fairly common to locally abundant summer and fall visitor on the ocean coast. Rare in winter and spring. Rare to locally uncommon on inland marine waters. Casual on freshwater.

Subspecies: *P. c. californicus*.

Habitat: Embayments, channels, and nearshore littoral waters, occasionally offshore waters. Roosts on islands and accreted islands and spits.

Occurrence: There were "large numbers" from the Columbia R. mouth to Willapa Bay from 1825 to about 1860 and birds were "common" in about 1890 at Willapa Bay and Dawson (1908, in Jewett et al. 1953) reported 100 or more off Cape Johnson. Numbers declined in the early 1900s, with records after that of singles or a very few birds except possibly from Willapa Bay s. Later attention focused on the collapse of the s. population, the source of birds dispersing n. to Washington, which was linked to pesticides in the 1960s (Schreiber and DeLong 1969). Even single birds were noteworthy in Washington by the 1960s.

Surveys from Grays Harbor from 1971 to 2000 (TRW) had single birds in 1977 and 1982. During the El Niño event of 1983 hundreds came n. Numbers were similar for several years, then dramatically increased in 1989 and remained at variably high levels through 1998 (Wahl and Tweit 2000a). Over 97% of 32,533 birds surveyed were in the channel or in littoral waters offshore. Small numbers occurred offshore in 1990 and to 50-100 m depth by 1993. One about 45 km off Westport on 29 Aug 1997 was farthest offshore (TRW).

The most important roost n. of the Farallon Is., California, in 1987-97 was in Willapa Bay where an average of 2178 birds were present during aerial surveys (Lowe 1997b). Birds commuted between there and Grays Harbor, where Whitcomb I. was another important roost prior to its disappearance in the 1990s subsequent to channel dredging. R. Lowe (in Briggs et al. 1992) estimated up to 7000 birds along the Washington-Oregon coast in late summer since 1985, and shore counts in the early 1990s peaked at 1000 birds each in Grays Harbor and Willapa Bay. In 1999 up to 6000 birds roosted on a sand island in Willapa Bay. Erosion and disturbance there resulted in relocation to surrounding estuaries, with over 9000 present at E. Sand I. in the Columbia R. in 2002 (Jaques et al. 2003). The first birds usually appear in late Apr or early May, and numbers increase through early summer and peak from Jul to Sep.

One or two birds were reported in inland waters annually in the early 1980s, with several per fall in the mid-1990s and occasional larger incursions subsequently. Most reports from the Strait of Juan de Fuca s. to Point No Point, less frequently in the San Juans, s. Georgia Strait, Port Susan, and the Central Basin off Seattle. Pelicans are rare in n. estuaries and s. of the Tacoma Narrows. Typically, few remain into Nov and birds are very rare by Dec. Occurrence on inland waters later than on the coast, with most records between mid-Aug and early Dec and a peak from mid-Sep to late Nov.

Changes in abundance of several marine species off the w. coast in the early 1990s were associated with changes in ocean productivity (see Ainley et al. 1995, Veit et al. 1997, Wahl and Tweit 2000a). Record numbers of pelicans appeared in the fall of 1997, with over 300 along the Strait of Juan de Fuca and 90 birds estimated in Hood Canal and Puget Sound s. to Olympia. The influx began on 15 Sep, peaked in early Oct, and the last report of any number was 13 at Protection I. on 19 Oct. One imm. at Bellingham on 28-30 Mar 1998 (TRW) likely wintered. An eastside record occurred in Oct 1997.

High proportions of imms. in flocks at Grays Harbor—up to 80%—noted particularly in 1985 and 1995 (TRW; see Briggs et al. 1992).

Remarks: The cleanup of toxic chemical discharge (e.g., DDT) in California was followed by increased breeding success, and expansion within traditional range. Wright et al. (2002) suggested the species may be on the verge of nesting in the Columbia R. estuary. Classed SE, FE, PHS.

Noteworthy Records: *Outer coast, Winter*: 1 on 18 Dec 1982 on Grays Harbor CBC; 1 on 18 Dec 1983 at n. jetty of Columbia R; several in early Jan 1993 at Tokeland; 1 on 19 Feb 1996 at Ocean Shores. *High counts*: 3000 on 28 Aug-10 Sep 1987 at Willapa Bay; 1150 on 22 Aug 1999 at Westport; 1000 on 27 Aug 1993 at Westport (TRW). *Inland waters*, Dec-Jul: 1 collected 10 Jan 1942 at Pt. Defiance, Tacoma (Jewett et al. 1953); 1 on 10 Dec 1983 at Ediz Hook; 1 on 20 Dec 1987 at Fox I.; 3 1st-year birds wintered in 1987-88 at Port Angeles; 2 ads. on 11 Dec 1987 at Lopez I.; 1 on 31 Dec 1988 on Olympia CBC; 1 on 1 Jan-4 Feb 1989 in "Pierce Co."; 1 on 17 Mar 1990 off Pt. Roberts (Wahl 1995); 2 on 2 May 1997 at Tacoma. *Freshwater*: 1 collected on 30 Oct 1921 at L. Kapowsin, Pierce Co. (Jewett et al. 1953); 1 on 16 Jan-3 Feb 1933 at Gravelly L., nr. Tacoma (Jewett et al. 1953); 1 on 13 Oct 1997 at McNary Dam.

Terence R. Wahl

Brandt's Cormorant *Phalacrocorax penicillatus*

Fairly common to locally abundant winter resident in marine waters; locally fairly common to common summer visitor; local breeder on the outer coast.

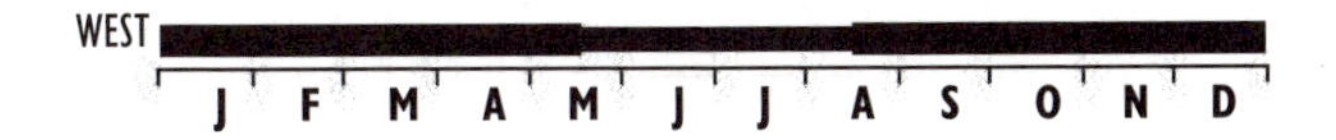

Habitat: Forages primarily in depths >20 m, often in passages and channels with strong tidal currents. Roosts and nests often with other cormorants, on cliffs, islands, and rocks.

Occurrence: Like Brown Pelicans, Heermann's Gulls, and Common Murres, many Brandt's Cormorants migrate n. in the fall from Oregon and California (see Briggs et al. 1987). Banding returns suggest that California was the origin of birds wintering in B.C. (Campbell et al. 1990a) and numbers wintering in Washington and B.C. are greater than numbers breeding in Washington, B.C., and Alaska (Briggs et al. 1992).

Though some birds return s. in late fall, many winter and return s. in the spring. Movements are imperfectly known: a bird banded as a nestling in Oregon was recovered at Bainbridge I. in Jan (Jewett et al. 1953), and birds banded in B.C. have been recovered in Washington and California (Campbell et al. 1990a). Winter range of Washington breeders is unknown.

Censuses in 1978-79 (MESA) found hundreds wintering along e. Juan de Fuca Strait and Rosario

Strait. Whale and Mummy Rks., Hall and Deadman Is., and Bird Rks. had from 150 to 2000 birds present in midday at roosting rocks, with many more probably there at night. Peak foraging abundance was at Active Pass, in the Canadian Gulf Is., where as many as 7000 have been reported (Campbell et al. 1990a). Other large concentrations occur in tidal fronts in Rosario Strait, nw. of Patos I. in Haro Strait, the e. Strait of Juan de Fuca, and Admiralty Inlet. Local and uncommon during winter s. of Admiralty Inlet. Relatively small numbers winter along the outer coast.

In inland waters, summering non-breeders are widespread, with 75 roosting at midday on Whale Rks. on 14 Jun 1979 (MESA). Very few are present in summer in inland marine waters s. of Admiralty Inlet: only 12 birds were noted in surveys of all suitable habitat there in summer 1982 (Wahl and Speich 1984).

Birds have bred in recent decades at four outer coast sites, with about 550 estimated at three islands off the n. coast and at cliffs at Cape Disappointment (Speich and Wahl 1989). Breeding numbers declined following the 1980s ENSO episode (Wilson 1991). Historic reports of breeding in inland waters are questioned (e.g. Lewis and Sharpe 1987) and breeding status is uncertain. Described as nesting on Matia I. and at Lopez Sound by Cantwell (Jewett et al. 1953). Edson (unpubl.) reported three nests among Pelagic Cormorant nests on Bare I. in 1937 and additional, untraceable reports exist. Birds have nested on Race Rks., nr. Victoria, B.C. (three nests in 1987: Campbell et al. 1990a).

Birds occur more often in flocks than other cormorants, often foraging on fish schools with Pacific Loons, alcids, and gulls. Seldom noted in shallow, soft-sediment estuaries. Peak numbers correspond with herring spawning in Feb and Mar (Campbell et al. 1990a). Long-term CBC data, which include large numbers of unidentified cormorants, indicate no pattern.

Remarks: Classed SC, PHS, Blue List.

Noteworthy Records: West, *Spring*: 1 on 25 May 1987, 64 km off Grays Harbor at 125° W (TRW). *Fall*: 2000 on 26 Sep 1978 roosting on Whale Rks., off Cattle Pt. (MESA); 500 on 19 Sep at Cattle Pass, San Juan I., and 3100 there Sep 30 1987. Winter: 1000 on 15 Jan 2000 at w. end of Shaw I.; 500 on Whale Rks. and 300 on Goose I. on 4 Nov 1989 (TRW); 512 on 21 Dec 1994 off Green Pt., Anacortes (TRW). CBC: 4413 in 1982 at San Juans Ferry; 1255 in 1989 at San Juan Is.; 906 in 1993 at Padilla Bay; 173 in 1996 at Tacoma; 133 in 1981 at Olympia; 110 in 1977 at Bellingham.

Terence R. Wahl

Double-crested Cormorant

Phalacrocorax auritus

Widespread, common to locally abundant resident in marine habitats and fairly common to common visitor to freshwater in w. Washington. Locally uncommon to common eastside resident and visitor.

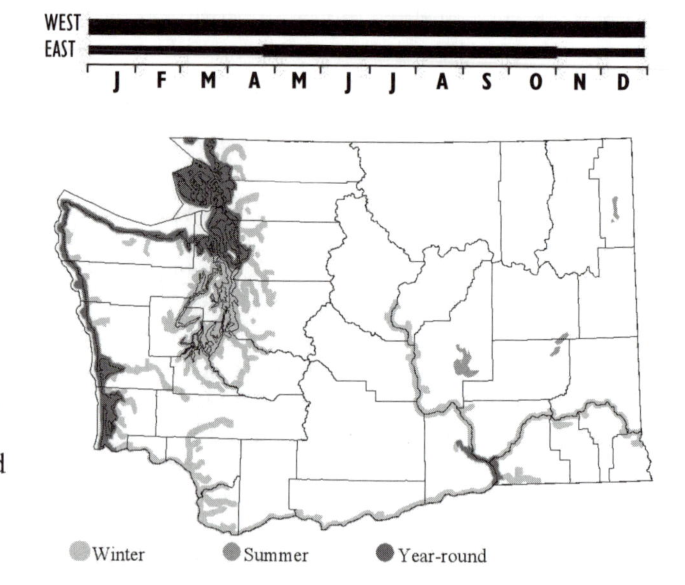

Subspecies: *P.a. albociliatus.*

Habitat: Breeds on open tops of offshore rocks and low islands, human-made objects, locally in trees. Forages primarily in shallow, soft-sediment marine embayments, lakes, and rivers. Roosts on islands, log-booms, piers. Small numbers on isolated lakes or ponds far from nest sites in summer and fall.

Occurrence: Reported nesting only on Washington's outer coast in the 1940s and early 1950s, described as migrants and winter residents in inland waters (Jewett et al.1953). In Jul 1995 10,000 were reported on the Columbia R. at Sand I., Oregon, and a winter roost of "as many as 70 birds" on pilings at the n. end of Bellingham Bay (Edson 1926) increased to 1000-1200 birds roosting all year on a pier there by 2000 (TRW). Birds commute to forage: hundreds roosting on islands in Rosario Strait pass through Deception Pass to forage in Skagit Bay or n. to Padilla and Samish bays. Others move to lowland lakes and rivers.

In e. Washington, the species was seen "commonly along the Columbia R. at all seasons" (Jewett et al. 1953). Local increases were apparent in the early 1970s. Most winter records from the Columbia R. from Richland to Wallula and on the Snake R. with 250 birds in Feb at Ice Harbor Dam. Birds reported from fall to spring at Chelan Falls, N. Potholes Res., Clarkston, Ellensburg, Malaga, O'Sullivan Dam, Snake R., and Tri-Cities area.

Double-crested Cormorant (G. Scott Mills)

An estimated 3300 nested at marine sites in the 1980s (Speich and Wahl 1989), and 4000 in 1993 (Siegel-Causey and Litvinenko 1993). About 1100 nested in n. inland waters, mostly along the e. Strait of Juan de Fuca and Rosario Strait. Colony locations and breeding success vary greatly. Birds nested at Colville and Hall Is. in 1966, then not again until 1970. Reports indicated an increase in 1984, mixed success in 1986, and none nested in the San Juan Is. in 1988 when outer-coast breeding numbers also were low (Wilson 1991). Nests were reported at about 14 outer-coast locations in the 1980s, from the Columbia R. n. to White Rk., with 610 nests in 1984 at Grays Harbor (Speich and Wahl 1989). At most outer-coast colonies breeding success was low in 1983 and in 1987 (Wilson 1991).

On the eastside, Jewett et al. (1953) reported one colony, at Goat I., Benton Co., in 1941. Nesting became more widespread, with 45 birds and 15 young at Umatilla NWR in 1972. At the N. Potholes, 38 birds with 9 young in 1980 had increased to 285 nests in 1989. About 30 pairs nested on pilings along the Pend Oreille R. nr. Usk in 1995, with fewer there in 1996 (A. Stepniewski p.c.).

Including foraging birds from Sand I., the state population in 1997 may have been as high as 20,000. Populations appear essentially non-migratory, though wandering: chicks banded at lower Columbia R. colonies in 1995 and 1996 were recovered in Puget Sound and as far s. as Los Angeles (Lowe 1997a). Though Nysewander et al. (2001) indicated a population decline from 1978-79 to 1994-95, long-term CBC data show significant winter increases statewide (P <0.01, Table 17), corresponding with continent-wide trends (see Hatch and Weseloh 1999).

Remarks: Populations increased after the elimination of predators and after egg collection ceased in mid-20th century (see Vermeer and Rankin 1984) and this increase may also be associated with availability of hatchery-raised prey. Due to believed impacts by birds on Columbia R. *salmonid* smolts (Lowe 1997a, Myers et al. 2002), USFWS proposed allowing federal, state, and tribal agencies to "take" as many as 200,000 Double-crested Cormorants at any season, without requiring annual permits (Ornithol. Newsl. 2002 146:3). In 1997-2001 an average of 857 per year were killed on the mid-Columbia R. (USDA 2003). Local observations suggest increased eastside numbers may have been due, however, to increased numbers of spiny-ray and other warm-water fishes (M. Denny p.c.). Though overall state population and possible shifts of breeding birds are unknown, numbers at E. Sand I. increased by 23% from 2002 to 2003, with 10,646 breeding pairs present on 11 Jun (B. Espenson, Columbia Basin Bulletin). No significant eggshell thinning was noted in the 1980s compared with 1947 in eggs from Grays Harbor, Protection I., or Colville I. (Speich et al. 1992). Classed PHS in Washington.

Noteworthy Records: *High counts,* **West**, *Summer*: 10,000 on 11 Jul 1995 at Sand I., Columbia R.; 6000 pairs there in 1997. *Fall*: 5000 on 5 Aug 2002 going s. at Ft. Canby; 3800+ on 14 Aug 2000 at Ilwaco; 1000 on 14 Oct 1995 at Anacortes. **East**, *Spring*: 20 in Mar 1971 at McNary

Table 17. Double-crested Cormorant: CBC average counts by decade.

	1960s	1970s	1980s	1990s
West				
Bellingham		78.8	238.1	526.2
Cowlitz-Columbia			198.9	124.2
Edmonds			121.8	151.7
Grays Harbor		42.9	65.5	87.7
Kent-Auburn			26.8	97.8
Kitsap Co.		100.8	265.6	349.8
Olympia			389.3	612.2
Padilla Bay			411.4	606.8
Port Townsend			56.8	106.1
San Juan Is.			368.8	225.1
San Juans Ferry		26.2	70.4	134.0
Seattle	26.7	100.7	338.3	523.2
Sequim-Dungeness			62.1	169.4
Tacoma		60.0	411.3	730.4
East				
Tri-Cities		1.8	23.0	158.7
Walla Walla		0.0	0.0	12.8
Wenatchee	0.0	0.0	0.4	12.6

NWR; 10 pairs on 2 May 1981 at Potholes Res.; 50-60 on 4 Mar 1996 at Almota, Franklin Co. *Summer*: 285 nests in 1989 at Potholes Res.; 45 including 15 young in 1972 at Umatilla NWR; up to 28 in 1988 at Yakima R. mouth. *Fall*: 30 in 1983 at N. Potholes. *Winter*: 150 (most imm.) on 6 Dec 1992 at the Dalles Dam; at least 80 on 22 Jan 1992 at Yakima R. mouth; 50 wintered in 1987-88 from Richland-Wallula; 21 on 5 Feb 1984 at Burbank.

Terence R. Wahl

Pelagic Cormorant *Phalacrocorax pelagicus*

Widespread, common year-round resident and breeder in marine habitats.

Subspecies: *P. p. resplendens*; *pelagicus* may occur in winter.

Habitat: Breeds on cliffs, rocks, human-made structures. Forages in nearshore waters, often with strong tidal action, along open shorelines, over kelp beds, infrequent in soft-sediment embayments.

Occurrence: Pelagic Cormorants nest over a wide range of habitats including navigation buoys with clanging bells in the San Juan Is., a navigation tower off Indian I., and abandoned piers at Dungeness and Port Angeles. In recent decades, about 4900 birds nested at about 63 locations in the state, often in small numbers. Colonies of 100 or more existed at 21 locations (Smith et al. 1997). About one-half nested at larger colonies at Cape Disappointment, and at Paahwoke-it, Tatoosh, Protection, Smith, and Colville and Castle Is. (Speich and Wahl 1989). Relatively few nest s. of Port Townsend in Puget Sound (Speich and Wahl 1989). Nesting was reported at Bremerton in 2003 and may occur at other scattered sites in Puget Sound (Smith et al. 1997). Use of nest sites and reproductive success varies, even in colonies relatively nr. each other, as in the e. Strait of Juan de Fuca in 1986. Some reproductive failure in the San Juan Is. attributed to boat traffic (Lewis and Sharpe 1987).

Migration incompletely understood, but in late Mar-Apr and Sep-Oct in B.C., most noticeably on the outer coast, with seasonal population shifts relative to prey availability (Campbell et al. 1990a). In 1933 a hatchling banded nr. Duncan, B.C., was retaken at Bellingham in Dec (Jewett et al. 1953). A first-winter bird from s. B.C. was shot in s. Puget Sound (Richardson 1997), and birds foraging in the San Juan Is. have been observed commuting to and from Mandarte I., B.C. (TRW). Distribution off Grays Harbor was similar to that of other cormorants: over 99% of birds were in the channel or immediately offshore (TRW).

Usually almost solitary, birds congregate locally, as at herring-spawn events (Campbell et al. 1990a). In 1978-79 surveys, birds were widespread throughout n. inland waters: 14 locations had at least one count of >100 birds (Wahl 1996). CBCs recorded 200 to 500 birds at locations such as Olympia, Tacoma, and in the San Juan Is. Roosts include nesting sites and locations such as cliffs on Willow I. in the San Juans and the se. tip of Guemes I. Summer distribution is less widespread than winter. In regions lacking nest sites, small numbers of non-breeders are noted (e.g. Wahl and Speich 1984, Richardson 1997). Contrasting with high CBC counts, relatively few birds are recorded in Puget Sound in summer: just 66 non-breeders were noted s. of Indian I. during an extensive survey in 1982 (TRW, S. Speich), but 264 pairs were reported nesting under a bridge at Bremerton in 2003 (*Bremerton Sun* 15 Jul 2003). Changes in winter populations are not evident on long-term CBC data.

Pelagic Cormorants seldom fly over land. Freshwater occurrence is extremely rare and limited to upper reaches of estuaries: not recorded away from saltwater in B.C. (Campbell et al. 1990a) or Oregon (Gilligan et al. 1994). A Quillayute R. record (Jewett et al. 1953) may have been in river-mouth tidal waters.

Remarks: Shoreline development and disturbance threaten nesting locations (Speich and Wahl 1989, Siegel-Causey and Litvinenko 1993). Classed PHS, GL.

Noteworthy Records: West, *Spring*: 225 on 30 Apr 1992 from Admiralty Inlet ferry (TRW). *High counts, 1978-79 MESA surveys*: 636 on 18 Sep 1979 at Protection I.; 239 on 30 Nov 1979 at Padilla Bay; 183 on 7 Aug 1979 at Port Angeles; 177 on 13 Oct 1978 at Voice of America; 151 on 22 Aug 1978 at Dungeness Spit; 124 on 14 Jun 1979 at Colville I; 100 on 2 Dec 1979 at Bird Rks. CBC: 514 in 1984 at San Juan Is.; 332 in 1995 at Olympia; 249 in 1994 at Tacoma; 218 in 1987 at Kitsap Co.; 205 in 1988 at Port Townsend; 204 in 1990 at Tacoma; 187 in 1989 at Sequim-Dungeness; 161 in 1989 at Padilla Bay. *Note*: Reports of birds on CBCs at Bellevue in 1983, Tri-Cities in 1994, and 35 on E. Lk. Washington on 26 Dec 1987 are suspect. A bird 40 km off Grays Harbor on 9 May 1982 may have been misidentified.

Terence R. Wahl

Red-faced Cormorant *Phalacrocorax urile*

Casual vagrant.

There is one record of this primarily Alaskan species from Washington: nr. Elwha R. mouth, on 8 May 1999 (Mlodinow and Pink 2000). Though there are no other records from the contiguous United States, there are two from B.C.: Masset Sound, Queen Charlotte Is., 10-11 Apr 1988 (Campbell et al. 1990a) and Dixon Entrance, 20 Jun 1999 (*Birders Journal* 8:160). Though the concentration of vagrant records from Apr to Jun might appear unusual for a bird more expected during winter, this may be due to this species' more recognizable alternate plumage.

Stephen G. Mlodinow

Magnificent Frigatebird *Fregata magnificens*

Casual vagrant.

At least two Magnificent Frigatebirds have occurred in Washington. The first was an imm. photographed over the Columbia R. at Umatilla NWR on 1 Jul 1975 (McCabe 1976). The second record consists of a series of sightings, presumably of the same imm., first noted over Commencement Bay on 7-8 Oct 1988, then at Pt. No Pt. on 11-17 Oct, at the Copalis R. mouth on 22 Oct, at Tokeland, 29 Oct, and finally on 31 Oct 1988 from the Astoria bridge over the Columbia R. (Tweit and Skriletz 1996).

This widespread tropical species wanders annually in small numbers to inland and coastal s. California. Northwest records, almost all along the coast, include five from Alaska, five from B.C., and 10 from Oregon. Most occurred between late Jun and late Oct, with the rest in winter and spring (Mlodinow 1998a).

The Great Frigatebird *F. minor* has been recorded twice in central California and once in Oklahoma (AOU 1998, McCaskie and San Miguel 1999, Howell 1994). The Lesser Frigatebird *F. ariel* has been seen once in Maine (Snyder 1961).

Steven G. Mlodinow

American Bittern *Botaurus lentiginosus*

Locally uncommon breeder in e. and w. Washington. Rare to locally uncommon in winter.

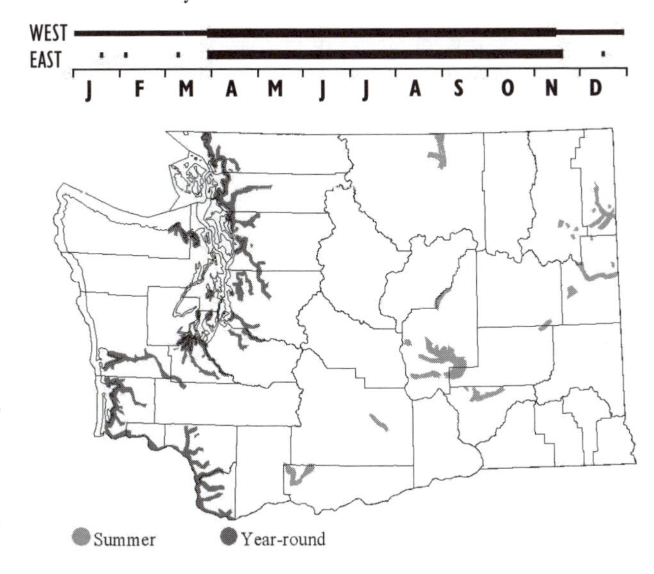

Habitat: Freshwater marshes and estuarine wetlands with dense vegetative cover. Rare in open habitats such as tidal flats or fields.

Occurrence: Widespread w. of Cascades, local due to habitat. Sometimes absent where suitable habitat appears present, including much of the Olympic Pen. (Smith et al. 1997). May nest in the San Juan Is. (Lewis and Sharpe 1987). In e. Washington nests locally at lower to mid elevations from Klickitat Co., Yakima Co., Columbia Basin, Spokane Co., to the n. Okanogan. Very local in se. counties (see Smith et. al. 1997).

Winter distribution appears to be determined by probability of severe freezes, which declines from n. to sw. CBCs show birds essentially absent n. of Skagit Co. in most winters, local in small numbers s. of that, mainly at Sequim-Dungeness, Seattle, Olympia, Grays Harbor, and Leadbetter Pt. More birds were found more consistently in the sw. (e.g. 1-5 birds on 24 out of 28 Grays Harbor CBCs, 1972-99). Recorded only once on CBCs e. of the Cascades, where there are a few other, scattered, winter records.

Numbers apparently have decreased with alteration of habitat and human activities in breeding habitats. Historical status (apparently never common: e.g., Jewett et al. 1953) somewhat uncertain. Due to secretive behavior and habitat associations, may be more widespread than described. Decline in numbers, based on lack of recent records in areas of known past occurrence in Washington and across N. America, however, appears certain (see Campbell et al. 1990a, Gibbs et al. 1992).

Remarks: It is unlikely that creation of secluded reservoirs has offset widespread loss of suitable situations throughout much of the state. Human presence in limited, previously undisturbed habitats including clearing of marsh edges around lakes through grazing, creation of lawns, dogs, disturbances from recreation like boating, water-skiing, and personal watercraft (e.g. Campbell et al. 1990a, Wahl 1995, Stepniewski 1999) are probable factors in status. A continent-wide decline for this and most other marsh dependent species placed the species on the Blue List (Tate 1986).

Noteworthy Records: West, *CBC*: 11 in 1986 at Skagit Bay, 6 in 1985 at Olympia, 6 in 1994 at Sequim-Dungeness, 5 in 1972 and 1978 at Grays Harbor. **East**, *Winter*: Birds at Toppenish NWR on 21 Feb 1977, and in Whitman Co. in mid-Mar 1976 may have been early migrants; one nr. Othello at Columbia NWR in Jan 1989 was present all winter.

Terence R. Wahl

Great Blue Heron *Ardea herodias*

Common resident statewide, especially in Puget Trough and lower Columbia R. Uncommon to rare in mountains and in arid uplands of e. Washington.

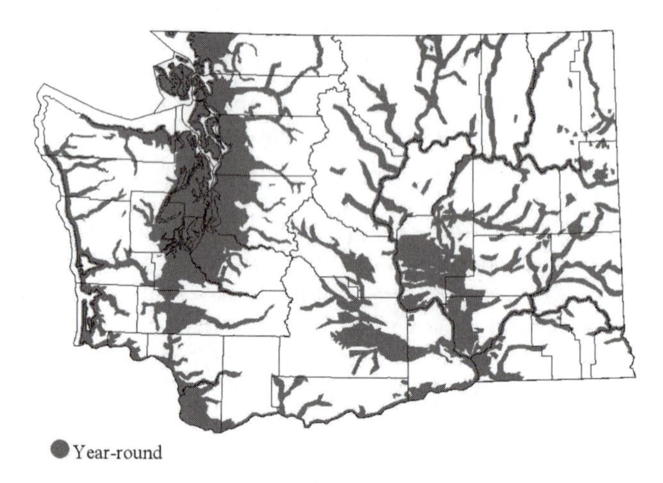

Subspecies: *A. h. fannini* in the Pacific coastal nw. (Payne 1979) and *herodias* in e. Washington (see Butler 1992).

Habitat: Marine shorelines, ponds, lakes, marshes, streams, ditches, sloughs, estuaries, intertidal areas, eelgrass meadows, and shorelines, also upland meadows, fallow fields and grassy margins including ditch edges. Nests in upland forests or isolated stands nr. productive feeding grounds and away from human intrusion. Nest sites in deciduous, conifer, and mixed stands of mature, large trees, and there are two accounts of nests on power transmission towers.

Occurrence: The extensive habitat associated with some of w. Washington's river deltas support the w. coast's largest concentrations of Great Blue Herons and some of the largest colonies in N. America. (Gibbs et al. 1987, Butler 1996).

Westside herons are largely non-migratory (Hancock and Kushlan 1984, R. Butler p.c.) but some wander or disperse, with some coastal birds recovered in the interior and the Columbia R. mouth (see Campbell et al. 1990a). Other occurrences include birds at 2100 m elevation in B.C. (Campbell et al. 1990a), nr. Mt. Baker (Wahl 1995), and birds at sea several times off Grays Harbor (TRW). Eastside birds likely "escape" to ice-free zones during harsh winter weather and freezes and some interior populations may be at least partially migratory (Campbell et al. 1990a).

Historical Washington breeding numbers uncertain. The 10 largest heronies—possibly 50% of breeders statewide—are in n. Puget Sound/s. Georgia Strait, the lower Columbia R. islands and the Potholes. Recent numbers at each active colony between 1990 and 2001 (WDFW data base 2001) resulted in an estimate of 6300 pairs. Along inland marine waters there were 3,027 pairs in 114 colonies, 1588 pairs in 78 colonies in the upper Columbia R. and e. Washington, 1340 pairs in 26 colonies on the lower Columbia and 348 in 22 on the Olympic Pen. and outer coast. However, 105 colonies had not been surveyed in the past five years and at least 28 were inactive or abandoned within the period. Abandonments resulted from logging, shooting, and damage or destruction in a 1996 ice storm. In 1999 extensive abandonment was reported throughout nw. Washington including the largest colony at the time, Birch Bay with 440 nests. Many colonies re-established in 2000 with, for example, 500 nests at March's Pt., 453 at Pt. Roberts (abandoned in 2004) and 261 at Birch Bay in 2002 (AME). Overall, Butler (1996) estimated 4000 pairs in the nw. breeding population, with the greater percentage within Puget Sound and Georgia Strait.

Overall nw. population trends are uncertain. Westside human development has had the greatest impact but effects of expansion of eastside wetlands through irrigation in the late 20th century are not certain. Increases over time are indicated: 200 birds nr. the mouth of the Columbia R. prior to 1860s were the most reported at one time by Jewett et al. (1953). CBC data indicate significant increases ($P < 0.01$) in overall winter numbers both e. and w. of the state.

Remarks: *A. a. treganzai* stated occurring nr. Moses L. (see Jewett et al. 1953, AOU 1957), but there is little support for three different subspecies breeding in Washington, and differentiation of subspecies and ranges is needed (D. Paulson p.c.). Though fledging success is greater in larger colonies (Butler 1992), which may contribute significantly to the genetic diversity and health of the regional population (DesGranges 1988), nesting concentrations create a high level of localized sensitivity and vulnerability and, if left unprotected, potentially place regional populations at risk. In 1997-2001, an average of 169 birds were killed per year under a juvenile salmonid protection program on the mid-Columbia R. (USDA 2003). Classed PHS.

Noteworthy records: *Highest colony count*: 500 nests at March Pt. in 2002. *Offshore*: single birds on 27 Aug 1983, 10 Sep 1989, and 1 Sep 1994 were 73-77 km off Grays Harbor (TRW).

Ann M. Eissinger

Great Egret *Ardea alba*

Locally uncommon non-breeding visitor e. and w. Washington. Local breeder.

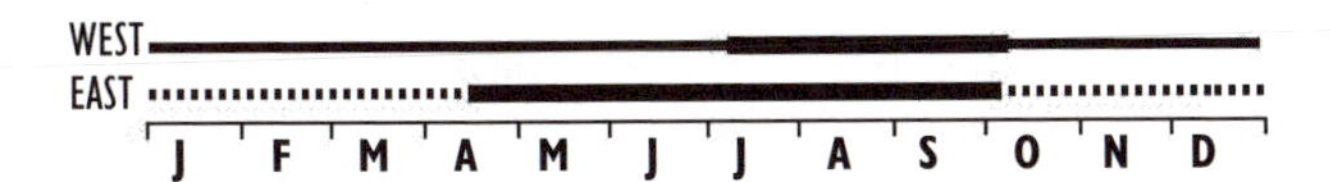

Subspecies: *A. a. egretta.*

Habitat: Wetlands, fields, estuaries, tidal flats; nests in willows.

Occurrence: Classed as hypothetical by Jewett et al. (1953), the range expansion of this conspicuous species during the last three decades was likely well documented due to the increase in field observers. Status in the state corresponds with similar, greater, increases in Oregon, and in B.C. (see Campbell et al. 1990a).

The first state report was from Turnbull NWR on 14 Jun 1949 (Canaris 1950, in Weber and Larrison 1977), followed by a bird at Crab Cr. on 10 and 16 Jul 1953 (Johnsgard 1954). Eastside occurrence increased rapidly after 1975, with non-breeders occurring widely s. from Reardan and the Potholes Res. Though recorded during all seasons, almost all records were from Apr to Sep. Most occurred from Spokane-Grant-Kittitas Cos. with many records from reservoirs and wetlands created by dams and irrigation projects such as Columbia Basin WMAs, Scootenay Res., Asotin-Clarkston and Walla Walla, Yakima R. delta, at McNary, Columbia and Umatilla NWRs, and at Lyle and Bingen on the Columbia R. Birds were almost annual in recent winters from Yakima to Walla Walla Cos.

Westside reports of seasonal numbers varied, with increases noted in 1973-74, 1986-86, and 1992. Birds occurred as far n. as Pt. Roberts and Drayton Harbor, w. to Dungeness, s. through Puget Sound and to Wahkiakum Co., sw. to Grays Harbor, Willapa Bay and Ilwaco. Not reported for the San Juan Is. (Lewis and Sharpe 1987). Birds now uncommon to locally fairly common during late summer-early fall in sw. Washington, likely representing post-breeding dispersal from farther s. Numbers appear in early to mid-Jul and start to decline in mid-late Sep. By early winter, birds are rare to uncommon in the sw. Farther n., birds are rare in late summer-early fall, very rare in winter. Rare to very rare on the westside, except in Clark-Cowlitz Co. lowlands, from Feb through Jun. All but 10 of 50 CBC records from 1973 to 1998 were s. and w. of Olympia.

The species breeds regularly only at Potholes Res. (75 pairs in 1999: Stepniewski 1999), at a small heronry along the Columbia R. between Benton and Franklin Cos. (Smith et al. 1997). Birds nested at Ridgefield NWR in 1998 and 1999, though success was unknown. Five apparent non-breeders were at a large heronry on Samish I. in 1987 and 1988.

Noteworthy Records: East, *high counts, Summer*: 75 pairs in 1999 at Potholes (Stepniewski 1999). *Fall*: 200 on 6 Oct 1998 at Potholes. *Winter*: 2 from Dec 1996 to 23 Feb 1997 at Walla Walla R. delta; 1 wintered 1996-97 at Lowden; 1 on 8 Jan 2000 at Selah; 1 on 22 Mar 1998 at Kahlotus L.; 1 on 24 Mar 1998 at Wahluke Slope WMA. **West**, *high counts, Fall*: Ridgefield, 125 on 18 Aug 1997; 124 on 12 Sep 2000; 90 in late Sep 1986; 89 on 31 Aug 1996; 62 on 8 Aug 1998. 8 on 16-19 Aug 1996 at Skagit WMA. *Winter*: 33 on 7 Dec 2000 at Ridgefield; 35 on 6 March 2002 at Vancouver lowlands. *CBC*: 50 birds statewide on CBCs 1973-98; first in 1973 at Leadbetter Pt.; 1 in 1993 at Toppenish first on eastside; 2 on 21 Dec 2002-10 Feb 2003 at Moses L.; 10 on 8-25 Jan 2003 at Ringold.

Terence R. Wahl

Snowy Egret *Egretta thula*

Very rare spring to late fall visitor, increasing in frequency in w. Washington.

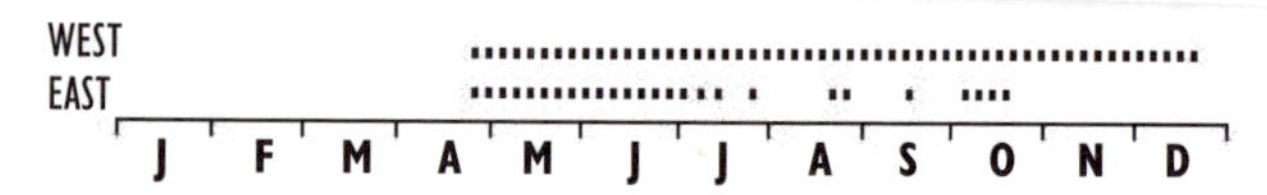

Subspecies: *E. t. brewsteri* (see Parsons and Master 2000).

Occurrence: Unrecorded in Washington until 1975, with the first westside records in 1984. Since 1993 Snowy Egrets have occurred almost annually in w. Washington, less than annually in e. Washington. Most records are from estuarine areas w. of the Cascades and along the Columbia R. in the interior. Though birds now breed in se. Oregon, a similar pattern of increasing abundance has been noted in w. Oregon. Prior to the late 1970s they did not appear regularly there (Gilligan et al. 1994). Of 36 Washington records from 1975 to 2002, most of the interior records are from May to Jul, while 12 of 20 westside records are from Apr to Jun and seven are from Jul to Oct. Thirteen of the westside records are from the sw. quarter, and all but two of the interior records are from the Columbia/Snake rivers basin.

Records: East: 11-12 May 1975 at Richland; 20 Jun 1976 at Tucannon R. mouth; 23 Jun-24 Aug 1976 at McNary NWR; 1-3 May 1977 at Clarkston; 20-27 May 1977 at Badger L., Spokane Co.; 3 Sep 1980 at McNary NWR; 7-21 Oct 1980 at Vernita Bridge, Saddle Mt. NWR; 3 Apr-25 May 1987 at Whitman Mission; 8 Jul, 18 Aug 1990 nr. Moses L.; 1-7 May 1997 at Crow Butte; 6 Jun 1999 at Rock I.; 9 Jun 2000 at Oroville; 13-15 Jul 2000 at Wilson Cr.; 2 on 9 May 2001 at Walla Walla R. delta; 20-25 May 2002 at Corfu; 2-27 Jul 2002 at Wallula. **West**: 22 Apr 1984 at Hoquaim; 2 on 9 Jun 1984 at Hoquiam; 29 Jul-9 Sep 1984 at Ocean Shores; 1-2 Oct 1985 at Washougal, 1-8 Nov 1986 at Crockett L.; 19-20 Dec 1986 w. of Raymond; 6-7 May 1987 at Ocean Shores; 25 Apr-3 May 1993 at Bay Center; 7-8 May 1993 at Lummi Flats; 26 Apr-14 May 1994 at Ocean Shores; 16 May 1994 at Dungeness NWR; 23-26 May 1994 at Shelton; 8 May 1996 at Olympia; 28 Sep-15 Oct 1997 at Bay Center; 23-25 Jun 1999 at Sequim Bay; 13 May 2000 at Tokeland; 7 Jul 2001 at Everett; 22 Aug-10 Sep 2001 at Blaine; 2 on 9-11 Oct 2001 at Ridgefield; 20-22 May 2002 at Edmonds.

Bill Tweit

Little Blue Heron *Egretta caerulea*

Casual vagrant.

Three records in Washington. Imms., documented by photographs, were at Judson L., Whatcom Co., 5 Oct 1974-5 Jan 1975 (Weber and Hunn 1978) and at Crockett L. 26-28 Oct 1989. An ad. was at Ellensburg on 8-9 Jun 2002. There are three records from Oregon (Marshall et al. 2003).

Bill Tweit

Cattle Egret *Bubulcus ibis*

Rare fall post-breeding dispersant and very rare spring and summer visitor.

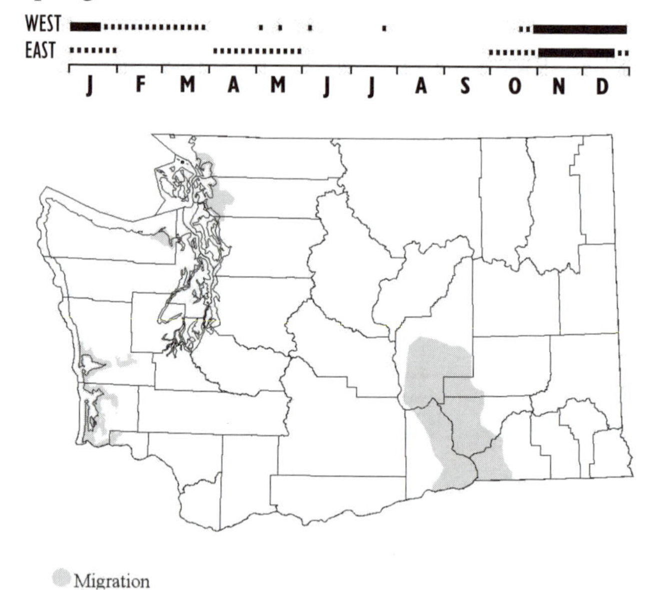

Subspecies: *B. i. ibis.*

Habitat: Open areas, primarily pasturelands.

Occurrence: Following rapid colonization of the New World (Telfair 1994), the first Cattle Egrets recorded in Washington appeared in Walla Walla Co. on 16 Oct 1967 and two years later on 30 Dec 1969 in Pacific Co. (Mattocks et al. 1976). Three more westside records followed: 30 Oct 1973 in Skagit Co., fall 1974 in Pacific Co., and fall 1975 in Whatcom Co. (Mattocks et al. 1976). The next eastside records were in fall 1977 when singles were found in Walla Walla and Asotin Cos. Since then, they have occurred almost annually in late fall and early winter on both sides of the state, but annual totals have varied greatly. Their seasonal pattern of occurrence is evidence of a consistent northward post-breeding dispersal in w. N. America, a phenomenon undescribed by Telfair (1994).

Of approximately 340 fall/winter records through Dec 1999, 62% were from the four highest seasons in 1984-85, 1992-93, 1994-95, 1997-98. Fall records have been relatively equally distributed between the westside (54%) and the interior (46%), though abundance trends in the two regions differ strongly; 49% of westside records but only 18% of the eastside records occurred prior to the 1990-91 season. Excluding the record incursion in 1992-93, the westside records have average four per year in the 1990s while the eastside records have averaged 13 per year. Cattle Egret numbers appear to have increased in the interior during the 1990s, while decreasing on the westside except during the 1992-93 incursion. Harris (1996) reports a similar decrease in nw. California after a peak in the late 1970s and early 1980s.

Westside fall-winter records are concentrated in several areas: the sw. coast (Pacific/Grays Harbor Cos.), the n. Olympic Pen. (Clallam Co.) and nw. Washington (Skagit/Whatcom Cos.). Eastside fall/winter records come primarily from the se. (Walla Walla/Benton Cos.) and the c. Columbia Basin (Adams/Grant Cos.).

The first spring/summer record was in 1980 when two were found in Grant Co. Since then there have been eight additional reports, ranging from 1 Apr to 20 Jul. Spring-summer reports are equally distributed between the eastside (five) and the westside (five).

Bill Tweit

Green Heron *Butorides virescens*

Uncommon summer resident, rare in winter at lower elevations in w. Washington. Very rare on eastside.

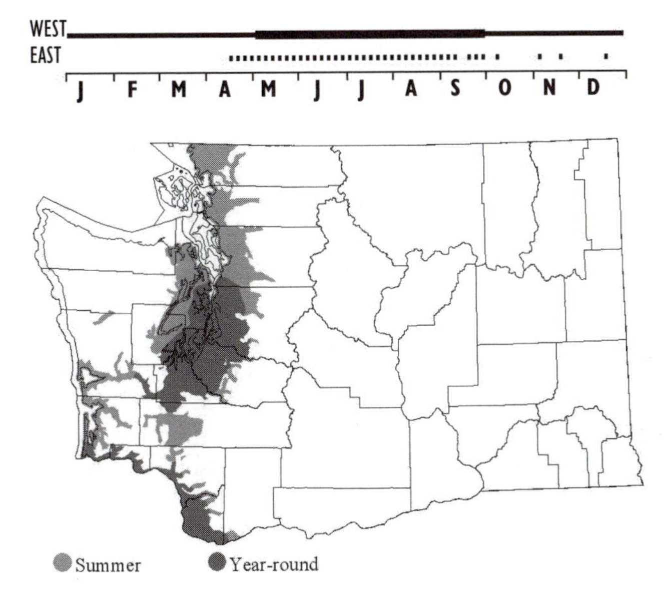

Subspecies: *B. v. anthonyi.*

Habitat: Forested freshwater wetlands, streams with heavy vegetative cover, noted on tidal flats or open freshwater edges nr. nest sites.

Occurrence: Obviously more numerous since Jewett et al. (1953) described it as "scattered and irregular" w. of the Cascades. Although Davis and Kushlan (1994) speculated that land-use changes, wetland drainage, and expansion of coastal recreational activities have likely reduced local populations, they reported that nesting and wintering seem to be undergoing long-term expansion to the n. along the Pacific coast.

Dawson and Bowles (1909) classed this species as hypothetical. Jewett et al. (1953) reported one bird nr. Kelso on 31 May 1938. Larrison (in Jewett et al. 1953) reported a breeding record nr. Camas. The first westside nesting record was at Union Bay, Seattle, in 1939 and 1940 (Hunn 1982). The first record in Whatcom Co. was in 1963, with frequency of reports increasing since then. Birds noted widely (Eissinger 1994, Wahl 1995), as far inland as Baker L. on the Skagit R. (USFS), with early reports of fledglings in 1980 and in 1985 (Wahl 1995). BBA surveys indicated nesting in the Puget Trough and lowlands s. to the Columbia R., with uncertain occurrence on the Olympic Pen. (Smith et al. 1997). The first B.C. record of the species was in 1953 (see Campbell et al. 1990a).

Birds are absent from the mountains, and eastside records are few (see Davis and Kushlan 1994). The first record in e. Washington was of a bird killed on the Walla Walla R. around 1914 (Jewett et al. 1953, USFW files). The species is a very rare breeder in e. Washington, with one recent record in Chelan Co. (Smith et al. 1997).

Fall migration generally late Aug-early Sep, but individuals winter as far n. as Vancouver I. (Davis and Kushlan 1994). CBC counts of wintering individuals show effects of weather on occurrence. Of 65 records through 1999, just five were in Whatcom and Skagit Cos. where severe freezes are more frequent, while 51 were from Edmonds to Olympia. Just two other winter records in Whatcom Co. (Wahl 1995) also reflect colder, more severe winter weather there. Washington CBC data from 1959 to 1988 indicated a significant increase in winter (Sauer et al. 1996) and BBS data also indicated at significant increase ($P < 0.05$) from 1966 to 2000 statewide (Sauer et al. 2001).

Noteworthy Records: East, *Spring*: 1 on 10 Apr 1970 at Sportsman's SP, Yakima Co.; 1 on 14 May 1987 at n. Yakima Canyon, Kittitas Co.; 1 on 20 May 1988 nr. Leavenworth; 1 on 18 May 1989 nr. Thorp. *Summer*: 1 in

1914 on lower Walla Walla R. (Jewett et al. 1953); 1 on 16 Jun 1973 at Sun Lks. SP; 1 on 7 Jun 1981 along Yakima R. w. of Cle Elum; 1 imm. in 1989 nr. Leavenworth; 1 on 9 Jul 1992 nr. Wapato; 1 on 16-17 Jun 2001 at Bingen; 1 on 6 Jul 2001 at Yakima R. delta. Fall: 1 on 19 Aug 1940 on Snake R. at Wawawai (Jewett et al. 1953); 1 on 3 Nov 1977 at McNary NWR; 1 on 29 Aug 1986 at L. Wenatchee; 1 on 6 Sep 1986 at College Place; 1 on 25 Aug 1986 nr. Yakima; 1 on 6 Sep 1987 at Ellensburg; 1 on 26 Sep-8 Oct 1988 at Yakima R. delta; 1 on 20 Aug 1994 at Crow Butte SP; 1 on 27 Aug 1994 at Vantage; 1 on 12 Sep at Hood Park, Walla Walla Co.; 1 on 24 Sep 1999 at Columbia NWR; 1 on 30 Sep-5 Oct 1999 at Richland; 1 on 21 Dec 1999 at Two R. WMA; 1 on 18 Sep 2000 at Yakima R. delta; 1 on 12-16 Nov 2001 at Yakima.

Ann M. Eissinger

Green Heron
(Shawneen Finnegan)

Black-crowned Night-Heron

Nycticorax nycticorax

Locally fairly common summer resident, locally uncommon to fairly common winter resident in e. Washington. Rare w.

Subspecies: *N. n. hoactli.*

Habitat: Forages widely in wetlands and fields, nests in dense riparian woods, often in willows, along watercourses.

Occurrence: Prior to the 1950s, nested at relatively few sites, all within the Columbia Basin (Jewett et al. 1953). Numbers increased and became more widespread, especially from the 1960s on (see Smith et al. 1997), as in B.C. in the 1970s and 1980s (Campbell et al. 1990a). This increase and shift followed creation of extensive habitat due to irrigation projects following construction of major dams in e. Washington, which provided wetlands in large areas of dry shrub-steppe habitats.

Recorded widely at low elevations, with numbers increasing noticeably in Apr. Nests in Lincoln, Adams, Douglas, Grant, Benton, Klickitat and s. Okanogan Cos., along Walla Walla and Touchet Rs., and at Toppenish (Smith et al. 1997, Stepniewski 1999). Formerly more widespread in se. counties (Weber and Larrison 1977). Colonies have varied in location and size, from a few pairs at St. Andrews and Banks L. to estimates of 1000-2000 pairs at the Potholes Res. in the 1980s. Birds occur n. of known nesting areas (e.g. Chewelah in 1981) and the increase in B.C. is associated with the increase in Washington's population (R. Friesz, in Campbell et al. 1990a). Many apparently leave after young fledge, but fall migration is seldom noted. In winter, largest numbers are recorded on CBCs at Moses L. (from 1989 to 1998), Tri-Cities (1967-98), and Toppenish.

Jewett et al. (1953) did not list occurrence in w. Washington but mention three "reports" in the late 1800s. The species now appears to be an uncommon, local winter resident on the westside, often at traditional roosts. Fall arrival

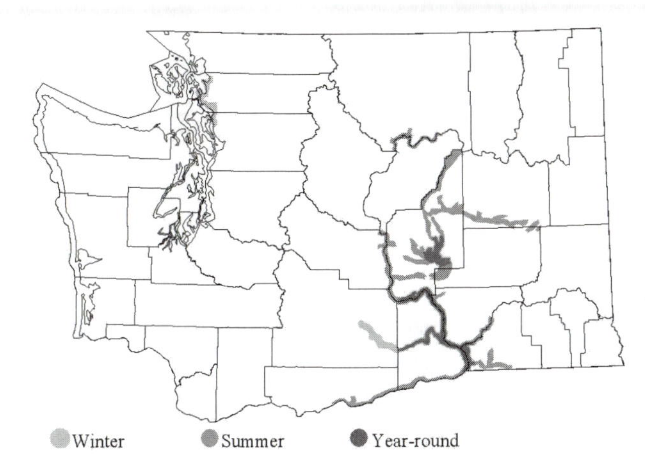

typically from mid-Sep to mid-Oct with birds leaving in Apr. Birds nested at least once in the 1990s nr. Stanwood (Smith et al. 1997), but current breeding activity there is unknown. There are a number of records from Jul and Aug nr. Ridgefield, possibly post-breeding wanderers from the s. or perhaps local breeders. Summer records from elsewhere on the westside are extremely scarce, and there are very few records from the Olympic Pen. at any season.

Remarks: A colony at Yakima R. delta was abandoned, believed due to human disturbance. Noted as "in trouble" in Idaho and Oregon with nesting failure associated with pesticides (Gilligan et al. 1994). In 1997-2001 an average of 164 birds/year (range 82-249) were killed under a juvenile salmonid-protection program on the mid-Columbia R. (USDA 2003). Over N. America, Blue-Listed from 1972 to 1982 due to extirpation in a number of areas, a Species of Concern in 1982 (Tate and Tate 1982). Classed PHS.

Noteworthy Records: West, *Summer*: 20 May 1999 at Skagit WMA; 1 on 20 Jun 1995 at Ridgefield NWR; 1 on 23 Jun 1999 at Port Townsend; 6 Jul 1998 at Ridgefield; 7 on 8 Jul 1997 at Ridgefield; 2 on 27 Jul 1995 nr. Stanwood; 2 at 3 Aug 1999 at Ridgefield; 4 on 4 Aug 1998 at Ridgefield; 14 Aug 1997 at Ft. Canby; 4 on 15-17 Aug 1994 at Ridgefield; 17 Aug 2000 at Ridgefield. *High counts, Winter*: 31 in 1986-87 nr. Stanwood. **East**, *high counts, Summer*: 1000-2000 pairs in 1982, 1500 pairs in 1983 at Potholes Res.; 100 pairs in 1983 at Moses L.; 100-200 pairs in 1983 at Frenchman Hills Wasteway; 100 pairs in 1983 at Wahluke HMA; 25 pairs in 1983 along Winchester Wasteway; 10+ pairs in 1982 at Sylvan L., nr. Odessa. *Winter*: 50 in 1982-83, 53 in 1983-84 at College Place. CBC high counts, East: 30 in 1992, 36 in 1994, 29 in 1997 at Moses L.; 17 in 1994 and 17 in 1997 at Tri-Cities; 15 in 1989 at Toppenish. West: 29 in 1987 and 1988 at Skagit Bay.

Terence R. Wahl

Yellow-crowned Night-Heron

Nycticorax violaceus

Casual vagrant.

Breeding in the c. and e. U.S. s. into Mexico, including Baja California and the Pacific coast, recorded once in Washington: an ad. photographed nr. Walla Walla 30 May-8 Jun 1993. Subspecies unknown: *N. v. violaceus* and *bancrofti* occur in N. America. Marshall et al. (2003) reported no records for Oregon.

Bill Tweit

White Ibis

Eudocimus albus

Casual vagrant.

The one Washington record is of a subad. at Bay Center on 30 Dec 2000 and subsequently near Menlo 8-21 Jan 2001 (Aanerud 2002). The White Ibis wanders widely but infrequently across the w. U.S., with two California, one Idaho, and one Oregon records. A subad. noted during Nov 2000 in Oregon was likely the same individual seen in Washington.

Steven G. Mlodinow

White-faced Ibis

Plegadis chihi

Rare spring visitor, casual in fall and winter.

Habitat: Open wetlands.

Occurrence: Breeding throughout much of the w. U.S., the Gulf Coast, portions of Mexico, and c. S. America, the White-faced Ibis has spread dramatically. Prior to 1981, there were only three records for Washington (Roberson 1980). From 1981 through 2001, ibis have occurred in varying numbers in 17 of 21 years, primarily in spring in wetlands or wet pasture in the Columbia Basin or the w. end of the Columbia Gorge. Years with more than five individuals reported include 1981 (6 birds), 1985 (20), 1987 (32), 1988 (91), 1992 (51), 1993 (10), 2000 (103), 2001 (295), 2002 (126). All but five of 757 individuals were in May or Jun. The geographical distribution is also tightly clustered. A majority (60%) of the records are from the eastside, and 84% of those are from Grant,

Adams, Franklin and Walla Walla Cos.; 65% (165/255) of the westside records are from Clark Co. at the w. end of the Columbia Gorge.

The increase in records in the late 1980s matches the peak in numbers of ibis breeding at Malheur NWR. in se. Oregon (Littlefield 1990) and a region-wide increase in numbers in the 1980s following the elimination of the use of DDT in the U.S. (Ryder and Manry 1994). The episodic nature of their appearance in Washington is driven primarily by drought conditions in their breeding range; the 2001 incursion was undoubtedly triggered by the most severe drought since 1977 (Tweit and Flores in press).

Noteworthy Records: *Early records*: 1 on 30 Oct 1909 at Clear L., Spokane Co.; 1 on 26 May-2 Jun 1951 at O'Sullivan Dam; 1 in mid-May 1974 at Reardan. *Recent Fall-Winter records*: 1 imm. on 20-23 Nov 1981 at Humptulips R. mouth; 1 imm. on 24 Jan 1982 at Nahcotta; 1 on Cowlitz-Columbia CBC in 1995; 1 on 19 Sep at Wallula.

Bill Tweit

Turkey Vulture *Cathartes aura*

Uncommon to fairly common migrant and summer resident, mostly in lower open forest habitats in most of the state. Local, fairly common fall migrant w., uncommon migrant e. of the Cascades. Rare in se. corner. Very rare in winter w., casual e. of the state.

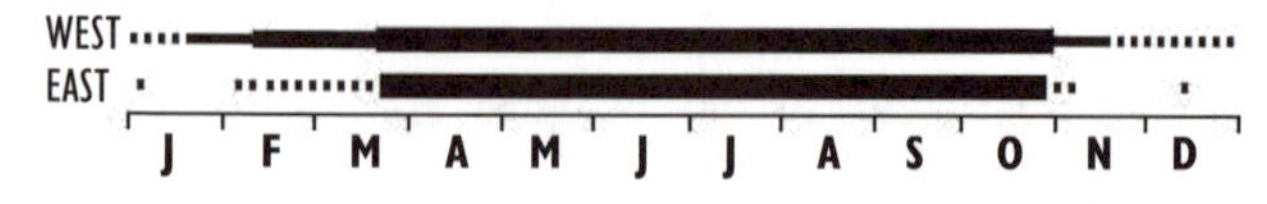

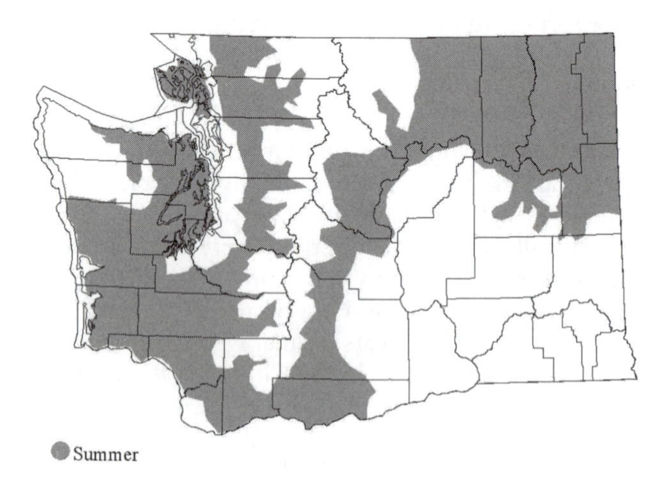

Subspecies: *C. a. meridionalis* in Washington.

Habitat: Nesting occurs up to subalpine forest, often in isolated rocky outcroppings or cliffs with caves, occasionally in hollow logs, abandoned buildings, brushy areas, often nr. dense stands of trees. Forages over open rangeland, agricultural fields, and similar areas, along shorelines scavenging for dead fish, seals, other animals, and at Columbia R. dams for salmon carcasses.

Occurrence: Observations suggest numbers have increased since Jewett et al (1953) noted up to 25 in late summer and fall: flocks of 100+ migrants are not unusual in the Puget Sound area. Some apparent increases likely due to increased observer effort, but range continent-wide has expanded northward, partly due to road expansion and habitat alteration resulting in road kills, domestic animal mortality, and industries such as fish-processing plants.

Monitoring of birds crossing the Strait of Juan de Fuca between Vancouver I. and the Olympic Pen. indicated up to 2000 per year in fall. A high count of 1138 on 26 Sep 1998 was half of that season's total at Salt Cr. where systematic censuses found large numbers (MacRae 1998). Observations on Vancouver I. in the 1990s confirmed the population size. Southbound routes include one along Hood Canal, across to the Kitsap Pen., then through the lowlands (MacRae 1999). Lack of systematic counts does not allow comparisons of numbers migrating on routes through the Puget Trough. A few westside migrants often noted as early as mid-late Feb with peak numbers from late Mar to early May. Fall migration peaks from mid-Sep to mid-Oct, with a few birds noted into early Nov.

In spring, vultures comprised just 3% of raptors counted in eight years of Cape Flattery censuses (Clark et al 1998). Numbers moving e. along Vancouver I. and congregating to cross at the narrower part of the Strait in the fall indicate minimal over-water migration farther w. Birds move along the outer coast at times, and migrating flocks are reported island-hopping in the San Juan Is. E. of the Cascades, migration is on a broad front. The majority of vultures there probably move along mountain slopes and the Columbia R. Few are reported from the Columbia Basin.

Breeders relatively widespread, occurring in undeveloped areas and foothills of the Puget Trough (Smith et al. 1997, C. Chappell p.c.), on e. and w. slopes of the Cascades, probably elsewhere. BBA surveys confirmed only four nests on the eastside, four on the w., with one on San Juan I. The Olympic Vulture Study recorded a nest in the n. Olympic foothills and two have been reported from the Chelan area. Sauer et al. (2001) indicate an increase in BBS numbers from 1966 to 2000.

Winter records primarily from the Puget Trough, with an unusual proportion from the San Juan Is. First CBC report in 1977, with nine reports from 1984 to 1989, and five in the 1990s. Reported on

18 counts or during Count Week: one to three birds on six CBCs in the San Juan Is., one three times at Bellingham, once each at E. Lk. Washington, Leadbetter Pt., Port Townsend, once in e. Washington at the Tri-Cities.

Remarks: Threats include habitat encroachment and shooting, poisoning of carcasses (for so-called predator management), being hit by vehicles while at road kills.

Noteworthy Records: West, *Winter*: 2 on 27 Nov 1981 at San Juan I. (Lewis and Sharpe 1987); 3 on 6 Feb 1983 at Shaw I. (Lewis and Sharpe 1987); 6 on 21 Dec 1986 on E. Lk. Washington CBC; 5 on 29 Jan 1995 at W. Dungeness; 2 on 3 Dec 1999 nr. Chehalis. *High counts*: 30 in Apr 1996 at Freeland (DM); 36 on 3 Oct 1976 at Deception Pass; 30 on 30 Jul 1997 at Everson; 17-24 at summer roost in 1999 nr. Enumclaw (DM); 31 on 30 Oct 1947 at Seattle (Eddy and Eddy 1948); 11 on 6 Oct 1967 roosting at Seattle; 50 on 2 Sep 1978 at Dungeness; 300 on 5 Oct 1980 at Silverdale; 126 on 27 Sep 1982 at Edmonds (MacRae 1983); 1138 on 26 Sep 1998 at Salt Cr. CP, Clallam Co. **East**, *Winter*: 7 on 7 Jan 1989 at Moses Coulee; 1 on 13 Dec 1999 at Cashmere. *High counts*: 13 on 1 Aug 1966 at Turnbull NWR; 8-10 on 24 Aug 1970 at Newport; 40 on 16 Sep 1981 roosting at Fish L., Chelan Co.; 15 on 23 Aug 1988 in the Teanaway Valley Kittitas Co. (DM); 30 on 18 Sep 1999 at Granger, Yakima Co. (DM).

Diann MacRae

California Condor *Gymnogyps californianus*

Extirpated.

Iten et al. (2001) provide a thorough description of the records in Washington, last recorded in 1897 (Jewett et al. 1953). Virtually all records were along the Columbia R., from nr. the mouth to 800 km upstream, apparently attracted by abundant salmon runs. Records were obtained in Jan, "spring," Sep and Oct. There is no evidence that they bred in Washington, although Wilbur (1973) argued for a small breeding population in the Pacific Northwest. An alternate explanation is that their occurrence in the state was apparently a post-breeding movement from California with some non-breeders possibly remaining year round.

Bill Tweit

Osprey *Pandion haliaetus*

Fairly common, local breeder and migrant statewide at many water bodies, mostly at lower elevations; very rare in winter.

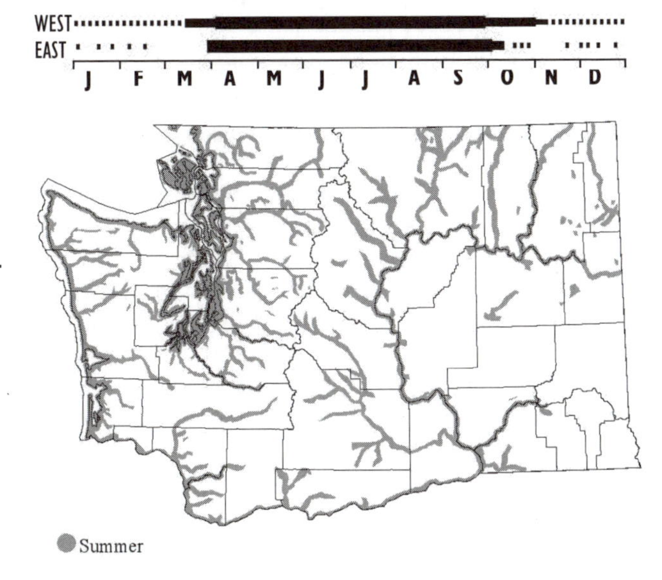

Subspecies: *P. h. carolinensis* (Smith et al. 1997).

Habitat: Nests near water bodies, often in lower-elevation forests and along marine shorelines. Nest sites typically in tall, emergent trees and readily used artificial nesting platforms. Forages on lakes, reservoirs, and marine habitats.

Occurrence: Wide-ranging, occurring statewide most often near large bodies of water in transition zones (Jewett et al. 1953, Smith et al. 1997). Uncommon nester at high elevations as at Ross L. (Wahl 1995) and L. Wenatchee and in Yakima Co. (Stepniewski 1999); noted foraging in many lakes in the Cascades (e.g. at Snoqualmie Pass and White Pass). In addition birds can be found rarely in shrub-steppe areas near Yakima, Klickitat, Benton, Franklin, Ferry, and Walla Walla Cos. (Knight et al. 1982, Smith et al. 1997). Absent from the Columbia Basin and Palouse, apparently nearly so in se. corner (Smith et al. 1997). Concentrations of nesting birds include the Pend Oreille R., with 22 nests visible from one location at Usk in 1995 (Smith et al. 1997), 23 active nests with 54 young in mid-Jul 2001 at Port Gardner (FN 55:475), sites on the Columbia R. from Lyle downstream to Washougal, and eight nests on outer coast in Clallam Co. in 1977.

Migration appears to occur on a broad front: there are no known concentration sites. Fall migration studies in the Cascades report small numbers (11-71) in a season (Hawk Watch International). The vast majority of birds leave for the winter, with records in the Cascades of birds passing migration-study sites like Hart's Pass (see Wahl 1995). Recent increasing records confirm the suggestion of Jewett et al. (1953) that some occur in winter, most regularly w. of the Cascades. Reported 33 times on CBCs from 1974 to 1999: 26 in the w. and 7 e. Birds were reported in consecutive years five times (e.g., birds at Olympia 1978-81), suggesting presence of the same individuals. Recent telemetry studies found that Ospreys from the Columbia R. wintered from n. Mexico s. to El Salvador, on the Gulf and Pacific coasts, moving s. using both inland and coastal routes (Martell et al. 2001).

Historically, this species has suffered population declines due to DDT poisoning, habitat degradation, and poaching. Jewett et al. (1953) noted the killing of Ospreys at fish hatcheries, and populations were noted to be declining during the 1950s. Banning of DDT and a decrease in persecution of Ospreys led to population increases. Increase of reports from field observers was noteworthy in 1981, and by the mid-1990s described as a rare but regular winter visitor. Like the Bald Eagle, has become to some extent locally urbanized, with a nest noted in Seattle in 1981 and two nests on light standards at Kamiak High School athletic fields near Edmonds in 2002 (E. Wahl p.c.). A statewide nesting survey in 1989 indicated a minimum of 339 occupied territories, up from 183 in 1984 (WDFW 1997b). BBS data from 1966 to 1998 indicate that nationwide population increased by over 15% (Sauer et al. 2000). Washington BBS data showed an increase of between 10.2 and 11.7% from 1966 to 1991 (Peterjohn 1991).

Noteworthy Records: *Late fall-winter*, **West**: 1 on 25 and 28 Jan 1919 on Duckabush R. (Jewett et al. 1953); 1 wintered 1967-68 at Clear L., Skagit Co.; 7-14 Nov 1975 at Seattle; 14 Jan 1978 at Ridgefield NWR; 29 Jan 1978 at Skagit Flats; 4 Jan, 4 Feb 1979 at Nisqually NWR; 22 Dec 1980 nr. Olympia and 12 Jan 1980 nr. Elma; 2 Jan 1982 at Grays R.; 18 Jan 1982 at Orcas I.; wintered 1987-88 and 1988-89 at Olympia; 18 Jan 1990 at Nolte SP, King Co.; 27 Jan 1991 at Bonneville Dam; 5 Jan 1992 nr. Tahuya and 29 Jan nr. Allyn, both Mason Co.; 5 Jan 1993 and 1 Dec 1994 nr. Bellingham; 15 Jan 2000 at Woodland; 29 Jan 2000 at Fir I.; 10 Feb 2000 at Kent; 25 Mar 2000 at Brady and Elma; 25 Feb 2001 at Issaquah; 19 Jan 2002 at Longview. **East**: 5 Nov 1972 n. of Richland; 24 Nov 1986 at Vantage; 16 Jan 1972 nr. Walla Walla; 24 Jan at Whitebluffs, Benton Co.; Jan and Feb 1983 in Yakima Co.; 9 Feb 1989 and 8 Dec 1991 at Lyons Ferry Fish Hatchery, Ferry Co.; 2 Dec 1998 at College Place; 5 in 2002-03 from Kittitas to Benton Cos.

Jennifer R. Seavey

White-tailed Kite *Elanus leucurus*

Uncommon resident in sw. Washington.

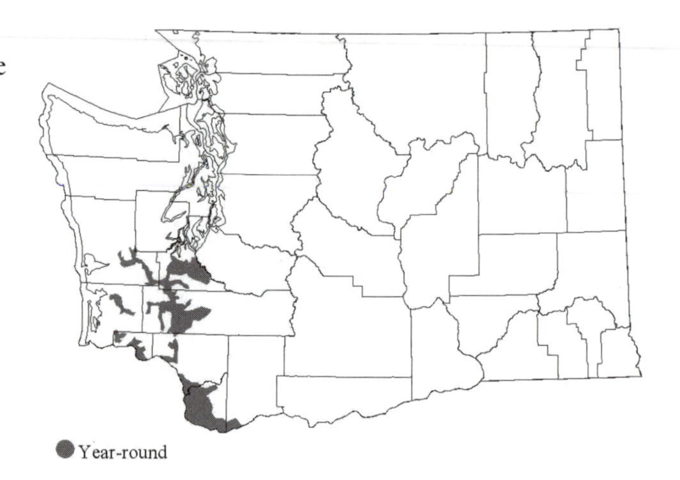

Subspecies: *E. l. majusculus.*

Habitat: Low-elevation grasslands.

Occurrence: Coincident with a dramatic range expansion throughout the New World beginning in the 1950s (Dunk 1995), the first White-tailed Kite recorded in Washington appeared on the Nisqually NWR on 10 Jul 1975 (Harrington-Tweit 1980). Nine more records accumulated during the 1970s, all from sw. Washington, primarily in Pacific Co. Kites appeared consistently near Raymond from 1977 on. Records increased dramatically during the 1980s, totaling 160. The first record from n. Puget Sound was in Skagit Co. on 3 Apr 1982. Beginning in 1985, apparently mated pairs on territory were noted in Grays Harbor and Pacific Cos., with the first breeding record near Raymond in 1988 (Anderson and Batchelder 1990).

During the 1990s the breeding distribution appeared to move inland, with apparent breeders consistently present at Francis in Pacific Co.; Julia Butler Hansen NWR in Wahkiakum Co.; Cowlitz Prairie, Hanaford Valley, Adna, Curtis, and Boistfort in Lewis Co. and near Rochester and Tenino in Thurston Co. At least 29 individuals were present at these locations in Feb 1999 and systematic surveys in fall 2001 of Thurston and Lewis Cos. found 33 ads. and 10 juvs. (RO). Probably due to a decline in observer interest and the population shift to inland areas with fewer observers, the 150 reports in the 1990s appeared equal to the 1980s reports. It is unclear whether the population continued to increase from 1980s levels, though five birds were at Julia Butler Hansen NWR on 1 Mar 2003.

Kites are still very rare (only about 12 reports/ records) w. of the Cascades n. of Grays Harbor, Thurston, and Pierce Cos.; these include 5 from

King, 4 from Skagit, 1 from Snohomish, 1 from Clallam, and 1 from Whatcom Cos. Although there have been at least three reports from e. Washington, none of them are documented.

Bill Tweit and Roger Orness

Bald Eagle *Haliaeetus leucocephalus*

Common breeder in w. Washington near lower-elevation water bodies, uncommon on e. slope of the Cascades and along rivers in ne. and Benton Co. Common in w., locally common e. in winter.

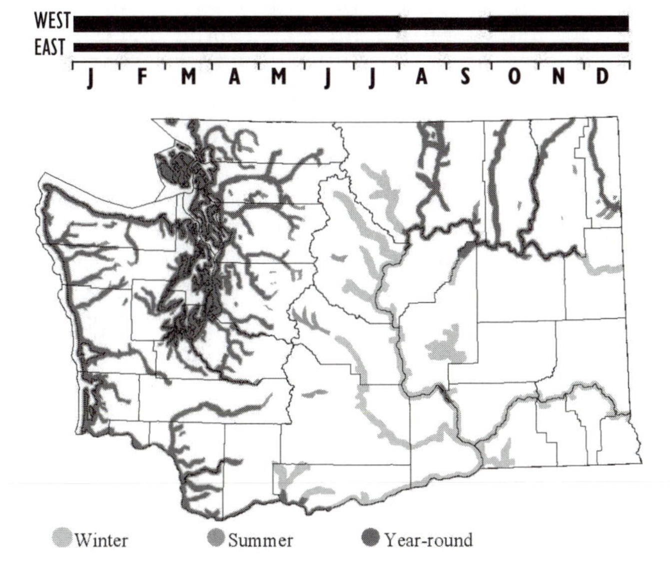

Subspecies: *H. l. alascansus.*

Habitat: Salt- and freshwater areas, uplands, mature coniferous forests. Nests commonly in coniferous forest with an uneven vertical structure and other old-growth characteristics (Anthony et al. 1982). Forages in coastal areas, wetlands, rivers, lakes, and fields. Birds often shift from marine shorelines to inland rivers while salmon carcasses are abundant.

Occurrence: Widespread and common in the Puget Trough, San Juan Is., on the Olympic Pen. including Hood Canal, and the s. coast including Grays Harbor. Common along most large rivers in w. Washington (Kitchin 1949, Jewett et al. 1953, Grubb et al. 1975, Wahl 1995, Smith et al. 1997), especially during salmon runs. Smaller numbers in e. Washington year-round, with nests noted along the Okanogan, Sanpoil, Kettle, Columbia, Colville, and Pend Oreille Rs. (Smith et al. 1997).

High breeding concentrations are in the San Juan Is. and Whatcom, Skagit, and Island Cos. and Olympic Pen. coastline. Small numbers nest in urban areas, including Seattle (Smith et al. 1997).

Table 18. Bald Eagle: CBC average counts by decade.

	1960s	1970s	1980s	1990s
West				
Bellingham		11.1	24.6	64.7
Edmonds			2.8	9.8
Grays Harbor		1.1	4.0	15.9
Kent-Auburn			3.0	14.1
Kitsap Co.		3.7	5.4	10.8
Leadbetter Pt		0.9	3.9	9.1
Olympia			10.1	29.5
Padilla Bay			71.4	137.0
Port Townsend			7.5	23.3
San Juan Is.			59.0	67.5
San Juans Ferry		14.6	16.6	27.5
Seattle	0.0	0.4	4.3	24.2
Sequim-Dungeness			14.4	46.0
Tacoma		4.3	12.1	30.0
East				
Spokane	1.0	4.9	7.9	18.0
Tri-Cities		0.2	2.2	10.5
Wenatchee	0.0	1.8	3.1	11.4
Yakima Valley		0.3	1.4	9.4

Territories in Washington have an average radius of 2.6 km (Grubb 1980) and, where water does not freeze in winter, may be held year-round (Rodrick and Miller 1996). The breeding season begins in Jan.

Though most nesting eagles are vulnerable to human disturbances, individual pairs have been known to habituate to disruptions (Rodrick and Miller 1996). There are several nests within the Seattle city limits and birds hunt regularly around human activity (Hunn 1982). Wintering areas have concentrations from late Oct through Jan, and birds concentrate in areas where forage is abundant and there is little disturbance (Rodrick and Miller 1996).

Eagle numbers declined nationally from half a million historically (e.g., see Dawson and Bowles 1909) to 417 in 1963 due to shooting, pesticide and direct poisoning, and habitat loss (USFWS 1999, Buehler 2000). Following the listing of this species in 1967 and the banning of DDT, numbers increased to over 5000 pairs (USFWS 1999). Numbers in Washington have recovered well under federal and state protection (see McAllister et al. 1986, Smith et al. 1997), which may lead to a change in state status from threatened to sensitive. Removal from the Federal Endangered Species Act is under consideration.

In 1982, 137 pairs were counted in Washington, increasing to 630 breeding pairs for 1998 (USFWS 1999), with a recent suggestion that this may be approaching carrying capacity (Buehler 2000). CBC data show significant increases ($P < 0.01$) at all long-term locations statewide (Table 18), though

Nysewander et al. (2002) indicate a decline in numbers from 1978-79 to 1994-95.

Remarks: The Pacific States Bald Eagle Recovery Plan advises that eagle nesting sites should be protected from human disturbances (see Grubb 1980, Anthony and Isaacs 1989, Rodrick and Miller 1996). The species has received more survey attention than most other species, with a number of volunteer efforts counting wintering eagles (e.g., for 10 years on the Skagit R. by North Cascades Institute) and agency surveys of nesting (along the lower Columbia R.; Oregon State Univ.). Impacts of increasing eagle populations on prey populations experiencing declines due to other environmental and human factors (e.g., Common Murres; J. Parrish) require study. Classed ST, FT, PHS.

Noteworthy Records: *Bald Eagle Survey totals,* Winter 1979-80: 935 ad. and 633 imm. (total: 1568), thought to be up from past years; Winter 1980-81: total 3197; 138 in winter 1978-79 in e. Washington; 250 in Jan 1986 count in e. Washington; 280 in winter 1971-72 in nw. Washington; 105 on 5 Jan 1975 along the Skagit R. near Rockport; 400 noted for 1991. *Summer*: 59 occupied territories and 67 young in 1986 in the San Juan Is. Fall: 75 on 24 Nov 1981 nr. Banks L.

Jennifer R. Seavey

Northern Harrier *Circus cyaneus*

Fairly common to common resident and migrant in open habitats in eastside, more localized in westside river deltas, adjacent areas. Uncommon to fairly common migrant in Olympics and Cascades. Uncommon to rare in ne. Washington.

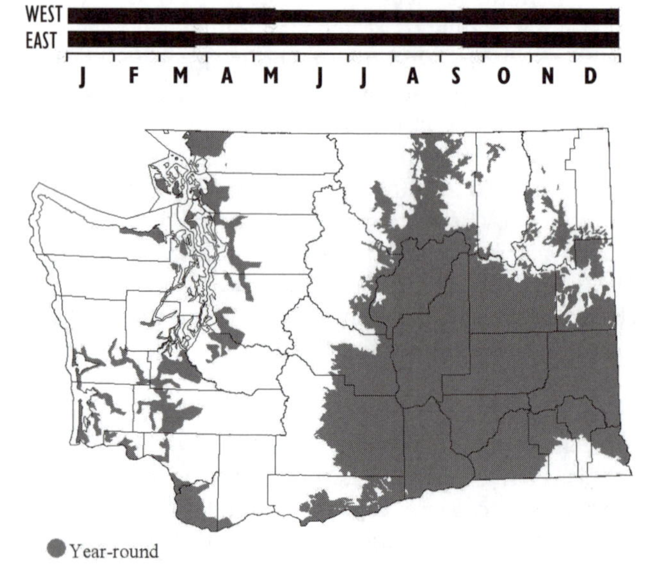

Subspecies: *C. c. hudsonius.*

Habitat: Breeds in primarily grassy marshes and wet prairie with tall grass marshes, frequently in cattail bulrush marshes, ponds, wet and dry meadows, scrub/shrub, grasses, and agricultural fields (Thompson-Hanson 1984). Forages in areas with thick grasses and shrubs, grasslands, cultivated fields, meadows or marshy wetlands, urban fringe habitats, and on tide flats and adjacent fields in Puget lowlands.

Occurrence: The Washington breeding population may be larger than generally believed. BBS data indicate an increasing trend from 1966 to 2000 in Washington (Sauer et al. 2001), similar to a winter trend over the U.S. (Root and Weckstein 1994; but see MacWhirter and Bildstein 1996). Nesting is widespread in suitable habitats statewide (see Thompson and McDermond 1983, Thompson-Hanson 1984, Smith et al. 1997). The harrier is highly adaptable in selection of nest sites and habitats (Hammerstrom and Kopeny 1981) and is possibly the second-most common breeding raptor in the open habitats of e. Washington up to the ponderosa zone (Smith et al. 1997). Nesting birds arrive in e. Washington in Feb (Thompson-Hanson 1984). Nests are placed anywhere from dry ground in dense shrubs or directly on water where the nests are supported by submerged shrubs or anchored vegetation (Thompson-Hanson 1984). Small numbers nest in isolated areas (e.g., narrow-margin rye-grass fields along streams) in Klickitat Co., such as Rock Cr. and Alder Cr. (M. Denny p.c.). Because preferred nesting habitat tends to be clumped, harriers can nest in close together in loose aggregations (but see Balfour and Cadbury 1979). Westside nesting has not been well documented (Thompson and McDermond 1983) but Whidbey I. supports relatively large numbers of nesting birds, with up to 30 nesting attempts per year from 1994 to 2003 (J. Bettesworth p.c.).

Fall migration in the Cascades, from late Aug to late Oct or early Nov, is dispersed, with concentrations on some ridgelines e. of the crest. Jewett et al. (1953) noted early records at 1800 m on Mt. Rainier, and one over Chopaka Mt. at 2000 m, and records at 1965 m at Table Rock, Columbia Co. (M. Denny p.c.) suggest some occurrences. Recent surveys at Slate Pk. in Sep 1991 had 66 birds in 27 days (Bettesworth 1991), at Diamond Head from 1993 to 1998 had 18-45 per season (Smith and Grindrod 1998), at Chelan Ridge from 1998 to 2001 averaged 129 per season, with 148 in 2002 (Smith 2002). In spring, mid-Mar to late Apr, at Cape Flattery, 48 birds occurred in 46 days in 1985 (C.M. Anderson p.c.) and from 1990 to 1997 averaged 17 per season (Clark et al. 1998). Most Whidbey I. nesting birds disperse in late summer/ fall in all compass directions while some overwinter to renest the following year (J. Bettesworth p.c.).

One of the most abundant westside wintering raptors. Birds are widespread in winter. Temperatures affect distribution: Contreras (1997) states that harriers usually vacate the Okanogan Valley when daytime temperatures drop below freezing (see Root 1988). Harriers roost communally in winter in flat, open terrain with low vegetation profile, typically tall grasses sometimes interspersed with dense, low shrubs growing in slight depressions in the landscape (PAT; see Root 1988).

Indications of local winter concentrations include 66-147 birds wintering on the Skagit/Samish flats (FRG unpubl.), 50 at Lummi Flats and 30 near Lynden in the 1990s (Wahl 1995) and 35-40 in the Walla Walla Valley (M. Denny p.c.). CBC data from 1960 to 2000 indicate a significant increase (P <0.05) in e. Washington with a slight positive percent annual change trend of 1.2 for the period of 1959-88 in Washington and a steady overall increase from 1959 to 2000.

Remarks: Habitats have no formal protection and human development often eliminates prime hunting and nesting habitat and crop harvesting may destroy nests, eggs, and young. Changes in plowing of fields from spring to fall have been suggested as affecting numbers wintering in the Skagit Flats (SGM). Continued residential and commercial development in the Kent Valley appears to be negatively affecting nesting attempts: down from an average of 2.7 per year (1995-2000) to only two attempts, total, in 2001-2003 (J. Bettesworth p.c.). Formerly known as Marsh Hawk.

Noteworthy Records: *Southerly migration limit*: fledgling from Whidbey I. to just n. of San Francisco. *Oldest breeding ad.*: female at 10 years (J. Bettesworth p.c.).

Patricia A. Thompson

Sharp-shinned Hawk *Accipiter striatus*

Fairly common winter visitor, fairly common breeder in forested regions statewide. Fairly common migrant, locally common in fall along ridges in the Cascade Mts. and in coastal areas.

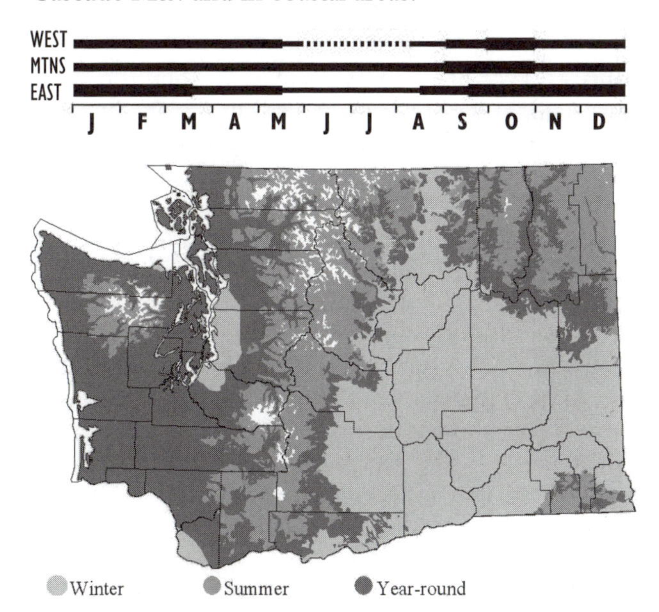

Subspecies: *A. s. velox* (see Bildstein and Meyer 2000).

Habitat: Breeds in many coniferous forest habitats, particularly those with an abundance of smaller-diameter trees that create dense areas of closed canopy (Reynolds et al. 1982); such areas may be younger stands of trees (e.g. 25-50 years) or dense areas within older stands. Uses lowland urban and suburban areas.

Occurrence: Jewett et al. (1953) considered this species "fairly common in summer, especially in the mountains." Recent authors state that it is uncommon or rare (Smith et al. 1997, Stepniewski 1999), but intensive searches, perhaps using taped recordings of vocalizations, may indicate greater abundance. Weber and Larrison (1977) reported only a single record from the Blue Mts., but intensive search in similar habitats nearby in ne. Oregon resulted in discovery of numerous sites (Henny et al. 1985; see Reynolds and Wight 1978).

Sharp-shinned Hawks can be quite common locally during migration. Between 1 Apr and 15 May 1985, 1828 birds were counted migrating past Cape Flattery (C. M. Anderson p.c.; Anderson et al. 1983). During fall migration between 27 Aug and late Oct 1999, 932 were observed at Chelan Ridge and 623 were at Diamond Head in the e. Cascades (www.hawkwatch.org) and a similar count of 1050 in 2000 was at Chelan Ridge.

Birds winter at elevations below 450 m in w. Washington, with occurrence particularly noticeable at urban bird feeders. Annual counts on long-term CBCs indicate significant increases (P <0.01) statewide, but questions of identification and numbers of unidentified accipiters suggest caution in use of CBC data (SGM).

Remarks: Breeding subspecies in w. Washington uncertain. *A. s. perobscurus* breeds in the Queen Charlotte Is. and may occur as a breeder, migrant, and winter resident on the Olympic Pen. (AOU 1957). Jewett et al. (1953) considered the entire westside population to be *perobscurus.* Smith et al. (1997) suggested that the range of the species did not include the coastal Sitka spruce zone. Bildstein and Meyer (2000) concurred, though the species occurs in Sitka spruce forest throughout coastal B.C. and Alaska (Bildstein and Meyer 2000) and further surveys may show it occurs along the outer coast (K. Bildstein p.c.). The dearth of records from the Sitka spruce zone may reflect a comparative lack of search effort or the species' secretive nature.

Noteworthy Records: *Late Spring-Summer*: 8 May 1980 (possibly a migrant) at Leadbetter Pt. (WDFW); 25-26 Jun 1977 at Ocean Shores (WDFW).

Joseph B. Buchanan

Cooper's Hawk *Accipiter cooperii*

Uncommon resident in coniferous and mixed-deciduous forests; rare in sw. lowlands. Rare to uncommon in lowland forests and valleys in winter, including the Columbia Basin; very rare in more-developed areas.

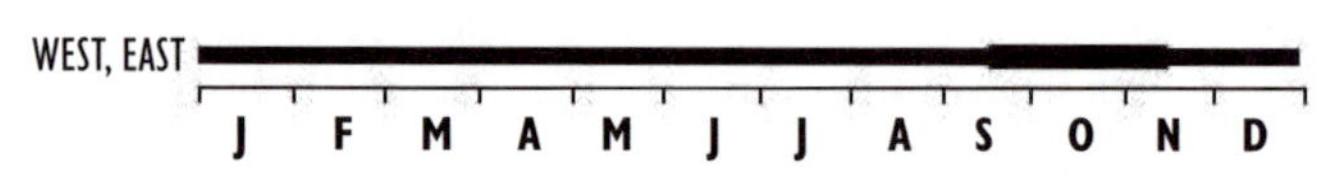

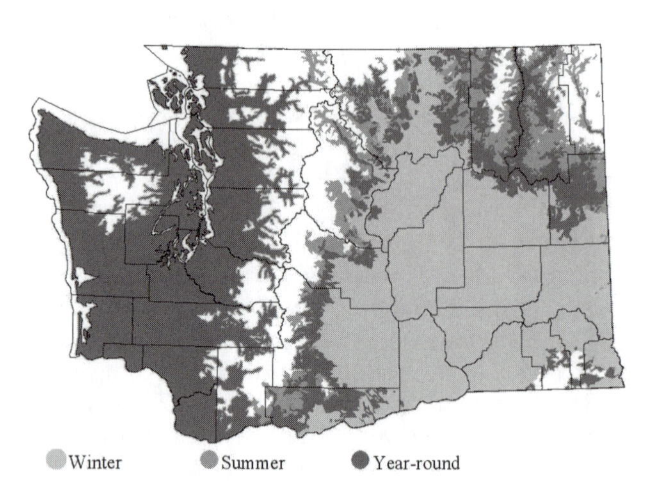

Habitat: Coniferous and mature deciduous stands. Nests typically in mid-aged, mature or older conifer or deciduous forest habitat in areas dominated by Douglas-fir, western hemlock, and bigleaf maple in low to mid-elevation areas w. of the Cascade crest, and fir-pine forests in e. Washington (WRDS 2001). May nest in urban parks and suburban woodlots. Winters in forested, edge, and urban habitats.

Occurrence: Cooper's Hawk is an uncommon to rare but regular breeder across forested areas of the state. Birds are likely in suitable habitat from coastal to higher-elevation forest below subalpine (Cassidy 1997a) zones. It is not known what proportion of birds occurring in Washington are residents or from Canada, and BBS methods may underestimate populations as birds may be overlooked in the breeding season.

Migrants occur regularly along Pacific Flyway routes, with Cape Flattery a staging area in spring for hawks crossing into Canada. Fall movements were noted at observation points at Entiat Ridge, Chelan Ridge (FRG 2001) and Slate Pk., Redtop Mt. in Kittitas Co., and Willapa Hills. Though less often reported than Sharp-shinned Hawks, numbers of birds winter throughout the w. lowlands to 450 m elevation. Regionally, locally nesting birds may winter statewide and there is some apparent winter migration downslope from the Cascades into the Puget lowlands.

Although riparian deciduous areas are often used, nests have been located in mature and semi-mature coniferous and pine-oak and pine-fir forests away from water (Wagenknecht et al. 1998, Finn et al. 1999, T. Fleming p.c., M. McGrath p.c., R. Rodgers p.c.). Active nests have been located in 50-80-year and older coniferous forests above 900 m (*contra* Smith et al. 1997), and some utilized mistletoe clumps for nesting in the Blue Mts. and in Oregon (Reynolds et al. 1982, Moore and Henney 1983, 1984, SMD).

Birds usually nest in crotches of older maple or cottonwood; red alder apparently usually lacks suitable structure unless the tree has a suppressed or deformed top. Nesting Cooper's Hawks are amazingly tolerant of human activity in urban areas in Vancouver, B.C. (Stewart et al. 1996) and Bellingham (TRW).

Population trends are uncertain. BBS data, perhaps not representative of breeding numbers, show varying trends with increases shown from 1966 to 2000 and 1966 to 79 and a decrease from 1980 to 2000 (Sauer et al. 2001). Long-term CBC data indicate fairly widespread local increases and significant increases (P <0.01) statewide, but accipter identification problems may pose questions in interpretation.

Remarks: Population reduction was considered desirable until the 1960s because of predation on other birds (Jewett et al.1953). Susceptible to *trichomoniasis* carried by avian prey associated with bird feeders in urban/suburban areas, causing mortality of chicks and fledglings and reduced multi-season productivity (Boal and Mannan 1999). Regularly detected in the lower elevations of the e. Cascades, but estimated breeding density is lacking. Habitat use and home range of both breeding and winter residents in Washington are unknown. Silvicultural options in managed forests will likely determine future distribution in extensive commercial forests in Washington. Formerly "Blue listed" because of pesticide contamination and heavy hunter persecution (Tate 1986).

Noteworthy Records: *Spring migration*: 50 on 6 Apr 1985 at Cape Flattery (FRG 1986). *Fall*: imm. banded on 16 Sep 2000 at Chelan Ridge recovered 4 Oct 2000 at Edwards AFB, CA: 1500 km in 19 days, average of 79 km per day (FRG 2001).

Steven M. Desimone

Northern Goshawk *Accipiter gentilis*

Uncommon to rare breeder in suitable coniferous forests; rare in sw. lowlands and apparently extirpated from the Puget Trough. Uncommon to rare in winter in lowland forests and valleys, including the Columbia Basin; very rare in more-developed areas.

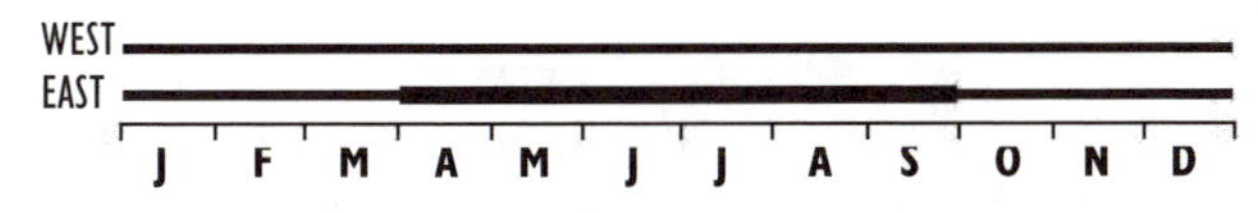

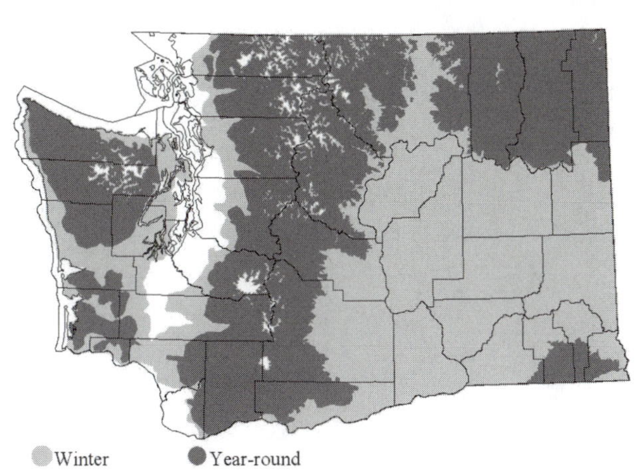

Subspecies: *A. g. atricapillus*, with Queen Charlotte Goshawk, *laingi*, possible but not substantiated (Palmer 1988, Johnsgard 1990, USFWS 1995, E. McClaren p.c.).

Table 19. Distribution of Northern Goshawk breeding birds (WDFW 2003).

Region	*Percent of breeding records*
East Cascades	50
West Cascades	27
Okanogan Highlands	11
Olympic Peninsula	10
Southwest	<2
Blue Mountains	<1
Puget Trough	-

Habitat: Mid-aged, mature, or older conifer forest habitat with complex stand structure (Fleming 1987, Finn 1994, McGrath 1997, Finn et al. 2002a, b), dominated by Douglas-fir and western hemlock w. of the Cascade crest, and a mix of Douglas-fir, true firs, pine, and larch e. of the Cascades, and also mature lodgepole pine stands in Okanogan Co. (SMD). Rarely nests in deciduous trees in w. Washington; utilizes some quaking aspen stands in e. Washington (WDFW, Wildlife Info. Systems). Forages also in the forest/shrubsteppe ecotone in winter (B. Behan p.c.).

Occurrence: Likely occurs in suitable habitat in all forested zones (Cassidy 1997a) from coastal to higher-elevation forest. Intensive though incomplete survey efforts indicated at least 340 reproductive territories in recent years, though some no longer support breeding birds. The greatest number—at least half of the known breeding sites—occurred along the e. Cascades in mid- to high-elevation mature forests (Table 19). Another 10% were in the Okanogan Highlands, and just under 40% on the Cascades w. slope, Olympic Pen., and sw., and <1% were in the Blue Mts. (WDFW, Wildl. Info. Systems, Smith et al. 1997). Historically, birds nested in w. lowlands from Semiahmoo s. to Ft. Steilacoom, w. to Cosmopolis and Shoalwater Bay (Jewett et al. 1953). Present-day absence in lowlands is undoubtedly due to habitat alteration and development.

Small numbers occur in fall migration in the Cascades, where counts at Chelan Ridge and Diamond Head in 1999 and 2000 ranged from 16 to 50 (HWI). Little is known about winter habitat use and migration patterns of Washington breeders. There are apparent elevational and seasonal shifts in winter from higher-elevation breeding ranges by both sexes; females may remain close to or occupy the breeding territory in mild winters, and males in w. Washington make limited movements to a separate winter range (Finn 2000, Bloxton 2002). Roughly two-thirds of wintering birds reported in the Okanogan Valley and Lower Columbia Basin are imms. (Contreras 1997). Goshawks are noted foraging in the forest/shrub-steppe ecotone in winter (B. Behan p.c.).

Jewett et al (1953) cited a major irruption in 1916-17 and Campbell et al. (1990b) cited a major invasion of s. B.C. in 1954. In Washington neither seasonal reports nor CBC data indicate evidence of a major irruption from 1970 to present. CBC data from 1959 to 1999 indicate widespread winter occurrence, more irregularly w. of the Cascades than e. Spokane had up to five birds per count, while other e. locations had smaller, fairly consistent occurrences. Minor incursions apparent in 1984, 1992, and 1994.

No reliable population estimate. Breeding-season survey efforts have had limited success in providing estimates of crude density (Wagenknecht et al. 1998, WDFW 2003), and demography studies and seasonal habitat use and home-range information are lacking.

Remarks: Several studies suggested that timber harvest impacts selection of nest sites (Reynolds 1989, Crocker-Bedford 1990, Ward et al. 1992, Woodbridge and Detrich 1994, Desimone 1997, Widen 1997). Managed forest techniques promoting and sustaining mature forest attributes (e.g., longer rotation length), research on major prey species and forest structure, coordinated multiple-breeding-season surveys to estimate territory spacing and density in various managed-forest landscapes (e.g., DeStefano et al. 1994) are likely necessary to maintain future distribution of territories in commercial forests. Analyses of recent DNA samples from Vancouver and Queen Charlotte Is. and the Olympic Pen. should clarify subspecies occurrence. A USFWS status review could not determine whether the population was increasing, declining, or stable (Kennedy 1997, USFWS 1997a). Classed SC, FSC, PHS.

Steven M. Desimone

Red-shouldered Hawk *Buteo lineatus*

Rare, but increasing, winter visitor along the lower Columbia R., very rare elsewhere.

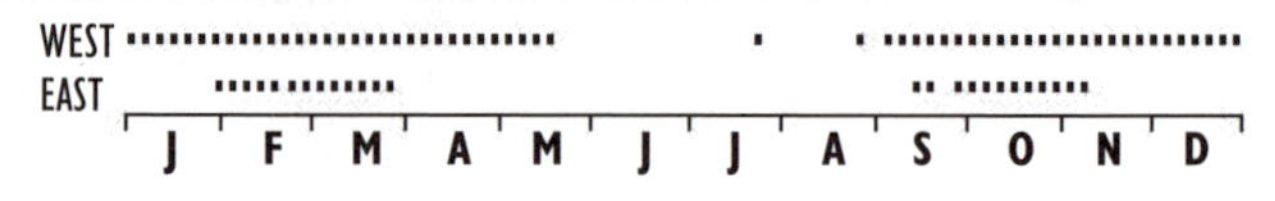

Subspecies: *B. l. elegans.*

Occurrence: Resident in the w. U.S. from s. California to sw. Oregon, spread into Washington in the late 20th century. An ad. found 20 Dec 1979-23 Feb 1980 at Nisqually NWR was the first state record. The second was a specimen, an imm. *B. l. elegans* at Cathlamet on 9 Sep 1988, and the third was an imm. in Everett from 24 Apr to 14 May 1992 (Tweit and Paulson 1994). Since 1993, Red-shouldered Hawks have been reported annually, primarily from the Columbia R. bottomlands at Ridgefield NWR. Their increased regularity is undoubtedly due to increasing numbers in w. Oregon, where there has been a dramatic increase in reports since 1971 (Henny and Cornely 1985). There have been 10 additional records away from the lower Columbia, including an ad. at Dungeness on 2 Jan-18 Feb 1997 (Aanerud and Mattocks 2000), an imm. at Spencer I. from 1 Nov 2000 to 21 Mar 2001 and an imm. at Bay Center on 4 Oct and 3 Nov 2002, and four interior records. An ad. was at Madame Dorian Park, Walla Walla Co., on 26 Sept-8 Nov 1997 and one imm. (B. and N. LaFramboise) wintered in W. Richland from 31 Jan to 22 Mar 1998. Birds were at Conboy NWR on 16 Sep 2000 and 12 Sep 2001. Six were reported from Skagit, Grays Harbor, Wahkiakum, and Clark Cos. in winter 2002-2003.

Remarks: Records from Naselle in May 2001 and Skagit in Sept 2001 were rejected by WBRC.

Bill Tweit

Broad-winged Hawk *Buteo platypterus*

Rare fall and spring migrant statewide.

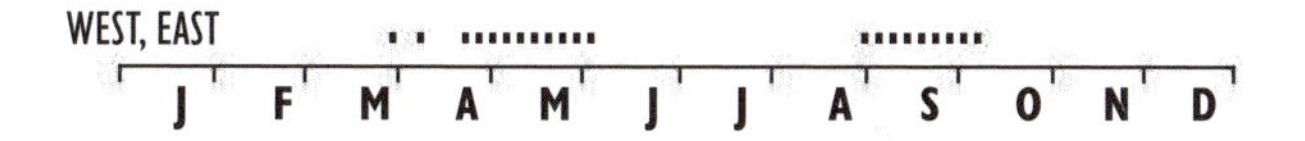

Subspecies: *B.p. platypterus.*

Occurrence: One of the rarest raptors occurring in the state, first recorded in 1970 with about 25 fall and eight spring records subsequently (Aanerud and Mattocks 1997). Birds may appear anywhere statewide, with records from Cape Flattery (W. Clark p.c.) e. to near Sullivan L. The first record of mountain migrant was in Sep 1995 (J. Bettesworth p.c.).

During a migration study in 1997-98, the USFS and HWI discovered a small but remarkable fall movement at Cooper Mt. and Chelan Ridge (K. Woodruff p.c.). There were two individuals in Sep 1997, seven in 1998 (S. Crampton p.c.), and five in 1999 (Hawk Watch International data). In 2000, five birds were seen from 14 to 17 Sep at Chelan Ridge (CMA). Nearby, "at least 25" in fall 1999 migrated past Rocky Pt. on s. Vancouver Is., to cross the Strait of Juan de Fuca (NAB 54:96) and "an amazing kettle of 60+" on 29 Sep 1999 migrated over Bonney Butte on Mt. Hood, Oregon.

It is generally assumed that records are of birds dispersing w. from Alberta. Farther s. on the Pacific coast and in the Great Basin, fall migrants were seen annually at several w. hawk banding stations, such as the Marin Co. headlands in coastal California (Golden Gate Raptor Observatory) and the Goshute Mts. in Nevada (HWI). It is possible that some Washington birds may pass these sites.

Two birds have been recovered. An imm. male, the first state record, was found dead beneath a plate-glass door in Tacoma on 1 Jun 1975 (Clark and Anderson 1984; Slater Mus. UPS, no. 2367). An imm. with a broken wing recovered n. of Pasco on 1 Oct 1982 (T. Mercer p.c.) was later transported to Texas and released.

Remarks: Ages given for only five records: three imms. and two ads. It is possible that birds seen in Washington are primarily imms. far from main migration routes. The recent discovery of fall migrants in the Cascades, however, supports the possibility of a range expansion and small breeding population in central B.C. (see Campbell et al. 1990b).

Noteworthy Records: *Accepted by WBRC*: 11 Sep 1970 at Spokane; 1 Jun 1975 at Tacoma (Clark and Anderson 1984); 1 Oct 1982 n. of Pasco (T. Mercer p.c.); 6 Oct 1990 nr. Kent (E. Hunn p.c.). *Other records*: 31 Aug 1980 at Sullivan L. (J. Acton p.c., R. Wilson p.c.); 17 Sep 1995 at Diamond Head, Kittitas Co. (M. Gleason p.c.); 22 Apr 1995 at Bahokus Mt., Clallam Co. (S. Salesky p.c.), 2 in 1997 at Cooper Mt., Chelan Co. (S. Crampton p.c.); 23 Apr 1998 at Umtanum Ridge (K. McBride p.c.); 1 May 1998 at Bahokus Mt. (A. Barna p.c.); 2 on 14 Sep, 3 on 15 Sep, 1 on 17 Sep, 1 on 18 Sep 1998 at Cooper Mt. (S. Crampton p.c.); "a few" on Sep 5-Oct 11 1998 at Chelan Ridge; 30 May 1999 at Wenas Cr.; 10 Sep at Spokane, 11 Sep and 14 Sep at Cooper Mt.; 1 or more 14-17 Sep 2000 at Chelan Ridge (S. Crampton p.c.); 11 Apr 2000 at Stanwood; 1 Apr 2002 at Phileo L., Spokane Co.; 4 Jun 2001 at Joyce. *Not accepted by WBRC*: a record from 5 Aug 1976 in Pend Oreille (Larrison 1977) due to insufficient documentation, particularly for the early date.

C. M. Anderson

Swainson's Hawk *Buteo swainsoni*

Fairly common summer resident e. of the Cascades in open range shrub-steppe habitat below tree-line elevation, rare migrant w.

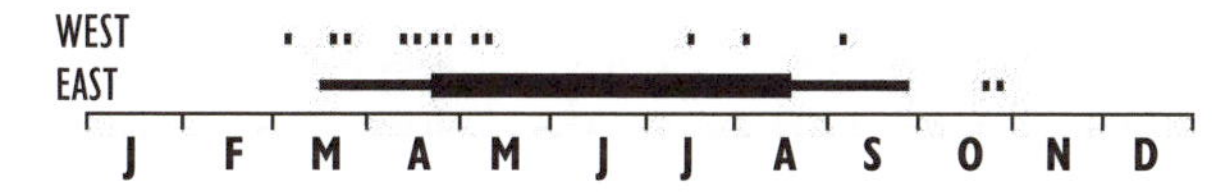

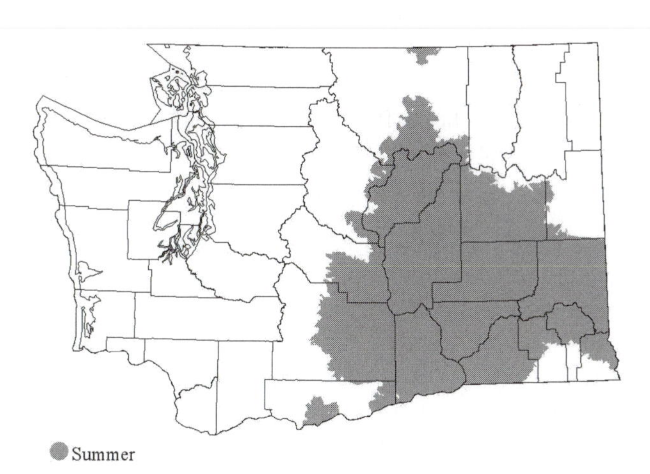

Habitat: Grasslands, shrub-steppe, nesting in isolated scrub trees or small woodland stands. May nest near isolated farms, but not a hawk of settlements. When coexisting with Red-tailed Hawk, Swainson's chooses smaller trees in smaller clumps (England et al. 1997).

Occurrence: The Swainson's Hawk is one of the world's longest-distance migrant raptors, with the estimated N. American breeding population of 500,000+ traveling up to 10,000 km to winter in Argentina (Woodbridge 1997). Studies since status was described by Jewett et al (1953) have clarified status of the relatively small Washington population.

Between 1980 and 2000 there were 456 territories recorded in Washington (J. Brookshier, WDFW). Most breeding records are from the Columbia Basin shrubsteppe habitat with a tiny isolated pocket near the Canadian border in Okanogan Co. and another pocket near the Oregon border in Klickitat Co. (Smith et al. 1997). Diet in Washington primarily pocket gophers and snakes (Bechard 1980).

Spring migration on the eastside is generally unremarked, with no concentrations reported. The earliest spring migrants are seen in Mar with the majority arriving Apr-May. In recent years, a few individuals are reported nearly annually on the westside from the Puget Trough and the Cape Flattery hawkwatch site. Recorded most months, but extremely rare in winter and rare w. of the Cascades at all times.

Fall migration begins in mid-Aug and is over by the end of Sep. It is primarily through low-elevation areas on the eastside; counts at fall raptor migration sites in the Cascades range from two to 17 (HWI) while migratory concentrations of 50-100 are routinely reported from the Columbia Basin and the Palouse. Aug records from high elevations in the Cascades (e.g. Jewett et al 1953) may indicate post-breeding dispersal to feed above timberline, similar to Prairie Falcon, rather than direct migration. In contrast to spring, fall migrants are very rarely reported from w. lowlands. Winter reports are inadequately documented, and not included here.

Remarks: The reason for an escalating, extreme decline in the 1990s of numbers of returning birds in spring was determined in 1997 when it was found that thousands were dying due to the misuse of *monocrotophos*, a grasshopper-control pesticide, in Argentina (Woodbridge 1997). International cooperation, educational campaigns, training in the detection of pesticides, and survey monitoring of dead birds helped to curtail mortalities. The work is ongoing with the cooperation of the Argentine wildlife agencies, the agricultural community, and American biologists. Classed GL.

Noteworthy Records: West, *Spring*: 1 on 15 Feb 1920 at Renton and 1 on 7 Mar 1892 at L. Washington (see Hunn 1982); 1 on 27 Apr 1969 at L. Terrell; 1 on 28 Apr 1975 at Seattle; 1 on 26 Apr 1981 at Whidbey I.; 1 on 25 Apr 1986 at Sequim; 1 on 1 May 1988 at Nisqually NWR; 1 on 15 May 1989 at Randle; 1 on 17 Apr 1992 at Bellingham; 1 on 21 Mar 1997 at Duvall; 2 on 22 Mar 1997 at Cape Flattery; 2 on 24 Apr 1997 at Cape Flattery; 1 in Apr 1979 on San Juan I. (Lewis and Sharpe 1987); 1 on 14 Apr 2001 at Ocean Shores; 1 on 3 May 1997 in w. Skagit Co.; 1 on 10 May 1997 nr. Bayview; 1 on 11 May 2001 at Crocker L., Jefferson Co.; 1 on 12 May 2001 at Scatter Cr.; 1 on 30 Apr 1998 at Cape Flattery; 1 on 18 Apr 2000 at Nisqually NWR; 1 on 14 May 2000 at Carnation; 1 light morph on 5 May 2000 at Everett STP (D. Beaudette p.c.). *Summer*: 2 on 13 Jul 1968 at Lopez I. Fall: 2 on 11 Aug 1973 at Mt. Rainier; 1 on 24 Sep 1974 nr. Ashford, Pierce Co.; 1 ad. on 6 Sep 1979 at San Juan I.; 1 on 4 Aug 1983 at 2000 m on Mt. Rainier; 1 on 22 Aug 1999 at Edmonds. **East**, *Spring, early*: 6 Mar 1975 at Sunnyside. *Fall, late*: 1 on 15 Oct 1996 at College Place; 1 on 18 Oct 1895 at Pullman (Jewett et al. 1953).

Diann MacRae

Red-tailed Hawk *Buteo jamaicensis*

Common resident, migrant, and winter visitor statewide.

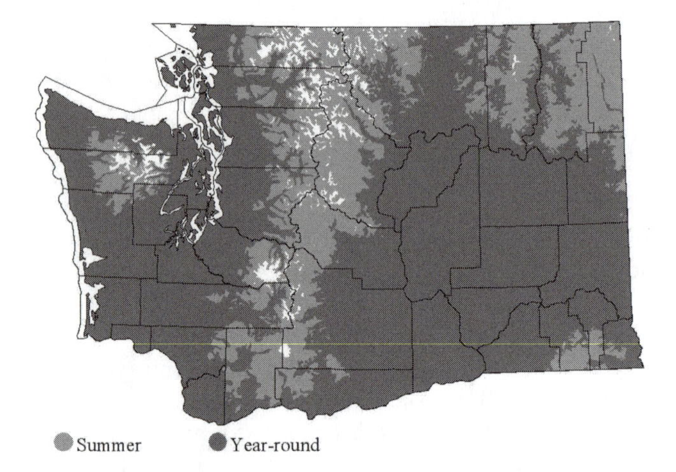

Subspecies: *B. j. calourous*, the Western Red-tail, is resident. Harlan's Hawk, *harlani,* and the Alaskan Red-tail, *alacensis,* occur in migration and winter (Jewett 1953).

Habitat: Farmlands, prairies, and other open areas, semi-open urban fringes. Nests in riparian areas, deciduous and mixed forests and, particularly e. of the Cascades, on cliffs.

Occurrence: A widespread species that has both benefitted and suffered from human-caused environmental changes, nests in all Washington habitats except dense coastal rainforests, high-elevation forests, alpine areas, and inner cities (Smith et al. 1997). Nest densities in lowland w. Washington are as high as any known for the species (Fitch et al. 1946, PDeB), but are lower in the mountains and e. Washington. Typical westside nest is in a large black cottonwood, usually in a woodlot. Birds also nest in Douglas-fir, big-leaf maple, paper birch, and western hemlock. Cliff nesting is rare in the Cascade foothills but frequent in e. Washington where birds also use ponderosa pine, quaking aspen, Lombardy poplar, and western larch as well as most trees used w. of the Cascades.

Migrant and winter populations include birds from Alaska and Canada (Preston and Beane 1993). Migration from late Aug to Nov peaks in late Sep-early Oct (Hoffman et al. 1992). Though migrants occur anywhere in the state, mountain ridges and coastal estuaries concentrate birds. Fall raptor counts from the Cascades indicate sizeable numbers—counts ranged from 225 to 450 in 1999 and 2000 (HWI).

Most wintering birds are on territory by late Oct. Distinctive individuals have been known to return to the same area every winter for many years. Locally raised young and winter visitors tend to occur away from established breeding territories. Wintering juvs. have been observed concentrating in hybrid poplar/cottonwood plantations in w. river valleys, a habitat used by few other species.

The species is polymorphic, with individuals ranging from pure white to almost black recorded in Washington, and variation noted in resident, migrant, and wintering birds. Albinism is relatively common in the w. subspecies (H. Kendall p.c.). Harlan's Hawk—a distinct species, subspecies, or color phase (Taverner 1936, Mindel 1983, Palmer 1988)—occurs uncommonly, primarily in the Puget Trough and Clark Co. It is rare on the Olympic Pen. and outer coast (Lavers 1975a). Harlan's is rare in late Oct, uncommon from Nov to mid-Mar, rare in late Mar (SGM). Most are dark-phase, but a few light-phase Harlan's reported regularly from Whatcom, Grant, and Benton Cos. Reports of Krider's Red-tails (*krideri*) probably refer to light-phase Harlan's or albinistic w. Red-tails.

In w. Washington, the Red-tail benefitted from the clearing of forest and the creation of intermixed woodlots and open-space habitats. Edson (unpubl.) implied in the early 1900s that it was rarely seen in nw. Washington before widespread logging. A steady increase in breeding pairs from 1978 to 2000 in Whatcom Co. is probably typical of w. Washington (PDeB). One of the most important westside prey items, the e. cottontail rabbit, was introduced in the 1920s and 1930s, but was not widespread until the 1940s (Dahlquist 1948). Jewett et al. (1953) reported that prior to 1950 Red-tails were more common in e. than in w. Washington. For breeding birds that situation is reversed today. Except for plantings of deciduous trees around homesteads, habitat changes have been mostly negative for eastside Red-tails. Agriculture in the Columbia Basin has negatively impacted ground squirrels, a major prey item, and monocultures such as wheat fields and orchards are probably not as suitable as native vegetation.

Robbins et al. (1986) show increases of breeding populations during the period 1965-79 in nearly all regions of N. America, and Sauer et al. (2001) show this for Washington. CBC data also reflect increases of wintering birds (Root 1988); Washington counts show a significant increase ($P < 0.01$), particularly on the eastside.

Remarks: Reports of *harlani* increased essentially statewide in the late 1990s, suggesting possible winter range expansion. Taxonomic status of *harlani* and *alacensis* is uncertain.

Noteworthy Records: *Harlan's Hawk, early Fall*: 27 Sep 2000 at McNary NWR. *Late Spring*: 13 May 2000 at March's Point. *High count*, **West**: 4 on 31 Jan 2001 at Ridgefield. **East**: 6 on 1 Nov 2001 at Kittitas Valley.

Paul DeBruyn

Ferruginous Hawk *Buteo regalis*

Local, uncommon breeder in Columbia Basin. Rare winter visitor in se. and s.c. Washington. Very rare spring migrant in w.

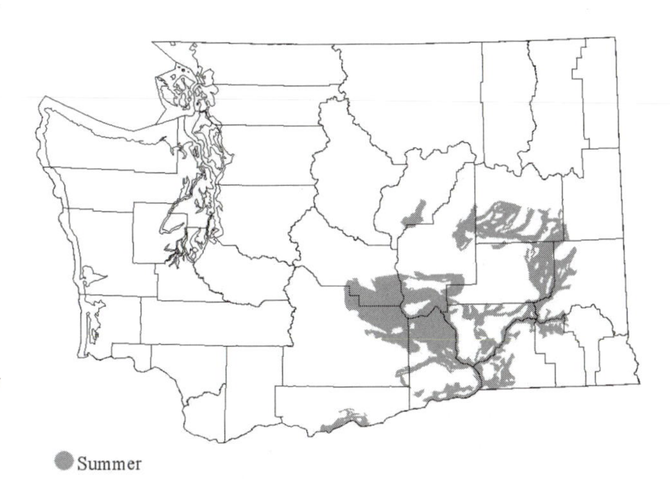

Habitat: Breeds in arid grasslands and shrub-steppe habitats using a variety of nest substrates, including cliffs, outcrops, trees (especially juniper and black locust), and artificial structures such as powerline towers and, infrequently, on pole-mounted platforms.

Occurrence: Known range includes Benton, Franklin, Douglas, Grant, Adams, Walla Walla, Whitman, Columbia, and Klickitat Cos., though usage fluctuates (Smith et al. 1997). At present, only 25% of the species' 200 territories are occupied in most years (Watson and Pierce 2000). No reliable abundance estimates were made prior to mid-1970s. Fitzner et al. (1977) found 71 nests on 31 sites in 1974 and 1975. They documented nesting by at least 15 pairs over two years and estimated the state population to be about 20 pairs. However, Friesz and Allen (1981) found 152 nests at 62 sites during a comprehensive search in 1981 and estimated the state population to be about 40 pairs. To date, over 200 territories in 12 counties have been documented, with more than 60% of nests in Benton and Franklin Cos. An average of 56 pairs nested annually from 1992 to 1996 (WDFW). Much of this apparent increase is almost certainly due to increased survey efforts.

Birds winter primarily in Mexico and the sw. and s.c. U.S. Numbers increase in Feb and decline in Sep. Satellite monitoring of 10 radio-tagged birds in 1999 documented a post-breeding movement across the Continental Divide in early fall to nw. Montana/ se. Alberta, with a minor additional movement further e. as three birds reached N. Dakota, Nebraska, and Oklahoma (Watson and Pierce 2000). Four of the 10 subsequently moved to California. Infrequent in winter; reported 19 times on CBCs from 1967 to 1999 at Tri-Cities, Two Rivers, Wenatchee, and Columbia Hills-Klickitat Valley.

Recently, specific localities have experienced clear population trends. The Hanford Site had no nests in 1973, presumably due to a lack of prey (Olendorff 1973), but in the late 1970s and early 1980s, one to three cliff nests were found. Construction of powerline towers during the mid 1970s created additional nesting opportunities; by 1991 the Hanford Site held 10 occupied nests, eight of them on towers (WDFW unpubl.). Although nest sites are not limited at Hanford, prey apparently is scarce. Birds often travel several kilometers, and cross the Columbia R., to forage on northern pocket gophers in alfalfa fields. Overall, state numbers are declining (Smith et al. 1997). In Esquatzel Coulee, Franklin Co., three or four nests were active yearly during the late 1970s and early 1980s, but were reduced to two in 1987 and one or none during the 1990s. Similarly, five nests in Yakima Co. in 1985 declined to one in 1995. In the 1920s and 1930s, Decker and Bowles (1933) noted a decline in the population in Benton Co. and placed the blame on sheepherders. Although such directed persecution is now rare, this species continues to be influenced by human activities. Birds are thought to be susceptible to human disturbance, especially early in their nesting cycle. The conversion and fragmentation of Washington's native shrub-steppe habitats have probably affected distribution and abundance. A widespread decline in the state's populations of jackrabbit and Washington ground squirrel may have altered the species' diet, leading to lower productivity and recruitment.

Remarks: More than 75% of known territories are on private land, with most others on federal holdings. Concerns raised recently regarding creation of wind turbine power-generation projects with respect to construction of extensive gravel access roads and mortality due to spinning blades (M. Denny p.c.). Classed ST, FC, PHS. A state recovery plan has been issued (WDFW 1996).

Noteworthy Records: West: 6 Apr 1985, 14 Apr 1993 and 7 Apr 1996 at Cape Flattery; 15 Jan 2001 at Ridgefield. **East**, *Summer, reports e. of current known range:* "several" in Jun-Jul 1968 nr. Clarkston, Steptoe, and Colton, Whitman Co.; 27 Jul 1972 at Clarkston; 6 Aug 1974 nr. Spokane; 30 Aug 1978 at Sherman Pass. Fall: 24 Sep 1974 at Pasco/ Prosser; 14 Sep, 16 Oct 1975 (1 light, 1 dark) at Kamiak Butte; 29 Nov 1975 at Warwick. *Winter*: 3 pairs remained in winter 1981-82 nr. Peola; 7 Feb 1982 at Orondo; 1 Jan 1984 nr. Lowden; 15 Dec 1988 at Waitsburg; 28 Dec 1989 nr. Clyde, Walla Walla Co.; 26 Dec 1995 at Burbank; 2000-2001 nr. Kahlotus; 27 Jan 2002 nr. Hanford; 9 Feb 2002 nr. Prosser; 1 in Franklin Co. and 1 in Walla Walla Co. in 2002-2003.

Scott A. Richardson

Rough-legged Hawk *Buteo lagopus*

Fairly common to common migrant and winter resident statewide.

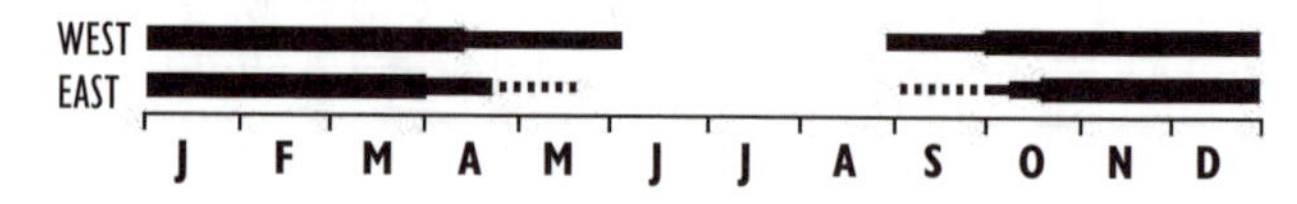

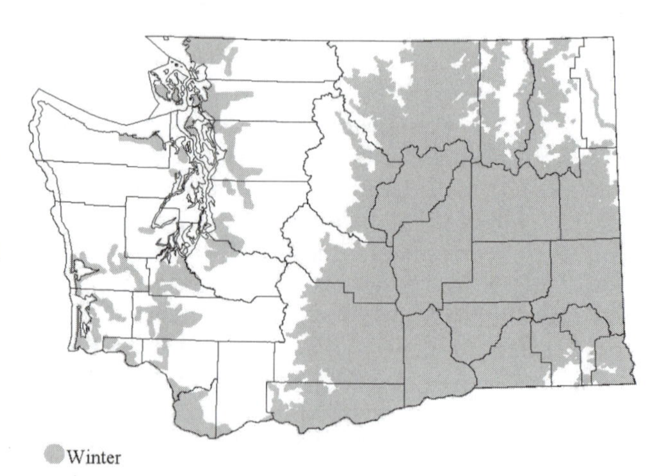

Subspecies: *B. l. sanctijohannis.*

Habitat: Lowland open fields, coastal shoreline, estuarine and freshwater marsh edges, irrigated valleys, and shrub-steppe areas. Noted in mountains in migration.

Occurrence: The Rough-legged Hawk is more common e. of the Cascades, including the Columbia Basin, than in the w. It is relatively widespread in open lowlands in the e. and the westside, though these are limited in developed Puget Trough counties (e.g. Kent-Auburn Valley, Nisqually delta, and Skokomish R. valley). Occurs in San Juan Is. (Lewis and Sharpe 1987) and in migration in small numbers along the n. outer coast. Like other raptors, observed migrating at high elevations. (C.M. Anderson, p.c,; Stepniewski 1999). Birds banded in Alaska have been recovered in B.C. (Campbell et al. 1990b) and Washington (Bechard and Swem 2002). Dates of peak migration for Washington are undescribed.

Abundance varies not only between years but also e. and w. Peak numbers mentioned in field reports in 1973-74, 1974-75, 1976-77 (see Campbell et al. 1990b), 1983-84, 1984-85, 1988-89; low numbers in 1965-66, 1967-68, 1968-69, 1978-79, 1986-87, 1987-88, 1990-91, 1994-95. CBC data from 1986, when effort was fairly consistent, show a peak in 1990-91, and lows in 1994-95, 1995-96, and 1999-2000. Relatively largest numbers (birds per party hour) recorded on CBCs at Bellingham, Ellensburg, Grays Harbor, Moses L., Padilla Bay, Toppenish, Two Rivers, and Walla Walla. Though cyclical numbers and relatively low numbers make trends difficult to determine, numbers decreased significantly long-term ($P < 0.01$). Declines may relate to loss of winter habitat to development and recent agricultural practices like planting of cottonwood trees for pulp production in previously open fields. Snow depths also may affect local winter distribution (Stepniewski 1999).

Remarks: Identification may be complicated by occurrence of light and dark color morphs and differences in appearance between sexes. Light-phase birds more prevalent than dark phase (2.7:1 in B.C. [Campbell et al. 1990b] and 5:1 in most winters in the lower Columbia Basin [M. Denny p.c.]), and sexes and age classes may vary in abundance from year to year (C. M. Anderson, in Wahl 1995).

Noteworthy Records: West, *Spring*: 1 on 21 Mar 1996 at Cape Flattery. *Fall*: 2 on 15 Oct 1967 on Mt. Baker. *Winter*: 72 censused in 1988-89 and 39 in 1990-91 in Skagit/Samish flats. **East**, *Summer-early Fall*: 1 on 19 Jun 1969 nr. Slate Mt., Hart's Pass; 1 on 5 Jun 2001 at Ritzville; 1 on 25 Aug 2001 nr. Gordon, Whitman Co. *Fall*: 1 on 18 Oct 1998 at Chopaka Mt. *CBC high counts*: Peak counts at Ellensburg were 2-3 times those elsewhere in the state; with 97 in 1987, 148 in 1990, 101 in 1993, 80 in 1996.

Terence R. Wahl

Golden Eagle *Aquila chrysaetos*

Locally fairly common permanent resident, rare migrant and winter visitor.

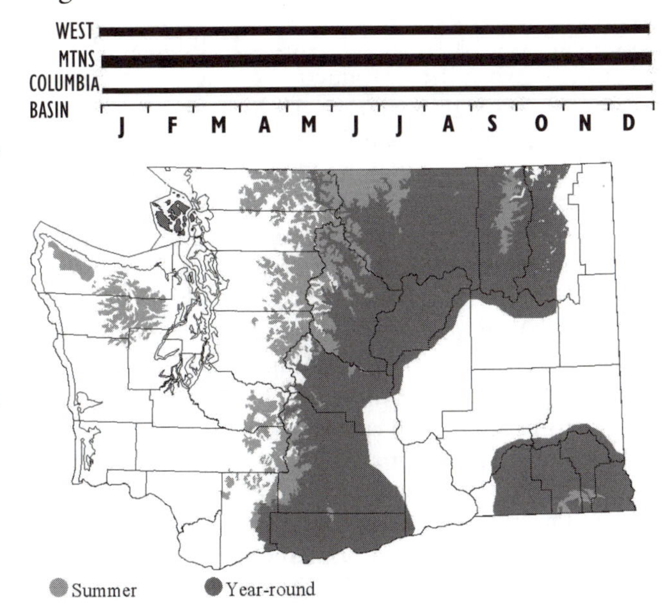

Subspecies: *A. c. canadensis.*

Habitat: Mostly in dry open forests of the eastside, shrub-steppe, canyonlands, and in high-elevation alpine zones of all regions. Associated with steep terrain, which very often includes cliff habitats. Nests on cliff ledges, rocky outcrops, large trees, power poles, transmission towers, or on the ground (Olendorff et al. 1981, Bruce et al. 1982, Knight et al. 1982). Most eastside nests on cliffs; westside nests are above timberline or in extensive clear-cuts. Open meadows, avalanche chutes, talus fields, rock outcrops, balds, bogs, burns, and clear-cuts used as hunting sites.

Occurrence: Breeding limited primarily to mountainous regions of the Okanogan highlands, the rain-shadows of the Olympics and Cascades, the Blue Mts. along the Snake and Grande Ronde Rs. and the San Juan Is. (Smith et al. 1997). Some sites have historic use: a nest in the Aeneas Valley in Okanogan Co. reportedly occupied for 100 years (R. Knight, p.c.; P. DeBruyn, p.c.).

Migration infrequently mentioned but noted along the Cascade crest in Yakima Co., for example (Stepniewski 1999). Moderate numbers (46-174 per season) noted at Cascade hawkwatch sites (HWI). Birds are fairly widely distributed in winter. CBC records are widespread in the e. with most consistent occurrence at Toppenish, Twisp, Walla Walla, Wenatchee, and Yakima Valley. Birds regular on CBCs in the San Juan Is., but irregular elsewhere w. of the Cascades.

Feed mainly on medium to large-sized mammals and birds including hares, rabbits, ground squirrels, and marmots (McGahan 1967, Snow 1973) with snowshoe hare and mountain beaver providing prey in w. Washington forests (Bruce et al. 1982). A population increase on Vancouver I., B.C., appeared correlated with the introduction of the e. cottontail rabbit there (Campbell et al. 1990b), and eagles resident in the San Juan Is. similarly occur where there is a large population of introduced European rabbits.

Nesting success varies by year and region. In n.c. Washington there were 20 successful nests out of 60 potential nest territories in 1982. Surveys in n. Chelan, Okanogan, and Douglas Cos. found about one-third of 115 historic nesting areas were active and 50% produced young in 1983. Twenty-four of 42 nesting attempts were successful in central Washington, producing 35 young in summer of 1984. Only two of seven nests were successful in Pend Oreille and Stevens Cos. in summer 1986. Of 68 nesting territories in Okanagon, 32 were occupied with 16 producing 21 young summer 1988. Westside nesting not well documented except in the San Juan Is. where up to five nests have been reported. Imm. birds reported essentially statewide year-round in summer and fall, and like ads., may be found with Bald Eagles at foraging opportunities.

Remarks: A state candidate species under review for listing as endangered, threatened, or sensitive (1998), never federally listed. Protected by the Bald Eagle and Golden Eagle Protection Act of 1940, this species was included in order to limit shooting of similar-appearing juv. Bald Eagles. Early declines in the W. resulted from shooting, trapping, and poisoning by ranchers and farmers who attributed livestock losses to depredation by eagles. More recent impacts include lead poisoning, electrocution on power lines, human disturbance around nest sites, overgrazing, and conversion of rangelands to agricultural, industrial, and residential uses (Olendorff et al. 1981, Harlow and Bloom 1989, Kochert 1989). Classed SC, PHS.

Noteworthy Records: West, *nesting*: 5 nests in 1979 on San Juan I. and James I.; birds seen in 1980 at breeding sites on San Juan I.; 3-4 pairs nesting in 1985 on San Juan I.; 2 pairs in May 1986 nesting w. side of San Juan I.; pair in 1988 nested at San Juan I.; nests in 1979 in Grays Harbor and Mason Cos., and Mt. Baker NF; two nests, one successful, in 1997 nr. Storm King Mt. and Kiona Peak, Lewis Co. (TB). **East**: pair nested in Apr 1968 at Banks L.; 2 birds in 1970 noted on cliffs by Alkali L., Grant Co.; pair in 1975 at nest in Ellensburg Canyon; 3 nests in 1978 in Ferry Co.; nests in 1979 in Ferry and Stevens Cos.; 16 active nests in 1981 in Okanagon Co.; incubating birds in 1984 at Naches and Oak Cr. WMA nr. Yakima.

Thomas Bosakowski

Common Kestrel *Falco tinnunculus*

Casual vagrant.

"Probably the most numerous species of *Falco* in the world" (Cade 1982), and described as the most common diurnal raptor in the w. Palearctic (Cramp and Simmons 1980). A rare vagrant to N. America, with only three certain records from the conterminous U.S.: Massachusetts (Cory 1888), New Jersey (Clark 1974) and Washington. On 31 Oct 1999, an imm. hunting in fields near Bow, Skagit Co. (CMA, M. Gleason, M. Muller and E. Deal), was captured, banded, and released (#1253-77695) on 3 Nov. It was observed eating voles, mice, and a small frog, and a duck killed by a Peregrine. It disappeared after Christmas.

Recorded also in B.C. in 1946 (Campbell 1985), in Nova Scotia and New Brunswick (AOU 1998). Seven reports have been published for the Aleutian Is. (Gibson and Kessel 1992), with at least two more for the Aleutians and a specimen from Little Diomede I. more recently (D. Gibson, p.c.). Most of the N. American records are from the fall. Also known as the Kestrel or European Kestrel.

C.M. Anderson

American Kestrel *Falco sparverius*

Common year-round resident in open habitats in e. Washington, uncommon on westside.

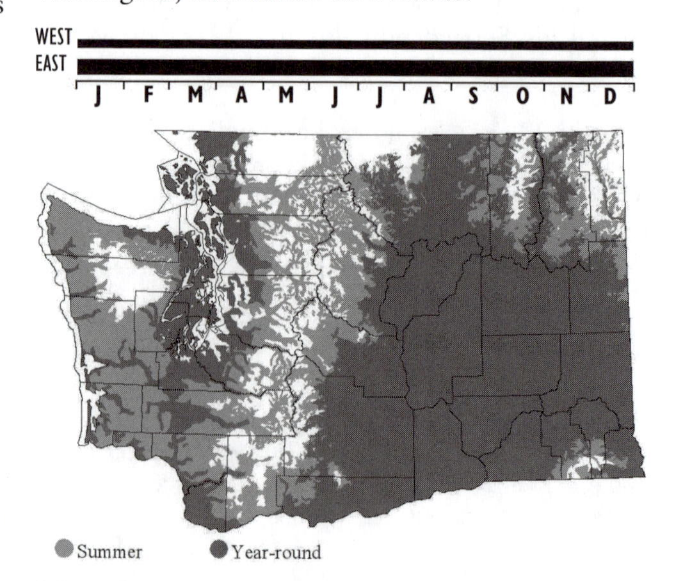

Subspecies: *F. s. sparverius.*

Habitat: Open habitats including agricultural areas, shrub-steppe, grasslands, meadows, burned-over forest, and clear-cuts (Runde et al. 1999). Occurs from low-elevation habitats to over 1500 m in the Blue Mts. and to 1800 m in Kittitas Co. (Smith et al. 1997).

Table 20. Average of American Kestrels counted at selected CBCs between 1990 and 1999. Data from Cornell Laboratory of Ornithology (birds.cornell.edu).

	Avg
West	
Bellingham	5.5
Columbia River Estuary	2.3
Grays Harbor	0.5
Kent/Auburn	5.2
Kitsap	0.1
Leadbetter Point	0.4
Olympia	3.6
Padilla Bay	3.5
Port Townsend	1.0
Seattle	0.7
Sequim/Dungeness	3.9
Tacoma	1.2
East	
Ellensburg	32.9
Moscow/Pullman	34.9
Spokane	6.7
Toppenish	90.5
Tri-Cities	34.6
Walla Walla	23.1
Wenatchee	24.5
Yakima	56.6

Occurrence: One of the more abundant raptors in e. Washington, where it has been recorded as a confirmed or probable breeder in nearly every county. It is uncommon in w. Washington, becoming more numerous, at least in winter, s. of the Puget Trough (SGM). Smith et al. (1997) described the range as the Puget Trough, the w. Cascades, and the ne. Olympic Pen., with no breeding records or suitable breeding habitats from much of the Olympic Pen. and sw. Washington. Jewett et al. (1953) reported breeding records from Aberdeen, Ozette L., Quinault, and Westport, and observations in the 1990s found kestrels w. of the range indicated by Smith et al (1997), in Grays Harbor, Mason, and Pacific Cos., and the w. portions of Clallam and Jefferson Cos. (T. Fleming p.c., D. Varland p.c., JBB).

Migration timing incompletely known. Mean spring arrival date in the Okanagan Valley in B.C. was 19 Apr (range 27 Mar-9 May; Cannings et al. 1987) and fall migration peaked in early Sep, much earlier than suggested by Jewett et al. (1953). Largest westside numbers are present from early Sep to mid-May, with peaks in Apr to mid-May and Sep-Oct (SGM). Stepniewski (1999) gives Mar-Apr as spring migration in Yakima Co. Kestrels are detected at hawk migration lookouts, but at Chelan Ridge and Diamond Head in 1999 it was seen less frequently than the Merlin (39 and 35 times, respectively; HWI). Kestrels are commonly encountered above timberline in late summer/early fall; some individuals appear to reside for periods of time there.

Birds are far more abundant during winter in eastside lowlands than on the westside (Table 20). Since 1974, counts of >50 kestrels at eastside CBCs were at Ellensburg (1 year), Toppenish (14), Tri-Cities (1), and Yakima (8). Three counts, all at Toppenish, exceeded 100 birds, with the highest 125 in 1990. During the same period, highest westside counts were 19 at Sequim/Dungeness in 1983 and 17 at Olympia in 1979 (birds.cornell.edu). Apparently only a small proportion of breeding birds from e. Washington over-winter locally: about 90% of the returns from birds banded on territory in e. Washington or Oregon were from Mexico (Henny and Brady 1994).

BBS data indicate changes over the past few decades. The population appears stable for the Columbia Basin from 1966-79, 1980-99, and 1966-99, but on the westside the population has fallen at a rate of -7.3% since 1966 ($P < 0.05$), and substantially (-9.6%: $P < 0.05$) since 1980 (Sauer et al. 2000). It seems likely that the negative trend is the result of habitat loss and degradation (e.g. due to the lack of cavity trees or snags). CBC data trends indicate an overall significant winter increase e. of the Cascades ($P < 0.05$) and a decrease w. ($P < 0.01$).

Remarks: Called Sparrow Hawk in the past.

Joseph B. Buchanan

Merlin *Falco columbarius*

Uncommon to fairly common winter resident at lower elevations statewide. Rare and local breeder, primarily on the westside in Puget lowlands and Olympic Pen.

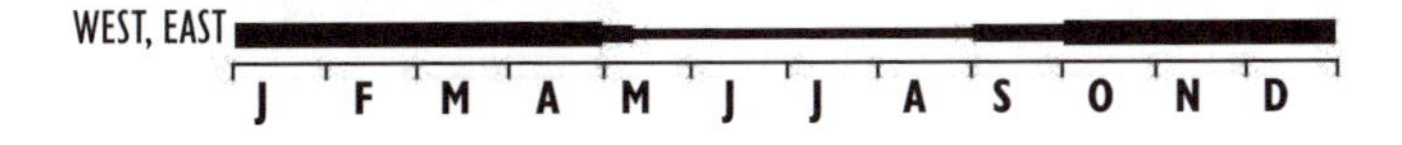

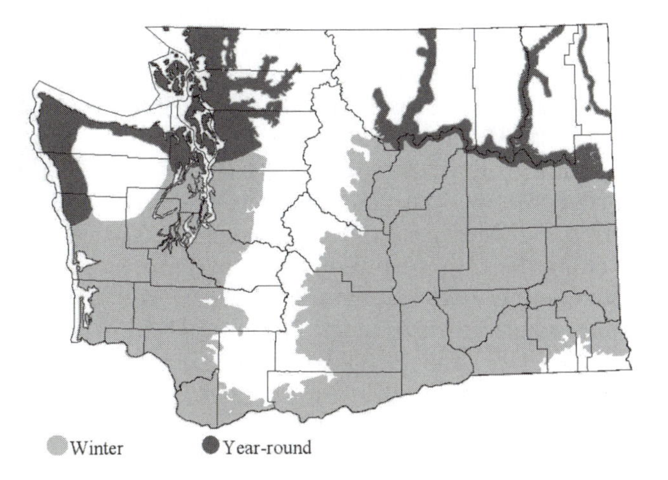

Subspecies: *F. c. columbarius* and *suckleyi* in Washington. Richardson's Merlin, *richardsoni*, likely very rare to rare migrant and winter resident in e. Washington (Stepniewski 1999). Westside breeders primarily *suckleyi*, and eastside primarily *columbarius* (TG).

Habitat: Nests in a range of habitats, from forests to riparian thickets (Palmer 1988, Johnsgard 1990, Sodhi et al. 1993). Many habitats during migration and winter, tending to follow flyways used by prey species (Aborn 1994). Foraging habitats including coastal beaches, estuaries, agricultural areas, orchards, cattle feedlots.

Occurrence: Apparently breeds primarily on the Olympic Pen., in the w. Cascades, the Puget Trough, occasionally near or in urban centers, and in the n.c. Cascades and in ne. Washington between e. Okanogan Co. and the Idaho border. Increasing in Puget Trough, with nests at Anacortes, Bellingham, Lk. Stevens, Marysville, Mt. Vernon, and Seattle in late 1990s-early 2000s (C. M. Anderson p.c.).

Found in many areas in winter and migrations (Cade 1982), including open areas where shorebirds and other small birds provide prey. Wintering Merlins occur below about 300 m elevation in most open habitats and in urban/suburban areas where waxwings, sparrows, and starlings are abundant. Birds are generally more common at large estuaries than at other sites. CBC data from 1987-88 to 1996-97 indicate birds were most abundant at westside sites such as Skagit Bay, Bellingham, and Padilla Bay and in urban areas such as Seattle, Wenatchee, and Kent-Auburn (Table 21).

Neither the timing nor magnitude of migration are well understood (Cannings et al. 1987). Local breeders are on territory when migrants occur during Apr-early May. Except at large estuaries where numbers may be observed, Merlins occur in small numbers and migration peaks are difficult to discern (Beebe 1974). At Cape Flattery, an average of 5.5 Merlins (0.03 per hour) were observed each spring between 1990 and 1995, substantially lower than the average total of 21.7 per year (0.11 per hour) between 1996 and 1998 (Van Der Geld 1998a, 1998b). Fall migration peaks in Sep-Oct (Cannings et al. 1987), with routes similar to those used in spring, but Merlins also occur at other areas, such as along the Columbia R. in e. Washington (Stepniewski 1999) and in the Cascades. The average number observed at Diamond Head in fall between 1991 and 1998 was 23; the fall migration high count was 55 at Chelan Ridge in 1998 (Van Der Geld 1997, McDermott 1998).

Table 21. Mean abundance of Merlins at CBC sites in Washington for the period 1987-88 to 1996-97.

Location	Mean	Range
West		
Bellingham	5.9	1-11
Grays Harbor	4.8	0-10
Kent-Auburn	4.9	3-7
Leadbetter Pt.	1.3	0-3
Olympia	3.1	0-5
Padilla Bay	5.9	0-9
Seattle	6.1	2-11
Sequim-Dungeness	3.4	1-7
Skagit Bay	6.6	4-11
Tacoma	4.1	1-7
East		
Ellensburg	1.3	1-2
Spokane	4.7	3-8
Tri-Cities	1.6	0-6
Toppenish	2.7	0-5
Walla Walla	1.7	0-4
Wenatchee	5.0	0-10
Yakima	2.7	1-6

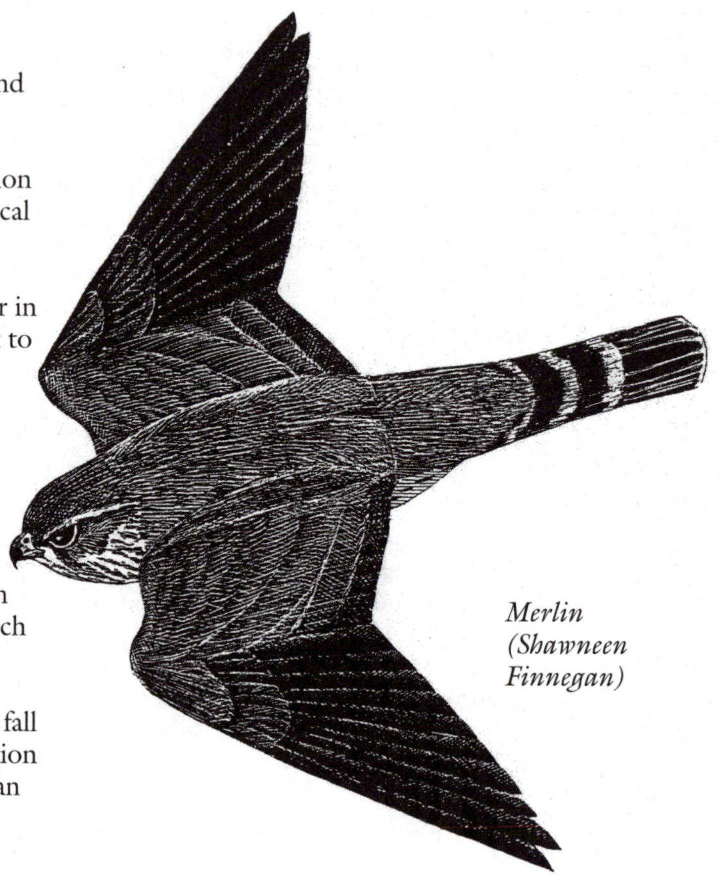

Merlin (Shawneen Finnegan)

The trend of Washington numbers over the last century is unknown, though CBC counts from 1960 to 1999 indicated increases statewide (P <0.01). Although the status of various subspecies populations is unknown, most wintering populations in N. America appear to be stable (Buchanan 1988a).

Remarks: Mixed pairs of *suckleyi* and *columbarius* are known on the westside (TG). Black Merlins *suckleyi* nest in interior B.C. (Cannings et al. 1987, Campbell et al. 1990b) and may be occasional breeders in e. Washington. Some N. American populations may have been impacted by the use of toxic chemicals in forestry and agriculture (Fox 1971a, 1971b, Fyfe et al. 1976, Fox and Donald 1980), but have apparently recovered. Classed SC, PHS.

Noteworthy Records: *Spring, late*: 2 May 1971 nr. Tower Mt., se. of Spokane; 6 May 1988 at Ocean Shores; 9 May 1974 at Ocean Shores; 11 May 1986 at Ocean Shores (JBB); 12 May 1990 at Leadbetter Pt.; 18 May 1969 at Leadbetter Pt. *Summer, not associated with known nesting sites*: 30 Jun 1988 in Seattle; late Jun 1977 in Okanogan Co.; 7 Jul 1984 atop Eagle Cliff, San Juan Co. (Lewis and Sharpe 1987); 7 Jul 1990 in s. Cascades; 9 Jul 1988 in Edmonds; 22 Jul 1990 in Olympia (Buchanan and Horn 1992); late Jul 1986 at Dungeness. Richardson's Merlin: 9 Apr 1988 at Yelm; 29 Dec 1990, e. of Moxee (Stepniewski 1999).

Thomas Gleason, Tracy L. Fleming, Joseph B. Buchanan

Eurasian Hobby *Falco subbuteo*

Casual vagrant.

An ad. was seen eating dragonflies and v.t. in Seattle on 20 Oct 2001 (K. Aanerud, P. Cozens). There are about 10 records from Alaska and one at sea 515 km e. of St. John's, Newfoundland (ABA 2002). Falconers believed there was very little likelihood that the Washington bird was an escape.

Terence R. Wahl

Gyrfalcon *Falco rusticolus*

Rare migrant and winter resident in lowlands primarily in nw. Washington and the Columbia Basin.

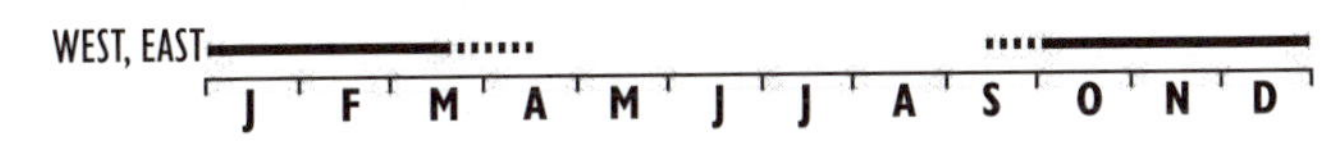

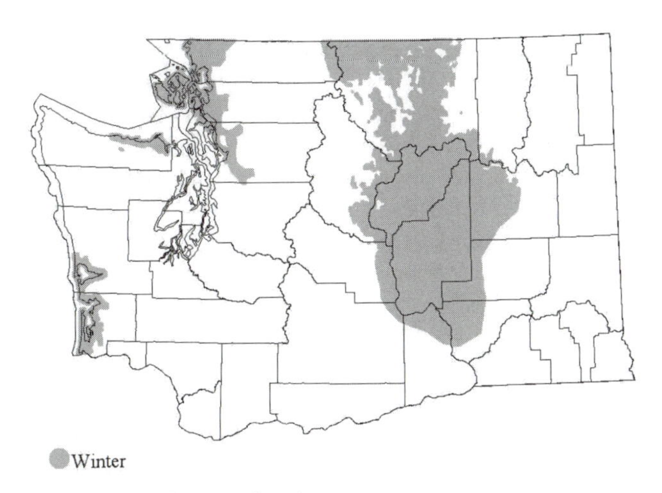

Subspecies: *F. r. uralensis.*

Habitat: Open lowland agricultural areas, river deltas. Roosts include utility poles and power pylons, cliffs on an offshore island.

Occurrence: Breeding in the holarctic arctic and subarctic, Gyrfalcons begin to arrive in Washington in mid-Sep, with the majority appearing in Oct or later. Wintering birds are annual both e. and w. Westside birds inhabit the agricultural lowlands of Whatcom, Skagit, Snohomish, and, more rarely, King Cos. They also occur near Sequim and along the outer coast in Gray's Harbor and Pacific Cos. (D.Varland p.c.). Birds are regular in the Columbia Basin, particularly Grant, Adams, Franklin, Lincoln, and Douglas Cos. Dobler (p.c.) reported sightings from virtually all e. Washington counties. Apparently restricted to lowlands: migrants have never been reported at fall hawk-banding sites in the Cascades. Birds remain into Mar. A radio-tagged ad. female departed the Samish Flats on 26 Mar. Imms. may remain in Washington until at least mid-Apr (CMA).

Prey includes waterfowl, upland game birds, doves, shorebirds, and gulls. A radio-tagged bird on the Samish Flats regularly pirated voles from N. Harriers. M. Perry (p.c.) also reports several instances of Gyrfalcons eating voles in e. Washington. F. Dobler (p.c.) observed frequent kleptoparasitism by Bald Eagles and Rough-legged Hawks.

Observations of eight birds banded in Skagit Co. showed that, like peregrines, they returned to the same locations for a number of years, with the wintering area selected during the first year. They often utilized the same fencepost or tree for a series of weeks throughout the winter, disappearing and then reappearing at random intervals. One bird foraging in the Samish Flats roosted on the same cliff on a nearby island every night for several months.

The majority of Washington birds are brown imms. and gray ads. White and black morphs are rare. "Silver" morphs with light gray backs and pure white undersides are seen occasionally (and see Friedmann 1950, Clum and Cade 1994). In w. Washington, white birds have been reported only three times: a juv. shot near La Conner in 1958, another seen in Snohomish Co. in 2000-01 (M. Prostor p.c.), and a third at Lummi Bay (R. Knapp p.c.). Black Gyrfalcons have been observed three times in w. Washington: one on the Samish Flats (CMA) and two in Whatcom Co. (R. Knapp p.c.).

In-hand records suggest different winter ranges of the sexes: eight Gyrfalcons banded in w. Washington by Anderson were all females. Five were brown juvs. and three were gray second-year birds. Two banded by D. Varland (p.c.) on the sw. coast were also brown juv. females. Eight Washington specimens—four from the Conner Museum (D. Johnson p.c.), three from the UWBM and one from the UPS (D. Paulson p.c.)—were also all females.

First recorded on a Washington CBC in 1972, then not from 1975 to 1981, and annually since somewhere except in 1987. Seven records were from Grays Harbor and Leadbetter Pt., one from Olympia, and 40 other westside CBC records were from Sequim-Dungeness, San Juan Is., Skagit and Whatcom Cos. Nine eastside occurrences were from Grand Coulee, Wenatchee, and Yakima Valley.

Remarks: A bird at Bellingham on 18 Aug is the earliest seasonal report: the possibility of an escaped bird cannot be ruled out.

C. M. Anderson

Peregrine Falcon *Falco peregrinus*

Fairly common resident, migrant, and wintering species in w. lowlands, rare to uncommon summer resident and migrant in the mountains, rare in e. lowlands.

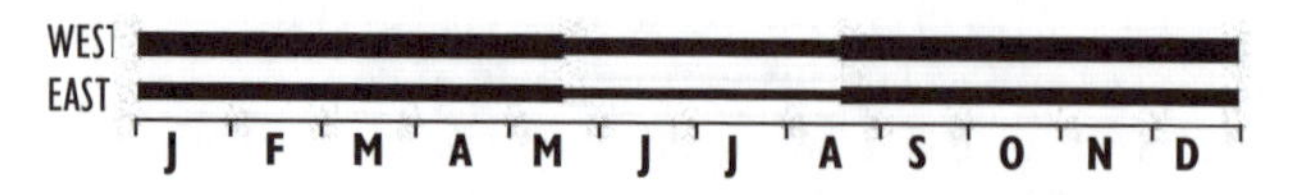

Subspecies: *F. p. anatum* and *pealei* nest; small numbers of *tundrius* in migration and winter. Coastal breeders considered *pealei,* inland birds *anatum*, with some mixing coastally and in the San Juan Is.

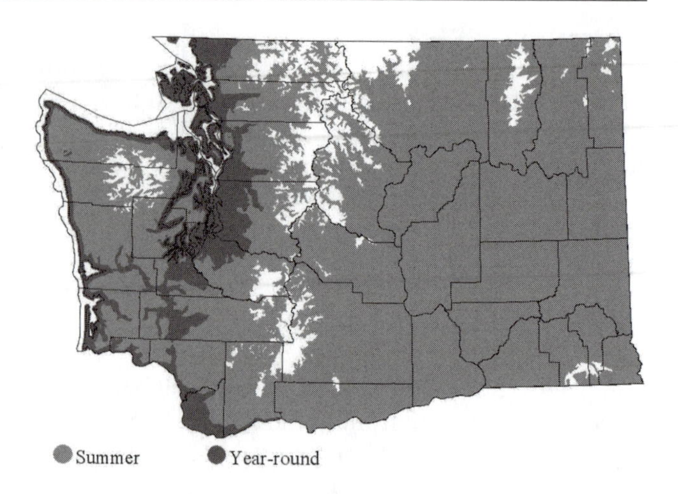

Habitat: Nest on cliffs, bridges, and tall buildings. Forage and winter on large river deltas, estuaries, agricultural fields, coastal beaches preying on shorebirds and waterfowl, and on starlings and pigeons in major cities.

Occurrence: The peregrine suffered a severe population decline in N. America during the 1950s and 1960s as a result of DDE-induced eggshell thinning (Hickey 1969, Cade et al. 1988). Numbers increased in Washington following the ban on DDT use in 1972. None of the 14 known historic Washington nest sites was occupied in 1976 when a nesting pair was found on the coast (CMA, J. Fackler); it was the only active peregrine nest known in Washington throughout the 1970s.

Since the mid-1980s, additional nests have been discovered on the outer coast (Wilson et al. 2000), the San Juan Is., the Columbia Gorge, and the Snake R. Canyon. Peregrines also breed in urban Seattle, Spokane, and Tacoma. Most recently, pairs have established nests in the Columbia Basin and on both sides of the Cascades. A recent record in Asotin Co. (Smith et al. 1997) follows an old record from the area (Weber and Larrison 1977). By 1990 birds were breeding successfully in several areas, and numbers subsequently increased. Nest sites range from near sea level to 900 m, though recorded to 1700 m. In 2001, about 75 breeding pairs were known (Hayes and Buchanan 2002). In addition, 145 captive-bred birds were released in the state between 1982 and 1997 (Allen 1991) to augment recovery.

Peregrines are primarily lowland migrants (Anderson et al. 1988). In fall, they move along the outer coast (D. Varland p.c.), through Puget Sound, and across the Columbia Basin. In spring, northbound migrants are most frequently seen along the coast (Herman and Bulger 1981, Clark and Clark 1998). Peregrines were rare fall migrants in the Cascades during a five-year study at Diamond Head, Kittitas Co. (J. Bettesworth p.c.) and at

Chelan Ridge, Okanogan Co., in 1999 and 2000, with less than 10 each year (HWI).

Most lowland breeders in Washington winter near nest sites. Banding data have shown that wintering resident population is supplemented by northern birds, including at least some breeding birds from the Queen Charlotte Is. (D. Varland p.c.) and Alaska (Anderson et al. 1988). Peregrines winter notably in the Lummi Bay and Samish/Skagit bays area, at Grays Harbor and Willapa Bay, and in many cities, including Seattle, Tacoma, Bellevue, Spokane, Olympia, Everett, and Bellingham. At least nine individuals wintered in Seattle in 2001 (R. Taylor p.c.). Some individual peregrines show remarkable winter philopatry (Anderson and DeBruyn 1979, Anderson et al. 1980, Dobler 1993).

Long-term CBC data show a significant increase w. of the Cascades (P <0.01) at locations where peregrines occurred relatively regularly.

Remarks: *F. p. anatum* and *tundrius* listed as endangered by the USFWS in 1973 and the species listed as endangered by WDFW in 1980 (Hayes and Buchanan 2002). The federal government has now delisted both subspecies; the species now classed SE, FSC, PHS.

C.M. Anderson and Steve G. Herman

Prairie Falcon *Falco mexicanus*

Uncommon resident in e. Washington, rare in winter w.

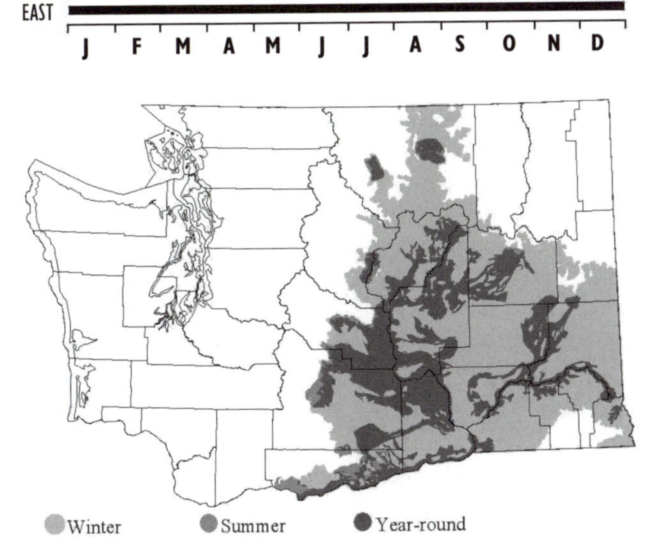

Habitat: Nests in canyon and coulee habitat in open areas and sagebrush country. Foraging habitat more diverse, can include agricultural lands (Smith et al. 1997). Winters also in open farmlands, marine shoreline habitats.

Occurrence: Fairly widespread, breeds along the e. slope of the Cascades, in upper sonoran and transitional zones below the ponderosa pine zone in the Columbia Basin, along the Columbia and Snake Rs., locally in the Methow and Okanogan valleys and in the Palouse and Whitman Co. (Smith et al. 1997).

Occurrence at high elevations noted by several authors, and thought to be primarily dispersing young (Smith et al. 1997, Stepniewski 1999). Above timberline in late summer/fall this is the second-most common falcon, after kestrel. Very low season totals (one to seven per year) at Cascades hawkwatch sites (HWI) indicate little actual migration in the mountains.

Wintering birds concentrate in the c. Columbia Basin (Jewett et al. 1953, Hays and Milner 1999), and birds that moved upslope following nesting shift to open habitats at lower elevations, foraging on prey like Horned Larks (Stepniewski 1999). Single birds recorded annually in winter w. of the Cascades (e.g., Larrison 1947a, Kitchin 1949, Lewis and Sharpe 1987, Wahl 1995) forage on waterfowl, shorebirds, and songbirds in open farmlands and along estuarine shorelines and tidelands in the n. Puget Trough. CBC data indicate fairly consistent winter presence at Ellensburg, Moses L., Tri-Cities, Toppenish, Two Rivers, Wenatchee, Walla Walla, and Yakima Valley.

Population trends unclear, but an overall decline suggested by historical assessments, from one of the more common birds in e. Washington (Dawson and Bowles 1909), with decreases noted in 1977 (Weber and Larrison 1977), and uncommon today (Smith et al. 1997). BBS data from 1966 to 2000 indicate a 4.4% annual decrease (Sauer et al. 2001). The decline of shrub-steppe habitat a concern in Washington (Hayes and Milner 1999), with much of this habitat being converted to agriculture and larger developments. Highly developed habitats not widely utilized by this species (Smith et al. 1997). Occurrence in w. lowlands fairly regular and apparently increased in recent decades, though the increase in numbers of observers is likely involved.

Remarks: Apparent decline attributed in part due to past DDT poisoning and the falconry trade (Weber and Larrison 1977). Recovery and current status since the banning of DDT are unknown. Prey populations in shrub-steppe habitats need to be maintained. WDFW has expressed concern over the limited nature of preferred nesting sites free from human disturbance which have abundant prey species available (Hayes and Milner 1999), and the species is classed PHS in Washington. Note: Two "breeds" records for Tacoma Flats and Nisqually (Kitchin 1923, and Decker and Bowles [1930], in Jewett et al. 1953) are questionable. Classed PHS.

Noteworthy Records: East, *CBC high counts*: 12 in 1998 at Columbia Hills-Klickitat Valley; 7 in 1993 at Toppenish; 7 in 1992 at Two Rivers; 6 in 1997 at Ellensburg; 5 in 1989 at Moses L.; 5 in 1988 at Yakima Valley. West, Spring: 2 on 5 Mar 1988 at Skagit Flats; 1 on 21 Mar 1999 at Lummi Flats; 1 May 1995 at Point No Point; 1 on 23 May 1993 at Stanwood (SGM). Fall: 15 Sep 1990 at Ocean Shores. Winter: 2 in 1999 at Padilla Bay CBC.

Jennifer R. Seavey

Yellow Rail *Coturnicops noveboracensis*

Casual vagrant.

Breeds across c. N. America w. to Alberta, with an isolated breeding population in the Klamath Basin of Oregon, winters in coastal areas of e. N. America. Two state records: an imm. collected on the Skagit flats on 16 Nov 1935 (Jewett et al. 1953) and 1 seen at Herman Slough, in Adams Co., on 30 Apr 1969 (Tweit and Skriletz 1996). Subspecies is *C. n. noveboracensis.*

Bill Tweit

Virginia Rail *Rallus limicola*

Fairly common summer resident statewide. Uncommon in winter.

Subspecies: *R. l. limicola.*

Habitat: Freshwater marshes and wetlands, less frequently brackish marshes. Will occupy small marshes created by agricultural runoff (Smith et al. 1997).

Occurrence: Though relatively common, the status of the Virginia Rail remains in some respects uncertain. Most conspicuous during spring and summer, birds breed locally throughout the Puget Trough extending inland up river valleys, the San Juan Is., w. to Grays Harbor and Willapa Bay. E. of the Cascades, found in the Columbia Basin, the Palouse, river valleys across from Okanogan to Pend Oreille and Spokane Cos., s. to Klickitat, Walla Walla, and the se. corner (Weber and Larrison 1977, Smith et al. 1997).

More numerous in Washington than the Sora, the Virginia Rail occurs on suitable lakes and marshes at moderate elevations, but essentially absent from higher-elevation meadows more often used by the Sora. Arrival dates on breeding grounds thought to be influenced by spring weather and emergent plant phenology and fall migration is variable, also influenced by weather conditions (Conway 1995). In addition to marshes, nesting birds occur in unexpected places: a pair nested on Tatoosh I. in 1998.

Though rails are present all year, some birds apparently leave Washington for the winter. Migration is little known, but evidenced by occurrence of this species in strange places like city streets (TRW).

Statewide, breeding numbers on BBS apparently increased (Sauer et al. 2000) and winter numbers recorded on CBCs increased from 1965 to -1999. Though the number of CBCs and effort increased, numbers of birds noted per count increased also. As noted by Stepniewski (1999), lower numbers or absence are noted in Yakima Co. following freezes, and mild winters lacking severe freezes in the 1990s may explain recent apparent increases in wintering numbers. Sensitivity to freezing is apparent also in the difference in occurrence on CBCs at Bellingham—one or two birds, nine times between 1973 and 1999—and locales less likely to be severely frozen such as Sequim-Dungeness—up to 20 birds, 19 times between 1980 and 1999—and Grays Harbor—up to 19 birds, 26 times between 1973 and 1999.

Remarks: Like other marsh birds, this species is habitat-critical. Local populations have probably decreased due to factors like loss of habitat including shoreline vegetation, and also to recreational disturbance, livestock grazing, predation and vastly increased numbers of Canada Geese (Wahl 1995).

Noteworthy Records: West, *Summer*: nest, 3 young on 28 Apr 1998 in a tiny, virtually dry sedge marsh on Tatoosh I. *CBC high counts*: 21 in 1998 at Everett, 19 in 1998 at Grays Harbor, 48 in 1999 at Kent-Auburn, 20 in 1999 at Sequim-Dungeness, 15 in 1996 at Toppenish, 14 in 1999 at Yakima Valley.

Ian Paulsen

Sora *Porzana carolina*

Fairly common summer resident in e. Washington, uncommon in w. Rare in winter.

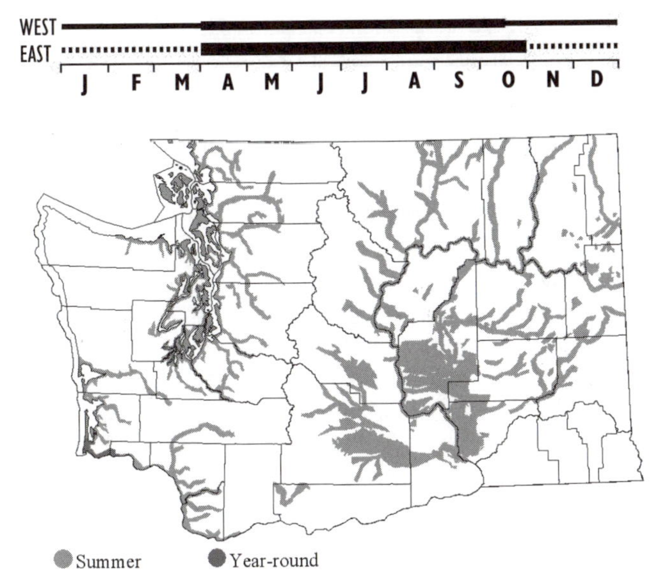

Habitat: Freshwater marshes, wet meadows and brackish marshes, wet areas in agricultural fields (Smith et al. 1997). Occurs in freshwater marshes and brackish marshes in winter.

Occurrence: Less numerous in Washington than the Virginia Rail, the Sora is more widespread, can occur in habitats with less water and at higher elevations—up to 1200 m (Smith et al. 1997). With a mean date for full sets of eggs of May 19 (Bowles in Jewett et al. 1953), most birds apparently arrive in early Apr. They are then widespread in habitat through the Puget Trough lowlands, nr. Sequim-Dungeness, the San Juan Is., and e. of the Cascades in Kittitas, Yakima, and Klickitat Cos., from Okanogan e. to Pend Oreille and Spokane Cos., and in the Columbia Basin. BBA surveys in the early 1990s did not find Soras along the outer coast from the Strait of Juan de Fuca to the Columbia R., in Wahkiakum, Clark, Skamania Cos., or in the se. corner of the state (Smith et al. 1997).

Fall migration usually occurs in Sep and Oct. Timing of migration, especially in the fall, may be determined in part by timing of frosts (Melvin and Gibbs 1996). The advent of portable tape players in the late 20th century probably resulted in better knowledge of winter presence of this species in the region (Contreras 1997). Soras were reported over 30 times on CBCs through 1999, almost all w. of the Cascades from Sequim-Dungeness and Skagit Bay south. Reported once each at Moses L. and Tri-Cities. Other field reports in the late 1990s suggested increased presence reflecting mild winters.

Remarks: Jewett et al. (1953) listed the Sora as a "common migrant and summer resident" but the species has declined, probably due to a loss of habitat in the state since the early 1950s. Marsh bird habitat has been negatively impacted in many localities through development, water-skiing and personal watercraft operation, and the spread of Canada Geese. Though not studied in Washington, effects on Sora populations are likely.

Noteworthy Records: West, *Winter high count*: 4 on the Seattle CBC in 1966. **East**, *Winter*: 1 on 28 Dec 1998 at Gloyd Seeps, Grant Co.; 1 on 18 Dec 1999 at Two Rivers CP; 1 on 2 Jan 2000 at Wallula; 1 on 15 Dec at Toppenish NWR.

Ian Paulsen

American Coot *Fulica americana*

Locally common year-round resident, mostly in lowlands, statewide. Uncommon in protected outer coastal and inland marine habitats in winter. Localized, concentrated in winter e. when water freezes. Locally abundant migrant e.

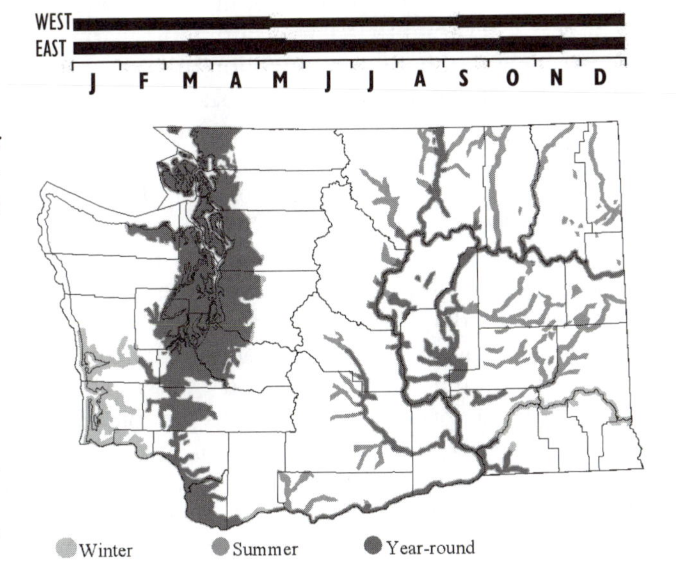

Subspecies: *F. a. americana* (see Taylor 1998).

Habitat: Ponds, lakes, slow-moving rivers at lower elevations, with emergent vegetation necessary for nesting (Smith et al. 1997). Occurs also in brackish and marine habitats in winter, particularly when freshwater freezes.

Occurrence: Common, but status and distribution less understood than that of many rare or threatened species. Widespread breeder throughout e. of the Cascades though not recorded during BBA surveys in three se. counties (Smith et al. 1997).

Uncommon on Cascade lakes to at least 1400 m elevation (e.g. Stepniewski 1999). Also local, widespread in the Puget Trough, w. to Sequim-Dungeness, the San Juan Is., and along the Columbia R.; relatively rare along outer coast, possibly only on sewage lagoons. Rare or absent in Grays Harbor, Pacific, Lewis and Skamania Cos. (Smith et al. 1997).

Migrant concentrations range from thousands in the e. and hundreds in the w. Peak numbers included 1000 on L. Washington in Oct 1919 (Jewett et al. 1953), 18,000 at Turnbull NWR in fall 1983, and 23,164 at Columbia NWR in fall 1984.

Part of the population moves south in winter (Jewett et al. 1953, Larrison and Sonnenberg 1968), and CBC totals show larger numbers w. of the Cascades. Eastside freeze-ups concentrate birds on open waters as at Tri-Cities and Wenatchee. Peaks of 3500-6000 reported Nov-Jan on the Yakima R., where numbers decrease over the winter and vary inter-annually due at least in part to severity of winter freezes (e.g. Stepniewski 1999). In the w. freezes on lakes drive some birds to marine waters and others to concentrate at open waters at parks and urban areas. Not described as occurring on saltwater by Jewett et al. (1953) or most previous authors (e.g. Edson [1919] does not report marine occurrence, though his notes mention this once, in Apr 1934 at the Nooksack delta). Local and usually uncommon, coots are widespread in estuaries and protected embayments: MESA surveys in 1978-79 recorded birds regularly at several bays in winter and, rarely, from mid-Apr to mid-Sep.

Population trends uncertain. Breeding numbers, particularly w. of the Cascades, appear subject to problems from development around lakes and marshes, clearing of lakeside vegetation, recreational activities like water-skiing and personal watercraft (Wahl 1995). Expanded reservoirs and pothole lakes and marshes over the past decades would appear to have favored increases e. of the Cascades. Winter trends based on long-term CBC data are uncertain, possibly due to changes in count coverage or changes in urban habitats where most counts are taken. Seattle and E. Lk. Washington showed increases, while Kitsap Co., Port Townsend, and Spokane showed decreases.

Remarks: Seasonal reports by field observers and publications declined over time (no AB reports after 1984). Coots are hunted in the state with an annual harvest of between 3000 and 6000 birds. Birds with white frontal shields suggesting the Caribbean Coot (*F. caribaea*) reported from the Pacific Northwest are aberrant American Coots (see Roberson and Baptista 1988).

Noteworthy Records: East, *Spring*: 10,000 in 1972, 11,000 in 1981 at Turnbull NWR. *Summer*: 500 pairs in 1973 at Turnbull. *Fall*: 5000 in 1972 at Turnbull and 10,000 estimated for Grant Co. coulee lakes; 20,000 on 24 Nov 1974 on Alkali L., Grant Co.; 13,000 in 1981 at Turnbull NWR; 1982 Washington aerial count of over 39,000 in the Columbia Basin with definite increase in mid-Oct, and 30,000 peak at Turnbull; 1983 peak at Turnbull was 18,060; 1984 peak 23,164 at Columbia NWR and 2000 at Turnbull. *Winter*: severe losses in 1968-69, McNary NWR population between 1000-1500; peak of 1020 in 1981-82 at Turnbull NWR, 500 at Columbia NWR. **West**, *Summer*: breeding at 8 localities in 1981, with 45 nests at Nisqually L. *CBC selected high counts*: 2137 in 1994 at Bellingham, 1742 in 1997 at Edmonds, 1349 in 1978 at Olympia, 1835 in 1997 at Padilla Bay, 16, 221 in 1993 and 10,216 in 1994 at Seattle, 1541 in 1976 at Tacoma, 1526 in 1987 at Tri-Cities, 4000 in 1971 at Wenatchee. (Lowest count at Seattle from 1959 to 1999 was 1500 in 1968).

Ian Paulsen and Terence R. Wahl

Sandhill Crane *Grus canadensis*

Regular fairly common to locally abundant migrant w. and e. Locally very rare to uncommon in winter w. Rare breeder e.

Subspecies: Greater, *G. c. tabida*, Canadian, *rowani* and Lesser Sandhill Crane, *canadensis*, in Washington.

Habitat: Open fields, river bottomlands, areas adjacent to estuaries, nests in wet meadows dominated by grasses, sedges, and willows or wet hay fields.

Occurrence: Status today differs from that generally described by Jewett et al. (1953) which in summary reported more widespread occurrence but with migrating flocks much smaller than known in recent decades.

The three subspecies evidence different occurrence patterns in Washington (Littlefield and Ivey 2002). Greaters are found in small numbers statewide, but most apparently overfly the state. Lessers migrate on both sides of the divide (Bettinger and Milner 2000) and comprise most of the 20,000+ cranes that migrate through e. Washington annually (Littlefield and Ivey 2002). Migration noted generally from Richland to Othello and from Ridgefield NWR n. to B.C. Peak numbers noted include 13,000 at Soap L. in Sep 1992, 3860 at Ridgefield NWR in Oct 1995, and 3500 in spring

Sandhill Crane
(G. Scott Mills)

1985 at Cape Flattery. Migratory patterns of the Canadian subspeices are poorly known.

Greater Sandhills are the only breeding race. Confirmed breeding within the Yakama Indian Reservation at two disjunct sites: Signal Pk. and Camas Patch, at Deer Cr., Yakima Co., and in Klickitat Co. at Panakanic Valley and at Conboy L. NWR (RHL, Engler and Anderson 1998b). No other breeding locations are currently known in Washington, but cranes were formerly thought to have nested on both sides of the Cascades (see Jewett et al. 1953). First described in 1941 (Jewett et al. 1953), ads. not observed at Signal Pk. again until 1991 due to restricted access. Ads. first noted at Conboy in 1974 (Engler and Anderson 1998a). Young first reported in 1984 at Conboy and 1994 at Signal Pk. (Leach 1995). Birds arrive on breeding grounds in late Feb-Mar (several weeks later at higher elevations). Young observed in mid-May and mid-late Jun, fledging from early to late Aug, and departure from breeding grounds by late Sep to mid-Oct (Engler and Anderson 1998a). Nesting birds total up to 19 pairs with 0-6 young raised annually (Engler and Anderson 1998a, Littlefield and Ivey 2002, RHL).

The only regular wintering area is the lower Columbia bottomlands of Clark and adjacent Cowlitz Cos., with up to 1000 wintering in recent years (Littlefield and Ivey 2002); these are thought to be the Canadian subspecies. Wintering was first noted in 1972-73, and birds were described as regularly wintering by 1985-86 (AB 40:318). Elsewhere, very rare in winter. Scattered individuals noted in summer since 1989 at Sauvie I., Oregon, and nearby Ridgefield NWR.

Remarks: Numbers believed to be increasing in w. N. America (see Tacha et al. 1992, 1994). Classed SE, PHS.

Noteworthy Records: *Early/late migration dates*, **East**: 4 on 2 Feb 1975 at Lowden; 1on 9 Jun 1982 at Nile, Yakima Co.; 9 on 17 Nov 1999 at Columbia NWR. **West**: 3 on 10 Jun 1980 at Sequim; 1 on 9 Sep 1973 at Willapa Bay; 3 on 18 Nov 1998 at Naselle. *Peak numbers*, **East**: 13,000+ on 15 Sep 1992 at Soap L.; 5000-6000 on 16 Apr 1984 at Conconully L.; 5000 on 21 Sep 1998 at Othello; 3800 in Oct 1996 at Columbia NWR. **West**: 3500 in spring 1985 at Cape Flattery; 3860 on 11 Oct 1995 at Sauvie I., Oregon, and Ridgefield NWR; 1700 on 10 Mar 1996 at Ridgefield NWR. *Summer, other than Ridgefield*: 9 Jun 1996 at Douglas and 6 Jul 1989 at Clallam. *Breeding early/late dates*: 8 Mar 1997 (Conboy; Engler and Anderson 1998b); 5 Apr 1993 (Signal Pk.; Leach 1995); 2 ads., 1 imm. on 26 Sep 1997 at Lateral A and Toppenish Cr., about 52 km from Signal Pk. (RHL). *Winter*: 300+ wintered in 1995-1996 Sauvie I., Oregon. 8 other winter sightings: 3 from Whatcom; 2 from Clallam; 1 each from Snohomish, Skagit, and Thurston Cos. Two young color-banded in 1996 at Conboy NWR seen nr. Glenn, California, 14 Dec 1997 (Engler and Anderson 1998a).

Rosemary H. Leach

Black-bellied Plover *Pluvialis squatarola*

Widespread, fairly common to common migrant and winter visitor at large estuaries in w. Washington. Local, rare in spring and uncommon in fall in e. Washington. Uncommon in summer at coastal estuaries.

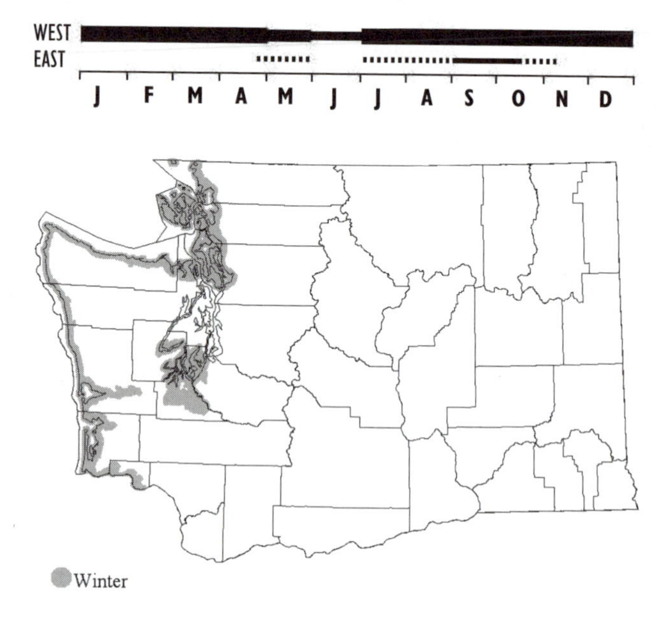

Habitat: Intertidal mud flats and sandy beaches; also uses freshwater habitats in small numbers; roosts in fields and on beaches (Paulson 1993).

Occurrence: The Black-bellied Plover has a circumpolar breeding distribution in arctic and subarctic tundra, winters s. to s. S. America, Africa, and Australia, and is common in n. temperate portions of its winter range, n. along the Pacific coast to s. B.C. Large numbers winter at Willapa Bay (Buchanan and Evenson 1997), Grays Harbor, and at a few scattered locations in Puget Sound (Evenson and Buchanan 1997). Over 1000 are regular at Willapa Bay and the adjacent coastal beaches. Several Puget Sound estuaries collectively support several hundred or more birds, although high counts occasionally exceed 1000. In 1978-79 MESA surveys of n. inland waters found highest numbers at the Dungeness area, Port Angeles, Sequim Bay, Samish Bay, Padilla Bay, and Lummi Bay. Birds move considerable distances up river valleys (e.g. to Satsop in the Chehalis Valley; C. Chappell p.c.).

This species is also abundant during spring migration with counts generally lower than those in winter. It is abundant at Willapa Bay (Buchanan and Evenson 1997), Grays Harbor (Herman and Bulger 1981), and Puget Sound (Evenson and Buchanan 1997). Migration peaks in mid-late Apr (Buchanan 1988c, see Herman and Bulger 1981). Migration extends well into May (Widrig 1979, Herman and Bulger 1981, Paulson 1993). Small numbers occasionally spend the summer (i.e., 5-25 June) in the region. Such birds are found at main wintering areas (Widrig 1979, Buchanan 1988c) and most are in imm. plumage (Paulson 1993).

Birds are abundant during fall migration. High counts for the period are primarily from Ocean Shores, Grays Harbor, and Leadbetter Pt. The highest Puget Sound count was at Dungeness Bay (Evenson and Buchanan 1997). Ad. fall migrants begin arriving in late Jun-early Jul whereas juvs. arrive in late Aug and Sep (Paulson 1993).

The species is a rare eastside migrant in spring, with most records coming from Reardan, Banks L., and the Yakima R. delta (Paulson 1993), from 29 Apr to 31 May. There are more fall eastside records (eight records of 1-10 birds), widely scattered at locations such as Reardan, Turnbull NWR, Richland, the Potholes Res. area, and Walla Walla R. delta. The species was considered hypothetical in the four se. counties by Weber and Larrison (1977). Late records are from mid- to late Oct, with one occurrence on 8 Nov. There are no winter records from the Columbia Basin.

The N. American population is thought to be about 200,000 (Morrison et al. 2000). Although adequate monitoring data are generally lacking, it appears that Washington populations are currently stable or perhaps increasing. Long-term CBC counts show no overall trend. Classed PHS.

Noteworthy Records: *High counts, Winter, outer coast*: 4049 on 21 Dec 1994 at Willapa Bay (Buchanan and Evenson 1997); 3195 on 11 Mar 2001 at N. Beach (JBB); 2700 on 7 Feb 2001 at Leadbetter Pt. (SGM). *Inland waters*: ca. 200 at Dungeness Bay, Sequim Bay, Samish Bay (Evenson and Buchanan 1997); ca. 1500 on 18 Dec 1982 at Padilla Bay (Paulson 1993); 502 on 14 Mar 2001 at Totten Inlet (JBB). *Spring, Willapa Bay*: 300 at Nemah R., 300 at Bear R., 280 at Leadbetter Pt. (Widrig 1979, Buchanan and Evenson 1997); Grays Harbor: 210 at Pt. New (Herman and Bulger 1981). *Inland waters*: 871 on 20 Apr 1998 at Totten Inlet (JBB), 200 at Samish Bay (Evenson and Buchanan 1997), 500 in late Apr at False Bay, San Juan I. (Lewis and Sharpe 1987). *Summer*: >100 in late Jun 1976 at Leadbetter Pt. (Widrig 1979); 70 on 17 Jun 1977 at Dungeness Bay. *Fall, outer coast*: Ocean Shores: 5000+ on 15 Sep 1996, 6-7000 on 3 Sep 1999. Leadbetter Pt.: 1400 on 21 Sep 1978 and 1500 on 16-17 Oct 1978 (Widrig 1979), 3000 on 2 Sep 1995 (SGM). Grays Harbor: 1300 on 20 Oct 1979 (Paulson 1993). Inland waters: 750 at Dungeness Bay (1991; Evenson and Buchanan 1997); 321 on 27 Jul 1997 at Drayton Harbor; 208 on 22 Oct 2000 at Totten Inlet (JBB). **East**, *Spring*: 8 on 14 May 1960 at Reardan (Paulson 1993); 7 on 16 May 1963 at Banks L.; 14 on 19 May 2000 at Swanson Lks., Lincoln Co. Fall: 20 on 21-22 Sep 1980 at N. Potholes

Res.; 25 in 1984 at Richland/Wallula area; 75 on 19 Oct 1986 at Walla Walla R. delta.

Joseph B. Buchanan

American Golden-Plover *Pluvialis dominica*

Uncommon to fairly common fall migrant and rare spring migrant along outer coast. Rare to uncommon fall migrant and very rare spring migrant elsewhere w. Rare fall migrant in e. Washington.

Habitat: Mudflats with nearby salicornia marshes or vegetated sand dunes. Also found on ocean beaches, plowed fields, freshwater marsh edges, and open tidal flats.

Occurrence: American Golden-Plovers occur in Washington primarily as southbound migrants along the outer coast and, to a lesser extent, in the Puget Trough. During fall, four or five are reported per year in e. Washington. Spring records average only one or two birds per year, all from the westside. There are no winter records from Washington or Oregon (Nehls 1994). Birds often associate with Pacific Golden-Plovers and Black-bellied Plovers and in Washington there is no clear difference in habitat preference between the two golden-plovers.

A decline in reports from an increased number of field observers at regular migration locations suggest a decline in numbers in the 1980s and 1990s, but systematic data are limited and trends uncertain.

Remarks: Until 1993, American and Pacific Golden-Plovers were considered conspecific as the Lesser Golden-Plover (AOU 1993). Because of the difficulty in distinguishing the two (Paulson 1993), differentiation can be a challenge, and some misidentification has likely occurred. The record listed for Kahlotus on 19 Dec 1924 (Sloanaker 1925, Paulson 1993) was based on supposition and could have been a Pacific Golden-Plover (D. Paulson p.c.). Campbell et al. (1990b) did not separate the two species and many Washington records cannot be attributed to either. Additional observations are required to understand the distribution, habitat use, and migration timing of both species. For details on taxonomic and winter occurrence see the Pacific Golden-Plover account. Classed PHS, GL.

Noteworthy Records: West, *Spring*: 21 Apr 1998 at Nisqually NWR; 3 on 25 Apr 1999 at Brady; 1 May 1998 at Leadbetter Pt.; 6 May 1998 at Ocean Shores; 2 on 9 May 1999 at Brady; 2 on 11 May 1987 at Ocean Shores (Paulson 1993); 22 May 1999 at Ocosta. *Fall, high counts*: up to 12 on 4-11 Aug 1994 at Ocean Shores; 4 on 22 Aug 1998 at Fort Lewis; 7 on 23-24 Aug 1991 at Ocean Shores; 22 on 2 Sep 1997 at Grays Harbor; 25 on 19 Sep 1997 at Grays Harbor; 53+ on 22 Sep 1984 at Ocean Shores; 45 on 28 Sep 1997 at Grays Harbor; 3 on 12 Oct 1997 at Fir I.; 4 on 16 Oct 1999 at Ediz Hook. Fall, late: 7 Nov 1999 at Ocean Shores. **East**, *Spring*: 1 on 29 Apr 2000 at Rock L., Whitman Co.; 1 on 10 Jun 1976 at Moses L. (Paulson 1993). *Fall*: 4 Sep 1976 at Turnbull NWR (Paulson 1993); 8 Sep 1995 at Atkins L., Douglas Co.; 8 Sep 1996 at Stallard L., Douglas Co.; 19 Sep 1999 at Wenas L.; 20 Sep 1997 at Yakima R. delta; 21 Sep 1996 at St. Andrews; 28 Sep 1997 at Swanson Lks.; 2 Oct 1999 at Othello; 3 Oct 1999 at Walla Walla R.; 2 on 5 Oct 1997 and 9 Oct 1998 at Yakima R. delta; 2 on 11 Oct 1996 and 1998, 25 Oct 1997 at Walla Walla R. delta; 11 Oct 1996 and 1998, 13 Oct 1996, 2 on 22 Oct 1998 at Scooteney Res.; 1-2 on 18-27 Oct 1996 at Yakima R. delta, 12 Nov 1980 at Walla Walla. *High count*: 17 in mid-Oct 1975 at Richland (Paulson 1993).

Joseph B. Buchanan

Pacific Golden-Plover *Pluvialis fulva*

Uncommon to fairly common fall migrant, very rare winter resident, and rare spring migrant along outer coast. Rare to uncommon fall migrant, very rare winter resident and spring migrant elsewhere in w. Washington. Casual in fall on eastside.

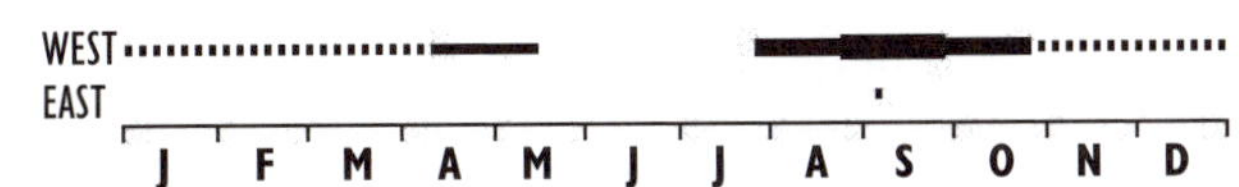

Habitat: Mudflats nr. *salicornia* marshes or vegetated sand dunes, ocean beaches, plowed fields, freshwater marsh edges, and open tidal flats.

Occurrence: Until recently considered conspecific with the American Golden-Plover. This, with problems of distinguishing the two species in the field, has resulted in incomplete knowledge of status in Washington.

Most birds occur along the outer coast during fall migration. They are uncommon to fairly common from late Jul to late Oct, with peak numbers from late Aug to early Oct. Ads. arrive first and are often present into Sep, while juvs. appear during late Aug or early Sep and are often found into Oct. Away from the outer coast, about three to four birds occur annually, mostly in the greater Puget Sound region.

Overall, appears to slightly outnumber American Golden-Plovers, with which it is often found. Pacific Golden-Plovers have been recorded only twice in e. Washington, both times in early Sep.

Of seven winter records, three have been identified as Pacifics and four remain unidentified. Of these three were found along the n. Olympic Pen., three (possibly the same birds) were in Skagit Co., and one in Spokane. The three unidentified birds from w. of the Cascades were likely this species: there are no confirmed midwinter records of American Golden-Plover from w. N. America (Mlodinow and O'Brien 1996) and few from e. N. America (Paulson and Lee 1992). A bird at Spokane on 19 Dec 1924 (Sloanaker 1925) was presumably either a very late American Golden-Plover or an unusual record of this species.

In spring, only one to three birds are seen most years, on the outer coast or in Puget Sound, primarily between mid-Apr and mid-May, but the species appears to outnumber American Golden-Plovers.

Numbers vary noticeably from year to year, though the magnitude of this is complicated by numbers identified simply as golden-plovers. In 1988 251 golden-plovers were reported in w. B.C., Washington, and Oregon. Of these, only 47% were identified to species, and over 60% of those were felt to be Pacifics. In 1996 over 105 Pacific Golden-Plovers were identified in w. Washington and Oregon, contrasting with fewer than 10 noted in 1994.

Remarks: This species and the American Golden-Plover were recognized as separate species in 1993 (Johnson and Connors 1996). Historical reports often included birds identified only as golden-plovers. Classed PHS, GL.

Noteworthy Records: West, *Fall, early*: Crockett L., 26 Jun 1999 (SGM, D. Duffy). *Winter*: 8 Feb 1986 and 12 Jan 1993 at Fort Flagler; 3 Feb 1993 at Dungeness Spit; 19-21 Feb 1995, 1 Jan-19 Feb 1996, 29 Dec 1996 at Samish Flats, (G. Toffic). *Spring, High count*: 4 on 4 May 2000 at Brady. *Fall, High count*: 41 on 28 Sep 1996 at Ocean Shores; 12 on 12 Oct 1997 at Fir I.. East: 8 Sep 1995 at Atkins L. Douglas Co., 9 Sep 1991 at Walla Walla R. delta.

Steven G. Mlodinow

Snowy Plover *Charadrius alexandrinus*

Local, uncommon summer resident, rare in winter on s. coast. Rare spring and summer visitor elsewhere in w. Washington. Very rare spring visitor in e.

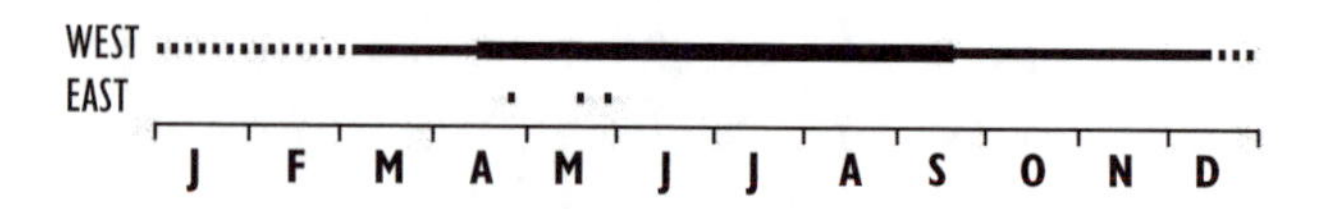

Subspecies: *C. a. nivosus*, breeds from Washington to s. Baja California and locally at interior sites in most states w. of the Rocky Mts.

Habitat: Coastal sand beaches and dunes, habitats typical of unstabilized sand spits and creek mouths. Nests seaward of the vegetated foredune or in flat, open areas with sparse vegetation. Salt flats, sand bars, or mud flats in e.

Occurrence: A species of range-wide concern. Restricted now to three breeding sites in Washington: Damon Pt./Oyhut Spit on the Ocean Shores Pen., Midway Beach, Pacific Co., and Leadbetter Pt./Gunpowder Sands on the Long Beach Pen. The number of nests located at Damon/Oyhut has been five or fewer during most years, with estimates of six in 1985 and eight in 1986. Leadbetter Pt. has supported similar numbers of nests, with high counts of seven in 1988 and 1989 and estimates of eight, nine, and 11 in 1997, 1979, and 1981, respectively. Other recent breeding-season sightings have come from Dungeness Spit, where up to six plovers have been seen during some summers in limited, marginal nesting habitat. Birds formerly bred at Copalis Spit in Grays Harbor Co., where an estimated six to 12 pairs nested during the late 1950s or early 1960s (G. D. Alcorn, letter dated 2 Mar 1983 to D. W. Heiser). Although Copalis Spit still holds suitable nesting habitat and has been surveyed regularly for plover activity, the last sighting was a single male in 1985.

Plovers nested regularly nr. Westport between 1915 and 1968, but the sand spit the plovers presumably inhabited has since eroded and the most recent report of nesting was in 1983. Other coastal sites probably supported nesting plovers before erosion. For example, the Cape Shoalwater vicinity on the n. side of Willapa Bay held approximately 339 ha of open dune habitat in 1937, but by 1992 had eroded to about 8 ha of poor plover-nesting habitat (J. Hidy, Willapa NWR p.c.).

Ads. reoccupy nesting areas in Mar or Apr and initiate most clutches between late Apr and late Jun. Chicks hatch a month after eggs are laid and fledging occurs from late Jun to Aug. These critical activities take place in conflict with peak human presence in time and place. Many color-banded plovers from coastal Oregon, and a few from coastal California, are observed at Leadbetter Pt. Banded plovers are rare at Damon Pt.

Surveys reveal evidence for a late-summer influx at Leadbetter Pt./Gunpowder Sands. In fall and winter, Snowy Plovers generally withdraw from Grays Harbor Co. Fall and winter abundance in Pacific Co. varies. CBCs at Leadbetter Pt. and Grays

Harbor have found birds six times between 1978 and 1995. Occurrence is very rare on the eastside.

Remarks: Disturbance in and nr. nesting areas has been partially allayed through placement of "No entry" signs, but maximum human presence occurs during the nesting season. Habitat is lost through erosion of unstabilized dune systems and by encroachment by European beachgrass. Two of the current nesting areas are in relatively secure habitat: Leadbetter Pt. within the Willapa NWR complex, and Damon Pt. and Oyhut Spit on state land managed for wildlife by the WDFW. The Pacific coastal population *(nivosus)* was listed as SE, FT, PHS, GL. A state recovery plan has been issued (WDFW 1995c). Known as Kentish Plover in the Old World.

Noteworthy Records: West, *Spring*: 1 on 6 May 1914 at Seattle area (Rathbun 1915). *High counts*: 35 on 11 Jun 1973, 30 on 1 May 1970, 16-20 ad. on 11 Jun 1993, 19 ad. on 11 Jun 1997 at Leadbetter Pt. *Summer*: 2 on 14 Sep 1975 at La Push. *Fall*: First state specimen on 3 Sep 1899 at "Grays Harbor"(UWBM 16519). *High counts*: 34 on 13 Sep 1995 at Gunpowder Sands; 20 on 20 Sep 1969 and 28 on 17 Dec 1978 at Leadbetter Pt. *Winter, high counts*: 21 on 24 Feb 1978 at Leadbetter Pt; 32 on 10 Feb 2001 at Midway Beach; 25 in 2001-02 at Midway Beach (SGM); 27 nests at Midway Beach and 6 in Damon Pt./Oyhut area, twice as many as in 2003, in summer 2004 (WDFW). *CBCs*: 28 in 1978, 9 in 1980, 2 in 1982, 6 in 1987 and 2 in 1988 at Leadbetter Pt.; 2 in 1994 at Grays Harbor. **East**, *Spring*: 1 on 28 May 1967 at Reardan; 2 prs in May 1985 at Goose I., Banks L.; 1 on 27-28 Apr 1987 at Wallula; male on 15 May 1993 se. of Soap L.; 1 on 25-26 May 2002 at Iowa Beef.

Scott A. Richardson

Semipalmated Plover *Charadrius semipalmatus*

Widespread, common migrant, locally rare to uncommon winter and summer visitor in w. Washington. Local, uncommon migrant, very rare summer resident in e.

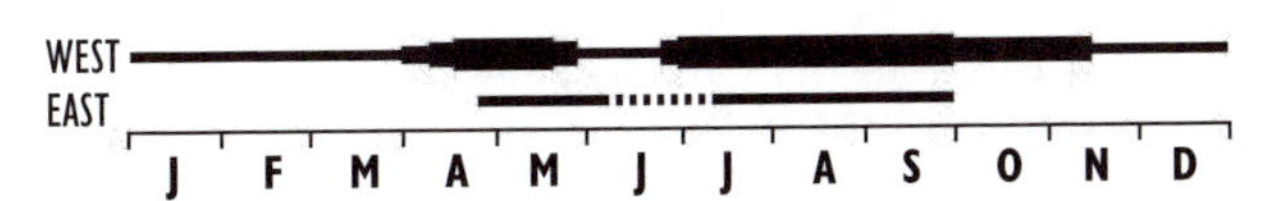

Habitat: Estuarine mudflats, coastal beaches, and freshwater shorelines.

Occurrence: This plover nests in arctic and subarctic regions (Paulson 1993, Nol and Blanken 1999) and as far s. as the Queen Charlotte Is. (Campbell et al. 1990b). It winters largely in coastal areas s. of Canada, to S. America (AOU 1998). In Washington

it is most abundant at Willapa Bay, Grays Harbor, and the adjacent coastal beaches.

First spring migrants arrive in w. Washington in early Apr (Widrig 1979) and numbers increase steadily (Widrig 1979, Herman and Bulger 1981). Birds are fairly abundant through the first week in May. Numbers drop off by mid-May. The species is far less abundant in spring at Puget Sound sites. Records of the first spring migrants in e. Washington occur later in the season and most records are from May. They are typically found in small numbers.

Fall migrants begin moving through in about mid-Jul (Widrig 1979) and, as in spring, are most abundant on the outer coast. The species is less common in Puget Sound, and even less so in e. Washington. Most migrants appear to move through the area by the end of Sep (Widrig 1979, Paulson 1993). Birds occur locally inland in migration, as at Elma in the Chehalis Valley (C. Chappell p.c.).

Winter high counts come from outer coastal beaches and estuaries, where dozens can be seen in some years (Buchanan 1992). Additional annual cumulative high counts there between 1986-87 and 1989-90 ranged from 14 to 36 (Buchanan 1992). Of about 40 CBC records in Washington, 27 are from Grays Harbor and Leadbetter Pt., with up to 34 at Grays Harbor in 1988. There are scattered records of wintering birds in Puget Sound (see Evenson and Buchanan 1997) with most records of single birds. There are no winter records from e. Washington.

Birds nested in at least two years at Ocean Shores, and displaying birds in other years suggests other attempts. A pair of territorial birds at Big Goose L., Okanogan Co., in 1991 suggests attempted nesting. There are a number of mid-Jun records at Willapa Bay and Grays Harbor, but no additional nesting records. Nesting has also been documented twice at Coos Bay, Oregon, in e. Oregon (Ivey and Baars 1990), and nr. Vancouver, B.C. (Campbell et al. 1990b).

The N. American population is estimated to be about 150,000 (Morrison et al. 2000).

Population status in the state is unknown but it appears that the winter range has expanded northward into Washington, as the species was not noted then by Jewett et al. (1953).

Remarks: Possibly conspecific with the Ringed Plover, *C. hiaticula* (Hayman et al. 1986), a species unrecorded in Washington (Paulson 1993). Classed PHS.

Noteworthy Records: *Outer coast, high counts, Spring*: 830 on 25 Apr 1981 at Grays Harbor sites (Herman and Bulger 1981); 264 in spring 1994 at Willapa Bay sites (Buchanan and Evenson 1997); 2000 on 3 May 1986 at Ocean Shores (Paulson 1993). *Summer*: 40 in Jun 1978 at Leadbetter Pt. (Widrig 1979); 30 at Grays Harbor in 1990. Fall: 1700 on 26 Jul 1978 at N. Beach (Paulson 1993); 525 on 25-26 Jul 1978 at Leadbetter Pt. (Widrig 1979), 1500 on 21 Jul 1998 at Midway Beach; 2320 on 1 Aug 1998 at Ocean Shores. *Winter*: 92 on outer beaches in 1988-89 (Buchanan 1992); 33 on 1 Dec 1978, at Leadbetter Pt. (Widrig 1979), 25 at Midway Beach on 20 Jan 1991. *Nesting*: 1973 and 1984 at Ocean Shores; territorial behavior at Ocean Shores in 1975, 1979 and 1980. *Inland marine waters, high counts, Summer*: 89 in 1992 at Dungeness Bay (Evenson and Buchanan 1997). Fall: 45 at Dungeness Bay in 1994 (Evenson and Buchanan 1997). Winter: 9 in 1987-88 in Puget Sound. **East**, *Spring, early*: 28 Apr 1987 at lower Crab Cr. and Soap L. (JBB); 30 Apr 1959 at Reardan (Johnson and Murray 1976). *Late*: 31 May 1975 at Richland; 4 Jun 1971 at Yakima R. delta. *High counts*: 30 in 1963 at Banks L. (Paulson 1993); 15 in 1977 at Yakima R. delta; 13 on 5 May 1968 at Reardan; 40 on 6 May 2000 at Toppenish. *Summer*: possibly nesting in 1991 at Big Goose L., Okanogan Co.; 1 there on 10 Jun 1995. *Fall*: 12 on 3 Sep 1951 at Twelve Mile Slough (Hudson and Yocom 1954); 10 in 1984 at Wallula. *Late*: 28 Sep 1974 at Richland.

Joseph B. Buchanan

Piping Plover *Charadrius melodus*

Casual vagrant.

One state record: an ad. ph. at Reardan on 13-16 Jul 1990 (Tweit and Paulson 1994). The only other record from the Pacific Northwest is one from coastal Oregon in Sep (Gilligan et al. 1994). Nearest regular occurrence is in s.c. Alberta. Classed GL.

Bill Tweit

Killdeer *Charadrius vociferus*

Locally widespread, fairly common to common year-round.

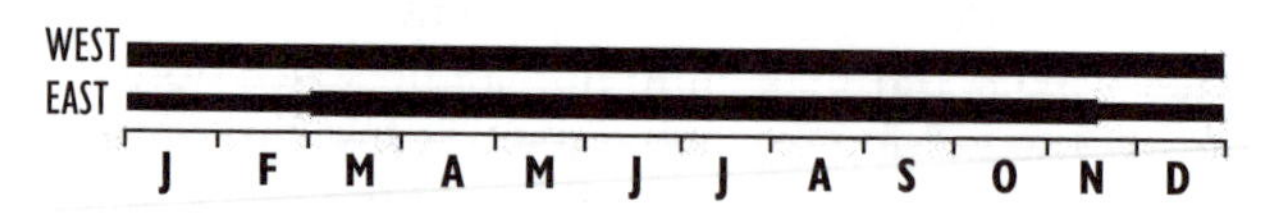

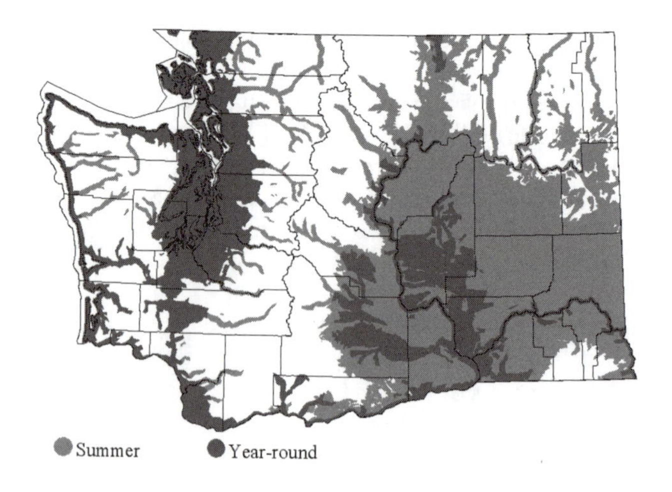

Subspecies: *C. v. vociferus.*

Habitat: Grass fields, pastures, meadows and wetlands, dry lakebeds and exposed shorelines, riverine gravel and sand bars, marine shorelines, gravel margins of unimproved and paved roads and airports.

Occurrence: Conspicuous vocalizations and behavior are likely responsible for the Killdeer to be considered more abundant than is warranted. It sometimes occurs in fairly large flocks but typically singly or in small groups. It occurs over most of N. America s. of the Arctic, to n. S. America (AOU 1998).

An early spring migrant in the region, although the timing of spring migration is not well documented due to the presence of a year-round, resident population and the lack of large concentrations of spring migrants. The breeding season commences long before most migrants have passed and broods of fledglings are present in Apr through Jun (Herman and Bulger 1981, Campbell et al. 1990b).

Movements of local nesters begin soon after the young fledge; local increases at estuaries are noted by Jun (Buchanan 1988c; Campbell et al. 1990b). Fall migration is from Jul-Aug through late Oct (Buchanan 1988c, Campbell et al. 1990b, Paulson 1993) although birds may relocate later due to weather conditions. High counts from e. Washington suggest that the species moves through the region in fairly large numbers.

Not abundant but widespread in winter at lower elevations throughout the state; recorded at every CBC location except Grand Coulee and at locations out of habitat (e.g., Mt. Rainier and N. Cascades). CBC totals in w. Washington averaged less than about 130 birds in 1974-1988 (see Paulson 1993), while much larger number occurred in Oregon's Willamette Valley, where three counts averaged 1240-4030 birds during 1974-1988.

The Killdeer is uncommon during winter in e. Washington (Harris and Yocom 1952, Johnsgard 1954) and often emigrates upon arrival of the first severe winter storms, possibly moving to the Willamette Valley. After these movements, most midwinter counts are low: the highest CBC average between 1968-69 and 1996-97 was 14.1 birds at Toppenish. CBC counts were consistently higher, and birds were present in more years at lower-elevation sites along the Columbia R. than at higher-elevation sites such as Ellensburg and Spokane.

BBS data indicate no population changes from 1966 to 2000 (Sauer et al. 2001) but with declines from 1980 to 1996 (-4.1%) over all physiographic regions statewide, ranging from -2.9% on the e. slope of the Cascades to -6.8% in the Columbia Plateau (Sauer et al. 1997). Reasons for the decline are uncertain and disconcerting as it is a generalist species. CBC data indicated no certain trend statewide, though there were apparent local declines at E. Lk. Washington and Tacoma. Classed PHS.

Noteworthy Records: West, *Winter*: 160-523 in winters between 1990-91 and 1995-96 at all inland estuarine sites combined (Evenson and Buchanan 1997). *CBCs, high counts*: 511 in 1976 at Tacoma, 339 in 1976 at Bellingham, and 312 in 1982 at Seattle. **East**, *Fall*: 2000 in early Sep 1969 at Turnbull NWR (Paulson 1993); 500 in fall 1973 at Turnbull NWR; 250 in late Oct 1969 at McNary NWR; 200 in late Jul, early Aug 1965 at Reardan. *CBCs*: 85 in 1999 at Toppenish, 57 in 1991 at Yakima, 114 in 1998 at Tri-Cities, 79 in 1981 at Walla Walla.

Joseph B. Buchanan

Mountain Plover *Charadrius montanus*

Casual vagrant.

A species that breeds on the Great Plains, and winters from c. California to e. Texas. Three Washington records: one collected at N. Cove on 28 Nov 1964 (Tweit and Skriletz 1996), one seen at Turnbull NWR. on 6 May 1968 (Aanerud and Mattocks 1997), and one at Ilwaco on 22 Dec 2000-6 Jan 2001. There are eight records, mostly from Nov to Mar, in w. Oregon. Classed GL.

Bill Tweit

Eurasian Dotterel *Charadrius morinellus*

Casual fall vagrant.

Eurasian Dotterels are scarce breeders in n. and w. Alaska, with other N. American records only from Washington and California (AOU 1998). California records were all coastal, with most between 6 and 20 Sep (Small 1994) except for one at Point Reyes from 17 Oct to 21 Nov 1992. Three records in Washington: an ad. female collected at Westport on 3 Sep 1934 (Brown 1934, 1935); an imm. (ph.) at Ocean Shores on 8 Sep 1979 (Paulson 1979); and one present (ph.) at Ocean Shores on 20 Oct-5 Nov 1999.

Steven G. Mlodinow

Black Oystercatcher *Haematopus bachmani*

Locally fairly common resident on exposed rocky shoreline and offshore islets on outer coast, San Juan Islands, and e. Strait of Juan de Fuca; often concentrate in small wintering flocks. Very rare in e. Washington.

Habitat: Feeding restricted to rocky marine intertidal shorelines; nests on open rock surfaces, mostly on non-forested islands, nearby rocks, or promontories.

Occurrence: A widespread, habitat-limited breeding resident along the Pacific coast s. to Baja California, local in Washington. Dawson and Bowles (1909) estimated a state population of 200, and Jewett et al. (1953) cited records that suggested at least this many or more frequented the outer coast. Nysewander (1977) counted 197 birds on the outer coast at Cape Disappointment and the n. rocky shoreline, and an additional 63-89 breeding in the inland marine waters, concentrated in the San Juan Is. and Smith/Protection Is. area. Rathbun (1915) reported oystercatchers as formerly not uncommon as summer visitors to s. Puget Sound, but this has not been seen in recent years.

Re-survey during May 2000 in the San Juan Is. found 79 birds, with at least 19 nests incubated, at 35 locations (DRN). Earlier surveys undoubtedly missed sites covered by the more intensive, standardized survey done in 2000, but in 1973-1980 the same area contained at least 90 oystercatchers (USFWS; Nysewander 1977) on 34 sites. During this 20-year period, many changes occurred in the San Juan Is., including increasing numbers of predators such as eagles, Peregrine Falcons, gulls, and river otters, and increasing numbers of people diving, fishing, boating in kayaks, or pursuing other activities close to breeding islands.

Breeding numbers in the inland marine waters have decreased slightly or remained relatively stable. Distribution may have changed somewhat; with larger concentrations, more have dispersed to smaller islands or rocks. It is likely that the remoteness and sea conditions have left populations on the outer coast relatively undisturbed.

Populations are essentially non-migratory. Groups of up to 75 non-breeding birds have been noted at times, generally in the proximity of their breeding grounds. Some local movement of non-breeders occurs: a bird banded in the Gulf Is., B.C., was sighted in the Deception Pass area in 2000 (S. Hazlitt p.c.), and small numbers on CBCs at locations such as Bellingham, Seattle, Skagit Bay and large numbers at Oak Harbor and Padilla Bay are often in habitats away from breeding areas.

Remarks: Classed GL, on NAS Watch List.

Noteworthy Records: *High counts, outer coast*: 51 in 1959 in Tatoosh I./Cape Flattery area, 40 in 1959 in Bodelteh/Ozette Is. area; 13 in 1974 in Norwegian Cr. area; 29 in 1974 at Cape Johnson vicinity; 36-40 in 1973-75 at Destruction I.; (all Nysewander 1977). *Inland marine waters*: 22 in 1988 at Protection I., 40 in Jun 1982 at Smith I. (WMNWR, USFWS); 79 in 2000 in San Juan Is./Deception Pass vicinity: Skipjack, Bare, Puffin, Little Sister, Chuckanut, Peapods, Flower, Cayou, Pointer, Bird, Small Hope, Williamson, Cone, Strawberry, Low (2), Ripple, Cactus, Gull, Barren, Sentinel, Goose, Deadman, Buck, Charles, Long, Hall Is., and Swirl Rk. (DRN). *MESA high counts*: 56 on 14 Jan 1979 at Clallam Bay; 41 on 4 Nov 1979 at Mummy Rks.; 35 on 13 Oct 1978 at Protection I.; 24 on 13 Mar 1979 nr. Friday Harbor; 16 on 9 May 1979 at Green Pt., Skagit Co. *CBC selected high counts*: 60 in 1994 and 61 in 1998 at Padilla Bay; 25 in 1980 and 1983 at Port Townsend; 23 in 1993 at Quillayute R.; 64 in 1995 in San Juan Is.; and 74 in 1998 at Oak Harbor. **East**: 1 on 8 Jan 1947 at Bumping L. Dam, Yakima Co., is the only Pacific Northwest inland record (Stepniewski 1999).

David R. Nysewander

Black-necked Stilt *Himantopus mexicanus*

Fairly common local summer resident in the Columbia Basin. Rare migrant, has nested in w. Washington.

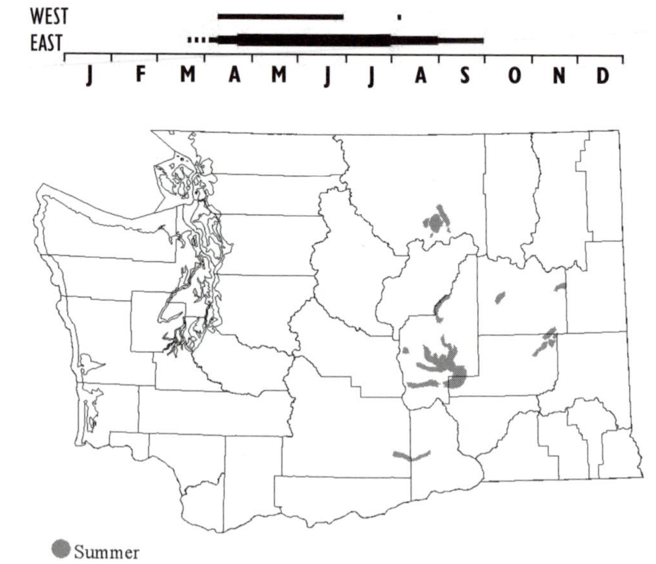

Subspecies: *H. m. mexicanus.*

Habitat: Breeds on lakes and ponds with marsh shorelines, often in association with American Avocets.

Occurrence: Breeding in scattered areas across the intermountain w., California, and e. areas, the estimated N. American population of 150,000 (Morrison et al. 2000) winters from the s. U.S. to n. S. America (AOU 1998). Stilts in Washington nest primarily in the Columbia Basin, where they have been recorded annually since 1977. Known areas include Potholes Res., Frenchman Hills Wasteway, nr. Othello, nr. George, Columbia NWR, Reardan, and Swanson Lks.

First documented nesting in Washington on 23 Jun 1973, with a single nest found sw. of Moses L. (Rohwer et al. 1979). Stilts were not reported again in the area until 1977, when nests were found at three localities (Rohwer et al. 1979). Birds have subsequently been reported nesting at several other locations and observed at numerous other sites, usually in small numbers. Rohwer et al. (1979) suggested that the colonization of these sites in the mid-1970s was the result of the presence of a dependable water source like the Winchester Wasteway during a period of extreme drought conditions throughout the intermountain w. Though numbers of breeding sites have increased, use of some of these sites has been intermittent. Population size and trends over recent decades are unknown, though nesting range expanded and numbers appeared much higher in the late 1990s (SGM). Stilts appear to be replacing avocets in some portions of the Columbia basin (BT).

Spring migrants arrive in late Mar-early Apr, and eastside numbers appear lower in Jun than in May (SGM). Fall migration is not obvious and the last migrants depart before the end of Aug. There are few records away from the known breeding locations during the migration period. There are nine records of this species in w. Washington, from San Juan I., Nisqually NWR, nr. Stanwood, at Bowerman Basin, Ridgefield NWR, Leadbetter Pt., and Seattle, and one nesting record at Ridgefield in 2001.

Remarks: Classed PHS in Washington.

Noteworthy records: West, *Spring*: 2 on 17 Apr 1981 at Nisqually NWR; 3 on 15 Apr 1988 at Nisqually NWR; 3 on 17 Apr 1984 nr. Stanwood: 4 on 24 Apr 1985 at Bowerman Basin; 1 on 21 Apr 1987 at Ridgefield NWR; 6 on 26 Apr 1987 at Leadbetter Pt.; 1 on 12 May 1988 at Montlake Fill, Seattle; 5 on 25 Apr 2002 Ridgefield NWR; 27 Apr 2002 at Stanwood; 7 on 9 Apr 2001 nr. Westport; 2 on 26 Apr 2001 at Crockett L.; up to 15 on 8-29 May 2001 at Ridgefield; 9 May 2001 nr. Sequim; 13 May 2001 at Redmond. *Summer*: 2 present most of summer 1988 at Nisqually NWR; 8 ad. hatched 3 young at Ridgefield in June 2001. *Fall*: 1 on 7 Aug 1977 at San Juan I. **East**, *early Spring*: 28 Mar 1994. *High count*: 33 on 10 May 1981 at Winchester Wasteway area, 47 there in 1984, 28 on 28 Apr 1987 (JBB), 45 on 16 Apr 1989 (Paulson 1993); 36 in 1985 at N. Potholes, 40 there in 1987: 21 in 1985 at McNary NWR; 16 in 1985 at Yakima R. delta, 28 there in 1987, 16 there on 13 Apr 1988; 12 in 1988 at Reardan; 100+ at Iowa Beef and 76 at Othello on 25 May 2001; 117 on 30 May 2002 at Othello. *Fall*: 6 Oct 2000 at Paterson Slough.

Joseph B. Buchanan

American Avocet *Recurvirostra americana*

Local, uncommon migrant and breeder in e. Washington. Local, rare migrant and very rare breeder in w.

Habitat: Alkaline ponds in summer. Ponds, intertidal flats, saltmarshes, flooded agricultural areas in migration.

Occurrence: Avocets are common only at several localities in the Columbia Basin, which is the nw. extent of their primary range (Paulson 1993). The first spring migrants in e. Washington typically arrive

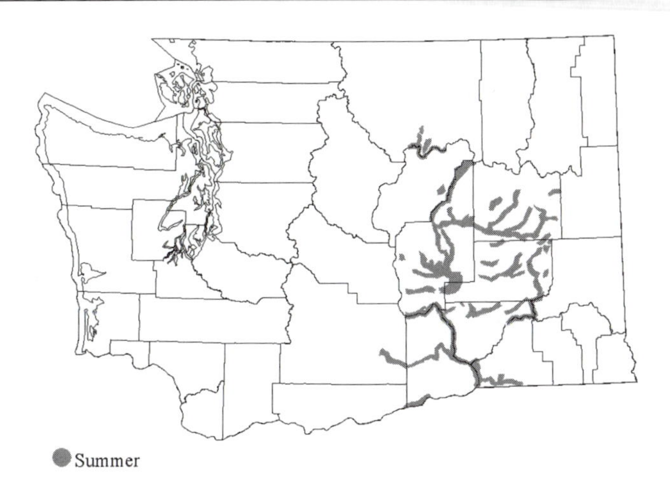

in Apr although there are records from Mar. Counts in e. Washington range as high as 80 birds. Regular sites include Yakima R. delta, ponds nr. St. Andrews, Alkali L. in Adams Co., and at Dodson Rd. and Frenchman Hills Rd. junction. They have been recorded in small numbers (<10) from numerous other sites including Moses L., Soap L., Duley L., L. Lenore, Turnbull NWR, Columbia NWR, and Reardan. They likely breed in most of the areas where they have been found. Fall migration takes place mainly in early to mid-Aug, and rarely into early Oct (Paulson 1993).

Vagrants, usually single birds or pairs, and rarely larger groups, occur in w. Washington and are found in ponds and on intertidal flats. Avocets have been recorded in every month (three to eight records each month) from Mar to Sep in w. Washington although there is only one record from Jul. Breeding was documented at Crockett L. in 2000.

The N. American population is estimated to be about 450,000 (Morrison et al. 2000). Based on limited BBS data, Sauer et al. (2000) indicate an increase in Washington. Extensive trend data are lacking, but it is possible that the population of this species has increased slightly in e. Washington due to the presence of a dependable water source associated with development of irrigation in the Columbia Basin. Data are required to assess whether water salinization is negatively influencing productivity or recruitment (Rubega and Robinson 1997). Classed PHS, GL.

Noteworthy Records: West, *Spring arrival*: 31 Mar 1988 at Seattle. *High count*: 39 on 28 Apr 1987 at Sequim Bay; 3 on 25 May 1987 at Lopez I.; 8-10 on 18 May 2000 at Crockett L. *Breeding*: 7 ads., 3 chicks on 6 Jul 2000 at Crockett L.; 2 pairs bred at Serpentine Fen, lower mainland B.C., in 1988 (Campbell et al. 1990b). *Fall, high count*: up to 15 on 13-20 Aug 1994 on Whidbey I. Late: 27 Sep 1986 at Nisqually NWR; 2 Oct 1997 at Steigerwald L. NWR. CBC: 1 in 1995 at Padilla Bay. **East**, *Spring arrival*: 5 Mar 1994, 19 Mar 2000, 4 on 20 Mar

1997, 2 on 20 Mar 1998 at Yakima R. delta; 21 Mar 1996 at Walla Walla R. delta. *High counts, Spring*: average of ca. 80 in 1991-95 at Columbia NWR (R. Hill p.c.); 80 in 1987 at Yakima R. delta; 50 on 19 Apr 1981 at Yakima R. delta; at least 50 on 13-14 Apr 1973 nr. St. Andrews; 48 on 25 Apr 1998 at Othello; 90 on 2 Jun 2001 at Wallula. *Summer*: 31 on 6 Jun 1997 at Walla Walla; 28 on 2 Jun 1992 at Alkali L. (JBB); estimated 35-40 pairs nested in 1950-51 at Potholes (Harris and Yocom 1952). *Fall*: 28 on 2 Aug 1998 at Iowa Beef; 25 on 17 Aug 1999 at Othello. Late fall: 11 Oct 1971 6.5 km w. of Othello (Verner 1974); 11 Oct 1999 at Yakima R. delta; 8 Oct 1989 at Benton; 4 on 8 Oct 1998 at Iowa Beef.

Joseph B. Buchanan

Greater Yellowlegs *Tringa melanoleuca*

Fairly common to common migrant, local uncommon winter visitor in w. Washington. Widespread, uncommon migrant e., and local, rare in the Columbia Basin in winter. Very rare in summer statewide.

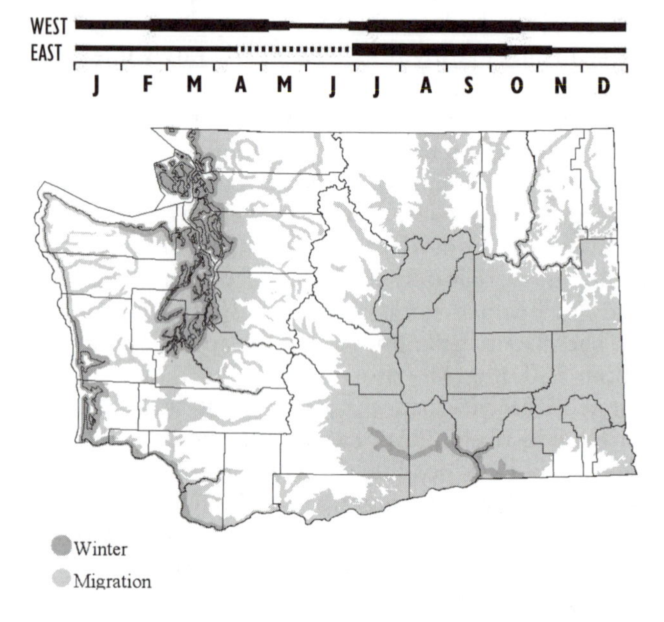

Habitat: Intertidal mudflats, freshwater habitats including small ponds, flooded fields, and riverine shorelines. Roosts in saltmarsh, on floating logs, docks and other structures (Paulson 1993).

Occurrence: First movement of spring migrants in w. Washington is in Feb-Mar (Buchanan 1988a), with peak passage in early to mid-Apr (Widrig 1979, Buchanan 1988b). Small numbers linger to mid-May, particularly at coastal estuaries (Widrig 1979, Herman and Bulger 1981). High regional counts of spring migrants come from Willapa Bay, Grays Harbor, and several sites in Puget Sound. The species is less abundant in e. Washington, where small flocks are most typical (Paulson 1993).

Table 22 . Greater Yellowlegs: CBC average counts by decade.

	1970s	*1980s*	*1990s*
Grays Harbor	12.1	20.4	32.6
Kitsap Co.	0.2	8.0	11.7
Leadbetter Pt	9.9	20.4	45.0
Olympia		30.8	33.2
San Juan Is.		18.4	12.3

Fall migrants begin arriving in late Jun-early Jul (Buchanan 1988b, Paulson 1993). Found in good-sized flocks, except in e. Washington, through the end of Oct. Fall migrants are generally found in the same areas used during spring.

In winter, the Greater Yellowlegs is most abundant at Willapa Bay and several sites in Puget Sound (Buchanan 1988b, Buchanan and Evenson 1997, Evenson and Buchanan 1997). It is generally uncommon throughout the remainder of w. Washington (Evenson and Buchanan 1997), with the northernmost "population" in the San Juan Is. (Lewis and Sharpe 1987). In e. Washington, it was recorded in 8 of 10 winters (13 records) between 1986-87 and 1995-96, mostly at Tri-Cities, Walla Walla area, and Yakima. Overall, there were at least 19 records in 14 of 26 years between 1970-71 and 1995-96, and all records were of single birds or small flocks.

Agitated birds during the breeding season at Muskrat L., Okanogan Co., suggested possible nesting there. The species breeds about 240 km n. of Washington, and an extralimital record is known from Oregon (Paulson 1993). Midsummer records come from Willapa Bay (Widrig 1979), Puget Sound (Buchanan 1988b), and several sites in e. Washington, including Muskrat L., where there are several records. Multiple records from 5 to 25 Jun in w. Washington involve non-breeders (Widrig 1979, Buchanan 1988b, Paulson 1993).

The N. American population is estimated at about 100,000 birds (Morrison et al. 2000). Long-term CBC data indicate increases at coastal locations, Olympia and Kitsap Co. (Table 22). Very small numbers on eastside CBCs increased slightly in the 1990s, perhaps indicating mild winter conditions. Classed PHS.

Noteworthy Records: *High counts,* **West**, *Winter*: 42-97 in 1992-93 and 1994-95 at Willapa Bay; ca. 35 at Eld Inlet; ca. 25 at Totten Inlet, 21 at Crockett L. (Buchanan 1988b, 1988c; Buchanan and Evenson 1997, Evenson and Buchanan 1997); 25 on 17 Jan 1971 at Westport; 88 at Nemah on 4 Jan 2001. *Spring, outer coast*: 117 on 27 Apr 1985 at Rialto Beach (JBB); 150 on 10 Apr 1987 at n. Willapa Bay; 262 on 29 Sep 1999 at Ocean Shores. *Inland waters*: 56 at Samish Bay (Buchanan 1988c); 154 on 9 Apr 1988 at Auburn; 79 on 20 Apr 1999 at Enumclaw.

Summer: 5 on 8 Jun 1974 and 17 Jun 1977 at Leadbetter Pt.; 2 on 23 Jun 1974 at Ocean Shores. *Fall, outer coast*: 285 on 3 Oct 1993 at Ocean Shores (SGM); 139 on 25 Oct 1980 at Ocean City SP (JBB). *Inland waters*: 74 at Drayton Harbor (Wahl 1995); 47 at Port Susan (Evenson and Buchanan 1997); 100 on 10 Aug 1996 at Deer Lagoon; 103 on 12 Sep 1996 at Olympia; 121 on 20 Aug 1998 at Eld Inlet; 115 on 5 Sep 1997 at Mud Bay. **East**, *Winter*: 5 on 3 Jan 1999 at Toppenish; 7 on 7 Jan 1984 at Asotin Cr. mouth; 7 on 28 Feb 1998 at Kennewick; 7 on 26 Feb 2001 at Yakima R. delta. *Spring*: 100 on 14 Apr 1957 at Four Lks. (Paulson 1993). *Summer*: 4 on 16 Jun 1969 at Columbia NWR; 4 on 21 Jun 1991 at Muskrat L., Okanogan Co.; 2 on 23 Jun 1998 at Yakima R. delta; 14 on 24 Jun 1998 at Sanpoil L. *Fall*: 19 on 26 Jun 1972 nr. Tri-Cities; 21 on 15 Jul 1999 at Columbia NWR; 25 on 16 Jul 1966 at Reardan (Paulson 1993); 23 on 17 Aug 1999 at Othello; 20 on 27 Oct 1998 at Yakima R. delta.

Joseph B. Buchanan

Lesser Yellowlegs *Tringa flavipes*

Widespread, uncommon spring migrant statewide, fairly common in fall migration. Very rare and irregular in winter, particularly in w. Washington.

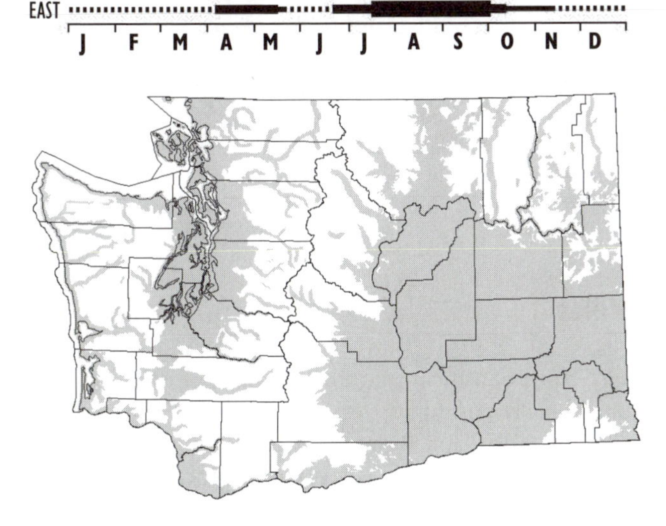

Migration

Habitat: Often in freshwater habitats—flooded fields and pastures, shorelines of shallow ponds and lakes, sewage-treatment lagoons—and estuarine areas though not often found on expansive tide flats (Paulson 1993).

Occurrence: Migration in Washington typically begins in mid-Apr and continues through about mid-May (Paulson 1993). There are several records of birds in early Apr. Small numbers generally occur in spring; flock size is typically less than about five birds although flocks of over 10 have been noted (Herman and Bulger 1981, Paulson 1993, Evenson and Buchanan 1997). Though the spring passage is later than that of the Greater Yellowlegs (Buchanan 1988a), the two species associate closely. Southbound migrant ads. generally begin arriving in late Jun-early Jul, followed by the arrival of juvs. in Jul through Oct (Paulson 1993). The species is more common in fall than in spring. Local concentrations of hundreds of birds have been noted in B.C. at sewage lagoons (Campbell et al. 1990b).

An irregular winter resident. Between 1967-68 and 1979-80 there were six reports of wintering birds (see Paulson 1993). Two midwinter records came from e. Washington, a region that supports very low numbers of shorebirds in winter. Subsequent winter reports included five during large-scale shorebird surveys in inland marine waters between 1990-91 and 1995-96 (Evenson and Buchanan 1997).

The N. American population is estimated at about 500,000 (Morrison et al. 2000). It is unlikely that the paucity of more recent field reports reflects a change in the abundance or occurrence of the species, but there is a lack of information necessary to adequately determine whether the status of the Lesser Yellowlegs in Washington, or elsewhere, has changed. Classed GL.

Noteworthy Records: West, *high count, Spring*: 20 on 3 May 2001 nr. Snohomish; 12 on 1 May 1999 at Skagit WMA; 10 on 24 Apr 1999 at Ebey I.; 10 on 16 May 1999 at Stanwood. *Fall*: 100 on 30-31 Aug 1988 at Kent sewage ponds; 65 on 11 Sep 1975 at Ocean Shores (Paulson 1993); 40 in 1991 at Port Susan (Evenson and Buchanan 1997); 30 in early Oct 1980 at Dungeness. *Winter*: 1 on 13 Dec 2002 nr. Sequim. **East**, *Winter*: 1 on 7 Feb1970 at Sportsman SP, Yakima Co. 1 on 31 Dec 1970 at Yakima R. delta; 2 in 1990-91 at Walla Walla (Cornell Lab. Ornith.); 18 Mar 1995 along Dodson Rd.; 22 Feb 2002 at Othello. *Spring, early*: 10 Apr 1954 at Potholes (Johnsgard 1954); 18 Apr 1999 at Dallesport; 9 Apr 1999 at Yakima R. delta; 24 Apr 1997 at Toppenish NWR (Stepniewski 1999). *High count*: 56 on 30 Apr 1993 at McCain Pond, Columbia NWR (R. Hill p.c.); 26 on 28 Apr 1994 at Columbia NWR (R. Hill); 22 on 5 May 1995 at Toppenish NWR (Stepniewski 1999). *Fall, early*: 20 Jun 1973 at Reardan; 23 Jun 1998 at Muskrat L., Okanogan Co.; 2 on 27 Jun 1998 at Atkins L. Late: 5 on 18 Nov 1998, 10-13 Nov 1995, 29 Oct 1997, 14 on 27 Oct 1998 at Yakima R. delta; 20 Oct 1957 at Reardan (Hall and LaFave 1958). *High count*: 150 on 17 Aug 1985 at Reardan; 65 on 22 Jul 1993 at Royal Slough, Grant Co.; 50 on 23 Aug 1989 at Yakima R. delta.

Joseph B. Buchanan

Solitary Sandpiper *Tringa solitaria*

Widespread uncommon spring and fall migrant in freshwater habitats statewide.

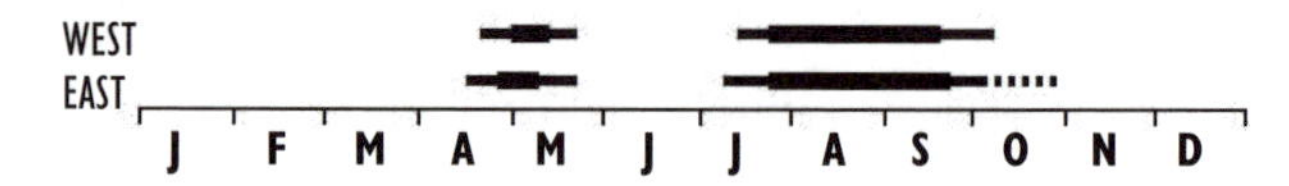

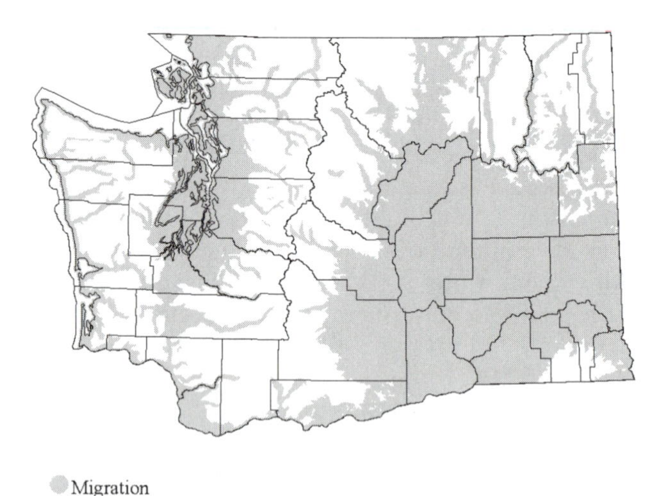

Migration

Subspecies: *T. s. cinnamomea* and *solitaria* in Washington (Paulson 1993).

Habitat: Many freshwater habitats, including sewage lagoons, flooded or muddy fields, and ponds in open or forested environments, often high in the mountains.

Occurrence: Spring migrants are present in Washington from about mid-Apr through May, typically moving through as singles or in small groups. The species has nested in both the c. Cascades in Oregon and in e. B.C. A record at Round L., Okanogan Co., on 24 Jun 1991 raised the possibility of nesting in the vicinity. This date is not extremely early for fall migrant shorebirds, especially after a failed nesting attempt, though the first migrant Solitary Sandpipers tend to arrive about mid-Jul. As suggested by Paulson (1993), observers should be on the alert for this species in small boggy lakes as it may eventually occur as a breeder in Washington.

As in other shorebird species, fall migration is much more protracted than the spring movement. Migrants typically begin moving through in mid-Jul (rarely earlier), peak in Aug, and appear to be scarce after Sep. The majority of records of multiple birds occur during Aug. Ads. generally migrate in Jul and juvs. follow later from late Jul through Sep (Paulson 1993).

The N. American population is estimated to be about 25,000 (Morrison et al. 2000). Trends in population size and distribution are unknown.

Remarks: Limited information suggests that both subspecies are equally abundant in the Pacific Northwest (Burleigh 1972, Paulson 1993). *Cinnamomea* breeds in Alaska and n. B.C., and might occur anywhere in Washington (Paulson 1993); *solitaria* breeds as far w. as e. B.C., and may be more likely in e. Washington (see Paulson 1993). Classed GL.

Noteworthy Records: West, *Spring, early*: 18 Apr 1998 at Ebey I.; 21 Apr 1998 at Nisqually NWR; 22 Apr 1998 at Olympia; 22 Apr 1998 at Woodland; 25 Apr 1997 nr. Elma. *Late*: 18 May 1975 at Burlington; 18 May 1999 at Ebey I.; 16-18 May 1993 at Seattle. *Fall, early*: 17 Jul 1999 at Bainbridge I.; 18 Jul 1999 at Seattle; 25 Jul 1977 nr. Monroe. *Late*: 8 Oct1995 at Spencer I.; 29 Sep 1999 at Edmonds; 27 Sep 1974 at McKenna; 2 Oct 2000 at Snohomish. High count: 5 on 25 Aug 1975 at Ridgefield. **East**, *Spring, early*: 15 Apr 1990 at W. Richland; 25 Apr 1998 at Royal City; 26 Apr 1998 at Toppenish NWR; 28 Apr 2000 at Columbia NWR; 30 Apr 1998 at Elk, Spokane Co. *Late*: 23 May 1999 at Ellensburg; 13 May 1999 at Cameron L., Okanogan Co. *High count*: 5 on 11 May 2002 at Rock L, Whitman Co. *Fall, early*: 25 Jun 2001 at Riverbend, Pend Oreille Co.; 1 Jul 1973 at Reardan; 5 Jul 1954 at Potholes (Johnsgard 1954); 11 Jul 1971 at Toppenish NWR (Stepniewski 1999); 12 Jul 1998 at Jameson L. *Late*: 26 Oct at Potholes (Harris and Yocom 1952); 4-7 Oct 1994 at Yakima R. Delta; 24 Sep 1988 at Grandview (Stepniewski 1999). *Fall, high counts*: 15 on 12 Aug 2001 at Lind Coulee, Grant Co.; 8 on 11 Aug 1999 at Tucannon R. Fish Hatchery, Columbia Co.; 8 on 18 Aug 1991 at Reardan; 4 on 17 Aug 1999 at Prosser; 4 on 1 Sep 1999 at Ellisford, Okanogan Co.

Joseph B. Buchanan

Willet *Catoptrophorus semipalmatus*

Rare to uncommon migrant in coastal estuaries, locally fairly common; very local, uncommon winter resident at Willapa Bay, very rare elsewhere in w. Washington. Very rare eastside migrant.

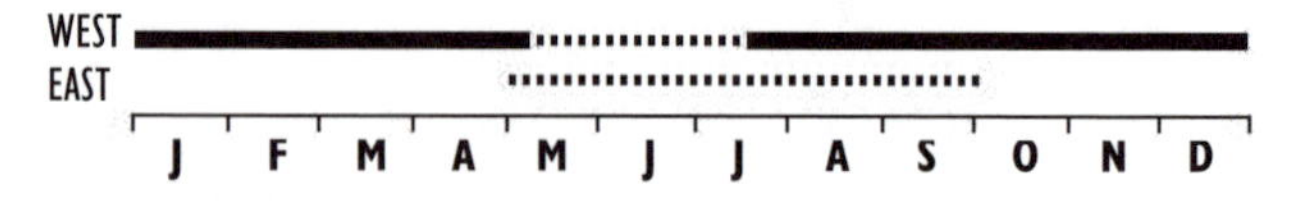

Subspecies: *C. s. inornatus* (Paulson 1993).

Habitat: Soft-sediment coastal estuaries and sandbars, shallow lakes and ponds.

Occurrence: Jewett et al. (1953) described the Willet as rare and with very few westside records and none from e. Washington. Birds are now regular in the w. and have been recorded in the e. in spring and fall and on three occasions in Jun. Westside seasonal records, including CBC occurrences, increased. In B.C., where the species was considered hypothetical in 1947, it is now rare to casual and has occurred as far n. as Fort St. John (Campbell et al. 1990b).

Primary winter distribution is between Tokeland and North R. on Willapa Bay with birds present at least 14 of 16 years between 1974-75 and 1989-90. The average count from years in that period with count data was 19, with a high count 35 in 1982-83. Single birds (two in 1982-83) have been reported at Ediz Hook in at least 10 winters between 1981-82 and 1995-96. There are few other winter records from w. Washington, and none from e. Washington.

Spring migration is in mid-Apr through May (Paulson 1993). Except at Willapa Bay migrants have been reported widely and are usually solitary. Most of the migrants in the interior have also been singles during May. There are no breeding records in Washington; there are few published records from the month of Jun. It is uncertain whether these were late spring or early fall migrants or non-breeding summer visitors. It is possible that Willets occasionally breed in e. Washington, particularly nr. small lakes and ponds in the Columbia Basin.

Fall records are numerous, with the species recorded annually in w. Washington. Early records range from early Jul, with most from Aug through Oct, and a peak in Sep. Many of these are from Willapa Bay; and several from nr. Port Angeles. Most of the remaining records are from Ocean Shores, and there are also records from Cape Flattery, Tatoosh I., Lummi Bay, and Budd Inlet. In addition, Widrig (1979) observed Willets at Leadbetter Pt. on eight dates between 26 Aug and 18 Sep 1978, with a high count of six on 10-11 Sep. Most of the few eastside records were in Sep.

Remarks: The origin of birds occurring in Washington may be from the ne. (e.g. Alberta or Saskatchewan), se. Oregon, where the species is common at Malheur NWR (Gilligan et al. 1994), or further e. The N. American population is estimated at about 250,000 birds (Morrison et al. 2000). Classed GL.

Noteworthy Records: West, *Spring*: 4 May 1968 at Bellingham; 30 Apr 1978 at Lopez I.; 10 May 1980 at Vancouver; 26 Apr 1984 at Bowerman Basin; 25 Apr 1981 at Ocosta and 30 Apr 1981 at Oyhut WMA (Herman and Bulger 1981); 2 Apr 1994 at Oak Bay; 26 May 2002 at Nisqually NWR. *Summer*: 11 Jun 1977 at Ocean Shores; 12 Jun 1976 at Leadbetter Pt. *Fall*: 1 Jul 1978 at Dungeness; 3 Jul 1986 at Tokeland; 1 on 18-25 July 1993 at Parego L., Whidbey I. *Winter, inland waters*: 22 Feb 1954 at Seattle (Eddy 1956); 25 Nov 1974 at Skagit Bay; 15 Dec 1995-13 Feb 1996 at Olympia; 9 Dec 1996 at Olympia. CBC: 1 in 1993 at Bellingham; 3 in 1961, 1 in 1995, 2 in 1998 at Olympia; 1 in 1989 at Sequim-Dungeness; CW in 1996 at San Juan Is. **East**, *Spring*: 4 May 1963, 5 May 1968, 24-25 May 1969, 21, 23 May 1972 at Banks L.; 15 May 1972 at Hanford AER; 12 May 1990 at Kahlotus L.; 22 May 1966, 8 May 1971, 3 May 1994 at Reardan; 11 May 1997 at Wallula; 31 May 1958 at Cow L., Adams Co. (LaFave 1959), 1986 at Yakima R. delta; 1987 at College Place; 1987 at Columbia NWR; 1988 at Othello; 26 May 2002 at Othello, 20 May 2000 at Swanson Lks., Lincoln Co.; 11 May 2000 at Sprague, 2 on 9 May 2001 at Walla Walla R. delta. *Summer*: 2 Jun 1951 nr. O'Sullivan Dam (Harris 1951); 4 Jun 1960 at Reardan (Johnson and Murray 1976); 17 Jun 1981 at Toppenish. Fall: 3 on 15 Aug 1979 at Yakima R. delta; 3 on 22 Sep 1976 at Turnbull NWR; 20-30 on 27 Sep 1973 nr. Wallula; 2 on 27 Sep 1990 at Walla Walla R. delta; 29 Sep 1976 at Wenas L.

Joseph B. Buchanan

Wandering Tattler *Heteroscelus incanus*

Uncommon to fairly common migrant, very rare winter visitor in w. Washington, very rare in e.

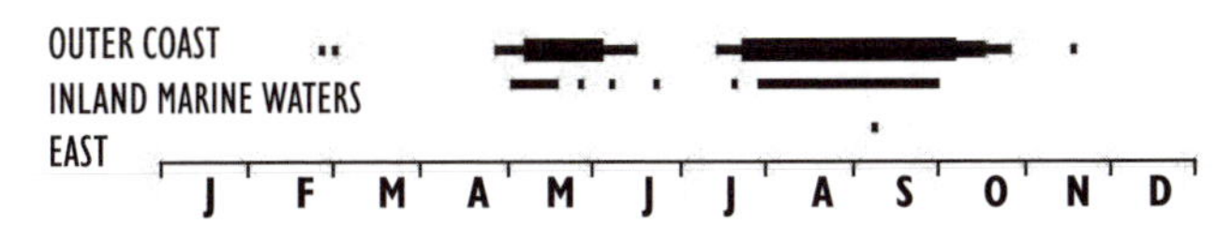

Habitat: Rocky shoreline areas and jetties; rarely uses other habitats (Paulson 1993).

Occurrence: Though widely spread during the year from s. Alaska s. through S. America and on w. Pacific islands (Paulson 1993, AOU 1998), the habitat-specific tattler is regular though local and thinly distributed in season in Washington. It is fairly common and occurs regularly along rocky shorelines and harbor jetties on the outer coast (Paulson 1993), less commonly along the e. Strait of Juan de Fuca, and it is rare in inland marine waters from Whatcom Co. (Wahl 1995) s. to Pierce Co. (e.g., Hunn 1982). During MESA surveys of n. inland marine waters in 1978-79 there were nine records, in May, Jul, and Aug, from Neah Bay e. to Dungeness, with two of these at Chuckanut Bay s. of Bellingham.

Spring migration is generally between late Apr and early Jun. Birds are usually encountered individually or in small flocks, but flocks of up to 30 have been reported. Although Jewett et al. (1953) considered Jun records from Flattery Rks. and Destruction I. to represent non-breeding birds (and see Paulson 1993), it is likely that many of these birds are late migrants (Nehls 1994).

Fall migrants typically begin arriving in coastal Washington in mid-Jul (ads.) and Aug (juvs.; Paulson 1993). The passage extends into Oct (Paulson 1993), occasionally into Nov. Given the winter distribution, it is not surprising that there are few records then. Single wintering birds have been reported from Ocean Shores, Tokeland, and Cape Flattery. Unusual fall records include one at Wenas Res. in e. Washington (Paulson 1993, Nehls 1994) and one along the Hamma Hamma R. in the se. Olympics.

The N. American population is estimated to be about 10,000 birds (Morrison et al. 2000), but little is known about status in Washington. Systematic surveys of the rocky shorelines of the outer coast, Strait of Juan de Fuca, and the San Juan Is. are desirable.

Noteworthy Records: West, *Spring, early*: 6 on 26 Apr 1986 at Ocean Shores. *Late*: 6 Jun 1986 at Iceberg Pt., Lopez I.; 11 Jun 1998 at Ocean Shores. *High count*: 30 on 11 May 1973 at Destruction I.; 25 on 16-17 May 1969 at Ilwaco; 15 in 1976 at Ocean Shores jetty. *Fall, early*: 7 Jul 1998 at Ocean Shores; mid-Jul 1974 at Strait of Juan de Fuca. *Late*: 12 Nov 1986 at Ocean Shores. *High count*: 15 on 7 Sep 1974 at Westport jetty; 10 on 16 Aug 1976 at Ocean Shores; Winter: 2 Feb 1964 at Birch Bay; 26 Feb 1977 at Ocean Shores; 23 Feb 1994 at Tokeland. 1 reported in 1966 on a CBC at Cape Flattery. *Mountains*: 1 about 15 km up the Hamma Hamma R. on the se. slope of the Olympics on 23 Aug. 1986. **East**, *Fall*: 3-5 Sep 1982 at Wenas Res.

Joseph B. Buchanan

Gray-tailed Tattler *Heteroscelus brevipes*

Casual vagrant.

One of several Palearctic shorebirds breeding in Siberia and wintering in Asia and Australia that are casual in Washington. There is one record—a juv. (ph.) at Leadbetter Pt. on 8 Sep 1975 (Tweit and Paulson 1994).

Bill Tweit

Spotted Sandpiper *Actitus macularia*

Fairly common migrant and summer resident statewide; uncommon in w. Washington and irregular in e. Washington in winter.

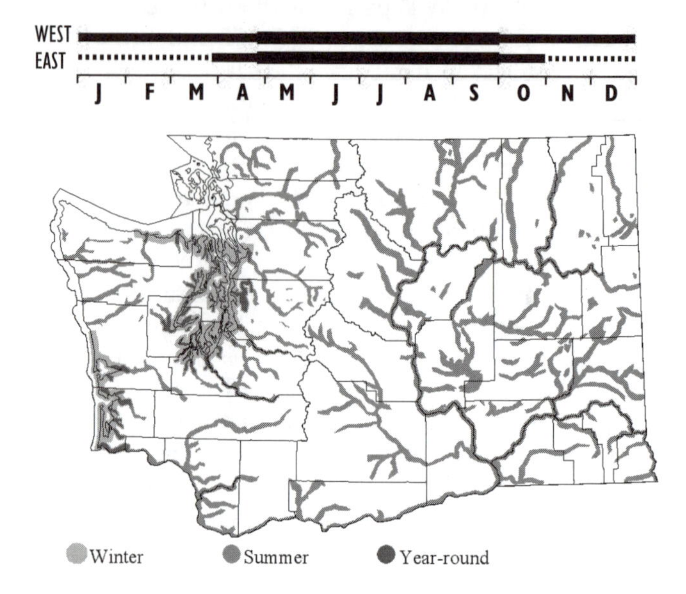

Habitat: Breeds in open areas along shorelines of lakes, ponds, rivers, streams, and freshwater wetlands in habitats ranging from shrub-steppe to coniferous forest (Paulson 1993). Nest substrate varies from low marsh vegetation to earth and cobble; presence of herbaceous cover appears to be essential (Paulson 1993, L. Oring p.c.). Migrants found in all breeding-season habitats, to lesser extent in tide flats and rocky shorelines (Paulson 1993).

Occurrence: Breeding across N. America, from the sub-Arctic s. over much of the U.S. and wintering from sw. B.C., s. to n. S. America (AOU 1998), a ubiquitous shorebird occurring essentially statewide in suitable habitat (Smith et al. 1997) from sea level to over 1800 m elevation (Cannings et al. 1987, Paulson 1993). Spring migrants begin arriving in late Apr or early May, fall migration is from Jul through Sep (Paulson 1993). Late migrants linger after that and small numbers winter in w. lowlands and protected estuaries (Buchanan 1988c). There are few recent winter records from e. Washington; similarly Cannings et al. (1987) reported only two records from the Okanagan Valley, B.C. Largest westside winter numbers evident in s. inland marine waters from CBCs in Kitsap Co., Seattle, Tacoma, and Olympia.

Although birds are widespread and relatively common, the population status is uncertain: numbers may be declining. The species may be locally impacted by changes in water management (Drut and Buchanan 2000), shoreline recreation, and perhaps, by grazing (Cannings et al. 1987). BBS data from 1966 to 2000 indicate conflicting trends at two different spatial scales: significantly increasing statewide (Sauer et al. 2001) but significantly decreasing in the Columbia Basin 1966-1996. No other trend data are available.

Noteworthy Records: West, *CBC high counts*: 45 in 1997 and 22 in 1985 at Olympia; 31 in 1991 at Kitsap Co.; 15 in 1978 at Port Townsend; 12 in 1967 and 1977 at Seattle; 19 in 1979, 17 in 1977, 16 in 1983, 14 in 1978 at Tacoma. **East**, *Winter*: 1 on 27-28 Jan 2001 and 3 Jan 2002 at Pasco, and 3 Feb 2001 at Prosser. CBC: 1 at Tri-Cities in 1976-77; 1 at Ellensburg in 1996-97. *High count*: 40 on 25 Jul at Prosser 1998.

Joseph B. Buchanan

Upland Sandpiper *Bartramia longicauda*

Very rare fall migrant, casual in spring. Formerly scarce and very local breeder.

Habitat: Nests in grasslands with areas of short sparse vegatation for feeding and areas of slightly denser and taller vegetation for nests. In Oregon, breeds in high-elevation (1000-1500 m) wet meadows within forest-dominated landscapes (McAllister 1995). During migration, found in a variety of open habitats with relatively short or sparse vegetation such as plowed fields, airports, golf courses, beach dunes, and sod farms.

Occurrence: The Upland Sandpiper was apparently never a well-established species in Washington. Breeding first noted at Stubblefield L., Turnbull NWR, in 1928 (Jewett et al. 1953). Almost all subsequent summer records have come from the Spokane Valley, with birds first found in 1929 and last seen in 1993 (McAllister 1995). Regular observations in this area from the mid-1950s on showed a very small but relatively stable population from the mid-1950s into the late 1980s, with annual totals varying from about two to 12 birds (McAllister 1995). During 1989-1993, only two to three birds were located annually, and none were found after that. Years with peak counts were 1980 and1955, with 11 and 12 birds, respectively (McAllister 1995). Small breeding populations persist in e. Oregon (Marshall et al. 2003), but this species is gone, or almost so, from Idaho. Declines have mostly been due to grazing and development, but the spread of spotted knapweed has also likely decreased the amount of suitable habitat (McAllister 1995).

Migrants are rare in Washington, with only about 17 records—nine from the westside and eight from the e. Thirteen were fall migrants and four were in spring. Fall dates range from 28 Jul to 6 Oct, with most between 18 Aug and 6 Sep. Spring records are from late Apr and late May. When the Spokane Co. colony was active, birds were reported there mostly from mid-May into Jul, but on one occasion, two ads. were seen as late as 24 Sep (McAllister 1995).

Westside records are largely from the outer coast, and all but three were during the 1990s. The reasons for this increase in reports is uncertain but is likely in part due to increased observer effort. The seven eastside migration records are widely scattered in both location and year.

Remarks: Classed SE, PHS, GL. Listed as Bird of Conservation Concern in 2002 (USFWS).

Noteworthy Records: *High counts*: 12 on 20 May 1956 and 9 on 18 May 1980 at e. Spokane Valley (McAllister 1995). *Migrants*, **West**: 6 Oct 1963 at Grays Harbor; 2 Sep 1977 at Ocean Shores; 29 May 1978 nr. Aberdeen; 19 Sep 1991 at Leadbetter Pt.; 27-30 Aug 1994 at Ocean Shores; 20 Aug 1995 at Leadbetter Pt.; 6 Sep 1997 at Ocean Shores; 18 Aug 1998 at Seattle; 22 Aug 1998 at Sequim. **East**: 22-23 Apr 1905 nr. Two Rivers (Dawson 1908); 29 Apr 1983 at Sprague L. (McAllister 1995); 28 May 2000 at Ephrata; 2 Aug 1929 at Turnbull NWR (Paulson 1993); 28 Jul 1948 nr. Touchet (Weber and Larrison 1977); 30 Jul 1980 nr. Ellensburg; 20 Aug 1988 at Reardan; 20 Aug 1998 at Loomis; 21 May 2003 nr. Lyons. Several other reports (McAllister 1995) are unreviewed and not included here.

Steven G. Mlodinow

Little Curlew *Numenius minutus*

Casual vagrant.

There are only six N. American records of this Asian species, four from California during fall-winter, one from Alaska during Jun, and one record in Washington, a bird at Leadbetter Pt., on 6 May 2001 (Mlodinow 2002).

Steven G. Mlodinow

Whimbrel *Numenius phaeopus*

Locally common spring migrant, fairly common to common fall migrant, uncommon local winter and non-breeding summer resident in w. Washington. Very rare migrant e.

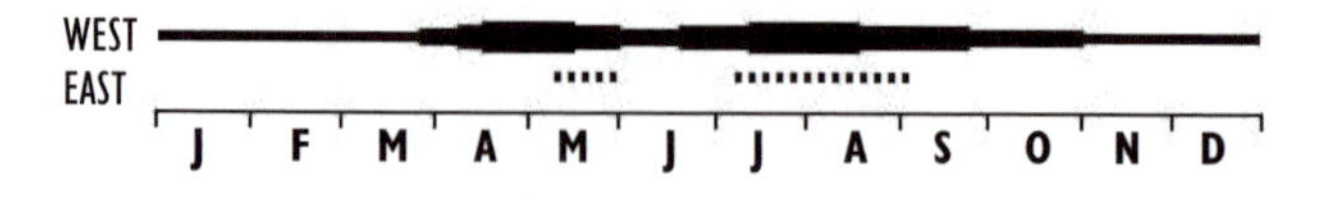

Subspecies: *N. p. hudsonicus.*

Habitat: Mud flats, sand beaches, dry or flooded fields, and even rocky shorelines (Paulson 1993).

Occurrence: Whimbrels breed in the Arctic and winter along coastal areas from the s. U.S. to n. S. America (AOU 1998). Though often seen in flocks of 10-15 birds, during eight years between 1990 and 1998 the average annual large flock size reported during spring migration was 448. Almost without exception, these flocks were in river valley agricultural areas (e.g. Chehalis R. Valley, fields near Mt. Vernon and in lowland Whatcom Co.) often removed from marine estuaries. Fall flocks appear to be slightly smaller and high counts are almost always from marine areas like Ocean Shores, Tokeland, and Leadbetter Pt., rather than agricultural areas and it is believed that Whimbrels use a more coastal migration route during fall (Paulson 1993).

Jewett et al. (1953) did not report Whimbrels as either summer or winter residents. Birds are now regularly encountered during winter (reported in at least 13 of 20 years between 1976-77 and 1995-96; Buchanan and Evenson 1997, Evenson and Buchanan 1997) at sites such as Tokeland and Ediz Hook and inland along the Chehalis R. From one to three birds occurred on 11 CBCs from 1976 to 1997 at Grays Harbor, Leadbetter Pt., Oak Harbor, Olympia, Sequim-Dungeness, and Kitsap Co. Birds are regularly encountered throughout Jun at locations such as Leadbetter Pt. and Ocean Shores (Widrig 1979, Paulson 1993).

Population trends unknown, but recent regular winter residency and occurrence of rather large flocks suggests that birds may now be more common in w. Washington than was indicated by Jewett et al. (1953). Scarcity in e. Washington reflects the strong marine nature of the species; it is rarely recorded in interior Oregon (Nehls 1994) and the Okanagan Valley in B.C. (Cannings et al. 1987) and Burleigh (1972) did not record it from Idaho. There were about 14 records from May to Aug 1951-2000 in e. Washington. Thirteen of these were of single birds, and one was of three. Classed PHS, GL.

Noteworthy Records: *High counts, Spring, outer coast*: 275 nr. Satsop, 5 May 1991; 450 nr. Elma, 17-18 May 1990 and 28 Apr 1996; 350, Ocean Shores, 26 Apr 1998. *Inland waters*: 832 on 3 May 1994 nr. Padilla Bay (Evenson and Buchanan 1997), 750 on 5 May 1990 at Skagit Flats. *Summer*: 60 on 8 Jun 1978 at Leadbetter Pt. (Widrig 1979), 38 on 8 Jun 1974 at Leadbetter Pt. *Fall, outer coast*: 950 on 22 Jul 2000 at Tokeland; 600 on 13 Jul 1988 at Ocean Shores. *Winter*: 9 in 1993-94 at n. Willapa Bay (Buchanan and Evenson 1997). **East**, *Spring*: 1 on 6 May 1973 at White Swan; 1 on 17 May 1986 at Thorpe; 1 on 25 May 1969 and 1 on 21-23 May 1972 at Banks L.; 1 on 21 May 1988 at Walla Walla R. delta; 1 on 25 May 2001 at Othello. *Summer*: 1 on 11 Jun 1951 at Potholes (Harris and Yocom1952); 1 on 21 Jun 1982 at Spokane Valley. *Fall*: "several" on 1 Jul 1973 nr. Touchet; 1 on 5 Jul 1975 at Dodson Rd.; 1 on 14 Jul 1997 atWalla Walla R. delta; 3 on 30 Aug 1987 at Walla Walla R. delta;1 on 11 Aug 2000 at Kahlotus L.; 1 on 29 Aug 1999 at Atkins L.

Joseph B. Buchanan

Bristle-thighed Curlew *Numenius tahitiensis*

Casual vagrant.

Breeding in w. Alaska and wintering in the tropical Pacific, this curlew is at best casual in Washington. The first accepted record was of one at Leadbetter Pt. on 1 May 1982 (Widrig 1983, Tweit and Paulson 1994). Widrig had reported an earlier sighting of two in the same locale in May 1980, but with insufficient substantiating details.

Subsequent state records occurred during an extraordinary landfall in May 1998 in w. N. America, with birds noted between c. California and n. Washington. Mlodinow et al. (1999) reported that in May 1998 between four and 10 individuals were found in Washington; two on Tatoosh I. and the remainder in the Westport/Ocean Shores area. The likely cause of the landfall was a climate anomaly associated with the W. Pacific Oscillation, creating extremely strong headwinds for northbound migrants over the nc. Pacific (see Mlodinow et al 1999). The 1998 records were from 8 to 24 May.

Bill Tweit

Long-billed Curlew *Numenius americanus*

Fairly common spring and summer resident in e. Washington. Local, fairly common winter resident at Grays Harbor and/or n. Willapa Bay; uncommon spring and fairly common autumn migrant and very rare summer resident in w.

Subspecies: *N. a. parvus* (Paulson 1993).

Habitat: Nest in a variety of grassland habitats (Allen 1980), and apparently capable of using certain degraded or exotic habitat conditions depending on the vegetation structure (Paulson 1993). They generally nest on level ground. Winter in soft-sediment estuaries and, in some cases, outer coastal beaches.

Occurrence: This species' regional status has changed substantially. Numbers declined rapidly in the vicinity of Prescott in the early 1900s (Dice 1918), and similar declines were later reported for much of e. Washington (Yocom 1956). Declines have been reported for much of the species' range (Page and Gill 1994); Brooks (1918) and Burleigh (1972) reported declines in the Okanagan Valley of B.C., and the Idaho panhandle region, respectively. These were attributed to conversion of grassland habitats to agricultural crops. Washington breeding distribution is centered in the Columbia Basin and adjacent areas including a small area of the Okanogan Valley near the Canadian border, e. portions of Kittitas and Yakima Cos., and much of the s. fringe of Klickitat Co. (Smith et al. 1997).

The species was apparently first reported as a winter resident in coastal Washington in 1970, when three were seen at Tokeland. It is now regular there and at Ocean Shores, with over 20 typically present at one of these locations. The establishment of the Tokeland area flock coincides with occurrence there of Marbled Godwits. This presence far n. of historical, contiguous winter range (from Humboldt Bay s. [Paulson 1993]) is unexplained, but may be due to localized prey abundance. Birds are rare there in summer, and are rare migrants and very rare summer and winter visitors elsewhere on the westside. Fall occurrence is generally greater than in spring. In the Puget Trough there were only about eight records from 12 Apr to 20 May, from 1971 to 1996, seven records from 9 Aug to 19 Sep, from 1969 to 1999, three winter records, and two in Jun. These were single birds except two and three birds on one occasion each.

Yocom (1956) suggested that curlews had recolonized historic areas, speculating that the rebound may have been in response to abandonment of agricultural practices at failed homesteads in favor of grazing. BBS data suggest stable or increasing numbers in the Columbia Plateau since 1966 (Sauer et al. 1997), and that the population has been stable for more than three decades (see Palmer 1967), although at a lower level of abundance. More recently, declines in abundance have been reported in the intermountain w. likely due to habitat loss (Paulson 1993, Oring et al. 2000).

Remarks: Occurrence of *N. a. americanus*, which occurrs s. of *parvus*, in Washington, is not documented (AOU 1957, Paulson 1993). The BBS technique may not be suitable for this species (Howe et al. 1989, Page and Gill 1994). Classed as PHS, in past a monitor endangered species in Washington, a candidate FE by USFWS ; GL and on NAS Watch List.

Noteworthy Records: West, *high counts, outer coast, Spring*: 17 on 12 Apr 1986 at Tokeland; 8 in spring 1993 in n. Willapa Bay (Buchanan and Evenson 1997). *Fall*: 110 on 30 Aug 1989 at Long Beach Pen. (Paulson 1993); 80 on 18 Aug 1987 at Tokeland (Paulson 1993); 80 on 3 Oct 1997 at Ocean Shores. *Winter*: 80 on 26 Dec 1995, 78 on 6 Feb 1983 at North R./Tokeland. *CBC*: Recorded 8 times between 1986 and 1999 at Grays Harbor, with 100 birds in 1992. **East**, *Spring, early*: 7 Mar 1997 at Walla Walla R. delta; 8 Mar 1999 at Columbia NWR. *High counts, Spring*: 60 on 10 May 1972 at Umatilla NWR; ca. 50 on 25 May 1976 s. of Ephrata; 150 on 30 May 1968 at McNary NWR: 50 pairs in 1968 at Columbia NWR. *Summer*: 80 on 19 Jun 1968 at McNary NWR; 30 on 24 Jun 1988 nr. Lowden; breeding population of about 300 birds at Hanford Site and adjoining Wahluke Slope in the late 1970s (Allen 1980). *CBC*: 5 in 1973 at Walla Walla.

Joseph B. Buchanan

Hudsonian Godwit *Limosa haemastica*

Very rare migrant.

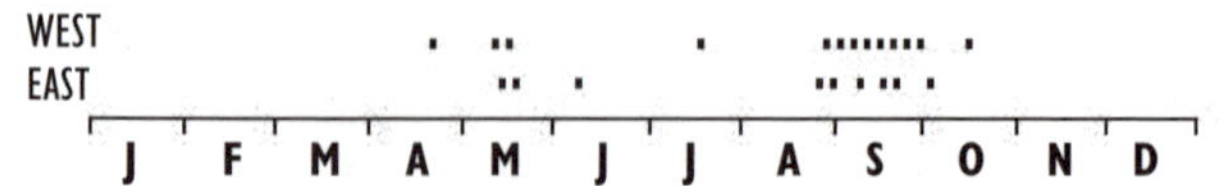

First recorded in Washington in Sep 1959 at Potholes Res. (LaFave 1960a). Since then, 24 more have been found, with sightings apparently increasing in frequency, possibly due to growing observer effort. Of the 26 birds seen, 17 have occurred between late Aug and mid-Oct. Most were juvs. The others were found during late Apr, two were in mid-late May, one each was in early Jun and late Jul.

Fourteen records were w. of and 10 were e. of the Cascades. On the westside, 10 were on the outer coast and four were in the Greater Puget Sound region. The pattern in Washington is similar, temporally and geographically, to that of the rest of the Pacific Northwest, though Washington has a higher proportion of inland records (Paulson 1993). Classed GL.

Records: West, *Spring*: 2 on 12 May 2002 at Bottle Beach; 14 May 1977 at Ocean Shores (Roberson 1980); 26-28 Apr 1992 at Ocean Shores. *Fall*: 1 on 26 Aug 2001 at Blaine; 3 on 24 Sep 1966 at Ocean Shores (Mattocks et al. 1976); 6-24 Sep 1975 at Aberdeen; 31 Aug 1983 at Samish Flats; 8 Sep 1990 at Ocean Shores; 8-13 Sep 1990 at Leadbetter Pt.; 12 Sep-4 Oct 1992 at Ocean Shores; ad. on 20 Jul 1996 at Crockett L. (SGM, R. Rogers, G. Toffic); 27 Sep-12 Oct 1996 at Blaine; 11-14 Sep 1997 at Blaine; 27 Aug 1999 at Ocean Shores; 26 Sep-10 Oct 1999 at Tokeland. **East**, *Spring*: 11 May 2002 at Texas L., Whitman Co., 15 May 2003 at Grandview; 19-21 May 1997 at Turnbull NWR; 8 Jun 1987 at Yakima R. delta. *Fall*: 12 Sep 1959 at Potholes Res. (LaFave 1960a); 15 Sep 1961 at Reardan; 2 Oct 1983 at Soap L.; 25 Aug 1984 at Reardan; 25-29 Aug 1992 at Othello; 31 Aug-6 Sep 1997 at Walla Walla R. delta; ad. on 21-22 Sep 1997 at Swanson Lks., Lincoln Co. Two published records not accepted by WBRC due to inconsistent or insufficient details: 3 on 24 Sep 1966 at Ocean Shores (Paulson 1993) and 9 Sep 1990 at Leadbetter Pt. (Tweit and Skriletz 1996).

Steven G. Mlodinow

Bar-tailed Godwit *Limosa lapponica*

Rare fall migrant and very rare spring migrant in w. Washington.

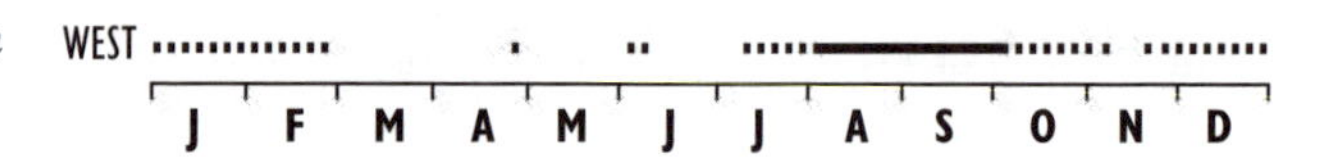

Subspecies: *L. l. baueri*. Reports of European, *lapponica*, probably refer to pale *baueri*.

Occurrence: First recorded in Washington on 4 Sep 1973 at Ocean Shores (Wahl 1973) with approximately 44 subsequent records through 2001. Most were during fall migration along the outer coast, especially on the n. shore of Willapa Bay. The 40 records of southbound birds span 11 Jul-21 Nov, peaking between early Aug-early Oct. Fall birds found only seven times found away from the outer coast: four times in Clallam Co., once in Tacoma, once at Kennedy Cr., in Mason Co., and once at Blaine. There have been only five spring records. The lack of pre-1973 records may be in part due to lack of awareness by observers of possible occurrence. Sightings have increased since then, from three records during the 1970s, to 12 in the 1980s, and 20 in the 1990s. One bird wintered at Tokeland in 2001-02.

Birds most often occur with Marbled Godwits, though some have been alone or with Black-bellied Plovers.

Remarks: The photograph of a bird at Neah Bay on 2 Jul 1974 was considered inadequate to establish identification (Paulson 1993). Classed GL.

Noteworthy Records: *Spring*: 21-28 Apr 1991 at Ocean Shores; 27 May 1998 at Ocean Shores; 8 Jun 1974 at Leadbetter Pt. (Paulson 1993); 10 Jun 1980 at Dungeness (Paulson 1993); 10 Jun 1983 at Bay Center. *Fall, early*: 11-12 Jul 1996 at Ocean Shores; 11 Jul- Aug 1998 at Tokeland. Late: until 19 Nov 1995 at Tokeland; until 21 Nov 1998 at Ocean Shores. Winter: 1 in 2001 at Tokeland. *High count*: 6 on 8 Jun 1974 at Leadbetter Pt. (Paulson 1993).

Steven G. Mlodinow

Marbled Godwit *Limosa fedoa*

Fairly common migrant and winter resident at Willapa Bay and Grays Harbor; rare elsewhere in w. Very rare migrant in e.

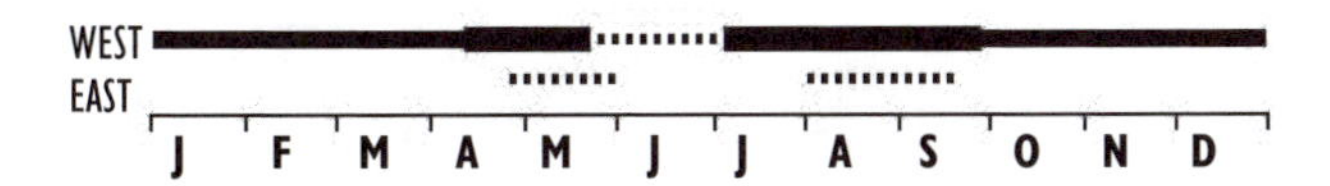

Subspecies: Birds wintering in Washington appear to be *L. f. beringiae* (Gibson and Kessel 1989); interior migrants may be nominate *fedoa* (Paulson 1993).

Habitat: Typically associated with tidal mud flats, but also use grass areas (Paulson 1993), coastal beaches (Buchanan 1992) and shorelines of shallow lakes, reservoirs, and river deltas in the Columbia Basin.

Occurrence: Sw. coastal Washington is the northernmost wintering area of this species which breeds in sw. Alaska ne. to James Bay (Gibson and Kessel 1989, Paulson 1993). Coastal Washington birds are thought to be from the small Alaskan breeding population. Large flocks are noted in spring, most often at Ocean Shores and Bottle Beach, less frequently at Willapa Bay. Migration is from Apr through May (Paulson 1993). A recent change in fall high counts at Willapa Bay—an increase from about 200 to 950 since 1990—mirrors the spring pattern. Most migrants move through between Jul and Oct, with ads. first and juvs. in Sep (Paulson 1993).

Birds present after Nov likely overwinter. Most winter records are from n. Willapa Bay between Tokeland and the mouth of the Willapa R., where numbers have increased dramatically from about 50 to 800 over the last three decades. Godwits typically roost at Tokeland, move e. on the falling tide, and return on the rising tide. These movements are usually made as a single large, sometimes loosely organized, group. Birds are fairly common winter residents at Willapa Bay and Grays Harbor, increasingly at Sequim/Dungeness, and rarely elsewhere at westside sites like Skagit Bay. There are no eastside winter records (Paulson 1993) and birds are rare in the Pacific Northwest interior (Nehls 1994).

Marbled Godwits occasionally noted in Jun along the outer coast are typically first-year birds that have ceased migration (Paulson 1993). Birds observed in very early or very late Jun, however, may be either unusually late spring or early fall migrants.

There are few spring records from e. Washington which lies slightly n. of the primary flyway linking coastal wintering areas and n. Great Plains breeding areas. Fall migrants are very rare, though more frequent than in spring. Birds are widely recorded, but averaged less than one record per year during the last four decades. Most records were from Reardan, and Yakima and the Walla Walla R. deltas.

Jewett et al. (1953) indicated the species had declined since the early 1900s. This may have been due to unregulated hunting and perhaps habitat loss on the breeding grounds (Paulson 1993, Page and Gill 1994). Apparent rarity or absence of a wintering flock at Willapa Bay before the 1960s may reflect low observer effort, but recent numbers likely represent a population increase, an expansion, or range shift of a portion of the regional population. Recent increases have also been noted in coastal California (Shuford et al. 1989). The continental population is estimated to be about 171,500 (Morrison et al. 2000). The magnitude of breeding habitat loss in the n. Great Plains may prevent the species from recovering its former abundance (Paulson 1993, Page and Gill 1994). Classed PHS, GL.

Noteworthy Records: *Outer coast, high counts, Winter*: 950 on 18 Jan 2002 at Tokeland; 462 in 1994-95 in n. Willapa Bay (Buchanan and Evenson 1997). *Spring*: 600 on 21-28 Apr 1991 at Ocean Shores. *Late Spring-early Fall*: 29 May 1970 at Tokeland; 17 Jun 1975 at Leadbetter Pt.; 10 on 28 Jun 1975 and 24 on 1 Jul 2000 at Ocean Shores. *Fall*: 850 on 7 Oct-15 Nov 2001 at Tokeland; 390 on 13 Sep 1997 at Raymond. *Inland marine waters, Spring, late*: 2 on 28 May 1988 at Sequim. *Fall, early*: 8 on 25 Jun 1999 at Sequim. *Late*: 30 on 17 Dec 2001 at Dungeness Spit. *High count*: 60 on 25 Sept 2001 nr. Sequim. **East**, *Spring*: 26 Apr 1997 and 6 on 3 May 1997 at Walla Walla R. delta; 9 May 2000 at Columbia NWR; 12 May 2000 at Swanson Lks. *Late*: 2 on 3 Jun 1951 at Potholes (Harris and Yocom 1952). *Fall, early*: 4 on 4 Jul 1979 and 6 on 11 Jul 1988 at Yakima R. delta; 2 on 5 Jul 1997 at Walla Walla R. delta. *Late*: 18 Sep 1976 at Calispell L.; 4 on 17 Sep 1997 at Yakima R. delta; 16 Sep 1969 at Reardan; 6 Sep 1958 at Cow L., Adams Co. (LaFave 1959). *High count*: 20 on 1 Sep 1997 at Yakima R. delta; 10 on 29 Jul 1984 at Walla Walla R.delta.

Joseph B. Buchanan

Ruddy Turnstone *Arenaria interpres*

Common spring, fairly common fall migrant in w. Washington; casual spring migrant and very rare fall migrant in e. Rare in w. during winter.

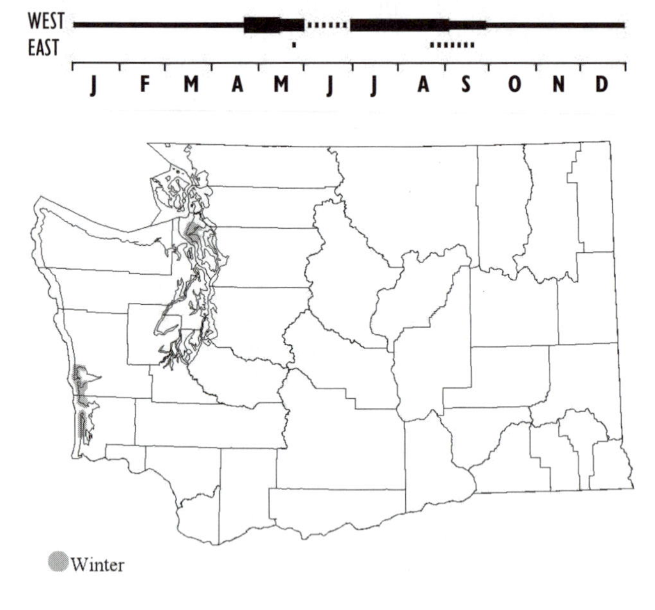

Subspecies: *A. i. interpres.*

Habitat: Rocky shorelines including jetties and bulkheads, coastal sand beaches, mud tide flats, and salt marsh (Paulson 1993).

Occurrence: Ruddy Turnstones are found in greatest numbers along the outer coast at sites such as Ocean Shores and Leadbetter Pt., and the Strait of Juan de Fuca. They are locally common in the San Juan Is. (Lewis and Sharpe 1987), less common in n. Puget Sound (Wahl 1995), and casual in s. Puget Sound (Buchanan 1988c). Most winter records are from more-protected areas away from the outer coast (e.g., Strait of Juan de Fuca, Whidbey I.). Winter scarcity is similar in coastal B.C. (Campbell et al. 1990b) and Oregon (Gilligan et al. 1994, Contreras 1997). They are usually found in small groups during migration but occasionally occur in flocks of 100 or more birds.

Most eastside records are of single birds, with one recent record of four. With the exception of one spring record, all eastside records are from 17 Aug to 20 Sep, similar to the range of dates reported from the Okanagan Valley, B.C. (Cannings et al. 1987). Records from e. Oregon span the period 3 Aug-10 Sep (Anderson 1988, Anderson and Bellin 1988, Nehls 1994) and there appear to be more spring records from that state (Gilligan et al. 1994).

Although population status is uncertain, it likely has changed little, if any, over the past 50 years in Washington. Between 1960 and 1999 there were 28 reports on CBCs, all in w. Washington, from Sequim-Dungeness s. to Tacoma and w. to Grays Harbor and Leadbetter Pt.

Remarks: Flocks should be scrutinized to determine whether, or to what extent, *morinella* occurs in the region (see Paulson 1993). In particular, records from interior areas of B.C., Washington and Oregon should be documented by photographs.

Noteworthy Records: West, *high counts, Spring*: 490 on 9-10 1981 at N. Bay (Paulson 1993); 387 on 11 May 1986 in 20 km of Ocean Shores beach (JBB); 200 on 13 May 1984 at Ocean Shores (Paulson 1993), 174 on 15 May 1983 at Willapa Bay and Leadbetter Pt. *Fall*: 400 on 21 Jul 1998 at Grayland Beach; 200 on 18 Jul 1998 at Ocean Shores; 150 on 31 Aug 1979 at Leadbetter Pt. (Paulson 1993); 60 on 2 Aug 1988 at Dungeness (Paulson 1993); 50 on 21 Jul 1986 at Dungeness; 42 on 10-11 Sep 1978 at Leadbetter Pt. (Widrig 1979). *Winter*: 25 in 1992-93 at Dungeness (Evenson and Buchanan 1997); 7 on 10 Jan 1993 at Penn Cove (SGM). **East**, *Spring*: 26 May 1957 at Cow L., Adams Co. (Hall and LaFave 1958). *Fall, early*: 17 Aug 1997 at West Medical L.; 2 on 20 Aug 2000 at Gap Rd. Pond, Benton Co.; ad. male on 26 Aug 1996 and ad. male on 29 Aug 1997 at Walla Wall R. delta (M. Denny p.c.). *Late*: 15 and 18 Sep 1957 at Reardan (Hall and LaFave 1958); 19-20 Sep 1965 at Reardan.

Joseph B. Buchanan

Black Turnstone *Arenaria melanocephala*

Regular, local, fairly common to common migrant and winter resident along marine shorelines. Rare in June. Very rare on eastside.

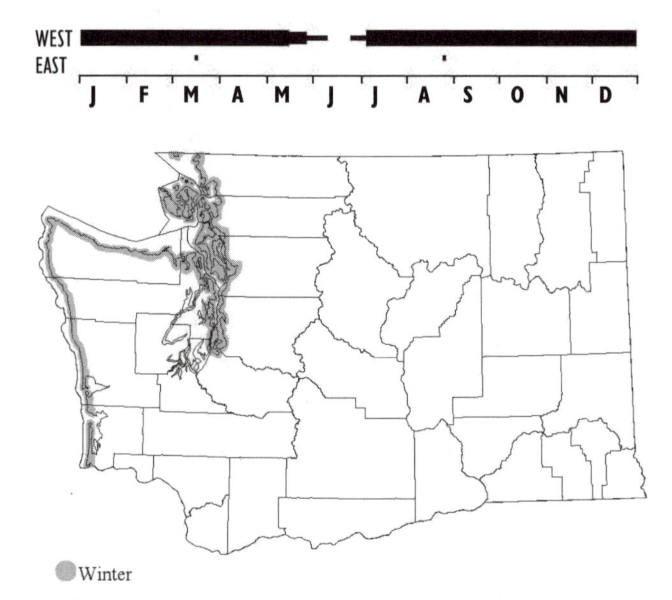

Habitat: Rocky shoreline habitats, including rock headlands and offshore rocks, gravel shorelines and jetties; occasionally sand beach habitats, mud flats, kelp beds.

Occurrence: The Black Turnstone breeds in w. Alaska and winters coastally s. to n. Mexico (Paulson 1993, AOU 1998). In Washington it is most common on CBCs at Leadbetter Pt., Seattle, and Sequim Bay/ Dungeness Bay, with average counts at those sites in the period 1984-1988 ranging between 22 and 27 birds (Paulson 1993). Recent data from Oak Harbor and Penn Cove, on Whidbey I., indicate that these areas may support the largest concentrations in Washington. Smaller numbers are reported at Oak Bay, Port Angeles, and Ocean Shores and several other CBCs in the state (Paulson 1993, Evenson and Buchanan 1997). The winter population in Washington appears to be numerically dominated by ads. (Paulson 1993).

Spring migration data are minimal, but it is believed that the migration is generally completed by mid-May (Paulson 1993). As in other seasons, flocks of hundreds of birds are occasionally encountered. Late migrants may be found into early Jun.

Although small numbers of fall migrants may begin arriving on the coast in late Jun through mid-Jul, largest numbers arrive between Sep and Nov (Paulson 1993). The species has rarely been reported away from the coast (e.g. interior and e. Oregon; Nehls 1995).

The N. American population is estimated to be about 80,000 (Morrison et al. 2000). Except for annual CBC efforts, there is very little information on the abundance of this species in Washington. A recent count of 1000 at Rat I., Jefferson Co., a site with no survey history, was the highest known for the species in Washington.

Paulson (1993) suggested that a decline of wintering birds has occurred in the nw. More comprehensive analysis might determine whether the reported changes reflect long-term change in abundance, or regular or periodic variation in abundance at CBC sites in the region, as might be indicated by the 8-15x increase in abundance (to an average of 1220 birds) at Comox, B.C., in 1979-83 compared to periods before and after those years (Paulson 1993). It is noteworthy that another inhabitant of rocky shorelines, the Rock Sandpiper, has experienced a substantial population decline in Washington (Paulson 1993) as part of a larger range contraction (Buchanan 1999).

CBC data indicate no overall trend over time; an apparent increase at Olympia may represent a change in census coverage.

Noteworthy Records: West, *high counts, Winter*: 585 on 28 Nov 1998 at Oak Harbor; 475 on 31 Dec 1999 at Penn Cove; 220 in 1990-91 at Oak Bay, Island Co. (Evenson and Buchanan 1997); 120 in 1994-95 at Port Angeles (Evenson and Buchanan 1997). *Spring*: 1000 on 16 Apr 2000 at Rat I., Jefferson Co.; 450 on 11 May 1968 passing the n. jetty of the Columbia R. (Paulson 1993); 250 on 7 May 1972 at Whidbey I.; 100 in 1994 at Kilisut Harbor (Evenson and Buchanan 1997). Late: 3 on 19 May 1974 at Kalaloch; 1 until at least 11 Jun 1977 at Ocean Shores (Paulson 1993). *Fall, high counts*: 410 on 17 Oct 1999 at Penn Cove; 100 on 25 Sep 1965 at Blaine. Early: 26 Jun 1988 at Smith I.(Paulson 1993); 29 Jun 1974 at Ocean Shores. **East**, *Spring*: 15-16 May 2001 at Yakima R. delta.

Joseph B. Buchanan

Surfbird *Aphriza virgata*

Local, common migrant and winter resident in w. Washington.

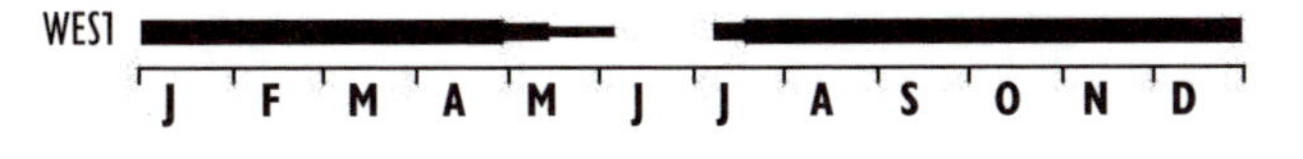

Habitat: Rocky shoreline habitats, including jetties and offshore rocks, log booms, occasionally foraging on sandy beaches and areas adjacent to rocky habitats.

Occurrence: Surfbirds breed in alpine tundra in Alaska and parts of Yukon Territory, and winter along the Pacific coast from s. Alaska to Chile (Senner and McCaffery 1997, AOU 1998). Found in Washington predominantly on outer coast, and more locally, in n. inland waters, often with Black Turnstones (Paulson 1993, Wahl 1995). Generally rare in c. and s. Puget Sound and Hood Canal.

Timing of migrations is not well known. The beginning of spring migration is not evident, nor is the peak, but most spring migrants appear to have moved through the region by the end of Apr (Paulson 1993, Nehls 1994), and a few stragglers occasionally linger into early Jun. The peak of fall migration appears to occur from mid-Jul to mid-Sep (Paulson 1993, Nehls 1994), with juvs. usually not arriving until early Aug. CBC data indicate no trend and low numbers at most locations.

Remarks: Identification of important staging areas and subsequent surveys would provide much needed insight on migration timing and population status. The N. American population of this monotypic species (AOU 1957) is estimated to be about 70,000 (Morrison et al. 2000).

Noteworthy Records: *Winter, high counts*: 130 on 1 Mar 1969 at Kalaloch; 120-150 in 1981-82 at Chuckanut Rk., Whatcom Co (Weisberg 1983); 120 at the Peapods, Rosario Strait, in 1978-79 (MESA); 50 on 13 Feb 1999 at Penn Cove. *Spring, late*: 7 on 4 Jun 1983 at Westport; 31 May 1985 at Sand I., Grays Harbor; 16-17 May 1969 at n. jetty of Columbia R. *Fall, high counts*: 250 on 15 Aug 1997 at Westport; 148 on 21 Jul 1982 at Ocean Shores (Paulson 1993); 125 on 18 Aug 1999 at Ocean Shores; 80 on 6 Nov 1998 at Penn Cove; 70 on 19 Aug 1997 at Port Townsend; 25 on 23-24 Jul 1963 at Destruction I.

Joseph B. Buchanan

Great Knot *Calidris tenuirostris*

Casual vagrant.

A Palearctic species, breeds in Siberia and winters in Austral-Asia, recorded once in Washington: a single-person sight record of an ad. at LaPush on 6 Sep 1979 (Tweit and Paulson 1994). This record was accepted by the WBRC for the Supplementary List. There are three other records s. of Alaska: one in Oregon in Sep 1990 (Lethaby and Gilligan 1992); one at Iona I., B.C., in Jan 1988 and one at Boundary Bay, B.C. in May 1987 (Campbell et al. 2000).

Bill Tweit

Red Knot *Calidris canutus*

Locally common to abundant spring migrant, uncommon to rare fall migrant, and very rare summer and winter visitor along the coast. Very rare fall and late spring migrant in e.

Subspecies: *C. c. roselaari*; possibly *rufa*.

Habitat: Primary foraging in estuarine intertidal flats; sand flats or sandy beaches less commonly used. Roosts include sand islands, sand spits, and sandy beaches in or near estuaries. Pastures adjacent to estuaries occasionally used for roosting or foraging.

Occurrence: Long known to be local and at times abundant in late spring at Grays Harbor and Willapa Bay (Jewett et al. 1953). Recent studies have shown the importance of Washington to the species. The only e. Pacific spring staging areas with numbers comparable to those in Washington are the Copper R. delta in s.c. Alaska (Isleib 1979, Paulson 1993) and the Yukon-Kuskokwim Delta in w. Alaska (Gill and Handel 1990). As elsewhere, most migrating knots occurring in Washington appear to fly great distances between stops.

Washington knots are a newly recognized subspecies, *roselaari* (Roselaar 1983, Piersma and Davidson 1992, Tomkovich 1992), breeding on Wrangel I. n. of Siberia and in nw. Alaska. The winter range may include the coast of California (a few hundred birds), w. Mexico, w. Florida, Texas and ne. S. America (Harrington 2001). Population is suggested to be about 20,000 (Piersma and Davidson 1992).

Washington's only concentrations occur in spring, very locally at Willapa Bay and Grays Harbor. Counts of over 1000 are typical only at extensive mudflats near river mouths at Bottle Beach/Ocosta and Bowerman Basin (Herman and Bulger 1981), and at the mouths of the North and Willapa Rs. (Herman et al. 1983). Knots are uncommon to fairly common in spring elsewhere in those embayments, and rare at other coastal locations. Large counts occur from about 22 Apr to 15 May, peaking during the first two weeks of May, with birds more widespread than earlier. Large late Apr counts are all from either Bottle Beach/Ocosta or at the Willapa R. mouth. Small summer flocks, up to 38 birds, are occasional there or in Grays Harbor.

Important high-tide spring roosting habitat appears limited to very few sites. In 1983 the majority in Willapa Bay apparently roosted on Ellen Sands, small unvegetated sand islands near major foraging sites w. of Bruceport (Herman et al. 1983). Most other Willapa birds roosted on a spit at Tokeland. Large numbers have been noted roosting at Bottle Beach and Bowerman Basin in Grays Harbor. On rainy days, knots may move into pastures adjacent to mudflats.

Fall migration is protracted, dispersed, and flocks seldom involve more than 20 birds. A few, mainly in Jul and Aug, are ads. but most are juvs. in Aug and Sep (Paulson 1993). Migration ceases by early Nov. Most records away from Willapa Bay and Grays Harbor are in the n. Puget Trough (e.g. Whidbey I., Skagit/Samish Flats) and at Dungeness, with very few s. of that. A few individuals rarely remain into winter: typically single birds during Dec.

Irregular in the Columbia Basin and e. counties, more frequent in the fall than the spring. Knots have appeared at the Walla Walla R. delta about every other year since the mid-1980s.

Remarks: Dependence on far-separated, localized staging areas and small population size make the species vulnerable (Harrington 2001). Grays Harbor and Willapa Bay are probably important to the survival of *roselaari*. Concerns include the spread of non-native smooth cordgrass, water pollution, pesticides and herbicides in estuaries, and sea-level rise. Disturbance from humans at roosting/resting sites (Pfister et al. 1992) should be considered in site management at Bottle Beach and Bowerman Basin. Maintenance of potentially ephemeral sand islands (e.g. Ellen Sands) as roosting habitat in strategic locations may also be important. Classed PHS, GL, WL.

Noteworthy Records: West, *Spring*: 6100 on 27 Apr 1981 in Grays Harbor, with 5000 at Bottle Beach/Ocosta; 3450 on 12 May 1983 at n. Willapa Bay, with 2570 at Willapa R. mouth; 3000 on 3 May 1986 at Bowerman Basin and 500+ at Grass Cr., Grays Harbor. *Summer*: 20 on 24 Jun 1977 at Ocean Shores; 38 on 4 Jun 1979 at Leadbetter Pt. *Fall*: 65 on 26 Aug 1979 at Leadbetter Pt.; 100 on 7 Sep 1982 at Ocean Shores; 120 on 13 Aug 2001 at Ocean Shores. *CBC*: 9 in 1976 at Leadbetter Pt., 7 in 1996 and 6 in 1982 at Grays Harbor; 1 each at Bellingham, Port Townsend, Oak Harbor and Sequim-Dungeness. **East**, *Spring*: 1 on 7 May 1972 at Banks L.; 1 on 11-12 May 1978 w. of Spokane; 1 on 20 May 1995 at Turnbull NWR; 2 on 19 May 2000 at Swanson Lks. *Fall*: 1 on 26 Sep 1965, 1 on 7 Sep 1969, 1 on 18 Aug 1973, 1 on 28 Jul 1991 at Reardan; 1 on 2 Sep 1970 at Potholes Res.; 1 on 30 Sep-1 Oct 1975 at Richland; 1 on 15 Sep 1978 nr. Davenport; 1 on 26 Aug 1995 at Atkins L.; 1 on 30 Aug 1987; 1 on 10 Oct 1987; 1 on 10 Sep 1990; 1 on 29 Sep 1992; 1 on 15 Aug 1995; 1 on 7 Sep 1995 at Walla Walla R. delta. 1 on 4 Sep 2001 at Yakima R.

Christopher B. Chappell

Sanderling *Calidris alba*

Common migrant and winter resident and fairly common non-breeding summer resident along outer coastal beaches. Local, fairly common in migration and winter in c. and n. inland marine waters. Local, rare to uncommon e.

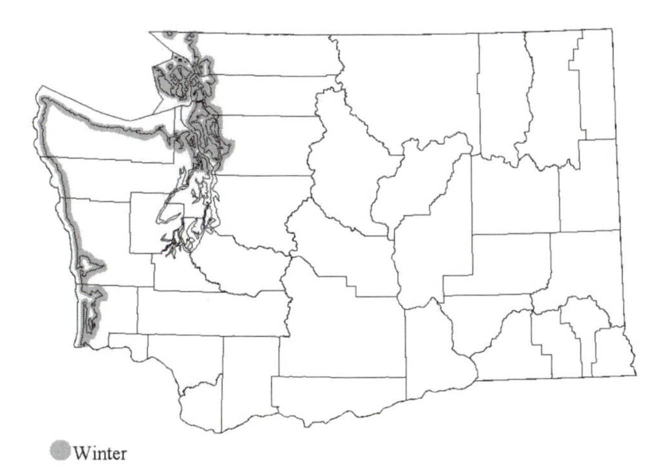

Habitat: Frequents hard-packed sandy beaches, and found there along the outer coast more often than other small shorebirds. Small numbers forage on intertidal mudflats, occasionally in rock/cobble habitats, and at margins of shallow ponds and lakes in e. Washington. Typically roost in foraging habitat, unlike many other sandpipers that move between distinct foraging and roosting sites.

Occurrence: N. American breeders have a vast winter coastal distribution from N. to S. America (AOU 1998). Sanderlings can be very abundant along coastal beaches during fall migration, with the first migrants arriving in mid-Jul (Widrig 1979, Paulson 1993). In winter, they are most abundant along Ocean Shores, Grayland Beach, and N. Beach on the outer coast. Over an eight-year period (1982-83 to 1989-90), the abundance of Sanderlings at these locations averaged 50.2, 43.3, and 46.6 per km, respectively, with associated high counts at each beach of 2312, 1604, and 2292 (Buchanan 1992). Three subsequent counts at N. Beach indicated a similar abundance of birds (average = 50.1 per km; Buchanan and Evenson 1997). Birds are much less abundant in the adjacent estuaries (Buchanan and Evenson 1997), occurring in sandy areas (e.g. Ocosta in Grays Harbor). Spring migrants begin moving through the area in Apr, and are most numerous in late Apr-late May (Widrig 1979, Paulson 1993). Sanderlings appear to be fairly regular summer residents along coastal beaches although they may be scarce or absent in some years (Paulson 1993).

Sanderlings are much less common in inland waters, though local counts can be substantial. The largest concentrations are apparently at Dungeness Bay, with birds uncommon to fairly common though local in the Puget Trough. Birds are occasionally noted in rock/cobble habitats in parts of Puget Sound (Paulson 1993).

In e. Washington, 78 Sanderlings were reported from 1996 to 2000, mostly in groups of <5 between mid-Aug to mid-Oct, mainly in Sep, with one Jul and two May records (SGM). Records are from ponds, lakes, and locations such as the Yakima R. delta. There are few eastside spring records and no summer records. There are no winter records from the Columbia Basin, although there is a single record from Okanagan Landing, B.C. (Paulson 1993). Comparatively higher eastside counts during the early to mid-1980s were unusual.

Winter densities on the three Washington coastal beaches are the highest reported in N. America (Myers et al. 1984). The species has experienced population declines at migration sites along the Atlantic coast (Howe et al. 1989), but it is unknown whether sub-populations occurring in Washington have experienced similar declines. Long-term CBC data indicate an increase at Leadbetter Pt. and Sequim-Dungeness but declines at other areas. The N. American population has been estimated at about 300,000 (Morrison et al. 2000).

Remarks: Potential impacts of human and vehicle use of beaches on populations include increased time in flight and reduction of time spent roosting (see MacWhirter et al. 2002). Classed PHS, GL.

Noteworthy Records: West, *Winter, high counts, outer coast*: 2973 on 4 Dec 1995, 2900 to 12-21 Jan 1983 (Myers et al. 1984), 4000 on 16 Nov 1996 at Ocean Shores. *Puget Trough*: 1312 in 1995-96, 280 in 1993-94 and 190 in 1992-93 at Dungeness Bay (Evenson and Buchanan 1997). 150 in 1990-91, 145 in 1994-95 nr. Foulweather Bluff (Evenson and Buchanan 1997); 150 in 1993-94 at Boz L. (Evenson and Buchanan 1997); 160 in 1992-93 and 125 in 1994-95 at Kilisut Harbor (Evenson and Buchanan 1997), and up to 440 at Cultus Bay (Van Velzen 1973). *Outer coast, Spring*: 27 Apr-5 May 1983: 7160 at Ocean Shores n. to Copalis, 3840 at N. Beach, and 2390 at Grayland Beach (Myers et al. 1984); 2908 at N. Beach (Buchanan and Evenson 1997). *Summer*: 104 avg. on 8-31 Jun 1978 on 6 counts at Leadbetter Pt. (Widrig 1979). *Fall, high counts*: 7900 on 30 Aug 1993 at N. Beach (Buchanan and Evenson 1997); 6 counts of >4000 on 24 Sep to 9 Oct 1978 at Leadbetter Pt. (Widrig 1979). **East**, *Spring*: 5 on 9 May 2001 at Walla Walla R. delta. *Fall*: 80 on 21-22 Sep 1980 at N. Potholes; up to 16 on 6, 23 Jul 1982 at Clarkston. *Early fall*: 6 Jul 1982 at Clarkston. *Fall, late*: 11 Oct 1969 at Stratford; 11 Oct 1996 at Walla Walla R. delta.

Joseph B. Buchanan

Semipalmated Sandpiper *Calidris pusilla*

Rare spring migrant; locally uncommon to fairly common fall migrant.

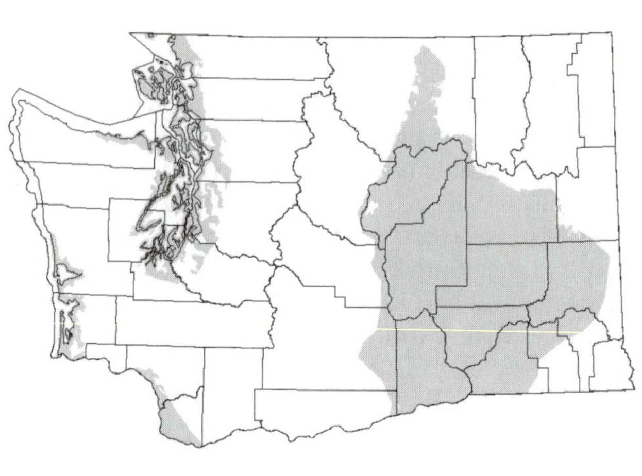

Habitat: Freshwater, brackish, and saltwater mudflats, often non-tidal habitats.

Occurrence: Usually found with Western Sandpipers, the Semipalmated Sandpiper has a complex pattern of occurrence in Washington, with migration and abundance apparently varying with latitude, longitude, and season. During spring migration, birds are rather rare throughout the state. On the westside, about two are seen annually, and on the eastside there were just three published spring reports from 1970 to 2000. All spring records have occurred between late Apr and early Jun (Paulson 1993).

In fall, birds appear to be more common in nw. Washington than elsewhere in the state. From Snohomish, Island, and Clallam Cos. n., this species is fairly common locally, with regular high counts of up to 15. Further s. on the westside, this species is less numerous, with only a handful found in Pacific and Grays Harbor Cos. each fall. Recent reports indicate lower numbers in e. Washington (contra Weber 1981). Migration often starts in late Jun when the first ad. or two arrives. Peak numbers of ads. move through in early and mid-Jul, and the last ads. usually leave by early Aug. Juvs. typically appear in late Jul, often in a fairly sudden burst, so that their peak occurrence is in late Jul-early Aug. By Sep, birds become scarce, and there are no records after Sep. This pattern likely results from a regular movement of birds from Alaska to the Frasier R. delta and nearby areas, with most of these birds apparently heading e. from there and rejoining the species' main migration routes (Paulson 1993). Classed GL.

Noteworthy Records: West, *high counts*: 28 on 4 Jul 1996 at Whidbey I. (SGM, S. Pink); 25 on 21 Jul 1968 at Protection I. (Paulson 1993); 24 at Everett 12 Aug 1997 (SGM). *Late*: 28 Sep 1977 at Ocean Shores (Paulson 1993). **East**, *high counts*: 10 on 17 Aug 1968 at Reardan; 14 on 15 Aug 2001 at Othello. *Late*: 2 on 29 Sep 2001 at Othello.

Steven G. Mlodinow

Western Sandpiper *Calidris mauri*

Abundant migrant, uncommon winter and summer resident in w. Washington, occasionally common in winter at Grays Harbor and Willapa Bay. Fairly common to common migrant and very rare winter visitor e.

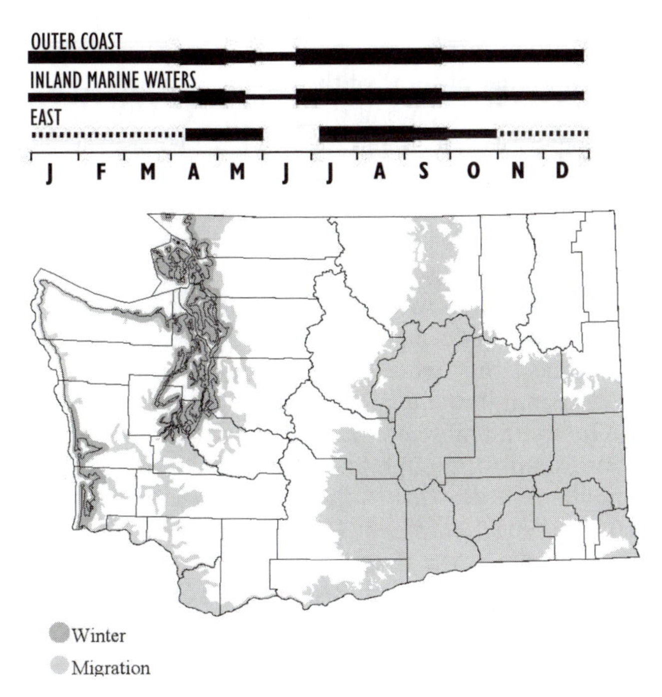

Habitat: Forages on estuarine tide flats, occasionally on beaches during high tides, upland flooded fields, pastures and on muddy margins of shallow ponds and lakes inland and in e. Washington. Roosts on beaches, jetties, log booms, golf courses, and rarely along grass meridians of roadways nr. major migration sites.

Occurrence: Campbell et al. (1990b) state that nearly the entire world population migrates through B.C., and the species was described as the most abundant shorebird in Washington by Jewett et al. (1953). This remains true during migrations (Herman and Bulger 1981, Paulson 1993). Possible population change is unknown because the first population estimates have only recently been made (JBB).

First fall migrants typically arrive in late Jun and numbers increase in early Jul. Juvs. appear Jun and increase in early Aug as ads. diminish. Numbers drop sharply after Sep. In winter, almost always found with Dunlins. Fairly common on the outer coast, where groups of 10 or more are regular. Uncommon in inland estuaries in winter, normally only as scattered individuals. Very rare in winter e., with most reports from the Columbia Basin.

Greatest abundance is at Grays Harbor and Willapa Bay. Spring counts at Bowerman Basin have ranged as high as 520,000 (Herman and Bulger 1981) and 460,000 birds (1982) but recent counts have been substantially lower (JBB). Recent spring counts at Willapa have ranged as high as 82,000 (Buchanan and Evenson 1997). Birds are locally abundant in Puget Sound, particularly in spring; high counts at Port Susan, Samish, Padilla, and Skagit bays exceeded 10,000 birds and three other sites—Drayton Harbor, Totten Inlet, Dungeness Bay—exceeded 3000 birds (Buchanan 1988c, Evenson and Buchanan 1997).

High counts at the few suitable sites in e. Washington—Columbia NWR, Turnbull NWR, Reardan Ponds, ponds nr. Othello, Sprague lagoons—during migration are typically <50, although there are several recent fall counts of >1000 birds, primarily at the Walla Walla R. delta (R. Hill p.c., M. Rule p.c., J. Acton p.c.). Although common as a spring migrant in e. Oregon (Paulson 1993, Nehls 1994), and regular, but uncommon in e. Washington, there are very few records of spring migrants in the Okanagan Valley, B.C. (Cannings et al. 1987), suggesting that interior migrants either move w. to the coast or make few landings in the region after departing Oregon. This regional pattern of abundance is generally true in fall as well.

CBC data include sizable numbers of unidentified small sandpipers ("peeps"). Long-term averages of these plus Western and Least sandpipers and Dunlins indicate no trends except at Grays Harbor where numbers in the 1980s and 1990s were less than half those of the 1970s. Only a small fraction (often <1%) of total peeps tabulated were identified as Western Sandpipers.

Remarks: A large portion of the species' Pacific Flyway population migrates through w. Washington and conservation of estuarine habitats is essential (Drut and Buchanan 2000). Birds are vulnerable to human disturbance while roosting on coastal beaches and possibly at foraging areas (Buchanan 2000). Classed PHS, GL.

Noteworthy Records: West, *high counts, Spring*: 520,000 on 25 Apr 1981 (Herman and Bulger 1981), 460,000 on 25 Apr 1982 (JBB) at Bowerman Basin. 82,575 on 27 Apr 1991 at Willapa Bay (Buchanan and Evenson 1997). 20,000 on 7 May 1994 at Crockett L.; 18,880 at Port Susan Bay and 18,637 in 1994 at Samish Bay (Evenson and Buchanan 1997). *Summer*: 20 on 8 and 11 Jun 1978; 70 on 21 Jun 1978 at Leadbetter Pt. (Widrig 1979); 6,000 on 2 Jun 1998 at Grays Harbor. *Fall*: 50,000 in 1991 at Port Susan Bay (Evenson and Buchanan 1997), 30,000 on 10 Jul 1979 at Grays Harbor (Paulson 1993); 22,000 on 27 Aug 1981 at Bowerman Basin (JBB); 20,000 on 13 Sep 1982 at Ocean Shores (Paulson 1993); 20,000 on 30 Aug 1999 at Crockett L. *Winter*: 2100 in 1982-83 at Grayland Beach (Buchanan 1992); 623 in 1988-89 at N. Beach, Pacific Co. (Buchanan 1992); 690 on 4 Jan 2001 at Leadbetter Pt. *CBCs, outer coast*: 1974-78 avg 863 per yr. at Grays Harbor; 1979-83 avg 755 per yr. at Leadbetter Pt. Puget Trough: 615 in 1993-94 at Padilla Bay, 224 in 1995-96 at Annas Bay (Evenson and Buchanan 1997); 100-180 in 5 of 8 years between 1980-81 and 1987-88 at Totten Inlet (Buchanan 1988c). **East**, *Spring*: 75 in 1963 at Banks L. (Paulson 1993). *Fall, high counts*: 1400 on 5 Sep 1999, 1100 on 26 Aug 1998 and 31 Aug 1999 at Walla Walla R. delta; ca. 2000 on 14 Aug 1988 at Reardan (J. Acton p.c.); 1000 on 9 Aug 1998 at Yakima R. delta; 650 on 7 Aug 1998 at Umatilla NWR; ca. 250 on 2 Sep 1992 at Othello ponds (J. Acton); 168 on 19-20 Aug 1998 at Columbia NWR (R. Hill); 132 in fall 1987 at Turnbull NWR (M. Rule p.c.); ca. 125 at on 15 Jul 1995 Sprague lagoons (J. Acton p.c.). *Winter*: small flock on 5 Dec 1980 at Vantage (Paulson 1993); 5 on 13 Feb 1988 at Walla Walla R. mouth; 1 on 25 Feb 1990 nr. Toppenish; 1 on Walla Walla CBC 1990-91.

Joseph B. Buchanan

Little Stint *Calidris minuta*

Casual vagrant

Washington's lone record of this Eurasian species was provided by an alternate-plumaged ad. (ph.) at the Yakima R. Delta 5-13 Aug. 2004. There are about 35 records from N. America, about half of which are from Alaska. The remainder are mostly split between the e. and w. coast; Washington's record is one of but a very few from the interior (Dunn et al. 2002).

Stebven G. Mlodinow

Least Sandpiper *Calidris minutilla*

Widespread and fairly common to common migrant. Uncommon, rather local winter resident w., rare to very rare e.

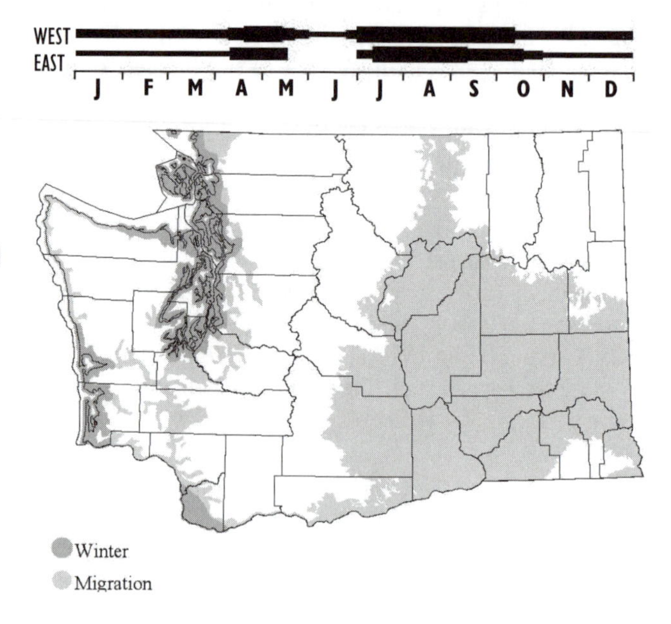

Habitat: Flooded fields and pastures, shorelines of lakes and ponds, tide flats, and beaches (Paulson 1993).

Occurrence: The Least Sandpiper winters along the w. coast s. to c. S. America (AOU 1998). Generally common during migration, its status is not as well known as that of some of the other sandpipers. It is often overlooked among the huge flocks of Western Sandpipers with which it migrates, and its preference for salt marsh and upper beach habitats probably make it slightly less detectable than other shorebirds and the habitat is less well covered by surveys and field observers. Least Sandpipers are most common in w. Washington, with hundreds often found in prime habitats during migration (e.g. Widrig 1979), and in B.C. they are also more abundant on the coast than in the interior (Campbell et al. 1990b).

One of the more common migrant shorebirds in e. Washington, though like all shorebirds there, its abundance is low due to the limited area of suitable habitats. For example, in 16 years (1984-1999) of regular visits to the Reardan ponds under a variety of site conditions, J. Acton observed an average seasonal peak count of about nine birds during fall migration.

Western, Least, and unidentified small sandpipers made up only a small fraction (1-5%) of sandpipers counted on long-term CBCs and Least Sandpipers only 5-10% of these. The highest average count was at Grays Harbor and averages there were 187 in the 1970s, 52 in the 1980s and 45 in the 1990s.

Joseph B. Buchanan

White-rumped Sandpiper *Calidris fusicollis*

Casual vagrant.

There are three Washington records of this Arctic breeder. All were ads.: one at Reardan on 20-21 May 1962, another at Reardan on 23 May 1964, and one at Dungeness on 7 Jul 1992 (Tweit and Paulson 1994). Other regional records are mostly from late May and late Jul to mid- Aug (Paulson 1993).

Bill Tweit

Baird's Sandpiper *Calidris bairdii*

Widespread, rare to uncommon spring and uncommon to fairly common fall migrant; more frequent and numerous e. of the Cascades.

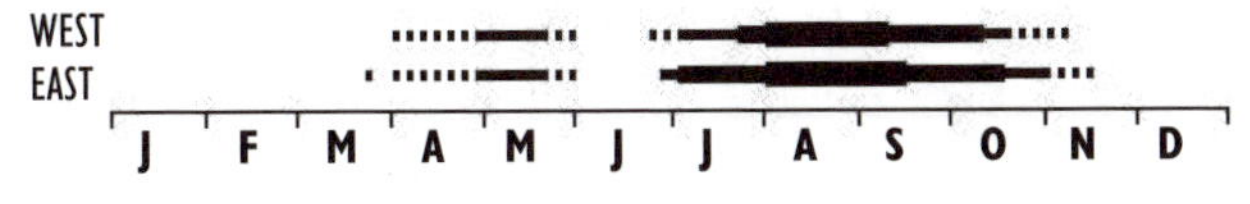

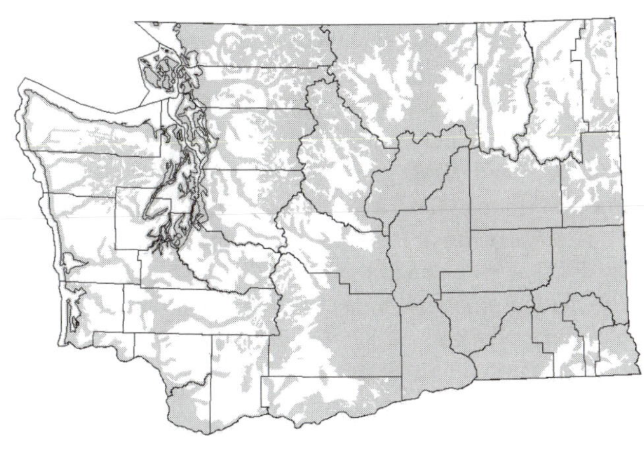

Migration

Habitat: Typically associated with drier habitats than used by most other shorebirds and most often observed on coastal or lakeshore dry beaches or grassy shoreline areas. Birds are known to migrate through high-elevation areas during fall and are often seen along the shores of alpine ponds (Paulson 1993).

Occurrence: Baird's Sandpiper migrates primarily through the Great Plains and relatively small numbers occur in Washington. According to Paulson (1993), most of the spring passage in the Pacific Northwest occurs in late Apr to mid-May; early migrants have been recorded in mid-Mar and late stragglers as late as early Jun. Perhaps due to its regular status, field reports have usually been generalized, resulting in a paucity of actual count data in the Pacific Northwest (see Paulson 1993, Nehls 1994). High spring counts in e. Washington generally do not exceed about 50, with counts in w. Washington much lower than this.

During fall migration, the species generally occurs in Washington from late Jun through late Oct with the peak movement in Aug-Sep in the w. and Jul-Sep in the e. (Paulson 1993). It is relatively more common in fall in e. than w. Washington. The mere three records described by Jewett et al. (1953) from e. Washington likely reflect low survey effort. There are few records from w. Washington between late Jun and mid-Jul, although this sandpiper is known to occur in the region at that time (Campbell et al. 1990b, Paulson 1993, Nehls 1994). The relative lack of records likely reflects the very rapid migration of ads. through the region in early fall (Jehl 1979); apparently most of the early migrants pass through the interior. Although Paulson (1993) doubted a record from mid-Nov at Reardan, it appears that occasional birds may pass through that late in the year (Campbell et al. 1990b, Paulson 1993, Nehls 1994). Migration in subalpine and alpine areas in the Cascades is known but data are minimal.

The population of this migrant, which winters s. to Tierra del Fuego, is estimated at 300,000 birds (Morrison et al. 2000).

Noteworthy Records: West, *Spring, early*: 1 on 2 Apr 2000 at Ocean Shores; 3 on 22-23 Apr 1979 at Leadbetter Pt. (Widrig 1979); 8 May 1997 at Steigerwald L. NWR. *Spring, high count*: 4 on 29 Apr-3 May 1976 at Seattle. *Fall, early*: 17 Jul 1993 at Yellow Aster Butte, Whatcom Co. (Wahl 1995); 6 on 25-26 Jul 1978 at Leadbetter Pt. (Widrig 1979). *Late*: 3 on 28 Sep 1996 at Ocean Shores; 23 Sep 1997 at Spencer I., Snohomish Co. *High counts*: 30 on 15 Sep 1989 at Ocean Shores; 20 on 19 Aug 1999 at Spencer I.; 12 on 8 Aug 1971 at Lummi Bay (Wahl 1995); 11 on 12 Sep 1978 at Leadbetter Pt. (Widrig 1979); 10 on 16 Sep 1982 at Kent. **East**, *Spring, early*: 21 Mar 1991 at Walla Walla R. delta; 5 Apr 1999 at Columbia NWR. *Spring, high counts*: 150 in 1963 at Banks L. (Paulson 1993); 27 on 29 Apr 1991 at Columbia NWR (R. Hill; 11 in 1987 at Reardan. *Fall, early*: 27 Jun 1998 at Atkins L., Douglas Co.; 29 Jun 1997 at Richland. *Fall, late*: 17 Nov 1962 at Reardan (Paulson 1993); 9-11 Nov 1996 at Yakima R. delta; 9 Oct 1999 at Atkins L. *Fall, high counts*: 105 on 27 Aug 1995 at Atkins L.; 66 on 26 Aug 1996 at Walla Walla R. delta (Paulson 1993); 56 in 1987 at Turnbull NWR (M. Rule p.c.); 50+ on 27 Aug 1989 nr. Banks L.; 40 on 6 Sep 1996 at Quincy STP; 30 on 18 Sep 1986 at Wenas L. (Stepniewski 1999); 150 on 26-29 Aug 2000 at Swanson Lks., Lincoln Co.

Joseph B. Buchanan

Pectoral Sandpiper *Calidris melanotos*

Uncommon to fairly common fall migrant in both e. and w. Washington; less common in spring. Occasional in summer.

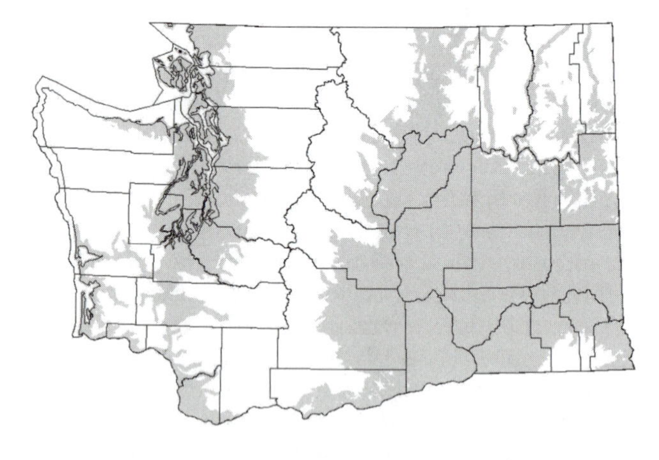

Migration

Habitat: Often in *Salicornia* marsh in w. Washington estuarine areas. Also in many freshwater habitats including flooded pastures and agricultural areas, and the shorelines of shallow ponds and lakes statewide (Paulson 1993).

Occurrence: The species breeds across most of the N. American tundra and winters in s. S. America (Paulson 1993, AOU 1998). During spring migration Pectoral Sandpipers are typically present during May although there are early records from late Apr and a late record from early Jun (Paulson 1993, Nehls 1994). Paulson (1993) suggested that a late Mar record from Victoria, B.C., was inadequately documented. Most published records of spring migrants consist of singles or small flocks (e.g. <6 birds), although in May 1992 there were three reports of flocks ranging in size from 14 to 50 in w. Washington. There are few records from the Pacific Northwest in mid-Jun (Paulson 1993) although the species should be expected on occasion due to its late spring migration and the early passage of ads. in fall.

Fall migration is much more protracted, beginning in late Jun or early Jul and continuing through Oct, with several records in late Nov (Paulson 1993). Birds are more common than in spring and occur in much larger flocks. There are at least seven published records of flocks of 15 or more birds in w. Washington (largest was about 150). In addition, Widrig (1979) reported relatively large flocks at Leadbetter Pt. during fall migration. There are also at least eight records of large flocks (up to 56) from e. Washington. Statewide, the largest flocks generally occur during late Aug through Oct. There are several records of birds present in the Pacific Northwest in Dec although none are from Washington (Paulson 1993).

The estimated N. American population is about 400,000 (Morrison et al. 2000). Data are limited, but as far as can be determined, there have been no changes in the status of this species in Washington since described by Jewett et al. (1953).

Remarks: A report of occurrence during a CBC count week at Kitsap Co. in 1977 is questionable.

Noteworthy Records: West, *Spring, early*: 19 Apr 1998 at Jensen Access, Skagit Co. *High counts*: 50 on 15-20 May 1992 at Lummi Bay (Wahl 1995); 30+ on 17 May 1992 at Port Angeles; 14 on 10-24 May 1992 at Seattle. *Late*: 8 Jun 1974 at Leadbetter Pt. *Fall, early*: 2 on 15 Jul 1968 at Bellingham. *Late*: 26 Nov 1982 at Duvall; 11 Nov 1999 at Fir I., Skagit Co.; 9 Nov 1996 at Samish Bay; 7 on 8 Nov 1996 at Nisqually NWR; 7 Nov 1999 at Ocean Shores; 29 Oct 1972 at Lummi Bay (Wahl 1995). *High counts*: 200+ on 14 Sep 1985 at Ocean Shores (Paulson 1993); 191 in 1978 at Leadbetter Pt. (Widrig 1979); 150 on 26 Sep 1976 at Leadbetter Pt.; 100+ on 11 Sep 1992 at Ridgefield NWR. **East**, *Spring, early*: 2 on 12 Apr 2002 at Lamont, Whitman Co.; 23-25 Apr 1985 at Walla Walla R. delta; 27 Apr 1954 at Potholes (Johnsgard 1954). *Late*: 20 May 2000 at Swanson Lks., Lincoln Co. *High count*: 6 in 1966 at Hauser L. *Fall, early*: 29 Jun 1953 at Potholes (Johnsgard 1954); 22 Jul 1964 at Reardan. *Late*: 17 Nov in se. Washington (Hudson and Yocom 1954); 16 Nov 1975 at Yakima R. delta; 6 Nov 1957 at Reardan (Hall and LaFave 1958). *High counts*: 56 on 24 Sep 1991 at Reardan; 53 in 1988 at Turnbull NWR (M. Rule); 50 in 1966 at Reardan (Paulson 1993); 40 on 11 Sep 1964 at Reardan; 40 in 1982 at Yakima R. delta.

Joseph B. Buchanan

Sharp-tailed Sandpiper *Calidris acuminata*

Rare fall and very rare spring migrant in w. Washington. Very rare fall eastside migrant.

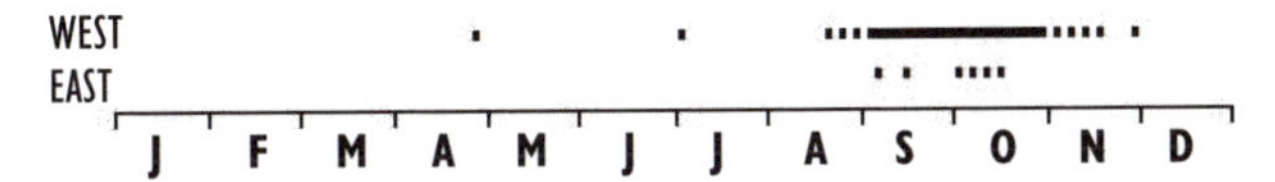

Habitat: Shallow-water salt and brackish marshes where *Salicornia* is dominant, and shallow freshwater marshes with abundant emergent vegetation, most often along the outer coast or in Puget Sound.

Occurrence: Usually accompanying Pectoral Sandpipers, this Asiatic shorebird was first reported in Washington on 2 Sep 1892, four shot at the Nooksack R. delta (Edson 1908), though this record is questionable (Wahl 1995). The next reports were of singles collected at Westport on 29 Oct 1927 and at Nisqually on 2 Nov 1927 (Jewett et al 1953). There was only one other record prior to 1963, when two were found at Willapa Bay, and one was seen s. of Birch Bay. Since then, Sharp-taileds have been recorded almost annually, with three to five per fall typical, and most often at Leadbetter Pt. and Ocean Shores. Annual totals were greatest from the late 1970s into the late 1980s.

The vast majority of Sharp-tailed Sandpipers in Washington have been juvs. on the westside. There are only 11 records from e. Washington, and only two records of ads.—one in late Apr and one in early Jul. Just nine other ads. have been found along the Pacific coast s. of Alaska, three between 8 Apr and 24 May and five between 21 and 27 Jul (Mlodinow 2001).

Sharp-tailed Sandpipers are somewhat more numerous just n. of the U.S.-Canada border in the Fraser R. delta, perhaps due to habitat being more accessible to observers, with high counts reaching 20 (Paulson 1993). A very late bird was seen there on 21 Dec 1976 (Paulson 1993).

Noteworthy Records: West, *Spring*: 26 Apr 1979 at Leadbetter Pt. (Paulson 1993). *Fall, early*: 2-5 Jul 1995 at Crockett's L.; 11 Aug 1940 at Tacoma 1940 (Slipp 1943). *Late*: 26 Nov 1977 at Ocean Shores; 28 Nov 1999 at Ridgefield NWR; 1 on 30 Nov 2001 at Brady, Grays Harbor Co. *Ads.*: 26 Apr 1979 at Leadbetter Pt. (Paulson 1993); 2-5 Jul 1995 at Crockett L. *High count*: 6 on 8-13 Oct 1978 at Leadbetter Pt. (Paulson 1993). **East**: 15 Sep 1972 at Soap L. (Meyer 1973); 13 Oct 1973 at Sunnyside; 28 Sep-5 Oct 1975 at Richland; 26 Sep 1982 at Othello; 12 Oct 1986 at Yakima R. delta; 26 Sep 1990 at Yakima R. delta; 2 on 29 Sep 1991 (1 to 6 Oct 1991) at Frenchman Hills Rd.; 7-8 Sep 1996 at Reardan; 26 Sep 1999 at Walla Walla R. delta (A. Stepniewski p.c.); 7 Oct 2000 at Othello (SGM).

Steven G. Mlodinow

Sharp-tailed Sandpiper (Shawneen Finnegan)

Rock Sandpiper *Calidris ptilocnemis*

Local, rare to uncommon migrant and winter resident in w. Washington.

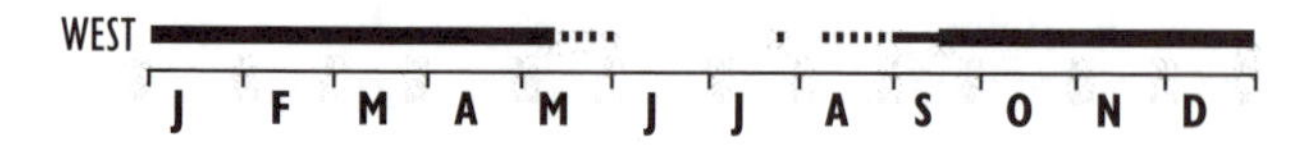

Subspecies: *C. p. couesi* and *tschuktschorum* apparently migrate through or winter in Washington (Conover 1944, AOU 1957, Paulson 1993).

Habitat: Primarily rocky shoreline habitats often at jetties on the outer coast. Also found on sand beaches, but not mud flats, a preferred winter habitat in Alaska (Gill and Tibbitts 1999).

Occurrence: The species breeds in tundra in w. Alaska (Gabrielson and Lincoln 1959, Gill and Handel 1981, AOU 1998), migrates and winters along the Pacific coast, rarely as far s. as c. California (Paulson 1993). The N. American population is estimated to be about 150,000 (Morrison et al. 2000). In Washington, Rock Sandpipers are very local from fall through spring, apparently most numerous in winter, often associated with turnstones, Surfbirds and Wandering Tattlers. Most high counts are from the Ocean Shores jetties. Other areas of fairly regular observations include Westport, Cape Flattery, and Crescent Bay, and scattered records of single birds have been reported from other sites in n. Puget Sound and the Strait of Juan de Fuca (e.g. Wahl 1995, Evenson and Buchanan 1997); there is a record as far s. as Tacoma.

Though spring migration has not been well described, there does not appear to be a substantial movement through the coastal region. Most migrants depart Washington by late Apr or early May, with records into late May. The timing of fall migration is also not well understood, but the primary movement appears to be in early Oct (Paulson 1993). There are a number of earlier records, including two extreme records from late Jul that were either very early migrants or birds that spent the summer locally.

Jewett et al. (1953) considered this a somewhat rare winter resident species, with high counts of 100 at Smith I., 50 at Destruction I., and 20 at Dungeness. In contrast, there have been few comparatively recent winter counts of >20 birds in Washington. Paulson (1993) indicated that winter abundance had declined along the nw. coast. This is evident in CBC data from Grays Harbor, the only location in Washington with consistent occurrence, where average counts declined drastically from 26.4 birds per year in the 1970s, to 13.6 in the 1980s to 5.8 in the 1990s. CBC numbers declined significantly ($P < 0.01$) from 1960 to 1999.

Regional CBC data also suggest that a massive range contraction, with a concurrent increase in abundance at certain major Alaskan wintering areas, occurred following the 1982-83 ENSO event (Buchanan 1999). There appears to have been no recolonization or redistribution of birds in the s. portion of the winter range (e.g. in Washington) in the years following the ENSO event. Recent counts are much lower than those prior to 1982-83.

Remarks: The relative abundance of the different subspecies is unknown. Overall population status indicates that surveys are needed at historically significant sites and also in other rocky shoreline areas where this species may occur. A bird of the nominate *ptilocnemis* was recorded at Ocean Shores in winter 2000-01 (Aversa 2001) and a bird was noted there on 20 Dec 2002 (NAB).

Noteworthy Records: West, *Winter, high counts (pre-1982-83)*: 50-60 in 1975-76 at Ocean Shores jetty; 25-30 after 6 Nov 1976 at Ocean Shores; ca. 50 on 26 Mar 1977 at Ocean Shores. *High counts (post- 1982-83)*: 20 in winter 1990-91 at Crescent Bay (Evenson and Buchanan 1997); 8 on 14 Mar 2000 at Ocean Shores; 7 on 25 Nov 1997 at Ocean Shores; 4 on 13-14 Feb 1998 at Ocean Shores jetty. *Spring, late*: 18 on 28 May 2000 at Cape Flattery; 24 May 1983 at Ocean Shores. *High counts*: 13 on 6 Apr 1997 at Ocean Shores; 12 on 3 May 1970 at Westport; 3 on 26 Apr 1998 at Ocean Shores; 3 on 18 Apr 1999 at

Rock Sandpiper (G. Scott Mills)

Penn Cove. *Fall, early*: 3 on 23 Jul 1997 at Jimmie-come-lately Cr., Clallam Co.; 3 on 23 Jul 1998 at Sequim; 11 Aug 1974 at Westport; 12 Aug 1972 at Ocean Shores; 28 Aug 1971 at Westport; 2 Sep 1963 at Cape Flattery. *High counts*: 5 on 26 Oct 1996 and 2 Nov 1996 at Westport; 3 on 27 Oct 1996 at Ocean Shores.

Joseph B. Buchanan

Dunlin *Calidris alpina*

Abundant migrant and winter resident in w. Washington. Locally fairly common to common migrant and rare to uncommon winter resident in Columbia Basin. Small numbers summer locally in w. Washington.

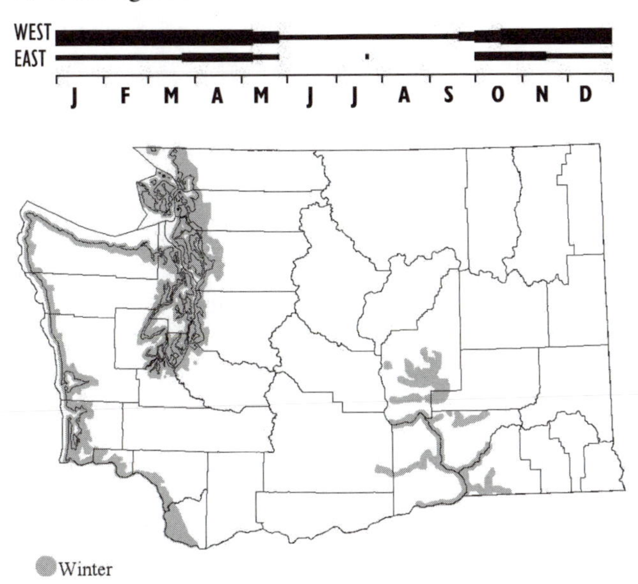

Subspecies: *C. a. pacifica*.

Habitat: Estuarine tidal flats, also uses freshwater habitats (Paulson 1993). Roosts on rocks, breakwaters, log booms.

Occurrence: The Dunlin breeds across much of the Arctic (Cramp et al. 1983), and overwinters primarily in the N. Hemisphere. It is found throughout Puget Sound, Willapa Bay, Grays Harbor, and Columbia R. estuaries. As many as 150,000-180,000 birds (>90% of the state's winter shorebird community) winter in w. Washington (Buchanan and Evenson 1997; Evenson and Buchanan 1997; G. Page p.c.), which is one of the most important wintering regions of the *pacifica* subspecies. Eastside winter records are scarce. Flocks of <10 birds have been recorded at several localities, but only at the Yakima R. delta are they fairly regular, occurring in 11 of 19 years between 1969-70 and 1987-88, with average flock size of about 25 birds.

Birds are widespread and abundant in w. Washington in spring. They are local and uncommon in e. Washington, where they have recently been reported annually, though there were no records from far e. Washington in any season between 1901 (Jewett et al. 1953) and 1947 (Booth 1957). Spring migration typically peaks in late Apr (Buchanan 1988c, Paulson 1993), with small numbers lingering to late May.

There are few westside summer records of Dunlins, and these have decreased in recent years (SGM). Most are from Leadbetter Pt. and Ocean Shores. It is likely that most of these birds spent the summer locally (Paulson 1993), though some may have been migrants (SGM).

Small numbers of fall migrants arrive in Washington as early as Aug, although most arrive after about mid-Oct (Paulson 1993). Subsequent high counts typically represent overwintering birds, although late migrants may remain into Nov (Buchanan 1988c). In e. Washington, the Dunlin is less common in fall than in other seasons.

The N. American population is estimated at about 1.5 million birds (Morrison et al. 2000). Willapa Bay, Grays Harbor, Padilla Bay, Port Susan, Samish Bay, and Skagit Bay qualify as internationally important sites (Western Hemisphere Shorebird Reserve Network) based on the abundance of overwintering Dunlins (Buchanan and Evenson 1997, Drut and Buchanan 2000).

Wintering numbers may have declined in Washington and elsewhere in w. N. America (Paulson 1993), but this is difficult to determine because of annual variability (>250% difference in winter counts at Willapa Bay [Buchanan and Evenson 1997]) and data presented by Paulson (1993) were not from major sites. Long-term CBC data indicate no overall trend although monitoring is advised.

Noteworthy Records: West, *Winter, high counts*: 27,260 at Bear R. and 13,500 at Willapa R. estuaries in Willapa Bay (Buchanan and Evenson 1997); 15,000 at Bowerman Basin (Brennan et al. 1985); 15,000 at Bellingham Bay; 10-20,000 at Drayton Harbor (Wahl 1995); 11,500 at Padilla Bay, 31,037 at Port Susan Bay, 15,000 at Samish Bay, 29,255 at Skagit Bay (Evenson and Buchanan 1997); 40,000 on 1 Jan 1998 at Port Susan Bay. 17,615 at Ocean Shores, 10,540 at Grayland and 33,424 at n. beaches were at high tide roosts (Buchanan 1992). *Regional totals*: 78,792 in "Puget Sound" (Evenson and Buchanan 1997), 69,850 at Willapa Bay (Buchanan and Evenson 1997), and 100,000 at Grays Harbor (Paulson 1993). *Spring, high counts*: 41,640 at Willapa Bay (Buchanan and Evenson 1997), 31,000 at Grays Harbor (Herman and Bulger 1981), 67,677 at Puget Sound estuaries with 35,000 at Port Susan, 12.339 at Padilla Bay, 12,973 at Samish Bay and 11,167 at Skagit Bay in Puget Sound (Evenson and Buchanan 1997). *Summer*: 50 on 8 Jun 1974 at Leadbetter Pt.; 3 on 25 Jun 1972 at Ocean Shores. **East**, *high counts, Winter*: 70 on 1 Jan. 1982 at Yakima R. delta; 300 on 10 Dec 1998 at Scooteney Res.; 200 on 14 Dec 1998 and 21 Feb 2000 at Walla Walla R. delta. Spring: 200+ on 29 Mar 1997 at Scooteney Res.; 400 on 30 Mar 1997 at Walla Walla R. delta; 100 on 20 Apr 1999 at Othello; 100 on 29 Apr 1991 at Columbia NWR (R. Hill p.c.). *Spring*, late: 4 on 22 May 1947 at Moses L. (Booth 1957); 1 on 26 May 1957 at Cow L. (Hall and LaFave 1958); 2 on 29 May 1921 nr. Grandview (Jewett et al. 1953). *Fall, high counts*: 46 on 31 Oct 1997, 120 on 5 Nov 1999 at Walla Walla R. delta; 60 on 14 Nov 1997 at McNary NWR; 46 on 24 Nov 1974 at the Tri-Cities.

Joseph B. Buchanan

Curlew Sandpiper *Calidris ferruginea*

Very rare spring and fall vagrant in w. Washington. Casual in e.

An Old World shorebird recorded across the breadth of N. America but barely annual along the Pacific coast s. of Alaska. Recorded six times in Washington: an ad. (ph.) 24 km s. of Ephrata, 10 May 1972 (Meyer 1973); an ad. seen at Ocean Shores, 5 Oct 1979; ad. at Leadbetter Pt., 17 May 1983; ad. at the Dungeness R. mouth, 29 Jul 1984; ad. (ph.) at Ocean Shores, 19 Sept 1990; a juv. at Ocean City, 7 Sep 1997; and an ad. at Long Beach, 5-11 Aug 2000. Other records from the Pacific Northwest include approximately 10 from B.C. and about 11 from Oregon. Most Pacific Northwest birds have been w. of the Cascades between mid-Jul and early Oct (Paulson 1993). The relative scarcity of records of juvs. is probably due to problems of identification.

Steven G. Mlodinow

Stilt Sandpiper *Calidris himantopus*

Widespread, rare to uncommon fall migrant in e. and w. Washington; irregular, rare spring migrant on eastside.

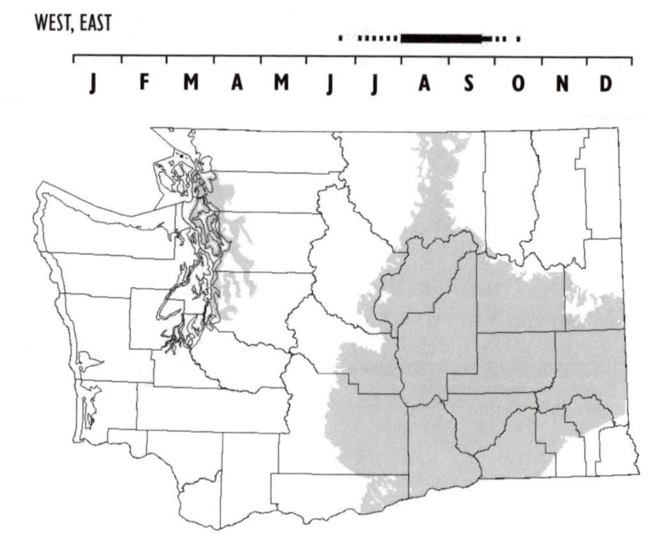

Habitat: Primarily in freshwater habitats including shallow ponds, flooded fields, brackish marshes, and sewage lagoons, rarely estuarine tide flats (Paulson 1993).

Occurrence: The Stilt Sandpiper breeds across much of the N. American arctic tundra and winters in w.c. and s.c. S. America (Paulson 1993, Klima and Jehl 1998), migrating primarily through the interior of N. America.

There are few records of spring migrants from the Pacific Northwest (Paulson 1993, Nehls 1994), with just three from e. and one from coastal Washington from mid-May through early Jun (Klima and Jehl 1998). Spring records are also scarce in s. B.C. (Cannings et al. 1987, Campbell et al. 1990b); there is one record in Oregon, a bird at Astoria on 20 Apr 2000.

Fall migrants are far more frequent and in greater numbers on the eastside though timing is similar statewide. Ads. occur from late Jun to about mid-Aug, and juvs. from about mid-Aug to early Oct (Paulson 1993). Fall migration typically consists of scattered individuals occurring at a few locations, with records mostly from Reardan, the Walla Walla and Yakima R. deltas and Island (Crockett L.), King, Snohomish, and Skagit Cos. on the westside, possibly reflecting observer effort at well-known locations. The highest counts are associated with the movement of juvs. through the region.

Total world population (AOU 1957, Klima and Jehl 1998) is estimated at about 200,000 (Morrison et al. 2000). Population change in Washington is uncertain. Although it is reported more often than indicated by Jewett et al. (1953) and other observers of that era (Harris and Yocom 1952, Hudson and Yocom 1954, Johnsgard 1954, Hall and LaFave 1958, LaFave 1959), this may at least in part be due to the increased field coverage. Classed GL.

Noteworthy Records: West, *Spring*: 11 Jun 1998 at Ocean Shores. *Fall, late*: 19 Oct 1996 at Sequim; 10 Oct 1971 at Whidbey I.; 5 Oct 1987 at Drayton Harbor (Wahl 1995); 29 Sep 1998 at Vancouver L.; 28 Sep 1998 at Ocean Shores. *High counts*: 23 and 15 on 20 Aug and 17 Aug 1997 at Everett STP; 13 on 18 Aug 1974 and 12 on 17 Sep 1987 at Crockett L. **East**, *Spring*: 24 May 1973 at Reardan; 3 on 12-22 May 1991 at Othello; 3 Jun 2000 at Othello. *Fall, late*: 4 on 17 Oct 1996 at Walla Walla R. delta; 4 on 17 Oct 1997 at Iowa Beef; 7 on 10 Oct 1997 at Walla Walla R. delta; 25 Sep 1974 at Richland. *High counts*: 45 in 1985 at Walla Walla R. delta; 15 on 19 Aug 1997 at W. Medical L.; 13 on 26 Aug 1989 at Reardan; 11 in 1987 at Reardan; 10 in fall 1987 at Walla Walla R. delta.

Joseph B. Buchanan

Buff-breasted Sandpiper *Tryngites subruficollis*

Rare fall migrant on the s. outer coast, very rare elsewhere.

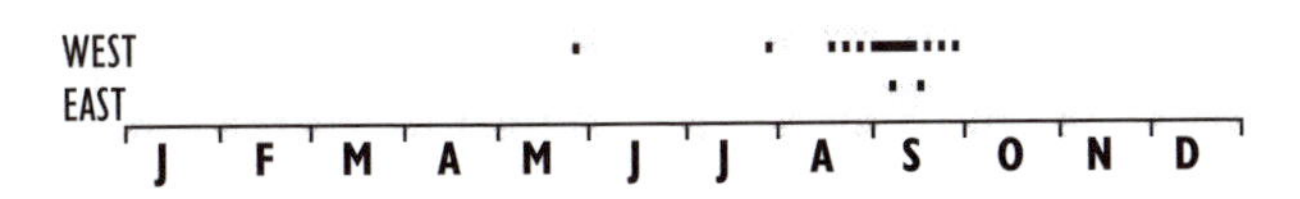

Habitat: Lake edges, mudflats, estuaries, dunes, short-grass fields.

Occurrence: This long-distance migrant has been recognized as a rare fall migrant in coastal Washington since the early 20th century, when two birds were collected by Lien in 1917 (Jewett et al. 1953). There were only 13 records prior to the 1970s, when regular coverage of the outer coast by increasing numbers of observers began. Since then, the species has been recorded almost annually in fall on the outer coast, in numbers that show strong annual variation (0-38), possibly related to strong annual fluctuations on the breeding grounds (Lanctot and Laredo 1994). Peak years included 1978 (20 birds), 1979 (38), 1985 (10), and 1993 (12). Numbers appear to be declining: there were about 66% the reports in the 1990s (41) than the previous two decades (about 65 per decade).

The vast majority (90%) of the 185 records through 1999 are from the outer coast, mainly at Ocean Shores and Leadbetter Pt., with another 6% from shoreline areas nr. Puget Sound. Only 4% are from inland areas, including at least three from the well-watched Reardan Ponds in e. Washington. Based on specimens and photographs, all fall records appear to be of imms. (Paulson 1993). Classed GL.

Noteworthy Records: *Outer coast, Spring*: 31 May 1984 at Leadbetter Pt. *Fall, early*: 30 Jul 1984 at Ocean Shores. *High counts*: 11 on 27 Aug 1978 at Ocean Shores and 20 on 5 Sep 1979 at Leadbetter Pt. *Fall, late*: 27 Sep 1980 at Ocean Shores. **West**, *inland*: 15 Sep 1991 at Lyman, Skagit Co. **East**, *Fall*: 1962 in "eastern Washington"; 7 Sep 1968 at Reardan; fall 1987 at Reardan; 18 Sep 1988 at Reardan.

Bill Tweit

Ruff *Philomachus pugnax*

Rare fall migrant. Casual spring vagrant.

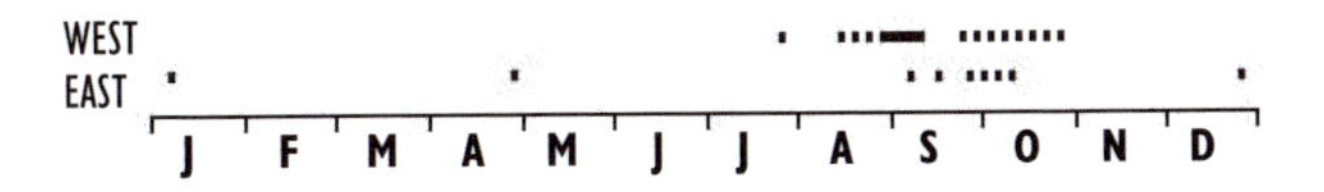

Habitat: Non-tidal wetlands with mud and emergent vegetation including saltwater marsh edges and flooded fields, occasionally in a wide variety of other shorebird habitats including exposed intertidal (e.g., Paulson 1993).

Occurrence: An Old World shorebird now found widely across N. America. The large number of records, including many from spring in e. N. America suggest breeding, though there is only one N. American nesting record (Gibson and Kessel 1977).

The first Washington record was of an imm. on Whidbey I., from 4 to 19 Sep 1971 (Binford and Perrone 1971). There have been at least 48 subsequent records, mostly on the outer coast during fall. There are two spring records, and only eight fall e. Most have been juvs., and no ad. has been seen after Aug (Paulson 1993). County record totals: Grays Harbor (19), Pacific (3), Clallam (4), Jefferson (1), Skagit (1), Island (3), King (3), Clark (1), Douglas (1), Lincoln (3), Benton (3), Yakima (3),and Walla Walla (4).

Records appear cyclical: 13 occurred between 1977 and 1980 and 17 were seen from 1994 to 1997, five were seen from 1981 to 1992, and only seven from 1994 to 2000. It is uncertain whether this pattern is similar elsewhere on the Pacific coast.

Noteworthy Records: West, *Spring*: 1 on 4-12 Apr 2003 at Brady Loop Rd. *Fall, early*: ad. on 27-28 Jul 1995 at Kent. *Late*: 1 on 25-26 Oct 1978 at Dungeness; 1 on 10-27 Oct 1976 at Ocean Shores; 1 on 28 Dec 2002-4 Jan 2003 nr. Satsop. *High counts*: up to 4 on 31 Aug-15 Sep 1979 at Ocean Shores; 3 on 8-20 Sep 1980 at Leadbetter Pt. **East**,

Spring: 1 on 27 Apr 1986 nr. St. Andrews. *Fall*: 1 on 22 Sep 1972 at Reardan (Paulson 1993); 1 on 27 Apr 1986 nr. St Andrews; 1 on 27 Sep 1987 at Reardan; 1 on 15-17 Aug 1994 at McNary; 26 Sep 1987 at Yakima R. delta; 1 on 7 Oct 1994 at Yakima R. Delta; 1 on 1 Oct 1997 at Walla Walla R. delta; 1 on 6-13 Sep 1999 at Yakima R. delta; 1 on 15-16 Sep 2000 at Sprague.

Steven G. Mlodinow

Short-billed Dowitcher *Limnodromus griseus*

Common to locally abundant spring migrant; rare summer visitor; common fall migrant; very rare winter resident in w. Washington. Very rare spring migrant, casual summer visitor, and rare fall migrant e.

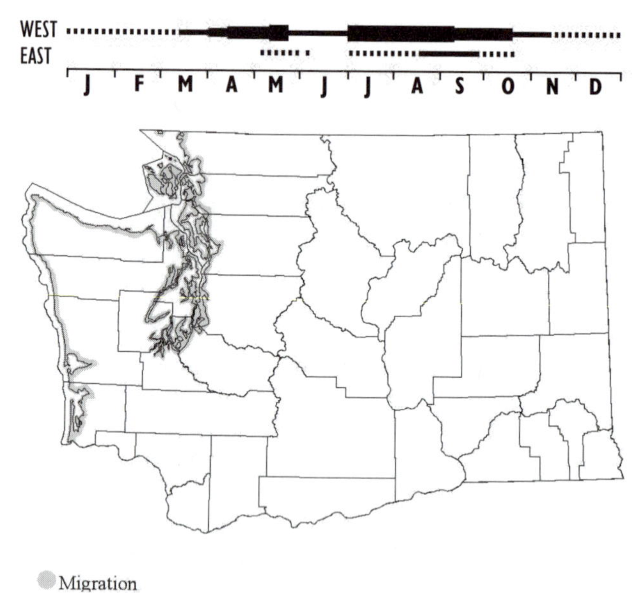

Subspecies: *L. g. caurinus* on the westside, occasionally in the interior; *hendersoni* appears to be found only in the interior (Paulson 1993).

Habitat: Strongly associated with estuarine mud flats; also uses a variety of other habitats opportunistically (Paulson 1993). Uncommon to rare away from tidal habitats and roosts.

Occurrence: The second or third most-abundant migrant shorebird in w. Washington, with largest flocks concentrating at specific localities (Paulson 1993). Flocks of thousands occur regularly, for example, at Grays Harbor (Herman and Bulger 1981) and Willapa Bay (Buchanan and Evenson 1997). High counts rarely exceed several hundred birds elsewhere in w. Washington (Paulson 1993, Evenson and Buchanan 1997). Smaller numbers are widespread, occurring from Mukkaw Bay, Hood Canal, and inland marine waters locations, where only occasionally more than 50 birds are noted. Ads. are present late Jun to mid-Jul. Juvs. arrive in late Jul, peaking in Aug-early Sept (Paulson 1993). In spring, birds are uncommon to fairly common in the Puget Trough (SGM).

The species is rare in e. Washington. There are relatively few spring records, but there are at least 35 fall records from the interior since 1954, with up to five birds occurring on 31 eastside records from 1997 to 2000, with 21 of these in late Aug-early Sep. It is a rare migrant in interior B.C. (Cannings et al. 1987), Idaho (Burleigh 1972), and e. Oregon (Gilligan et al. 1994), where Nehls (1994, 1989) considered it regular.

The Short-billed Dowitcher has recently experienced a significant population decline at migratory stopover sites along the e. coast of N. America (Howe et al. 1989, Morrison et al. 1994). It is possible that this decline is limited to *L. g. griseus*, the easternmost breeding subspecies, and may not apply to the w. subspecies. Data from w. N. America are lacking and are required to assess the status of populations that migrate through Washington.

Remarks: Based on the few eastside records it is not clear which subspecies is most common (Paulson 1993). Some of the specimens identified by Burleigh (1972) and Weber (1985) were thought to have been misidentified (Paulson 1993; see below). Population size is difficult to estimate due to overlap with Long-billed Dowitcher (Jehl et al. 2001). Classed GL.

Noteworthy Records: West, *Winter*: 16 Dec 1995-31 Jan 1996 at Dungeness (Evenson and Buchanan 1997); 25 Dec 1992 at Stanwood; 27 Dec 1992 at N. Cove; 2 on 8 Jan 1989, 3 on 30 Jan 1988 at Long Beach (Buchanan 1992); 20 Mar 1966 at Grays Harbor; 25 Mar 1976 at Dungeness. *CBC*: 4 reports from 1959 to 2000. *Spring, high counts, outer coast*: 23,000 on 23 Apr 1983 at Willapa Bay; 34,000 on 27 Apr 1981 at Grays Harbor (Herman and Bulger 1981); 26,000 on 26 Apr 1990 at one site in Grays Harbor; 10,000 on 6 May 2001 at Leadbetter Pt. *Inland marine waters*: 200 on 13 May 1981 at Seattle, 70 in 1994 at Crockett L., 46 in 1993 at Deer Lagoon, 281 dowitcher sp. in spring 1993 at Sequim Bay, 70 dowitcher sp. in 1993 at Port Susan Bay (Evenson and Buchanan 1997). *Summer*: 70 in 1974 at Leadbetter Pt. (Widrig 1979); 50 in 1990 at Grays Harbor. *Fall, high counts, outer coast*: 3000 on 29 Jun 1977 at Willapa Bay; 1300 on 9 Jul 1978 at Leadbetter Pt. (Widrig 1979); 12,000 on 10 Jul 1979 at Grays Harbor (Paulson 1993). *Inland marine waters*: 300 in 1991, 78 in 1993 at Dungeness Bay; 80 in 1991 at Samish Bay; 46 in 1991 at Crockett L. (Evenson and Buchanan 1997); 75 on 12 Aug 2000 at Boz L. (SGM). **East**, *Spring*: 7 May 1977, 14 km se. of George; 2

on 8 May 1998 at Toppenish NWR (Stepniewski 1999); 3 on 9 May 1997 at Folsom L., Whitman Co.; 9 May 1998 at Grandview Sewage Lagoons (Stepniewski 1999); 6 probable *hendersoni* on 12 May 1990 in Grant Co.; 12 May 1991 nr. Othello; 20 May 1996 at Creston; 24-26 May 1980 at Reardan; 30 May 1999 at Iowa Beef; spring 1985 at McNary NWR; spring 1985 at West Medical L. *Summer*: 3 on 5 Jun 1997 at Folsom L. *Fall*: *caurinus* on 3 Jul 1993 at Potholes Res.; 2 on 16 Jul 1990 at Reardan; 6 on 28 Jul 1994 at Wenas L. (Stepniewski 1999); 7 Aug 1991 at Potholes Res.; 2 on 19-20 Aug at Columbia NWR (R. Hill); 1-3 on 23-29 Aug 1987 at Reardan; 2 on 25 Aug 1991 at Walla Walla R. delta; *hendersoni* on 5 Sep 1980 in Whitman Co. (Weber 1985); about 8 widespread records from 1982 to 1987 and an additional 20 records from 1997 to 1999.

Joseph B. Buchanan

Long-billed Dowitcher

Limnodromus scolopaceus

Regular common migrant in w. Washington; fairly common on outer coast and uncommon to rare in winter in inland waters. Common migrant, very rare in winter on eastside.

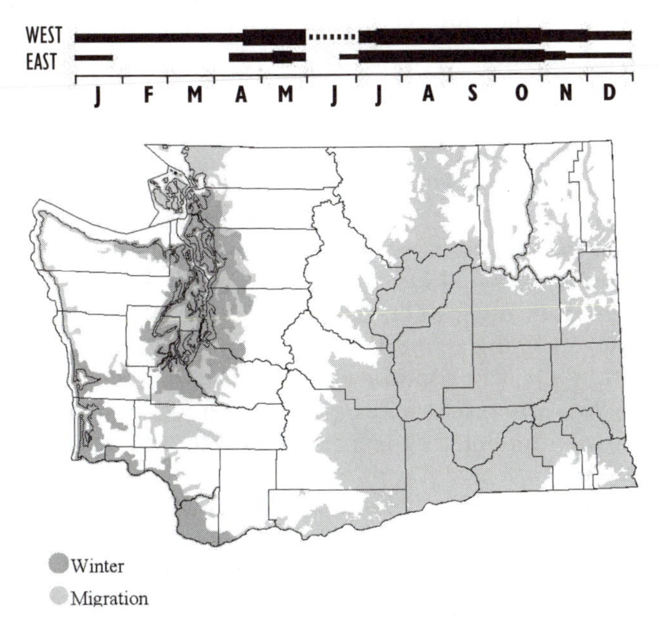

Habitat: Many habitats, including mud flats, freshwater wetlands, and flooded open fields or grasslands (Paulson 1993).

Occurrence: The Long-billed Dowitcher has a wide distribution but, as Paulson (1993) states, it is not abundant anywhere in Washington. Birds are common only in fall migration at Yakima and Walla Walla R. deltas in e. Washington. High counts of 100 or more birds have been made during migration at these sites and at westside locations (e.g. Spencer I., Crockett and Boz Lks [SGM]: and smaller numbers at Stanwood and Skagit WMA [SGM]). In winter it occurs regularly only at Grays Harbor and Willapa Bay. Small numbers are found at a number of other sites in Puget Sound (Buchanan 1988c, Evenson and Buchanan 1997). As elsewhere, this species occurs more frequently in freshwater habitats while the Short-billed Dowitcher occurs more in marine areas, though there is overlap (Paulson 1993).

As for several other migrating shorebirds, abundance in e. Washington (up to hundreds at Othello and Lind Coulee; SGM) is perhaps best described as intermediate between that noted in B.C. and that in Oregon. In spring, the Long-billed Dowitcher is abundant in e. Oregon (Paulson 1993, Nehls 1994), fairly common during the peak of spring migration in e. Washington, and casual in the Okanagan Valley of B.C. (Cannings et al. 1987). A similar pattern is noted in fall when it is considered uncommon in interior B.C. (Cannings et al. 1987), common in e. Washington, and abundant in e. Oregon (Nehls 1994, Paulson 1993). These differences likely reflect the availability of suitable habitats in the three areas.

Population trend of this species in Washington—or elsewhere—is unknown, but based on descriptions by Jewett et al. (1953) and recent records, it seems likely that it has not changed greatly, if at all, over the past 50 years. Including unidentified dowitchers presumed to be this species, long-term CBC data show low numbers.

Remarks: CBC reports of 2000 at Leadbetter Pt. in 1973 and a count of 1065 unidentified dowitchers at Everett in 1984 are suspect.

Noteworthy Records: *High counts*, **West**, *Winter*: 94 in 1993-94 at Willapa Bay (Buchanan and Evenson 1997); 26 on 1 Dec 1978 at Leadbetter Pt. (Widrig 1979); 116 on 10-16 Dec 1978 at Westport; 54 on 14 Dec 1996 at Spencer I.; 125 in 1992-93 at Deer Lagoon (Evenson and Buchanan 1997); 54 at Skagit Bay in 1994-95 (Evenson and Buchanan 1997); 28 in 1994-95 at Cultus Bay (Evenson and Buchanan 1997); 400 on 31 Jan 1984 nr. Woodland. *Spring*: 200 on 13 May 1981 at Seattle; 32 on 14 May 1979 at Leadbetter Pt. (Widrig 1979); 70 in 1994 at Crockett L. and 46 in 1993 at Deer Lagoon (Evenson and Buchanan 1997). *Fall*: 70 on 7 Sep 1978 at Leadbetter Pt. (Widrig 1979); 400 on 7 Aug 1998 at Ridgefield NWR; 163 on 4 Oct 1980 at Grays Harbor (Paulson 1993); 135 on 10 Oct 1999 at Skagit WMA. **East**, *Winter*: 10 on 2 Dec 1987, 2 on 28 Dec 1974 at Yakima R. delta; 8 on 6 Dec 1981 at Richland; 2 in early Jan 1971 at Tri-Cities. *Spring, high counts*: 10 on 8 May 1999 at Iowa Beef; 10 on 11 May 1999 at Philleo L.; ca. 100 on 12 May 1951 at Potholes (Harris and Yocom 1952); 500 on 12 May 1965 at Turnbull; 10 on 28 May 1971 at Reardan. *Summer*: 6 Jun 1950 at Moses L. (Harris and Yocom 1952); 14 Jun 1951 at Potholes (Harris and Yocom 1952); 2 nr. Moses L. and 1 at Othello on 3 Jun 2000 (SGM). *Fall, high counts*: 61 on 31 Jul 1999 at Iowa Beef; 100+ on 23 Sep 1957 at Reardan (Hall and LaFave 1958); 60 on 3 Oct 1988 at Grandview Sewage Lagoons (Stepniewski 1999); 72 on 4 Oct 1998 at Scooteney Res.; 160 on 17 Oct 1989 at Yakima R. delta; 165 on 17 Oct 1998 at Walla Walla R. delta; 200 on 19 Oct 1974 (first reported as *griseus*) at Yakima R. delta; 45-50 on 21 Oct 1972 at Reardan; 64 on 22 Oct 1998 at Scooteney Res.; 101 on 25 Oct 1997 at Wallula; 300 on 27 Oct 1976 (first reported as *griseus*) at Toppenish NWR; 60 on 13 Nov 1976 (first reported as *griseus*) at Toppenish NWR; 160 on 21 Dec 2002 at Moses L.

Joseph B. Buchanan

Jack Snipe *Lymnocryptes minimus*

Casual vagrant.

The Jack Snipe breeds across much of the far ne. Palearctic but is exceedingly rare in N. America. It has been recorded once in Washington: a single bird well seen and described by two experienced observers at the Skagit WMA on 9 Sep 1993.

The few other N. American records are specimens: St Paul I., Alaska, in 1919 (ABA 2002); Makkovik, Labrador (Austin 1929), 24 Dec 1927 (Hanna 1920); Gray Lodge WMA, Butte Co., California, 20 Nov 1938 (McLean 1939); and Colusa Co., California, 2 Dec 1990.

Steven G. Moldinow

Wilson's Snipe *Gallinago delicata*

Widespread, fairly common migrant, summer and winter resident in lower-elevation areas; locally common in winter.

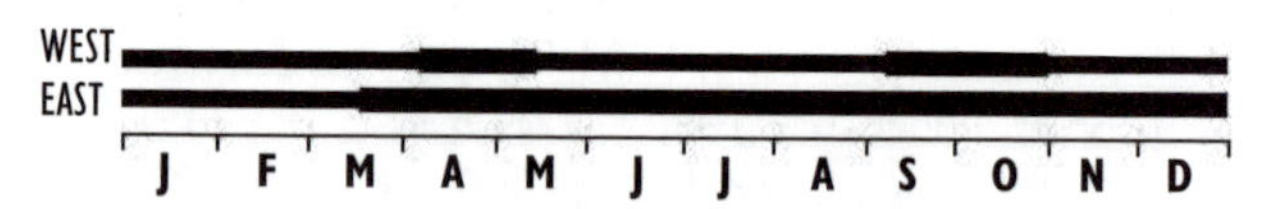

Habitat: Nests in open lowland wetlands but also found up to 1700 m elevation in subalpine wet meadows. Also in marine wetlands, flooded fields, and grassy areas in migration and winter (Jewett et al. 1953, Paulson 1993, Smith et al. 1997).

Occurrence: Widespread with a known or suspected breeding range in Washington primarily in the Columbia Basin e. of c. Kittitas and Yakima Cos., the Okanogan Highlands, localized areas in the Methow Valley, the Puget Trough, lowlands s. to Clark Co.

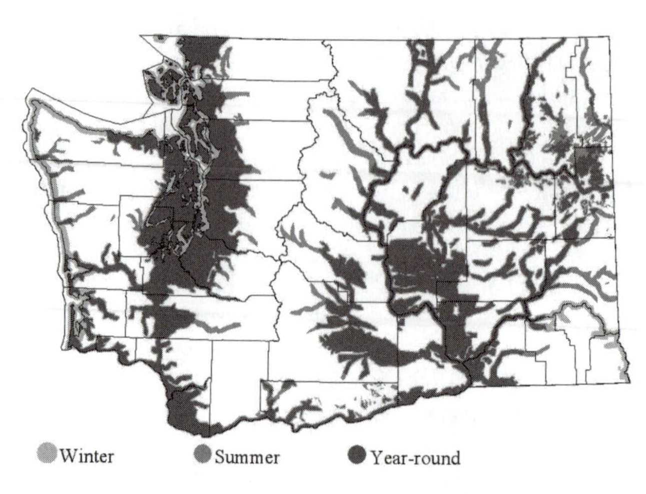

and w. to Grays Harbor, Willapa Bay, and the mouth of the Columbia R. Breeding is uncertain in se. counties (Weber and Larrison 1977) and not recorded during BBA surveys in Columbia, Garfield, and Wahkiakum Cos. (Smith et al. 1997).

Migration, winter movements, and ecology are not well known (Paulson 1993). Although difficult to detect in some years, spring migration peaks in late Mar to early or mid-May. Though first fall migrants may appear as early as Jul, most occur in late Aug-early Sep, with peak numbers through late Oct-early Nov (Paulson 1993, Nehls 1994, Stepniewski 1999, SGM). Some birds migrate at high elevations (Jewett et al. 1953, Wahl 1995), and birds are often found locally in flocks during migrations and winter (Campbell et al. 1990b).

Snipe reportedly leave interior areas during severe winters (Campbell et al. 1990b, Contreras 1997, Stepniewski 1999), and Paulson (1993) suggests that the species is found in low numbers in the interior after mid Oct. Few likely survive cold winters in Yakima Co. (Stepniewski 1999), and the few remaining after severe freezes in the n. Puget Trough may be found locally in urban ditches (Wahl 1995). The success of eastside snipe hunters suggests early-season hunting before the weather becomes severe.

Although this is a game species, its population status is not well known (Arnold 1994). Based on surveys of hunters, the harvest in 2002-03 was 3480, compared to over 40,000 in 1975, while the number of hunters declined about 50% (WDFW unpubl.). BBS data indicate that statistically significant declines in snipe abundance occurred in the Columbia Plateau Province between 1966 and 1996 and a decline statewide between 1966 and 1991 (Sauer et al. 1997, 2000), though the adequacy of the BBS for censusing the species is uncertain (Mueller 1999). Average CBC totals range from 7.3 per year (Moses L., high eastside site) to 67 (Bellingham 1974-78, high westside site; Paulson 1993). CBC counts show declines (Table 23).

Table 23. Wilson's Snipe - CBC average yearly counts by decade per 100 party hours.

	1960s	*1970s*	*1980s*	*1990s*
West				
Bellingham		22.1	9.9	12.6
Cowlitz-Columbia		25.1	13.0	
Edmonds			2.0	22.1
E. Lk. Washington			19.2	7.5
Grays Harbor		40.7	29.1	17.4
Kent-Auburn			23.0	15.1
Leadbetter Pt		16.9	13.6	11.1
Olympia			12.3	6.4
Padilla Bay			1.5	2.3
Port Townsend			2.6	2.2
Seattle	19.4	7.9	9.9	4.6
Sequim-Dungeness			22.0	9.2
Tacoma		42.3	15.4	3.8
East				
Ellensburg			17.9	16.6
Spokane	11.4	4.6	0.8	0.9
Toppenish			12.5	10.6
Tri-Cities		13.8	8.3	6.3
Walla Walla		15.4	13.4	5.2
Wenatchee	4.4	3.5	3.5	1.0
Yakima Valley		3.0	7.5	12.1

Remarks: Separated from Common Snipe *G. gallinago* in 2002 (AOU 2002). Species-specific surveys, banding, and telemetry may be required to monitor winter populations and movements. Classed GL.

Noteworthy Records: *High counts,* **West**: 50 on 6 Mar 1997 at Elma; 100 on 14 Nov 1999 at Dungeness; 50 on 28 Oct 1995 at Spencer I. (SGM). **East**: 9 on 14 Jan 1997 at College Place; 75 on 1 Oct 1996 at Ellensburg; 10 on 9 Oct 1999 at Yakima R. delta; 32 on 5 Nov 1999 at Yakima R. delta.

Joseph B. Buchanan

Wilson's Phalarope *Phalaropus tricolor*

Fairly common breeder in e. Washington; less common in migration. Rare on westside.

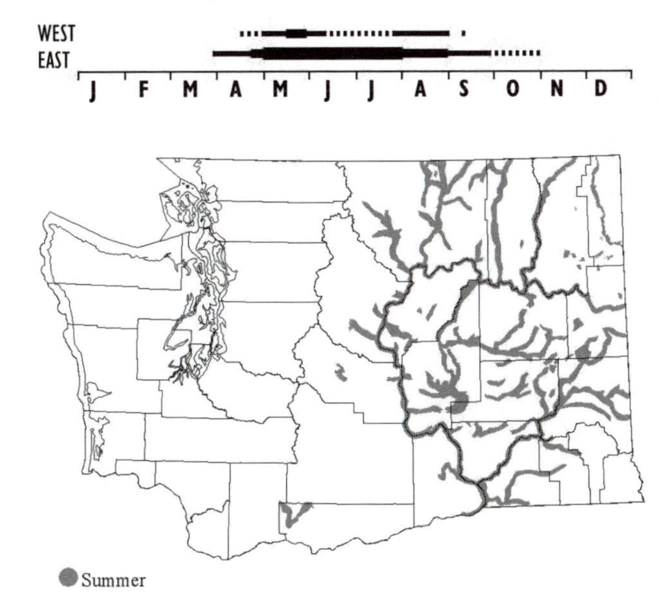

Habitat: Lakes, ponds, flooded fields, sedge meadows, and freshwater marshes.

Occurrence: Jewett et al. (1953) considered this a "regular summer resident" in e. Washington, and Smith et al. (1997) considered it to be fairly common there. Birds breed locally throughout the Columbia Basin and lower-elevation areas, primarily e. of the Columbia and Okanogan Rs. (Smith et al. 1997). Breeding varies between years, including locations at higher-elevation lakes in Kittitas Co. (Smith et al. 1997). Nesting distribution is more local in Yakima Co. than historically and birds are negatively affected by dry conditions (Stepniewski 1999).

Birds apparently nest rarely in w. Washington but there are virtually no recent nesting records w. of the Cascades in Washington or Oregon (Paulson 1993, Dowlan 1996). The species nested formerly at the Nisqually delta (Paulson 1993) and the Tacoma tideflats (Jewett et al. 1953). Recent breeding-season records nr. Elma (G. Schirato p.c.) suggest nesting there. Small numbers of migrants are widespread w. of the Cascades, utilizing flooded fields and sewage ponds in spring, and less common in fall at marshes and sewage ponds and occasionally tidal flats. Fall birds are almost all juvs., with most in early Aug. Unlike the other two phalaropes, birds are unrecorded at sea off Washington during migrations between N. and S. America. Birds occasionally winter as far n. as the s. U.S. (see Colwell and Jehl 1994).

Populations apparently respond numerically to local and regional changes in water conditions (Paulson 1993). BBS data from the Columbia Basin show a 10.9% decline in abundance between 1980 and 1996 (Sauer et al. 1997), a period during which conditions were drier than previously. There are no population estimates from Washington, but the N. American population is estimated at roughly 1.5 million birds (Morrison et al. 2000). Classed GL.

Noteworthy Records: West, *Spring, high counts*: 6 on 8 May 1996 at Stanwood STP; 10 on 9 May 1976 at Nisqually NWR; 10 on 11 May 1975 at Nisqually NWR; up to 6 after 13 May 1975 at Burlington; 34 on 30 May 1998 in w. Snohomish Co.; 7 on 31 May 1998 at Kent; 7 in 1969 at Leadbetter Pt.; 7 in 1969 at Bellingham. *Summer*: 3 on 6 Jun 1981 at Dungeness; 3 on 13 Jun 1998 at Nisqually NWR; 16 Jun 1999 at Monroe; 23 Jun 1968 at Leadbetter Pt.; Jun 1976 at Skagit Bay; Jun 1988 on Lopez I.; 2 young, ad. male on 9-14 Jul 1975 at Nisqually NWR; noted in 2000 nr. Elma (Greg Schirato p.c.); 7 Jul 2001 nr. Brady. *Fall, high count*: 12 on 14 Jul 1975 at Blaine (Paulson 1993). *Late*: 20 Aug 1966 at Leadbetter Pt.; 25 Aug 1968 at Conway; 25 Aug 1973 at Whidbey I.; 2 on 6 Sep 1972 at Pt. Roberts. **East**, *Spring, early*: 2 on 25 Mar 1999 at Hayford, Spokane Co. *Fall, late*: 23 Oct 1996 at Yakima R. delta. *High counts*: 87 on 1 Aug 1996 at Othello; 44 on 31 Jul 1999 at Iowa Beef; 22 on 28 Apr 1994 at Columbia NWR (R. Hill p.c.); 22 on 7 Aug 1998 at Umatilla NWR; 20 on 19, 20 Aug 1998 at Columbia NWR (R. Hill p.c.); 18 on 12 Jul 1998 at Othello; 14 on 3 Jul 1998 at Iowa Beef; 100 on 1 Jun 2002 at Othello and 100 on 29 Jun 2001 at Wallula (SGM).

Joseph B. Buchanan

Red-necked Phalarope *Phalaropus lobatus*

Common to abundant spring and fall migrant in continental shelf waters, fairly common to locally common in inland marine waters, uncommon on freshwater; locally uncommon to common migrant e.

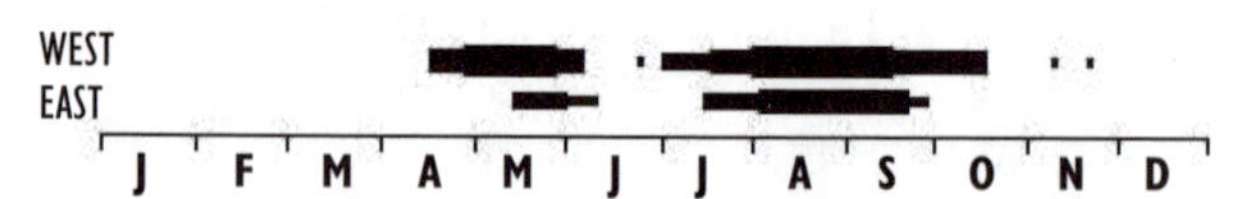

Habitat: Concentrates at current boundaries offshore and in areas of local upwelling and tidal convergences, inland lakes, reservoirs, and sewage-treatment lagoons.

Occurrence: This phalarope migrates between the low-arctic and winter range in the Humboldt Current and other S. Hemisphere areas. Surveys off Grays Harbor in Apr-Oct 1971-2000 had 1193

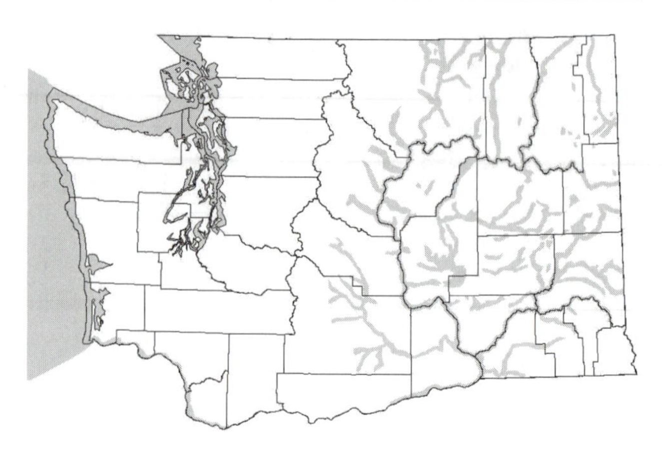

records totaling 34,768 birds from inside Grays Harbor to 2000 m depth offshore, indicating migration primarily over the continental shelf along the w. coast (Wahl et al. 1989). Over 86% of phalaropes specifically identified were Red-necked (TRW). Briggs et al. (1992) noted a ratio of 10:1 of this species vs. Red Phalaropes on shipboard censuses primarily within the 200 m depth contour, and Morgan et al. (1991) found 98% of phalaropes in the spring in the Queen Charlotte Is. were Red-necked Phalaropes.

Concentrations occur at productive areas: thousands forage at Swiftsure and LaPerouse banks off sw. Vancouver I. during migration (Morgan et al. 1991). During surveys in 1978-79 small numbers were noted in the Strait of Juan de Fuca, n. Admiralty Inlet, channels in the San Juan Is. (e.g. Cattle Pass) and Georgia Strait (MESA), and at sewage-treatment lagoons (e.g. Ocean Shores, Hoquiam, and Everett). Storm-driven birds are found in ditches and puddles along the coast.

Abundance varies, with high numbers offshore in fall migrations in 1980, 1984, and 1987-89, and low numbers in the 1990s' ENSO episode (Wahl and Tweit 2000a). In inland waters, birds are rare to uncommon in spring, peaking in mid-late May. Ads. are rare to uncommon in fall, with juvs. arriving in late Jul-early Aug. Variation is considerable inland also. High numbers were present in the San Juan Is. in 1981, for example, with high mortality evident in Aug of that year. In the ENSO year of 1983 there were very low numbers in the San Juans (Lewis and Sharpe 1987). Widespread reports of larger than normal numbers in inland waters (e.g. 1994) suggest a shift in migration routes (Wahl and Tweit 2000a).

Common, widely recorded, usually in small numbers in fall in e. Washington (Jewett et al. 1953). Records are from the L. Lenore, Tri-Cities area, Walla Walla R. delta (Ennor 1991), and Whitman Co. (Weber and Larrison (1977).

Remarks: Populations of planktivores like phalaropes are likely early indicators of changes in ocean productivity. Migration routes through e. Washington relative to concentrations at Mono L., California, should be studied (Paulson 1993). Classed PHS in Washington.

Noteworthy Records: West, *Spring, high counts, off Westport*: 1150 on 30 Apr 1994, 1531 on 4 May 1991. *Inland marine waters*: 215 at Swantown on 20 May 1995. *Summer*: singles at Leadbetter Pt. on 12 Jun 1971 and 17 Jun 1975 (Paulson 1993). *Fall, high counts off Westport*: 3129 on 26 Jul 1984, 1774 on 12 Aug 1989, 5600 on 8 Sep 1980. *Inland marine waters*: >1000 in Aug 1981 at Cattle Pass (Lewis and Sharpe 1987); >500 on 6 Sep 1993 at Edmonds; 500 on 16 and 28 Aug 1993, 4000-5000 on 3 Sep 1994, and 3000 on 11 Sep 1997 at Point No Point. **East**, *Spring*: 300 in 1963 at L. Lenore; 53 on 25 May 1995. *Fall, high counts*: 120 on 13 Aug 1995 at Othello STP; 120 on 27 Aug 1995 at Atkins L. Douglas Co.; 400-500 in late Aug-early Sep 1986 at Soap L.

Terence R. Wahl

Red Phalarope *Phalaropus fulicarius*

Fairly common to common migrant over continental shelf and oceanic waters; irregular, rare to occasionally abundant storm-driven migrant along the coast and in inland marine waters. Rare to uncommon in winter. Very rare migrant e. of the Cascades.

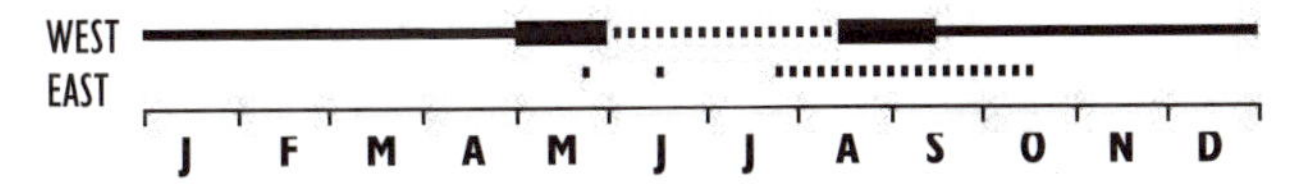

Habitat: Oceanic and continental shelf waters, deeper inland marine waters, foraging along convergence lines formed by currents and tidal action. Also lakes, puddles, ditches, and sewage ponds when storm blown.

Occurrence: Rarely encountered in most years except offshore, Red Phalaropes are less common than the Red-necked Phalarope. Surveys off Westport from 1971 to 2000 recorded only about 15% as many as *lobatus*. Migration is oceanic, with birds noted not only to the outer limit of Westport surveys but also across the N. Pacific (Hayman et al. 1986, Wahl et al. 1989): the outer continental shelf represents only the periphery of migration distribution. There were only three records, from the Strait of Juan de Fuca, during extensive MESA surveys of inland waters in 1978-79.

Recorded during all months, with peak numbers in mid-late May and from mid-Aug through mid-Sep. Spring migration is brief, apparently predominantly oceanic, and records of any number of birds are few. Offshore surveys and coastal reports indicated peak numbers in 1974, 1980, 1992, 1994, and 1996. Fall numbers are variable, with very high numbers in 1988, high numbers in 1976, 1980, and 1981, and lowest numbers in 1983-84 and the 1990s ENSO events (Wahl and Tweit 2000a). In inland waters, in contrast, thousands were noted off Victoria, B.C., in Nov 1982, at the beginning of the ENSO (Campbell et al. 1990b) and widespread numbers were observed in winter 1995-96 in the Strait of Juan de Fuca, s. to Olympia, and inland to Ridgefield NWR on the Columbia R.

Presumably because of this species' relatively late migration, storm-driven mortality is frequently reported (see Paulson 1993). Records in coastal and inland marine waters are attributed to these events (as in winters 1985-86, 1995-96, and see Bond 1971) with dead birds found on beaches and live birds in puddles in parking lots as well as in inland waters. Widespread storm-related mortality was evident in Dec 1995-early Jan 1996 with numbers of live birds found along the ocean coast and inland. Similar mortality of Red-necked Phalaropes is virtually unknown in Washington, likely reflecting their absence by the onset of severe late-fall storms.

Except for one record in May and one in Jun, all 20 or so widespread eastside records are during fall migration, peaking from mid-Aug to mid-Sep. These records, incidentally, are almost all of juvs. (Paulson 1993), strongly suggesting indeed that inland occurrence is of vagrants.

Remarks: Classed GL. The Gray Phalarope in Europe. Taxonomic name corrected in 2002 (AOU 2002).

Noteworthy Records: West, *outer coast, Spring*: 196 on 16 May 1976 off Westport. *High count*: ca. 3000 on 5 May 1994 passing Ocean Shores. *Summer*: 13 on 13 Jun 1982 off Westport. *Fall*: 880 on 11 Sep 1988 off Westport. *Inland waters*: 5000 on 11 Nov 1982 at Clover Pt.. Victoria, V.I. *Fall, late*: 1 on 16 Dec 1982 at Carson, Skamania Co.; 360 at Dungeness Spit and 150 at N. Beach, Jefferson Co., on 16 Dec 2002. **East**, *Spring*: 1 on 22 May 1996 at Rimrock L., Yakima Co. *Summer*: 1 in breeding plumage (ph.) on 14-15 Jun 1981 nr. Anatone. *Fall*: 1 on 19 Jul 1995 at Big Meadow L., Pend Oreille Co.; 1 on 6 Aug 2000 at Iowa Beef; Walla Walla R. delta: 1 on 16 Aug 1988, 2 on 13-14 1986, 1 imm. on 17 Sep 1989, 1 on 9-25 Sep 1991, 3 on 23 and 27 Sep 1983, 1 on 4 Oct 1985, 1 on 11 Oct 1980. Reardan: 1 on 14 Oct 1962, 1 on 7, 11 Sep 1973, 1 on 17 Aug 1974 and 1 on 29-31 Aug 1992. 2 on 19 Sep 1970 at Coulee City; 1 collected on 21 or 22 Sep 1980 at N. Potholes (specimen to WSU.); 1 on 5 Oct 2000 at Soap L.; 1 on 5-6 Oct 1994 at Richland; 1 on 7 Oct 1989 nr. Cle Elum.

Terence R. Wahl

South Polar Skua *Stercorarius maccormicki*

Uncommon to fairly common fall migrant, rare spring migrant and summer visitor in oceanic waters. Very rare to rare in nearshore and inland marine waters.

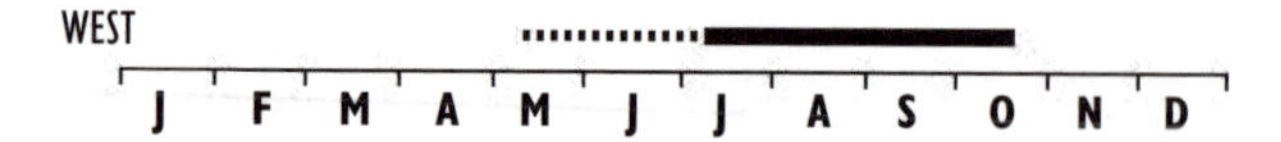

Habitat: Oceanic and continental shelf waters, occasionally nearshore and in Strait of Juan de Fuca and Georgia Strait.

Occurrence: Contrary to Jewett et al. (1953) it is now accepted that this species is the only skua recorded in Washington (Wahl 1975, Mattocks et al. 1976), if not the entire w. coast of N. America (see Devillers 1977). This species, with an estimated 5000-8000 pairs breeding in Antarctica and the S. Shetland Is. (Enticott and Tipling 1997), may migrate a greater distance than any other species other than the Arctic Tern. Jewett et al. (1953) give records of specimens and sightings to 1945. Boat trips and surveys off Westport began in 1966 and established that skuas were regular in varying numbers off Washington.

Surveys in 1971-2000 had 366 records of 437 birds. Some were migrating directly s., some were observed searching for possible food sources, and 165 were at fishing vessels or chums scavenging and parasitizing other seabirds (see Wahl 1977). Birds were recorded to the w. limit of Westport surveys, and seasonally over much of the N. Pacific (Wahl et al. 1989). Numbers vary from year to year. Years of highest abundances were 1975, 1976, 1977, 1986, 2001, and 2002 with low numbers from 1987 to 2000 (Wahl and Tweit 2000a). One-day high counts in 1976, 2001, and 2002 occurred during years of seabird abundance while low counts of occurrences and individuals were particularly noticeable in the 1990s (Wahl and Tweit 2000a).

Birds occur very infrequently in nearshore and inland marine waters. None were observed during about 6000 censuses in inland marine waters during MESA surveys in 1978-79.

Birds return to the Antarctic in Sep-Oct. Juvs. fledge in Feb-Apr, disperse farther than ads., which reportedly remain nr. breeding areas, and follow a clockwise route around the N. Pacific (Harrison 1983), reaching Japan in early May-late Jun (Olsen and Larsson 1997). Westport surveys support this, with just two records in May, five in Jun and main occurrence concentrated from early to mid-Jul through early Oct. Birds seen per trip increased from 1.3 in Jul and Aug to 2.3 in Sep to a peak of 2.9 in Oct. Records from B.C. range as late as 22 Nov (Campbell et al. 1990b).

Remarks: There is possible misidentification of apparent skuas nr. shore or in inland waters due to confusion with dark-phase Pomarine Jaegers.

Noteworthy Records: West, *outer coast/offshore, Spring*: 1 on 14 May and 1 on 16 May 1978; 1 in May 1989 at 46° 40' N over 994 m depth from aircraft (Briggs et al. 1992). *Summer*: ca. 6 off Grays Harbor and 2 off Jefferson Co. on 28 Jun 1917 (Jewett et al. 1953); 2 on 26 Jun 1979, 2 on 14 Jun 1987, 8 on 16 Jul 1976, 1 on 4 Jul 1987, 1 on 5 Jul 1988 off Grays Harbor, and 1 on 2 Jul and 2 on 7 Jul 1996 off Cape Flattery. *Fall, high counts*: 12 on 8 Aug 1969 at fishing fleet nw. of Cape Flattery; 14 on 26 Sep 1976 at research vessel off Grays Harbor, 17 on 3 Oct 1976, 21 on 22 Sep 2001 off Westport (TRW). *Inshore*: 1 on 8 Sep 1973 at Westport (Wahl 1975); 1 on 1 Oct 1976 at Ocean Shores; 1 on 24 Aug 1996 at Westport. *Inland marine waters*: 1 on 7 Oct 1982 and 1 on 25 Aug 1983 at Pt. Roberts (Wahl 1995); 1 in Skagit Bay on 16 and 18 Sept 1970; 1 on 8 Nov 1979 at Pt. Roberts; 1 on 26 Aug 1996 off Ediz Hook.

Terence R. Wahl

Pomarine Jaeger *Stercorarius pomarinus*

Fairly common spring and fall migrant and rare winter visitor over outer coastal and continental shelf waters. Rare migrant on inland marine waters. Rare in e. Washington.

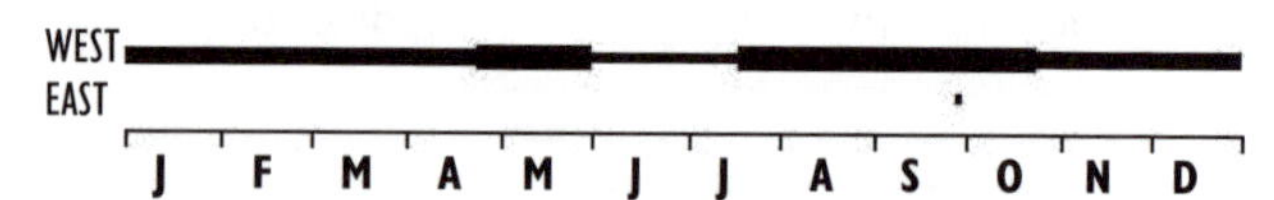

Habitat: Oceanic, outer continental shelf waters, also nearshore, inland marine waters, freshwater reservoirs.

Occurrence: A migrant noted all across the N. Pacific (Wahl et al. 1989) but relatively infrequently seen from shore. Off Westport from 1971 to 2000 over 90% of 1485 records of 3537 birds were over 50-1000 m depth. Pomarines are normally found farther offshore than the Parasitic Jaeger and were recorded w. to 126° 20' W. Birds occurred mainly during fall migration from Jul to early Nov, with 14.9 birds observed per trip. Peak numbers were from Aug to Oct, with lower numbers in the spring migration, mainly in Apr-early Jun. Variably present during winter off Washington, with birds noted on 11 of 19 one-day trips (1.2 per trip) during Dec-Mar. Birds are infrequent in inland waters (Lewis and Sharpe 1987, Campbell et al. 1990b, Wahl 1995). Though increasingly reported from mid-Sep to mid-Oct at foraging areas such as Point No Point with increasing observer effort, reports attributed to Wetmore by Jewett et al. (1953) of a number of birds seen together on several instances between Seattle and Port Townsend do not correspond with present status. E. Washington records are very limited, and just four records from interior B.C. (Campbell et al. 1990b) also indicate infrequent inland occurrence.

Attracted to fishing vessels and chums. Off Westport from 1971 to 2000 about 25% of over 3500 birds were associated directly with feeds at vessels, and about the same number diverted from migration to investigate chums away from vessels. Though jaegers often take prey directly, migrations coincide with those of suitable victims for robbing: kittiwakes, Sabine's Gulls, and Arctic Terns.

Like other jaegers, known to migrate in "waves," and even if this was not always directly visible off Westport it was implicit by varying numbers noted on surveys on consecutive days. Numbers off Westport varied between years. Densities were noticeably much higher than average in 1986 (including 190 on 23 Aug) and 1987. Numbers observed in fall 1997, 1998, and 1999 were low (Wahl and Tweit 2000a) but increased noticeably in 2000 and especially in 2001 (TRW). Birds may not breed at all during some years, and timing of and numbers noted in migration may vary accordingly (see Wiley and Lee 2000). Age class data off Washington are limited, but proportions of age classes appear to vary interannually in the fall, perhaps reflecting breeding success in the Arctic.

Remarks: Most birds seen off Washington are light or "medium" phase, with about 5% of birds "dark" phase. (Both Pomarine Jaegers and skuas vary in plumage characters and can be misidentified under some conditions.) Called Pomarine Skua in England.

Noteworthy Records: *Offshore, off Westport, Spring, high count*: 8 on 15 May 1977. *Summer*: 1 on 26 Jun 1977. *Fall, high counts*: 190 on 23 Aug 1986, 147 on 22 Sep 2001, 108 on 12 Sep 1992, 103 on 5 Sep 2002, 102 on 13 Sep 1992, 83 on 6 Oct 1985, 77 on 26 Aug 1995, 58 on 22 Jul 1989, 52 on 4 Oct 1987, and 34 on 7 Nov 1981. *Winter*: 8 on 6 Dec 1983, 2 on Jan 20 1990, 1 on 21 Jan 1995, 3 on 2 Feb 1997. *Coastal, Fall, high count*: 15 on 2 Sep 1995 at the n. jetty of the Columbia R. Winter: 1 on 15 Dec 1974 at Grays Harbor; 1 on 26 Feb 1994 at Ocean Shores; 1 ad. on 5 Feb 1995 at N. Head lighthouse. *Inland marine waters, Summer*: 1 on 11 Jun 1998 at Edmonds. *Late fall*: 1 on 6 Nov 1997 at Point No Point; 1 on 8 Nov 1980 at Pt. Roberts (Wahl 1995). **East,** *Fall*: 1 on 28 Sep 1997 on the Columbia R. at McNary Dam.

Terence R. Wahl

Parasitic Jaeger *Stercorarius parasiticus*

Fairly common spring migrant, fairly common to common fall migrant on oceanic and outer coastal waters; uncommon fall and rare spring migrant, very rare summer visitor on inland marine waters. Rare migrant in e. Washington.

Habitat: Oceanic, coastal, and inland marine waters, lakes and rivers in migration.

Occurrence: The jaeger likely to be seen from shore or in inland waters of Washington. Off B.C., Morgan et al. (1991) found that the majority of southbound Parasitic Jaegers were seaward of the continental shelf and Wahl et al. (1989) recorded birds migrating across the entire N. Pacific. Surveys from 1971 to 2000 off Grays Harbor found most birds over the shelf.

Surveys provided 539 records of 749 birds, most from May through Oct, with one record each in late Apr and early Nov and two in Jun. Morgan et al. (1991) suggest that non-breeders and failed breeders peak during the last half of Aug and breeders and juvs. peak in late Sep off B.C. Though 60% of records were from the outer shelf where fishing vessels and other seabirds concentrate, only about 17% of birds observed were directly associated with vessels. Abundances were high in 1973, 1977, and 1978 and numbers appeared to decline following the 1970s (Wahl and Tweit 2000a).

During MESA surveys in 1978-79 almost all records were from mid-Aug to early Nov, with 107 birds widespread in inland waters. The majority were in the Strait of Juan de Fuca, Admiralty Inlet, Rosario Strait, Haro Strait, Georgia Strait, and the

San Juan Is. Other records indicate seasonal occurrence over virtually all inland waters s. to Olympia, with relatively few in shallow estuarine embayments. There are a number of Jun records in inland waters. Fall records are later than those off the outer coast, with a number of Nov records (ca. 10 for Whatcom Co.; TRW) and a few in Dec.

Most inland waters records are of one or two birds, but with local aggregations of up to 37 noted at Point No Point, in the San Juan Is. (Lewis and Sharpe 1987), at Pt. Roberts, and Bellingham (TRW). This corresponds with reports of patterns of migrating, loose groups cited by others (see Cramp et al. 1985). Recent records suggest increased use of inland waters while coastal and offshore records declined (Wahl and Tweit 2000a).

Attracted to fishing vessels and chums. Occurrence is strongly associated with migrating Sabine's Gulls, Arctic and Common terns and, particularly in inland waters, Bonaparte's Gulls.

Eastside records are almost all from reservoirs on major rivers (see Denny 1995). Most jaegers there are this species, as is the situation elsewhere (e.g., Great Lakes, see Sherony and Brock 1997).

Remarks: Called the Arctic Skua in England (Cramp et al. 1985).

Noteworthy Records: *Offshore, high counts, Spring*: 18 on 19 May 1974 off Grays Harbor. Fall: 22 on 21 Sep 2002. *Inland marine waters, Spring*: ad. on 8 May 1994 at Mukilteo SP; 1 on 23 May 1995 from Mukilteo-Clinton Ferry; 1 on 1 Jun 1995 at Seattle. *Summer*: 1 on 15 Jun 1976 at Bellingham; 2 on 7 Jun 1979 at Boundary Bay. *Fall, late*: 16 Nov 1968, 23 Nov 1969, 22 Nov 1975, and 18 Nov 1976 at Bellingham. *Winter*: 1 on Tacoma CBC on 15 Dec 1990, 1 at Grays Harbor CBC on 20 Dec 1992. *High counts*: 37 on 19 Sep 1995, 23 on 24 Sep 19 1994 and 25 on 19 Sep 2001 at Point No Point; 26 on 10 Oct 1996 at Bellingham (TRW). **East**, *Spring*: 2 on 26 Apr 1985 at Walla Walla R. delta (Denny 1995). *Summer*: 1 on 30 Jun 1969 at Richland (Weber and Larrison 1977). Fall: 1 on 12 Oct 1973 at Richland (Denny 1995); 1 on 20 Oct 1973 at Yakima R. delta (Denny 1995); 1 on 28 Oct 1988 at McNary NWR (Denny 1995); 1 on 10 Nov 1996 at Yakima R. 1 on 14 Sep 1993 at 1800 m elevation at Diamond Head, Kittitas Co.

Terence R. Wahl

Long-tailed Jaeger
(G. Scott Mills)

Long-tailed Jaeger *Stercorarius longicaudus*

Uncommon to occasionally common fall migrant, rare to uncommon spring migrant over the outer continental shelf and oceanic waters. Rare fall migrant over coastal and inland marine waters. Very rare migrant e.

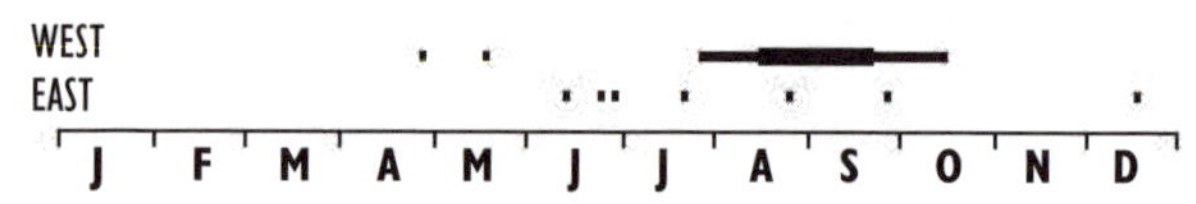

Subspecies: *S. l. longicaudus.*

Habitat: Oceanic in migrations, recorded in nearshore waters and inland water lakes and reservoirs.

Occurrence: The Long-tailed Jaeger was not listed for Washington by Jewett et al. (1953), probably reflecting limited observations offshore. The first record was from O'Sullivan Res. on 27 Jun 1953 (see Mattocks et al. 1976).

Off Westport, surveys in 1972-2000 had 422 records of 1387 birds. One bird was in Grays Harbor and 43 were over depths of <250 m. The rest were over the outer continental shelf and oceanic waters—the species migrates essentially across the entire N. Pacific (Wahl et al. 1989). Like other jaegers, noted on occasion migrating in "waves," with numbers of birds moving s. in the fall in loose aggregations. On consecutive weekends in Aug 1988, 89, 100, and 193 birds were observed, and on 12 and 13 Sep 1992, there were 99 and 102 birds, respectively. Reported to form flocks of hundreds on occasion during migrations in other parts of the world (Harrison 1983).

Harrison (1983) suggested that Long-tailed Jaegers are less piratical than other jaegers. Morgan et al. (1991) suggested that the apparent migration nearer the B.C. coast in fall than in spring may be due to greater fishing activity in the fall. About 26% of birds observed off Westport were associated directly with fishing vessels, a ratio similar to that of Pomarine Jaegers. There is an apparent association with migrating Sabine's Gulls and Arctic Terns (Wahl 1975).

In Washington, this is the most oceanic of the jaegers (and see Cramp et al. 1985, Wahl et al. 1989), with birds seldom seen from shore on the outer coast and only about 14 records from inland marine waters. Birds migrate through e. Washington in small numbers with six of nine records to date from Benton and Walla Walla Cos. (see Denny 1995). The Dec record is later than any to date for w. Washington. Largest seasonal numbers off Washington were in 1988, 1992, and 2001 (Wahl and Tweit 2000a, TRW).

Remarks: Almost all ads. are light phase: dark-phase birds are very rare (Harrison 1983) or questionable (Enticott and Tipling 1997, Olsen and Larsson 1997). Four dark-phase ads. were recorded singly on surveys off Grays Harbor (TRW). Called the Long-tailed Skua in Britain.

Noteworthy Records: West, *Spring, outer coast*: 1 on 26 Apr 1997 and 1 on 18 May 1996 off Westport; 1 on 20 May 1994 at Ocean Shores. *Inland marine waters*: 1 ad. on 13 May 1991 at Bellingham (Wahl 1995). *Fall, high counts*: 89 on 13 Aug, 100 on 20 Aug, and 193 on 27 Aug 1988; 99 on 12 Sep and 102 on 13 Sep 1992 off Westport. *Inland marine waters*: 1 on 25 Oct 1963 in e. Strait of Juan de Fuca; 1 on 18 Oct 1975 in Upright Channel; 1 on 8 Oct 1977 at Seattle; 1 ad. on 26 Sep 1979 at s. San Juan Channel (MESA); 1 in Sep 1984 off S. Beach, San Juan I. (Lewis and Sharpe 1987); 1 on 11 Sep 1981 at Blaine; 1 on 5-6 Sep 1983 and 2 on 17 Oct 1984 at Pt. Roberts (Wahl 1995); 1 on 3 Sep 1994 at Point No Point; 1 ad. on 13 Sep 1995 nr. Edmonds; 1 ad. on 7 Oct 1995 at Kingston; 1 juv. on 23 Oct 1995 and 1 ad. on 26 Jul 1996 at Point No Point; 1 ad. on 7 Jul 1995 between Port Angeles and Victoria. **East**, *Summer*: 1 on 12 Jun 1969 on the Columbia R., Benton Co. (Weber and Larrison 1977); 1 ad. on 27 Jun 1953 at Potholes Res. (Denny 1995; specimen WSU); 1 ad. on 30 Jun 1973 at Yakima R. delta (Denny 1995). *Fall*: 1 ad. on 22 Jun 1995 at the Walla Walla R. delta (Denny 1995); 1 on 25 Aug 2001 at L. Lenore; 1 ad. on 7-8 Sep 1995 at Richland (Denny 1995); 1 found dead on 28 Sep 1957 nr. Reardan (Weber and Larrison 1977); 1 on 17 Dec 1982 at Hanford Res. (Denny 1995).

Terence R. Wahl

Laughing Gull *Larus atricilla*

Casual vagrant.

Common in s. and e. N. America, the Laughing Gull has occurred three times in Washington. Two state records were during the 1982-83 El Niño event and suggest migration n. with Heermann's Gulls. A first-year bird was reported on 1 Sep 1975 at the n. jetty of the Columbia R. Onc (ph.) on 14 Aug 1982 off the Westport Jetty was the first documented for the state. An ad. on 18 May 1983 nr. S. Bend on Willapa Bay was the second documented report.

There are two records for Oregon (Gilligan et al. 1994), and one for B.C. Some birds migrate considerable distances: a bird banded in the e. U.S. was recovered in Panama (Cramp et al. 1983), suggesting that birds from the Atlantic may winter off c. America and nw. S. America, and birds occur in Hawaii, Clipperton, and other Pacific islands. The breeding population, estimated at about 400,000 pairs, increased during the 20th century (Enticott and Tipling 1997).

Terence R. Wahl

Franklin's Gull *Larus pipixcan*

Uncommon fall migrant, very rare spring migrant and summer visitor in w.; rare spring and fall migrant and very rare summer visitor in e. Washington. Casual in winter.

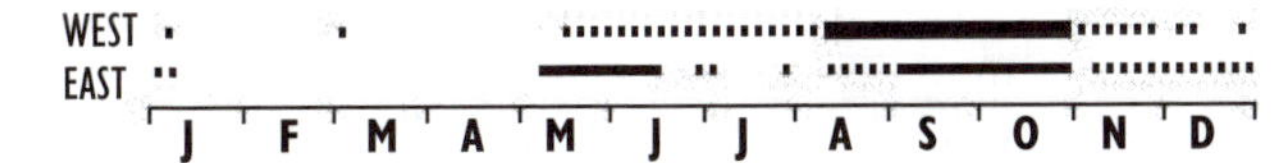

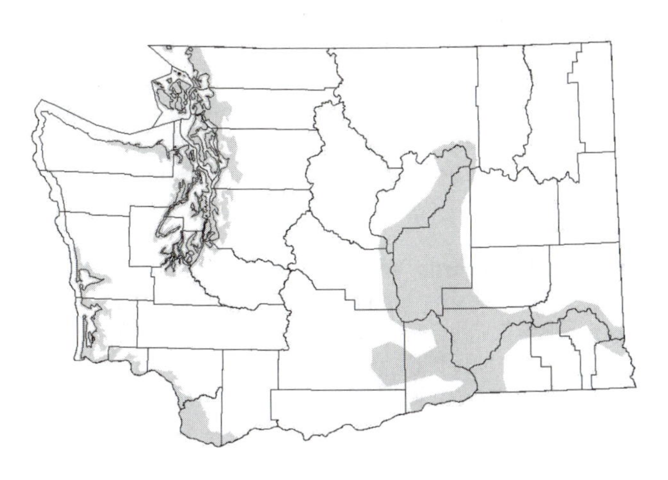

Habitat: Sewage ponds, sheltered bays, tidal flats, and river estuaries.

Occurrence: In 1950, Franklin's Gull was known only from a specimen collected at L. Waughof, Pierce Co., on 25 Oct 1941 (Slipp 1942b) and from two additional specimens collected at the same location in the following fall (Slipp 1943, Jewett et al. 1953). The species now occurs mainly as an uncommon fall migrant in the inland marine waters from King Co. n., and is a rare spring and fall migrant in the se.

By the 1960s two to six birds were found annually in w. Washington, with to five to 15 noted annually during the 1970s and 20 or more per year during the 1980s. Reports decreased to five to 15 per year during the 1990s, in spite of much greater observer effort. Birds are rare except in from the n. Puget Sound region and there are very few records from the outer coast. Single birds have been observed in flocks of gulls foraging at fishing vessels off Grays Harbor, 30-40 km offshore (TRW). Usually found with flocks of Bonaparte's Gulls and sometimes with Mew Gulls (Campbell et al. 1990b).

There were several e. Washington records during the 1960s and 1970s, but the species did not become regular there until the mid-1980s. From 1994 to 1999, the eastside averaged about four per year. In 2000, Franklin's Gulls irrupted into e. Washington, with 44 reported, mostly during late May-early Jun. Aggressive behavior displayed by two of these birds at L. Lenore on 8 Jun was suggestive of nesting (T. Aversa p.c.). Records have been concentrated around the Walla Walla R. delta but have occurred n. to Spokane, Reardan, and Atkin's L., Douglas Co. Almost all fall state records are of imms., with the earliest being found on 22 Jun.

Changes in Washington probably reflect an overall westward range expansion during the last 50 years (Burger and Gochfeld 1994, Johnson 1994a), and perhaps, more locally, an increase in the Oregon breeding population (Littlefield 1990). Fall records may not represent birds on active southbound migration, but rather a post-breeding dispersal: the species is known to disperse 500 km or more before engaging in a directed southward movement (Burger 1972).

Noteworthy Records: West, *high counts*: 11 on 21 Aug 1966 at Bellingham (Wahl 1995); 21 on 8 Aug 1971 at Lummi Bay (Wahl 1995); 13 on 22 Oct 1978 at Everett; 26 on 24-30 Sep 1981 at Snohomish and Everett sewage ponds; 20+ on 16 Sep 1984 at Everett; 20+ on 6 and 25 Oct 1987 at Everett. 15 on 5 Sep 1994 at Everett. *Spring*: 3 Mar 1995 nr. Monroe; 15-16 May 1988 at Ocean Shores; 20 May 1972 at Bellingham; 24 May 1998 at Point No Point; 6 on 30 May 1998 at Silvana; 30 May 1998 at Everett. *Summer*: 14 Jun 1980 at Point No Point; 18 Jun 1994 at Spencer I.; 2 on 20 Jun 1986 at Pt. Roberts; 3 on 22 Jun 1995 at Everett Sewage Ponds; 6 Jul 1995 at Blaine. *Fall*: 19 Aug 1979 at Grays Harbor; 12 Aug 1989, 18 Aug 1990, and 3 Aug 2000 off Grays Harbor (TRW). *Winter*: 1 Dec 1989 at Speiden I.; ad. on 4 Dec 1999 at Sequim; 27 Dec 1980 at Bellingham; 3 Jan 1982 at Kent. **East**, *high counts*: 30 on 25 Oct 1967 at Stubblefield L., Turnbull NWR; 14 on 5 Jun 1993 at Walla Walla R. delta. *Summer*: up to 10 on 8 Jun-2 Jul 2000 at L. Lenore; 25 Jun 1988 at Wallula; 27 Jun 1963 in Grant Co.; 29 Jun 1985 at College Place; 3 on 1-4 Jul 2000 at Wilson Cr. Winter: 10 Nov 1981-2 Jan 1992 at Spokane.

Steven G. Mlodinow

Little Gull *Larus minutus*

Rare spring and fall migrant, very rare in winter and very rare to casual during summer in w. Washington. Casual on eastside.

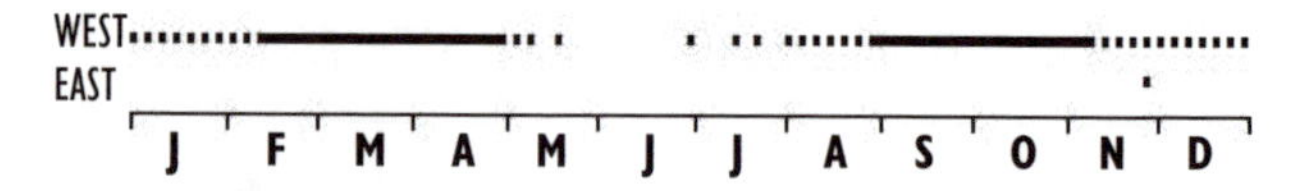

Habitat: Tidal convergences and other marine habitats, lakes, sewage-treatment ponds.

Occurrence: A relatively recent addition to the avifauna of w. N. America, noted first in 1968 in California (Garrett and Dunn 1981). Over 10 are reported annually along the Pacific coast, with about 100 records in Washington through 2000. There were 67 confirmed and probable N. American nesting records from 1962 to 1989, mostly from the Great Lakes and St. Lawrence R. basin (Ewins and Weseloh 1999). Birds in Washington almost certainly originate in N. America though there no recent breeding records known.

The first Washington record was of an ad. at Pt. Roberts on 5 Nov 1972 (Campbell et al. 1974, 1990b). Three were found in Washington during fall 1974. The remainder of the 1970s featured about two Little Gulls per year. In the 1980s, this number increased to about four per year, and in the 1990s, about six annually. Almost always with Bonaparte's Gulls.

Records are almost entirely from the Puget Tough including the San Juan Is. and w. Clallam Co. There is one record from the outer coast and one from e. Washington. Records by county are as follows: Kitsap (29 records), Snohomish (27), King (18), Whatcom (6), San Juan (40), Pierce (4), Island (2), Jefferson (2), Clallam (2), Skagit (1), Grays Harbor (1), and Klickitat (1).

Typically, two to four birds are found each fall, with occurrence generally from late Sep into mid Oct. One or two are noted in spring, mostly between mid-Mar and mid-Apr. Midwinter birds are very scarce, though at least one or two ads. were seen at American L., Pierce Co., for at least 10 consecutive seasons.

Noteworthy Records: West, *high counts*: 3 on 11 Oct 1986 at Everett sewage ponds; 3 on 10-11 Apr 1994 at Point No Point. *Outer coast*: ad. on 21 Sep 1975 at Ocean Shores. *Summer*: ad. on 24 Jun 1985 at Thatcher Pass; ad. on 15 Jul 1989 at Dungeness; 1 on 26-27 Jul 2000 at Point No Point; 1 mid Jul - 1 Aug 2001 at Everett. **East**: 21-26 Nov 1989 at John Day Dam.

Steven G. Mlodinow

Black-headed Gull *Larus ridibundus*

Very rare to rare westside fall migrant. Very rare in winter and spring.

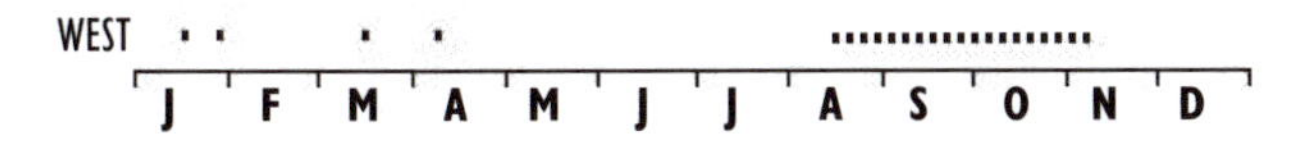

Occurrence: First recorded in the U.S. in 1930, when one was collected in Massachusetts (Emilio and Griscom 1930), Black-headed Gulls were uncommon in New England and the Maritime Provinces by the 1970s. As of 1994 this species had been seen in at least 34 states (Mlodinow and O'Brien 1996). Nonetheless, it still remains fairly rare along the Pacific coast and casual in the interior of w. N. America (Mlodinow and O'Brien 1996).

In Washington, Black-headed Gulls have been seen 15 times, almost always with Bonaparte's Gulls. The first record was of an imm. seen at Ocean Shores on 4 Nov 1972. All records since then have been of ads. in the more protected saltwater from Dungeness/San Juan I. e. Almost half have been found in the large Bonaparte's Gull flocks at the Everett sewage ponds and Point No Point. Dates range from 17 Aug to 6 Apr, with a peak from mid-Sep to early Oct.

The high ad.:imm. ratio of Washington records may well indicate returning individuals, though it may also be due to the difficulty in identifying young birds.

Records: imm. on 4 Nov 1972 at Ocean Shores; ad. on 27 Aug 1986 at Dungeness; ad. on 17 Sep 1987 at Orcas I.; ad. on 5 Oct 1987 at Green L., Seattle; ad. on 20 Dec 1987 at Crockett L.; ad. on 17-31 Jan 1993 at Nisqually NWR; ad. (ph.) on 17 Aug-10 Oct 1993 at Everett sewage ponds; ad. on 27 Oct 1994 at Seattle; ad. on 22 Dec 1994 at Point No Point; ad. on 27 Dec 1994 at Edmonds; ad. on 9-20 Mar 1995 at Point No Point; ad. on 6 Apr 1996 at Point No Point; ad. on 30 Sep 1996 at Everett sewage ponds; ad. on 28 Sep-11 Oct 1997 at Everett sewage ponds (D. Duffy p.c.); ad. on 8 and 17 Mar 1998 at Point No Point.

Steven G. Mlodinow

Bonaparte's Gull *Larus philadelphia*

Locally common to abundant spring and fall migrant on salt- and freshwater, locally uncommon to fairly common winter and summer visitor in w. Rare migrant in mountains. Locally rare to uncommon migrant, rare winter visitor e.

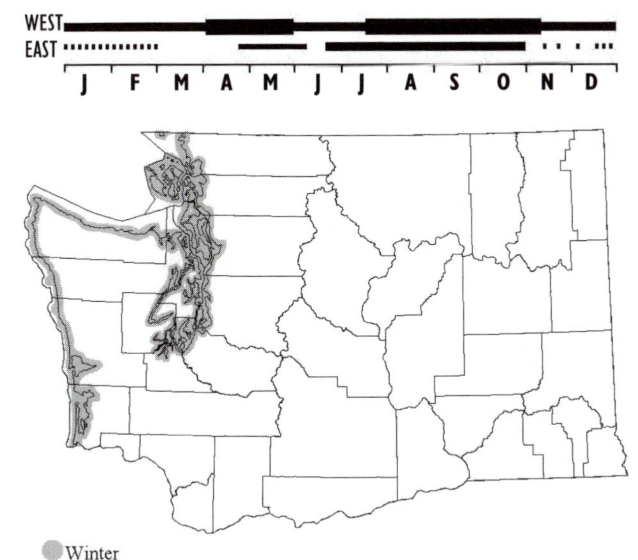

Habitat: Widespread in tidal convergences, shallow intertidal areas, sewage-treatment lagoons and other freshwater habitats.

Occurrence: A conspicuous migrant on the inland marine waters of Washington, widespread in many habitats. Large numbers are consistent at locations such as Deception Pass, Pt. Wilson, Point No Point, Tacoma Narrows, Cattle Pt. on San Juan I., n. Haro Strait, Pt. Roberts, Everett sewage-treatment lagoons, and at Active Pass, in B.C.

Noticeable numbers are absent only briefly in summer, with last flocks of spring migrants noted in late May. Ninety-five percent of 2000 birds at the Nooksack delta on 28 May 1972 were subadults (Wahl 1995), suggesting most northbound ads. had passed earlier. Numbers increase again when the first fall migrants reappear in late Jun-early Jul; early fall migrants may include failed breeders. Some non-breeders spend the brief summer period locally, relatively nr. closest breeding areas in s.c. B.C. (Campbell et al. 1990b).

Outer coastal and offshore waters migration not well known. Surveys off Grays Harbor recorded birds from 100 to 2000 m depth in Apr, with others close inshore in May and Sep-Nov. Briggs et al. (1992) reported the species as a sporadically common migrant over the continental shelf in Apr and Sep. Surveys off Grays Harbor found much lower numbers, however, than in inland waters and Briggs et al. (1992) reported abundances of

Table 24. Bonaparte's Gull: CBC average counts by decade.

	1960s	*1970s*	*1980s*	*1990s*
Bellingham		114.1	25.7	10.5
Edmonds			130.0	61.6
E. Lk. Washington			16.9	1.4
Kitsap Co.		34.7	44.1	57.4
Olympia			296.7	144.5
Padilla Bay			27.7	4.1
Port Townsend			67.9	12.8
San Juan Is.			56.4	17.9
San Juans Ferry		62.6	122.4	1.0
Seattle	216.9	125.4	487.1	98.8
Sequim-Dungeness			4.4	15.6
Tacoma		2343.6	1844.1	1462.0

Bonaparte's Gulls off California of as much as 10 times greater than off Washington. This suggests that many migrants move inland (e.g., see Wahl 1995), taking advantage of extensive, productive inland waters of Washington in migration. E. of the Cascades, birds occur locally in relatively very small numbers during spring migration and in somewhat widespread numbers in the fall.

Spring migrations appear to coincide with herring-spawning events, particularly in the Strait of Georgia. With scoters and other gulls, hundreds of Bonaparte's forage on herring eggs attached to eelgrass and rocks. In migration, this species is often accompanied by and preyed upon by Parasitic Jaegers. In Aug-Sep, flocks of Bonaparte's Gulls are often observed foraging on flying termites, sometimes feeding in company with larger gulls, terns, Common Nighthawks, swifts, or starlings.

Population trends are uncertain, but field observations and most long-term CBC data suggest an overall winter decline. Numbers of wintering birds are larger and more consistent in s. Puget Sound (Table 24) than they are in n. waters.

Noteworthy Records: West, *high counts, Spring*: 5000 on 25-27 Apr 1991 at Bellingham and 2000 on 2 May 1971 at L. Terrell (Wahl 1995). *MESA counts at herring spawn*: 1640 on 27 Apr 1979 at Cherry Pt., 1564 on 22 Apr 1978 at Birch Bay, 1220 on 1 May 1978 at Sandy Pt. *Other counts*: 1509 on 7 May 1979 in n. Rosario Strait and 1197 on 27 Apr 1979 nw. of Patos I. from aircraft (MESA). 2500 on 5 Apr 2000 at Point No Point. 378 on 17 Apr 1994 off Westport. *Fall*: 10,000 on 18 Oct 1994 at Everett STP; 3500 on 24 Oct 1999 at Point No Point; 2000 on 28 Jun 1970 at Bellingham. *MESA*: 1037 on 6 Aug 1978 at Samish Bay, 815 on 17 Sep 1978 in Speiden Channel, 672 on 19 Sep 1979 at Padilla Bay, 631 on 25 Aug 1978 at Dungeness Bay, 570 on 14 Aug 1979 at Port Angeles; from aircraft on 7 Dec 1978: 545 at Vendovi I., 645 at Sinclair I,, 500 n. of Cypress I., and 1200 at Peapods. *Winter*: 2900 on 17 Jan 1984 in Hale Passage, Pierce Co.; 527 on 19 Jan 1986 in Elliott Bay (WDFW aircraft survey); 3500 on 22 Feb 1996 at Point No Point. *CBC*: 7280 in 1979 at Tacoma, 1707 in 1988 at Seattle, 1201 in 1988 at Olympia. **East**, *high counts, Spring*: 50 on 10 May 1995 at Sprague L. *Fall*: 30 on 26 Oct 1996 at Yakima R. delta; 40 on 24 Oct 1998 at Lyons Ferry SP; 121 on 25 Oct 1998 at Walla Walla R. delta, 60 on 21 Nov 1998 at McNary Dam; 46 on 1 Nov at Yakima R. delta and 25 on 6 Nov 1999 at L. Lenore. *Winter*: 3 ad. on 5 Dec 1993 at Banks L.; 1 on 1 Jan 1994 at Kennewick STP; 2 on 1995 CBC at Wenatchee; 60 in mid-Dec 2002 at Moses L.

Terence R. Wahl

Heermann's Gull *Larus heermanni*

Common to locally abundant summer and fall visitor on the ocean coast, Strait of Juan de Fuca, Admiralty Inlet, Strait of Georgia, and adjacent waters. Rare to uncommon s. of Admiralty Inlet and in inland estuaries. Very rare to rare in winter. Very rare on the Columbia R. e. to Tri-cities.

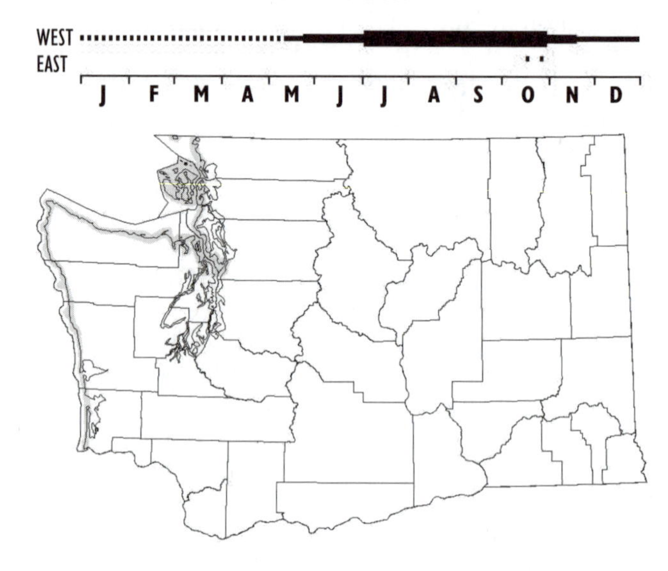

Habitat: Roosts on offshore rocks, islands, and kelp beds, adjacent to foraging areas in tidal convergences. Often bathe in nearby freshwater.

Occurrence: Numbers arrive on the outer coast beginning in Jun, spreading northward and inland in locally large numbers before leaving in late fall. On surveys off Westport records were concentrated from Jul to Oct, with very small numbers before and after that period. On surveys of inland waters in 1978-79 only three of 634 records were prior to Jun, 529 were from Aug to -Oct, 55 were in Nov, and three in Dec (MESA). Hundreds may be present in late Oct-early Nov in Rosario Strait and the e. Strait of Juan de Fuca (MESA). Variable small numbers over-winter, with CBC records of one or two birds reported at a number of locations in inland waters, irregularly at Grays Harbor and apparently at other coastal locations.

Birds occur almost exclusively in coastal nearshore waters and harbor entrances. Of over 100,000 birds noted on surveys in 1972-2000 from Westport, 99% were over water depths of <20 m; one was 64 km offshore. In inland waters, large flocks occur along the Strait of Juan de Fuca, in Admiralty Inlet, n. through Rosario and Haro straits and the San Juan Is., where flocks of 2000 are not uncommon (Lewis and Sharpe 1987). Surveys in 1978-79 (MESA) found lower numbers in the Strait of Georgia n. to Pt. Roberts. Low numbers occur in the large estuaries and in Puget Sound. There are very few records on the Columbia R., with one bird upstream at Richland.

Ads. migrate farther n. than young birds. Campbell et al. (1990b) reported an ad.:imm. age ratio of 91:9 from May to Nov in B.C. Age composition is variable, with a greater proportion of young birds apparently occurring during years of severe ENSO events along the s. Pacific coast (e.g.. in 1983; TRW).

Numbers on the sw. Washington coast increased from 1972 to 1996 (Wahl and Tweit 2000a), likely reflecting, in part, increased breeding success at Isla Raza, Mexico (Islam 2002). Numbers vary and during years of abundance birds are found farther inland, as in 1968 when hundreds were at Pt. Roberts and Sandy Pt. in Whatcom Co., and on Bellingham Bay (Wahl 1995). Classed GL.

Noteworthy Records: West, *high counts, outer coast*: 5000 on 7 Aug 1994 at Ocean Shores; 3010 on 10 Sep 1977 and 3737 on 22 Jul 1991 at Westport. *Inland marine waters*: 1700 on 14 Aug 1994 at Whidbey I.; 1455 on 7 Aug 1978 at Dungeness Spit, 873 on 24 Sep 1979 at Port Angeles, 525 on 5 Oct 1978 at Protection I., 602 on 4 Nov 1979 in s. Rosario Strait (MESA). *Winter-Spring*: 4 Mar 1971 at Gooseberry Pt. (Wahl 1995); 7 Apr 1978 and 22 Feb 1979 at Neah Bay (MESA); 12 Feb 1994 at Tacoma, possibly same 31 Mar 1994 at Olympia; 2 Mar 1995 and 18 Dec 1999-27 Feb 2000 at Tacoma; 16 Dec 1995 at Seattle; 15 Jan 2000 at Orcas and Shaw Is. 2 on 26 Nov 1994, 1 on 8 Jan 1995, 4 on 3 Dec 1995, 10 on 4 Dec 1995, 1 on 15 Jan 1996 and 1 on 1 Dec 1999-30 Jan 2000 at Edmonds. *CBC*: 1-2 each in 11 years from 1973 to 1996 at Grays Harbor, 13 records of 1-2 birds each from 1974 to 1999 in inland waters. *Freshwater*: 1 on 20 Jun 1916 on L. Crescent (Jewett et al. 1953); groups in Jun and Jul 1918 at L. Union, Seattle (Jewett et al. 1953); 1 on 30 Sep 1994 at Marysville STP. **East**: 1 1st-year on 15 Oct 1995 at Roosevelt, Klickitat Co.; 1 2nd-winter on 25 Oct 1995 at Richland.

Terence R. Wahl

Black-tailed Gull *Larus crassirostris*

Casual vagrant.

This ne. Asian species has appeared more than 30 times in N. America with records from as far afield as Newfoundland, Belize, and Sonora. Oddly, there are relatively few Alaskan records, only one from B.C., none from Oregon, and one from California (Lethaby and Bangma 1998, Dunn et al. 2002)

Washington's sole record was furnished by a well-photographed ad. at N. Cove from 3 Aug through at least 20 Oct 2004.

Steven G. Mlodinow

Mew Gull *Larus canus*

Common to locally abundant migrant and winter visitor in coastal and inland marine waters and freshwater habitats, fairly common in winter in offshore waters; locally uncommon in summer. Rare migrant and winter visitor in e.

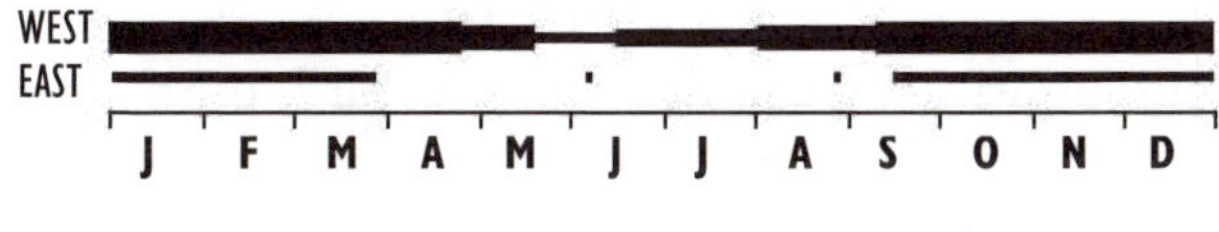

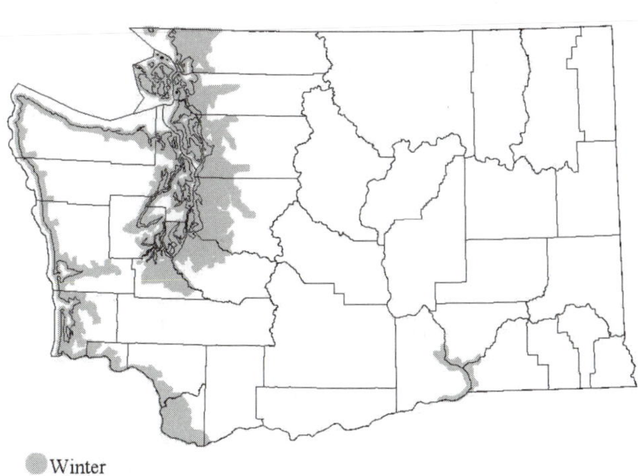

Subspecies: *L. c. brachyrhynchus* (see Moskoff and Bevier 2002).

Habitat: Marine and freshwater habitats, including tidal fronts, convergences, and narrow passages, exposed intertidal flats, sewage-treatment lagoons. Forages in inland fields in winter.

Occurrence: The second-most abundant gull in winter in inland marine waters. Concentrates especially in tidal convergences and fronts, at events such as herring spawns, and in exposed intertidal areas. During 1978-79 MESA surveys birds were widespread, with 2302 records of almost 75,000 birds during 7178 census units. Large numbers readily take advantage of human-provided foraging opportunities at sewage-treatment lagoons (e.g., Everett) and outfalls.

Hundreds forage in fields saturated or flooded during winter and this can result in underestimates of populations made from censuses in marine areas (Wahl 1995); hundreds of birds leave roosts on saltwater at first light, fly inland, and return at dark to roost on log booms, piers, and other situations. Concentrations in 1978-79 included over 1000 each in Speiden Channel, Drayton Harbor, Cherry Pt., Dungeness Bay, and Deep Cr. Other surveys and observations indicated concentrations at many estuaries and tidally active areas like the Narrows and Hale Passage, Pierce Co. (e.g. over 14,000 on the Tacoma CBC in 1979). Long-term CBCs showed sizable variations and no certain trends.

Mew Gulls are regular but less numerous in winter along the outer coast. On surveys off Westport in 1972-2000 there were 215 records with all but eight of these records from Oct to May. Most were inside Grays Harbor. From Jan to Mar small numbers were recorded to the 1000 m depth contour, with none farther offshore. Birds offshore were with Black-legged Kittiwakes. Noteworthy were 300 birds over 50-70 m depths on 11 Jan 1977. Relative scarcity of Mew Gulls offshore is suggested also since Briggs et al. (1992) did not report them from extensive aircraft surveys.

Mew Gulls breed on Vancouver I. (Campbell et al. 1990b) and post-breeders appear in n. inland waters in Aug. Some non-breeders are present each summer: 1978-79 data from Drayton Harbor and Bellingham included small numbers of subadults in Jun (Wahl 1995, MESA). In spring, the proportion of subadults to ads. noted at Drayton Harbor increased from Feb into late Apr (Wahl 1995), suggesting young birds had wintered farther s.

May be increasing in e. Washington at reservoirs and expanded foraging opportunities in recent decades. Most records are from the Columbia R. upstream to the Tri-Cities, and birds are locally uncommon from the Klickitat to the Tri-Cities. Records of one or two birds, however, are still considered worth reporting by field observers.

Noteworthy Records: West, *high counts, Spring*: 1300 on 25 Apr 1979 at Cherry Pt. herring spawn; 1280 on 13 Mar 1978 in Speiden Channel; 1182 on 30 Mar 1978 at Dungeness Bay; 1157 on 29 Mar 1978 at Drayton Harbor (MESA); 2500 on 17 Apr 1999 at Silvana; 5000 on 16 Mar 2001 nr. Woodland. *Fall*: 2007 on 29 Aug 1979 at Deep Cr. (MESA). *Winter*: 1300 on 2 Jan 1980 in flooded field 16-19 km inland in Whatcom Co. (Wahl 1995); 2500 on 26 Feb 1998 at Stanwood and 3370 on 21 Dec 1999 at Monroe 32 km inland. 800+ on 2 Jan 1995 at Point No Point; 2150 on 11 Nov 1996 at Stillaguamish flats. *CBC*: 14,227 in 1979 at Tacoma, 6487 in 1989 at Everett, 4249 in 1980 at Bellingham, 3000 in 1986 at Bellevue. **East**, *Summer*: 1 on 8 Jun 2000 at L. Lenore. *Fall*: 1 on 25 Aug 2001 at L. Lenore; 1 on 21 Sep 1995 at Kennewick. *Winter, high counts*: 30 on 22 Mar 1997 at White Salmon; 12 on 19 Nov 2000 at John Day Dam. *CBC*: 11 in 1997 at Bridgeport.

Terence R. Wahl

Ring-billed Gull *Larus delawarensis*

Common migrant and breeding summer resident, fairly common to locally common winter resident in e. Washington; locally common migrant, summer and winter resident, local breeder on westside.

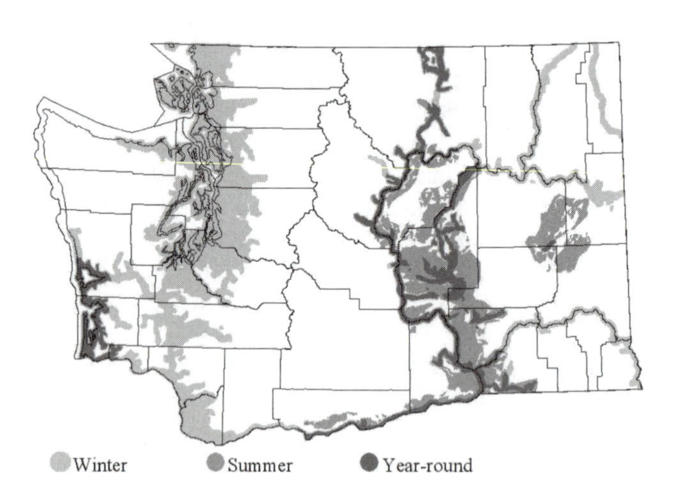

Habitat: Rivers, lakes, impoundments, farmlands, urban parks, parking lots, and fields and estuaries, particularly along the e. edge of the inland marine waters and major coastal embayments. Nests on unvegetated islands.

Occurrence: As it has over N. America, the Ring-billed Gull has increased in abundance in Washington over recent decades (Conover 1983, Ryder 1993). Described as a migrant, casual in winter by Jewett et al. (1953), with a single 1930 nesting record at Moses L. and nesting in the 1950s at Moses L., Sprague L. and Twelve-mile Slough in Adams Co., and on the Columbia R. nr. Pasco. New colonies were subsequently established, some abandoned, some relocated over time (see Conover et al. 1979). Breeders increased to over 17,000 by 1977 with major colonies nr. Richland, at Banks L., Sprague L., and Potholes Res. (Conover et al.

1979). Locations varied into the 1980s, with birds taking advantage of newly created habitat along the Columbia R. (e.g. Crescent I.; Ackerman 1994), and 7000 pairs nested in the s. Columbia Basin in 1996 (Ackerman in Smith et al. 1997). Status change in e. Washington parallels that in B.C., where breeding was not documented until 1966 (Campbell et al. 1990b). Migrants now occur across much of the region.

In w. Washington summering non-breeders and winter residents increased particularly in inland estuaries and adjacent farmlands. Based on 1978-79 data and subsequent censuses (Wahl and Speich 1984, Wahl 1995, 1996), both summer and winter populations increased over two decades. Birds are now widespread in estuarine and waterfront habitats, inland parks, and flooded or freshly plowed fields, where flocks of hundreds are locally common. Low numbers in the San Juan Is. (Lewis and Sharpe 1987) likely reflect less available habitat.

Two nests were found in a Caspian Tern colony in Willapa Bay, and two nests in a gull colony in Grays Harbor in 1976 and subsequently (Penland and Jeffries 1977, Speich and Wahl 1989). In 1995 two pairs nested at Everett: nesting may occur wherever habitat is available.

Distribution on the outer coast is mainly inland. Though birds are numerous in fields, mall parking lots, and park areas along the Chehalis R. at Aberdeen, few are noted 15-20 km to the w. at Westport. During 1972-2001 surveys from Westport, there were only 59 records, of 349 birds, with just 20 nr. fishing vessels over 100-1000 m depth in Jul-Oct. These probably followed California Gulls offshore.

Ring-billed Gulls feed on flying termites and perch on branches to snatch berries off trees (Ennor 1991). This versatility, decreased predation on gulls and eggs by humans, increased availability of food resources due to activities like farming and garbage availability, foraging at picnic areas, and creation of nesting habitats at impoundments and reservoirs and irrigated areas (Conover et al. 1979, Conover 1983, Smith et al. 1997) likely contribute to population increases over N. America. BBS data indicate a 1.8% annual increase in Washington breeding birds (Sauer et al. 2001) and long-term CBC data (Table 25) illustrate widespread winter increases especially on the w. side of the Cascades.

Remarks: Data are minimal, but Ring-billed Gulls were among the most numerous birds killed or "hazed" in Washington for agricultural "damage caused by migratory birds." In 1996, for example, 4066 Ring-bills were killed and 84,770 were hazed (USDA 2001). The average number lethally removed during 1997-2001 was 7367 per year (USDA 2003).

Table 25. Ring-billed Gull: CBC average counts by decade.

	1960s	*1970s*	*1980s*	*1990s*
West				
Bellingham		43.1	71.8	262.3
Edmonds			123.7	424.2
Everett		34.7	186.7	288.6
E. Lk. Washington			49.2	263.9
Grays Harbor		8.5	18.0	21.2
Leadbetter Pt		13.6	63.0	20.5
Olympia			76.0	137.8
Padilla Bay			315.4	201.8
Seattle	86.4	40.8	227.2	461.0
Sequim-Dungeness			7.1	13.2
Tacoma		33.2	50.2	124.9
East				
Spokane	3.1	13.7	80.0	142.4
Tri-Cities		567.4	651.2	911.8
Walla Walla		164.9	0.2	16.0
Wenatchee	75.7	19.2	25.4	30.0

Noteworthy Records: West, *Summer*: 40 on 9 Jun 1976 nesting at Willapa Bay (Penland and Jeffries 1977); 106 on 7 Jun 1982 on Gunpowder I. (Speich and Wahl 1989); 18 on 14 Jul 1977 on Whitcomb I. (Speich and Wahl 1989); 106 on 30 May 1981 on Sand I. (Speich and Wahl 1989); at least 2 pairs nested at Everett in 1995; 1 ad. on 3 Jul 1995 at Diablo L., Whatcom Co. *Fall*: 400 on 29 Jul 1979 at Padilla Bay, 242 on 3 Oct 1979 at Lummi Bay, and 563 on 14 Aug 1979 at Drayton Harbor (MESA); up to 300 non-breeders in summer in 1990s at Bellingham Bay (TRW). *Winter, CBCs*: 2070 in 1999 at Edmonds, 1726 in 1982 at Padilla Bay, 1581 in 1994 at E. Lk. Washington, 1420 in 1991 at Skagit Bay, 891 in 1999 at Bellingham, 826 in 1992 at Seattle, 651 in 1983 at Everett. **East**, *Spring-Summer, numbers at colonies in 1977*: 678 nr. Richland; 1726 at Island 18, Benton/Franklin Cos., 250 at Cabin I., Yakima Co., 5436 at Banks L.; 1702 at Sprague L.; 2292 at Potholes Res. (Conover et al. 1979). *Winter, CBCs*: 1292 in 1994 at Tri-Cities, 1451 in 1995 at Two Rivers, 125 in 1994 at Spokane.

Terence R. Wahl

California Gull *Larus californicus*

Common to locally abundant summer resident and breeder, uncommon to locally common winter visitor in e. Washington. Common to abundant fall migrant, common spring migrant and local summer visitor; uncommon in winter w.

Habitat: Lakes, estuaries, coastal and offshore waters; fields, airports, garbage dumps, sewage lagoons, and outfalls. Breeds on rocks and islands in inland lakes and reservoirs.

Occurrence: The California Gull has undergone a range and population expansion in w. N. America. First Washington colony known in 1932 (Jewett et al. 1953), eight documented by 1977 (Conover 1983) and additional sites later (e.g. Ackerman 1994). About 7000 birds nested in 1996 at a number of sites along the Columbia R. and locally in the Columbia Basin (Smith et al. 1997), and abundance is apparent along marine shorelines and offshore in the fall.

Spring numbers suggest most migrants occur inland. Counts during 1978-79 MESA surveys peaked in the low hundreds. Briggs et al. (1992) reported movement offshore during Mar, and surveys noted many fewer than in fall at "feeds" off Grays Harbor (TRW). Large numbers reported in Apr-May on the Lewis and Cowlitz Rs. during smelt runs (Jewett et al. 1953). Small numbers of non-breeders noted locally in summer in Puget Sound, Hood Canal, and Whidbey I. (Wahl and Speich 1984). MESA surveys noted flocks at a number of locations including Drayton Harbor and Bellingham.

Though some birds cross the Cascades, major rivers appear to funnel fall migrants to the coast (Briggs et al. 1992) where birds linger before moving s. Ads. and first-year birds appear in late Jun (Hunn 1982, Ratoosh 1995, Wahl 1995). Patterns of age classes are variable, however, with subadults often present in numbers. Observations in sw. B.C. in the 1950s and 1960s found birds hatched and marked in Saskatchewan, Alberta, N. Dakota, Montana, Wyoming, Idaho, Washington, Oregon, and California (Wahl 1975, 1995, Campbell et al. 1990b). Flocks of hundreds to thousands widespread along shorelines and estuaries of the Strait of Georgia, the Strait of Juan de Fuca, and the ocean coast. One aerial census along the s. shore of Strait of Juan de Fuca in late Aug 1979 recorded over 8000, along with large numbers of unidentified gulls (MESA).

Surveys during Jul-Oct 1971-2001 found these the most numerous gull over the continental shelf and adjacent oceanic waters. About 80% of 250,000 recorded over the outer continental shelf were at fishing vessels (TRW). Vermeer et al. (1989) suggested that nocturnal feeding under vessel worklights (Wahl 1977) and daytime scavenging could explain the dominance of California Gulls far offshore. First-year birds increased with distance from shore and, at farthest-offshore points, outnumbered other age classes, possibly due to competition nr. shore with ads. and other larger gulls (e.g., 40 of 44 birds at a research vessel stopped-on-station 135 km offshore on 17 Sep 1976 were birds hatched that summer [TRW]).

Population increase followed creation of major reservoirs in e. Washington, expansion of irrigation and farming, increased garbage and other urban food sources, and reduced predation from shooting, egg taking, and colony disturbance. The introduction of industrial fishing by foreign vessels in the 1960s with its huge amount of bycatch/waste likely increased the survival of first-year birds, which normally suffer high mortality.

Recent trends uncertain. Off Grays Harbor from 1971 to 1998, no overall trend was evident (Wahl and Tweit 2000a), and "average" numbers were present from 1999 to 2001 (TRW). CBC data, complicated by numbers of unidentified gulls, indicate no certain trend.

Remarks: Total population estimated at 200,000 by Enticott and Tipling (1997). The phase-out of foreign fishing fleets off Washington in the mid-1970s and expansion of fishing/processing off B.C. may have affected regional offshore distribution. From 1997 to 2001 an average of 2224 per year were killed in a program to protect juvenile salmonids on the Columbia R. (USDA 2003).

Noteworthy Records: West, *Spring*: 341 on 23 May 1979 at Bellingham Bay (MESA); 550 on 16 Mar 2001 at Woodland (SGM); 415 on 2 Apr 1994 off Grays Harbor. *Summer*: 125 on 1 Jun 1979 at Bellingham Bay; 104 on 7 Jul 1979 at Drayton Harbor (MESA);140 on 24 Jun 1982 at Commencement Bay. *Fall, off Grays Harbor*: 3411 on 31 Jul 1990, 2966 on 17 Aug 1991, 6093 on 12 Sep 1989; 11,396 on 15 Oct 1972 (Wahl 1975). *Outer coast*: 1407 on 7 Nov 1981; 15,000 on 1 Aug 2001 and 10,000 on 28 Oct 2001 at Ocean Shores (SGM); 5000 on 19 Sep 1998 and 7000 on 22 Jul 2001 off Cultus Bay (SGM); 2298 on 15 Aug 1978 at Port Angeles, 3436 on 30 Aug 1979 at Neah Bay, 1800 on 15 Aug 1978 at Jamestown, 2006 on 29 Aug 1979 at Deep Cr., 2683 on 15 Aug 1978 at Dungeness Bay, 1284 on 30 Aug 1978 at Clallam Bay (MESA). *Winter*: 2 on 3 Dec 1983, 1 on 16 Jan 1993, 4 on 28 Mar 1982 off Grays Harbor. **East**, *Spring*: 1000 on 30 Mar 1996 at Wallula. *Summer, estimated nesting in 1977*: 1690 at Banks L., 4 at Cabin I.,Grant Co., 5910 at Island 18, Franklin Co., 856 at Little Memaloose I., Klickitat Co., 960 at Miller Rks., Klickitat Co., 436 at Potholes Res., 3600 at Richland , 428 at Sprague L., 4380 at Three Mile Canyon across Columbia R. in Oregon (Conover 1983). Crescent I., estimated 1000 nests of Ring-billed and California gulls in 1994 (Ackerman 1994). *Winter, CBC*: 321 in 1994 at Tri-Cities.

Terence R. Wahl

Herring Gull *Larus argentatus*

Uncommon to locally common migrant and winter visitor in w. Washington, fairly common to common migrant and winter visitor in e. Very rare summer visitor w. and e.

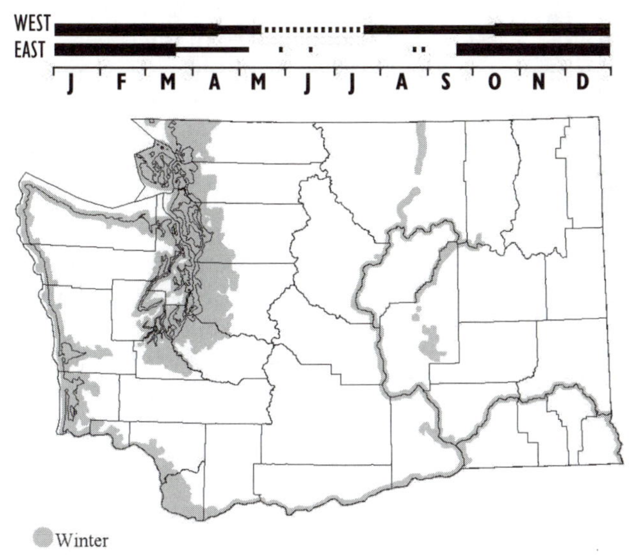

Subspecies: *L. a. smithsonianus.*

Habitat: Oceanic and nearshore marine waters, lakes and rivers, farmlands, dumps, playfields, newly plowed fields.

Occurrence: Published historical status is confused. Jewett et al. (1953) listed Thayer's Gull as a race, and field observers apparently seldom distinguished between that and the Herring Gull until the species were split in 1972 (AOU 1973). As in B.C. (Campbell et al. 1990b), many records attributed to the Herring Gull prior to at least 1973 may pertain to either species. The Herring Gull is predominant along the coast and offshore, and Thayer's Gull is more widely distributed in inland marine waters (e.g. MESA, Wahl 1975).

Primary distribution in inland waters along the Strait of Juan de Fuca: almost all records of more than a few birds during 1978-79 MESA surveys were from the Port Angeles/Ediz Hook area, plus a few counts of flocks at Deception Pass and in Georgia Strait. Though small numbers breed in s. coastal B.C. (Campbell et al. 1990b), only small numbers of non-breeders are reported in Washington in summer.

On 1972-2000 surveys off Grays Harbor relatively few birds were in nearshore waters; most winter birds were over depths of >100 m to the limit of surveys over oceanic waters. During research-vessel cruises birds were recorded to 170 km offshore, dispersing offshore in fall, being widespread in mid-ocean in winter, and returning to land in spring (Sanger 1970, 1973, Morgan et al. 1991, Briggs et al. 1992). Off B.C., Morgan et al. (1991) found Thayer's vastly outnumbered Herring Gulls off sw. Vancouver I. in Nov. Off Washington birds concentrated at fishing vessels, though there were few surveys during the species' main winter season.

This gull is noted feeding in flooded or saturated fields, golf courses, and parks in winter. Like other species, noted feeding on spawning salmon in B.C. (Campbell et al. 1990b), and large numbers are noted at dams and at fish runs in migration and winter on the lower Columbia R. (Jewett et al. 1953) and upstream to the Tri-Cities (SGM).

CBC data suffer statewide from identification confusion and are unsuitable to determine population trends. Many historical published field reports are also uncertain.

Remarks: The identification of large, black-wing-tipped *Larus* gulls may be as difficult in Washington as anywhere. Not only do Ring-billed, California, Herring and Thayer's gulls occur regularly in many plumages, but confusion is also possible due in extensive and variable intergradation of Western and Glaucous-winged gulls (see Scott 1971), and hybridization between Herring and Glaucous-winged gulls in Alaska (e.g., Patten and Weisbrod 1974). A reported 614 Herring Gulls were reported killed during salmonid predation-control actions in 2001 (USDA 2003), though identification may be questionable.

Noteworthy Records: West, *Spring*: 102 on 28 Mar 1982, 654 on 2 Apr 1994, 520 on 22 Apr 1995 and 132 on 13 May 1979 off Westport; 3300 on 16 Mar 2001 at Woodland. *High counts, Fall*: 289 on 29 Oct 1978, 101 on 3 Oct 1979 at Ediz Hook, 96 on 6 Nov 1978 at Deception Pass, 80 on 19 Sep 1978 over Alden Bank (MESA); 517 on 3 Oct 1976, 145 on 13 Nov 1982 off Westport. *Winter*: 135 on 15 Jan 1994 and 2 Feb 1992 off Westport; 146 on 1 Dec 1978 at Port Angeles (MESA). CBC: 1311 on 1 Jan 1986 at Cowlitz-Columbia. **East**, *Fall*: 1 on 26 Aug 1995 at Cusick; 11 on 1 Oct 1995 at the Walla Walla R. delta. *Note*: 2000 reported on 8 Jul 2000 at the Walla Walla R. delta require further documentation.

Terence R. Wahl

Thayer's Gull *Larus thayeri*

Common migrant and winter visitor in w. Washington, casual in summer. Rare migrant and winter visitor e.

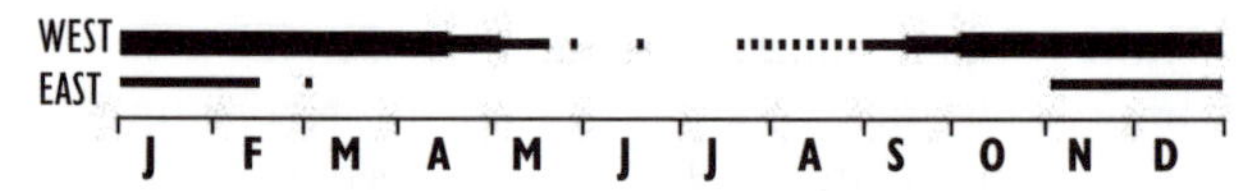

Habitat: Offshore and nearshore waters, shorelines, bays, sewage-treatment ponds, farm fields, athletic fields, garbage dumps. Roosts on islands, spits, log booms, waterfront buildings, fields.

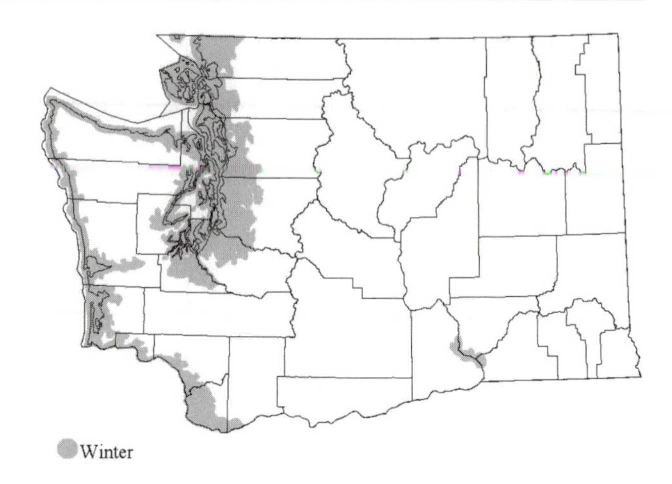

Occurrence: One of the three or four most numerous gulls wintering in nw. Washington and sw. B.C. Long-term status and abundance of Thayer's Gull uncertain due to taxonomic changes: once a species, then a subspecies of the Herring Gull ("Thayer Herring Gull" in Jewett et al. 1953) and a full species again in 1972 (AOU 1973). This confused field identification, counts, and knowledge of the status of the two species in the ne. Pacific region, the Thayer's primary winter range.

Seasonally widespread, generally uncommon but locally common to abundant in inland marine waters; fairly common to common on the outer coast. From Oct to Apr, ads. appear to outnumber ad. Herring Gulls at Puget Sound sewage-treatment ponds by a wide margin (Mattocks et al. 1976). Thayer's probably is more numerous than CBC records indicate (Lewis and Sharpe 1987, Wahl 1995). Flocks often roost on small rocks and islands remote from observers and forage more often in tidal convergences and deeper waters than other larger gulls.

On 1978-79 MESA surveys Thayer's occurred along the Strait of Juan de Fuca and n. in the San Juan Is., Whatcom Co., and adjacent B.C. waters while Herring Gulls concentrated along the Strait of Juan de Fuca. CBC data and other reports show numbers of Thayer's regularly at Ediz Hook and the mouth of the Elwha R. on the Strait of Juan de Fuca and locally in s. Puget Sound at Tacoma. Thayer's may be relatively less common at garbage dumps and more frequent at sewage-treatment

Thayer's Gull (G. Scott Mills)

ponds than larger gulls (TRW), due to type of food or competition. Birds forage in inland fields, though in smaller numbers than other species.

Off Grays Harbor from 1972 to 2000 there were 98 records of 277 birds from mid-Sep to 1 May, mostly nr. fishing vessels, to 180 km offshore (TRW). Herring Gulls were noted more frequently and in larger numbers though Morgan et al. (1991) reported Thayer's Gulls outnumbered Herring Gulls by 9:1 off Vancouver I. Similar to the more inland, nearshore distribution of Thayer's Gull in Washington, Vermeer and Morgan (1997) found they vastly outnumbered Herring Gulls in a major inlet on the w. coast of Vancouver I. during Nov-Feb surveys and described large concentrations at herring spawns, and 7000 were reported at the Delta Dump, lower mainland B.C. on 29 Oct 1989. Migration essentially undescribed. Birds appear in mid-late Sep, with records from Jul to Aug. Numbers decrease in late Apr, with small numbers noted into May, and records in Jun.

Jewett et al. (1953) did not note eastside occurrence. Records now indicate that Thayer's Gull may be a regular migrant, with reports from fall into midwinter in very small numbers: one or two birds, mainly along the Columbia R. and lower Snake R. where gulls and observers concentrate, upstream to the Tri-Cities.

Remarks: Thayer's may be reclassified, again, as a subspecies of the Iceland Gull (Godfrey 1986, Campbell et al. 1990b).

Noteworthy Records: West, *Spring, late*: 1 ad. on 31 May 1982 at Seattle (Ratoosh 1995). *Fall, early*: 1 on 16 Jul 2000 at Dungeness; 2 on 24 Sep 1995 at Anacortes. *Winter, high counts*: 1950 on 6 Nov 1999 at Ediz Hook; 500 in 1993-1994 at Tacoma; 148 on 11 Jan 1979 on Bellingham Bay (MESA). CBCs: 332 in 1992 at Tacoma, 285 in 1985 at Edmonds, 159 in 1980 at Bellingham. **East**, *Spring*: 2 on 1 Mar 1998 at Wallula. *Fall, early*: 1 ad. on 28 Oct 1995 at Roosevelt, Klickitat Co. *Winter, late*: 1 on 17 Feb 1996 at Richland; 1 on 16 Feb 1997 at Yakima R. delta.

Terence R. Wahl

Iceland Gull *Larus glaucoides*

Very rare winter visitor.

Casual in winter in the ne. Pacific s. to California. There are eight records of the Iceland Gull accepted by the WBRC and several unaccepted reports. Listed as casual in B.C., with one record from Shuswap L. in Jun 1971: several reports from nr. Vancouver, B.C., attributed to variants of *thayeri* by Campbell et al. (1990b) who list *kumlieni* as likely in the region. The ultimate status of the species may depend not only on documentation and but also taxonomy of *thayeri*, *glaucoides*, Kumlien's Gull, *kumlieni* (see Cramp et al. 1983, Harrison 1983, Campbell et al. 1990b, Enticott and Tipling 1997, Snell 2002). Washington records have been attributed simply to *L. glaucoides.*

Records accepted by WBRC: 1st-year bird on 16 Apr 1977 offshore sw. of Grays Harbor; 1st-year each on 17 Mar 1986 and 20 Mar 1989 (both ph.) at Port Angeles (Tweit and Skriletz 1996); 1st-year on 25 Nov 1990 at Wallula (Tweit and Skriletz 1996); 1 1st-winter on 7 Dec 1991 at Banks L.; 1 on 18 Dec-8 Jan 1995 at Clarkston (Aanerud and Mattocks 1997); 1 1st-year at Samish Flats on 6 Apr 1997; 1 ad. at Tacoma on 8-24 Jan 2000; 1 ad. at Clarkston on 2-21 Mar 2002.

The specimen cited in Jewett et al. (1953) has been determined to be a pale California Gull.

Terence R. Wahl

Lesser Black-backed Gull *Larus fuscus*

Casual vagrant.

First record, from the Walla Walla R. delta, 6 Feb-3 Mar 2000. Apparently the same ad. was also there during the next three winters. This species was first recorded in the U.S. during 1934 (Edwards 1935), and records slowly accumulated thereafter until the mid-1970s when the number of reports from N. America began to rise rapidly (Post and Lewis 1995). Birds are now recorded annually in California. Subspecies likely *graellsii.*

Steven G. Mlodinow

Slaty-backed Gull *Larus schistisagus*

Very rare winter visitor in w. Washington.

The Slaty-backed Gull was unrecorded s. of Alaska prior to 1974 when one was found in B.C. (Roberson 1980). Washington's first, in 1986, represented only the fourth regional record. Since then, apparently increasing in N. America, with the species now a locally uncommon resident in w. Alaska (Heinl 1997), where breeding was first recorded in Jul 1996. Since 1993, Slaty-backed Gulls have been recorded annually in se. Alaska and nr. Vancouver, B.C.

There are five Washington records, all of ads.: 1 (ph.) at Elwha R. mouth, 31 Dec 1986-4 Jan 1987; 1 (ph.) at Tacoma 1 Jan-28 Apr 1994 (Tweit and Skriletz 1996); 1 seen at Nisqually NWR and Thurston Co. dump, 30 Dec 1995-22 Feb 1996; 1 at Tacoma 3-6 Jan 1998 (likely same bird found in 1994); 1 at Tacoma 7-28 Nov 1999, which was likely a different bird than those in Tacoma in 1994 and 1998.

Most Slaty-backed Gulls recorded in the Pacific Northwest have occurred in flocks of large gulls including also the uncommon Glaucous Gull. Regional records have been from late Dec to mid-Mar.

Steven G. Mlodinow

Western Gull *Larus occidentalis*

Common to abundant resident and breeder on outer coast; fairly common on outer Strait of Juan de Fuca; rare to uncommon in inland marine waters; rare in e. Washington.

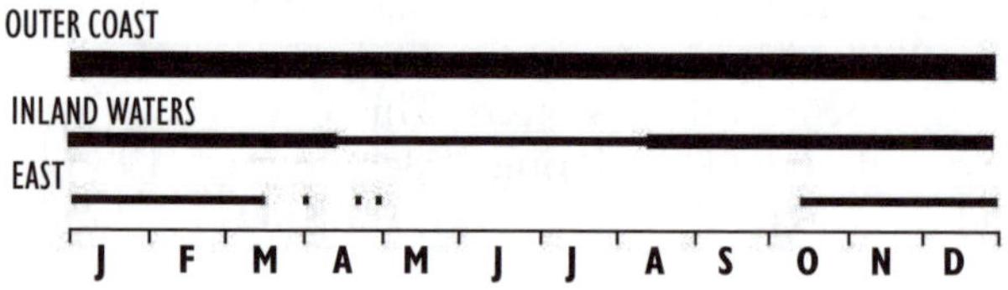

OUTER COAST

INLAND WATERS

EAST

J F M A M J J A S O N D

Winter Year-round

Subspecies: *L. o. occidentalis.*

Habitat: Nests on offshore rocks, islands along the outer coast; forages widely in marine habitats, at fishing vessels, processing operations; very local on freshwater.

Occurrence: The Western Gull, resident on the outer coast, has perhaps the smallest breeding population of any *Larus* gull in w. N. America (see Vermeer et al. 1993). Birds breed primarily along the s. Washington coast, with about 4000 pairs nesting from Destruction I. s. to Oregon (Speich and Wahl 1989). This is essentially the n. limit of the nesting range and overlaps that of the Glaucous-winged Gull. Non-breeding Westerns occur in inland marine waters, concentrating in the w. Strait of Juan de Fuca but individuals or small numbers are found s. to Tacoma on CBCs. Non-breeders are reported along sw. Vancouver I., B.C. (Campbell et al. 1990b), and some are reported in se. Alaska (see Heinl 1997). Birds have been reported fairly regularly in recent years from fall into spring inland in sw. Washington along the Columbia R. as far n. as Bridgeport, and e. to Clarkston.

The "northern" Western Gull exemplifies the challenges of gull identification in Washington and of interpretation of historic literature and even recent data of a common species. Early observers suggested hybridization with other species (see Jewett et al. 1953) and recent authors (e.g. Hoffman et al. 1978, Bell 1996) documented hybridization with the Glaucous-winged Gull. The area of hybridization in Washington and Oregon has increased over time (Bell 1996), with hybrids difficult to identify and many reports (e.g., CBC records) and data from field surveys must be interpreted conservatively (e.g., Wahl and Tweit 2000a). A recent proposal to name these hybrids "Olympic Gulls" (Boekelheide 1998) indicates the extent of hybridization, even in inland marine waters. "Western Gulls" nesting in B.C. are probably hybrids (Campbell et al. 1990b) as are most non-breeders occurring in se. Alaska (Heinl 1997). Most descriptions of Western Gulls occurring in e. Washington have not established that these were not hybrids—most noted e. of Klickitat probably are hybrids.

Birds are attracted to fishing vessels, though apparently less so than Glaucous-winged Gulls and much less so than California Gulls (Wahl and Heinemann 1979, Wahl and Tweit 2000a) and concentrate nearer shore (Wahl 1975). Reproductive success appears related to oceanographic conditions. Though data for Washington are lacking, Briggs et al. (1987) described declines in California populations in 1977-78 and 1982, which were periods of ocean warming.

Remarks: Pierotti and Annett (1995) raised conservation issues relating to the species' relatively small population and restricted distribution: pesticide contamination, hybridization, ocean productivity declines, concentration of 30% of the breeding population at Se. Farallon I., off San Francisco, and that the species is viewed as a predator on other protected seabird species. Study may determine that Western and Glaucous-winged gulls are one species.

Terence R. Wahl

Glaucous-winged Gull *Larus glaucescens*

Common to abundant resident on marine waters, shorelines, and westside lowlands. Fairly common along larger rivers, rare to locally uncommon winter visitor elsewhere in e. Washington; has bred locally inland along the Columbia R.

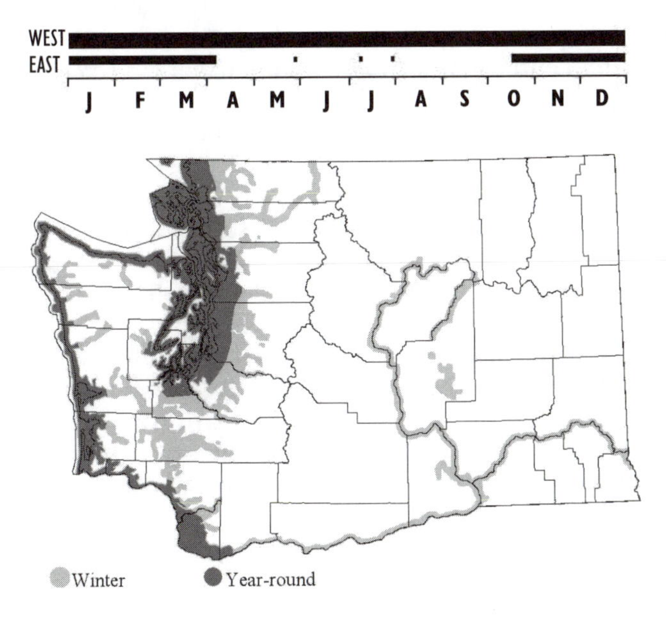

Habitat: Nests on offshore rocks and islands, breakwaters, dredge-spoil islands, roofs, pilings. Roosts also on log booms and piers. Forages widely in marine habitats, at fishing vessels and processing plants, locally on freshwater, soaked/flooded fields, garbage fills, parks.

Occurrence: The status of the Glaucous-winged Gull involves extensive hybridization (or intergradation) with the Western Gull from mid-Oregon n. to the Strait of Juan de Fuca (Hoffman et al. 1978, Bell 1996). Intergrades are numerous, even in inland waters, and often only a minority have a typical Glaucous-winged wing-tip pattern. Some CBCs and surveys (e.g., Wahl and Tweit 2000a) have "lumped" the species.

Birds range farther offshore in fall and winter (Sanger 1972, Briggs et al. 1992) with some, mostly first-winter birds, shifting s. (Campbell et al. 1990b). First-winter birds from inland waters colonies have dispersed into interior B.C., the Columbia Basin, and Baja California, and some older birds may continue wandering: a fledgling banded on Protection I. was found 18 years later in Eureka, California (TRW). Post-breeding ads. from the San Juans are widespread in Puget Sound; some beg on window ledges in downtown Seattle. Flocks occur at seasonal events such as smelt runs on the Lewis and Cowlitz Rs. (Jewett et al. 1953) and herring spawns (Wahl et al. 1981, Wahl 1995). Birds winter well inland along almost all major rivers. On Jul-Oct surveys off Grays Harbor about 67% of over 161,000 Glaucous-winged/Western gulls were within 6-7 km of shore, 19% were offshore at fishing vessels (compared with 83% of California Gulls; TRW). In summer non-breeders are dispersed: an estimated 14,000 in May-Jun 1982 were far from colonies in Hood Canal and Puget Sound (Wahl and Speich 1984).

In the late 1980s, some 37,000 birds bred at over 100 Washington sites, with about two-thirds in the San Juan Is., the e. Strait of Juan de Fuca and Georgia Strait including Protection, Colville, Smith, and Minor Is. Large colonies are at Tatoosh, Carroll, and Destruction Is. on the n. outer coast. S. of Pt. Grenville, colonies are in Grays Harbor and Willapa Bay at Gunpowder and E. Sand Is. (Speich and Wahl 1989). S. of Admiralty Inlet and in urban areas most nests are on human-provided situations (Eddy 1982, Speich and Wahl 1989) with, for example, 500-1000 nests on roofs at Bellingham in 2003 (TRW).

The B.C. population increased by 3.5 times in 50 years and numbers in Strait of Georgia colonies doubled in about 30 years (Campbell et al. 1990b, see Hatch 1993b). A large colony at Willapa Bay doubled from 1977 to 1981 to over 3000 pairs (Speich and Wahl 1989). Increases are attributed to garbage, sewage, fisheries discards, plowed or flooded fields, hatcheries, bird feeding, cessation of shooting and egg taking, and provision of nest sites (see Campbell et al. 1990b, Paulson 1992, Vermeer et al. 1993, Wahl 1995). Reports in the late 1990s indicated declines in numbers at a number of sites in the Strait of Juan de Fuca/San Juan I. area (TRW).

Breeding success varies greatly with natural food availability and other factors. Production at the Swinomish colony nr. Anacortes dropped about 50% in 1977, following closure of an adjacent landfill, while coastal colonies were apparently normally productive. Seattle CBC numbers declined after 1964, likely reflecting changes in garbage-disposal practices. Birds on Protection I. showed normal productivity while near-total nesting failure in San

Juan Is. colonies in 1964 and 1967 was attributed to visits by humans, often with dogs, into colonies.

Eastside occurrence noted in 1954 (Weber and Fitzner 1986, see Binford and Johnson 1995), and in winter 1962-63 one in Spokane was "noteworthy." Numbers increased at Banks L. and on the Columbia and Snake Rs. (e.g., six birds per year in the 1970s, 17 in the 1980s, 31 in the 1990s on Tri-Cities CBC). Birds now common at dams on the lower Snake R. and downstream from the Tri-Cities. Breeding reported on Miller I. and on Island 18 nr. Richland in 1981, and subsequently in Klickitat Co., at Priest Rapids, nr. Richland, and possibly at Walla Walla R. delta (Smith et al. 1997). Though the Island 18 bird was shown to be a hybrid Glaucous-wingedXWestern gull (Weber and Fitzner 1986), most inland breeders appear to be Glaucous-winged Gulls.

Remarks: Hybrids with Herring and Glaucous gulls from Alaska occur. Garbage dumps, airports, fields, and productive intertidal areas and roosts are often relatively close together, nr. urban centers, and conflicts with human activities often result (see Vermeer et al. 1993). Impacts on other species need to be monitored.

Noteworthy Records: *High counts, Summer*: 8000 on 26 Jul 1997 at Cultus Bay (SGM); 4043 on 5 Jul 1978 at Tatoosh I., 2413 on 23 May 1978 at Swinomish Channel and 2374 on 3 Aug 1979 at Protection I. (MESA). *Fall*: 6628 on 15 Sep 1978 at Port Angeles (MESA). *Winter, CBCs*: 13,809 in 1961 at Seattle; 11,493 in 1979 at Tacoma.

Terence R. Wahl

Glaucous Gull *Larus hyperboreus*

Rare to locally uncommon migrant and winter visitor.

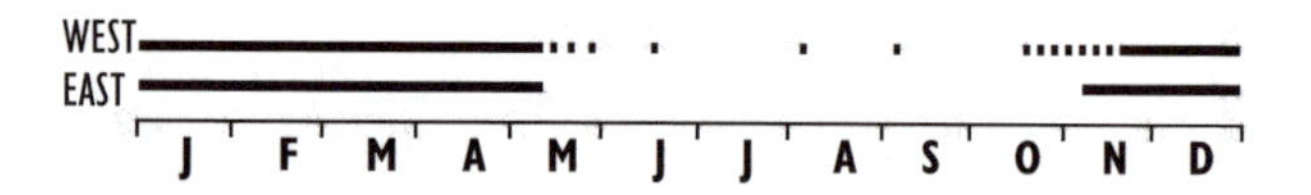

Subspecies: *L. h. barrovianus.*

Habitats: Open water, bays, beaches, fields, garbage dumps. Offshore waters during migration.

Occurrence: Glaucous Gull reports have increased over time. Jewett et al. (1953) listed only three specimens and four sight reports, all in w. Washington. Known status today may reflect the increased number and capability of observers in the 1980s and 1990s.

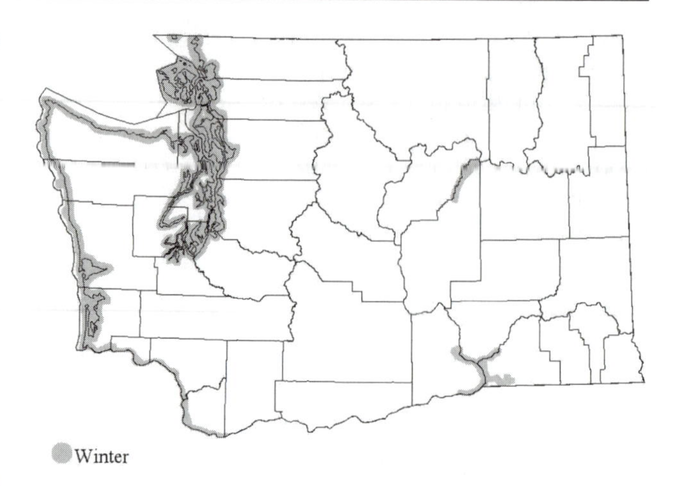

Records in w. lowlands are widespread, and the creation of reservoirs behind dams and of pothole lakes in e. Washington almost certainly has attracted more birds. Westside records average about eight per year, with four on the eastside (SGM). Four birds in Apr and six in May on surveys off Grays Harbor were apparently moving n., with most seen over the outer continental shelf 35-55 km offshore (TRW), suggesting migration.

Almost all birds noted are first-winter birds: Campbell et al. (1990b) reported that 87% of birds aged on CBCs in B.C. were first- or second-year birds. Though the conspicuous first-winter plumage may more readily draw observers' attention, this may also be because young gulls wander more frequently and farther than ads..

Variable numbers winter statewide, with the number of records influenced by CBC efforts. Number of occurrences per year on CBCs in the 1990s appear to correspond with other reports of abundance. Reports usually of single birds, though reports of two to six birds increased in the 1990s. Though fewer reports of individual numbers of birds were published in the late 1980s and 1990s, noticeably above-"normal" numbers occurred in fall-through-spring in 1970-71, 1975-76, 1981-82, 1983-84, 1986-87, 1989-90, 1990-91, 1992-93, 1994-95, and 1995-96. Counts and observers are numerous and more suitable gull habitat exists in w. Washington but reports from e. CBCs increased in the 1980s and 1990s.

Remarks: Hybridizes with Herring Gull ("Nelson's Gull") and Glaucous-winged Gull in Alaska. Confusion with "pure" Glacous-winged Gulls is suggested by the report of 50 to 500 *hyperboreus* at a cannery in Westport in May 1920 (Jewett et al. 1953). Observers may also confuse albinistic or bleached-out spring and summer Glaucous-winged Gulls with first-winter Glaucous Gulls. Jewett et al. (1953) listed the nominate subspecies *hyperboreus* as present.

Noteworthy Records: *Late Spring-Summer*: 1 on 18 May 1968 at Leadbetter Pt.; 2 on 28 May 1971 at Ocean Shores; 1 on 14 Jun 1991 at Ocean Shores. *Fall, early*: 1 (ph.) on 5 Aug 2001 at Ocean City; 1 on 3 Sep 2000 at Neah Bay. *Winter, high counts*, **West**: 6 on 13 Mar 1994 at Woodland; 3 on 16 Jan 1971 at Drayton Harbor (Wahl 1995); 3 on 1 Jan 1978 at Tacoma; 3 in 1999 at Grays Harbor CBC; 2 on 17 Dec 1994 at Grays Harbor; 2 on 21 Dec 1996 at Grays Harbor. **East**: up to 6 in Dec 1985-Feb 1986 at Tri Cities; 5 on 28 Jan 1993 at Walla Walla R. delta; up to 4, including 1 ad., in Jan-Feb 1982 at Tri-Cities; 3 1st-year birds on 14 Dec 1980 at Tri-Cities area; 6 in Jan 1997 and 5 in Jan 1998 at Two Rivers CBCs; 2 on 23 Dec 1990 at Spokane; 2 on 2 Jan 1995 at Two Rivers.

Terence R. Wahl

Great Black-backed Gull *Larus marinus*

Very rare vagrant

Washington's first Great Black-backed Gull, apparently a second-year bird, was extensively photographed at Renton 12 Jan. through 16 Feb. 2004. The only previous records w. of the Continental divide are from Kodiak I., Alaska, Feb. to Apr. 1995 and Kamloops, B.C., 18 to 23 Dec. 1988 (Campbell et al. 1990b, Gibson and Kessel 1997). This species has been undergoing a substantial range expansion in N. America. In 1931, it bred only on Canada's e. coast and was a casual winter vagrant w. to the w. Great Lakes (AOU 1931), but now it has bred as far w. as Wisconsin, is uncommon during winter on the w. Great Lakes, and there are multiple records from as far w. as Colorado (Sibley 2003).

Steven G. Mlodinow

Sabine's Gull *Larus sabini*

Fairly common to common migrant over offshore waters; rare to uncommon migrant in inland marine waters, very rare in winter. Very rare spring, rare fall migrant in e.

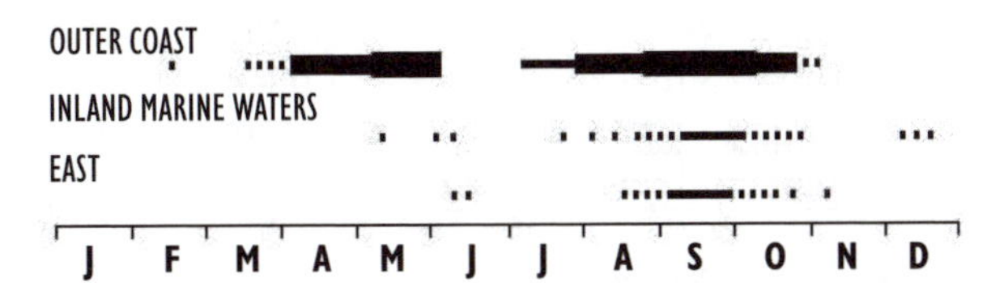

Habitat: Outer continental shelf during migrations, inland marine waters and reservoirs and lakes in w. and e. Washington.

Occurrence: Increased observer effort, especially offshore, explains the difference between status of Sabine's Gull known today compared to 1953 when Jewett et al. stated "definite records are few" and described the species as an apparently rare coastal migrant. It is a regular migrant on the ocean, occurring much less frequently inland.

Spring migration is brief, with most birds passing in May. Failed breeders may depart from Arctic nesting areas in early Jul (Morgan et al. 1991) and successful breeders and young occur later during the protracted fall movement. Numbers observed in fall are much greater than those in spring. Infrequent Jun records represent either late or early migrants. Winter records are few, with several records for Dec, Feb, and Mar from B.C. (Campbell et al. 1990b). One CBC report and an 1899 winter report for Washington could have been misidentifications.

Offshore abundance varies considerably, with apparent peaks in 1971, 1986, 1987, 1992, 1995, 1996, 2000, and 2001 (Wahl and Tweit 2000a, TRW). A record of 300 in 1963 in e. Washington represents an extremely unusual occurrence. Apparent increases in records in inland waters of w. and e. Washington (two and seven birds per year, respectively) likely resulted at least in part from increased observer effort, though eastside reports appeared to indicate increases in 1995 and 2000.

Birds migrate widely over the continental shelf. Competition with larger species and food-item size likely explain departure after a brief investigation of feeds at fishing vessels. On trips off Grays Harbor in late 1971-2000 a total of 6634 birds were recorded. Only 62 were first-year, noted as early as 12 Aug, though most were from late Aug into Oct.

Flocks concentrate at productive areas off the Strait of Juan de Fuca, particularly at Swiftsure Bank, with the largest number recorded in Washington just s. of there. Though over 2800 birds were noted at Swiftsure during the fall migration in 1989 (M. Force p.c.), very few were noted off Grays Harbor, suggesting that flocks disperse when they leave areas of high food concentration.

Birds have been noted to 180 km off Grays Harbor to survey limits at 126° 30' W, with concentrations over the outer continental shelf and shelf break. Over 80% of birds were noted over depths of 200-1000 m. Sabine's Gulls appear much less frequently over mid-ocean than Arctic Terns (Wahl et al. 1989).

Reports of fall migrants increased in the 1980s and 1990s in e. Washington, usually of single birds or small numbers in the Columbia Basin, at potholes and along major rivers.

Remarks: No subspecies (Cramp et al. 1983). As many as four given by others, with *wosnesenski* and *sabini* occurring in Washington (Jewett et al. 1953), and with *wosnesenski* and *tschuktschorum* given by Campbell et al. (1990b) and Morgan et al. (1991).

Noteworthy Records: *Outer coast, Spring*: 172 on 18 May 1996, 89 on 16 May 1992 off Grays Harbor. *Summer*: 3 on 17 Jun 1973, 2 on 11 Jun 2002 off Grays Harbor. *Fall*: 566 on 13 Sep 1992, 270 on 5 Oct 1986 off Grays Harbor; 1353 on 18 Sep 1979 from aircraft nr. Swiftsure Bank, B.C. (MESA). *Inland waters, Spring*: 1 on 8 May 1968 at Bellingham; 1 on 31 May 1971 at Blaine; 2 on 21 May 2003 in Rosario Strait. *Fall, late*: 1 imm. on 17 Oct at Blaine; 1 on 22 Oct 1971, 15 Oct 1972, and 17-19 Oct 1982 at Pt. Roberts; 1 on 21 Sep 1982 s. of Whidbey I.; 1 on 24 Sep and 7 Oct 1989 in Puget Sound. **East**, *Spring, late*: 300 on 9 Jun 1963 at Blue L., Grant Co., and ca. 4 hrs later about 130 km to the e. at Reardan; 1 ad. on 12 Jun 1964 at O'Sullivan Dam. *Fall, late*: at least 7 on 1 Sep-10 Oct 1997 on the Columbia R. between Walla Walla R. delta and Yakima R. delta; 1 imm. on 6 Oct 1962 at Soap L. 1 on 11-12 Oct 1969 at Banks L. 1 on 9 Nov 1991 at John Day Dam.

Terence R. Wahl

Black-legged Kittiwake *Rissa tridactyla*

Fairly common to common migrant and winter resident offshore and along the outer coast; irregular, locally uncommon to common summer resident coastally; rare to uncommon in inland marine waters. Very rare e.

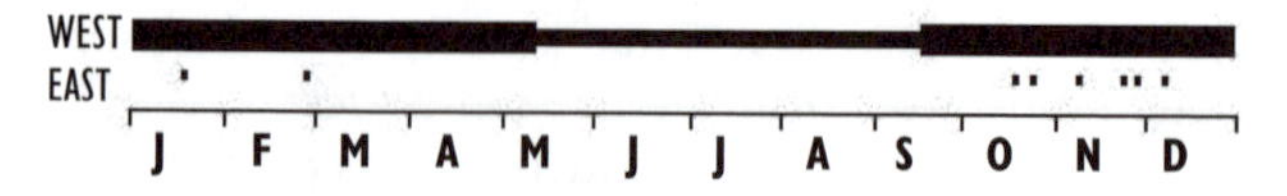

Subspecies: *R. t. pollicaris.*

Habitat: Continental shelf and oceanic waters in winter, nearshore coastal waters and embayments in summer.

Occurrence: Status is better understood since Jewett et al. (1953), likely reflecting offshore surveys and observer effort, described the Black-legged Kittiwake as an occasional spring and probable fall migrant and winter visitor.

Habitat use changes seasonally. Off Grays Harbor 75-100% of birds from Nov to Apr were offshore (>20 m depth). Birds in Jan 1977 occurred to at least 170 km offshore (TRW), Briggs et al. (1992) reported similarly, and birds winter across the ocean (TRW). Much smaller and variable numbers in May-Oct seldom offshore: 94-100% were over nearshore waters (depth <20 m) and inside Grays Harbor (TRW). Migration along the outer coast and offshore peaks in Apr-May, though varying from year to year (Briggs et al. [1992] reported no kittiwakes seen during aircraft surveys in Apr 1989). Smaller fall movements peak in Sep-Oct.

Many occurrence patterns may be associated with breeding success or failure (Hatch 1987), relating to broad-scale oceanographic conditions. Noteworthy numbers occurred in Washington in 1965, 1969, 1972, 1975-77, 1982-83, 1985-87, 2001, and 2003. Large numbers may be storm blown inshore and into inland waters (Campbell et al. 1990b), with mortality in 1976 notable (e.g., Harrington-Tweit 1979a). Inland records in 1988 and the 1990s contrasted with low numbers coastally and likely reflected poor ocean-feeding conditions (Wahl and Tweit 2000a). Summer reports range from none to hundreds of birds. E. Washington records corresponded with at least some of peak coastal years. On CBCs, 14 records were from the 1970s, two were from the 1980s and just six from the 1990s, suggesting changes in long-term occurrence.

Offshore from the Strait of Juan de Fuca to s. California in Jan 1977 and 1979, numbers of imms. increased with decreasing latitude, suggesting ads. wintered farther n. (TRW), agreeing with reports of large numbers of imms. along the Washington coast during spring. Age classes present in summer vary: in some years almost all the small numbers at Grays Harbor in that season are in ad. plumage (TRW) as in other locations (e.g. Wahl 1995), while in 2001 almost all were imms. (SGM).

Occurrence in inland waters is predominantly during migrations, occasional in winter and infrequent in summer. Most records are in tidal convergences or "fronts" from the Strait of Juan de Fuca, from Point No Point n. into the Strait of Georgia, and are of single birds or small numbers though large numbers have been recorded in peak years at Victoria, B.C.

Kittiwakes have been reported inland along the Columbia and Snake Rs. in recent decades. This likely reflects creation of reservoirs and increased observer effort. Beginning in 1975 there were about nine records through 2000 with only one, at Reardan, away from river habitats.

Remarks: Vulnerable to poor ocean productivity: Hatch (1987) describes total breeding failure at six colonies in Alaska in 1983, a year of a major El Niño event. N. Pacific breeding population estimated at 2.6 million birds (Hatch et al. 1993).

Noteworthy Records: West, *Fall-Spring, outer coast*: 1000 on 16 Sep 1972 at Ocean Shores; 400, mostly ad., on 6 Oct 1974 at Westport; 300 on 3 Dec 2000 at Cape Flattery (SGM); 2500-3000 on 21 Feb 1986 at Ocean Shores; ca. 1000, most imm., on 16-18 May along the sw. Washington coast; 800 imm. on 31 May 1976 at Ocean

Shores. *Off Grays Harbor*: 295 on 6 Sep 1983, 321 on 9 Oct 1977, 326 on 7 Nov 1981, 481 on 20 Jan 1990, 356 on 27 Mar 1982, 2063 on 2 Apr 1994, 800 on 20 May 1970 (TRW). *CBC*: Grays Harbor: 40 on 19 Dec 1976. *Summer*: 100 at LaPush on Jul 15 1969; >100 in 1972, 200 in 1977 at Ocean Shores. *Inland waters, Fall-Spring, high count*: 100 on 4 Sep 1972 at Kingston. CBC: 2 in 1970, 1 in 1979, CW in 1995 at Tacoma; 1 each in 1973, 1977, 1976 on San Juans Ferry; 1 in 1976 at Sequim-Dungeness. *Summer*: 60 in 1965 in Bellingham Bay. **East**: 1 on 29 Feb 1976 at Clarkston,; imm. on 20 Nov 1983 at Clarkston; imm. on 3 Sep 1989 at Reardan; 1 on 10 Nov 1990 at Asotin; 1 imm. on 22 Oct at White Bluffs, Grant Co., and 1 on 24-26 Nov 1995 at John Day and Dalles dams; 1 on 24 Nov-5 Dec 1996, 1 on 16 Jan 1998, 1 Dec 2001 at John Day Dam ; 1 on 8 Oct 2002 at Priest Rapids L.; 1 on 13-14 Jan 2002 at Walla Walla R. delta.

Terence R. Wahl

Red-legged Kittiwake *Rissa brevirostris*

Very rare migrant and winter visitor on the outer coast and oceanic waters.

Listed as hypothetical by Jewett et al. (1953), now known from very few records. The Red-legged Kittiwake probably is a sporadic winter visitor in offshore waters, where winter observations have been limited. Six recent records from offshore and nearshore waters, along with records from Oregon, suggest more frequent occurrence, and more offshore surveys in winter may well result in more frequent records.

Like other marine species (e.g., see Hatch 1987, 1993b), breeding success, population size, and extra-limital occurrence likely affected by oceanographic variations, which were especially apparent in the mid-1980s and the 1990s. Populations have declined in recent decades, with fewer colony sites being used and numbers decreased at the major colony, on St. George I. in the Pribiloffs (Byrd and Williams 1993a, Hatch 1993b). Gilligan et al. (1994) list seven records for Oregon. Recorded in California, Nevada, Yukon (as in Enticott and Tipling 1997). Not given for B.C. by Campbell et al. (1990b). Classed GL.

Records: 1 on 27 Jan 1974 at Leadbetter Pt. (Tweit and Skriletz 1996); 1 found dead on 1 Dec 1978 at L. Ozette (Tweit and Paulson 1994); 1 on 28 Jun-5 Jul 1999 at Tatoosh I.; 1 on 19 Jan 1991 off Grays Harbor; 1 on 21 Mar 1998 and 1 on 19 Aug 2000 were up to 60 km offshore. Rejected by WBRC: 1 on 2 Apr 1987 and 1 on 26 Apr 1987 at Ocean Shores (Aanerud and Mattocks 1997).

Terence R. Wahl

Ross' Gull *Rhodostethia rosea*

Casual vagrant.

The one Washington record was of an ad. (ph.) below McNary Dam, on 27 Nov 1994. The bird remained there until 1 Dec. This was only the 19th record for the contiguous U.S., with most of the previous 18 records coming from the ne. and midwest between early Nov and early May (Mlodinow and O'Brien 1996). In the nw., there was one previous late winter record from Oregon and one late fall record from B.C. (Campbell et al. 1990b).

Steven G. Mlodinow

Ivory Gull *Pagophila eburnea*

Casual vagrant.

This Arctic gull has been identified in Washington once: a bird seen at Grays Harbor on 20 Dec 1975. The species rarely ranges s. of the ice pack, and in w. N. America it is very rare s. of sw. Alaska. It has occurred in B.C. six times between Sep to mid-Feb 2001, with three of these prior to 1925 (Campbell et al. 1990b). Recorded in w. interior N. America in B.C. and Montana (Campbell et al. 1990b, Skaar 1992).

Steven G. Mlodinow

Caspian Tern *Sterna caspia*

Common to abundant summer resident and breeder on the lower Columbia R., fairly common to common locally on the outer coast and in inland marine waters. Locally common breeder and migrant in e. Washington.

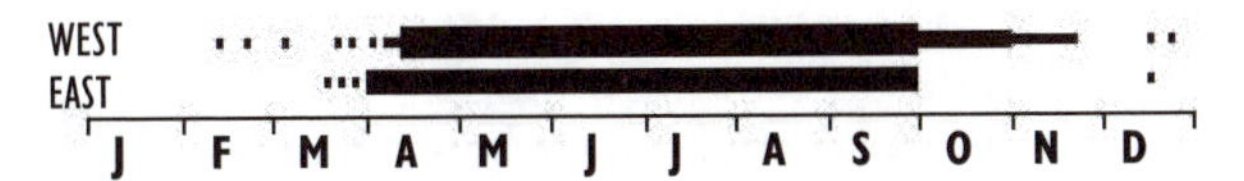

Habitat: Nests on unvegetated, predator-free sand/gravel islands created by wave action and dredging operations. Forages in estuaries, embayments, nearshore coastal waters, lakes, and reservoirs; roosts on islands, log booms, and roofs.

Occurrence: Dramatic changes in the late 20th century reflected adaptability to habitat alteration, human-provided food opportunities, possibly to climate change, and to conflicts with human interests (see Cuthbert and Wires 1999). From small numbers in the mid-1900s the nw. breeding

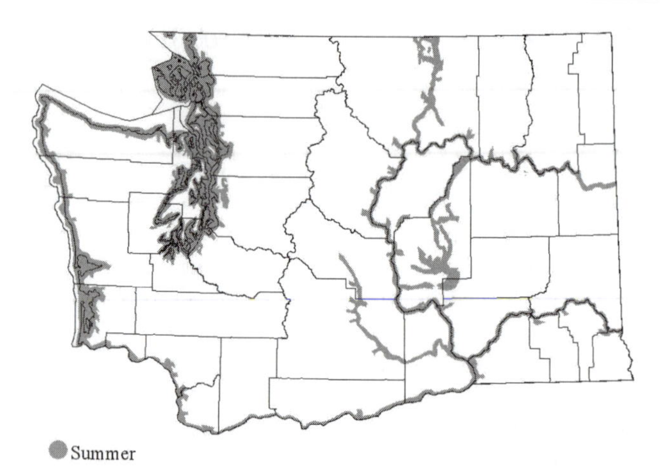

population grew to over 30% of the N. American total by 1998 (Lowe 1998)—the largest regional concentration of the estimated world population (Enticott and Tipling 1997) and potentially more vulnerable to anthropogenic events in Washington than anywhere else in its worldwide range (Thompson et al. 2002).

Casual migrants appeared on the coast in 1930, birds nested in 1932 at Moses L., and with 50 pairs on a Columbia R. island, Benton Co. (Jewett et al. 1953). Status varied by years and location (Penland 1982), ranging to about 300 pairs, with migrants or non-breeders reported e. to Turnbull NWR. In 2001 none nested in the Hanford Reach, while about 1200 birds were at three locations e. of the Columbia R. (www.columbiabirdresearch.org). In 2003 terns were noted nesting at Goose I., in the Potholes for the first time, and at Solstice I. and Banks L. (www.columbiabirdresearch.org).

W. Coast distribution shifted n., to coastal colonies (Gill and Mewaldt 1979). Birds first nested in w. Washington at Grays Harbor in 1957 (Penland 1981, 1982, Smith et al. 1997). Numbers there, at Willapa Bay and E. Sand I., nr. the mouth of the Columbia R., reached thousands in the 1980s and 1990s (see Penland 1981, 1982, Speich and Wahl 1989), with the species one of the most abundant and conspicuous in summer on the s. coast.

Inland, birds occurred on dredge-spoil islands at Swinomish Slough, nr. Anacortes, as early as 1972 (TRW). Hundreds were there in 1990, and small numbers nested in 1991. Breeders reached Everett in 1990, with 2600 ads. and 2500 chicks in 1994 on a site destined to become Naval Station Everett (Bird 1994). In 1995 an attempt to relocate the Everett colony to nearby Jetty I. failed (Bird 1995, Smith et al. 1997). Though 500 birds were reported in 1997 the colony was essentially eliminated. Also in the mid-1990s, dredging operations in Grays Harbor resulted in virtual elimination of accreted islands by ocean swells and nesting terns virtually disappeared in 1995. The two largest w. coast colonies were thus eliminated. Terns nesting at the ASARCO site at Commencement Bay in 1999 were enticed to relocate on a barge 7 km away in 2001. This site was closed in May 2001 (Shugart and Tirhi 2001, Thompson et al. 2002), effectively eliminating breeding terns in inland marine waters.

Many breeders apparently relocated to Rice I., 34 km inland up the Columbia R. By 1998 there were 10,000 pairs plus non-breeders, with fledglings increasing to 4000 in 1998 (Lowe 1998). Studies showed that Rice I. birds were eating some salmon smolts. Agencies eliminated Rice I. habitat and encouraged birds to relocate downstream at E. Sand I., where the diet of terns included fewer salmonids (Roby et al. 1998, Harrison 2002, Lyons et al. 2002, Roby et al. 2002): Rice I. diet included 77-90% salmonids, E. Sand I. terns had 31-47% (www.columbiabirdresearch.org). Presently, about two-thirds of Caspian Terns breeding on the w. coast nest at E. Sand I., the largest colony in the world. Aerial photo counts showed 9933 breeding pairs there in 2002, with a decrease to 8325 pairs in 2003 (B. Espenson, Columbia Basin *Bulletin*).

Non-breeders increased in inland waters. Counts in 1978-82 were usually <10 birds (MESA, Wahl and Speich 1984). Birds were widespread in nearly all estuaries by the 1990s, with parents and fledglings noted far from colonies. Counts in the 1990s included, for example, 350 non-breeders in summers 1995-2004 at Bellingham (Wahl 1995, TRW) and 500 at Tacoma in 2001 (G. Shugart p.c.). Birds reported in coastal B.C. in 1959, the Queen Charlotte Is. in the 1980s, nested nr. Vancouver in 1984 (Campbell et al. 1990b). Recorded in Alaska by 1981, with a bird at the Yukon-Kuskokwim delta in 1994 and up to three nests nr. there in 1996 and 1997 (see McCaffrey et al. 1997).

Remarks: Destination or purpose of nocturnal flights of large, vociferous flocks up to 25 km inland from Bellingham roost unknown (see Wahl 1995). USFWS decided in 2001 not to create a refuge at Sand I. (Harrison 2001, 2002). In 1997-2001 an average of 476 birds were killed per year as part of a management program to protect juvenile salmonids on the mid-Columbia R. (USDA 2003). Classed PHS in Washington.

Noteworthy Records: West, *Spring, early*: 1 on 19 Feb 1994 at Naselle, 1 on 11 and 19 Feb 1978 nr. Conway. *Nesting, Rice I.*: 1100 pairs in 1984, none in 2003 (www.columbiabirdresearch.org); *E. Sand I.*: 9000 pairs in 2001, 9900 pairs in 2002 with 15,000 birds there. *Grays Harbor*: 300 chicks in 1972; 2040 nests in 1977; 1900 nests in 1979; 2775 nests in 1984; colony almost unoccupied in 1995; no birds observed in 2003 (www.columbiabirdresearch.org). *Willapa Bay*: 30 in 1977, 500-800 pairs in 1980, none in 1984. *Both areas*: 4000 pairs along the coast in 1982; near-total nesting failure in 1983; 5450 pairs at Grays Harbor and Willapa Bay in 1985. *Offshore*: 1 on 27 Jul 1984 ca. 22 km off Grays Harbor. *Inland marine waters, Swinomish*: 300 on 14 Jul 1990; nests, eggs for first time in 1991; no birds in 1992; 800 on 10 Jul 1996. *Everett*: 600 birds with young on 16 Jul 1992; 2600 ad., 1500 young in 1994; 500 on 6 Jul 1997. May have nested at Bellingham in 2002 and 2003, and 2 fledglings were present there on 5 Jul 2004 (TRW). *Fall, late*: 1 on 19 Dec 1975 at Ocean Shores. **East**, *Spring, early*: 1 on 19 Mar 1996 at Richland, 1 on 24 Mar 1998 at Wahluke Slope WMA. *Summer*: 150 young in 1972 at Umatilla Refuge; 35 pairs nr. Richland in 1974; 220 pairs, 150 young in 1981, 268 ad. in 1991 at two colonies at Potholes; 100 birds at Crescent I., Walla Walla Co., in 1993 (Ackerman 1994). 30 birds at Goose I., Banks L.; 200-300 at Solstice I., Potholes Res, 30 at Harper I., Sprague L. in 2001. 300-400 birds at Crescent I., none elsewhere in Jun 2003 (Columbiabirdresearch). *Fall, late*: 1 on 20 Dec. 1966 at McNary NWR.

Terence R. Wahl

Elegant Tern *Sterna elegans*

Irregular, rare to locally common summer visitor on outer coast, inland marine waters.

Habitat: Coastal sandspits and beaches, roosts on occasion on breakwaters and jetties, forages in nearshore waters and estuarine embayments.

Occurrence: Though most birds winter from Guatemala s. to c. Chile (Clapp et al. 1993), northward post-breeding dispersal was evident when birds were first recorded in Oregon and Washington in 1983. The species became an indicator of SSTs, upwelling, and productivity along the w. coast: occurrence was coincident with ENSO and other warm-water episodes in the 1980s and 1990s (Wahl and Tweit 2000a).

Major incursions occurred in 1983, 1992, and 1997. After the ENSO of 1983, there were no records from 1984 to 1989. Small numbers were reported in 1990, 1991, 1993, 1994, 1996. 1998, 1999, and 2000. The highest count was 1000 birds at Willapa Bay in 1992. Birds occurred in inland marine waters during the main coastal occurrences in 1983, 1992, and 1997, with a few birds found in 1993 and 1996 and just one report in 1998. Oregon had small incursions also in 1987, 1989, and 1995.

During incursions, birds concentrated at Willapa Bay and Grays Harbor, generally from mid-Jul to early Sep. None were noted outside of the Grays Harbor channel entrance during surveys offshore (TRW). In inland waters highest numbers were along the e. Strait of Juan de Fuca shoreline, and birds were occasional s. into Admiralty Inlet with individuals or small numbers at Pt. Roberts, Boundary Bay, and Tacoma.

Remarks: Breeds at very few locations, with 97% at Isla Raza, in the Sea of Cortez. Clapp et al. (1993) and Enticott and Tipling (1997) detail population trends and threats, including possible declines in a major prey species, northern anchovies (see Burness et al. 1999, Wahl and Tweit 2000a). Recent expansion, increases, and decreases in populations at s. California colonies are documented (Collins 1997). Classed GL.

Noteworthy Records: West, *Summer, early*: 2 on 4 Jul 1992 and 1 on 9 Jul 2000 at Ocean Shores. *High counts, Summer/Fall, outer coast*: 100 on 4 Sep 1983 at Grays Harbor; 104 in 1990 at Tokeland; 300 on 23 Jul 1992 at Ocean Shores; 1000 on 21 Aug 1992 at Willapa Bay; 100-150 on 16 Jul-23 Aug 1997 on coast. *Inland marine waters*: 40 on 7 Aug 1992 at Dungeness, and 52 on 3 Sept 1992 at Oak Bay Jefferson Co.; 23 on 17 Aug- Sep along the e. Strait of Juan de Fuca. *Fall, late*: 4 on 19 Oct 1990 at Tokeland.

Terence R. Wahl

Common Tern *Sterna hirundo*

Common to abundant spring and fall migrant w.; very rare in spring, uncommon fall migrant e. of the Cascades.

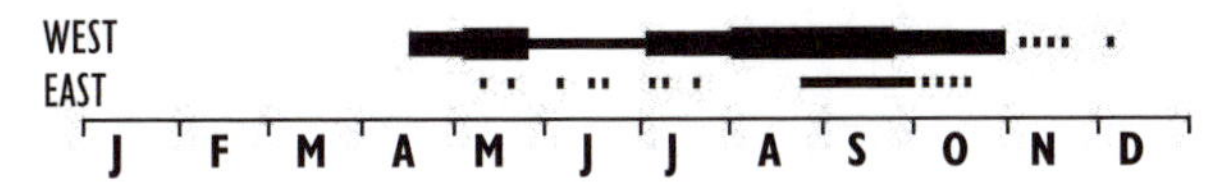

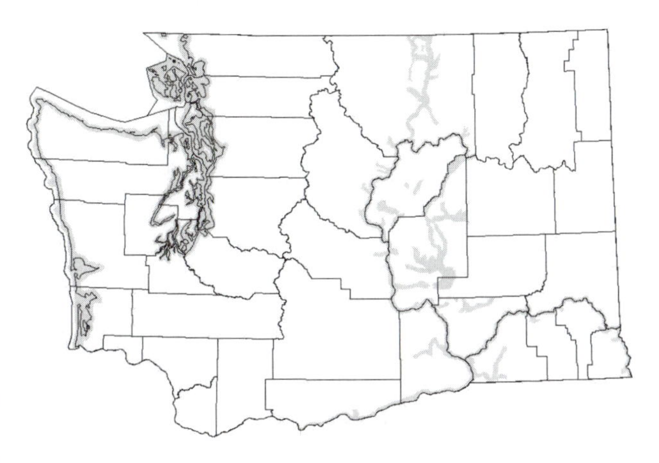

Migration

Subspecies: *S. h. hirundo.*

Habitat: Nearshore marine waters, tidal convergences, shallow embayments, lakes, and reservoirs.

Occurrence: Migrants are relatively widespread in w. Washington, and large feeding flocks occur along the outer coast, the Strait of Juan de Fuca, and in inland waters. Uncommon to common in spring, with highest numbers at Grays Harbor. Highest numbers in inland marine waters are in fall. The four highest 1978-79 MESA counts were from Port Angeles, with other high counts along the Strait of Juan de Fuca. Other concentrations peaked from mid-Aug into late Sep at Point No Point, Rosario Strait, Speiden Channel, and other locations with active tidal fronts. Small numbers observed in Jun and early Jul, particularly at areas where large flocks of migrants occur (e.g., Drayton Harbor). Whether these are late spring or early fall migrants or non-breeders not completing migration into the interior is uncertain, though Campbell et al. (1990b) state that Jun birds are imms.

Though there is some overlap, most Common Terns migrate nearshore while the Arctic Tern migrates mainly from the outer continental shelf across the breadth of the Pacific Ocean (Wahl et al. 1989). On surveys in 1971-2000 off Grays Harbor 93% of Common Terns were within 6 km of the shoreline or in Grays Harbor, with almost all of the rest within about 10 km farther offshore.

In e. Washington small numbers are reported sporadically along the Columbia R. e. to the Columbia Basin and Turnbull NWR, almost all in the fall. Numbers noted are relatively low but reports were more frequent in the 1960s.

Migration routes to and from interior N. American nesting areas and the coast are uncertain. The relatively few reports from interior Washington and B.C. (Campbell et al. 1990b) do not indicate large enough numbers to explain the origins of the thousands of birds noted w. of the Cascades, but Campbell et al. (1990b) suggest migrants use the Columbia and Frasier Rs. as routes between the interior and the coast.

Remarks: Some reports of breeding in the state (e.g. Weber and Larrison 1977) probably refer to Forster's Tern, *S. forsteri*. A reported sighting of the Asian subspecies, *longipennis*, at Vancouver, B.C., in 1975 suggests "different looking" Common Terns should be carefully identified.

Noteworthy Records: West, *Spring, early*: "few" on 24 Apr 1985 at Stanwood; 100 on 29 Apr 1977 at Ocean Shores. *High counts, Spring*: 300-400 on 1-14 May 1974 at Samish I.; 500 on 9 May 1981 and 800 on 20 May 1999 at Ocean Shores; 1000+ on 6-14 May 1989 in Grays Harbor channel; 1125 on 16 May 1992, 1169 on 18 May 2002 off/nr. Grays Harbor (TRW). *Summer*: 50 on 4 Jul 1971 at Bellingham; 50 on 5 Jul 1969 at Blaine; 42 on 5 Jun at Everett; up to 25 on 12 and 17 Jun 1976 at mouth of Columbia R. *Fall*: 2000 on 1-3 Sep 1994 and 27 Aug 2000 at Point No Point; 1000 on 17 Sep 1987 from San Juan Is. ferry, 3000+ at S. Beach, San Juan I. and 500+ in Speiden Channel on 5 Oct 1987 (Lewis and Sharpe 1987); 1010 on 17 Sep 1979 at Port Angeles, 1000 at Dungeness Spit on 13 Sep 1979 (MESA). *Late Fall*: 1 on 2 Dec 1981 off Nisqually R. delta; 1 on 7 Nov 1992 at Willapa Bay; 1 on 12 Nov 1977 at Dungeness Spit; 1 on 3 Nov 1979 at Drayton Harbor (MESA). **East**, *Spring*: 1 on 13 May 1999 at Cassimer Bar, Okanogan Co.; 14 on 24 May 2002 at L. Lenore and 1 on 24 May 2002 at Wilson Cr. (SGM). *Summer*: 2 on 9 Jun 1965 at Reardan; 2 on 1 Jul 2001 at McNary NWR (SGM). *Fall*: 28 on 2 Oct and 2 on 11 Oct at McNary NWR, 2 on 9-10 Oct at Yakima R. delta, 2 on 17 Oct 1998 at Kennewick; 10 on 29 Aug at Walla Walla R. delta, 4 on 9 Sep and 1 on 28 Sep at Yakima R. delta, 1 on 14 Sep at Lyle and 1 on 28 Sep 1999 at Richland. *High counts*: 28 on 31 Aug at Potholes Res. and 21 on 2 Sep 1965 at L. Lenore; 31 on 29 Aug 1968 at Medical L. 15 on 30 Aug 1969 at Turnbull NWR.

Terence R. Wahl

Arctic Tern *Sterna paradisaea*

Fairly common to common spring and fall migrant offshore, uncommon to locally common inland in w. Washington; rare, local breeder at Everett. Very rare to rare migrant in e.

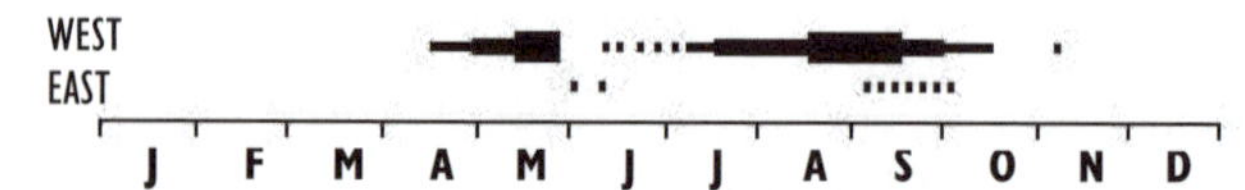

Habitat: Continental shelf and oceanic waters. Inland marine waters, lakes, and reservoirs. Breeds on dredge-spoil and waterfront open space in Everett.

Occurrence: Knowledge of the Arctic Tern's status in recent decades has resulted from surveys off the outer coast and from increased observer identification skills and effort statewide. Though westside status has remained the same, eastside status probably changed, with migrants attracted to human-created reservoirs that did not exist or were not investigated when, for example, Jewett et al. (1953) did not list occurrence there. A very small, unusual, and marginally established nesting "colony" at Everett was first noted in 1977 (Manuwal et al. 1979).

Oceanic migrations peak in May and in mid-Aug to mid-Sep. Surveys in 1971-2000 off Grays Harbor found that the number of birds noted per day in Sep was twice that of May (TRW; and see Campbell et al. 1990b). Usually seen in small migrating groups, but flocks (e.g. 150 birds on 7 Sep 1975) are noted on occasion. Numbers observed over the continental shelf apparently declined after the mid-1980s (Wahl and Tweit 2000a). Like other migrants, birds noted in Jun may be late spring or early fall migrants.

Arctic Terns migrate across the breadth of the Pacific Ocean (Wahl et al. 1989), and censuses off Grays Harbor recorded birds to the limit of surveys at 126° 30' W (TRW). That offshore distribution overlaps only slightly with that of the Common Tern: 98% of Arctic Terns occurred over waters >20 m depth, while 93% of Common Terns were noted in nearshore waters of <20 m depth (TRW). Arctic Terns do occur, however, at many coastal locations.

In inland marine and freshwater habitats birds are infrequent and usually with Common Terns. Surveys in 1978-79 recorded only two birds, both in the San Juan Is. (MESA). Other records are from Bellingham, Pt. Roberts, Dungeness, Whidbey I., and Ridgefield NWR on the Columbia R. Except for one specimen collected in the spring, the few records from e. of the Cascades are from the fall migration period.

Following the discovery of seven pairs nesting at Jetty I. nr. Everett in 1977 (Manuwal et al. 1979), this "colony" was monitored carefully as birds moved to the Everett waterfront, then moved around the area during construction of the naval base and other developments. The number of pairs varied from one to five, with pairs relocating and volunteer observers protecting nest sites and attempting to encourage birds to locate back on the dredge-spoils at Jetty I. (see Bird 1994, 1995). Though birds were noted at the Everett site through 2000, there were few records in the adjacent inland waters during the nesting and post-nesting season that might have been attributed to birds from Everett. The future of this unfortunately located population is uncertain but it is a unique case history in avian tenacity, if not of gene-pool considerations.

Noteworthy Records: West, *Spring, early*: 7 Mar 1995 at Everett; 14 Apr 2001 at Ocean Shores. *Summer/early Fall*: 2 on 27 Jun 1992 off Grays Harbor; 26 in Jun 2000 off the Olympic Pen; 1 ad. on 21 Jun 1995 at Bellingham roosting with Caspian Terns (Wahl 1995); 7 on 6 Jun 1999 at Everett; 2 on 5 May-21 Jun 2002 at Everett, without sign of successful nesting; 1 on 2 Jul 1995 at Deer Lagoon; 2 on 30 Jun 2000 at Useless Bay. *Fall, late*: 1 on 25 Sep 1992 at Ridgefield NWR; 1 on 8 Oct 1978, 1 on 5 Oct 1980, 3 on 10 Oct 1981 off Grays Harbor; 1 juv. on 10-14 Oct 1988 at Edmonds; 1 on 6-10 Nov 1999 at Pt. Wilson. *High counts*: 382 on 10 Sep 1978, 216 on 12 May 1985 off Grays Harbor. **East**, *Spring/Summer*: 1 collected on 21 May 1975 at Ringold, Franklin Co. (Weber and Larrison 1977); 1 in Jun 2002 at Othello; 1 on 13 Jun 2001 at Confluence SP. *Fall*: 1 on 2 Oct 1977 at the Yakima R. delta; 1 collected in 1978 at Snake-Clearwater confluence; 1 on 12 Sep 1993 at Walla Walla R.; 1 on 5 Sep 1994 at Walla Walla R. delta; 1 on 15 Sep at Soap L. and 3 on 19-21 Sep 1996 at Banks L.; 1 on 10 Sep 2000 at Priest Rapids. *Note*: the record of "about 15" at Crescent L. on 15 Apr 1916 (Jewett et al. 1953) probably refers to Common Terns.

Terence R. Wahl

Forster's Tern *Sterna forsteri*

Local, fairly common to common summer resident in lower elevations of e. Washington; rare migrant west.

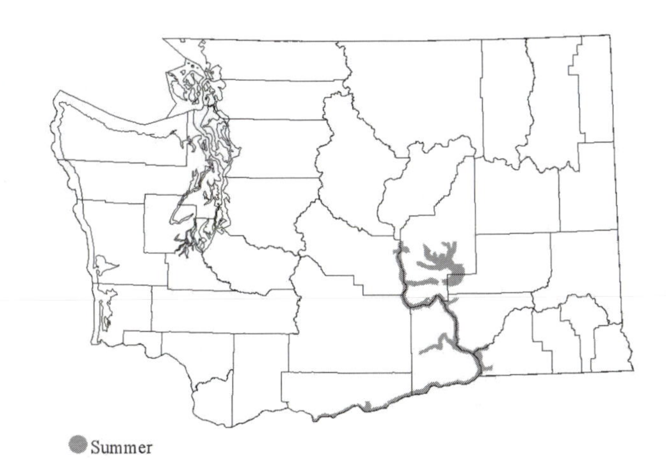

Habitat: Nests at ponds, lakes, marshes, forages widely including reservoirs behind dams.

Occurrence: Long known as a summer resident, Forster's Tern remains less understood as to status and distribution than many less conspicuous species. During the summer Forster's Terns are fairly common in the Potholes region, and s. along the Columbia R. to the Tri-Cities, with smaller numbers nesting along the Columbia R. w. to Klickitat Co. (Smith et al. 1997). Outside those areas birds are widespread in low numbers in suitable ponds, lakes, and marshes in e. Washington in summer and non-breeders and migrants are seen in many locations (Jewett et al. 1953, Weber and Larrison 1977, Ennor 1991). Recent nesting at some historical locations (Jewett et al. 1953) and at other later locations like Crescent I. (Ackerman 1994) is uncertain. Jewett et al. (1953) believed the state lacked suitable nesting habitat due to lack of cover and alteration for farming, though reservoirs behind dams and irrigation projects have increased habitat subsequently.

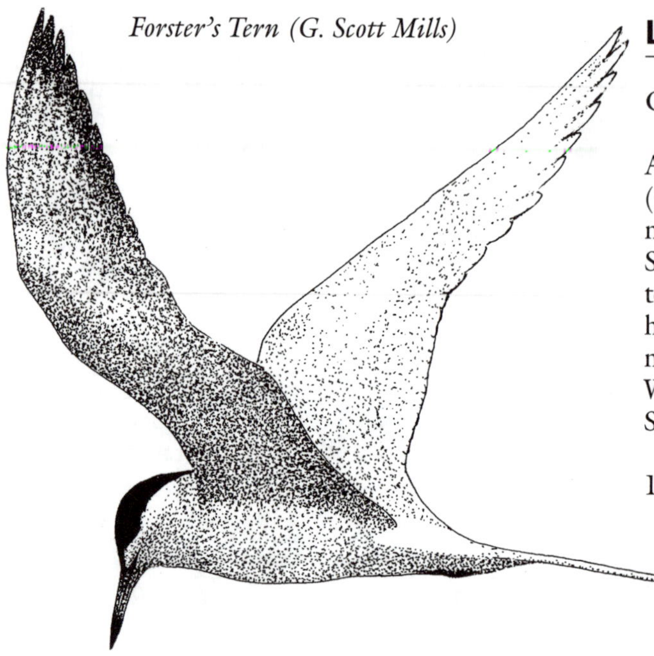
Forster's Tern (G. Scott Mills)

W. of the Cascades, there have been about 30-35 records, including several in early to mid-Jul, mostly from the n.c. Puget Trough, concentrated from late Sep to early Nov. The number of reports increased during the 1980s and 1990s, likely due in part to the increasing number of observers. The species is casual on the outer coast. Like the situation regarding a number of other inland-nesting migrants, the origin of birds migrating w. of the Cascades in particular is of interest. Few birds breed in B.C.: most of the Forster's Tern population in w. Canada breeds e. of the Rocky Mts. in Alberta.

The population trend in Washington is uncertain. A 38% increasing trend from 1966 to 2000 (Sauer et al. 2001) is based on very limited BBS samples that likely do not suitably represent the distribution (see McNicholl et al. 2001). Paulson (1992) stated that the population declined in recent decades.

Remarks: Conversion of habitat to farmland (Jewett et al. 1953) and effects of pesticides (see Paulson 1992) are mentioned as possible reasons for long-term decline in population. Classed PHS.

Noteworthy Records: East, *Summer, high counts*: 300 on 29 Aug 1988 at N. Potholes; 69 on 15 May 2000 at Yakima R. delta; 93 on 30 Jun 2001 at Paterson Slough. **West**, *inland waters, Spring*: 1 on 17 Mar 1990 at Everett; 1 on 30 May 1998 at Skagit WMA; 3 on 24-27 May 2001 at Camas; 2 on 28 May 2001 nr. Snohomish; 1 on 9 Jun 2000 at Monroe; 1 on 7 Jun and 9 Jul 2000 at Everett. *Summer-early Fall*: 1 ad. on 4 Jul 1976 at Bellingham; 1 on 7 Jul 1992 at Discovery Bay. *Fall, late*: 1 on 9 Nov 1986 at Everett; 1 on 8 Nov 1994 at Everett. Outer coast: 1 on 13 Sep 1987 at Grays Harbor (TRW).

Terence R. Wahl

Least Tern *Sterna antillarium*

Casual vagrant.

A Least Tern on 26-31 Aug 1978 at Ocean Shores (ph.) was the first record for Washington and the northernmost on the Pacific coast (Tweit and Skriletz 1996). It was with Caspian and Common terns. With a small species population, threatened by habitat loss and beach disturbance during the nesting season, the occurrence of even one bird in Washington may be unrepeated for years to come. Subspecies was *S. a. browni.*

The species has been recorded in Oregon between 1964 and 1999 (Johnson 1998, OBRC) and once at L. Osoyoos, B.C., in 1998. An estimated 800 pairs nest in California and w. Mexico (Enticott and Tipling 1997). Habitat loss is a threat, though most nesting areas are protected and managed (Clapp et al. 1993). Classed GL.

Terence R. Wahl

Black Tern *Chlidonias niger*

Locally fairly common to common breeder, widespread uncommon migrant in e. Washington; rare migrant and breeder west.

Subspecies: *C. n. surinamensis.*

Habitat: Marshes, sloughs, and small lakes; lakes and westside marine waters in migration.

Occurrence: Widespread in migration in much of e. Washington, the Black Tern is a characteristic species of marshes and small lakes. Uncommon through most of its nesting range, greatest abundance in

Washington is from Okanogan Co. e., with hundreds reported at Turnbull NWR (Smith et al. 1997). Birds "probably" breed up to about 900 m elevation in the ponderosa pine zone, are less common in the Potholes area and largely absent from the hotter, arid regions of the Columbia Basin (Smith et al. 1997). Though Weber and Larrison (1977) suggested the species was a probable summer resident in se. counties, BBA fieldwork in the late 1980s-early 1990s (Smith et al. 1997) apparently did not confirm this. Birds breed in some years at Conboy L. NWR in Klickitat Co. (Smith et al. 1997) and bred at Ridgefield NWR in 2001 and 2002 (SGM).

Presence in the region is relatively brief. Arrival in B.C. is mostly in mid-late May, and departure from breeding areas is early, with birds mostly gone by mid-Aug (Campbell et al. 1990b). Though fall reports from e. Washington are few, there are records from 14 and 30 Sep.

Data on interannual variability and trends in Washington are minimal, though numbers reported at Turnbull NWR since the early 1970s may suggest a decline. In the Okanagan Valley, B.C., numbers fluctuate noticeably over time (Cannings et al. 1987). This complicates determination of trends but, as in the species overall distribution in N. America and Eurasia (see Cramp et al. 1985), Cannings et al. (1987) and Campbell et al. (1990b) state that B.C. populations have declined over the long term due to habitat loss.

Though Black Terns have nested at Pitt L. in sw. B.C. (Campbell et al. 1990b) and birds are seen in w. Washington in Jun, there are are just two state breeding records w. of the Cascades. A few birds are reported in spring and fall in the lowlands, with records from Dungeness e. and s. to Vancouver and along the outer coast from Grays Harbor s.

Remarks: Population declines worldwide due to habitat loss cited by Campbell et al. (1990b), Enticott and Tipling (1997), Dunn and Agro (1995). Classed PHS.

Noteworthy Records: East, *Spring, early*: 2 on 22 Apr 2000 at Cheney. *Summer*: 300 breeding at Brook L. in 1908 (Jewett et al. 1953); 50 at Hauser L. in 1922 (Jewett et al. 1953); about 600 in summer 1970 at Turnbull NWR; ca. 230 in summer 1972 at Turnbull NWR; 40 pairs in summer 1988 at Goose L., se. of Omak. *Late Summer-Fall*: 1 on 30 Sep 1985 at Walla Walla R. delta; 1 on 30 Aug 2000 at McNary Dam; 1 on 4 Sep 2000 at Dodson Rd; 7 on 19 Aug and 3 on 25 Aug 2001 at Sprague L. **West**, *Spring*: 1 on 30 Apr 1989 at Snohomish. *Fall*: 1 on 7 Oct at Port Ludlow, and 1 on 23 Oct-6 Nov 1976 at Ocean Shores.

Terence R. Wahl

Common Murre *Uria aalge*

Common to abundant resident along the outer coast; fairly common to common winter visitor, locally fairly common to common in summer in inland marine waters.

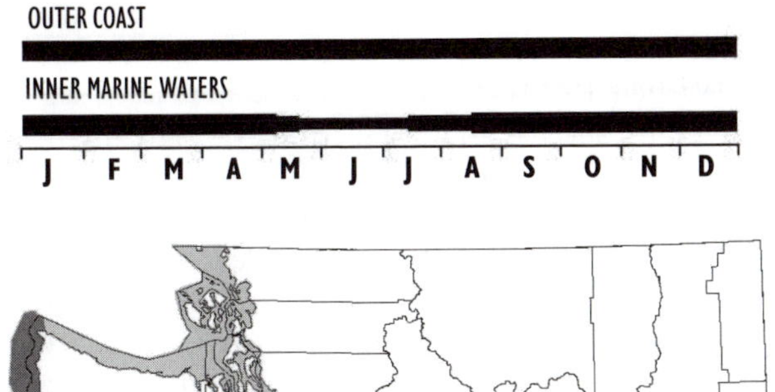

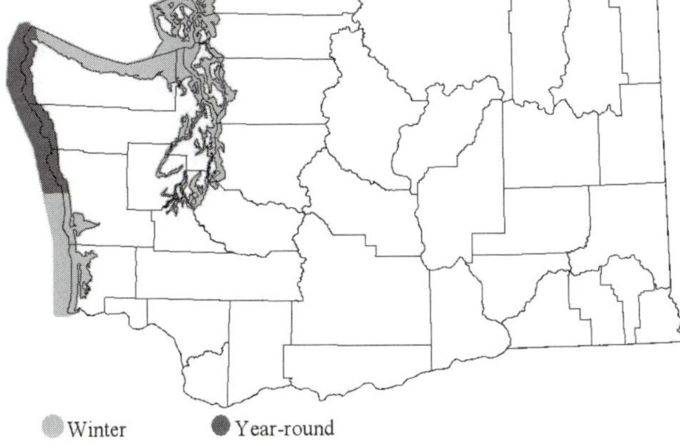

Subspecies: *U. c. inornata* breeds; *californica* also in winter.

Habitat: Breeds at colonies on rocks, islands, and cliffs along the outer coast; forages in nearshore continental shelf waters, deeper inland waters.

Occurrence: One of the most abundant N. Pacific seabirds, with about five million breeding in Alaska, 30,000 in B.C. and Washington, and one million in Oregon and California (Briggs et al. 1992). Relatively small numbers have nested at some 15 state locations from Tatoosh I. s. to Pt. Grenville (Smith et al. 1997, Warheit and Thompson 2002).

Surveys off Grays Harbor in Jul-Oct 1972-2000, found 96% of murres in <100 m depth (TRW). In inland waters, birds concentrate in the Strait of Juan de Fuca, and deeper channels and embayments such as Admiralty Inlet, Rosario Strait, Deception Pass, n. Haro Strait, Tacoma Narrows, Case and Carr inlets (MESA, Wahl and Speich 1983) and at seasonal events like spring herring spawns (Campbell et al. 1990b, Wahl 1995). Though most leave inland waters by early Apr, sizable numbers may remain into Jul (Wahl and Speich 1984, Lewis and Sharpe 1987).

Seasonal movements incompletely understood. Northward coastal flights—presumably mostly females—notable in Jul-Aug while males accompany flightless chicks. Birds banded in Oregon and Alaska have been recovered in winter in B.C. (Campbell et al. 1990b) and birds from Oregon recovered in Washington coastal and inland waters (Jewett et al. 1953). Numbers off Oregon decline in early fall and highest densities are off Washington and n. Oregon in fall and winter (Briggs et al. 1992). Washington waters support an estimated 50,000-200,000 in winter—10-20 times summer numbers—and may include birds breeding from Alaska s. to California (see Warheit 1997).

Breeding success and populations vary greatly. Reports indicated good nesting success in Washington and Oregon in 1965 and 1984; poor in 1967, 1969, 1980-83, 1987, 1989, and 1991-98. Washington breeding numbers ranged from 18,355 in 1979 to 31,520 in 1982, crashed to 3190 in 1983 (Wilson, in Smith et al. 1997) and 860 in 1993 (Manuwal et al. 1996). In 1994, the Tatoosh I. colony was the only active site in the state (Parrish 1995) and was reported to have been expanding rapidly prior to that (Takekawa 1992). In Oregon murres nested in 1988 but numbers declined in 1993 (Manuwal et al. 1996). Range-wide, reproductive failure was relatively consistent in the 1980s and 1990s and the population declined overall (Byrd et al. 1993).

Low numbers of breeders in Washington following the ENSO of 1982-83 correlated with the intensity and occurrence of such events as early as 1905 and 1907 (Wilson 1991). These conditions occurred again in the 1990s (Wahl and Tweit 2000a), and nesting failure in California in the 1992-93 ENSO was as severe as in 1983 (Sydeman et al. 1994). An example of ocean-wide impacts: an estimated 120,000 murres died in a "wreck" in the Gulf of Alaska in early 1993 (Van Pelt and Piatt 1995). During the 1983 ENSO birds off Grays Harbor foraged on anchovies close inshore. They showed a similar shift in the 1990s but nearshore water temperatures remained high and murre numbers declined overall (Wahl and Tweit 2000a; see Morgan et al. 1991, Ainley et al. 1996). Nesting failure results in ads. leaving colonies earlier and moving into inland waters because birds are unencumbered by flightless young and ocean prey is scarce, as in 1983 and the 1990s (e.g. Hodder and Graybill 1985, Lowe 1996, 1997a). CBC data (Table 26) indicated widespread long-term winter declines in inland waters. Wilson (2003) concluded the population in Washington was stable in the 1990s-early 2000s (see Warheit and Thompson 2002).

Following an oil spill off Grays Harbor in winter 1988-89, 72% of oiled birds recovered in Washington and B.C. were murres, with an estimated 33,000 killed (Campbell et al. 1990b, Ford et al. 1991). During the 1991 breeding season, 3100 of 4300 birds killed by a spill off Cape Flattery were murres (Divoky 1992); 900 of these were young birds just off nest ledges. Warheit (1997) estimated up to 19,500 were killed, though no drastic change was reported in ads. nesting at Tatoosh I. in 1992 (Takekawa 1992). In the early 1990s, 45%-96% of birds killed in gillnets in inland waters and coastal estuaries were murres (see Lewis and Sharpe 1987, Campbell et al. 1990b, Kaiser 1993, Warheit 1997). On colonies, gulls and crows take eggs and young, and locally there is predation by eagles and Peregrine Falcons (Takekawa 1992).

Table 26. Common Murre: CBC average counts per year by decade.

	1960s	*1970s*	*1980s*	*1990s*
Bellingham		326.8	53.5	22.2
Grays Harbor		1521.1	2515.5	29.9
Olympia			225.3	30.3
Padilla Bay			963.3	858.1
Port Townsend			57.9	40.1
San Juan Is.			738.8	280.7
San Juans Ferry		966.4	1741.6	754.3
Seattle	7.9	34.8	55.2	46.5
Sequim-Dungeness			168.1	271.1
Tacoma	10.0	224.1	693.6	390.0

Remarks: Classed SC, PHS.

Noteworthy Records: *Summer*: 2000 on 26 Jul 1997 "off Whidbey I." *Fall*: 1390 on 14 Sep 1979 at Pt. Partridge (MESA). *Winter*: 31,263 on 16 Jan 1993, 10,576 on 20 Jan 1990 off Grays Harbor; 4100 on 28 Feb 1979 (aircraft) in mid-Strait of Juan de Fuca, 3014 on 16 Dec 1978 in s. Rosario Strait, 2595 on 5 Nov 1979 at Green Pt., Anacortes, 1244 on 12 Feb 1978 in n. Haro Strait (MESA); 893 on 29 Dec 1979 in s. Admiralty Inlet; 882 on 14 Dec 1983 (aircraft) in s. Carr Inlet; 867 on 10 Dec 1982 (aircraft) in Case Inlet; 633 on 2 Feb 1983 at The Narrows (TRW). *CBC*: 10,000 in 1976 and 1988 and 8000 in 1985 at Grays Harbor; 4450 on 1 Jan 1980, 2998 on 4 Jan 1981 and 2745 on 30 Dec 1989 at Padilla Bay; 3371 on 2 Jan 1983 and 1270 on 16 Dec 1995 at Tacoma; 3307 in 1978, 1304 in 1980, 1259 in 1973 on San Juans Ferry; 1825 in 1993 at San Juan Is.; 1564 on 2 Jan 1977, 1011 in 1969 at Bellingham.

Terence R. Wahl

Thick-billed Murre *Uria lomvia*

Very rare winter visitor offshore and in inland marine waters.

Subspecies: *U. l. arra.*

Occurrence: Probably more frequent than recorded to date. Accepted sight records starting in the mid-1970s resulted from observer skill and increased observations offshore. The relative proportion of inshore and inland records demonstrates greater field effort there than offshore. Birds may be present in large flocks of Common Murres but conditions may not permit identification of this species. Except for one late Sep record, accepted Washington records are from early Dec to mid-Feb.

Irregular occurrence s. of the normal winter range may depend on oceanographic conditions (see Roberson 1980), and a decline in breeding populations in the Bering Sea in the late 1970s to 1980s (Gaston and Jones 1998) may also have affected dispersal.

Remarks: A small colony discovered in 1981 on Triangle I., nw. of Vancouver I., had 70 ads. there in 1982 (Campbell et al. 1990b). This is about 800 km from the nearest known colony. Called Brünnich's Guillemot in Europe.

Records: *Fall*: 1 on 22 Sep 1976 off Grays Harbor (Tweit and Paulson 1994). *Winter*: 1 collected on 19 Feb 1933 at Westport (Tweit and Skriletz 1996); 1 on 6 Dec 1979 nr. Friday Harbor (Tweit and Paulson 1994) 2 on 15 Dec 1979 at Ocean Shores (Tweit and Paulson 1994); 1 on 31 Dec 1986 at Drayton Harbor (Tweit and Paulson 1994); 1 killed by an oil spill off the outer coast in Dec 1988 (Tweit and Skriletz 1996); 1 on 8 Jan 1977 (TRW), 1 on 20 Jan 1990 (Tweit and Paulson 1994), 1 on 17 Feb 2002 off Grays Harbor, 1 on 17 Dec 2002 w. of Port Townsend. *Note*: a report of 1 on 21 Sep 1976 at Ediz Hook was rejected by WBRC (Tweit and Paulson 1994).

Terence R. Wahl

Pigeon Guillemot *Cepphus carbo*

Locally fairly common to common resident in inland marine waters. Locally common summer resident, rare in winter along outer coast.

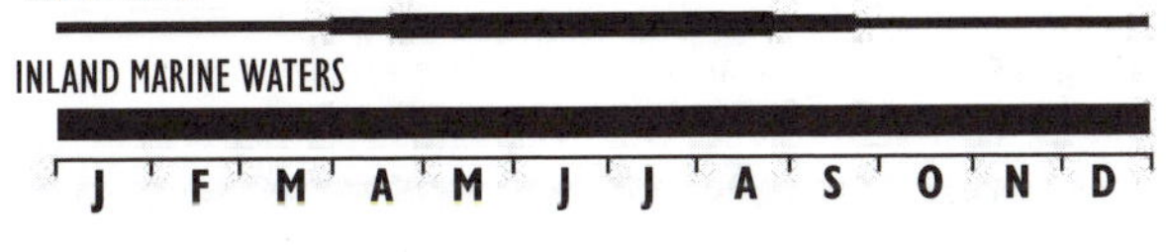

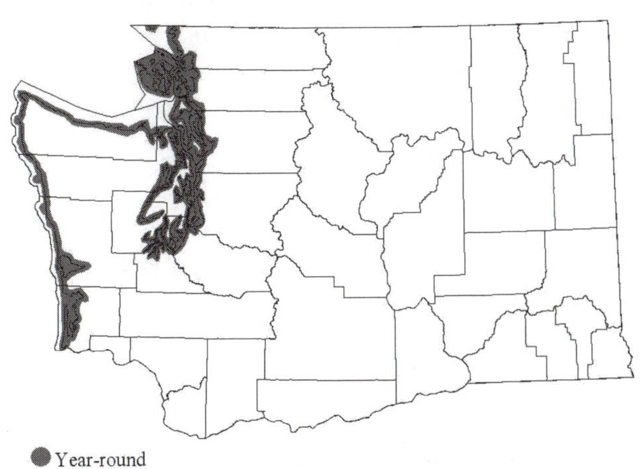

Subspecies: *C. c. adianta* breeds s. to Washington; *aurica* in Oregon and California (Ewins et al. 1993).

Habitat: Nearshore marine waters along rocky shorelines, in passages and shallow embayments,

Pigeon Guillemot (G. Scott Mills)

productive feeding areas such as tidal convergences. Nests in crevices, cliffs, and talus slopes, under beach logs and other natural situations, under piers, in drain pipes, jetties, and nest boxes, often in scattered pairs but in colony-like aggregations where sites are available (Speich and Wahl 1989).

Occurrence: Widespread and relatively adapted to humans. In the early 1980s, birds were known to nest at about 140 sites, with about 52% of these in the San Juan Is. and adjacent areas and 30% of the population at Protection I. (Speich and Wahl 1989). In 1978-79 MESA surveys most guillemots were nr. major colonies at Protection I. and the San Juan Is. (Speich and Wahl 1989). In Jul 1986, 645 guillemots were found in the San Juan Is. with 139 of these at Sucia I. Comparatively small numbers nest along the s. outer coast where birds nest in limited habitats at jetties and rocky headlands (Speich and Wahl 1989).

Almost all outer coast guillemots move into protected waters of se. Alaska (Ewins 1993), B.C. (Campbell et al. 1990b, Vermeer and Morgan 1997), and Washington (Wahl and Tweit 2000b). CBC totals, especially in Washington's s. inland waters, are noticeably higher than summer populations (Speich and Wahl 1989). Briggs et al. (1987) found most birds left California from Sep to Feb, with fledglings banded in California recovered in Oregon and Washington within weeks of departure from the Farallones. Ainley et al. (1990a) stated that 30% of winter band recoveries of Farallones imms. were from Washington and B.C.

About 95% of birds off Grays Harbor from 1972 to 1998 were within a short distance of nest sites and adjacent foraging areas in summer (Wahl and Tweit 2000b). Offshore migration was suggested in Aug and Sep when birds in breeding plumage were flying n., over 65 km offshore, almost certainly migrating from California and/or Oregon (Wahl and Tweit 2000b). Ewins (1993) suggested this migration is made before the full post-breeding molt. There are no offshore records in Oct and winter.

Nesting numbers vary, and estimates range from 4270 birds (Speich and Wahl 1989) to 3000 pairs at Protection I. alone in 1993 (U. Wilson, in Smith et al. 1997), and to 16,000 in inland marine waters alone in 2002 (Hodder 2002). Reproductive success varies with oceanographic conditions (see Ewins et al. 1993): a breeding population of 3000 pairs on Protection I. in 1993 dropped to 1967 pairs in 1995 (U. Wilson, in Smith et al. 1997). Long-term CBC data showed an increase (P <0.01) while Nysewander et al. (2001) indicated a decline from 1978-79 to 1994-95.

Remarks: Scattered nesting and non-breeding distribution believed to limit vulnerability, but seasonal concentration in inland waters may increase vulnerability to gillnet entanglement and oil spills. Low mortality in the winter 1988-89 oil spill off Grays Harbor probably due to virtual absence of birds in winter (Wahl and Tweit 2000b). Differences in population estimates may represent natural variation but also census methods and estimation assumptions (see Galusha et al. 1987). Historic numbers probably cannot be reconciled with the species-specific study estimate of about 16,000 birds breeding in inland marine waters in 2002 (Hodder 2002), which included numbers at over 300 colonies previously not documented. Classed PHS, GL.

Noteworthy Records: *CBC, selected records:* 1221 on 22 Dec 1989 at Sequim-Dungeness, 249 on 21 Dec 1996 at Tacoma, 243 on 1 Jan 1994 at Skagit Bay, 239 at Padilla Bay on 31 Dec 1994, 217 on 20 Dec 1995 at San Juans Ferry, 164 at Olympia on 22 Dec 1979, 129 on 17 Dec 1988 at Port Townsend, 150 on 17 Dec 1988 at San Juan Is.

Terence R. Wahl

Long-billed Murrelet *Brachyramphus perdix*

Very rare fall vagrant in w. Washington. Casual in e. Washington.

Status and distribution of the Long-billed Murrelet in N. America are uncertain, almost certainly due to previous conspecific status with Marbled Murrelet. As of Nov 1997, there were only 36 records from N. America, including a number from the interior, with four from Oregon, but none from B.C. (Mlodinow 1997).

There are six Washington records: one (ph.) sw. of Lopez I. on 12 Aug 1993 (Skriletz 1996); one (v.t.) at Olympia, 12 Aug 1994 (Mlodinow 1997); one at the Elwha R. mouth on 16 Aug 1995; one (ph.) at Edmonds,16 Nov 1995 (Mlodinow 1997); two at Ocean Shores, 6 Aug 1999; one 15 km w. of Pomeroy, Garfield Co., on 13 Aug 2001.

Along the Pacific coast and in Puget Sound, Long-billed Murrelets have been observed mostly at locations favored by Marbled Murrelets. between 25 Jun to 9 Dec (Mlodinow 1997). Pacific coast records appear coincident with observer effort and preparedness.

Steven G. Mlodinow

Marbled Murrelet *Brachyramphus marmoratus*

Uncommon to fairly common resident in marine waters, very rare on freshwater; nests in low to mid-elevation coniferous forests w. of the Cascade crest.

OUTER COAST

INLAND WATERS

J F M A M J J A S O N D

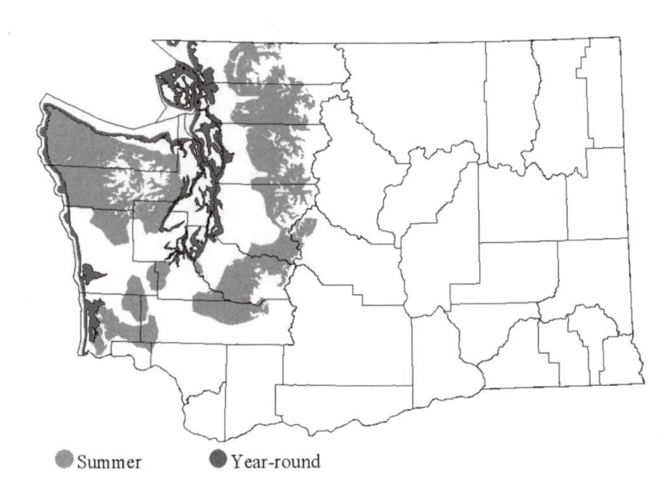

Habitat: Most commonly in shallow (<20 m depth) nearshore habitats, within 200-400 m of shore, areas of tidal activity (Speich and Wahl 1995, Thompson 1997a, b) and rocky, complex coastlines and habitats with kelp or cobble substrates. Rare on inland lakes (Carter and Sealy 1986). Nests in live canopies of conifers typically in old-growth forests or younger forests with old remnant trees (Hamer and Nelson 1995, Nelson 1997).

Occurrence: Marbled Murrelets are most numerous year-round in inland waters n. of Everett-Kitsap Co., fewer in s. Puget Sound and n. outer coast, rare along outer coast s. of Grays Harbor (Speich et al. 1992, Thompson 1997a, b). Local differences in abundance likely reflect differences in marine habitat suitability. Seasonal numbers vary widely, but causes of large changes in numbers (e.g. Ralph et al. 1996, Sustainable Ecosystems Institute 1997) are unclear.

Numbers decrease in winter coastally from Port Angeles to Willapa Bay, except for increased densities at the mouth of Gray's Harbor (Speich and Wahl 1995, Thompson 1997a, b). The marked winter increase in n. inland waters (Speich et al. 1992) apparently results from immigration from B.C. and the outer coast. Two records far offshore in Mar and May suggest offshore migration, similar to that of the Pigeon Guillemot (Wahl and Tweit 2000b).

Nesting status uncertain in the 1950s (Jewett et al. 1953). Because of its association with old-growth forests, species has been severely affected by loss of nesting habitat. Surveys of potential habitat during the early to mid-1990s (WDFW) indicated highest nesting or presence on the Olympic Pen., the n. Cascades, and in limited remaining habitat in sw. Washington. None were found in the San Juan Is. or Puget Trough lowlands. Though apparently suitable habitat was available, relatively few were located in the s. Cascades and none in Clark or w. Cowlitz Cos. Most detections were on federal lands. Nesting was detected as far as 84 km inland in the n. Cascades, while in the sw. region all sites were within 40 km of marine waters (WDFW). Most inland records are from Apr to late Aug. The species is not recorded occurring e. of the breeding range.

Table 27. Marbled Murrelet: CBC average counts by decade.

	1960s	1970s	1980s	1990s
Bellingham		69.5	71.7	5.6
Edmonds			5.2	1.5
Grays Harbor		5.1	6.0	2.4
Kitsap Co.		26.0	15.6	6.6
Olympia			30.4	9.0
Padilla Bay			56.3	38.6
Port Townsend			16.3	3.3
San Juan Is.			230.0	96.8
San Juans Ferry		19.4	48.0	36.7
Seattle	5.8	7.3	2.5	0.8
Sequim-Dungeness			33.8	13.4
Tacoma		42.1	7.8	8.5

Based on qualitative assessments of numbers, Speich et al. (1992) concluded numbers had declined in the Puget Sound region. Wahl and Tweit (2000a) found a significant decline at Grays Harbor. Numbers decreased on all CBCs (Table 27) and Nysewander et al. (2001) indicated a large decrease from 1978-79 through 1994-95. Locations with high counts since the mid-1990s (e.g. Point No Point) lack earlier data for comparison. Historical data from Alaska and B.C. (e.g., Campbell et al. 1990b) and modelling (Beissinger 1995) also indicate long-term population declines rangewide. Nesting habitat in the Puget lowlands has been virtually eliminated, and little remains in sw. Washington. Logging of nesting habitat is considered the most significant cause of rangewide decline in numbers, though oil spills, other pollution, and fish-net entanglement also contributed. At least two of nine significant oil spills in Washington from 1971 to 1992 caused mortalities significant to local murrelet populations (Carter and Kuletz 1995).

Remarks: Ralph et al. (1995) estimated Washington breeders at 5500, of a world total of 270,000. Classed ST, FT, PHS, GL. The 1994 Northwest Forest Plan (USDA and USDI 1994) was planned to protect most existing habitat on federal lands for 50-100 years and restore habitat. Forest practice rules enacted in 1997 may slow loss of habitat on non-federal lands—considered critical, particularly in areas where there is little federal land (USFWS 1997b). At-sea murrelet surveys on the outer coast by WDFW resulted in population estimates of 1963 birds in 2001, 2301 in 2002, and 3390 in 2003. Conclusions drawn from these data are uncertain. The Long-billed Murrelet was formerly considered subspecific.

Noteworthy Records: *Nesting*: 103 on 31 Jul 2002 noted by sight, sound at Heart of the Hills CG, Olympic NP. *High counts*: 5200 passed to the s. at first light at Pt. Roberts on 28 Jan 1978; 662 on 11 Apr 1993 off Mukilteo; 400 on 2 Mar 1995 at Point No Point were peak of 700 total there on 28 Feb-2 Mar; 800 on 6 Aug 1996 between Port Angeles and Pillar Pt. Offshore counts: 2 on 19 May 1979 ca. 47 km offshore and 2 on 20 Apr 1991 ca. 70 km off Grays Harbor (TRW).

Richard Fredrickson and Terence R. Wahl

Kittlitz's Murrelet *Brachyramphus brevirostris*

Casual.

The single Washington record of Kittlitz's Murrelet was of a bird at Friday Harbor on 4 Jan 1974 (ph., Tweit and Paulson 1994). This Alaskan/Siberian species has been found s. of Alaska only three other times, between 16 Aug and 12 Apr, in B.C., s. California, and Indiana (Devillers 1972, Mlodinow 1984, and Campbell et al 1990b). Classed GL.

Steven G. Mlodinow

Xantus' Murrelet *Synthliboramphus hypoleucus*

Very rare to rare visitor to outer continental shelf and oceanic waters.

Subspecies: Both *S. h. scrippsi* and *hypoleucus* occur in Washington (see Jehl and Bond 1975, Drost and Lewis 1995).

Occurrence: Though specimens had been obtained earlier, Xantus' Murrelets were essentially unknown in Washington until offshore boat trips began in the 1970s and later when observers accompanied research cruises. About 40 records were acquired off Grays Harbor, usually of pairs or two birds together. From 1974 to 2002, birds were recorded offshore during 18 years and, due to observation conditions, may have been missed during other years. Birds were noted from Jul to Oct (once in May), with numbers relatively concentrated in Aug-early Oct. Later occurrence and winter range are unknown.

Occurrence likely influenced by oceanographic conditions along the Pacific coast. More birds were noted prior to a general "regime shift" of sea temperatures and productivity conditions in the mid-late 1970s (see Ebbesmeyer et al. 1991). Most occurred in relatively warm SSTs during all years, with many recorded in SSTs of 16° C or higher (see Roberson 1980). In many years, high SSTs are far offshore and this may affect whether or not birds are noted during one-day surveys. Birds were noted annually in the late 1990s, though actual numbers present were likely relatively lower because of increased survey effort. Birds were seen to the outer limits of surveys at about 125° 36' W.

Both subspecies have been collected (Roberson 1980), and sight indentifications vary. Most of 106 birds seen off Oregon-Washington in Aug-Sep 2001 were *hypoleucus*; most were beyond 90 km offshore, with *scrippsi* predominating to the s. and *hypoleucus* predominating to the n., particularly far offshore (M. Force, NAB 56:99). Though subspecies were not often recorded, almost all birds noted on surveys off Grays Harbor to about 65 km offshore were believed to be *scrippsi* (TRW). On 19 Jul 2003 two *hypoleucus* were recorded about 80 km offshore, over SST of 17° C.

Remarks: Classed hypothetical by Jewett et al. (1953) who stated that a specimen of 23 Jan 1936 was later identified as a Marbled Murrelet. Though the similar-appearing Craveri's Murrelet, *S. craveri*, was not suggested as occurring to date, the WBRC (Aanerud and Mattocks 1997) decided to question specific identification and to classify records lacking descriptions of definitive field marks as Xantus's/Craveri's, on the assumption that Craveri's could occur, though it is not recorded n. of Sonoma Co., California (Small 1994). Classed GL. Designated a Bird of Conservation Concern in 2002 by USFWS.

Noteworthy Records: *Early*: 1 m. collected off Copalis Beach on 6 Dec 1941 and 2 collected about 200 km ssw. of Cape Flattery on 7 Aug 1947 (Mattocks et al. 1976). *Off Grays Harbor*: 2 on 16 May 1976, 1 on 18 Jul 1976, 1 on 27 Jul 1991, 1 on 15 Aug 1984, 2 on 17 Aug 2002, 20 on 24 Aug 1975, 1 on 26 Aug 2000, 1 on 5 Sep 2002, 2 on 6 Sep 1997, 3 on 8 Sep 1974, 1 on 9 Sep 1984, 3 on 10 Sep 1978, 2 on 11 Sep 1983, 1 on 12 Sep 1987, 2 on 13 Sep 1998, 3 on 20 Sep 1992, 2 on 21 Sep 1996, 1 on 4 Oct 1987, 2 on 7 Oct 1979, 2 on 9 Oct 1977, 1 on 9 Oct 1999, 6 on 11 Oct 1970 (TRW). *Other offshore*: 8 on

13 Aug-5 Sep 2001 off Washington (M. Force p.c.); 1 on 1 Aug 1981 off Cape Flattery (TRW); noted in Aug and Oct 1988 sw. of Cape Flattery at about 48° 05' N, 126° 40' W and attributed to B.C. (Morgan et al. 1991); 2 on 13 Sep 2001 155 km off Grays Harbor. *S. h. scrippsi*: 3 on 28 Jul 1983, 2 on 4 Aug 2001, 2 on 14 Aug 1999, 3 on 20 Aug 1977, 18 on 8 Oct 1978 off Grays Harbor (TRW); 2 on 5 Sep 2001 108 km off Destruction I. *S. h. hypoleucus*: 2 on 19 Jul 2003 82 km off Cape Shoalwater; 2 on 6 Sep 2001 64 km off Leadbetter Pt.; 6 on 11 Sep 1978 off Grays Harbor (TRW).

Terence R. Wahl

Ancient Murrelet *Synthliboramphus antiquus*

Locally fairly common to abundant migrant and winter visitor offshore and inland marine waters, probably rare breeder. Very rare in fall and winter in e. Washington.

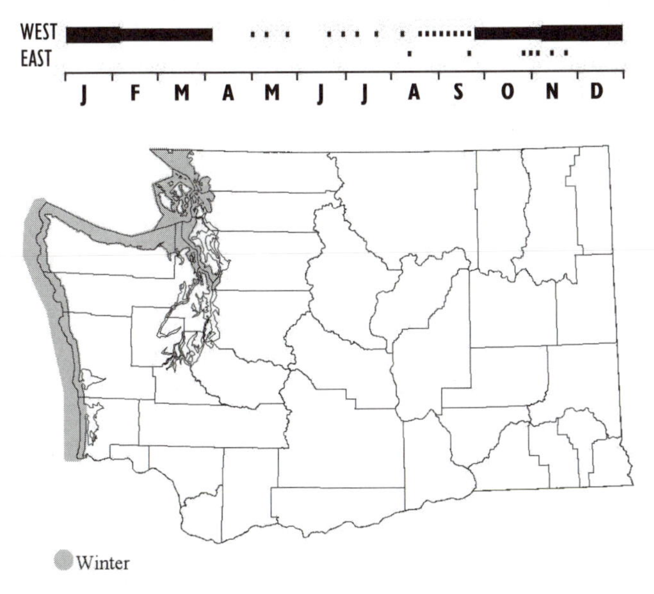

Habitat: Continental shelf waters; deeper offshore waters, tidal fronts in inland marine waters.

Occurrence: Large concentrations reported especially from Nov to Jan, notably along the e. Strait of Juan and adjacent deep waters. In inland marine waters birds forage in tidal convergences in the Strait of Juan de Fuca, at Pt. Wilson and Admiralty Inlet, Point No Point, and in Rosario and Haro straits, the Strait of Georgia (as at Pt. Roberts and Alden Bank), in San Juan and Speiden channels (MESA, TRW), and occasionally in similar habitats in s. Puget Sound.

Status offshore and on the outer coast less well understood. Some birds occur in winter (TRW), and birds are seen from shore on occasion (e.g. Hoge and Hoge 1991). Many of the estimated 400,000 nesting in Alaska and 500,000 nesting in B.C. (Briggs et al. 1992) migrate off Washington to wintering areas s. to Baja California. Most occur nr. the shelf break in B.C. (Morgan et al. 1991, Gaston 1992), and the few seen during aircraft surveys in 1989-90 were over the outer slope off n. Washington (Briggs et al. 1992). On one-day surveys 1971-2000 off Grays Harbor, 82% of 333 birds were over depths of 100-1000 m (TRW).

There is one Washington breeding record. R. Hoffman collected a female and two eggs at Carroll I. in 1924 (Jewett et al. 1953). Subsequent observers have reported single birds or pairs in May, Jun, and Jul off Grays Harbor, Destruction I., and Alexander I. Some were in breeding plumage, others in juv. plumage. Dead birds have been recovered at Ocean Shores in summer, and live birds recorded in inland waters then. Recent nesting is likely—birds were off the Olympic Pen. in Jun of 2000 and 2002 and a small juv. was noted ca. 45 km offshore on 23 May 2004 (BT). A pair with a downy young was seen off Westport on 17 May 2003 (BT).

The Ancient Murrelet is notorious for storm-blown occurrence far from its normal range and, as in B.C. (Campbell et al. 1990b), Montana, and Oregon (Gilligan et al. 1994), there are a number of such records in Washington. Birds have been found at Spokane, on the Snake R., the Columbia R., and nr. the Skagit R.

Seasonal occurrence and abundance probably relate to oceanographic variations. In inland waters, reports of 7000 and 13,000 at Victoria in Nov 1983, and of 5000+ on 21 Nov 1997 between Victoria and Race Rocks, B.C., occurred during years of ENSO conditions of low ocean productivity. Similarly, reports of birds in summer (e.g., "inexplicable numbers" in the region in mid-Jul 1984, "unusual numbers" in Jul 1993) may have represented dislocated birds. Long-term CBC averages indicate increases in the Strait of Juan de Fuca at Sequim-Dungeness and in the San Juan Is.

Remarks: Like other seabirds, vulnerable to oil spills. Ancient Murrelets go to sea when only a few days old, and observations of small, downy young would more likely indicate local breeding than would reports of adult size juvs. Classed GL.

Noteworthy Records: West, *Spring*: 216 on 13 Mar 1998 at Point No Point; 1 on 10 May 1997 at Sequim Bay. *Summer*: pair, downy young off Westport on 17 May 2003; ads., small juv. ca. 45 km offshore on 23 May 2004; 12-15 on 31 May 2003 nr. Smith I. (K. Wiggers p.c.); 10+ 12-18 Jun 2002 off the Olympic Pen.; 2 on 30 Jun 1998 w. of Destruction I.; about 10, some pairs, in Jun 2000 off the Olympic Pen; 1 on 6 Jul 1993 at Pt. Wilson; 1 on 18 Jul 1993 at Whidbey I. *High counts, Fall*: 165 on 11 Nov 1998 at Point No Point; 1878 on 18 Dec 2000 nr.

Sequim. *Winter*: 1000 on 3 Feb 1999 at Port Angeles; 100 on 15 Jan 2000 at Cattle Pt. *Outer coast*: 122 on 15 Jan 1994 off Grays Harbor (TRW). *CBC high counts*: 453 in Jan 1997 from San Juans Ferry; 272 in 1997 at Sequim-Dungeness; 311 in 1988 at San Juan Is.; 193 in 1979 at Tacoma; 80 in 1979 at Olympia; 75 in 1979 at Kitsap Co.; 68 in 1988 at Seattle; 63 in 1997 at Port Gamble. Inland Puget Trough: 1 dead on 11 Nov 1988 at Marblemount. **East**: 1 in late Oct 1963 in Spokane Co.; 1 on 11 Nov 1965 at Long L. Dam; 1 on 31 Oct 1988 and 1 on 23 Sep 1994 at Vantage; 1 on Nov 2 1989 on the Grande Ronde R.; 1 on 23 Nov-3 Dec 1989 at John Day Dam; 1 on 10-14 Aug 2001 at Chief Joseph Dam.

Terence R. Wahl

Cassin's Auklet *Ptychoramphus aleuticus*

Variably common to abundant breeder on outer coast, resident and migrant offshore; rare to uncommon visitor to inland marine waters.

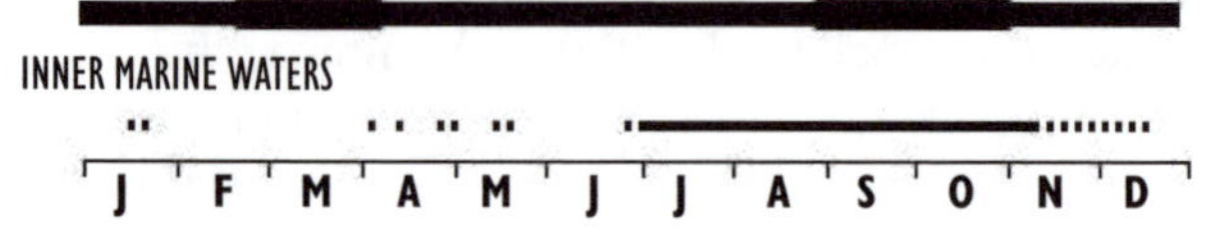

Subspecies: *P. a. aleuticus* (Manuwal and Thoreson 1993).

Habitat: Nests on forested offshore rocks on the outer coast; forages over the outer continental shelf and slope, also in the Strait of Juan de Fuca and adjacent deeper waters of inland channels.

Occurrence: The most abundant breeding seabird in Washington, though few are seen except offshore or during nocturnal visits to colonies. In 1978-82, major colonies were Alexander I. (54,600 birds), Carroll I. (15,400), Jagged I. (8000), middle and e. Bodelteh (8600), with smaller numbers on Petrel Rock and Tatoosh I. (Speich and Wahl 1989). The estimated 90,000 breeders are, however, a small part of the species' total population estimate of 3.6 million birds (Enticott and Tipling 1997). About 80% of all Cassin's Auklets nest in B.C. (Campbell et al. 1990b).

Birds forage offshore in summer and more inshore in fall (Vermeer and Morgan 1997). Briggs et al. (1992) reported greatest densities from mid-slope waters (>200 m depth) offshore with concentrations where upwelling occurs along the edges of underwater canyons.

Large numbers migrate offshore. Briggs et al. (1987) reported a 100% increase in Sep and Oct off California, with numbers reaching 500,000 to one million in Jan or Feb. Off Washington and Oregon in Sep 1989 and 1990, Briggs et al. (1992) found highest numbers s. of the Columbia R., with numbers n. of that only seaward of the shelf break. From 1972 to 1998 one-day surveys off Grays Harbor, Wahl and Tweit (2000a) found Jul-Aug numbers increased by 50-500% in Sep-Oct, suggesting movements of local breeders and young and migrants from the n. passing through. Spring migration is less well known. Birds arrive between late Feb and Apr in B.C. (Campbell et al. 1990b). Numbers off Grays Harbor in spring were much smaller than in summer or fall (TRW) though Briggs et al. (1992) reported birds were concentrated more inshore in Apr and May than later.

Birds winter from s. B.C. s. to Baja California (Campbell et al. 1990b), with some birds from Alaska and B.C. moving s. (Enticott and Tipling 1997). The winter range of Washington breeders is unknown.

For inland waters, Jewett et al. (1953) noted Nov records in 1925 and 1941 at Tacoma, and about 40 reports from Pt. Roberts s. through the e. Strait of Juan de Fuca to c. Puget Sound have followed (e.g. Lewis and Sharpe 1987, Wahl 1995, MESA). Reports suggest increased occurrence in the 1990s coincident with widespread reports of decreases along the w. coast. Most records are from Jul to Dec, suggesting dispersal of failed breeders, young birds, and migrants dislocated due to poor ocean productivity.

Population declines in recent decades were reported due to a number of causes, including predation on colonies, pollution, and fisheries conflicts (Enticott and Tipling 1997). The most important recent cause is likely a decline in ocean productivity (Ainley et al. 1990b). Ainley et al. (1994) give a 5% decrease in numbers at the Farallon Is. between 1971 and 1989. From 1972 to 1998, numbers off Grays Harbor declined significantly (Wahl and Tweit 2000a), with effects of persistent ENSO conditions especially noticeable in the 1990s. About 500-1000 were found beached on the Long Beach Pen. in Jul 1992.

Remarks: Following an oil spill in winter 1988-89, 72 Cassin's Auklets, 0.8% of identified dead birds, were recovered (Ford et al. 1991). Some records prior to the late 1970s, particularly those of groups of birds, may have been misidentifications. Classed SE, FSC, PHS.

Noteworthy Records: *Inland marine waters, Spring*: 15 on 22 Mar 1985 in Upright Channel and 1 on 4 Apr 1979 at Friday Harbor (Lewis and Sharpe 1987). *Summer*: 1 nr. Williamson Rocks and 2 nr. Bird Rocks in Rosario Strait on 21 Jun 1967 (Lewis and Sharpe 1987). *Fall*: reported on 1 Nov 1925 and on 21 Nov 1941 nr. Tacoma (Jewett et al.

1953). 1 seen on 5 Sep 1988 and 1 beached on 17 Aug 1994 at Pt. Roberts (Wahl 1995). *High counts off Grays Harbor*: 500 on 21 Jan 1995, 3135 on 7 Feb 2003, 486 on 27 Jul 1991, 379 on 5 Aug 1979, 404 on 6 Aug 1994, 333 on 13 Aug 1988, 342 on 22 Aug 1976, 355 on 8 Sep 1974, 575 on 24 Sep 1972, 763 on 5 Oct 2002, 812 on 7 Oct 1990, 2265 on 8 Oct 1978 (TRW).

Terence R. Wahl

Parakeet Auklet *Cyclorrhynchus psittacula*

Very rare, possibly rare winter visitor and migrant over the continental shelf.

Habitat: Outer continental shelf and oceanic waters.

Occurrence: The only live birds recorded in Washington were offshore in outer continental shelf waters, and specimens recovered during winter and early spring are all from the outer coastal beaches. Occurrence is likely influenced by variations in oceanographic conditions (Harrison 1983). Offshore winter surveys from research vessels and from Grays Harbor have been limited and long-term patterns are uncertain. That birds may be more numerous is indicated by recovery of beached birds. In winter 1988-89, for example, five specimens were recovered in Washington following an oil spill off the Columbia R. and 15 more, believed oiled off Washington, were found in B.C. (Campbell et al. 1990b). Roberson (1980) reported 29 specimens recovered in c. and n. California prior to 1945, with 14 of those at Monterey Bay in Jan 1908. Through 1994 there were about 55 records from the lower 48 states, with 51 between 20 Nov and 25 Apr (Mlodinow and O'Brien 1996).

Noteworthy Records: *Specimens not located by WBRC*: 1863 (Jewett et al. 1953) and 1 on 18 Dec 1934 from Westport (Jewett et al. 1953). *Accepted by WBRC* (see Tweit and Skriletz 1996): 1 on 21 Feb 1937 and 1 on 13 Feb 1959 from Grayland, 1 on 21 Apr 1937 and 1 on 25 Feb 1990 from Westport, 1 on 11 Apr 1944 from Copalis. *Other specimens reported*: 1 on 18 Jul 1976 at Westport (Roberson 1980); 1 on 17 Jan 1983 and 4 in the last week of Dec 1988 from Ocean Shores, 5 in late Dec 1988-early Jan 1989 from the Nestucca oil spill. 2 oiled specimens recovered in Mar 1999 at Long Beach (J. Skriletz). *Live birds*: 1 on 20 Apr 1991 and 1 on 25 Apr 1992 (Tweit and Paulson 1994), 3 on 2 Feb 1997, 1 on 14 Mar 1999, and 1 on 24 Apr 1999 off Grays Harbor (TRW; see Aanerud and Mattocks 2000). *Not addressed by WBRC*: 1 in the Strait of Juan de Fuca on 15 Jan 1907 (Jewett et al. 1953). 1 on 25 May 2004 off LaPush (BT). *Rejected by WBRC* (Tweit and Skriletz 1996, Aanerud and Mattocks 1997): 1 on 15 Apr 1973, 1 on 15 Apr 1978 from the Port Angeles-Victoria ferry nr. Dungeness, 12-15 on 12 May 1978 from the Edmonds-Kingston ferry.

Terence R. Wahl

Whiskered Auklet *Aethia pygmaea*

Casual vagrant.

The single Washington record of this Bering Sea breeder, a bird noted at Penn Cove on 16 May 1999, represents the only N. American record away from Alaska (Mlodinow and Duffy 2000). Whiskered Auklets are generally relatively sedentary (Byrd and Williams 1993b, Gaston and Jones 1998), but there are at least three records of vagrants in Japan (Brazil 1991). Though the species is presumably more likely in winter, the apparently out-of-season date is similar to Jun and Jul records of Least and Crested auklets, two other Beringian species, in California (Mlodinow and Duffy 2000). Classed GL.

Steven G. Mlodinow

Rhinoceros Auklet *Cerorhinca monocerata*

Common to locally abundant summer resident on the outer coast, offshore waters and n. inland waters; fairly common to common winter resident on the outer coast and s. inland waters.

WEST
J F M A M J J A S O N D

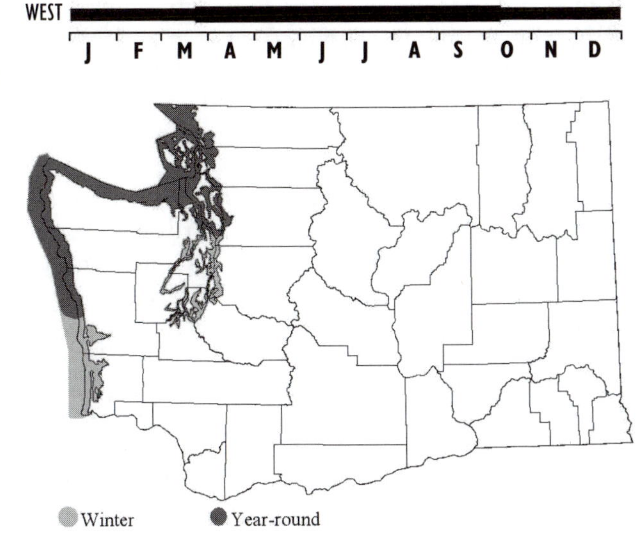

Habitat: Nests in burrows in suitable soils on islands; forages over the continental shelf, in nearshore waters, deeper channels, and embayments.

Occurrence: The Rhinoceros Auklet breeds at a few large colonies from n. Japan to California (Byrd et al. 1993). Over half of the N. American population of 600,000-900,000 nests in B.C. (Campbell et al. 1990b). Washington's two main colonies are the southernmost major w. coast sites—only about 1000 birds breed in Oregon and California (Briggs et al. 1992).

Auklets nested historically on Whidbey I. (Richardson 1961), but for decades nesting in Washington has been concentrated at Protection I. in the e. Strait of Juan de Fuca and at Destruction I. off the outer coast (Speich and Wahl 1989). In 1980 the Protection I. nesting population was estimated at 34,000 birds, Smith I. at 2600, and in 1982 Destruction I. had 23,000 (see Smith et al. 1997). Small numbers nested also at Tatoosh, Alexander, and E. Bodelteh Is.

Rhinos forage widely in n. inland waters in summer, congregating in channels and areas of the e. Strait of Juan de Fuca, Admiralty Inlet, and adjacent waters. Counts of 150 or more birds in 1978-79 were all between 15 May and 27 Sep, within foraging range of breeding birds at Protection and Smith Is., and a survey while Rhinos were feeding young on Protection I. in Jul 1979 showed a restricted foraging range, with over 90% of birds in the e. Strait of Juan de Fuca and n. Admiralty Inlet, and 80-85% within 37 km of the island (Wahl and Speich 1994). A survey in summer 1982 found numbers of ads. s. to Oak Bay, a few non-breeders s. of that, and none on Hood Canal (Wahl and Speich 1984). Outer coast birds feed from nearshore to the outer shelf and shelf break areas. Birds noted off Grays Harbor in 1972-98 concentrated in the channel and near shore in May-Aug, with numbers offshore increasing in Sep-Oct (Briggs et al. 1992, TRW). Numbers of ads. equivalent to about 10% of the population on Destruction I., about 75 km away, may occur in the Grays Harbor channel in summer (TRW).

Following nesting, most birds move s. offshore in Sep-Oct (Briggs et al. 1992, TRW). Birds banded in B.C. have been recovered in winter in California (Campbell et al. 1990b): there are few birds off California in summer and numbers increase to 100,000-300,000 in winter (Briggs et al. 1987). In Washington, relatively low numbers winter offshore (Briggs et al. 1992, TRW). Concentrations occur in s. Puget Sound, mostly from Seattle, Tacoma, and Olympia (Wahl and Speich 1983, Table 28), small numbers in the San Juan Is., and scattered numbers elsewhere in n. inland waters.

In spring, groups of 15-20 migrating n. 180 km off Grays Harbor in April 1977 and counts of 100-200 in April 1988 (TRW) were likely enroute to colonies in B.C. and Alaska. In early spring, newly arrived birds or "honeymooning" pairs are seen farther away from colonies than they are later on during the nesting season.

Table 28. Rhinoceros Auklet: CBC average counts by decade.

	1960s	*1970s*	*1980s*	*1990s*
Edmonds			7.3	13.8
Grays Harbor		2.4	1.1	0.6
Kent-Auburn			4.1	6.0
Kitsap Co.		3.0	1.3	4.0
Olympia			93.2	76.7
Port Townsend			0.1	6.1
San Juan Is.			7.5	7.2
San Juans Ferry		3.8	4.8	41.5
Seattle	31.5	35.8	45.0	51.2
Sequim-Dungeness			0.3	2.7
Tacoma		110.6	149.5	401.4

Numbers off Grays Harbor from 1972 to 1998 increased while other breeding alcids decreased (Wahl and Tweit 2000a). Numbers nearshore in Sep-Oct 1988 increased greatly while offshore numbers declined from 1972-87 patterns (TRW), reflecting lower productivity offshore and attraction to anchovy schools inshore (Wahl and Tweit 2000a). Inland, numbers of nesting birds decreased at Protection and Smith Is. in the 1990s (e.g. Smith et al. 1997). Numbers foraging in inland waters were larger than normal, however, likely indicating a shift to inside waters. Above-average numbers were reported in 1979-80, and CBC data (Table 28) indicate increases in s. inland waters.

Oil spills and colony predation are threats (Enticott and Tipling 1997). Reports of relatively low mortality following winter oil spills (Harrington-Tweit 1979a, Ford et al. 1991) reflected the seasonal absence of auklets. Concentrations at a few large colonies, many adjacent to major shipping routes, increase the potential for mass mortality, and one major foraging area, Salmon Banks off San Juan I. (Wahl and Speich 1994), is also a major gillnet fishery area.

Remarks: Establishment of the Zella M. Schultz NWR on Protection I. in the 1980s preserved that major colony from real-estate development. Classed PHS.

Noteworthy Records: *High counts, inland waters, Spring-Fall*: 550 on 7 Apr 1990 in n. Upright Channel, San Juan Is. (TRW); 450 at Deception Pass on 22 Aug 1971, 3000 on 18 Jul 1998 at Cultus Bay, 500 on 10 Aug 1998 at Point No Point. 1269 on 13 Jun 1979 and 1036 on 15 May 1979 from Port Townsend-Keystone ferry, 1100 on 27 Sep 1979 at Voice of America, 814 on 13 Jun 1978 in San Juan Channel nr. Friday Harbor (MESA). *Off Grays Harbor*: 2539 on 27 Jul 1991, 3770 on 6 Aug 1994, 3288 on 20 Aug 1988, 3453 on 13 Aug 1994 (TRW). CBC: 1050 in

1995, 646 in 1991, 614 in 1990, and 511 in 1979 at Tacoma. 918 in 1979, 352 in 1978, 315 in 1981, and 253 in 1993 at Olympia. 244 in 1995 at Seattle.

Terence R. Wahl

Horned Puffin *Fratercula corniculata*

Very rare to rare visitor off the outer coast and in inland marine waters, very rare e.

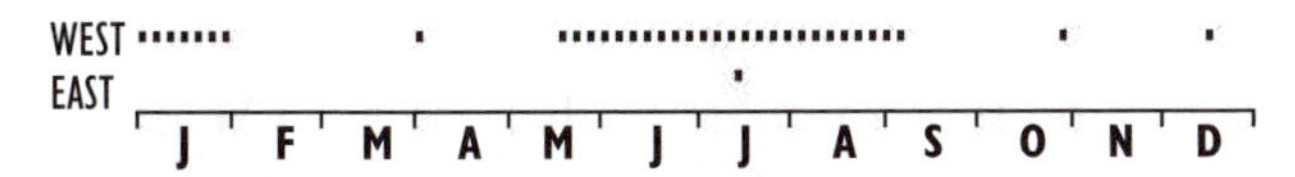

Habitat: Continental shelf and oceanic waters, Strait of Juan de Fuca and adjacent waters.

Occurrence: Since 1953 birds have been reported more frequently, though variably, particularly in summer, in nearshore and inland marine waters along the Pacific coast, from B.C. to California (see Hoffman et al. 1975, Roberson 1980). High seas surveys starting in the mid-1970s showed that wintering and non-breeding Horned Puffins, like Tufted Puffins, occur widely over much of the N. Pacific (Wahl et al. 1989).

In 1977 birds were found nesting in the Queen Charlotte Is. and breeding was probable at 10 more sites s. to Vancouver I., B.C. (Campbell et al. 1990b). As in Oregon (Pitman and Graybill 1985) and California (e.g. Roberson 1980), birds occurred nr. or at auklet/puffin colonies in Washington, both on the outer coast in 1981 and 1993-94 and at inland locations in the e. Strait of Juan de Fuca at Protection I., Smith I., Williamson Rock and Pt. Partridge in 1977-81, though nesting was not documented.

In early 1984 Pitman and Graybill (1985) found numbers over 185 km off California in February and 630 km off Oregon, with a few far off Washington in May. Reports in late 1980s-early 1990s showed numbers 100-200 km off California in late Apr to mid-May (Mlodinow and O'Brien 1996). Hoffman et al. (1975) suggested that Horned Puffin distribution may have switched seasonally as well as geographically in response to large-scale oceanographic changes and Washington records of Horned Puffins may well reflect these changes (see Wahl and Tweit 2000a).

Remarks: Large numbers noted in seabird die-offs. Hundreds washed up in Oregon in 1933 (Gabrielson and Jewett 1970), and Alcorn (1959a) reported 70 dead, along with about twice as many Tufted Puffins, in a 1.6 km stretch of beach nr. Grayland in Apr 1959. Beached birds and other reports indicate apparent occurrence is more frequent in Oregon. One noteworthy record from Coulee City in 1967 is the only inland record away from Alaska.

Noteworthy Records: West, *Spring-Fall*: 1 on 19 Jul 1975 off Grays Harbor; 1 on 2 Jul 1977 at Cape Flattery; 1 on 24 Jul 1977 at Williamson Rocks (Thoresen 1981); 1 ad. 5-26 Aug 1979 on Protection I; 1 in late Jun through Jul 1980 at Protection I.; 1 on 30 May 1981 nr. Smith I. (Aanerud and Mattocks 1997); 1 on 3 Sep 1981 off Pt. Partridge (Aanerud and Mattocks 1997); 1 on 15 May 1993 off Grays Harbor; 1 on 26 Oct 1993 at Sekiu (Tweit and Skriletz 1996); 1 on 11 Jun 1994 at Pt. Grenville and 1 ad. on 7 Aug 1994 at Ocean Shores; 1 on 21 Aug 2000 at Dungeness Spit; 1 on 5 Sep 2002 off Grays Harbor. *Winter*: 1 on 9 Jan 1977, 124 km offshore; 1 on 15 Dec 1978, 134 km offshore; 1 on 15 Dec 1991 off Kingston (Tweit and Paulson 1994). **East**: 1 live bird at reported at Coulee City in Jun 1967 (Larrison and Sonnenberg 1968). *Rejected by WBRC*: 1 on 15 Apr 1978 at Dungeness (Tweit and Skriletz 1996). *Specimens examined by WBRC* (Tweit and Skriletz 1996): 5 from 1933 from Westport (2 from 10 Jan, 2 from 27 Jan, 1 from 19 Feb); 1 from Apr 1959 at Grayland; 1 from 16 Dec 1973 at Long Beach; 1 from 28 Jan 1979 at Dungeness. *Note*: 1919 specimen from Jewett et al. (1953) not seen by WBRC; 8 reported oiled and 3 oiled specimens from *Nestucca* spill in winter 1988-89 (Ford et al. 1991).

Terence R. Wahl

Tufted Puffin *Fratercula cirrhata*

Locally common breeder on the n. outer coast, uncommon elsewhere in marine waters, rare s. of Admiralty Inlet. Very rare in winter.

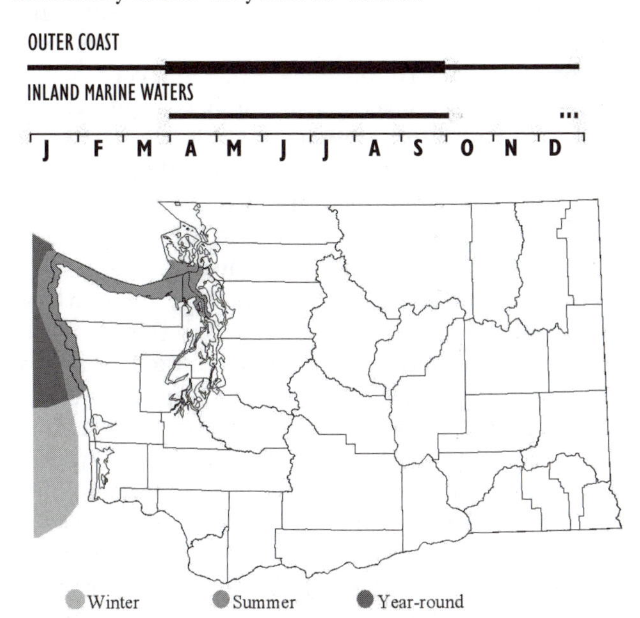

Habitat: Continental shelf, deeper outer coastal and n. inland marine waters. Nests in burrows on suitable islands.

Occurrence: Nests locally on rocks, islands, and mainland cliffs along the coast from Tatoosh I. to Pt. Grenville and along the Strait of Juan de Fuca (Speich and Wahl 1989). Of an estimated 1989 state population of 23,000, largest numbers were at Jagged I., with 7800 birds, and Alexander I., with 4000 (Speich and Wahl 1989). During inland waters surveys n. of c. Admiralty Inlet in 1978-79 all but three of 63 puffins noted were nr. Protection I., Smith I., and the s. San Juan Is., with larger numbers in the outer strait nr. Seal and Sail Rocks and Tatoosh I. (MESA, Speich and Wahl 1989).

Occurrence is seasonal, with birds moving to oceanic waters in winter (Briggs et al. 1992, see Wahl et al. 1989) and nearshore and continental shelf records are infrequent. Briggs et al. (1987) attributed increased winter numbers off California to a southward shift of n. breeders. Beached birds following winter storms and oil spills show presence of puffins offshore (e.g. Jewett et al. 1953, Alcorn 1959a), but relatively low numbers found after recent winter spills confirm predominant mid-ocean distribution (see Harrington-Tweit 1979a, Ford et al. 1991).

On surveys from Grays Harbor, puffins in Jul-Oct were farther offshore than other alcids, with 72% of 549 birds found over depths of >100 m. Distribution was farther offshore in Apr and Aug-Oct than it was during the nesting season, and on surveys to 185 km offshore in Jan 1977 all 21 birds were 64-155 km offshore (TRW).

A long-term decline in numbers and nesting locations was shown by Speich and Wahl (1989). In inland waters known or probable breeding sites were historically at 12-16 locations, while in recent decades there were very small numbers at just four sites. At Smith I. for example, 500 birds in 1914 decreased to 44 in 1979 and eight in 1982. At Protection and Smith Is. combined, estimates of 1000 pairs in the 1950s declined to 800 pairs in the 1970s and to 13 pairs at Protection I. in 1995 (Smith et al. 1997), with 18 birds there and 6 at Smith I. in 2001 (U. Wilson). Though this e. population was reported extirpated in the late 1990s (Wilson, in Piatt and Kitaysky 2002), up to 20 birds were reported nr. traditional nesting areas there in 2002-03 (D. Given-Seymour p.c., K. and J. Wiggers p.c.). In the early 1900s puffins were reported nesting in the San Juan Is. at Puffin, Bare, Skipjack, and Flattop Is. Though some sites remained active into the 1940s and 1950s, birds disappeared from these locations. Suggested causes include predation by eagles, increasing gull populations, human disturbance, and gill nets (e.g., Jewett et al. 1953).

During mid-1980s and late 1980s-90s ENSOs regional populations declined. Colonies in Oregon and at Pt. Grenville showed irregular but obvious declines (e.g., Speich and Wahl 1989, Lowe 1993), and much lower than normal breeding success in summer 1993. In inland waters, puffins were noted only rarely along the s. shore of Lopez I. in the late 1990s. Off Grays Harbor, 60-80 km from the nearest nesting colony, average numbers seen per trip in May-Sep declined from seven in the 1970s to four in the 1980s and one in the 1990s (TRW), and none were seen on May-Sep trips in 1997. This significant decline (Wahl and Tweit 2000a) corresponded with the trend in nesting birds on Tatoosh I. (R. T. Paine in Piatt and Kitaysky 2002). Numbers off Grays Harbor increased in spring-summer 2002 (TRW) when birds were also noted at Smith I. (D. Given-Seymour p.c.) and from three to 10 birds reported nr. Protection I. (J. d'Amore p.c.).

Remarks: Classed SC, FSC, PHS.

Noteworthy Records: West, *outer coast, Spring/Summer*: ca. 150 on 7-9 May 1971, several hundred in spring 1973 on Destruction I.; 80 on 26 Apr 1974 at Pt. Grenville; 320 nr. Tatoosh I. on 1 May 1992 (TRW); 40-50 on 26 Jun 1974 at Pt. Grenville. 150 on 6 Sep 1978 at Tatoosh I. *High counts off Grays Harbor*: 44 on 16 May 1976, 35 on 15 May 1977, 32 on 12 May 1985, 17 on 19 May 1974, 14 on 4 May 2002 (TRW). *Late Fall/Winter*: 2 on 13 Nov 1982, 6 on 15 Dec 1978 off Westport (TRW). *Spring, early*: 75 on 14 Feb 1975 at Cape Flattery. *Inland waters*: 2 on 11 May 1979 between Edmonds and Point No Point were southernmost noted during 1978-79 MESA surveys; 10 on 30 Jul 2002 at Smith I. (D. Given-Seymour p.c.); 18-19 on 31 May 2003 nr. Protection I. (K. and J. Wiggers p.c.). *Fall, late*: 1 on 18 Sep 1985 off Protection I. CBC: 1 in 1980 and 1 in 1986 at Port Townsend, 1 in 1993 at Sequim-Dungeness.

Terence R. Wahl

Rock Pigeon *Columba livia*

Locally common to abundant resident in unforested urban, agricultural and some natural areas.

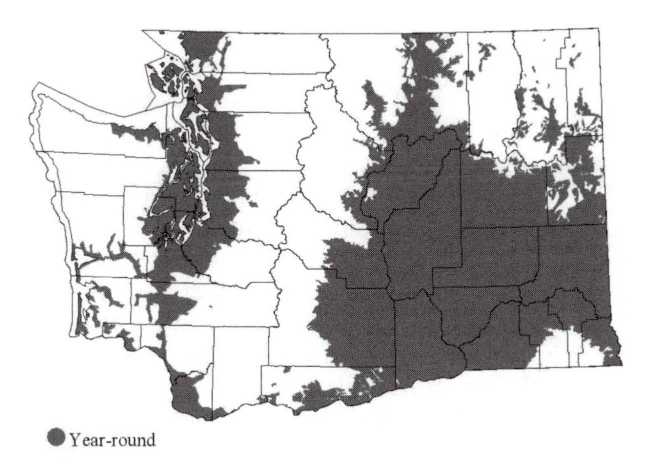

Habitat: Urban and agricultural areas with available nesting habitat, forages widely.

Occurrence: Native to s. Europe and introduced by early emigrants to much of the rest of the world—one of the most successful species at adapting to human situations worldwide. Introduced in N. America in the early 1600s (Johnston 1992), with expansion from the e. coast, including arrival in Washington, undocumented.

Great majority of population is in urban areas, though numbers also associate with farms in agricultural areas. Feral birds appear to be local all year, though birds may commute distances from roosts or nest sites to food sources. Nest sites include "any flat surface under cover" (Smith et al. 1997). These and roost sites include buildings, bridges, barns, silos, docks, and pilings. As elsewhere in the world, feral birds have probably nested on cliffs away from human settlements including sites at Palouse Falls, Alta L. (see Smith et al. 1997), the Snake and Grande Ronde R. valleys (Weber and Larrison 1977), Skagit R. (Wahl 1995), Yakima R., and Priest Rapids L. (Stepniewski 1999). Other sites are likely.

Attracted to feedlots, poultry farms, and other agricultural opportunities, grain spilled along roads, and to urban feedstores. Along with gulls and House Sparrows readily accepts handouts from people feeding birds at parks and waterfronts. Like other pigeons, occurs at mineral springs and along marine shorelines, presumably for sodium and potassium (see Sanders and Jarvis 2000).

The species serves as an important dietary component of Peregrine falcons re-establishing in urban areas (e.g. Columbia Tower in downtown Seattle). And, in a large old barn in Whatcom Co. unused for 10 or more years and occupied by Rock Pigeons and nesting Barn Owls, an estimated 50 cu. m of accumulated pigeon feathers presumably represented food packaging discarded by the owls (TRW).

Though historical state population size was undescribed (e.g. Jewett et al. 1953), BBS data show significant statewide population increases from 1966 to 2000 (Sauer et al. 2001). Continent-wide CBC data 1987-91 showed little increase (Johnston and Garrett 1994). Highest CBC counts in Washington were in urban circles and likely related in part to priorities of censusing coverage.

Remarks: Flocks usually show considerable plumage variation, reflecting interbreeding of "wild" birds with escapes and releases from "tame" populations (see Cramp et al. 1985). Attempts to reduce populations elsewhere have proved generally ineffective. Species not included on CBCs until 1973. Formerly Rock Dove.

Noteworthy Records: *CBC high counts*: 3083 in 1982 at Tacoma; 2896 in 1997 at Seattle; 2326 in 1995 at Spokane; 2065 in 1985 at Yakima Valley; 2009 in 1983 at Wenatchee.

Terence R. Wahl

Band-tailed Pigeon *Patagioenas fasciata*

Fairly common to locally common in summer in w. lowlands, uncommon to locally common in winter. Very rare to rare e. of the Cascades.

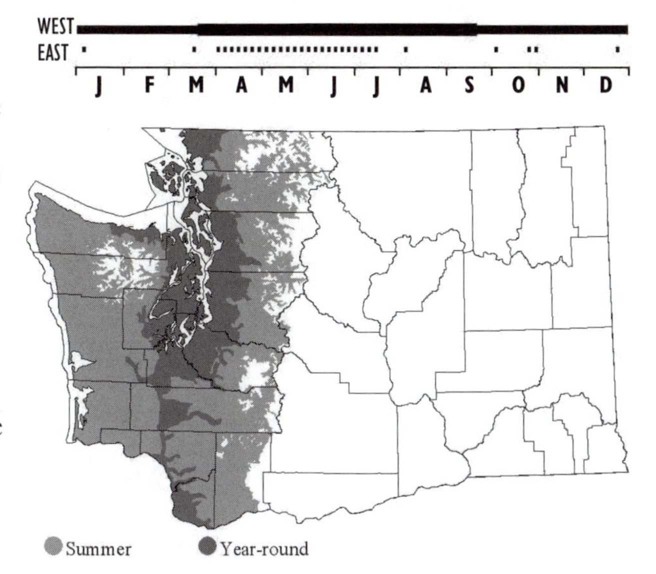

Subspecies: *P. f. monolis.*

Habitat: Coniferous and deciduous woodlands and edges in low- and mid-elevation forests in Humid Transition and Canadian zone, suitable parks and urban residential areas. In spring and fall visits cultivated fields.

Occurrence: A species of concern, with numbers declining from hunting in the w. lowlands by the early 1900s. Protected by treaty in 1916 and "barely holding its own" by the 1950s (Jewett et al. 1953). Hunting in state closed in 1991 through 2001 (WDFW 2000b) due to rapid population decline. Nests w. of the Cascades, particularly in "gulches tributary to Puget Sound" (Jewett et al. 1953), in the Columbia R. Gorge to just w. of the Cascade crest, and scarce but regular from there to White Salmon (Stepniewski 1999). Proximity of mineral springs to food sources important for nesting success (WDFW 1997c, Sanders and Jarvis 2000). Foods include elderberries, Pacific madrone, cherries, cascara, and acorns (Jewett et al. 1953, Godfrey 1986).

After decline in lowland food supplies birds move upslope to forage and may occur then and in migration at high elevations (Jewett et al. 1953). Many birds move s. to winter along the Pacific coast (see Campbell et al. 1990b).

Following earlier records (e.g Cooper 1860 in Jewett et al. 1953), there have been widespread reports of small numbers e. of the Cascades, particularly since the 1960s. Reported in Spokane, Asotin, Benton, Columbia, Yakima, Chelan, Easton, Kittitas, Whitman Cos. (AFN-NAB reports, 1960-99). There has been a northward and eastward range expansion into B.C. in recent decades (Campbell et al. 1990b).

Variable winter numbers concentrate in the s. Puget Sound area. The 42 highest CBC counts (1959-98) were in Tacoma or Seattle, but birds recorded on almost all lowland counts in w. Washington. CBC numbers at the few locations with long-time counts show little trend overall, but a decrease is suggested. Kent-Auburn, Olympia, and Tacoma numbers declined. Kitsap Co. apparently showed a very slight increase. Seattle, with longest count history (1959-98), declined. Recent BBS data, from 1966 to 2000, suggested a decline of 3.0% per year (Sauer et al. 2001; see Keppie and Braun 2000).

Remarks: A State and Federal game species, classed PHS, GL, on NAS Watchlist, and a priority species in assessment of land bird populations (Pashley et al. 2000). Protozoan disease, possibly transmitted at bird feeders, suspected in large-scale mortality (WDFW 1997c). Preservation of mineral springs habitat (Sanders and Jarvis 2000), nesting habitat, and suitable forest and urban vegetation practices needed for maintenance of populations (WDFW 1997c).

Terence R. Wahl

Eurasian Collared-Dove *Streptopelia decaocto*

Rare visitor.

The occurrence of Eurasian Collared-Doves in N. America appears to be primarily from escapes and releases in Nassau, Bahamas, during the early 1970s (Green 1977, Smith 1987). This species probably first appeared in Florida during the early 1980s (Smith 1987, and see Romagosa and McEneaney 1999). The species inhabits suburban and agricultural areas. Washington's first record was of a bird at Spokane on 2 Jan 2000. The second was at Wenatchee 9-17 Jul 2002, paired with a Turtle Dove. An apparent hybrid of the two species was at Othello on 7 May 2002. Birds were reported heard at Rock Is. on 30 Apr 2003 and seen at Pt. Whitney on 13 May 2003. A photograph of a bird at College Place, 11 Jan 1996, is suggestive of this species but is not definitive.

Steven G. Mlodinow

White-winged Dove *Zenaida asiatica*

Very rare visitor.

Six records: one in the Puyallup R. valley on 7 Nov 1907 (Jewett et al. 1953), one at Cypress I. on 19 Jul 1997, one at Redmond on 19 May 1999, one at Tokeland on 8 Oct 1999, one at Tatoosh I. on 12 Jun 2000, and one nr. Ellensburg on 8-10 Jun 2002. Another report, one at Ocean Shores on 19 Jul 1982, was rejected by WBRC. Recent Washington records match increasing vagrancy across N. America. The Washington specimen identified as *Z. a. mearnsi*.

Bill Tweit

Mourning Dove *Zenaida macroura*

Widespread common summer resident, fairly common to common winter visitor e. Locally fairly common summer resident and uncommon to fairly common winter visitor west, mainly in lowlands from s. Puget Sound to the Columbia R.

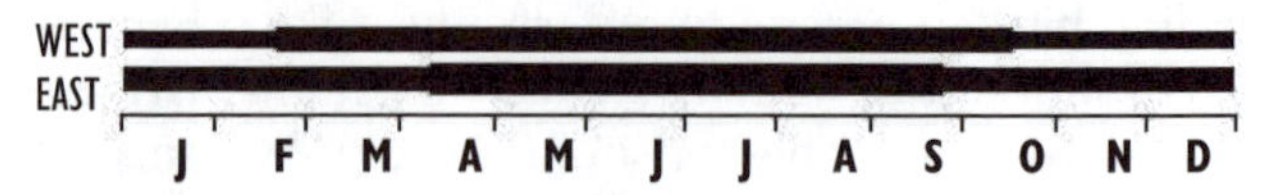

Subspecies: *Z. m. marginella*.

Habitat: Eastside lowlands to open lower Ponderosa forests (Larrison and Sonnenberg 1968). Breeds locally, most often on agricultural land, including

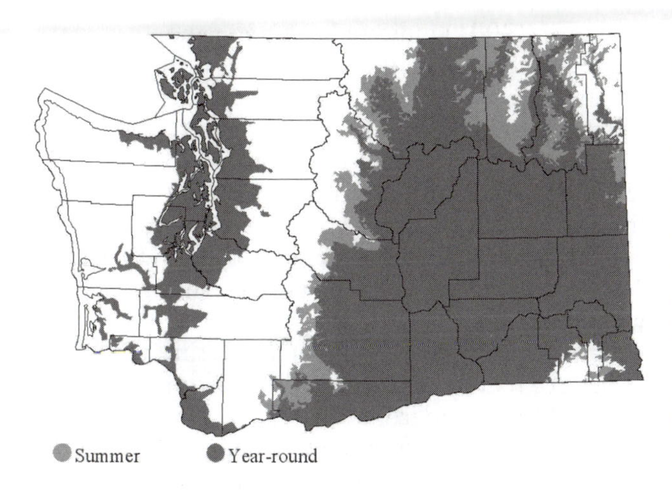

orchards, also rangeland, pastureland, farmland, riparian woodland, open forest, adjacent to water sources (Campbell et al. 1990b), poplar windbreaks (Larrison and Sonnenberg 1968), tree plantings, and structures (Smith et al. 1997).

Occurrence: Characteristic of the dry, open lowlands e. of the Cascades, Mourning Doves are found to 1500 m elevation (Jewett et al. 1953). Formerly restricted in w. to prairies; clearing for agriculture, tree planting, increased number of structures for nesting have increased available habitat, and distribution is now essentially statewide in summer (Smith et al. 1997).

Irregular, scattered distribution w. of Cascades, in drier cultivated habitats, logged and burned-over areas, prairies from Puget Trough s. to the Columbia R. (Smith et al. 1997). Range expanding in w. in logged-off areas (Smith et al. 1997). Noted increasingly common on prairies nr. Ft. Lewis in summer 1988. An uncommon breeder in the San Juan Is., where it was formerly much more common (Lewis and Sharpe 1987) and has declined as breeder in King Co. (Hunn 1982). Not recorded during BBA surveys w. of Port Angeles or coastally s. to the Columbia R. (Smith et al. 1997).

Largely migratory, wintering s. of Washington: band returns show that some breeding in B.C. winter in Nevada and Arizona (Campbell et al. 1990b). Fairly common, irregular in more open country in winter in protected low-elevation areas (Larrison and Sonnenberg 1968). In winter, when snow covers food, in interior B.C. and in Yakima Co. Most common at cattle feedlots (Campbell et al. 1990b, Stepniewski 1999). On the westside, usually found nr. stored seed or feeders in winter (SGM). Away from CBCs, winter flocks of 50-120 are considered noteworthy by field observers. Much larger numbers are often found within count circles e. of the Cascades.

Flocks form after breeding (Bent 1932), often feeding on waste grain in harvested wheat fields

Table 29. Mourning Dove: CBC average yearly counts per 100 party hours by decade.

	1960s	*1970s*	*1980s*	*1990s*
West				
Bellingham		3.1	0.2	2.5
Cowlitz-Columbia			18.6	13.3
Kent-Auburn			1.4	4.4
Olympia			1.9	8.6
Padilla Bay			12.0	13.5
Sequim-Dungeness			2.6	7.5
Tacoma		8.1	21.7	10.3
East				
Ellensburg			215.7	574.7
Spokane	25.6	6.5	12.4	60.8
Toppenish			471.1	400.5
Tri-Cities		29.7	77.0	240.9
Walla Walla		14.1	12.5	21.6
Wenatchee	50.8	11.9	3.1	84.1
Yakima Valley		21.0	14.9	64.7

(Jewett et al. 1953) just before and during hunting season (Stepniewski 1999). Obtains dietary minerals on beaches (Bent 1932) and at mineral springs (Wahl 1995; see Sanders and Jarvis 2000), and birds may nest in proximity to estuarine marshes (Campbell et al. 1990b). On balance, the species benefits from availability of feedlots in winter, suffers from vegetative succession and hunting (Paulson 1992b) and hunting pressure tends to keep numbers below carrying capacity (Smith et al. 1997). Based on surveys, the hunting harvest in 2000-01 was 97,529 in e. Washington, 2202 on the westside (WDFW 2001) during season 1-15 Sep (WDFW 2000b).

Though long-term declines are indicated for the w. U.S. (Mirarchi and Baskett 1994), BBS data for Washington suggest a slight increase (Sauer et al. 2001). Long-term CBC counts indicate increases on long-established eastside CBCs (Table 29).

Noteworthy Records: West, *Spring*: 1 on 14 May 1972 at Cape Flattery. *Summer*: 2 on 22 Jul 1970 at 2100 m elevation on Mt. Rainier. *High counts, Winter*, **East**: 600 on 10 Feb 1969 at McNary NWR; 70 on 2 Jan 1999 at Goldendale. **West**: 60 on 26 Oct 1994 nr. Puyallup; 200 on 30 Oct 1994 nr. Coupeville; 120 on 5 Dec 1999 at Elma; 50 on 31 Dec 1998 at Auburn; 50 on 3 Jan 1999 at Samish Flats; 68 on 5 Jan 1999 at Elma; 80 on 19 Jan 2002 nr. Toledo. *CBC*, **East**: 744 in 1990 at Moses L., 713 in 1994 at Ellensburg, 528 in 1992 at Toppenish, 452 in 1994 at Tri-Cities, 206 in 1995 at Wenatchee, 182 in 1998 at Spokane, 137 in 1993 at Twisp, 124 in 1994 at Yakima Valley. **West**: 137 in 1983 at Tacoma, 129 in 1996 at Skagit Bay, 87 in 1989 at Oak Harbor, 81 in 1992 at Padilla Bay, 45 in 1997 at Sequim-Dungeness, 40 in 1991 at Olympia.

Terence R. Wahl

Black-billed Cuckoo *Coccyzus erythropthalmus*

Casual vagrant.

Breeding throughout most of e. N. America, this is a very rare vagrant in the w. There are four records for Washington, all from a relatively narrow time period from mid-Jun to early Jul: one collected at Kamiak, Whitman Co., on 1 Jul 1952; one calling male ne. of Albion, Whitman Co., on 22 Jun 1958; a male specimen from Bremerton on 26 Jun 1978 and one at Davis L., Pend Oreille Co., on 19 Jun 1988 (Tweit and Skriletz 1996; Aanerud and Mattocks 1997; Aanerud and Mattocks 2000).

Bill Tweit

Yellow-billed Cuckoo *Coccyzus americanus*

Formerly an uncommon westside breeder, now very rare visitant statewide.

Subspecies: Unknown whether recent records are *C. a. occidentalis* or e. race, *americanus.*

Habitat: Primarily riparian woodlands, although Bent (1940) cites nests in fir woodlands, Jewett et al. (1953) cite a nest on an open brushy hillside, and Wahl (1995) notes that Edson reported use of semi-rural areas.

Occurrence: Although treated as a breeding species by Jewett et al (1953) in the Puget Trough and by Gabrielson and Jewett (1940) in the Columbia R. bottomlands, the last known Washington breeding records were from the 1930s (Layman and Halterman 1987). This extirpation is part of a broad decline throughout the w. (Layman and Halterman 1987) and the species' breeding range in w. N. America is currently limited to California and the desert sw. Spring arrival appeared to be in early to mid-May, as Jewett et al. cite a 3 May record in Tacoma and a nest with two incubated eggs on 1 Jun at Tacoma, and Gabrielson and Jewett (1940) cite a 19 May arrival date in Multnomah Co., Oregon. Typical departure dates uncertain: Gabrielson and Jewett suggest early Sep while Jewett et al. cite a late Sep and Nov records.

Accounts of abundance prior to extirpation are mixed (Iten et al 2001). Birds were variously described as abundant in the 1830s nr. Vancouver, fairly plentiful nr. L. Washington in the 1920s and 1930s, regular but not numerous in Whatcom Co. in the same time period, and Jewett et al (1953) described them as rare in w. Washington. While these may describe a trend, Oregon reports cited by Iten et al. (2001) indicate large population fluctuations between years.

The primary factor in disappearance appears to be loss of suitable riparian habitat (Iten et al. 2001), though they were reported to breed in a broader range of habitats. Broadscale application of pesticides from the 1940s on may also have contributed to their decline throughout the w. (Gaines and Laymon 1984), decreasing overall range and causing withdrawal from peripheries such as w. Washington.

There have been only nine records since 1941, the last of regular occurrence in Whatcom Co. (Wahl 1995). Most range from early Jun to early Aug, with one Nov specimen. Although Roberson (1980) suggests that most recent records originate from e. populations, recent w. Washington records are from areas that historically held breeding birds, raising a glimmer of hope that a vestigial breeding population may still exist. Classed SC, FC, PHS.

Records: *post 1941*: 1 on 10 Jul 1974 at Beaux Arts, King Co. (Tweit and Skriletz 1996); 1 on 11 Jun 1978 at George; 1 calling male on 26 Jul-1 Aug 1979 nr. Sultan; 1 on 20 Sep 1987 at Sacajawea SP; 1 on 5 Jun 1990 along the Milton-Freewater Hwy, Walla Walla Co.; 1 found dead on 5 Nov 1990 nr. Omak; 1 on 20 Jul 1991 at Tonasket (Aanerud and Mattocks 1997); 1 on 3 Aug 1996 at Elma; the remains of 1 (ph.) in a Peregrine nest in downtown Seattle in mid-Jun 1997 (Aanerud and Mattocks 2000).

Bill Tweit

Barn Owl *lyto alba*

Widely distributed throughout non-forested lowlands statewide.

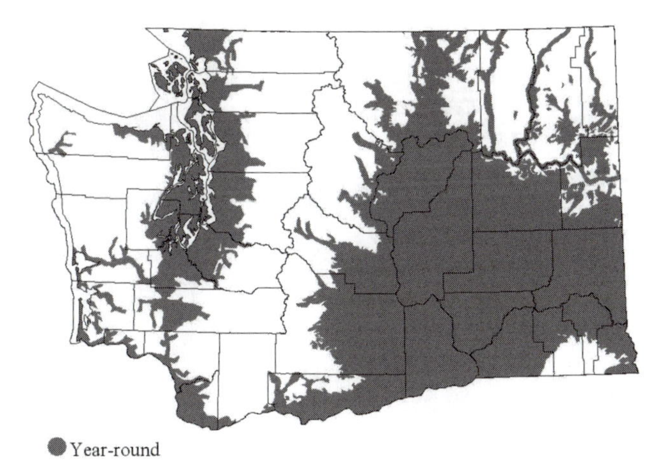

Subspecies: *T. a. pratincola.*

Habitat: Nearly all open habitats including agricultural areas, shrub-steppe and other arid habitats of the Columbia Basin, wetlands and riparian areas, and clear-cuts (Marti 1992, Smith et al. 1997). Nests in human-provided structures, cavities in cliffs, and in hollow trees or snags (Marti 1992).

Occurrence: Found or expected to occur in lowland areas of the Puget Trough s. to Vancouver, sw. Washington, the extreme ne. part of the Olympic Pen., the Columbia Basin, and sw. Klickitat Co. (Smith et al. 1997). Westside range shown by Smith et al. (1997) is supplemented by seasonal detections from sw. Mason Co. and adjacent Grays Harbor Co., s.c. Lewis Co., ne. Clark Co. (Irwin et al. 1991), and w. in Clallam Co. to Crescent Bay, and possibly to L. Ozette (H. Opperman p.c.). Additionally, 27 records between 1972 and 1987 in the Okanagan Valley of B.C., where birds were first recorded in 1972 (Cannings et al. 1987), suggest that records in lowland areas of Okanogan Co. should be expected. Occurrence n. of Spokane Co. in Stevens and Pend Oreille Cos. is possible (K. Knittle p.c.).

Relatively little is known about this species in Washington. Strong association with agriculture widely noted: fairly recent appearance in Yakima Co. suggested with irrigation and farming (Stepniewski 1999), more common than formerly in se. counties (Weber and Larrison 1977), including areas along the Snake R. and Walla Walla Valley (K. Knittle p.c.). Not recorded in the San Juan Is. until 1947, first reported breeding there in 1960 (Lewis and Sharpe 1987). First noted in B.C. in 1909, and probably established there with conversion of forest to agriculture (Campbell et al. 1990b).

The species has adapted to nest in many situations (see Campbell et al. 1990b), including urban situations like vacant rooms above a tavern, under a water tower atop a building in downtown Bellingham (TRW), under bridges, in Wood Duck boxes, drain pipes, holes in dirt banks, cliffs, canal banks, and underneath an Osprey nest. Noted expansion in interior s. B.C. suggested due to nest-box programs in Idaho and Washington.

Additional study needed to better determine range and population levels. Population in sw. B.C., at the n. limit of the breeding range, is declining due to loss of foraging habitat to development (Campbell et al. 1990b). Stepniewski (1999) suggested that mild winters increase survival and expansion. Relatively low numbers are found on CBCs, depending on observer effort and efficiency. CBC trends mixed in w. Washington, apparent increases on the eastside, possibly due to less severe winter weather. The species is generally thought to be resident in most areas although there is evidence of migration in some populations (Marti 1992).

Noteworthy Record: *Winter roost count*: 19 on 26 Dec 1990 nr. Moxee (Stepniewski 1999).

Joseph B. Buchanan

Flammulated Owl *Otus flammeolus*

Uncommon to fairly common summer resident in ponderosa pine zone in e. slope Cascades, Kettle Range, Selkirk Mts., and Blue Mts.

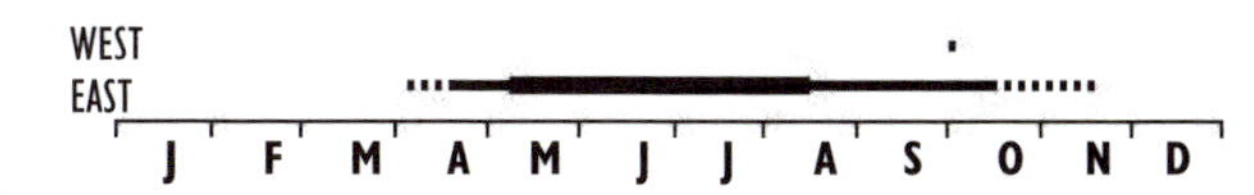

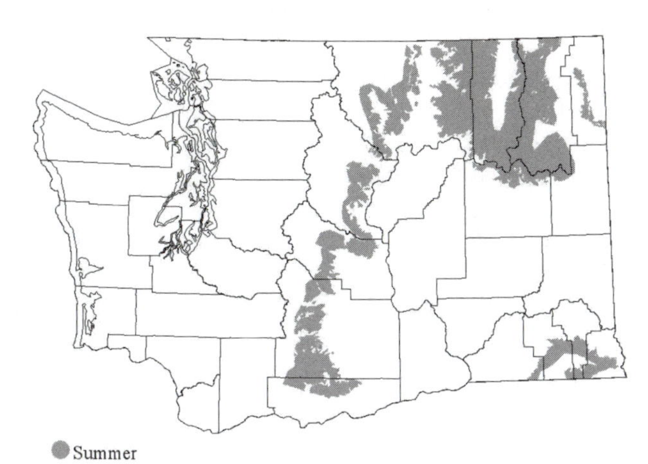

Subspecies: *O. f. flammeolus.*

Habitat: Mid- and late-seral ponderosa pine forests with sparse canopy cover, cavity trees or snags, and generally open understory.

Occurrence: Breeding from s. B.C., e. Washington, n. Idaho, and w. Montana s. in dry forests to n. Mexico, birds winter from c. Mexico s. to El Salvador (AOU 1998). Jewett et al. (1957) considered this a rare species and gave only five specimen records, all from spring or summer. Though 50 years later there are only 16 "probable" or "possible" breeding records reported by Smith et al. (1997), the species is much more common than suggested by the published records. Knowledge of occurrence in Washington really began in the 1970s and 1980s with the increase of numbers of mobile field observers. It is uncommon in many ponderosa pine forests and may be fairly common in c. and n. Klickitat Co. (WDFW database).

A late spring migrant and, except for rare arrivals in Apr, most birds appear to arrive in early May. They cease to vocalize after young have fledged and are less likely to be detected after about mid-Aug (Cannings et al. 1987). Birds are occasionally encountered during migration, particularly when they appear in low elevations. Late migrants (late Oct to mid-Nov) appear to be rare.

Population status unknown, but habitat protection and enhancement activities will likely be important for the conservation of this species. Unfortunately, older ponderosa pine forests are now diminished due to timber harvest and invasion of Douglas-fir and grand fir during a century of fire suppression in e. Washington forests (Agee 1993).

Remarks: Classed SC, PHS, GL. Two reports from winter nr. Newport—5 Dec 1984 and 4 Jan 1985—appear doubtful given the highly migratory nature of this insectivorous owl.

Noteworthy Records: East, *Spring, early*: 10 Apr 1990 at 10 km ne. of Swauk Pass (JBB); 4 May 1981 and 1982 at Penticton, B.C. (Cannings et al. 1987); 8 May 1977 10.5 km w. of Satus Pass; 12-13 May 1989 at Blewett Pass. *Fall, late*: 1 found dead on 17 Nov 1979 between Walla Walla and Dixie, Walla Walla Co.; 1 found dead on 22 Oct 1902 at Penticton, B.C. (Cannings et al. 1987); injured bird on 10 Oct 1971 at St. John, Whitman Co. (Weber and Larrison 1977); 24 Sep 1989 at Indian Canyon, Spokane Co. (J. Acton); 21 Sep 1980 at Davenport Cemetery, Lincoln Co. (J. Acton). **West**: 2 Oct 1980 at Mercer I.

Joseph B. Buchanan

Western Screech-Owl *Megascops kennicottii*

Fairly common resident statewide except in the Columbia Basin.

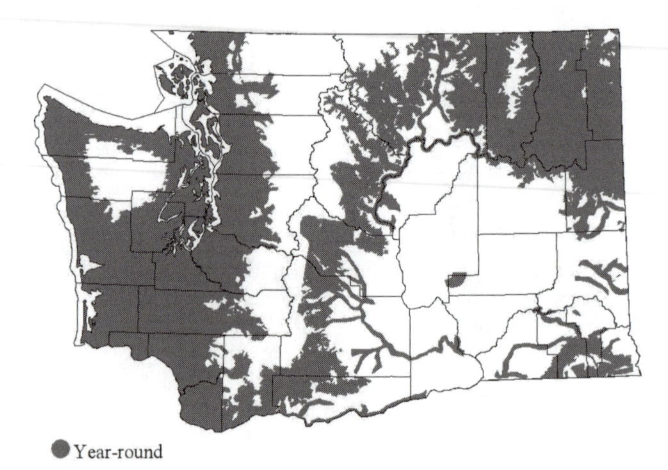

Subspecies: *M. k. kennicottii* on westside, *macfarlanei* on eastside.

Habitat: Riparian areas, many habitat types containing hardwoods (alder and maple patches in conifer zones [Irwin et al. 1991]), cottonwood groves along mainstem water courses, conifer-dominated forests, city parks (Stepniewski 1999), cemetery borders, residential areas.

Occurrence: Like other nocturnal owls with described distribution often based on minimal detections, screech-owls are one of the most common owls in Washington. They are widespread at low to mid elevations in many habitats from urban situations to riparian zones and westside mid-elevation Douglas-fir, western hemlock, and Sitka spruce zones and eastside grand fir zones. The w. Washington range is slightly greater than that suggested by Smith et al. (1997), as there are many detections in breeding season from s.c. Lewis Co., adjacent Cowlitz Co., and ne. Clark Co. (Irwin et al. 1991). Birds are virtually absent from the Columbia Basin (Smith et al. 1997), but should be expected to occur there locally if suitable conditions develop along riparian zones or in small woodlots nr. human dwellings.

As in the case of other owls, detectibility varies seasonally (Irwin et al. 1991). Irwin et al. (1991), conducting owl surveys in sw. Washington, encountered screech-owls at a rate of 0.7 detections per hour in Mar 1988 and over one detection per survey hour in early Aug 1988. Other information regarding the abundance or distribution of this species is lacking and population status is unknown. Cannings and Angell (2001) cite mortality due to collisions with vehicles and state anecdotal evidence of negative impacts of local populations with arrival of Barred Owls.

Remarks: Jewett et al. (1953) believed that *O. k. brewsteri* occurred along the Columbia R. as far e. as Walla Walla, but this distribution was not recognized by the AOU (1957). Nocturnal owls are at best minimally sampled on CBCs, with some long-term data likely indicative of effort put into finding the species rather than systematic sampling or censusing.

Joseph B. Buchanan

Great Horned Owl *Bubo virginianus*

Fairly common resident statewide.

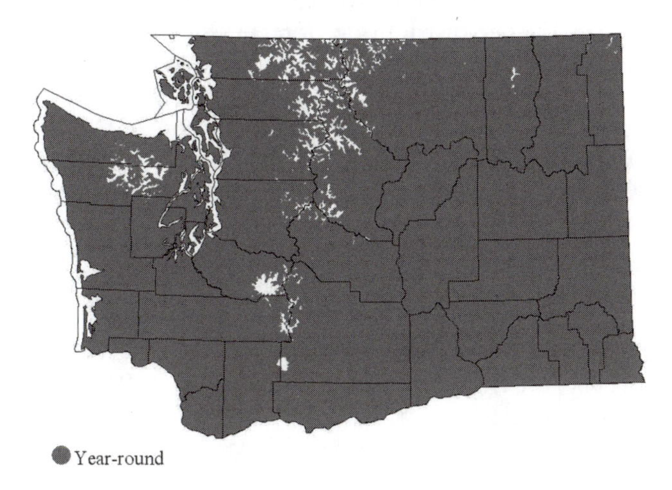

Subspecies: *B. v. saturatus* breeds in w. Washington, *lagophonus* in e. (AOU 1957). The northern, *subarcticus* (Houston et al. (1998), occurs in winter.

Habitat: Open habitats including agricultural areas, shrub-steppe, meadows, open areas within forested landscapes or in open forest, urban areas.

Occurrence: Perhaps the most common owl in Washington, with surveys in westside forests indicating it was at least as common as the screech owl or saw-whet owl (Irwin et al. 1991). It is an "unspecialized carnivore" preying on almost anything (Smith et al. 1997), including Red-tailed Hawks, barn and Short-eared owls (Lewis and Sharpe 1987), and poultry (especially in severe winters: Jewett et al. 1953). It may be as abundant as the Red-tailed Hawk though much less frequently detected (Stepniewski 1999).

Breeding range includes most of N. America except for far n. areas above tree line (AOU 1998, Houston et al. 1998). Resident in many parts of its range, but n. owls migrate s. during winter to c. and n. states (AOU 1998). In Washington birds occur from sea level to the Cascade crest at 1670 m in Okanogan Co. (Smith et al. 1997), with some downslope movements noted in winter (Jewett et al. 1953).

The species occurs in almost every habitat except lowland mature rain forest and alpine parkland/tundra/ice (Smith et al. 1997). It generally uses open habitats including agricultural areas, shrub-steppe, meadows, and open areas within forested landscapes (or in open forest), and in recent decades in urban areas with forested areas, parks (e.g. Hunn 1982, Wahl 1995).

Although specific population monitoring is lacking, the Great Horned Owl is likely doing well, as substantial areas of forest are cleared annually for timber production (Johnson 1993, Stepniewski 1999) and it is able to exist in rather close proximity to human dwellings. It has adapted to results of irrigation and agriculture in the Columbia Basin, usurping barns formerly used by nesting barn owls (Smith et al. 1997).

Remarks: Jewett et al. (1953) indicated that *lagophonus* ranges w. of the Cascade crest and may breed in parts of Lewis, Cowlitz, and Clark Cos. Recent records do not support this. *Lagophonus* apparently occurs in nw. Washington outside the breeding season (Jewett et al. 1953) and *saturatus* wanders during the non-breeding period with records in the Okanagan in B.C. (Cannings et al. 1987) and from Moscow, Idaho (Burleigh 1972). *Subarcticus* (Houston et al. 1998), formerly referred to as *wapacuthu* and *occidentalis* (Jewett et al. 1953), migrates to Washington during winter (AOU 1957, Houston et al. 1998). The winter range of this subspecies is not well documented.

Joseph B. Buchanan

Snowy Owl *Bubo scandiacus*

Irregular uncommon to rare winter visitor statewide.

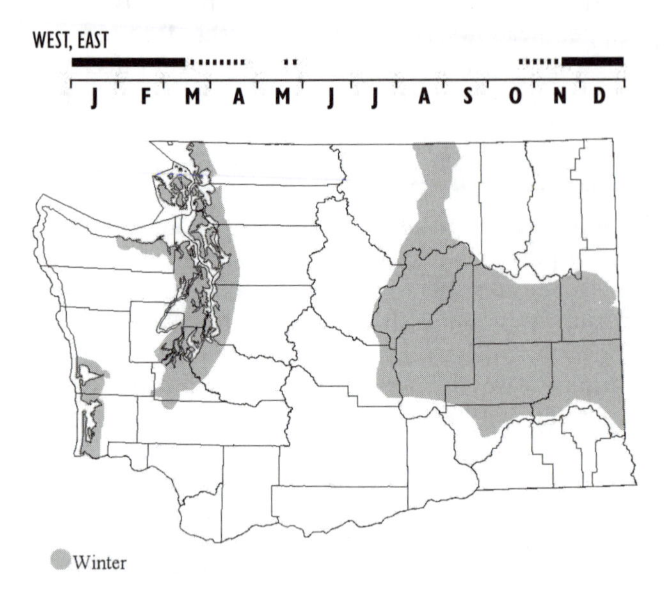

Habitat: Open country.

Occurrence: An almost-annual wintering species in Washington, ranging in abundance from hundreds during invasion winters to rarely absent. In the last 110 years, there have been seven major invasions noted and three echo flights (Table 30). The single largest invasion was the winter of 1916-17, with a very large echo flight the following winter (Jewett et al. 1953). Apparently, the next invasion was not until 50 years later, as this flight was reported to constitute the "greatest invasion ... since the winter of 1917-18" and Hanson (1971) could only find a record of one apparently minor flight, winter 1950-51, in the intervening time period. Three additional invasions have occurred on decade intervals, with all but the most recent having an echo flight.

Distribution during the invasion years is very tightly clustered; the 1996-97 pattern described by Cassidy (1997b) appears very similar to distributions noted in 1916-17 and 1973-74, with the largest concentrations found in the river deltas of n. Puget Sound (Nooksack, Samish, Skagit, and Stillaguamish). Smaller numbers were found along the Strait of Juan de Fuca and the outer coast, and in Grant, Lincoln, and Douglas Cos. in the interior. Other areas generally supported less than five birds. The 1984-85 invasion was apparently limited to the westside, as only "a few" were noted in e. Washington.

In non-invasion years, records average 12 per winter, five on the eastside and seven on the westside. Since 1963-64, there has been only one winter with no records (1969-70) and only three with more than 20 records (1977-78, 1993-94 and 2000-01). The distribution of records in non-invasion years is even more tightly clustered than in invasion years, with the majority of the records from the n. Puget Sound river deltas, coastlines, and Douglas, Lincoln, Grant, e. Kittitas, Spokane, and e. Whitman Cos. in the interior. Very irregular s. of Moses L. and Pullman.

Remarks: Formerly *Nyctea scandiaca* (AOU 2003).

Noteworthy Records: *Early date, invasion year*: 20 Oct 1996 at Port Angeles; non-invasion year 30 Oct 1975 at Bellingham. *Late date, invasion year*: 29 May 1974 at McNary NWR. *Non-invasion year*: 12 Apr 1975 at Davenport.

Bill Tweit

Table 30. Snowy Owl irruptions (> 50 birds) and echo flights into Washington.

Winter	*Magnitude*	*Extent*	*Comments*	*Reference*
1889-90	Prob. > 50	NW Wash		Jewett et al. 1953
1896-97	Uncertain	Statewide		Jewett et al. 1953
1916-17	Prob. > 1000	Statewide		Jewett et al. 1953
1917-18	Prob. > 250	Coastal	Echo flight	Jewett et al, 1953
1966-67	> 200	Statewide		AFN 21:441, 450; Hanson 1971
1973-74	> 200	Statewide		AB 28:666, 843
1974-75	50-100	Statewide	Echo flight	AB 29:718, 733
1984-85	50-75	Mostly West		AB 39:203
1985-86	22	West	Echo flight	AB 40:318
1996-97	> 100	Statewide	No echo flight in 97-98	FN 51:795, Cassidy 1997

Northern Hawk Owl *Surnia ulula*

Rare winter visitor.

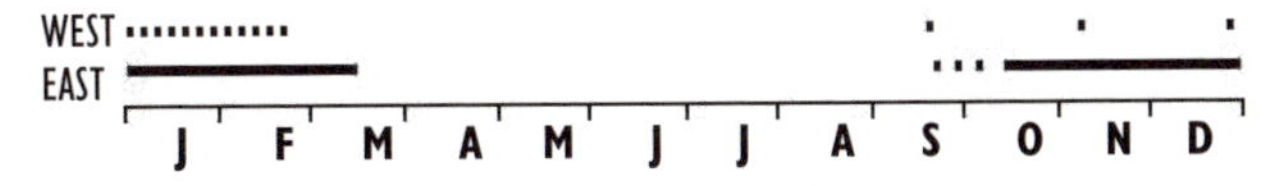

Subspecies: *S. u. caparoch.*

Habitat: Open country with tall perches, often nr. wooded areas; several Washington records from areas of recent burns.

Occurrence: Breeding in the Arctic s. to c. B.C. and Alberta, the hawk owl ranges irregularly in winter. There are about 31 Washington records through 2001, primarily from the n. counties (61%). Fourteen of these were after 1990, with four records from the winter of 1992-93 and five from the winter of 2000-01. Only nine records have been from the westside, and until the wintering bird nr. Custer, Whatcom Co., in 2001, there had been no westside records since 1922. Despite recent breeding records at Manning Provincial Park., B.C., 1980-84 (Campbell et al. 1990b), there have been no indications of breeding in Washington. Though Cannings et al. (1987) cite two invasion years in the Okanagan Valley, 1913 and 1926, those did not appear to extend into Washington. Just three state records occurred in years of N. American irruptions noted by Duncan and Duncan (1998).

Noteworthy Records: *Pre-1953*: 3 undated records from Whatcom Co.; 2 on 15 Sep 1897 at Martin, Grant Co.; 2 in Sep 1905 nr. Glacier; 1 on 24 Sep 1914 at Kachess; 1 on 21 Dec 1914 at Pullman; 2 in winter 1919-20 nr. Boundary RS, Whatcom Co.; 1 on 12 Nov 1922 at Camano I.; 1 on 5 Dec 1926 nr. Yakima (Jewett et al. 1953). *Subsequent records*: 1 on 1 Dec 1959 at Grassy Top Mt., Pend Oreille Co.; 1 on 26 Oct 1966 nr. Tiger; 1 on 20 Jan-14 Feb 1982 at Bridgeport; 1 on 6 Jan 1983 nr. Twisp; 1 on 14-24 Nov 1992 at Sherman Pass; 1 on 27-28 Nov 1992 e. of Oroville; 1 on 11-29 Dec 1992 nr. Spokane; 1 in late Jan-17 Feb 1993 nr. Pearygin L. SP; 1 on 28 Dec 1995 at Winthrop; 1 from mid-Jan to 17 Mar 1997 at Cheney; 1 on 4 Oct 1999 at Cooper Mt.; 1 on 16 Oct 1999 in Stevens Co.; 1 on 30 Dec 1999-5 Jan 2000 at Chelan; 1 on 13 Oct 2000 at Chelan Ridge; 1 on 14 Oct 2000 at Snow Peak, Ferry Co.; 1 on 12-25 Nov 2000 at Colville; 1 on 30 Dec 2000-24 Feb 2001 at Custer; 1 on 2 Feb 2001 nr. Bickleton.

Bill Tweit

Northern Pygmy-Owl *Glaucidium gnoma*

Uncommon in forested areas statewide.

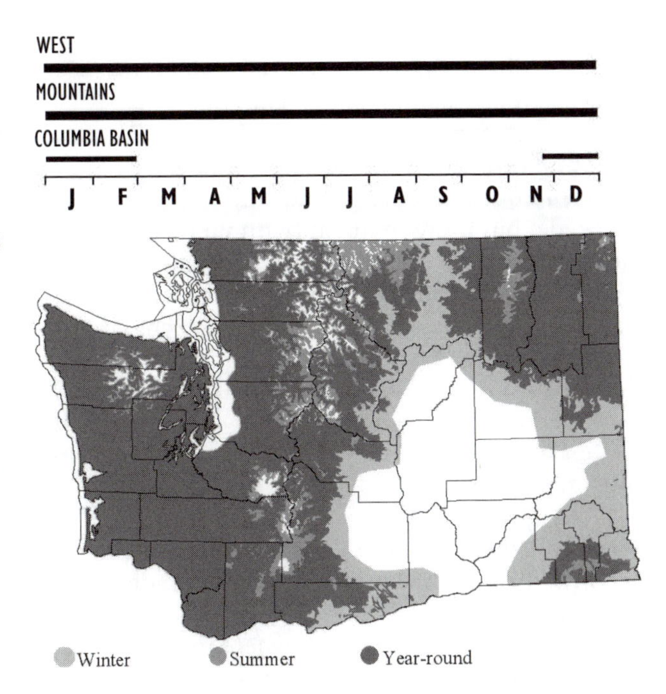

Subspecies: *G. g. grinnelli* and *californicum.*

Habitat: Coniferous woodlands, often in Douglas-fir forest, also in ponderosa pine, mixed conifer-hardwood, mixed conifer, high-elevation, deciduous, and western juniper forests with larger trees and moderately sized snags.

Occurrence: The pygmy-owl's range in Washington covers all coniferous forest types in w. Washington, the e. slope of the Cascades, the Selkirks, and the Blue Mts. (Smith et al. 1997). Its abundance is no better understood today than it was by Jewett et al. (1953). Although it is detected far less frequently than the two other small forest owls in low to mid elevations (Irwin et al. 1991), it may be more common than current knowledge indicates (Smith et al. 1997, Stepniewski 1999). Acoustic lure surveys in Oregon resulted in an average detection rate of 0.91 pygmy-owl per hour of survey (Sater 1999), equivalent to the highest detection rates for saw-whet owl and screech owl reported by Irwin et al. (1991).

In the recent Oregon surveys the species was encountered three times more often in Douglas-fir forests than in western juniper areas (Sater 1999). Although found in many habitats, the pygmy-owl was found on transects supporting larger trees than those trees on transects with no owl detections (Sater 1999). Pygmy-owls use moderately sized snags for nesting (Giese 1999), a resource found in diminishing supply in managed forests, and the availability of suitable snags may influence local distribution and abundance of this species.

The annual abundance of wintering pygmy-owls in the Okanagan Valley of B.C. occasionally fluctuates substantially. Pygmy-owls have gone undetected there in some years (Cannings et al. 1987), and population-assessment efforts may require long-term surveys. CBC occurrences, often reflecting downslope movements (see Holt and Petersen 2000), are widespread and irregular statewide, but more frequent (with up to seven birds) at Spokane, Twisp, Wenatchee, and Walla Walla on the eastside than on the westside.

Pygmy-owls are very infrequently detected in the Columbia Basin. The vast majority of records from that area are from Asotin, Walla Walla, and Whitman Cos., perhaps due to the availability of habitat along the Snake R. Records in that area were between Nov and late Feb.

Remarks: Jewett et al. (1957) stated that the range of *grinnelli* included the e. slope of the Cascades and the Okanogan Mts.; their map showed no occurrence in the se. Cascades where they now occur (Smith et al. 1997). The range of *californicum* (referred to as *pinicola*) was defined as ne. and extreme e. Washington and the Blue Mts. (Jewett et al. 1953). The AOU (1957) separated the range of the two subspecies along the crest of the Cascades.

Noteworthy Records: East, *Winter*: 24 Nov 1976 at Asotin and 2 on 4 Dec 1976 nr. Asotin (Weber and Larrison 1977). *WDFW records*: 1 Nov 1977 (injured) at Walla Walla; 21 and 23 Nov 1977 at Pullman; 23 Nov 1977 at Asotin; 23 Nov 1979 at Chief Joseph WRA; 25 Nov 1977 at T26N R21E S7; 12 Dec 1977 at Rocky Reach Dam; 17 Dec 1977 n. of Pullman; 1 flushed from sage on talus slope on 17 Dec 1981 at T23N R24E S33; 19 Dec 1977 at Asotin; 12 Jan 1972 at Walla Walla; 14 Jan 1982 at Walla Walla; 17 Jan 1979 on Waluke Slope; 20 Jan 1980 at Patit Cr., Columbia Co. *Specimens*: 4 on 15 Nov-27 Feb from Pullman (WSU Connor Mus. collection; Hudson and Yocom 1954); winter 1973-74 at Walla Walla (AB 28:3).

Joseph B. Buchanan

Burrowing Owl *Speotyto cunicularia*

Locally fairly common to uncommon breeder in shrub-steppe in e. Washington. Very rare in winter w. and e.

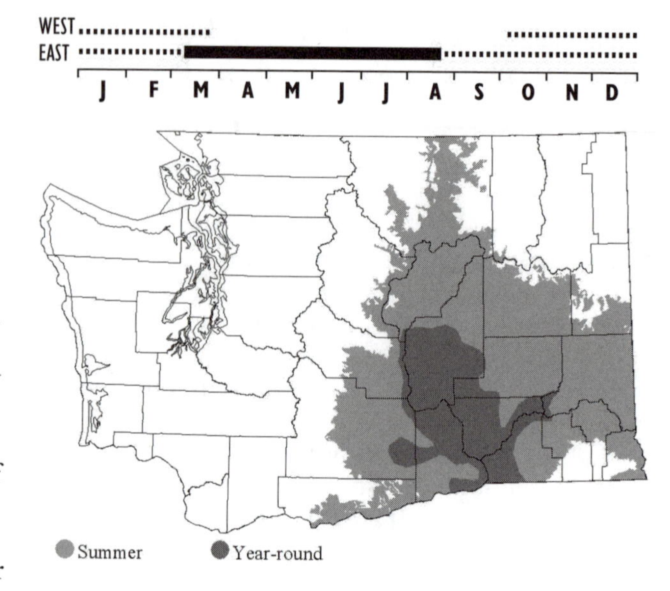

Subspecies: *S. c. hypugaea.*

Habitat: Breeds in the open grassland and shrub-steppe regions. Low vegetation stature, presence of elevated perches, and high percentages of bare ground appear important (Green and Anthony 1989).

Occurrence: While breeding populations can be found throughout the Columbia Basin where suitable habitat occurs, the bulk of the Washington nesting population is in the Moses L., Royal City, George, Ephrata, Frenchman Hills, Saddle Mt., and Columbia NWR regions of Grant Co., and nearby Ringold and Mesa areas of Franklin Co. (Friesz 1979). Significant populations have also been found nr. Kiona, the Davenport area, the McNary NWR and nr. Lowden, the Wapato-Toppenish region, and nr. Brewster. Approximately 150 Burrowing Owls were estimated present on the McNary NWR in the summer of 1974.

The species utilizes burrows excavated by badgers, yellow-bellied marmots, and California ground squirrels nr. Walla Walla, but also uses artificial structures and burrows. Twenty-eight active nests were documented in the Ephrata-Frenchman Hills area in 1984, and over 100 throughout the Columbia Basin in 1988. Present status of populations at these locations is uncertain, prompting the WDFW to begin a monitoring program in the Columbia Basin. Possibly nested in w. Washington historically (Jewett et al. 1953, Smith et al. 1997).

Winter records are irregular and occurrence apparently decreasing w. of the Cascades, with the last record 1989. Winter field reports from 1963 to 1998 included about 16 widespread on the eastside. Seven westside records were from Sandy Pt., the lower Skagit R. valley, Ocean Shores, Seattle, Ediz Hook, and Dungeness Spit. CBC records statewide are few.

Some populations, especially nr. Richland and Walla Walla, and in Yakima Co. (see Stepniewski 1999), have experienced declines due to habitat loss since the late 1970s. A recently invaded population of California ground squirrels nr. Walla Walla is providing nesting habitat, but presents a management conflict since the ground squirrel is an undesirable species targeted for control.

Remarks: Green and Anthony (1997) provided six recommendations for managing Burrowing Owls in the Columbia Basin, dealing with nesting habitats and sites, foraging areas, and habitat management. Further considerations include whether population declines are directly related to declines in badger populations, whether new populations be propagated or existing habitat be enhanced, and how livestock grazing influences habitat. A candidate for Washington State listing as a threatened or endangered species. Classed SC, FSC, PHS.

Noteworthy Records: East: 1-10 Mar 1881 arrivals at Ft. Walla Walla (Bendire 1881); 12 Mar 1921 at Wallula (Jewett et al. 1953); 373 records in 15 cos., 1894-1979 (Friesz 1979); 30 Jan 1976 at Othello (Woodby 1976); 50 in early Dec 1967 at McNary NWR; 140-150 in summer 1975 at McNary NWR. **West**: 23 Sep 1899 at Tacoma (Bowles 1906); 10 Feb 1915 at unknown island, e. Strait of Juan de Fuca (Brown 1924); 10 Apr 1917 at Bellingham (Brown 1924); 17-18 Apr 2003 at LaConner; 12 May 1926 at Mora (Bowles 1926); 18 Oct 1942 at Battle Ground (Beer 1944); numerous records, 1938-74 at Moon I., Grays Harbor Co. (G. Alcorn in Brown 1976). 1 banded in Jan 1964 nr. Clear L., Skagit Co.; 1 on 24 Nov 1980 at San Juan I.; 1 on 20-21 Mar 1999 at Skagit WMA; 1 on 24-25 Oct 2000 nr. Penn Cove. CBC: 1 in 1969, 1970 and 1983 and 1-3 birds/year from 1989 to -1999 at Tri-Cities; 1 at Yakima Valley in 1998; 1 at Grays Harbor in 1976; 1 in 1987 and noted in CW 1989 at Sequim-Dungeness.

Gregory A. Green

Spotted Owl *Strix occidentalis*

Widespread, uncommon resident on the Olympic Pen. and in the Cascade Mts.; rare in sw. Washington and very rare elsewhere away from the Cascade foothills.

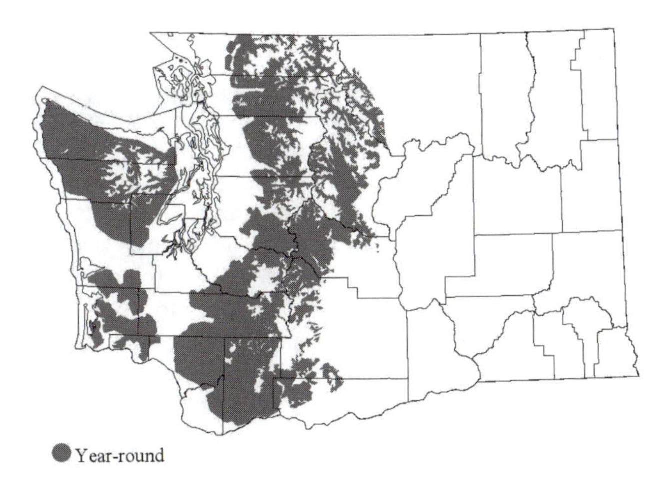

Subspecies: *S. o. caurina.*

Habitat: Mature and old-growth Douglas-fir and western hemlock forests, also second-growth stands with residual attributes, such as large-diameter snags, usually nesting in cavities and broken tops of large diameter trees or snags, to about 1100 m elevation in the Olympics, 1400-1600 m in the Cascades and Okanogan Co. In the e. Cascades mistletoe often provides nesting structures.

Occurrence: The Spotted Owl is strongly associated with old-growth Douglas-fir and western hemlock forests in w. Washington (Forsman et al. 1984, Forsman and Giese 1997). On the eastside they nest in conifer and mixed-conifer forests (see Buchanan et al. 1993) with about half of the known nest sites in stands between 60-130 years of age (Buchanan et al. 1995). Its forest-type dependence has led to dramatic population changes since three birds were collected on the e. shore of L. Washington in 1911 (Jewett et al. 1953). Mark-recapture data indicate that populations are declining statewide (Forsman et al. 1996). Although anthropogenic habitat loss can be traced back to Native Americans and European settlers, much of the change can be attributed to forest harvest practices of the last 50-100 years (USDI 1992).

As of March 2001 WDFW recognized 1223 site centers including those no longer active (Table 31). Most sites are not regularly monitored and it is likely that the number of active sites is less than indicated. Of the sites, 81% are centered on federal lands, 9% on state-managed lands, 9% on private lands, and 1%

Table 31. Status of Spotted Owl site centers in Washington in 2001 (WDFW).

Type of site	*Number*
Pair or reproductive site	899
Two birds, pair status unknown	22
Territorial single	121
Site status unknown	157
"Historic" or inactive site	24

on tribal lands (WDFW). Despite the high association with federal lands, about two-thirds of the known sites occur on or adjacent to nearby state or private lands (Buchanan 1997).

Spotted Owls have huge home ranges. On the Olympic Pen. the median annual home-range size of resident, reproductive pairs was 5760 ha; the median size in the Cascades was 2880 ha (see Hanson et al. 1993). Spotted Owls do not migrate, but juvs. and nonterritorial ads. may disperse 16-24 km, and rarely up to 113 km (E. Forsman p.c.). Dispersing juvs. are occasionally encountered in unusual locations (e.g., one found dead on a roadside in nw. Olympia; WDFW files). Breeding birds often leave nest territories in fall and may move 16 km or more from the nest during the winter.

The loss of Spotted Owls from the Puget Trough and sw. Washington, although undocumented, has likely been dramatic. An estimated 516-1160 site centers may have existed in these lowlands at some point in the past. The remaining 12 sites in the region are essentially unprotected under current state forest-practices regulations (WAC 222-16-086). The Barred Owl, which expanded into Washington in the mid 1960s, may compete with and hybridize with the Spotted Owl (Hamer et al. 1994). Systematic monitoring of the two owls and their relationships is essential.

Remarks: Recent demography research indicates continuing population declines statewide (Anthony et al. 2004) and declines of 60-70% in a number of occupied sites were reported in several areas (Hicks and Herter 2003). Voluntary timber harvest strategies using alternative silvicultural practices will be needed to maintain the owl's distribution in sw. Washington. Successful conservation largely hinges on management of federal lands (USDA and USDI 1994), and conservation efforts in key non-federal landscapes (WAC 222-16-086). See Gutierrez et al. (1996) for description of management and political aspects of conservation of this species. Populations of *occidentalis* occur in California and *lucida* in the s. Rockies (AOU 1957). Classed SE, FT, PHS, GL.

Joseph B. Buchanan

Barred Owl *Strix varia*

Widespread, uncommon to fairly common year-round resident in forested regions statewide.

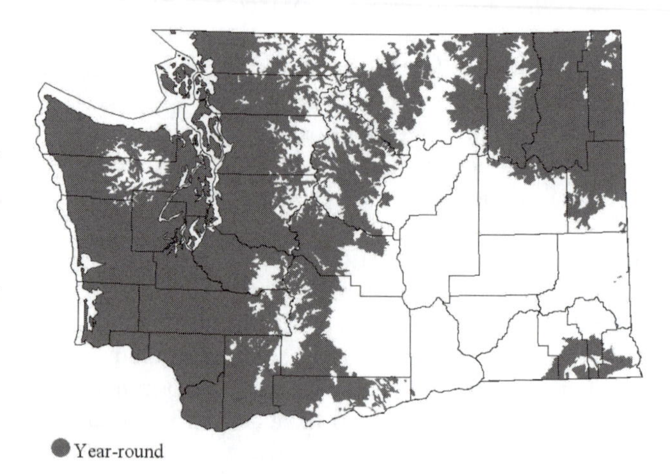

Subspecies: *S. v. varia.*

Habitat: Hardwood and old-growth and second-growth conifer forests nr. sea-level to conifer and mixed-conifer forest up to about 1900 m in the Blue Mts.

Occurrence: The Barred Owl's range enlarged dramatically in the late 20th century and it is now found in essentially all forested areas of the Pacific Northwest (Smith et al. 1997, see Wright and Hayward 1998). It is non-migratory throughout its range but has strong dispersal capability and has expanded its range into and throughout the Pacific Northwest since first recorded in Washington in 1965 (Dark et al. 1998, Mazur and James 2000).

In e. N. America, the species appears to prefer older, mixed conifer/hardwood forests that often contain forested wetlands (e.g. Nicholls and Warner 1972), and resides in boreal forest in parts of Canada. In Washington, it commonly nests also in old-growth forest (WDFW). It has adapted to using habitats in close proximity to human structures and activity and is found in small urban and suburban woodlots and forested parks as well as in wilderness areas far from human activity.

This species was first recorded in Washington in 1965 in Pend Oreille Co. though it is likely that birds first entered the state earlier. There were additional records from ne. Washington by 1974, and the species was first recorded in w. Washington in 1972. Birds were then reported regularly, primarily from nw. and ne. Washington. Reproduction was first confirmed, in Skagit Co., in 1975. By 1978 Barred Owls were found in the Cascades within 16 km of Oregon with nesting reported regularly by 1977. After the late 1970s the species began to expand into westside lowlands with numerous detections by 1980, including in Seattle's Discovery Park, where owls soon nested. A Barred Owl was recorded on San Juan I. in December 1981 and by 1985 they nested along the Queets R. on the w. Olympic Pen. (Sharp 1989). Birds expanded through sw. Washington, including Long I. in Willapa Bay, in the mid-late 1980s (Irwin et al. 1991). Areas in sw. Washington and the interior of Olympic NP supported fewer birds in that period (WDFW files) and were obviously colonized more recently (Irwin et al. 1991; E. Seamans p.c.). The population is apparently still expanding (see Mazur and James 2000).

Although this is now one of the more common forest owls, little is known about population density. Spotted Owl surveys primarily from the late-1980s to the mid-1990s showed that Barred Owls occur in proximity to hundreds of Spotted Owl locations in the Cascades and the Olympic Pen. (WDFW files). In the c. Cascades it is about as numerous as the Spotted Owl (Herter and Hicks 2000). Birds were relatively widespread on CBCs in w. Washington from 1980 to 2000.

Remarks: Significant issues associated with the species' "invasion" are 1) that it may compete for resources with the threatened Spotted Owl, 2) hybridization between the two may compromise the Spotted Owl's genetic integrity, 3) the Barred Owl is a habitat and prey generalist, and 4) the Barred Owl appears to have greater reproductive output. Evidence of resource competition is lacking, but other information suggests that competition between the two species may be occurring (Pearson and Livezey 2003, Kelly et al. 2003) and the two species use similar habitats and the Barred Owl is more aggressive. Hybridization has been documented in Washington (Hamer et al. 1994) and hybrid offspring are known to be reproductively capable. The frequency and ecological significance of hybridization is unknown and should be monitored.

Noteworthy Records: East, *early records*: 2 Oct 1965 nr. Ruby, 3 km n. of Blueslide, Pend Oreille Co.; 27 Oct 1967 at Spokane, (Johnson and Murray 1976), 15 Oct 1968 on Mica Peak, Spokane Co. (found shot; Johnson and Murray 1976), 23-24 Oct 1971 nr. Albion, Whitman Co. (Johnson and Murray 1976), 15 Oct 1972 nr. Pullman, (Johnson and Murray 1976). Columbia Basin: 26 Jan 1982 at Pasco, (WDFW files); Feb 1978 at Hanford Res.; 7 Oct 1995 at Palouse Falls SP; 21 Nov 1982 nr. Walla Walla; 30 Nov-3 Dec 1982 nr. Moses L.; fall 1986 e. of Walla Walla,; several fall, winter records at or nr. Asotin. **West**, *early records*: 1972 at Ross L. (Wahl 1995), road kill in Dec 1973 at Skykomish, (Mattocks et al. 1976), Sep 1974 at Diablo L. (Wahl 1995); 15 May 1975 in Skagit Co. (Leder and Walters 1980) was first breeding record for state.

Joseph B. Buchanan

Great Gray Owl *Strix nebulosa*

Rare local breeder in n.c. Washington, very rare winter visitor across n. counties.

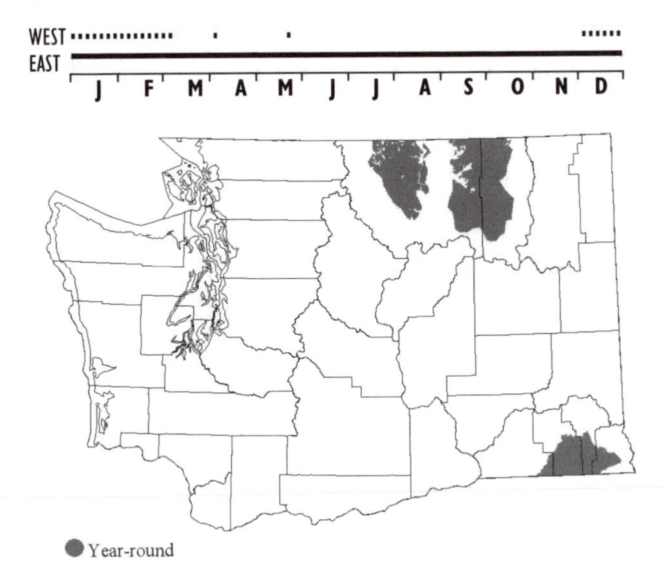

Subspecies: *S. n. nebulosa.*

Habitat: Mature mixed conifer forests of Douglas-fir, western larch, and ponderosa pine or mature subalpine forests of Englemann spruce, subalpine fir, and lodgepole pine adjacent to forage habitats in forest openings and meadows, with quaking aspen often present. Nests in broken-top snags, mistletoe brooms, nests of other species.

Occurrence: Recent studies (Winter 1986, Bryan and Forsman 1987, Franklin 1987, Bull and Henjum 1990) have improved understanding of Great Gray Owl ecology, but distribution, habitat use, and general ecology in Washington remain incompletely known. A rare resident, with a nesting pair found in a broken-top ponderosa pine snag in Okanogan Co. in 1991 (Stepniewski and Woodruff 1997), the first confirmed breeding record. Ten pairs found subsequently indicate a breeding range within e. Okanogan and w. Ferry Cos. (Smith et al. 1997).

Historic w. Washington breeding records from the mid- to late 1800s (Jewett et al. 1953) remain unconfirmed. The total of 52 records (16 given by Jewett et al. 1953 and 36 later non-breeding records in WDFW files) were from 20 counties, with 10 records in Whatcom, eight in Skagit, and five in Okanogan Cos. Other records are from Pend Oreille, Klickitat, and Skamania Cos. and n. Oregon very close to Walla Walla Co. (K. Knittle p.c.). The earliest reported records from Jewett et al. (1953) were from Pacific, Grays Harbor, King, Pierce, Whatcom, and Kittitas Cos. While changes in land use due to logging and urbanization may explain apparent absences there, many questions remain about historical status.

Small winter numbers occur irregularly, presumably during years of low prey abundance in Canada. Minor, consecutive irruptions into nw. Washington occurred in 1996 and 1997. None of four birds found in 1996 returned to the same locations in 1997, when at least six birds wintered—three in Whatcom, one in Skagit, one in Snohomish and one in King Co. Five of these were discovered in late Jan to mid-Feb, suggesting a synchronous incursion (CMAn). Two birds wintered within 16 m of each other in Skagit Co. in 1996, as two did in Whatcom Co. in 1997 (CMAn). In 1996 and 1997, two birds radio-tagged in winter in nw. Washington remained locally until early Apr and were tracked to interior B.C., with the first bird remaining there until at least 15 Sep 1996.

Remarks: Washington status description based primarily on chance encounters when birds are seen hunting during daylight hours. Surveys since 1995 for Great Gray Owls within Spotted Owl range on USFS and BLM land located one new territory and other sites may be undisclosed. While the species' range-wide population trend appears stable and gene flow does not appear restricted (Hayward 1994), the status of the small Washington population is unknown and systematic surveys farther e. in Washington are needed to determine distribution and population trends. Another obvious need is for long-term investigation of the relationship of forest management with the owls (see Bull and Henjum 1990, Habeck 1994, Hayward 1994, Hayward and Verner 1994).

C. M. Anderson and Kent Woodruff

Long-eared Owl *Asio otus*

Uncommon resident generally in lower elevations in e. Washington, less frequently observed in winter. Rare, seldom detected, in lower elevations in w. Washington, has nested several times in recent decades.

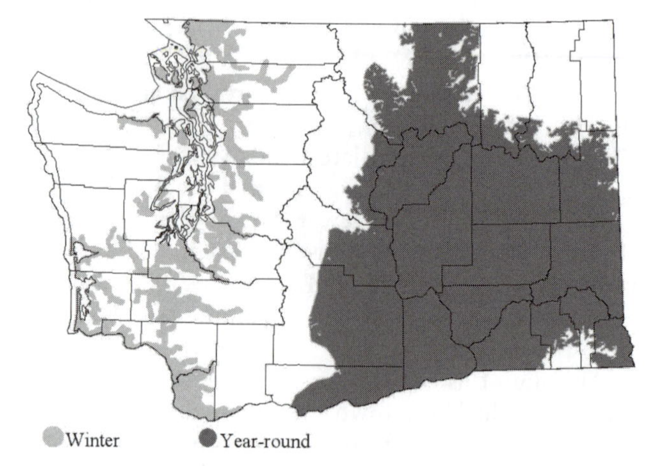

Subspecies: *A. o. tuftsi.*

Habitat: Nests in riparian areas, using abandoned corvid nests, and in rock formations, and in winter often found in communal roosts of several or more birds located in dense thickets.

Occurrence: Widespread resident across much of N. America, wintering s. to mid-Mexico (AOU 1998). In Washington found chiefly in the Columbia Basin and other eastside low-elevation areas, where Jewett et al. (1953) considered it a common breeder and a less common winter resident. Birds are noted occasionally to mid-elevations in both e. and w. Washington. Status appears to have changed little, if any, over the last several decades. Though Hudson and Yocom (1954) reported no winter records of the species in se. Washington, Weber and Larrison (1977) reported it there and K. Knittle (p.c.) reported both summer and winter records, especially in the Snake R. drainage.

The Long-eared Owl has long been considered rare to uncommon in w. Washington (e.g., Dawson and Bowles 1909, Jewett et al. 1953, Hunn 1982, Lewis and Sharpe 1987, Wahl 1995), in w. B.C. (Campbell et al. 1990b), and w. Oregon (Gabrielson and Jewett 1940, Gilligan et al. 1994). Jewett et al. (1953) listed only a handful of observations and one breeding record. Since then, the number of reports has increased dramatically, with annual reports since 1992. There are over 70 records since 1954 and from all but four w. Washington counties (Kitsap, Mason, Wahkiakum, and Island). There are at least seven confirmed breeding records and numerous records from the breeding season. Given the secretive nature of this owl and the difficulty of separating it in the field from the Short-eared Owl under some conditions (Wahl 1995, D. Paulson p.c, C. M. Anderson p.c.), its abundance is likely greater than records indicate.

Most westside records are from the e. Puget Trough. While they show no clear seasonal pattern, there are more records from Nov to Apr. Long-eared Owls have been recorded in small numbers in recent years at tree line in the Cascade Mts. (e.g. Hart's Pass; Mt. Rainier) in the early fall, perhaps indicating seasonal movements. Despite the fact that lengthy migrations have been documented in other parts of N. America (Marks et al. 1994, Houston 1997) more information is needed to determine the seasonal movements of this species in Washington. Knittle (p.c.) reported wintering birds at the n. tip of the Long Beach Pen.

Noteworthy Records: East, *Winter roost high counts*: 13 in mid-Nov 1996 nr. Moxee (Stepniewski 1999); 6 on 20 Dec 1975-28 Feb 1976 at Whelan, Whitman Co. (Woodby 1976); 6 on 28 Dec 1980, 10 in Jan 1979 at Asotin Slough (WDFW): 7 on 28 Jan 1981 nr. Rose Cr., Whitman Co. (WDFW); 13 on 4 Feb 1979 at Asotin (WDFW); 22 on 4 Feb 1988 nr. Wapato (Stepniewski 1999); 6 on 5 Feb 1999 at Rice Rd. HMU, Walla Walla Co.; 12+ on 8 Feb 1997 at Big Flats HMU, Franklin Co.; 12 on 1 Mar 1980 at Experiment Station nr. Moxee (WDFW), 7 there on 18 Mar 1979, 10 there on 26 Mar 1983. Fall, high roost counts: about 10 on 13 Oct in Douglas Co. (WDFW). **West**, *breeding/Summer:* May 1987 on San Juan I. (Lewis and Sharpe 1987); 3 Apr 1992 at Blaine; spring 1993 at Rockport; 29 Mar 1993 at Ferndale; 26 May 1998 at Glacial Heritage, Thurston Co. (RR); spring 2001 at Seattle.

Joseph B. Buchanan and Russell E. Rogers

Short-eared Owl *Asio flammeus*

Locally common winter visitor and migrant, uncommon summer resident in open lowland habitats e., more local w.

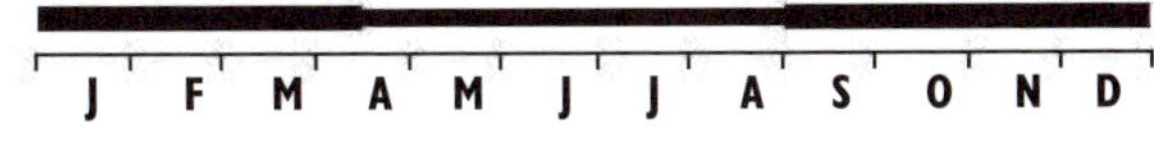

Subspecies: *A. f. flammeus.*

Habitat: Open-country habitats at moderate altitudes statewide, including grasslands, rangelands, marshes, shrub-steppe, meadows, fields, estuaries, and airport edges.

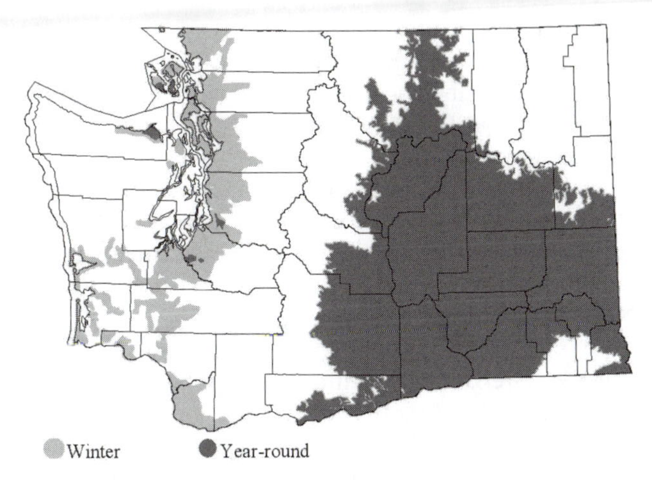

Occurrence: A species that has declined considerably since 1950s, now absent from much of the w. lowlands due to land-use changes and in parts of the e. due to intensive agricultural activities that have destroyed nesting habitat (Smith et al. 1997). Knowledge of status changes is complicated by presence of both breeding and wintering populations, and by often very localized distribution.

Described by Jewett et al. (1953) more thoroughly for e. (where it was "common") than for w. Washington. Though commented upon infrequently in breeding season e. of the Cascades (e.g. Stepniewski 1999), the species is still fairly widespread there, noted especially in the Columbia Basin (see Smith et al. 1997). W. of the Cascades, nests locally in Whatcom Co., Nisqually delta, Kent-Auburn Valley, Ocean Shores, Leadbetter Pt. (Smith et al. 1997) and, occasionally, on s. San Juan I. (Lewis and Sharpe 1987). Probably nested in 1977 at Sand Pt. in Seattle and nr. Everett but seldom reported since then. Reproductive success low in some agricultural areas due to nest/nestling loss to farm equipment (Campbell et al. 1990b).

An irruptive species, variable from year to year, often due to local prey populations as well as prey availability during nesting season elsewhere. CBC data and field reports showed low numbers statewide in winters of 1974-75 and in the west in 1991-92. Local and regional populations vary (see Stepniewski 1999), with low numbers on one side of the Cascades often noted while high numbers are reported on the other side. Foraging and, consequently, presence is affected by snow depths (see Stepniewski 1999).

Migrations not well described, but apparent in Mar-Apr, and late Oct-early Nov. Noted on occasion migrating in the mountains (e.g. Wahl 1995). As in other parts of the world, birds have been noted at sea during fall migration, with records up to 73 km offshore (TRW).

Whether breeding birds remain to winter is uncertain, but birds banded from late Jan to mid-May 1964-67 at Vancouver, B.C., were recovered in e. Washington, Oregon, and California as much as two years later (Campbell et al. 1990b).

Long-term CBC data indicate significant declines (P <0.01) in both w. and e. Washington. Numbers in sw. B.C., a main wintering area, have declined with habitat loss, and possibly due to competition with Northern Harriers (Campbell et al. 1990b).

Remarks: Breeding and wintering populations have declined over substantial parts of the species' range in the nw. (Tate 1986). Destruction of grasslands habitats noted as a probable cause of long-term decline (DeSante and George 1994). On NAS Blue List 1982-86 and Watch List since then due to apparent continent-wide decline.

Noteworthy Records: West, *Summer*: up to 7 in 1975 at Nisqually NWR. *Fall*: 30 in late Oct 1975 at Nisqually NWR; 1 on 19 Aug 1981 and 2 on 6 Oct 1985 ca. 37 km offshore and 1 on 20 Sep 1997 ca. 73 km off Grays Harbor (TRW). *Winter*: 25 on 21 Jan 1973 at Lummi Flats. **East**, *Fall*: 31 in 2.6 km transect in 1964 nr. Spangle. *Winter*: 34 on 29 Dec 1969 nr. Othello. *CBC high counts*: 22 in 1970, 24 in 1976, 35 in 1977, 31 in 1983, 15 in 1998 and 20 in 2000 at Bellingham, 19 in 1970 at Spokane, 14 in 1995 and 19 in 1999 at Skagit Bay.

Terence R. Wahl

Boreal Owl *Aegolius funereus*

Uncommon resident in subalpine forest areas in e. slope of n. Cascades, the Selkirks, and Blue Mts. Occasional vagrant to lower elevations.

MOUNTAINS EAST

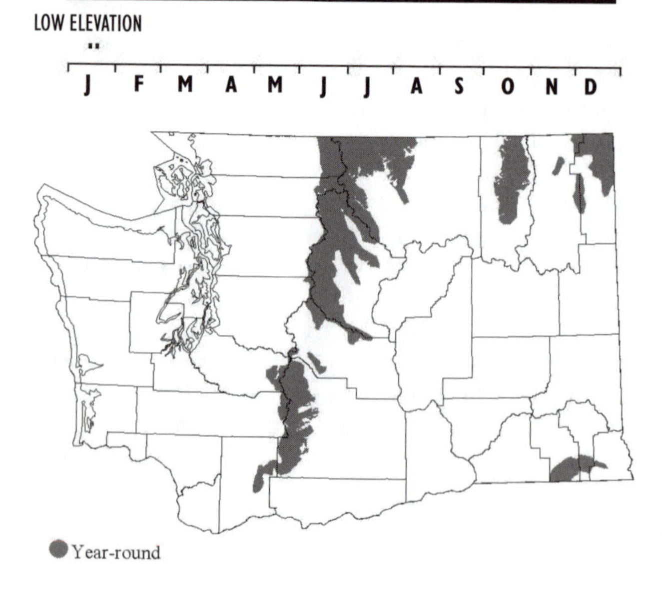

Subspecies: *A. f. richardsoni*.

Habitat: Subalpine conifer forests.

Occurrence: Jewett et al. (1953) considered this owl "hypothetical" in the state based on a Jan 1905 record from Whatcom Co. reported by Dawson (1908). Distribution remained unclear until extensive surveys in 1985 and 1987 showed the species was fairly widespread and regularly encountered in suitable high-elevation habitat. In 1985 Whelton (1989) made 33 detections of at least 23 individuals along 35 km of transects in Engelmann spruce, lodgepole pine, and subalpine fir forests in Ferry, Pend Oreille, and Stevens Cos., and in 1987 he made 12 detections of an unreported number of Boreal Owls in Engelmann spruce, grand fir, lodgepole pine, and subalpine fir forests in Columbia Co. All survey routes were between 1525-1950 m elevation. Boreal Owls have since been reported numerous times in the n. tier of counties, and less frequently in the Cascades and Blue Mts., with most records in fall. Only two nests are documented for the state (the first in 1992 at Thirtymile Meadows, Okanogan Co.; Stepniewski 1996); and a previous report, which is doubtful given its low elevation and location far from suitable habitat in Pullman (Batey et al. 1980, and see Smith et al. 1997).

Detections reported by Whelton (1989) and other sources, including WDFW data, in Washington were made in Jan (3 detections), Mar (3), Apr (9), May (1), Jun (10), Jul (4), Aug (7), Sep (38), Oct (23), Nov (1), and Dec (3); the greater number of detections in autumn likely reflect the fact that in this season the higher elevations are most accessible to observers. The seasonal pattern may also indicate a seasonal change in calling rates by the owl (Stepniewski 1996).

Much remains to be learned about this species in Washington. Smith et al. (1997) reported only seven "confirmed" or "possible" breeding records. Smith et al. (1997) also indicate "core zones" of the species only in the n. tier of counties; extremely small areas of "peripheral zone" habitats were thought to occur in the Cascade and Blue Mts. Despite this, it seems likely that the range extends through much of the Cascade Mts. high country; it is known to occur in the e.c. Cascades of Oregon (Gilligan et al. 1994). The species is thought to be nomadic in other parts of its range, moving from the traditional home range when prey populations are too low (Hayward and Hayward 1993). Given the dearth of information relating to the species' distribution, it is not surprising little or nothing is known about its population status or habitat associations in Washington.

Noteworthy Records: East, *low elevation*: 10 Jan 1974 at Pullman (Johnson and Hudson 1976); Sep-Oct 1988 nr. Cle Elum (WDFW). *Blue Mts.*: 7-14 Jun 1987; 27 Aug 1995 at Tablerock Lookout, Columbia Co.; 9 reports from summer 1988 from Columbia Co.; 1 Oct 1995 at Mt. Misery, Garfield Co.; 4 Oct 1996 and 4 on 11 Oct 1991 in Columbia Co. **West**, *low elevation*: injured bird on 14 Jan 1992 nr. Snohomish. *Cascades*: 6 Jan 1977 in Chumstick Valley, Chelan Co.; 11 Sep 1992, 5 on 28 Sep 1998, 2 Oct 1993 at Sunrise, Mt. Rainier NP. 20 Oct 1998 at White Pass; Oct 1992 on N. Fork of Ahtanum Cr., Yakima Co. (Stepniewski 1999); fall 1988 from Kittitas Co.; fall 1992 at Darland Mt., Yakima Co.

Joseph B. Buchanan

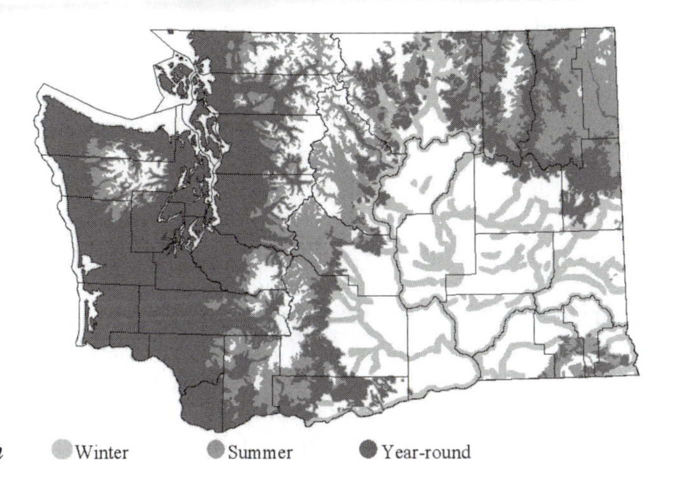

Northern Saw-whet Owl *Aegolius acadicus*

Fairly common resident in forests statewide.

WEST

EAST FORESTS

COLUMBIA BASIN

J F M A M J J A S O N D

Subspecies: *A. a. acadicus.*

Habitat: Nests in all forest types, including riparian areas, below the subalpine (Cannings 1993, Smith et al. 1997). Migrants and wintering birds in open areas associated with dense conifer roosting opportunites (K. Knittle p.c.).

Occurrence: One of the more common owls in westside forests through the lowlands, San Juan Is. and Cascades, and in the e. Cascades, ne., and Blue Mts. Almost strictly nocturnal, it occurs in a wide range of situations including urban parks, residential areas, and orchards, and may roost in buildings and nest in Wood Duck boxes (Wahl 1995). Nesting noted from sea level to high elevations—1600 to1850 m in the Blue Mts. in June 1987.

Migration and/or altitudinal movements are inconsistently described but seasonal shifts are apparent (e.g. Stepniewski 1999) and some portions of the population may be nomadic (see Marks 1997) and involve individual birds occurring in unusual locations to take advantage of uniquely available resources. A southward movement takes place in mid-Oct to early Nov in se. B.C. (Campbell et al. 1990b), and birds present in the Columbia Basin in winter may have migrated s. from n. Washington or Canada (Cannings 1993) although some likely moved down from higher elevations.

Saw-whet owls occur widely in forested portions of the state although detectability varies seasonally (Irwin et al. 1991). Peak detection rates of this owl exceeded one per survey hour in both years of an intensive survey effort in managed forest landscapes in sw. Washington; detection rates varied among years, and rates were lowest in Jul in 1987 and Jun-Aug in 1988 (Irwin et al. 1991). Saw-

Saw-whet Owl (G. Scott Mills)

whet owls were less frequently detected in 1987 (one per 18.3 km surveyed) compared to 1988, when 240 independent detections were made (one per 7.1 km, Irwin et al. 1991). It is not known whether the differences in detection rates among years reflected changes in the abundance of the species.

Birds are irregular and infrequently detected in the Columbia Basin. With the exception of three or four breeding records, observations there were primarily from fall and winter (at least 40 winter records since 1950; 23 from 1970-89). Unlike pygmy-owl winter distribution in the Columbia Basin, which appears, based on the available records, concentrated in Asotin, Walla Walla, and Whitman Cos., about two-thirds of the winter saw-whet records (19) come from other areas, notably Grant and Douglas Cos. There are no midsummer records, and it is not clear whether this early-nesting species departed early in the few seasons when nesting was documented or whether they simply became difficult to detect.

Noteworthy Records: East, *Fall*: 3 Aug records and 1 Nov record (no year given) in potholes region (Harris and Yocom 1952); 11 Aug 1950 nr. O'Sullivan Dam (Hudson and Yocom 1954); 14 Sep 1973 5 km sse. of Pullman (Conner Museum #74-161); 22 Oct 1988 at Fields Springs SP (WDFW); Oct 1983 in Moses Coulee; 12 Nov 1949 on Cabin Ridge, Garfield Co. (Hudson and Yocom 1954); fall 1972 in Pasco; fall 1980 at Ephrata and Quincy. *Spring*: 24 Apr 1954 in Potholes area (Johnsgard 1954); 1 May 1977 at Field Springs SP (WDFW); 5 May 1973 n. of Coulee City (WDFW); 17 May 1964 at Sprague L. (WDFW); 20 May 1978 at Davenport Cemetery (WDFW); spring 1985 at Fishhook SP nr. Pasco and nr. Newman L. *Breeding*: 15 Mar 1999 pair at George; 26 Mar 1948 (nest with 3 eggs) about 8 km w. of Dayton (Lyman and Dumas 1951); 18 May 1987 (pair with 3-4 young) at Lower Granite Dam Campsite; spring 1989 suspected of nesting at Rose Cr. Preserve n. of Pullman; 21 Jun 1970 (details lacking) at Rose Cr. Preserve (WDFW).

Joseph B. Buchanan

Common Nighthawk *Chordeiles minor*

Common migrant and summer resident in e. Washington. Fairly common migrant and summer resident in w. Washington except uncommon in much of the Puget Sound lowlands.

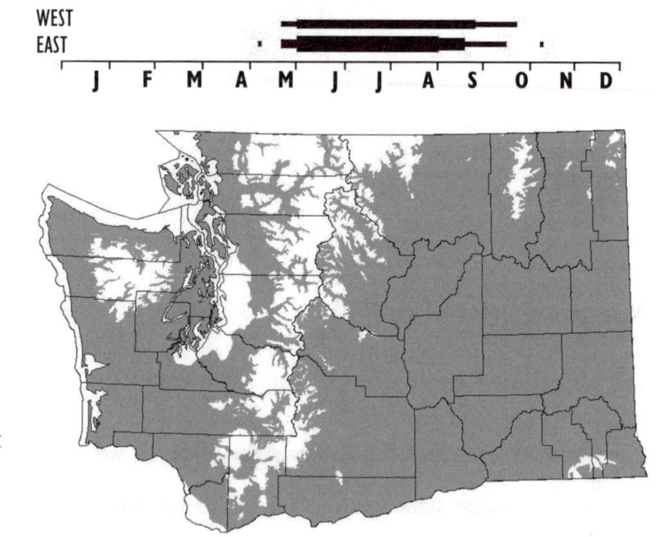

Subspecies: *C. m. minor* in w., *hesperis* in e. Washington.

Habitat: Nests on ground in open terrain including fields, fresh burns, and clear-cuts, and various gravel surfaces, including gravel roofs in urban settings (Poulin et al. 1996).

Occurrence: Breeds throughout most of Washington; not found in areas of extensive forest (except when foraging or traveling) or at higher elevations. The species nests up to about 1250 m in B.C. (Campbell et al. 1990b), but occurs at elevations to 2100 m in Idaho (Burleigh 1972). The elevational limit of the species in Washington is not known.

Often found in flocks of hundreds during fall migration (Poulin et al. 1996) and at prime foraging sites during the breeding season in e. Washington. It appears to travel in much smaller flocks in w. Washington (see Campbell et al. 1990b).

The status of the species is of concern as population declines have been reported from various regions in N. America (Poulin et al. 1996) and BBS data indicate significant declines over the w. BBS survey area, over the contiguous U.S., and over N. America (Sauer et al. 1999a). Sauer et al. (2001) indicated a decline of 1.1% per year in Washington from 1966 to 2000. Ministry of Environment and Parks (1998) cautions that BBS data may not be adequate to assess population trends of this species. Although the species appears to be reasonably common in e. Washington (see Stepniewski 1999), there is concern that lowland Puget Trough populations have declined (Wahl 1995, C. Chappell p.c.) though apparently not the San Juan Is. (Lewis and Sharpe 1987). One explanation may be that preferred nest locations on flat gravel roofs are less common, partly due to competition with roof-nesting gulls in urban areas of the Puget Sound lowlands (TRW). A decline in the Willamette Valley, Oregon, however (Gilligan et al. 1994, Sauer et al. 1999a), suggests that other factors are involved.

Remarks: The species is not known to enter torpor (M. Brigham p.c.): reports prior to May are of doubtful validity. More effective survey methods are required to monitor nighthawk populations throughout their range. Intergrades from the Tri-Cities and Sprague make exact ranges of the subspecies uncertain to date (see Selander 1954).

Noteworthy Records: West, *Spring arrival*: 1 May 1979 at Seattle; 27-30 May 1969 at Seattle; 28 May 1998 at McChord AFB; 30 May 1996 at Naselle. *Fall, late*: 11 Oct 1981 at Sequim, 19 Oct 1978 at Olympia. *High counts*: 40+ on 24 Aug 1996 at Whidbey I.; 30 on 23 Sep 1966 at Bellingham (Wahl 1995); 25 on 3 Aug 1997 at Shelton; 24 on 3 Aug 1999 at Silverdale. **East**, *Spring arrival*: 5 May 1973 at Clarkston; 15 May 1970 at Turnbull NWR; 18 May 1964 at Columbia NWR; 18 May 1996 at George; 21 May 1970 at Tri-Cities; 21 May 1998 at Umatilla NWR; 22 May 1999 at Yakima. *Fall, late*: 27 Sep 1972 at Spokane; 29 Sep 1989 at N. Richland; 4 Oct 1972 at Spokane; 7 Oct 1970 at Clarkston; 7 Oct 1989 at Walla Walla R. delta; 11 Oct 1972 at Ellensburg (Verner 1974); 9 Nov 1989 at Spokane. *High counts*: 500+ Priest Rapids 14 August 1993 (Stepniewski 1999); 120 on 26 Jun 1998 at Vernita; 100+ on 27 Aug 1989 7 km nw. of Harrah (Stepniewski 1999); 100 on 3 Sep 1964 at Newport; 100 on 25 Jun 1972 at Spokane; 100 in Jul 1980 at Winchester; 100 on 22 Aug 1997 at Naneum Meadows, Chelan Co.; 75 in Jul 1980 at Quincy; 53 on 6 Sep 1996 at College Place; 50 on 10 Jun 1970 at Clarkston.

Joseph B. Buchanan

Common Poorwill *Phalaenoptilus nuttallii*

Uncommon migrant and summer resident in e. Washington. Very rare spring and autumn migrant in w., very rare in summer.

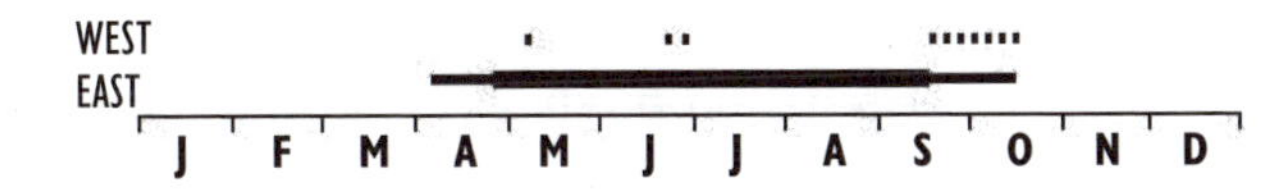

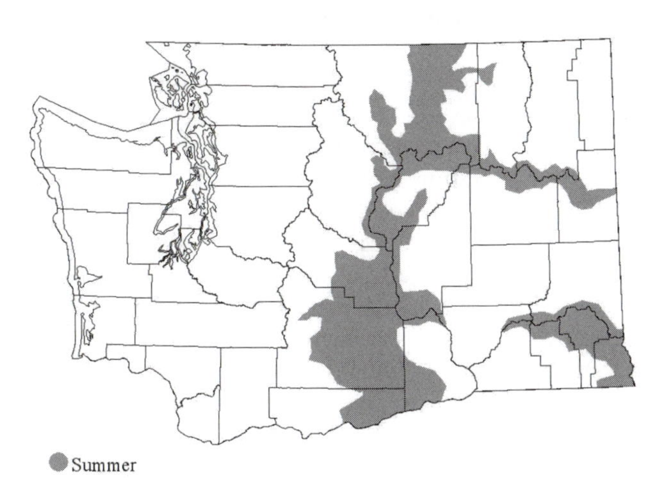

Subspecies: *P. n. nuttallii.*

Habitat: Nests in dry, open, grassy or shrubby areas (Swisher 1978) in canyons and the lower slopes of the Cascade foothills, Okanogan Highlands, and the Blue Mts., generally in the transition zone between ponderosa pine or dry Douglas-fir forest and shrub-steppe (Jewett et al. 1953, Cannings et al. 1987, Smith et al. 1997), but also in areas at or above 1200 m elevation (e.g. at Blewett Pass; JBB).

Occurrence: Though long known in the state (Jewett et al. 1953), the poor-will's distribution remains incompletely understood. BBA surveys in 1987-96 were unable to confirm occurrence in some areas along the Columbia and Snake Rs., in Channeled Scablands and other locales where birds had been reported historically (see Smith et al. 1997). Changes in status over time are unknown—monitoring data are lacking. Migration routes and timing are poorly understood (Csada and Brigham 1992). Records from the Okanagan Valley, B.C., indicate that spring arrival dates are quite variable, ranging between 21 Apr and 18 May in one 10-year period (Cannings et al. 1987).

Eight records from w. Washington since 1964 are of interest. Six are from mid-Sep to mid-Oct and, with the exception of a Tacoma record, are from Whatcom and Skagit Cos. Jewett et al. (1953) also cite an Oct record from Bellingham (see Wahl 1995). The other two records are from Jun, at Enumclaw and the Mt. Hardy burn, Skagit Co.

These two records occurred during the presumed peak of the nesting season (Csada and Brigham 1992). Although these may represent off-course migrants (the Jun records perhaps were birds that abandoned distant territories following failed nesting attempts), it is possible that they were locally nesting birds. Poorwills are thought to be site-faithful to breeding sites (Csada and Brigham 1992) and this would be consistent with the rather clustered nature of the records referred to above.

Remarks: It has been hypothesized that timber harvest practices may create suitable habitat and may have somewhat extended the range of the species in Oregon by creating more open habitats for foraging and nesting (Horn and Marshall 1975); it seems possible that this may have occurred in Whatcom and/or Skagit Cos. and in Yakima Co. on the s. slope of Mt. Adams in 1994-95 (Smith et al. 1997). The effects of forest fires may similarly influence habitat suitability in certain circumstances. Dedicated surveys in harvested or burned areas may clarify the situation.

Noteworthy Records: East, *Spring, early*: 6 Mar 1992 at Rattlesnake Hills, Yakima Co. (Stepniewski 1999). *Fall, late*: 12 Oct 1972 at McNary NWR; 12 Oct 1972, se. Washington (Hudson and Yocom 1954); 30 Oct 1990 at Rattlesnake Hills, (Stepniewski 1999); 6 Nov 1994 at Satus Cr., s. of Toppenish (Stepniewski 1999). **West**, *Spring*: 3 May 1984 at Marblemount. *Summer*: 19 Jun 1964 nr. Enumclaw; 26 Jun 1999 at Mt. Hardy burn, Skagit Co. *Fall*: 6 Oct 1941, n. of Bellingham, (Wahl 1995); 16 Sep 1976 at Burlington; 11 Oct 1983 at Burlington; 2 Oct 1989 in Skagit Co.; 4 Oct 1991 at Tacoma; 14 Oct 1994, Bellingham (Wahl 1995); 22 Sep 1996 at Hart's Pass.

Joseph B. Buchanan

Black Swift *Cypseloides niger*

Uncommon summer resident and migrant on the westside, locally common in the Cascades. Rare e. of the Cascades and on the outer coast.

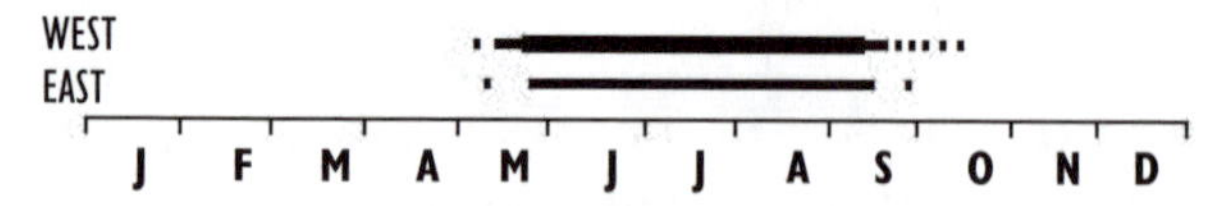

Subspecies: *C. n. borealis.*

Habitat: Nests behind waterfalls and on steep cliffs, mostly in montane forests. Forages over riparian areas, open lowlands, rivers and ponds.

Occurrence: The Black Swift is one of the least understood breeding species in Washington. "A denizen of mountain canyons and rugged seacoasts" (Cannings et al 1987), its status is somewhat enigmatic. It is uncommon within most of its range but BBS data indicate the highest densities in N. America in sw. B.C. and w. Washington (Price 1995). Population size remains uncertain, but BBS data from 1966 to 2000 indicate declines in both Washington and B.C. (Price 1995, Sauer et al 2001).

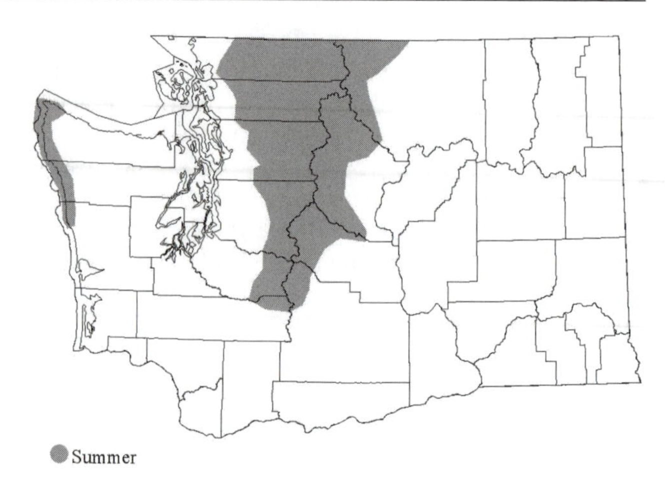

Breeds up to 2600 m on both sides of the Cascades, mainly n. of Snoqualmie Pass (Smith et al. 1997). Nesting birds forage down to sea level, especially during rain and low overcast, when large flocks are sometimes noted in Puget Trough lowlands. This is recorded s. to Mt. Adams, across Puget lowlands, w. to the San Juan Is., and e. to Ferry Co. (Smith et al. 1997). Summer records from the Olympic Pen. suggest nesting there, though it is unconfirmed.

Due to inaccessibility, reports of nesting are scarce but sites were found in Douglas Co. and nr. L. Chelan (Dawson and Bowles 1909). Subsequently, nesting was confirmed nr. Copper Mt., Whatcom Co., in the N. Cascades in 1988 (Smith et al. 1997) and very probable sites were nr. Stehekin in 1994 (D. Drummond p.c., AME) and at Goodell Cr., n. of Newhalem (Wahl 1995).

First spring migrants typically in mid-May, with peak migration mainly from late May through early to mid-Jun. Fall migration mostly from mid-Aug through early to mid-Sep. Migrants in presumed non-breeding habitats include flocks at 1500 m elevation at Mt. Baker and on the Olympic Pen. at Sequim, Cape Flattery. and Tatoosh I.

Remarks: Nest sites in Washington have been in montane habitats, but nesting has occurred on a sea cliff in s. California (Vrooman 1901), in a sea cave (Legg 1956), and in a limestone cave in Colorado (Davis 1964 in Foerster and Collins 1990). Classed GL.

Noteworthy Records: *High counts, Spring-Summer*: 550 on 30 Jul 1994 at Everett (SGM); 500 on 17 Jun 1979 at Wells Dam Res.; 500 on 8 Jun 1996 at Spencer I.; 400 on 18 Jun 1981 nr. Bridgeport; 300 on 3 Jun 1978 at Newhalem. *Fall*: 270 on 30 Aug 1980 at Brewster; 200 on 24 Aug 1978 nr. Cashmere; 180 on 15 Aug 1995 at Everett (SGM); flock on 18 Aug 1965 at 1500 m at Mt. Baker; hundreds on 6 Sept 1976 nr. Sequim; ca. 30 on 12 Sept 1965 at Cape Flattery; "a few" on 21 Sept 1979 at Tatoosh I. *Late fall*: 13 Oct 1990 nr. Snohomish; 7 Oct 1999 at Marymoor Park (M. Hobbs p.c.).

Ann M. Eissinger

Vaux' Swift *Chaetura vauxi*

Fairly common summer resident and migrant in w., uncommon in e. Widespread spring and fall migrant, locally abundant during migration.

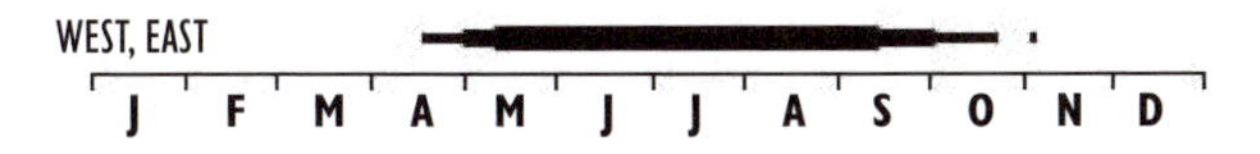

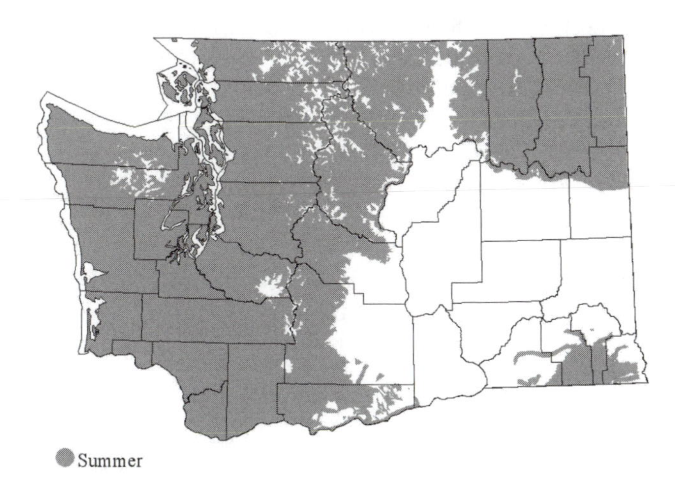

Subspecies: *C. v. vauxi.*

Habitat: Breeds mainly in old-growth forests, foraging over forests, grasslands, and water (Bull and Collins 1993), nesting and roosting in cavities within live trees, large, hollow snags (Bull and Hohmann 1993), and urban chimneys.

Occurrence: Breeding range includes w. Washington, the e. slope of the Cascades, and forests of ne. Washington and the Blue Mts., at elevations from sea level to 1524 m (Smith et al. 1997). Birds typically use trees and snags in old-growth forests and riparian areas (Summers and Gebauer 1995), and remnant trees and snags in logged stands (Bull and Hohmann 1993). Trees used in the Pacific Northwest include hollow conifers, cottonwood, and bigleaf maple.

Spring migrants appear in late Apr, peak in mid- and late May. Fall migration appears to begin in mid-Aug and end by mid-Sep, with a few birds lingering into early to mid-Oct. Compared with B.C. and Oregon, statewide field reports have been inconsistent seasonally, geographically, and numerically. Roosts in particular have not been consistently monitored. Eastside occurrence along the e. slope of the Cascades from Chelan to Klickitat Co. Other reports are scattered from Richland, Pend Oreille Co., Spokane, Pullman, Walla Walla, and Asotin.

Following human settlement, birds have been frequently observed nesting in chimneys (Bull and Cooper 1991), with older brick chimneys apparently preferred (e.g., Thompson 1977c). Large flocks (from 100 to >30,000) sometimes use large residential and industrial chimneys as night roosts during migration, and may attract onlookers at chimneys from Sumas s. to Vancouver, and in Klickitat and Walla Walla Cos. e. of the Cascades. Between 1996 and 1998, for example, flocks of up to 1000 roosted in chimneys in the S. Capitol neighborhood of Olympia in the spring and the fall, with larger flocks in the fall. About six pairs were seen there in the summer.

BBS data indicate that numbers in Washington declined significantly from 1966 to 2000 (3.32% per year; Sauer et al. [2001]). The extensive loss of old-growth forests and their conversion to second-growth forests (Bolsinger and Waddell 1993, Bolsinger et al. 1997) appear to have dramatically reduced the availability of nesting and roosting habitat. This decline may continue as older forests are logged and replaced with young stands that contain few, if any, large, hollow snags or live trees with cavities (Bolsinger and Waddell 1993, Bolsinger et al. 1997). The availability of suitable chimneys may be a significant factor affecting long-term status. These provide important temporary shelter for migrating swifts and might determine some migration routes. It is unlikely, however, that chimneys can replace older forests as primary habitat.

Remarks: Roosting congregations have prompted the capping of chimneys by some home owners due to smell, noise, or flights into the living room. Others mistake the swifts for bats, but homeowners are more likely to feel uncomfortable with numerous spectators crowded nearby at sunset during migration, which can last several weeks. Vaux' Swifts successfully nested in artificial nest boxes in ne. Oregon (E. Bull, USFS, PNW Res. Station, La Grande, OR). Designated SC, PHS.

Noteworthy Records: *Spring, early*: 1 on 20 Apr 1966 at Spokane. *Migration roosts*: 1100 on 1 Sep 1973; 1500 on 27 Aug 1977, 1500-2000 on 24 Aug 1980 at Klickitat; 400-500 on 19 May 1974, 3000 in fall 1986 at Walla Walla; 1500 on 10 May 1990 at Sumas (A. Eissinger p.c.). *Flocks*: "thousands" on 20 Sep 1988 at Bellingham, 1000 on 8 May 1999 at Enumclaw.

Jeff C. Lewis and Noelle Nordstrom

White-throated Swift *Aeronautes saxatalis*

Locally common migrant and summer resident nr. cliffs in e. Washington. Very rare migrant in w.

Subspecies: *A. s. saxatalis.*

Habitat: Ponderosa pine zone and below, nr. suitable nesting habitat along major rivers (Smith et al. 1997), strongly associated with basalt cliffs used for nesting and adjacent rivers, lakes, and lowlands for foraging.

Occurrence: Washington colonies are located along the Columbia R. from Wallula Gap to Stevens Co., the Spokane R., the lower Snake R. (in portions of Franklin, Walla Walla, and Whitman Cos.), the Okanogan R., the lower reaches of the Methow R., a portion of the Yakima R. w. of the Yakima area, the Tieton R. drainage, the lower Grande Ronde R., and along the Grand Coulee. Isolated breeding areas are located in sw. Spokane, nw. Whitman, n. Pend Oreille Cos. and possibly in Asotin Co. (Smith et al. 1997). Most common at coulees and cliffs in the Columbia Basin (Smith et al. 1997).

Numbers at nesting colonies vary. Most colonies in the Okanagan Valley in B.C. range between eight and 20 birds although numerous smaller colonies are also known (Cannings et al. 1987). Numbers in Washington appear similar although larger colonies are known: Weber and Larrison (1977) reported a colony of about 50 at Banks L. It is unknown whether larger counts (1000 nr. L. Lenore, 4 May 1963; 350 in Grand Coulee area, Grant County, 10 Apr 1969) represented individuals from local colonies, aggregated numbers from multiple colonies, or migrants. Some nesting cliffs are over 300 m high, although some are scarcely over 30 m (e.g., nr. Cashmere, JBB). Birds apparently attempted to nest under the freeway bridge over Latah Cr. at Spokane in 1986, and have nested under Interstate 82 bridges in Kittitas and Yakima Cos. in 1980, and at least one has nested under a house roof (Yocom 1966).

Jewett et al. (1953) considered this species a rare migrant and summer resident in e. Washington and similar, early accounts for Okanagan Valley, B.C. (Cannings et al. 1987), suggest that the species is now more common in comparison. It is thought to be more common in this region than in se. Oregon, which surprised Gabrielson and Jewett (1940), given the abundance of apparently suitable habitat there. Philopatry to traditional breeding sites (Dobkin et al. 1986, Campbell et al. 1990b) may discourage range expansion, although the swift has expanded its range northward in B.C. (Cannings et al. 1987). Migration dates uncertain but reportedly vary between colonies in the fall (SGM). There are three w. Washington records, from 27 Apr to 8 Aug.

Sauer et al. (2001) indicate a significant decline in numbers from 1966 to 2000 in Washington, though BBS methodology may not sample swifts adequately (see Ryan and Collins 2000).

Noteworthy Records: *Spring arrival*: at 2 sites in e. Washington before Apr in 1990; 3 on 17 Mar 1996 at Wallula Gap; 1 on 20 Mar 1996 at Yakima Canyon; 30+ on 29 Mar 1997 at Vernita Bridge; 10+ on 30 Mar 1997 at Port Kelly, Walla Walla Co.; 3 on 24 Mar 1998 at Wallula Gap; 1 collected on 27 Mar 1959 at Park L. (Johnson and Murray 1976); 5 Mar 1999 at Vantage (Stepniewski 1999). *Spring and breeding season high counts*: 1000 on 4 May 1963 nr. L. Lenore; 350 on 10 Apr 1969 nr. Grand Coulee. *Late fall*: 3 Nov 1980 at Alkali L. **West**, *Spring*: 27 Apr 1983 at L. Louise, Bellingham. *Fall*: 2 on 8 Aug 1980 at Stevenson. 21 Jul 1999 at Owyhigh L. trailhead, Mt. Rainier NP. *Note*: reports inadequately documented: 9000 on 12 Jun 1990 at Ephrata (NAB editorial comment: "which seems like more than the entire number known from all colonies in the state!") and a report of a bird, heard only, on 16 Feb 1987 at Yakima R. Canyon.

Joseph B. Buchanan

Ruby-throated Hummingbird

Archilochus colubris

Casual vagrant.

Breeds throughout most of e. N. America and winters in C. America: *the* hummingbird for much of its extensive range. Commensurate with its extreme rarity in w. N. America and the relative difficulty of field identification, there is only one Washington record—a sight record of a male ne. of Liberty, Kittitas Co., on 28 Jun 1992 (Tweit and Skriletz 1996).

Bill Tweit

Black-chinned Hummingbird

Archilochus alexandri

Uncommon to locally common summer resident in low-elevation riparian areas in e. Washington. Very rare w.

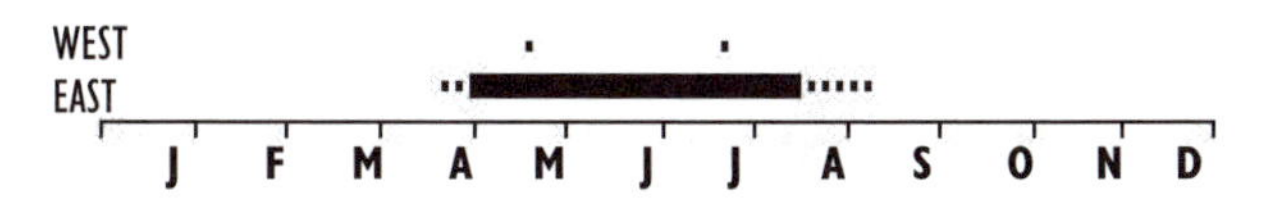

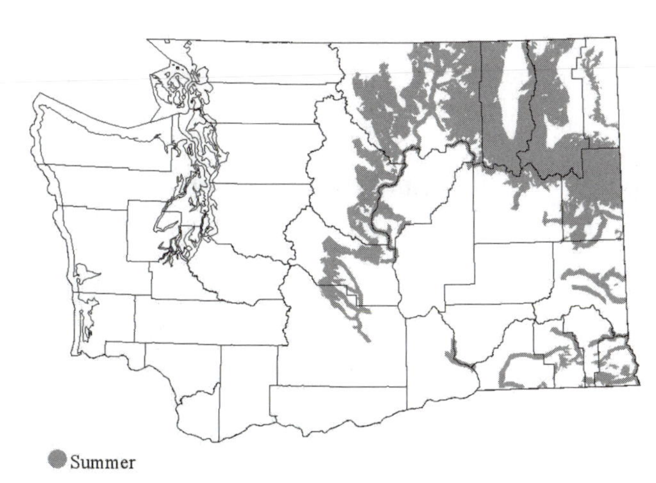

Summer

Habitat: Open riparian woodlands, particularly deciduous groves, low-density developed areas in upper Sonoran and lower transition zones with conifer forests secondary. Attracted to urban plantings and feeders (Campbell et al. 1990b). Interior Douglas-fir at the edge of core zones, Grand fir in the Blue Mts. are peripheral (Smith et al. 1997).

Occurrence: Fieldwork in recent decades added details about distribution of a species considered "relatively rare" by Jewett et al. (1953); it breeds throughout much of e. Washington. Recorded on the lower e. slopes of the Cascades from Yakima Co. n. and e. to Pend Oreille and Spokane Cos. and in the se. corner of the state (see Smith et al.1997), at lower elevations than the Calliope Hummingbird. Uncommon in the lower Columbia Basin, with abundance varying locally (see Weber and Larrison 1977, Ennor 1991, Stepniewski 1999). Black-chinned Hummingbirds are more common in ne. counties than along the e. slope of the Cascades and are recorded in the Blue Mts. (Smith et al. 1997).

Timing of migration is incompletely known. A bird in late Jul at Othello (B. Flores p.c.), a bird at Blewett Pass on 10 Jul and one at Mt. Pleasant, Skamania Co. on 21 Jul were apparent early migrants. Two that summered at Richland until 6 Sep were the latest recorded in Washington.

The species is casual w. of the Cascades in B.C. (Campbell et al. 1990b). It has been recorded in w. Washington at Everett and in Skamania Co.

Population trends are uncertain though, based on BBS data, Sauer et al. (2001) suggested an increase from 1966 to 2000. The species' range has apparently expanded in the U.S. (Baltosser and Russell 2000).

Noteworthy Records: West, *Spring*: 1 on 19-20 May 2000 at Everett; 1 on 21 Jul 2000 at Mt. Pleasant, Skamania Co. *Summer*: 1 on 10 Jul 1979 at Blewett Pass on 10 July 1979.

Terence R. Wahl

Anna's Hummingbird

Calypte anna

Rare to locally fairly common westside resident. Very rare year-round e.

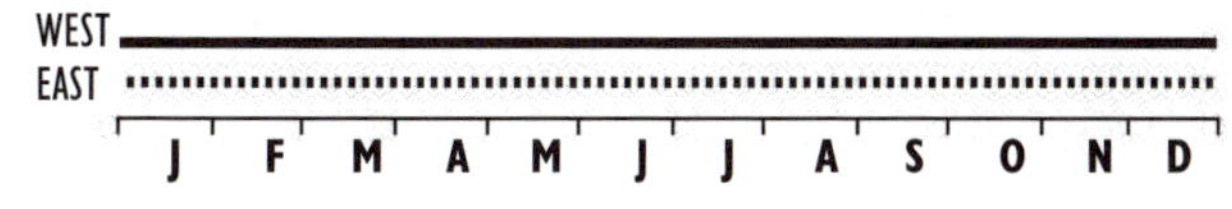

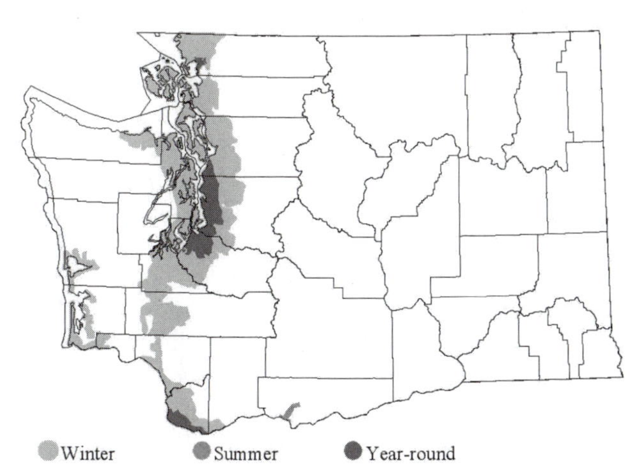

Winter Summer Year-round

Habitat: Urban gardens, parks, shrubs, less frequently in riparian habitats away from human habitation (e.g. Smith et al. 1997).

Occurrence: A sw., coastally occurring species with a noteworthy northward range expansion in the last 50 years. The availability of cover and plants such as winter jasmine in human-associated habitats, along with the increase in bird feeders, apparently explains the dramatic range expansion (see Zimmerman 1973, DeSante and George 1994, Russell 1996). First recorded in 1964 in Seattle, documented nesting in 1976 in Tacoma (Smith et al. 1997). Earlier nesting records at Victoria and Vancouver, B.C., in 1958 and 1959, respectively (Campbell et al. 1990b). First regional occurrence possibly as early as 1944 (see Campbell et al. 1990b, Wahl 1995). Core breeding range apparently from Edmonds to Tacoma, with records of wintering and confirmed but inconsistent records of possible breeders from Pt. Roberts through the lowlands s. to Olympia, in Lewis Co. and Ridgefield, w. to Clallam and Pacific Cos. (Smith et al. 1997), and nr. Lyle, along the Columbia R.

CBC records extend from Bellingham and Sequim-Dungeness s. in the Puget lowlands to Olympia. Abundance appears to be increasing. CBCs in Seattle, where bird feeders are numerous and well counted, found the species on all but four counts 1965-99 and counts increased steadily over time, to 54 in 1998 and 1999. Tacoma reported birds almost annually 1970-99, with a peak of 14 in 1998. The Victoria, B.C., CBC recorded 198 birds in 1999.

First recorded e. of the Cascades at Yakima in 1972 (Stepniewski 1999). Widespread, scarce records subsequently n. to Tonasket, e. to Spokane, Walla Walla, and Asotin, with most w. to White Salmon and Lyle during all months. CBCs recorded small numbers of birds at Spokane, Wenatchee, Yakima, Tri-cities, and Lyle. Breeding suspected nr. Klickitat, but first recorded at Ephrata in Mar 2002.

Noteworthy Records: East, *Winter*: 1 on 9-11 Jan 1975 at Yakima; 1 until 8 Mar 1975 each at Wenatchee and Cashmere; 2 in winter 1975-76 at Wenatchee and 1-4 until 4 Jan 1976 at Yakima; 1 on 13 Feb 1977 at Yakima; 1 on 10 Jan 1977 at Spokane; 1 wintered in 1989-1990 at Wenatchee; 1 until 8 Jan 1993 at Richland; 1 wintered at Kennewick, 1 nr. Yakima on 28 Dec-30 Jan, 1 at Yakima 10 Dec 2001-10 Feb. 2002, 1 at Cashmere in 2002-2003, 1 on 16 Feb 2003 at White Salmon. *Summer*: noted in 1974 at Cle Elum; 1 in 1978 at Husum; 1 on 5 Jun 1982 and 15 and 28 May 1987 at Tieton; reported in 1989 in the upper Wenatchee Valley; 1 on 20 Jul 1998 at Spokane. Nested in 2002 at Ephrata.

Terence R. Wahl

Costa's Hummingbird *Calypte costae*

Casual visitor.

Costa's Hummingbird has been recorded over 40 times in Oregon and at least eight times in B.C., but there are only five Washington records: a male (ph) nr. Frederickson, Pierce Co., early Aug-3 Oct 1998 ; a male (ph.) in Richmond Beach, 27 Apr-15 May 2000, and a male (ph.) in Redmond 18-24 May 2002. One bird (v.t.) was nr. Mt. Vernon on 15 May 2003. A report from Shelton on 14 Apr 1989, originally rejected by the WBRC (Tweit and Paulson 1994), was recently reviewed and accepted.

The first nw. record was in 1972 with birds noted in Oregon and B.C. (Campbell et al. 1990b, Gilligan et al. 1994). Following that, records appear to have steadily increased, with birds occurring annually in Oregon. Records there are year-round and are concentrated in the s.c. portion of the state. Peak period of Pacific Northwest records has been in Apr and May. Classed GL.

Steven G. Moldinow

Calliope Hummingbird *Stellula calliope*

Common summer resident and migrant on the e. slope of the Cascades, ne., and Blue Mts.; very rare in w. Washington.

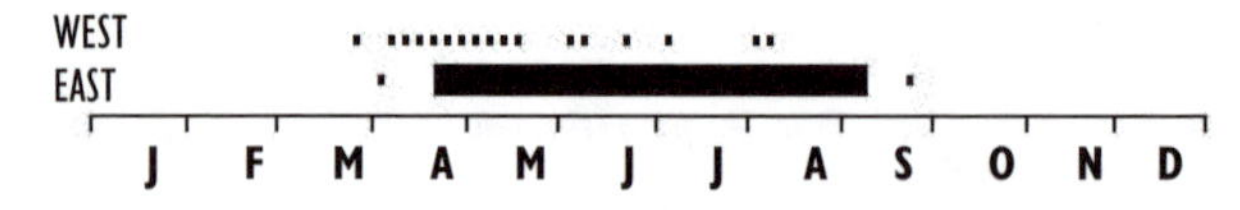

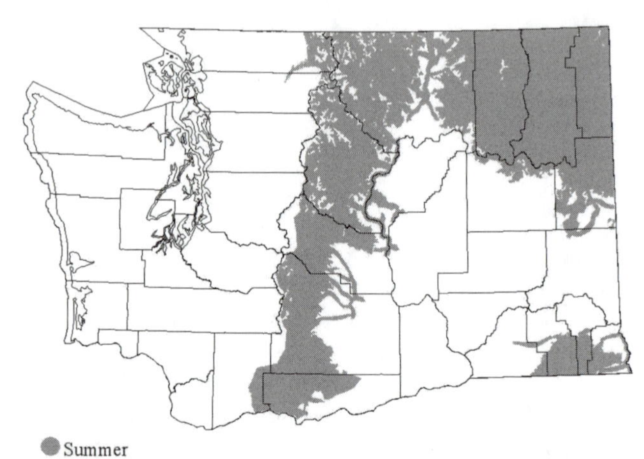

Habitat: Open montane forest, mountain meadows, and riparian thickets of aspen, willow, and alder in all forested zones below the subalpine fir zone (Smith et al. 1997), sometimes in shrubby habitats at low to medium elevations (Stepniewski 1999). Chaparral, lowland brushy areas, deserts, and semi-arid regions during migration and in winter.

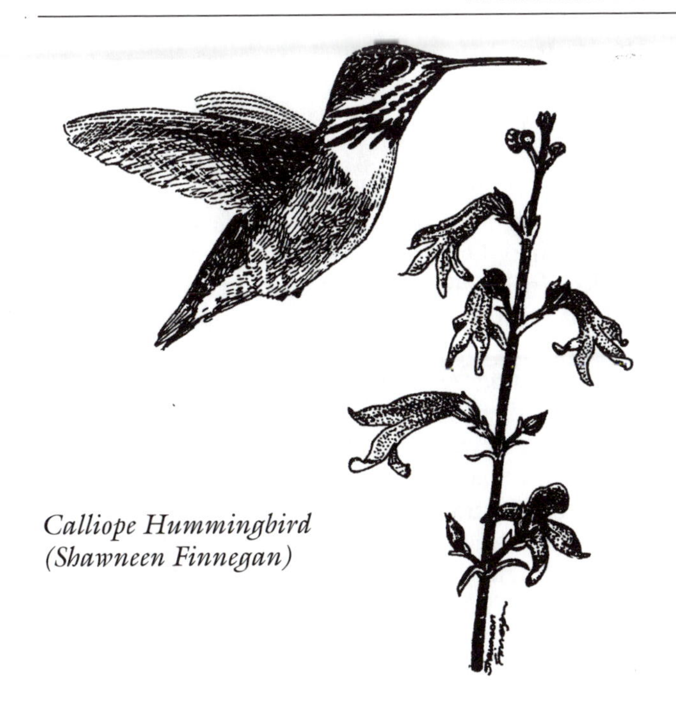

Calliope Hummingbird (Shawneen Finnegan)

Occurrence: The Calliope Hummingbird breeds mainly above 1200 m, from the U.S.-Canada boundary on the e. slope of the Cascades, e. to Pend Oreille Co. and in the Blue Mts. Occurs regularly in migration in the Columbia Basin. Essentially absent from w. Washington.

Dates of arrivals, departures, and peak migration periods in the state are poorly known, with remarkably few reports for a regularly occurring species. Reports indicate most arrive in mid-Apr e. of the Cascades, and there is a report of 65-70 at a feeder in Dixie, Walla Walla Co., on 24 May 1998. Reports of apparent migrants are scarce after mid-May. Males apparently leave breeding territories in Jun or Jul while females are still incubating, with few reports of any birds after early to mid-Sep, and with a late date of 22 Sep 1968 from Spokane considered noteworthy.

There are several reports of possible westside breeding, though migrant post-breeders are likely. A young bird was netted and examined in hand at Schrieber's Meadow, Mt. Baker, on 1 Aug 1986, and another seen 2 Aug in Mt. Rainier NP. Males displayed on territory at the same site in a pine microhabitat at the NP campground at Newhalem, Whatcom Co., in early Jun 1981-89, and a bird was noted at Colonial Cr. campground in Jul 1970 (Wahl 1995). This species typically migrates southward along the mountains and winters from Baja, California to s. Mexico (AOU 1983).

BBS data from 1966 to 2000 indicated a statewide decline (Sauer et al. 2001).

Remarks: Classed GL. Formerly; *Trochila (Calothorax) Calliope.*

Noteworthy Records: West, *Spring*: 1 on 26 Mar 1997 at Mt. Pleasant, Skamania Co.; 1 on 9 Apr at Washougal and 1 on 9 Apr 2000 at Olympia. *Summer*: 1 in Jul 1970, 1 on 2 Jun 1979, and territorial males in Jun 1981-1989 at Colonial Cr. campground, Whatcom Co. (Wahl 1995); 1 on 18 Aug 1981 nr. Enumclaw; 1 on 21 Jun 1986 along Swift Res., e. of Cougar, Skamania Co.; 1 on 8 Jun 1998 at Skykomish. *Fall*: young netted on 1 Aug 1986 at Schrieber's Meadow, Mt. Baker; 1 on 2 Aug 1986 in Mt. Rainier NP. **East**, *Spring, early*: 1 on 22 Apr 1963 and 1 on 20 Apr 1999 at Spokane. *High count*: 65-70 on 24 May 1998 at a feeder in Dixie, Walla Walla Co. *Fall, late*: 1 on 22 Sep 1968 at Spokane; 1 on 5 Sep 1999 at Richland.

Devorah A. N. Bennu

Broad-tailed Hummingbird *Selasphorus platycercus*

Casual vagrant.

There is one accepted Washington record of this interior N. American desert species: a female at Asotin, 25-27 Aug 2000. Several previous reports were rejected by the WBRC (Tweit and Paulson 1994, Tweit and Skriletz 1996, Aanerud and Mattocks 1997).

Broad-tailed Hummingbirds breed close to se. Washington in c. Idaho, and the bird in Asotin may have been dislocated by extensive forest fires in Idaho that summer.

Steven G. Mlodinow

Rufous Hummingbird *Selasphorus rufus*

Common migrant and summer resident in w. Washington, e. slope of the Cascades; fairly common in ne. and Blue Mts. Very rare in winter.

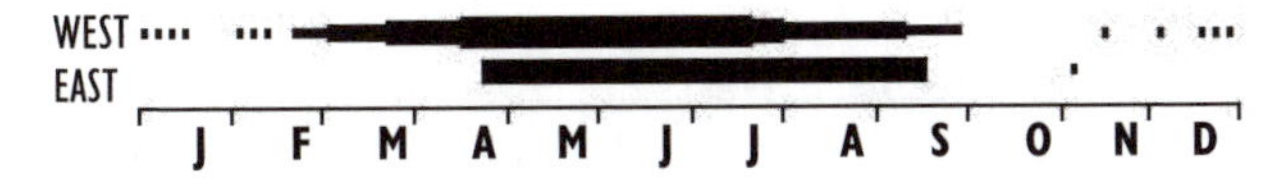

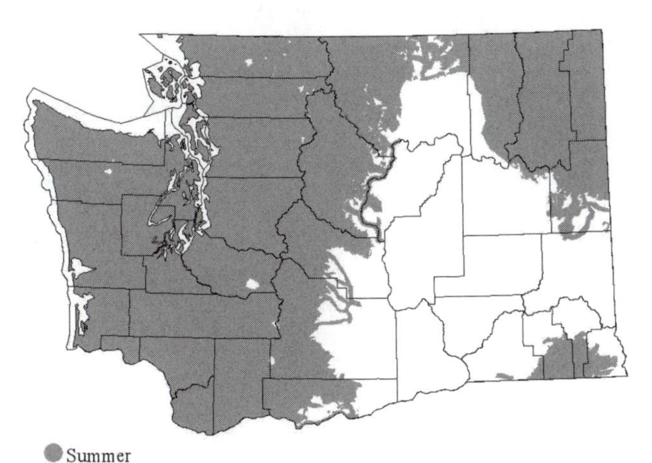

Habitat: Edge habitats nr. coniferous forests and second-growth, including thickets and brushy areas; also frequent found in meadows, by stream banks, chaparral, wetlands, and residential areas (Smith et al. 1997).

Occurrence: The northern-most occurring hummingbird, with the longest migration of any hummingbird (Terres 1980, AOU 1998). Birds are most common w. of the Cascades and found throughout much of the e.-slope Cascades, ne., and Blue Mts. forests in summer (see Smith et al. 1997; it was considered an "uncommon summer resident" in se. mountains by Weber and Larrison 1977). Very rarely recorded in winter on the westside.

On the westside, first ad. males usually appear in early Mar, but most arrive in late Mar with females appearing a week or two later. On the eastside, males have been reported in late Apr. Males depart the lowlands soon after breeding and are often noted in higher elevations while females raise the young.

Numbers decline sharply in Aug and birds are uncommon by Sep. Spring migrants tend to move mostly through the lowlands, and are noted especially foraging on early-blooming shrubs like the crimson-flowered currant and later on introduced plants and flowers of madrona trees and salmonberry (SGM), and also tree sap (Terres 1980). Fall migrants tend to migrate through mountain meadows where flowers are still blooming (e.g. Stepniewski 1999). Birds were noted 60-70 km off Grays Harbor in late Aug and Sep (TRW).

BBS data indicate a significant decrease in numbers between 1966 and 2000, with a decrease of 1.5% annually (Sauer et al. 2001).

Remarks: Winter records attributed to this species, especially those before the known occurrence of Anna's Hummingbird (see that account), may be questionable. Classed GL. Formerly *Trochilus rufus.*

Noteworthy Records: West, *late Fall-early Spring*: 1 male on 3-4 Nov 1977 on Mercer I.; 2 separately on 14 Nov 1968 in Bellingham; 1 on 12 Dec 1978 in Seattle; 1 on14 Dec 1979-29 Feb 1980 at a feeder with a pair of Anna's Hummingbirds in Seattle; female or imm. on 3 Dec 1992–31 Jan 1993 at Burien; 1998; 1 for several days in Jan 1974 at Seattle; 1 on 1 Feb at Bellevue 1998; 1 on 9 Feb 1996 at Aberdeen; 1 arrived on 11 Feb 1979 at Seattle. *Offshore*: 1 on 18 Aug 1970 about 72 km w. of the entrance to Juan de Fuca strait; 1 on 29 Aug 1971 64 km off Westport. **East**, *Spring, early*: 1 on 23 Apr 1976 at Spokane. *Summer*: birds in 1988 in Othello and at Columbia NWR led to suspicion of breeding. *Fall, late*: 21 Sep 1968 at Spokane; 1 until 18 Oct 1979 at Yakima.

Devorah A. N. Bennu

Allen's Hummingbird *Selasphorus sasin*

Casual vagrant.

There is one known Washington record of this species, which breeds along the Pacific coast s. of Cape Blanco, Oregon, and winters primarily in w. Mexico. Its relatively short migration range and difficult identification are factors that probably contribute to the scarcity of records n. of the regular breeding range. A male was collected at Seattle 27 May 1894 (Tweit and Skriletz 1996), and identified as *S. s. sasin*. Jewett et al (1953) refer to a second specimen, apparently destroyed in 1885.

Allen's constitutes a superspecies with Rufous Hummingbird, *S. rufus*, and green-backed males of that species need careful distinction. Field reports from Apr 1979 nr. Dungeness, in 1983 at Poulsbo, Redmond, and Ocean Shores were uncertainly identified and one from Willapa Bay was *rufus*. A *rufus* specimen with a solid green back was found in late Apr 1987 on Stuart I., San Juan Co. (to Burke Museum, UW). Classed GL.

Bill Tweit

Belted Kingfisher *Ceryle alcyon*

Fairly common breeder in w., uncommon breeder in e. Uncommon on outer coast and at high elevations. Fairly common breeder and migrant, uncommon in winter in e.

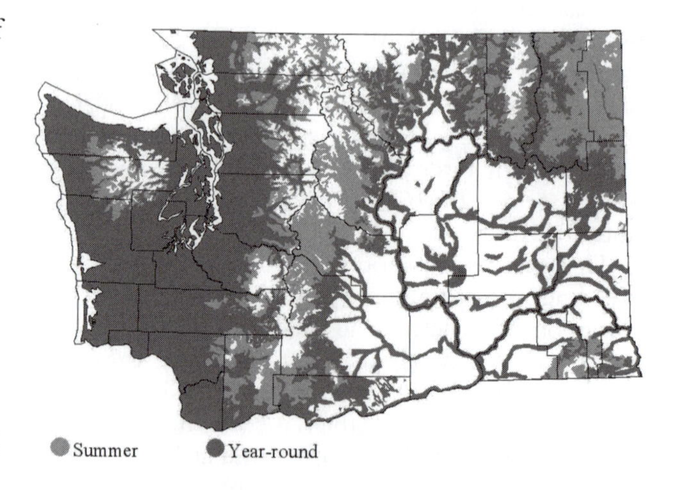

Subspecies: *C. a. caurina.*

Habitat: Clear, calm streams, rivers, ponds, lakes, estuaries, and marine shorelines where prey can be seen (Hamas 1994). Uses sand, clay, or gravel banks for nesting. Winters in ice-free areas, avoiding turbid and fast-flowing waters. May rely upon human-altered landscapes where few natural nest sites are available; gravel pits, landfills, and road cuts supported most nests in a n.c. Minnesota study (Hamas 1974).

Occurrence: Most frequent along streams, rivers, ponds, and inland marine habitats, but less numerous along outer coast embayments and waterways. Not normally found at high-elevation lakes, occurring below subalpine fir and mountain hemlock zones (Smith et al. 1997). Distribution limited primarily by nest sites in summer and prey availability in winter (Davis 1980).

Less common in winter, particularly in the e., where the bulk of the population migrates (Jewett et al. 1953). Post-breeding and post-fledging movements begin in midsummer, with obviously increased numbers along streams.

During the 1920s, kingfishers usually were present in the vicinity of Washington fish hatcheries, but numbers were kept low (Guberlet 1930). They continue to be named as a predator at many hatcheries in the w. U.S. They are shot or trapped at some (Parkhurst et al. 1987, Pitt and Conover 1996), and 21 were killed in 2001 during a salmonid-protection program on the Columbia R. (USDA 2003).

No significant trend apparent in BBS data between 1966 and 2000 (Sauer et al. 2001). Kingfishers are reported from most CBCs across the state, with annual totals regularly reaching 400 or 500. The highest counts come from circles encompassing inland marine waters. Consistently low counts are made at coastal locations and at high elevations. CBC data indicate an increase ($P < 0.01$) in winter numbers e. of the Cascades from 1970 to 1999.

Remarks: Kingfishers are sensitive to human activity and may avoid or vacate disturbed areas, particularly during the nesting season (Hamas 1994). They apparently have not been negatively affected by organochlorine contaminants, such as DDT and PCBs (Hamas 1994). Effects of increasing development on distribution and abundance, patterns of migration, and origins of winter populations require study.

Noteworthy Records: 200 noted on 3 Aug 1972 along the Cowlitz R. from the hatchery to Toledo, a stretch of about 13 km (B. Smith p.c.); 1 ca. 42 km off Grays Harbor on 31 Jul 2004 (BT).

Scott A. Richardson

Lewis' Woodpecker *Melanerpes lewis*

Locally common to uncommon summer resident, rare to locally common winter resident in e. Washington. Rare migrant and very rare winter visitor w.

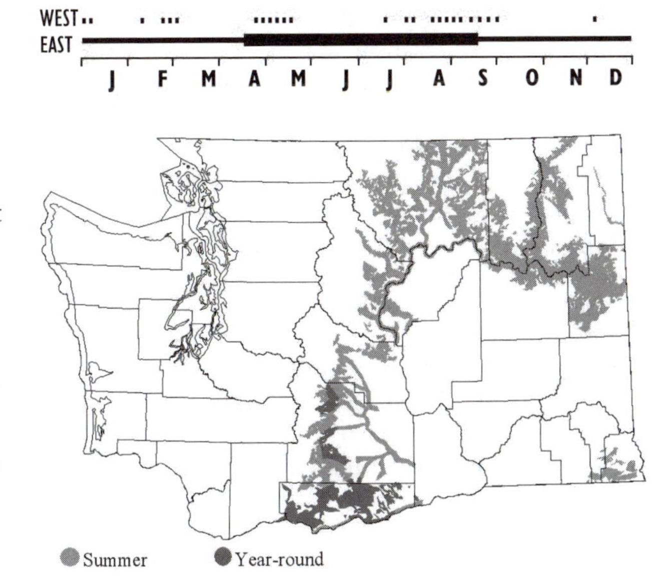

Habitat: Breeds in open canopied forests or woodlands with a brushy understory and large snags. Ponderosa pine forests are typically preferred habitat at higher elevations, while riparian woodlands dominated by cottonwoods, oak woodlands, cultivated orchards, field edges, urban habitats and recently burned pine stands are selected at lower elevations. Winters in oak woodlands, particularly in s.c. Washington, and commercial orchards (Bock 1970, Tobalske 1997, Stepniewski 1999).

Occurrence: Spring migration typically from mid-Apr to early May (Bock 1970). Early arrival dates reported as 5 Mar in se. (Hudson and Yocom 1954, Weber and Larrison 1977) and 18 Apr in Okanogan Valley in 1990. Birds then found in forests and steppe/shrub-steppe mosaics along the edges of the pondersosa pine zone of the e. flanks of the Cascades, in the Blue Mts. and the Okanogan Highlands (Smith et al. 1997) and oak woodlands of Klickitat and Yakima Cos. (Stepniewski 1999).

Breeds in Chelan, Klickitat, Lincoln, Okanogan, and Yakima Cos.; probably in Grant Co.; possibly in Douglas, Ferry, Kittitas, Pend Oreille, and Spokane Cos., with breeding likely in Stevens Co. (Smith et al. 1997). Formerly common in summer in the Blue Mts. (Booth 1952, Hudson and Yocom 1954, Weber and Larrison 1977); no longer found in Columbia and Walla Walla Cos. Nesting birds in Asotin Co. represent the last lingering population in the Blue Mts. region (Smith et al. 1997).

Breeding in w. Washington was likely always sporadic (Alcorn 1959b), but decreased following the 1950s. Recent records, mainly in early-late May and with most from mid-Aug to late Sep, average about 1.5 per year (SGM). Jewett et al. (1953) listed westside breeding/summer records for Clark, Gray's Harbor, Pacific, Skamania, Wahkiukum, and Whatcom Cos. (e.g., see Wahl 1995). An "abundant" breeder in sw. B.C. 1920-40 (Campbell et al. 1990b), last nested at Victoria in 1963. Nearby, it was once an uncommon summer resident in the San Juan Is., but now strictly a rare winter visitor and migrant (Lewis and Sharpe 1987).

Timing of fall migration varies locally, but birds generally depart in Aug or early Sep (Bock 1970). Jewett et al. (1953) observed that migration began as early as mid-Aug and peaked in Sep. The latest fall record listed in se. Washington is 6 Dec (Hudson and Yocom 1954, Weber and Larrison 1977).

During winter, food supply is apparently the most important aspect of habitat selection. Local occurrence is dependent upon irregular yearly mast production, with virtual absence in winter 1991-92, for example (and see Bock 1970, Winkler et al. 1995, Tobalske 1997, Stepniewski 1999). CBC records in the late 1990s show concentrations in the oak area at Lyle and Columbia Hills-Klickitat.

BBS data suggest a slight decline over N. America from 1966 to 1999 (Sauer et al. 2000). Based on Washington routes, a decline of 9.5% per year is indicated (Sauer et al. 2001). Irregular occurrence and insufficient CBC records prevent analysis of winter trends in Washington. Range-wide there has been a non-significant 1.0 % winter decrease annually between 1959 and 1988 (Sauer et al. 1996).

Remarks: Much of the westside decline may be attributed to habitat alterations (Smith et al. 1997). Modern forest-management practices have replaced natural fire cycles, thereby reducing suitable habitat, especially on the westside. Local declines may also be due to competition with starlings for nest cavities (Smith et al. 1997). Classed SC, PHS, GL.

Noteworthy Records: *CBC high counts*: 158 in 1999, 81 in 1997, and 58 in 1998 at Lyle; 35 in 1998 and 21 in 1999 at Columbia Hills-Klickitat. **West**, *reports since 1980*: 25 Sep 1981 at Redmond; spring 1991 at "Puget Sound"; 9 May 1992 at Nisqually NWR; 14 Aug 1993 in San Juan Is.; 18 Sep 1993 at Granite Falls; 29 Aug 1996 at Port Angeles; 30 Aug 1996 at Nisqually NWR; 8 Sep 1996 at Kent; 12 May 1998 in Skagit Co.; 20 Aug 1998 at Seattle; 29 Aug 1998 on San Juan I.; 12 Sep 1998 at Kirkland; 6 May 1999 at Little Rock, Thurston Co.; 8 May 1999 at Swauk Prairie, Skagit Co.; July 2000 at Little Rock; 17 Sep 2000 at Bellingham; 6-15 May 2001 at Nisqually NWR; 27 Sep 2001 at Brady; 1 on 10-24 Feb 2002 nr. Joyce, Clallam Co.

Michael S. Husak

Acorn Woodpecker *Melanerpes formicivorus*

Very localized, uncommon resident in Klickitat Co.

Subspecies: *M. f. bairdi.*

Habitat: Foothill and montane woodlands, closely associated with oak trees.

Occurrence: The first accepted occurrence of Acorn Woodpeckers in Washington came in Mar 1979 nr. Ft. Simcoe, with all subsequent records from s. of that. Members of a flock of nine individuals found nr. Lyle in October 1989 remained in the area to nest the following spring, the first breeding record for the state. Birds were thought to have abandoned the Lyle area in 1992 (Smith et al. 1997) but the small population was later relocated. Probable nesters have also been observed nr. Balch L. (Smith et al. 1997).

Washington birds occur in a timbered, arid transition zone e. of the Cascades, nr. and along the Klickitat and Columbia Rs. This xeric region of the Columbia Gorge is dominated by ponderosa pine-garry oak forests and scattered prairies (Franklin and Dyrness 1973). In Washington, nesting behavior is poorly described but probably comparable to other N. America populations.

In the nw. U.S., this species appears to be expanding its range northward with recently established, isolated populations in Klickitat Co. and nr. Hillsboro and Yamhill, Oregon (FN 49:303). BBS data for 1966-96 indicate a slight continent-wide increase (0.5% per year) in populations (Sauer et al. 1997). CBC data indicate this is often the most abundant woodpecker species within its range of distribution, with a maximum density of 7.96 individuals per hr (Root 1988), though Washington occurrence is infrequent and in small numbers.

Noteworthy Records: sighting on 29 Sep 1978 at Washougal probably 1st state record; 25 Mar 1979 at Ft. Simcoe; single sighting 3 May 1979 at Lyle; 1st winter record in 1989-90 at Lyle; 1st breeding record for state in 1990 at Lyle; reappearance of Klickitat population 1995-96 at Lyle; pair in Jun 1995 at Bickleton (Smith et al. 1997). *CBC*: 3 in 1997, 2 each in 1998 and 1999 at Lyle.

Michael S. Husak

Williamson's Sapsucker *Sphyrapicus thyroideus*

Fairly common summer resident on the e. slopes of the Cascades, the Okanogan highlands, and in Blue Mts. Very rare in winter.

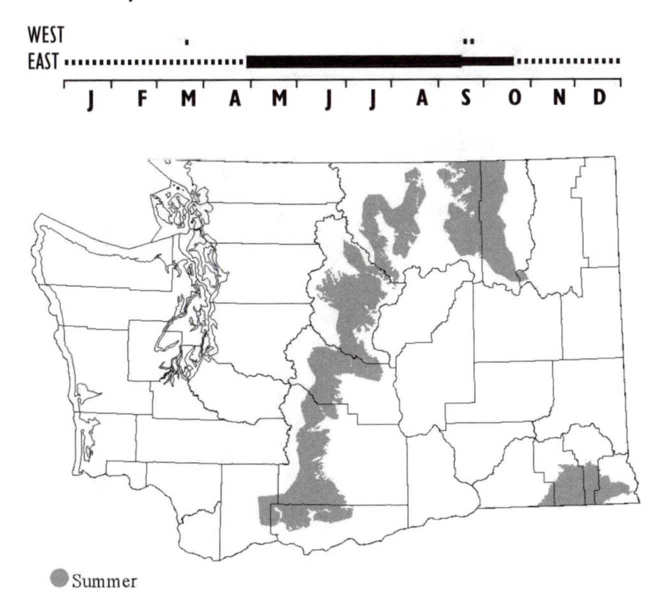

Subspecies: *S. t. thyroideus.*

Habitat: Open coniferous and mixed-coniferous subalpine forests (Smith et al. 1997), at low to mid elevations in the Cascades, at all elevations in the Okanogan and Blue Mts. In winter and migration, also in pines and pine-oak woodlands. Considered an indicator species because of its specific habitat requirements (Dobbs et al. 1997): western larch must be a component of the forest (Smith et al. 1997).

Occurrence: Males arrive and establish territories one to two weeks before females arrive (Dobbs et al. 1997). Birds are then common in summer in grand fir and Douglas-fir zones, and ponderosa pine zones in the e. Blue Mts. and Okanogan Highlands and relatively uncommon along the e. slope of the Cascades, Okanogan and Methow valleys, and considered an uncommon summer resident on the w. Blue Mts. (Hudson and Yocum 1954, Weber and Larrison 1977, Smith et al. 1997). Confirmed breeding in Asotin, Chelan, Garfield, Kittitas, Okanogan, and Yakima Cos. and probably breeding in Columbia Co. (Smith et al. 1997). Very rare in winter in Washington—most move s., likely to c. California. Females are more nomadic than males and tend to move further s. (Short 1982).

Williamson's Sapsuckers have been declining range-wide, with the strongest declines occurring in the Pacific Northwest (Dobbs et al. 1997). From 1966 to 2000, populations of Williamson's Sapsuckers in Washington have exhibited a negative trend (-0.5% per year; Sauer et al. 2001). Though over all of N. America there was a positive trend from 1966 to 1979 (35.2%), populations declined from 1980 to 1999 (-1.8% per year; Sauer et al. 2000).

Remarks: Because of declines in population numbers, Williamson's Sapsuckers are considered sensitive species by wildlife agencies in Oregon and Utah (Dobbs et al. 1997).

Noteworthy Records: West: 1 dead on 15 Mar 1981 at Seattle; 1 on 21 Sep 1982 at Hurricane Ridge, Olympic NP; 1 on 17 Sep 1994 on Bainbridge I. **East**, *Winter*: 1 dead on 30 Dec 1987 nr. Battle Ground; 1 on 23 Nov and 1 Dec 1985 nr. Sacheen L., Pend Oreille Co.; 1 on 15 Dec 1987 nr. Leavenworth; 1 on 19 Dec 1998 at Kennewick; 1 on 19 Dec 1998, 1 on 28 Dec 2001 and 1 on 11 Jan-11 Feb 2003 at Trout L.

H. Dawn Wilkins

Yellow-bellied Sapsucker *Sphyrapicus varius*

Casual visitor.

There are few records for this species in Washington, though it breeds in n. B.C. and is a regular fall-winter vagrant in California. The first record was an ad. at Ellensburg 16 Dec 1989-18 Feb 1990 (Tweit and Paulson 1994) and the second was one at Pe Ell 28 Feb-15 Mar 1997. The third was one at Sacagawea SP on 1 Oct 1999 and fourth was one at Kent 30 Dec 2001-5 Jan 2002 (SGM).

Sapsucker species were separated in the 1980s (AOU 1983, 1985) and field reports previous to that may be confusing (see Hamilton and Dunn 2002). A bird on 13 Oct 2001 at Hood Park, Walla Walla Co., was probably a Yellow-belliedXRed-naped sapsucker (SGM).

Bill Tweit

Red-naped Sapsucker *Sphyrapicus nuchalis*

Fairly common summer resident in e. Washington, very rare in winter. Rare in summer along w. side of the Cascades; very rare in w. lowlands in winter.

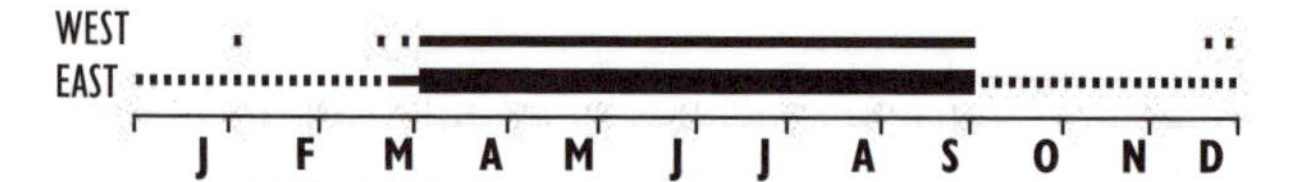

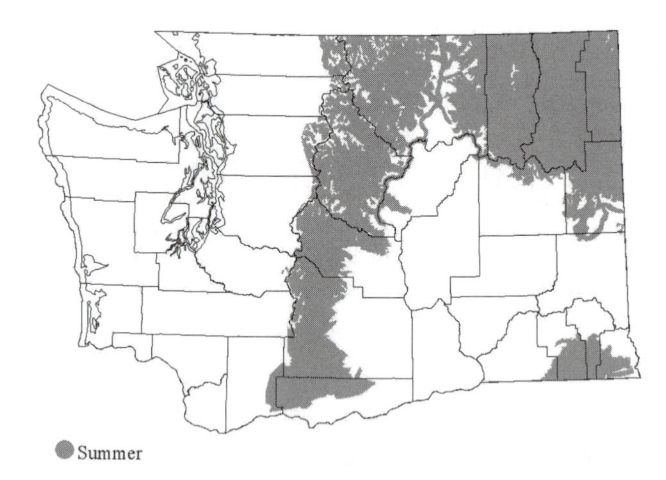

Red-naped Sapsucker (Shawneen Finnegan)

Habitat: All forested zones of e. Washington and coniferous forests e. of the Cascade crest, in high elevations w. of the crest (Smith et al. 1997), also in deciduous forests, including aspens (Stepniewski 1999). Broadleaf forests, riparian woodlands, and parks and gardens in winter.

Occurrence: Variously described as a summer resident and very rare winter resident by Jewett et al. (1953), common during early winter in the n. Cascades by Burdick (1944), rare in summer in se. Washington (Hudson and Yocum 1954), where Weber and Larrison (1977) classed them uncommon migrants and very uncommon summer residents. It now is described as occurring widely on the e. slope of the Cascades, ne. counties, and Blue Mts., and locally w. of the Cascade crest (Smith et al. 1997). Distribution today is somewhat uncertain due to hybridization with the Red-breasted Sapsucker, probably throughout the Cascades where both species occur. Hybrids have been found widely in winter and migration, w. at least to Ridgefield and e. to Douglas Co. (SGM) and some birds identified as Red-naped may have been hybrids.

Most birds arrive in spring between the last week in Mar and late Apr (Howell 1952) and depart during the early to mid-Sep (M. Denny p.c.). Most westside records are from Apr to mid-May.

From 1966 to 1999, populations of Red-naped and Red-breasted sapsuckers have increased in Washington (+1.9%; Sauer et al. 2000). Across N. America, sapsucker populations have increased (0.3%; Sauer et al. 2000). One to three Red-naped Sapsuckers were reported eight times on CBCs between 1985 and 1997, with six reports in westside lowlands and at Chelan and Twisp in the e.

Remarks: Formerly a subspecies of the Yellow-bellied Sapsucker, the Red-naped Sapsucker is now considered a separate species (Johnson and Zink 1983, Johnson and Johnson 1985, AOU 1998) genetically more closely related to Red-breasted Sapsuckers than Yellow-bellied Sapsuckers (Cicero and Johnson 1995; but see Hamilton and Dunn 2002).

Noteworthy Records: West, *lowlands*: 9 Apr 1985 at Seattle; 28 Sept 1985 on San Juan I.; 12 Apr 1987 at Seattle; 19 Mar 1988 at Carnation; 20-27 Apr 1988 at Waldron I.; 10 Apr 1989 at Seattle; 20 Apr 1991 nr. Elma; 9 Jun 1991 at Big Four Marsh, Snohomish Co.; 4 Jul 1992 e. of Mt. Vernon; pair on 25 Jun 1995 nr. Rockport; 16 Jun 1996 at Beckler R., Snohomish Co.; 1 on 5 Feb 1997 at Kirkland; 1 on 28 Mar 1997 at Johns R, Grays Harbor; 1 on 19 Apr 1997 at Seattle; 1 on 4 Jul 1998 at Joyce, Clallam Co.; 1 on 20 Apr 1999 at McCleary; hybrid on 19 Dec 1999 at Ridgefield; 1 on 8 Apr 2000 at Olympia; 1 on 16 May 2000 at Ridgefield; 1 on 24 Sep 2000 at Tokeland; 1 on 17 Dec 2000 at Ridgefield; 1 on 25 Apr 2001 at Lummi I.

H. Dawn Wilkins

Red-breasted Sapsucker *Sphyrapicus ruber*

Fairly common summer resident w. of the Cascade crest, uncommon in winter. Local, uncommon in summer, very rare in winter on e. slope of the Cascades; very rare elsewhere e.

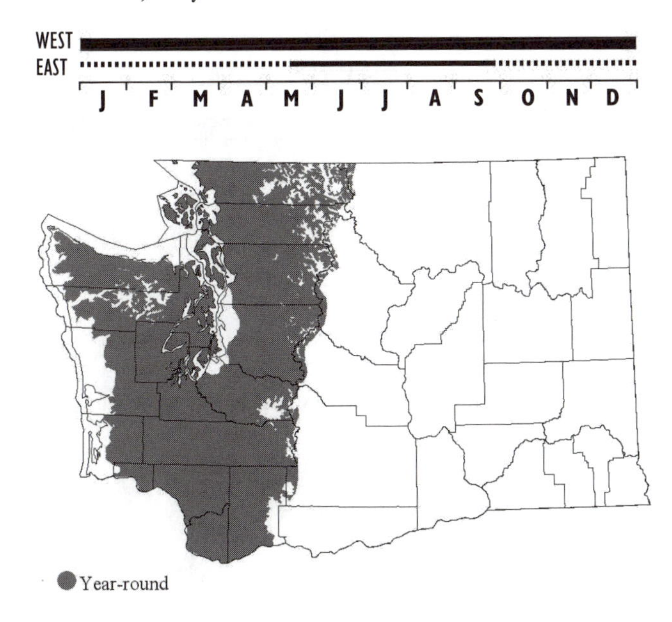

Subspecies: *S. r. ruber*.

Habitat: Wet coastal coniferous forests in the Puget Trough and the subalpine forest zones of w. Washington, uncommon breeder in broadleaf forests and riparian woodlands, and uncommon in winter in urban habitats. Coniferous forest, aspen parklands, aspen-ponderosa, and yellow pine woodlands of the e. Cascades (Smith et al. 1997).

Occurrence: Though range-wide some birds migrate for the winter (AOU 1998), many westside Red-breasted Sapsuckers are year-round residents (Jewett et al. 1953). Spring migration begins around the first week of Apr, and birds are widespread in westside forests in the breeding season. Recorded or probable throughout w. Washington, they are local breeders just e. of the Cascade crest from May to Oct, fairly common in sw. Yakima Co. in summer (Stepniewski 1999), and there are "wandering" breeders in s.c. Okanogan Co., Blewett Pass, and Brooks Memorial SP (Smith et al. 1997).

The beginning of fall migration is unknown. Winter numbers are variable and irregular (e.g. Wahl 1995), perhaps due to downslope migration during severe winters, with more reportedly occurring on Seattle CBCs then (Hunn 1982), and populations of sapsuckers fluctuate dramatically in B.C. (Campbell et al. 1990b). Variations may reflect winter survival and subsequent breeding populations.

From 1996 to 2000, populations of Red-naped and Red-breasted sapsuckers reportedly increased in Washington (+1.96%; Sauer et al. 2001). Across N. America, populations of Yellow-bellied, Red-naped, and Red-breasted sapsuckers increased (0.3 %; Sauer et al. 2000). CBC data for Washington suggest no trend.

Remarks: Formerly considered a subspecies of the Yellow-bellied Sapsucker, now a separate species (Johnson and Zink 1983, Johnson and Johnson 1985, AOU 1998) genetically more closely related to the Red-naped Sapsucker (Cicero and Johnson 1995). Eastside species occurrence remains uncertain in some areas—this species is believed to hybridize with Red-naped in Yakima Co. (Stepniewski 1999). Washington status of the s. subspecies, *S. r. daggetti*, reported on the e. slopes of the Cascades by Jewett et al. (1953), is uncertain.

Noteworthy Records: East: 1 on 30 Mar 1969 nr. Cle Elum; 1 on 14 Jun 1981 nr. White Pass; 1 on 24 Jun 1981 at L. Kachess; 1 on 8 Oct 1981 36 km e. of Wenatchee; 2 on 19 Jun 1982 at Rimrock L.; 1 in late Jun 1982 sw. of Republic; 1 on 27 May 1985 at Wenas Cr.; 1 in fall 1987 at Fields Spring SP; 1 on 28 May 1988 at Ahtanum Cr.; 1 on 7 Oct 1992 at Liberty L., Spokane Co.; 1 on 5 Dec 1992 at Windust Park, Franklin Co.; 1 on 17 Dec 1994 at Pasco; 1 on 28 Jun at Rimrock L. and 1 on 3 Jul 1998 at Leech L., Yakima Co.; 1 on 9 May at Rattlesnake Springs and 1 on 9 May nr. Rimrock L., Yakima Co., and 2 on 25 Jun 1999 at Elk Meadows CG, Klickitat Co.; 2 on 4 Jun 2000 nesting at Conboy L. NWR; 2 on 25 Jun 2000 at Rimrock L.; 1 on 3 Dec 2000 and on 27 Jan 2001 at Bingen; 3 on 24 Sep 2001 at Lower Trout L. Valley, Klickitat Co.; 2 in winter 2001-02 in Walla Walla Co.

H. Dawn Wilkins

Downy Woodpecker *Picoides pubescens*

Common resident statewide in lower elevations, rare or absent above 900 m elevation.

Subspecies: *P. p. gairdneri* in coastal forested areas; *turati* of the Cascades e. slope intergrades with *leucurus* of inland e. Washington. The northern, *nelsoni,* reported once w. of the Cascades (Jewett et al. 1953).

Habitat: Riparian woodlands and lowland broadleaf forests throughout, uncommon in shrub and urban habitats. Mostly in riparian corridors in lowland e. Washington. Noted foraging in winter at non-woody plants in scrub and fields (Godfrey 1986).

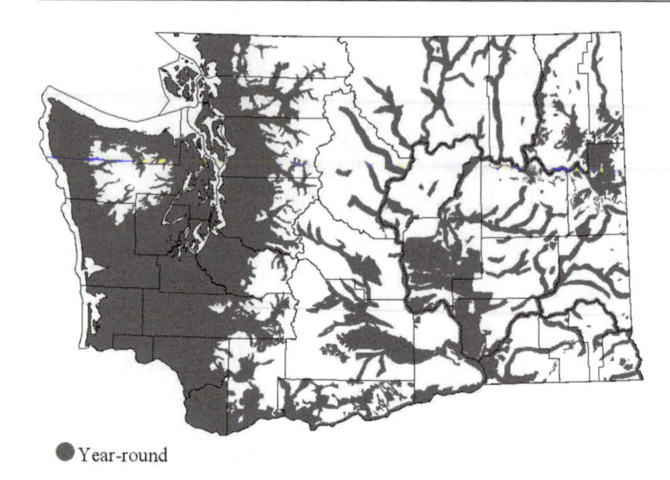

Occurrence: Noted in near-wilderness but predominantly in lowlands including urban areas. A species that benefitted from habitat changes resulting from human settlement in w. lowlands due to replacement of softwood forest by hardwoods (Smith et al. 1997). Downy Woodpeckers are common permanent residents throughout, though more common in e. Puget Sound lowlands than on the Olympic Pen. Occasional to upper limits of the Canadian zone (Jewett et al. 1953). Uncommon in se. Washington (Hudson and Yocom 1954, Weber and Larrison 1977, Smith et al. 1997).

Surveys from 1982 to 1997 confirmed or found probable breeding widely across the state with evidence of nesting lacking only in the se. corner in Adams, Benton, Franklin, and Garfield Cos. Local and uncommon in open, nearly treeless areas in the e. Columbia Basin and the Palouse (Smith et al. 1997).

BBS trends in Washington and N. America are divergent (Sauer et al. 2000). In Washington, from 1966 to 1999, Downy Woodpeckers exhibited a positive trend (+2.1 % per year), while the 30-year trend across N. America was a decrease (-0.2 % per year). From 1982 to 1996, BBS data indicated an increase (> +0.25 % per year) in urban areas, the Olympic Pen., and w., s.c., and c. Washington, and declines (-0.25 % per year) in e., ne., and n.c. Washington, as well as the Upper Cascades (Sauer et al. 2000). Few Downy Woodpeckers (0–1) were observed per BBS route throughout c. and e. Washington, as well as the Olympic Pen. while more (2–3) were observed per route in w. and sw. Washington and in urban areas (Sauer et al. 2000).

Though wandering or down-slope winter movements occur (e.g. Stepniewski 1999), the species is resident. CBC data indicated very low abundance (<1 per party hr) in se. Washington and the tip of the Olympic Pen. and relatively low abundance (2–3 per party hr) elsewhere in the state (Sauer et al. 1996). High counts were usually from urban-centered circles with the most intensive coverage (e.g., 86 in 1998 at Seattle) and long-term CBC data indicate a significant increase (P <0.01) statewide.

Noteworthy Records: *CBC high counts*: 74 in 1992 at Bellingham; 23 in 1983 at E. Lk. Washington; 22 in 1996 at Edmonds; 70 in 1990 at Ellensburg; 36 in 1997 at Everett; 33 in 1983 at Grays Harbor; 62 in 1993 at Kent-Auburn; 43 in 1997 at Olympia; 44 in 1993 at Padilla Bay; 23 in 1999 at Port Gamble; 50 in 1989 at Skagit Bay; 25 in 1998 at Sequim-Dungeness; 86 in 1998 at Seattle; 27 in 1963 at Spokane; 54 in 1993 at Tacoma; 22 in 1998 at Wahkiakum; 69 in 1993 at Yakima Valley.

Douglas R. Wood

Hairy Woodpecker *Picoides villosus*

Fairly common to common permanent resident.

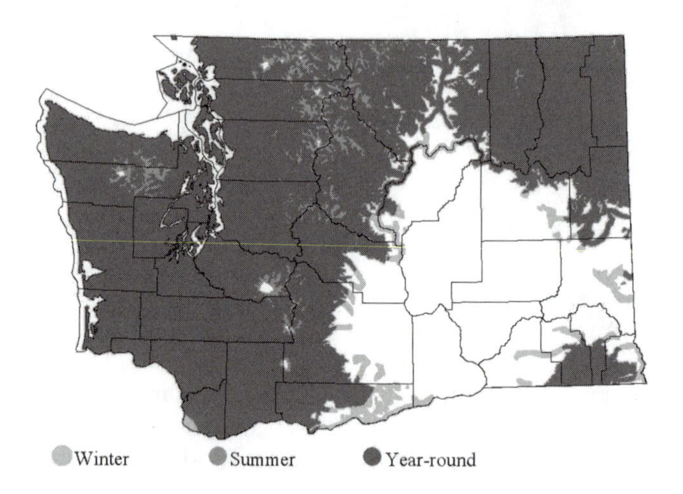

Subspecies: *P. v. harrisi* in coastal areas and *orius* in the Cascades.

Habitat: Commonly inhabit low- and mid-elevation dry and wet coniferous forests, less commonly subalpine forests, uncommon to locally fairly common seasonally in broadleaf forest and riparian areas (Franklin and Dyrness 1973, Smith et al. 1997, SGM). Uncommon in urban/suburban habitats statewide, rare or absent from grassland, desert, and agricultural habitats in e.c. Washington.

Occurrence: Hairy Woodpeckers are resident across N. America (Sauer et al. 2000), from c. Canada and Alaska s. into Mexico and C. America (Short 1982). They are resident from sea level to 2700 m in the Pacific Northwest, with altitudinal migration from higher to lower elevations occuring in the w. and n. range (Jewett et al. 1953, Short 1982, Stepniewski 1999). In surveys from 1982 to 1995, the species was a confirmed, probable, or possible breeder statewide except in the Columbia Basin with no breeding evidence documented in Adams, Benton, Douglas, Franklin, and Grant Cos. (Smith et al. 1997). Hairy Woodpeckers are generally separated by habitat from Downies which most often frequent deciduous and riparian habitats (Lewis and Sharpe 1987, Smith et al. 1997, Stepniewski 1999).

Jewett et al. (1953) classified Hairy Woodpeckers as common or fairly common residents of w., ne., and se. Washington. Hudson and Yocom (1954) reported birds were uncommon residents in se. Washington, though Weber and Larrison (1977) classified the species as a fairly common resident of forested areas there. Downslope movements in winter noted in some regions (Jackson et al. 2002).

BBS data indicated an increasing trend for Hairy Woodpeckers in Washington (Sauer et al. 2000) with an increase (+1.0 % per year) 1966–99, including a sharp increase (+19.1 % per year) 1966-79. The trend moderated from 1980-99 (+1.3 % per year). These parallel trends across N. America, though the national trend 1966-79 (+4.2 % per year) was markedly lower than that for Washington. Trends varied between areas, with declines noted in the upper Cascades and in c. Washington (Sauer et al. 2000). Birds are widespread in low abundances on CBCs (Sauer et al. 1996). Long-term CBC data suggest a westside increase.

Remarks: Short (1982) recognized *harrisi*, *orius*, and *septentrionalis*, a n. race that extends southward into n. and e. Washington (see Campbell et al. 1990b, Smith et al. 1997).

Douglas R. Wood

White-headed Woodpecker

Picoides albolarvatus

Uncommon to locally fairly common resident in ponderosa pine forest on e. slope of the Cascades, ne. mountains and Blue Mts. Very rare in w. Washington.

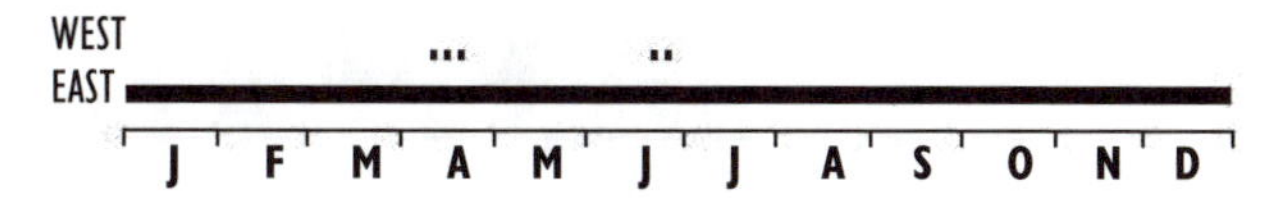

Subspecies: *P. a. albolarvatus.*

Habitat: Coniferous forests in ponderosa pine zones, occurs in adjacent open forest in Douglas-fir zone in ne. mountains and occasionally to 1000 m elevation (Smith et al. 1997). Infrequent in other habitats.

Occurrence: A species probably as closely associated with one vegetation type as any in the state, the White-headed Woodpecker is locally fairly common on the e. slopes of the Cascades, particularly the Yakama Indian Reservation in Yakima and Klickitat

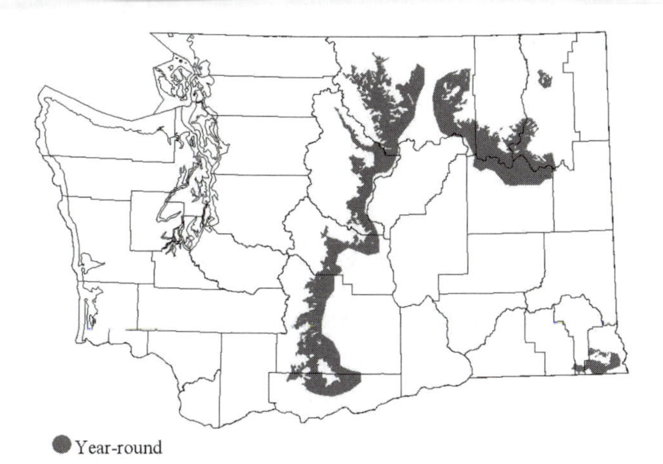

Cos. (RHL, Marshall 1997). Regular at several Yakima (Stepniewski 1999) and Chelan (Marshall 1997) Cos. sites. Locally uncommon in Kittitas and Okanogan, very local in the Blue Mts. (Smith et al. 1997). Reported from Ferry, Stevens, and Spokane Cos. (Smith et al. 1997).

Washington is essentially the n. limit of the species' range, which extends to s. California. A peripheral population is local in interior B.C. (Campbell et al. 1990b), and abundance may increase from n. to s. through its range (Marshall 1997).

Seasonal numbers and distribution are incompletely known. Records reflect survey effort, which has been local. Summer concentrations are known from the Yakama Indian Reservation (Marshall 1997), where birds were first recorded in 1977. Surveys from 1991 to 1998 totaled 49 records, including some nests (RHL) with locations fairly widely distributed across the Yakama Indian Reservation and dates ranging from Mar to Oct. Winter status there is uncertain.

Migration, dispersal, and altitudinal shifts are undescribed, though Stepniewski (1999) notes rare downslope movements in winter. Birds were recorded infrequently on CBCs from 1959 to 2000, with five occurrences at Twisp, three each at Walla Walla and Camas Prairie-Trout L., and nine elsewhere in e. Washington. Otherwise, there have been fairly widespread reports from low elevations nr. breeding habitats. The lack of reports from the Columbia Basin and just four westside records also suggest migration is minimal if it occurs at all. Fewer winter reports are likely due to lower observer effort (Marshall 1997).

The current range is similar to historic descriptions (Marshall 1997), though numbers and local distribution appear to have declined with harvest in ponderosa pine (Smith et al. 1997), particularly of large-sized trees (Marshall 1997). Decreased numbers on the Spokane Indian Reservation were attributed to loss of big pines (Acton, in Stepniewski 1999). Classed SC, PHS, GL.

Noteworthy Records: West: 1 on 9 and 20 Apr 1988 at Weir Prairie nr. Rainier, Thurston Co.; 1 on 25 Jun 1999 nr. Nisqually; 1 on 23 Jun 1989 nr. Glacier, Whatcom Co.; 1 on 15 Apr 1995 at Orcas I.

Rosemary H. Leach

Three-toed Woodpecker *Picoides dorsalis*

Uncommon, locally fairly common, resident in high-elevation coniferous forests in Cascades, mountains across n. counties to ne. corner, and in Blue Mts. Local and rare in Olympics. Rare on w. side of the Cascade crest.

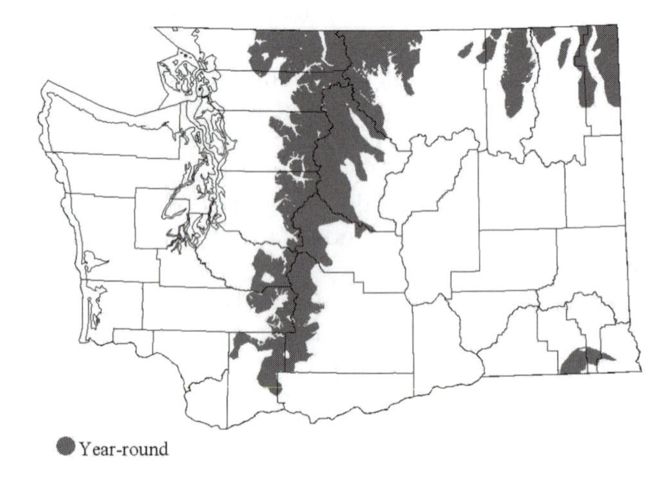

Subspecies: *P.d. fasciatus.*

Habitat: Coniferous forests from mid-elevations to subalpine zone, with some found in adjacent lower zones on Cascades e. slope (Smith et al. 1997).

Occurrence: Probably the state's least-known woodpecker due to its quiet vocalizations, behavior, and its distribution at high elevations in interior western hemlock, mountain hemlock, alpine/parkland and subalpine fir zones (Smith et al. 1997). The preponderant distribution of bird observers and censusing efforts in lowlands has almost certainly resulted in apparent, if mistaken, absence or undercounting in censuses.

Described nests show high-elevation distribution: 1700 m in the Blue Mts. (where it is rare) and the Olympics (Jewett et al. 1953), and ca. 1500 m in Kittitas Co. (Smith et al. 1997). Present in lodgepole pine but not ponderosa pine in Yakima Co. (Stepniewski 1999). E. Washington distribution overlaps to some extent that of Black-backed Woodpecker, which occurs generally at lower elevations. Wood-boring beetles are important prey throughout its range (Bent 1939, Cramp et al. 1985) and occurrence reported related to insect outbreaks, especially in burns resulting from forest fires, though less frequently reported doing this in Washington than the Black-backed Woodpecker.

Distribution generally described by Jewett et al. (1953) but became more detailed with greatly increased bird-watching effort starting in the late 1960s. Reports in e. Washington documented occurrence in Stevens, Pend Oreille, Yakima, Okanogan, and Kittitas Cos. Some westside report locations: Mt. Baker, White Pass, Mt. Rainier NP, Rainy Pass, Hart's Pass, Mt. Adams, Skamania Co., and Mt. Hardy burn, Skagit Co.

Birds reportedly move downhill in winter, at least on occasion (Campbell et al. 1990b, Smith et al. 1997). In e. N. America some move s. "nearly every winter" with "several large-scale movements" evident there (Bent 1939; see Leonard 2001). In Eurasia, where the species is resident, dispersive, and irruptive to a limited extent, some move lower and some to higher elevations in winter (Cramp et al. 1985). There are essentially no reports of the species in lowlands away from nesting range in Washington.

Remarks: Formerly known as Northern Three-toed Woodpecker.

Noteworthy Records: East, *Summer*: nest in Blue Mts. (Jewett et al. 1953). **West**: nest in 1921 on Happy L. Ridge, Olympics (Jewett et al. 1953), and a bird reported on 21 Jan 1985 at Hurricane Ridge. *CBC*: 1 at Chelan in 1970 and in 1995; 2 at Chewelah in 1978, 2 at Twisp in 1995. *Note*: 1 at Kitsap Co. in 1983 and 1 at Leadbetter Pt. in 1973 are questionable.

Terence R. Wahl

Black-backed Woodpecker *Picoides arcticus*

Rare to locally uncommon resident in mid- to high-elevation coniferous forests e. of the Cascade crest, rare w. of the crest.

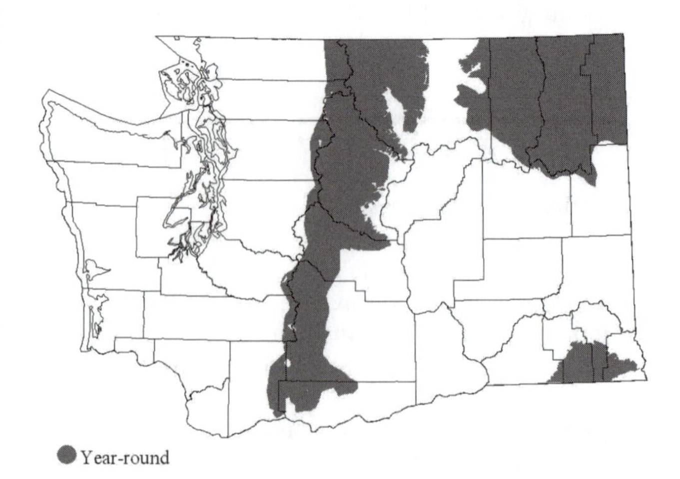

Habitat: Coniferous forests, open meadows in forested areas, often with diseased trees associated with forest fire burns.

Occurrence: Occurring essentially on the e. slope of the Cascades in the e. ponderosa pine zone, across the ne. forests, Spokane Co., and the Blue Mts., the relatively inconspicuous Black-backed Woodpecker is one of the least-known of its family in Washington. Described as "very localized" by Jewett et al. (1953), known distribution remains that today. Non-migratory, but wandering widely, it is closely associated with recent fire-killed dead or insect-infested old-growth stands (Powell 2000). Bark beetle populations are extremely variable, and the species' wanderings find new food sources after forest fires (Stepniewski 1999). Very local w. of the Cascades in high-elevation mountain hemlock, subalpine fir and alpine parkland zones (Smith et al. 1997).

Birds may be more noticeable in burns and anecdotal field reports may have been biased by observers attracted to burned areas where birds were previously reported by others, but state records generally correspond with observations elsewhere, coming mostly from habitats burned within the previous five to six years (Hutto 1995, see Dixon and Saab 2000). Almost half of 19 sightings from field observers from 1963 to 1999 were reported as from burned habitat, with sightings continuing in some areas for three to four years after the burn. These locations were in Spokane and Stevens Cos.

Twenty reports away from burns (or not described as in burned habitats) were in the Blue Mts., in Stevens, Pend Oreille, Spokane Cos., and on e. slopes of the Cascades in Okanogan, Chelan, Kittitas, Yakima, and Klickitat Cos. Five westside sightings from Mt. St. Helens included several from post-eruption burned habitat. Two more were from Mt. Rainier. Ten sightings 1991-98 by biologists on the Yakama Indian Reservation included three from two burned sites and others from old growth or insect-infested stands (RHL). In winter, 12 CBC records from 1978 to 1996 were limited to Chewelah, Spokane, and Twisp.

Remarks: Classed SC, PHS. Known previously as Arctic Three-toed or Black-backed Three-toed Woodpecker.

Rosemary H. Leach

Northern Flicker *Colaptes auratus*

Common resident in lowland and mountainous areas statewide. Uncommon in sagebrush desert and agricultural areas, absent from high-elevation meadows in winter. Rare or absent from parts of se. and c. Washington.

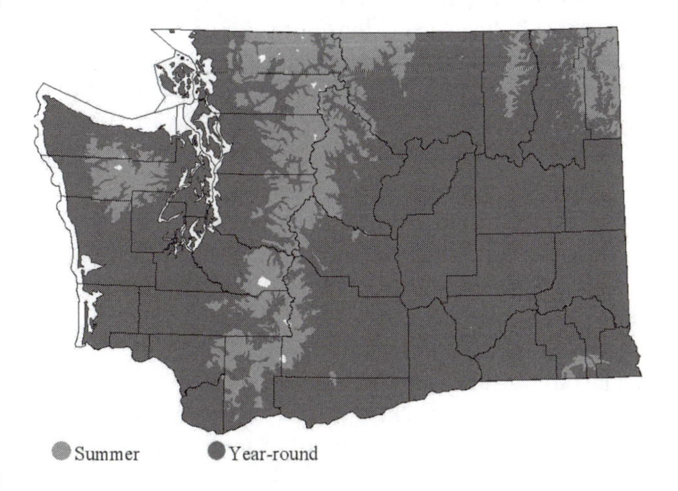

Subspecies: Red-shafted, *C. a. cafer*, and Yellow-shafted Flicker, *auratus*.

Habitat: Commonly breed in coniferous and broadleaf forests, as well as riparian woodlands, parks, and gardens (Smith et al. 1997); uncommon in wet meadows, sagebrush desert, lowland rain forest, and agricultural areas during the non-breeding season, and rare or absent in dry grasslands.

Occurrence: Common statewide (e.g. Jewett et al. 1953, Stepniewski 1999), flickers have been locally reduced in numbers from nest-cavity competition with the arrival of starlings (e.g., Wahl 1995). Nevertheless, birds are common in all major forested zones, in residential areas, city parks, and gardens. Uncommon in dense, lowland rain forests and absent from alpine tundra, dense urban areas, and dryland wheat farms (Smith et al. 1997), rare or absent from desert shrub, shrub-steppe, and alpine zones (Franklin and Dyrness 1973, Smith et al. 1997). BBA surveys in the 1990s indicated almost certain to confirmed breeding in all counties (Smith et al. 1997).

BBS data from 1982 to 1996 (Sauer et al. 2000) indicated variability in mean numbers of flickers observed per route varies throughout Washington. Few (0–1) were observed in urban areas and the se. portion of Washington, more (2–3) were on the Olympic Pen. and in sw. and s. Washington. Birds were moderately abundant (4–10) in c., n., and ne. areas. CBC data indicate moderate abundance (4–10 Red-shafted Flickers per count) in the n. half and parts of e. and w. Washington and moderately high abundance (11–30 birds per count) in the sw. and s.c. areas (Sauer et al. 1996).

Though flickers are considered resident, they "expand their range" during the non-breeding season (Root 1988). Altitudinal migration from higher to lower elevations often occurs in winter (Moore 1995), and the incursion of Yellow-shafted birds and hybrids is evidence of migration of boreal populations. A bird on a ship off Cape Flattery in fall 1898 (Jewett et al. 1953) preceded others noted up to 60 km off Grays Harbor in Sep 1976 and 1993 (TRW), presumably migrating.

Yellow-shafted Flickers are uncommon and hybrids common in winter in Washington; no nesting records exist for these morphs (Smith et al.1997). Hudson and Yocom (1954) and Jewett et al. (1953) earlier documented records of Yellow-shafted Flickers and hybrids in the state and reports today are regular.

BBS trends for flickers in Washington and N. America are markedly different (Sauer et al. 2000). From 1966 to 1999, flickers in Washington exhibited a positive trend (+0.3 % per year), while continent-wide numbers declined (-2.3 %/year). Long-term CBC data indicate a significant increase ($P < 0.01$) in e. Washington and some local increases in winter.

Remarks: Jewett et al. (1953) listed *cafer* as occurring w. and *collaris* e. of the Cascades. Recent lumping of forms may reduce the frequency and accuracy of field reports of the forms occurring here.

Douglas R. Wood

Pileated Woodpecker *Dryocopus pileatus*

Fairly common resident in coniferous, deciduous, and mixed forests over a wide elevation range statewide.

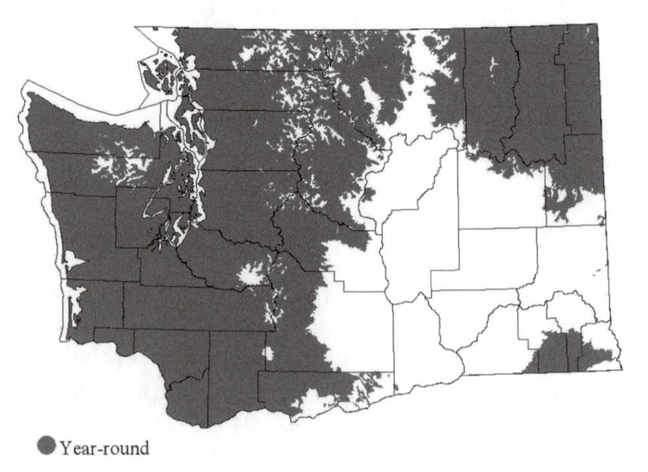

Subspecies: *D. p. picinus.*

Habitat: Mid- and late-seral forests with large, decadent live trees and snags for nesting, roosting, and foraging; also forage on logs and stumps. Can occur in younger forests with large remnant trees and snags. Less common at high elevations; probably due to low density of large trees and snags in high-elevation habitats.

Occurrence: Like some other resident species, the Pileated Woodpecker appears to have been infrequently reported until habitat concerns arose in the 1960s. In w. Washington, birds are locally common in mature and old stands of Douglas-fir, western hemlock, Pacific silver fir, and western redcedar. Most documented nests have been in western hemlock and Pacific silver fir, and most roosts in western hemlock and western redcedar (Aubry and Raley 2002a). Birds forage in a broader range of habitats including young forests, thinned stands, and along edges of openings and clear-cuts. Birds have been observed up to 1524 m on the w. slope of the Cascades (T. Kogut p.c., CMR). In the Puget Trough, birds have been found nesting in forested suburban habitats and large urban parks (Wahl 1995, Smith et al. 1997, A. Carey p.c., CMR).

On the e. side of the Cascades, birds are fairly common in a variety of low- to mid-elevation mixed conifer and conifer-hardwood forests that have large-diameter live and dead trees (Dumas 1950, Hudson and Yocom 1954, Weber and Larrison 1977, R. Leach p.c., K. Woodruff p.c.). Nesting documented in ponderosa pine, western larch, quaking aspen, and western redcedar (Madsen 1985, Bevis 1994, H. Murphy p.c.). Similarly to the westside, foraging habitats include young and mature forests and edges along pastures and agricultural lands.

Birds maintain territories year-round in Washington (Aubry and Raley 1996). Migratory movements by ads. have not been documented, and reports of increases in numbers during fall and winter are probably related to dispersal of juvs. (Bull and Jackson 1995).

Observers have long commented on the potential impacts, including land development and forest practices, on this species (e.g. Edson 1926, Jewett et al. 1953, Stepniewski 1999). In Kittitas Co., Bevis (1994) found birds only in unmanaged and shelterwood stands and not in stands where most of the overstory had been removed. Field observations before and after logging in Yakima Co. indicated decreased abundance after harvest and, when detected, birds were typically associated with remnant patches of old growth (R. Leach p.c.). While smaller woodpeckers have been documented using snags in clear-cuts, there is no evidence that Pileated Woodpeckers will nest or roost in remnant live or dead trees in clear-cuts (Aubry and Raley 2002b). Birds in suburban areas may decline locally with development and conversion of wooded areas for commercial or residential purposes.

Pileated Woodpeckers are detected irregularly and in very low numbers on almost all state BBS routes. The species has been detected fairly regularly (≥69% of the years surveyed) nr. Newhalem, Carnation, Mt. Rainier, and Cusick, and occasionally elsewhere. Indications of a decline from 1966 to 2000 (Sauer et al. 2001) are questioned because detections are sporadic and by Jun this species is fairly quiet. While birds are detected regularly on some westside CBCs, data are insufficient to determine population trends, and winter counts may be influenced by dispersing juvs. During the late 1980s and 1990s the highest number of CBC detections were in the San Juan Is., followed by the Olympia area, Kitsap Co., Padilla Bay, and the Kent-Auburn area. Though there is little information on abundance in Walla Walla Co., birds were detected there during the 1984, 1994-96 CBCs. BBS and CBC data are probably best used simply to document presence of the species in Washington.

Remarks: Smith et al. (1997) list *abieticola* for Washington. Classed SC, PHS.

Catherine M. Raley and Keith B. Aubry

Olive-sided Flycatcher *Contopus borealis*

Fairly common summer breeding resident in forested habitats, spring and fall migrant statewide.

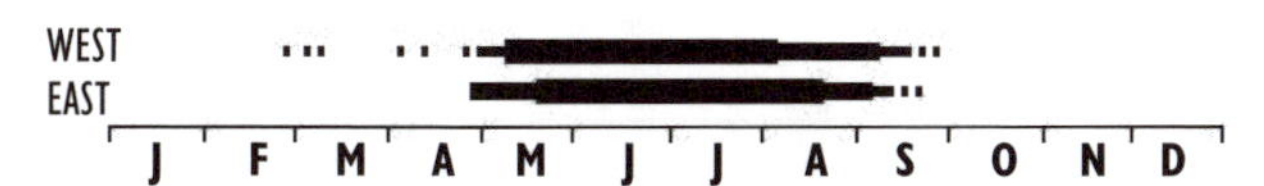

Habitat: Open mature stands of conifers or forest stands with high perches in tall trees and snags along the edges of clearings (Brown 1985, Sharp 1992) around lakes, marshes, beaver ponds, bogs, avalanche chutes, talus fields, burns, clear-cuts, and windthrown forest patches. Also breeds sparingly in city parks and residential areas (Smith et al. 1997; TB).

Occurrence: Closely aligned to the distribution of forested areas, this flycatcher appears to be very rare along the w. side of the Olympic Pen. and rare in the Blue Mts. (Smith et al. 1997). Sharp (1992) concluded that the species is apparently reliant on mature stands. Other studies showed correlation with recent clear-cuts, sapling conifers, and mature conifers though not with pole conifers 27-44 years

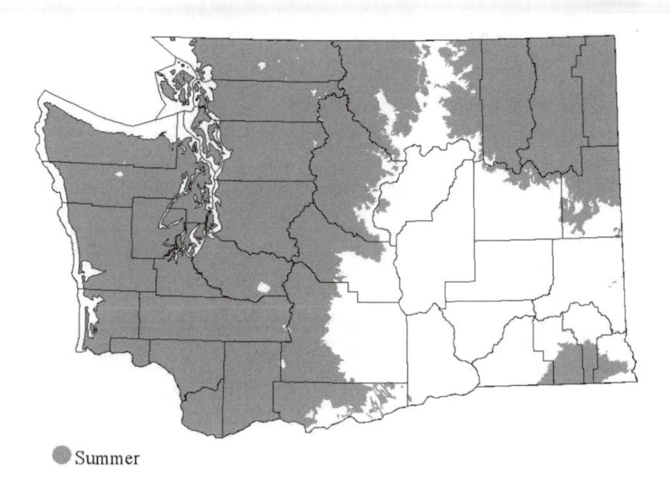

old (Bosakowski 1997). Hutto (1995) considered the species to be relatively restricted to early post-fire conditions in Montana. A study in California showed over half of over 400 detections of the species were in forest edges (Rosenberg and Raphael 1986). Associations need further clarification.

As for some other summer-only residents, migration knowledge consists of dates of seasonal appearance and disappearance. Migration is mostly from mid-May to early Jun, and from early to mid-Aug to early Sep statewide. The species is considered uncommon to fairly common away from breeding habitat on the westside and an uncommon migrant away from breeding locations on the eastside and such occurrence, as in the Columbia Basin, is noteworthy.

Population trends are apparent. Noted as declining in Washington (Sauer et al. 2001), the Northwest (Paulson 1992b), in N. America by Peterjohn and Sauer (1993), and locally (e.g., Wahl 1995). Smith et al. (1997) do not mention trends. Sharp (1992) noted that this flycatcher was widespread on all national forests in Oregon and Washington, but found declines (26 routes) significantly outnumbered increases (12 routes) on breeding bird survey routes conducted from 1968 to 1989, though Washington declines were not statistically significant. Effects of logging suggested as a cause for decline (Hejl 1994), but long-term decline noted on a BBS route along the Skagit R. where habitats remained undisturbed (Wahl 1995). Marshall (1988) noted disappearance from undisturbed sequoia forest in California and suggested habitat loss in the wintering grounds as a cause. Classed GL.

Noteworthy Records: *Spring, early*: 24 Feb 1977, 1 Mar 1980, 5 Mar-5 Apr 1976 at Seattle (Hunn 1982). *Summer*: 1 on 9 Jun 1963 at Columbia NWR was unusual. *Fall, late*: 1 on 22 Sep 1981 at Dungeness.

Thomas Bosakowski

Western Wood-Pewee *Contopus sordidulus*

Common spring migrant. Fairly common to common summer resident and fall migrant.

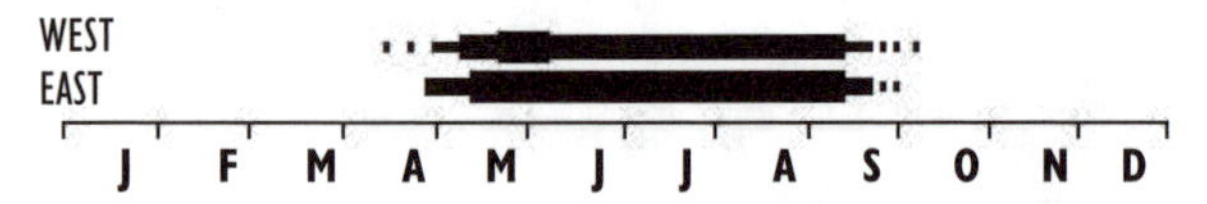

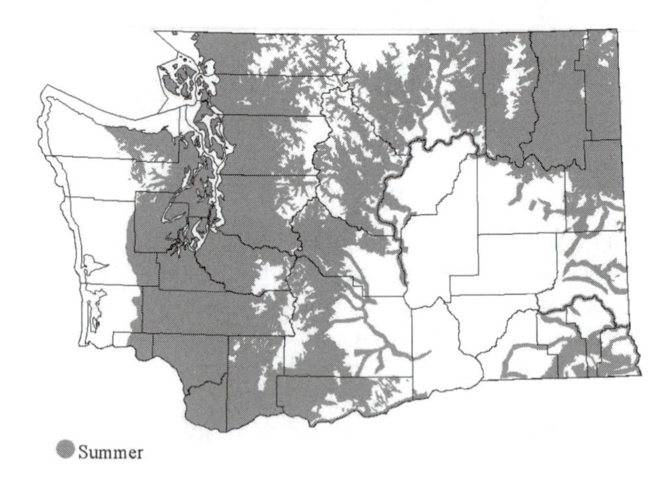

Subspecies: On westside, *C. s. saturatus.* On eastside, chiefly *veliei* (a.k.a. *richardsonii*), though *saturatus* and intergrades occur (Jewett et al. 1953, Smith et al. 1997).

Habitat: Primarily dry coniferous forests and lower-elevation riparian woods in e. Washington, mostly in riparian areas on the westside (Smith et al. 1997). Migrants occur in a wider variety of wooded habitats but still largely avoid dense coniferous woodlands.

Occurrence: Apparently taken for granted, little known in status and distribution or at least little commented upon by many Washington authors, the pewee is relatively local though widespread in w. Washington's wetter woodlands. It is really much more at home and more frequently heard and seen in the drier wooded habitats of e. Washington. Less common at high elevations than the Olive-sided Flycatcher, though reported at 2100 m nr. Hart's Pass (J. Duemmel p.c.).

Spring migration peaks during late May-early Jun, while fall migration peaks during mid-late Aug. Birds were widely detected in habitats statewide during BBA surveys in the 1980s and early 1990s (Smith et al. 1997). Birds were least common on the w. Olympic Pen., where the species is a rare breeder (Smith et al. 1997), as they are along the s. coast (BT).

Counter to earlier estimates (Bemis and Rising 1999), BBS data show an increase of 1.6% per year 1966-2000 (Sauer et al. 2001). Some local populations, particularly on the westside, may have declined due to habitat loss (TRW).

Remarks: Pyle (1997) considers differentiation within this species weak and clinal, and field identification to race is very unlikely.

Noteworthy Records: East, *high counts*: 45 on 24 Jun 2000 at Teanaway/Liberty, Kittitas Co.; 30 on 1 Aug 1996 at Cooke Canyon, Kittitas Co.; 30 on 30 Jun 2001 at Coppei Cr.; 30 on 28 Jun 2000 at Washtucna; 29 on 16 Aug 1998 at Naches. **West**, *high counts*: 20 on 3 Aug 1999 at Ridgefield NWR. Spring, early: 15 Apr 1968 at Seattle; 19 Apr 1994 at Seattle; 19 Apr 1981 nr. Anacortes; 23 Apr 1981 at Seattle. *Fall, late*: 2 Oct 1987 at Ridgefield NWR; 10 Oct 1997 at W. Seattle.

Steven G. Mlodinow

Alder Flycatcher *Empidonax alnorum*

Very rare vagrant.

Prior to 1973, the Alder Flycatcher and the Willow Flycatcher were considered subspecies of Traill's Flycatcher (AOU 1973). Alder Flycatchers breed as far w. as Alaska and as close as se. B.C. (see Campbell et al. 1997) and are vagrants in Washington. Identification by sight alone is not reliable, but vocalizations can allow accurate differentiation. Experienced observers have heard birds sounding like Alder Flycatchers in Washington, typically during Jun. Sonograms of vocalizations of a bird in Okanogan Co. in 2002 confirmed the occurrence of this species in the state (D. Paulson p.c., SGM), and another was identified by sound and recordings by a number of observers nr. Marblemount in Jun 2004 (S. Atkinson p.c., G. Bletsch, p.c., W. Weber p.c.).

Steven G. Mlodinow

Willow Flycatcher *Empidonax traillii*

Common migrant and summer resident in w. Washington. Fairly common migrant and summer resident in e.

Subspecies: *E .t. brewsteri* breeds (Smith et al. 1997). Some birds in e. Washington possibly *adastus* (see Jewett et al. 1953, Pyle 1997).

Habitat: Moist habitats with dense deciduous thickets, riparian areas, marsh and pond edges, and swamps with thick growth of alder, willow, dogwood, or wild rose. Recent clear-cuts with thick

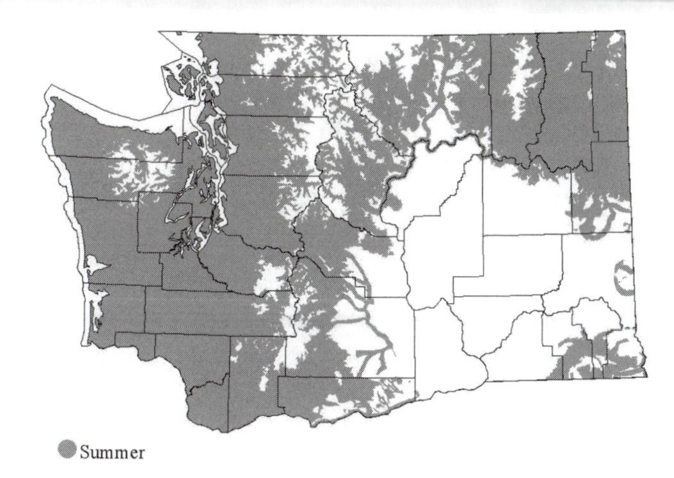

brush also can have high breeding densities. Migrants frequent similar habitats.

Occurrence: The Willow Flycatcher is a common breeder in Puget Trough river valleys and clear-cuts, less common along the outer coast and the w. Olympic Pen., where it is an uncommon to rare breeder and migrant (Smith et al. 1997). On the eastside, Willow Flycatchers are locally common breeders, though uncommon in the Columbia Basin (Smith et al. 1997). Birds in the s. Cascades are recorded breeding in clear-cuts up into the silver fir zone (BT).

Spring migrants may be arriving earlier. Historically, they arrived in Washington in the last few days of May or the first week of Jun, but beginning in about 1992, birds were reported annually during mid-May, with good numbers present during the last week of May. In 1999, 2000, and 2001 arrival dates were even earlier, with the first individuals being found during late Apr and early May. The reason for this change is unclear, but it does not seem to be occurring in s. California, where an early May report is still considered exceptional (M. A. Patten p.c.).

Observers in Spokane and Yakima Cos. (e.g., Stepniewski 1999) commented on a possible long-term decline locally. BBS data from 1966 to 2000 indicate a statewide decline (Sauer et al. 2001) as in much of N. America (Sedgwick 2000). Loss of riparian habitat from causes including cattle grazing, and cowbird parasitism are cited as causes of decline in B.C. (Campbell et al. 1997).

Remarks: Prior to 1973, the Alder Flycatcher and the Willow Flycatcher were considered subspecies of Traill's Flycatcher (AOU 1973). Classed GL.

Noteworthy Records: *High counts, Summer*: 79 on 15 Jun 1975, 73 on 28 Jun 1997 at Spencer I. *Spring*: 50 on 29 May 1975 at Camano I. *Fall*: 45 on 15 Aug 1996 at Nisqually NWR. *Spring, early*: 14 Apr 1968 at Bellingham (Wahl 1995); 28 Apr 2000 at Spanaway; 2 May 2000 at Skagit WMA; 4 May 2000 at Washtucna; 6 May 2001 at Othello; 8 May 1999 nr. Sequim. Fall, late: 1 Oct 1995 at ALE, Benton Co. (LaFramboise and LaFramboise 1999); 13 Oct 1981 at Leadbetter Pt.

Steven G. Mlodinow

Least Flycatcher *Empidonax minimus*

Rare spring migrant and summer resident and very rare fall migrant in e. Washington. Very rare migrant and summer resident in w.

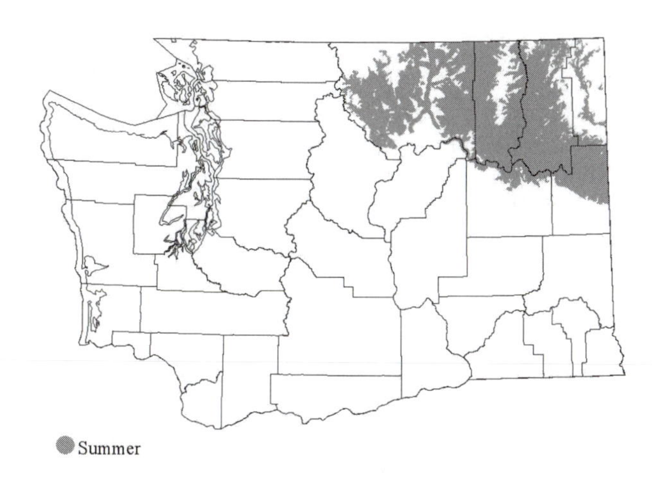

Habitat: Mostly semi-open stands of deciduous trees, especially aspen, often in riparian areas. Migrants recorded in a wide variety of open and semi-open brushy and wooded locations.

Occurrence: The Least Flycatcher is a fairly recent addition to Washington's avifauna. The first state record is of a bird collected nr. Anacortes on 23 Jun 1958 (Stein and Michener 1961). The next record was 10 years later at Turnbull NWR on 1 Jun 1968. Since then, the frequency of reports has steadily increased with at least 97 records through 1999. Records by five-year period 1970-74 (1), 1975-79 (7), 1980-84 (8), 1984-89 (12), 1990-94 (26), 1995-99 (41). Though some of the increase was likely due to greater numbers of observers, much of it was probably due to range expansion. Historically, in B.C. the species was probably limited mostly to the Peace R. area in the ne. part of the province (Munro and Cowan 1947). By the late 1950s, the range extended into the province's central interior region, and it had reached the southern interior, including the Okanagan Valley, by the early to mid 1970s (Cannings et al. 1987, Campbell et al. 1997), and probably into Washington.

Most records are of singing males, between late May and early Jul, peaking from mid to late Jun. Records from late summer and fall are quite scarce. Eastside records outnumber those from the westside by about 4:1, with nearly 50 records in Okanogan, Yakima, Pend Oreille, and Spokane Cos. Multiple records of migrants or nesting birds have come from the Wenas area, from nr. Winthrop, Sullivan L., and at Eloika L. in Spokane Co., at Davenport, Washtucna, and Vantage. At least four pairs were found in summer 2002 at the Kalispell Indian Reservation, Pend Oreille Co. The only confirmed nest to date was nr. Monroe in 1990.

Noteworthy Records: *Spring, early*: 27-30 Apr 1994 at L. Terrell (Wahl 1995); 28 May 1997 at Davenport (J. Acton); 2 on 10-31 May 1981 nr. Winthrop; 12 May 1999 at Oroville. *Summer*: 1 singing on 20 Jun 1976 nr. Maple Falls, Whatcom Co. (Wahl 1995). *Fall*: 17 Aug 1998 at Seattle; 28 Aug 1994 at Washtucna; 28 Aug 1989 at Vantage; 28 Aug 2000 at Washtucna; 7 Sep 1987 at Vantage; 19 Sep 1998 at Skagit WMA; 29 Sep 1995 at Washtucna; 12 Sept 2000 at Oroville; 2 on 4 Sep 2000 at Washtucna; 1 on 8 Sep, 3 on 9 Sep and 2 on 15 Sep 2001 at Washtucna; 1 on 30 Sep 2001 at Hooper, Whitman Co.

Steven G. Mlodinow

Hammond's Flycatcher *Empidonax hammondii*

Common summer resident and migrant in most of westside and forested areas of e. Washington. Widespread, uncommon migrant elsewhere.

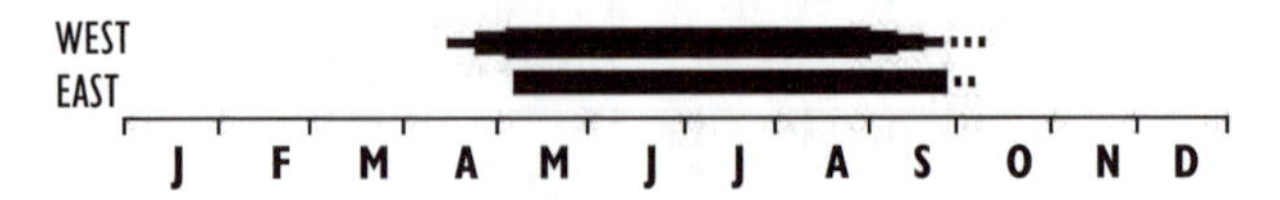

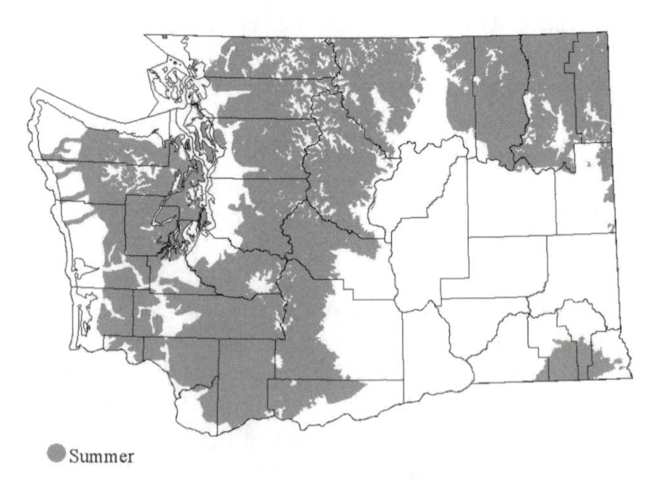

Habitat: Breeds in most forested areas and riparian habitats, eastside pine forests. Absent from urbanized areas in the Puget Sound lowlands (Smith et al. 1997).

Occurrence: Hammond's Flycatcher has been noted as a relatively widespread summer resident in Washington since early in the 20th century (e.g., Jewett et al. 1953). It is fairly common to locally common above about 150 m elevation, uncommon below in westside lowland forests, fairly common in montane mixed forests, uncommon in higher-elevation subalpine parklands and subalpine areas of the Olympics and Cascades (Jewett et al. 1953, CBCh) and common to fairly common in Puget Sound lowlands, though it is absent from urbanized areas.

Typically considered a species of conifer forests, it is locally fairly common to common in westside riparian broadleaf or mixed forests also (e.g. Jewett et al. 1953, Sharpe 1993, Stepniewski 1999). Birds have also been noted on territory in riparian wetlands dominated by tall shrubs on the w. side of the Olympic Pen. but are rare to absent elsewhere in lowlands there (Sharpe 1993, CBCh).

Fairly common to common in eastside conifer forests, birds breed uncommonly in deciduous broadleaf riparian groves within forested landscapes (Stepniewski 1999). Stands dominated or co-dominated by oak or madrone are avoided (Manuwal 1989, CBCh).

In conifer forests, Hammond's Flycatcher occurs mainly in relatively closed-canopy forests, and is much less numerous in open woodlands (Smith et al. 1997, CBCh). Where it occurs in open-canopy woodlands, such as ponderosa pine, it is usually in a small patch with a more closed canopy. Very young stands are shunned and westside birds typically do not appear in significant numbers until stand age is 50-60 years (Sakai 1987, Bryant et al. 1993, Hansen et al. 1995, CBCh). Thinning of very dense Douglas-fir plantations can create conditions more favorable to the species (Hagar et al. 1996). Birds are more numerous in mature and old-growth conifer forests than in young forests in the Cascades in s. Washington (Manuwal 1991).

Migrants are uncommon away from westside breeding areas but more widespread on the eastside. Migrants are noted until early Jun; fall movements peak in mid-Aug to early Sep.

Though Dawson and Bowles (1909) described the species as a fairly abundant summer resident on Capitol Hill in Seattle in Jun 1895, where they were singing actively and "twenty or thirty might have been seen in the course of a morning's walk," the species has disappeared as a summer resident from the urbanized areas of the Puget Trough (Smith et al. 1997). Washington BBS data, sampling mainly non-urbanized areas, indicate a significant increase from 1966 to 2000 (Sauer et al. 2001).

Noteworthy Records: *Spring, early*: 1 on 14 Apr 1985 at Carnation; 15 Apr 1985 at Seattle; 15 Apr 1990 nr. Tacoma. *Fall, late*: 30 Sep at Kamiak Butte; 30 Sep 1986 at Seattle; 5 Oct 1997 at Everett; 10 Oct 1997 at Two Rivers Park, Benton Co. *High counts*: 65 on 31 Aug 1997 at Yakima Training Ctr.; 35 on 28 Aug 1994 at Washtucna; 20+ on 11 May 1996 at Hog Ranch Butte, Yakima Training Ctr., with 15 there on 12 May 1996 (A. Stepniewski).

Christopher B. Chappell

Gray Flycatcher *Empidonax wrightii*

Rare migrant and locally fairly common breeder in e. Washington. Casual migrant w.

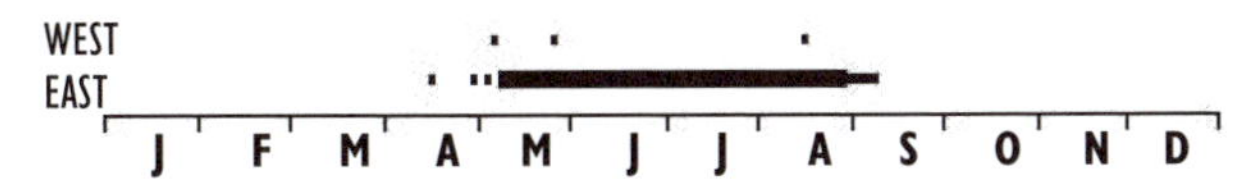

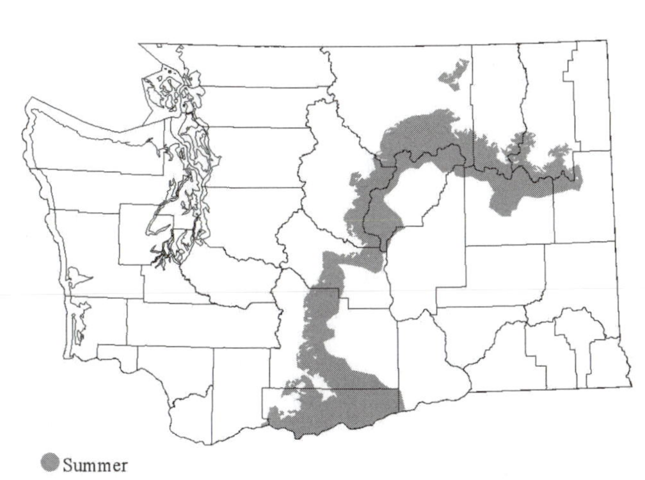

Habitat: Open pine forests with no understory (Smith et al. 1997). Most Washington sites have a understory of pine grass and, unlike breeding locations elsewhere, lack sage (Lavers 1975b). Migrants can occur in a wide variety of open woodlands or riparian areas.

Occurrence: First recorded in the state on 31 May 1970, when Larrison (1971) found one singing along Wenas Cr. The first nest was found there in late May 1972 (Yaich and Larrison 1973). The following summer, 10 were found in that area and others were found nesting in Klickitat Co. (Yaich and Larrison 1973, Lavers 1975b). Over the next 25 years, further range expansion was noted. The first breeding record in the Okanagan Valley, B.C., occurred during 1980 (Cannings et al. 1987), and the first singing birds in ne. Washington were found during Jul 1985 along the Little Pend Oreille R. and nr. Newport. Found nesting in Canada in 1986, with 13 singing birds located in the Okanagan Valley (Cannings 1987). Nests were recorded in 1990 and 1991, in Lincoln, Stevens, and Spokane Cos., and in 1992 at the Hanford Site. There is also at least one record for the Blue Mts.: three singing males nr. Godman Springs on 2 Jul 1978.

Colonization of e. Washington is part of a larger range expansion (Cannings 1987, Johnson 1994a) though some of the increase in reports may be due to observer effort. Stepniewski (1999) speculates that "extensive, selected logging of ponderosa pine stands ... has allowed the Gray Flycatcher to colonize the e. slopes of the Cascades," creating a new, open parkland-like habitat. Current distribution is apparently along the e. slope of the Cascades from Klickitat Co. n. to s. Okanogan Co. and e. along the Columbia and Spokane R. drainages into Spokane Co. (Smith et al. 1997). Records from Twisp, the Pend Oreille and the Blue Mts., indicates that breeding may occur there, at least sporadically.

On the eastside, migrants are uncommon away from breeding areas, with passages occurring from late Apr to late May and from early Aug to mid-Sep. The highest count away from breeding distribution was four, at Washtucna, on 25 May 2002. Gray Flycatchers are extremely rare in w. Washington, with only one fall and six spring records.

BBS data indicate a rangewide increase of 11.6% per year 1966-1995 (Sterling 1999), though BBS data are likely insufficient to determine trends in Washington.

Noteworthy Records: West: Aug 1982 at Fort Casey; 28 May 1984 at Seattle; 9 May 1993 at Seattle; 1-2 May 2000 nr. Arlington; 6 May 2000 at Steigerwald L.; 15 May 2003 at Seattle. **East**, *Spring, early*: 19 Apr 1994 at Wenas Cr.; 22 Apr 1989 at Turnbull NWR; 30 Apr 1994 nr. Bickleton; 30 Apr 1995 at Richland. *Fall, late*: 12 Sep 1998 at Snively Gulch, Benton Co.; (LaFramboise and LaFramboise 1999); 12 Sep 1998 at Lewis and Clark SP; 20 Sep 1998 at Bickleton.

Steven G. Mlodinow

Dusky Flycatcher *Empidonax oberholseri*

Fairly common to common summer resident and fairly common migrant in e. Washington. Rare and local breeder and very rare migrant in w. Washington.

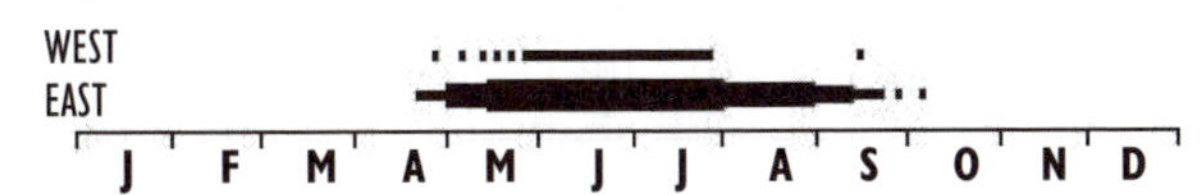

Habitat: Breeds in lower-elevation open dry coniferous woodlands with shrubby understory, but is also found in similar habitats at higher elevations (Smith et al. 1997). Brushy habitats, streamsides,

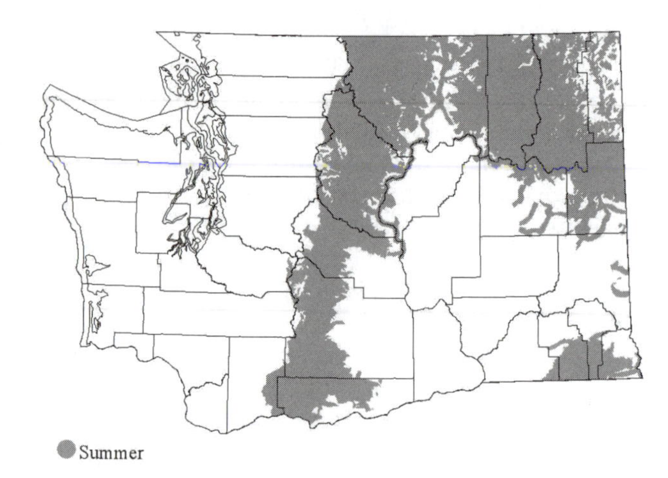

and woodlands, especially with shady, broad-leafed cover in migration (Sedgwick 1993).

Occurrence: Perhaps the most characteristic *Empidonax* of dry, open eastside coniferous forests, found mainly at lower elevations than Hammond's Flycatcher, typically in ponderosa and lodgepole pine stands, burns, and south-facing, forested slopes of the Cascades, from Okanogan e. to Pend Oreille, s. to Badger Mt. in Douglas Co. (D. Stephens p.c.), Lincoln and Spokane Cos., and in the Blue Mts. (Smith et al. 1997). In Jul, post-breeders disperse downslope, as to Douglas Cr., Douglas Co. (D. Stephens p.c.).

A few birds are found where drier microclimates exist in similar forests on the Cascade's w. slope. Local, apparently breeding populations have been found in e. Skagit, Snohomish, King, Pierce, and Skamania Cos. Mostly these are in localized patches of open coniferous woodlands that resemble typical eastside habitat. A singing bird was also found on the dry ne. portion of the Olympic Mts. at Royal Cr. Basin, Jefferson Co., on 19-20 Jul 1986.

Migration periods have been difficult to ascertain due to flycatcher identification problems, but northbound birds are detected mostly during mid-late May, with some movement undoubtedly occurring into early Jun. Fall migration seems to occur largely from early to late Aug. W. of the Cascades, spring migrant Dusky Flycatchers are less than annual, occurring almost entirely from 10 to 22 May, and there is just one fall record.

BBS data show an apparent increase in Washington of 4.9% per year between 1966 and 1991 (Smith et al. 1997), although a more recent analysis indicates a possible slight decline from 1966 to 2000 (Sauer et al. 2001). Throughout its range, this species seems to be holding its own if not increasing in numbers, with a rangewide increase of 3.6% per year from 1966 to 1991, possibly due to forestry practices that thin dense coniferous stands and leave small openings (Sedgwick 1993).

Remarks: Rathbun (in Jewett et al. 1953) reported birds in May 1913 in mountains above L. Crescent, and Jewett et al. (1953) stated the species was a summer resident throughout the state. See Smith et al. (1997) for a discussion comparing habitat separation and overlaps with Hammond's Flycatcher.

Noteworthy Records: West, *migration*: 29 Apr 2001 at Rockport; 10 May 1987 at Kent; 11 May 1993 at Seattle (K. Aanerud); 13 May 1997 at Spencer I.; 14 May 1974 at Seattle; 15 May 2000 at Skagit WMA (SGM); 18 May 1993 at Olympic Hot Springs; 2 on 18 May 1991 at Pt. Roberts (Wahl 1995); 18 May 2002 at Seattle; 19 May 1999 at Seattle; 20 May 1998 at Olympia; 22 May 1983 at Leadbetter Pt; 11 Sep 1999 at Vancouver L. **East**, *Fall, late*: 23 Sep 1998 at Robinson Canyon, Kittitas Co.; 2 Oct 1999 at Wahluke Slope WMA.

Steven G. Mlodinow

Pacific-slope Flycatcher *Empidonax difficilis*

Common summer resident and migrant

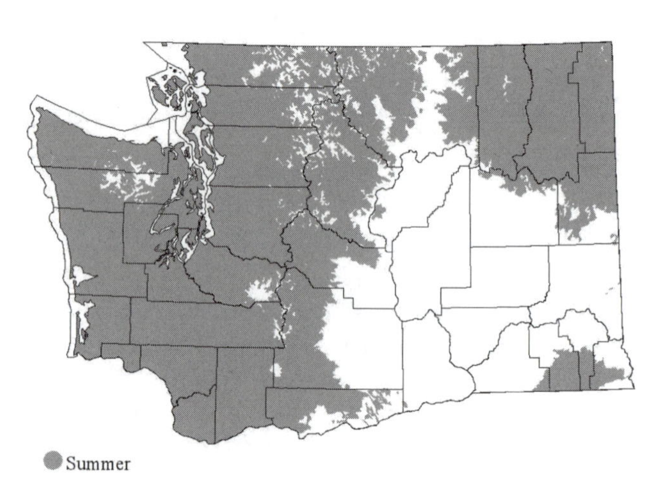

Subspecies: *E. d. difficilis.*

Habitat: Breeds in coniferous forests with closed canopy and open understory (Wahl 1995), and also in deciduous riparian areas, especially in e. Washington (Jewett et al. 1953). Wider variety of wooded locations including riparian scrub in migration.

Occurrence: Widespread statewide in suitable habitat, this flycatcher's status and distribution is somewhat clouded since it and the Cordilleran Flycatcher were considered subspecies of the Western Flycatcher and there are difficulties in differentiating Pacific-slope and Cordilleran flycatchers in the field. Johnson (1980) states that the ranges of Cordilleran and Pacific-slope Flycatchers are split in two by the Cascade crest in Oregon and California, but for most of e. Washington and se. B.C., the divide between these species' breeding ranges is uncertain. Cannings et al. (1987) determined that "Western Flycatchers" in the Canadian Okanagan were Pacific-slopes, and field observations seem to indicate that the great majority of "Western Flycatchers" in e. Washington are also Pacific-slopes, with the exception of those in the Blue Mts. There, some birds may be Cordilleran Flycatchers, but intermediates may well also be possible. At present, birds away from the Blue Mts. are assumed to be Pacific-slope Flycatchers.

The only *Empidonax* occurring in dense, wet rain forests along the outer coast (Smith et al. 1997), this species breeds more commonly in w. Washington's moist forests than in e. Washington's drier woodlands, where it is more localized in riparian areas. Where it occurs nr. Hammond's Flycatcher it is more often at lower elevations.

Migrants in eastside lowlands are fairly common. Spring migration throughout the state lasts into late May-early Jun and fall migration may start in early Aug. Though currently considered common on the eastside, Jewett et al. (1953) considered Western Flycatchers to be uncommon in e. Washington. There has been a dramatic increase in the Western Flycatcher breeding populations of e. B.C. and Alberta (Semenchuk 1992, Campbell et al. 1997), and an increase in abundance in e. Washington may occur. BBS data, however, indicate a statewide decline from 1966 to 2000 (Sauer et al. 2001).

Noteworthy Records: East, *Spring, early*: 19 Apr 1994 at Wenas Cr. *Fall, late*: 1 Oct 1995 at FEALE, Benton Co. (LaFramboise and LaFramboise 1999). **West**, *Spring, early*: 3 Apr 1980 at Seattle. *Fall, late*: 13 Oct 1997 at Elma.

Steven G. Mlodinow

Cordilleran Flycatcher *Empidonax occidentalis*

Possibly breeds in Blue Mts.

Occurrence: Status in Washington is uncertain. Johnson (1980) and Johnson and Marten (1988) show the species occurring with certainty only in se. Washington, but they also leave much of e. Washington as a question regarding Cordilleran vs. Pacific-slope flycatchers. To date, separation in the field has been by vocalizations only, and that is uncertain. Birds sounding like Cordilleran Flycatchers have been reliably reported only from the Blue Mts. (e.g. K. Aanerud p.c., SGM), though birds vocalizing like Pacific-slope Flycatchers and birds giving intermediate vocalizations also occur there. Elsewhere on the eastside, the great majority of "Western Flycatchers" appear to be Pacific-slope. Jewett et al. (1953) reported a Cordilleran Flycatcher specimen from Yakima Co. in May, but Johnson (1980) reviewed that specimen and, based on measurements, felt it more likely a Pacific-slope. Weber and Larrison (1977) list the Western Flycatcher as occurring from mid-May to early Sep in se. Washington, and this probably reflects the seasonal occurrence there of Cordilleran Flycatchers.

Remarks: The uncertain status and distribution of the Cordilleran Flycatcher is due to the difficulty of identifying the two closely related forms, even by vocalizations (see Johnson 1994b, Lowther 2000). Cannings and Hunn (unpubl.) show a complete integradation of vocalizations across the Pacific Northwest. Field identification by non-vocal characteristics is not reliable. Therefore, identification of any "Western" Flycatcher is somewhat tenuous.

Steven G. Mlodinow

Black Phoebe *Sayornis nigricans*

Very rare visitor.

A regular resident of the sw. U.S. n. to sw. Oregon, recorded six times in Washington. One record was at Moclips, Grays Harbor Co., 27 Feb 1980, one at Clear L., Yakima Co., 21-26 May 1989, one at Washougal from 20 Nov 1997 to 2 Jan 1998 (Tweit and Paulson 1994), one nr. Cathlamet from Oct 2001 to Feb 2002 and in 2002-03, one at Julia Butler Hansen NWR 16 Oct-30 Nov 2002 and on 1 Mar 2003, and one at Ridgefield NWR Sep 2002-17 Mar 2003. Additional records are expected, as this species is expanding its range northward in Oregon (Gilligan et al 1994). The subspecies possibly *S. n. semiatra*.

Bill Tweit

Eastern Phoebe *Sayornis phoebe*

Very rare visitor.

More scarce than would be predicted in Washington, based on the w. extent of their breeding range—into ne. B.C.—and regular occurrence in California. The first record was a winter bird in Bay City 16-23 Dec 1989 and the second a singing male nr. Chillowist, Okanogan Co., 22 Jun-3 Jul 1991 (Paulson and Mattocks 1992). Four subsequent records were at

Washtucna on 27 May 2000 (Aanerud 2002), L. Ozette on 3 Jun 2000 (Aanerud 2002), nr. Havillah on 10 Jun 2000 (R. Flores), and Leavenworth on 24 May-26 Jun 2001 (Aanerud 2002).

Bill Tweit

Say's Phoebe *Sayornis saya*

Common migrant and summer resident and very rare winter resident on eastside. Rare spring migrant, very rare fall migrant, and casual during winter on westside.

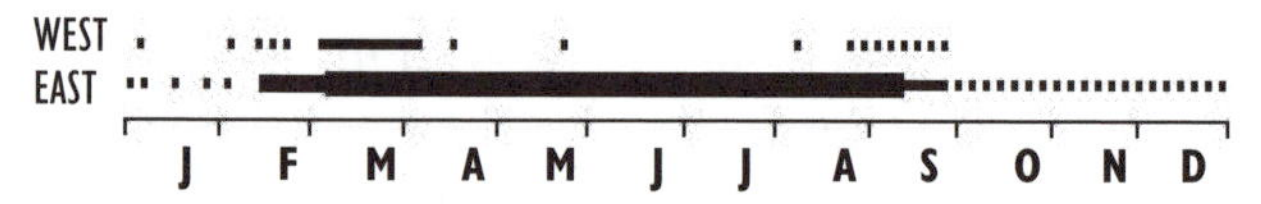

Subspecies: *S. s. saya*.

Habitat: Treeless habitats, particularly shrub-steppe and agricultural areas. Solid structure such as a rocky outcropping or building for nesting (Smith et al. 1997).

Say's Phoebe (G. Scott Mills)

Occurrence: A widespread species of dry areas of e. Washington, less numerous in ne. counties and Blue Mts., where it is absent from forested areas except in the foothills (Smith et al. 1997). Spring migrants arrive exceptionally early for an insectivore, with the first couple typically noted in mid-late Feb, making it the earliest-arriving flycatcher in Washington (Campbell et al. 1997). In the fall virtually all birds leave by the end of Sep, but in most years one or two linger into Dec or early Jan from Yakima s. The species is casual in midwinter, and recorded 13 times at scattered locations on CBCs, once in 1966 and the rest from 1984 to 1999.

Say's Phoebes are rare in w. Washington, averaging about two per spring and with a total of only about 10 in fall. Jewett et al. (1953) suggested the species may be a summer resident in sw. Washington but Smith et al. (1997) stated there were no records on BBA surveys. The timing of spring records mirrors the eastside migratory period closely. Occasional irruptions do occur, as happened during the spring of 1999, when eight Say's Phoebes were found on the westside. Notably, that was an exceptional spring for several other eastside species in w. Washington as well. Westside spring records are concentrated in the Puget Trough, with very few farther w. Birds in fall have been more widely distributed.

BBS data indicate a long-term increase in Washington (Sauer et al. 2001, see Schukman and Wolft 1998).

Remarks: Jewett et al. (1953) states that Washington breeders and migrants are *S. s. yukonensis*, but this race is not recognized by most recent authorities (Schukman and Wolf 1998).

Noteworthy Records: East, *Winter*: 4 Jan 2002 at Chelan Falls; 14 Jan 1995 nr. Walla Walla; 23 Jan 1975 nr. Yakima; 9 Feb 1924 at Grand Dalles (Jewett et al. 1953). **West**, *Winter*: 2 Jan 1971 at Lummi Flats. *Spring, early*: 8 Feb-5 Mar 1975 at Ridgefield NWR. *Spring, late*: 23-24 May

1975 at Camano I. *Fall, late*: 8 Aug 1996 at Point No Point; 27 Aug 1987 at McCleary; 30 Aug 1991 at Lummi Flats (Wahl 1995); 31 Aug 1979 at Orcas I.; 8 Sep 1980 at Ocean Shores; 13 Sep 1996 at Copalis; 18 Sep 1986 at Seattle; 21 Sep 1993 at Ocean Shores; 22 Sep 1990 at Sumas Mt., Whatcom Co. (Wahl 1995); 24 Sep 1973 at Seattle. *High count*: 50 on 6 Mar 1995 at Webber Canyon, Benton Co.

Steven G. Mlodinow

Vermilion Flycatcher *Pyrocephalus rubinus*

Casual vagrant.

A regular resident in s. N. America and most of S. America, the Vermilion Flycatcher has been recorded in Washington four times, in fall or winter. An ad. male occurred at Redmond, 25 Jan-17 Mar 1988 (Tweit and Skriletz 1996), a female at Ridgefield NWR, 31 Dec 1995-25 Jan 1996 (Aanerud and Mattocks 1997), an ad. female on Fir I., Skagit Co., 24 Oct-10 Nov 1999, and an imm. male nr. Stanwood on 1 Nov 2002.

Bill Tweit

Ash-throated Flycatcher *Myiarchus cinerascens*

Locally common breeder in oak woodlands and riparian habitat of s.c. and e. Washington from May through July. Vagrant on w. side of Cascade range.

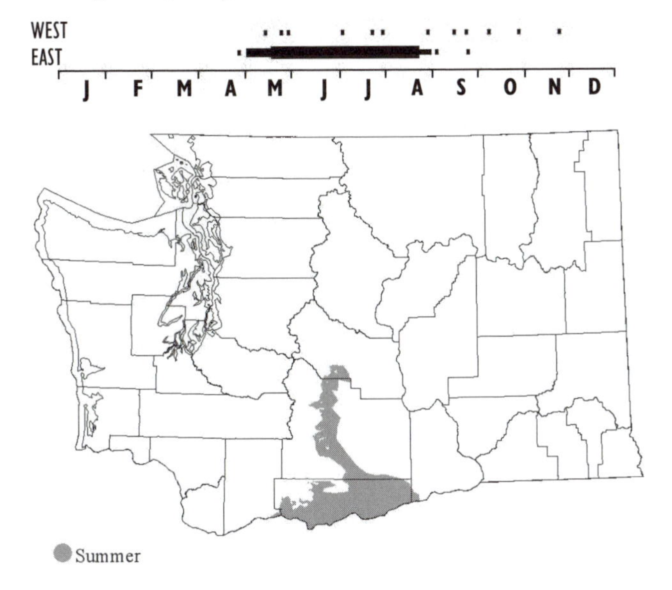

Subspecies: *M. c. cinerascens* in Washington.

Habitat: Breeds in pure, mature stands of Oregon white oak, stands with mature oak mixed with ponderosa pine, and riparian zones (JRS), peripheral in central arid steppe zone (Smith et al. 1997). Natural tree cavities and woodpecker-created holes are used for nesting.

Occurrence: Historically considered an irregular or uncommon Washington breeder (Dawson and Bowles 1909, Jewett et al. 1953); there are mixed reports on the distribution and abundance in Washington. Kitchin (1934) describes an increase in abundance in e. Washington. Larrison and Sonnenberg (1968) reported a decreasing and restricted abundance around the c. part of the state. Weber and Larrison (1977) listed it as a summer visitor and rare migrant in se. Washington. Main breeding range is the oak woodlands of Klickitat and Yakima Cos., where this species is common (JRS). It is a local, scarce breeder in Chelan, Grant, and Adams Cos., possibly nests in Douglas and Lincoln Cos. (Smith et al. 1997). Summer records from Okanogan Co. likely pertain to non-breeding vagrants (SGM). Migrants arrive at the end of Apr to mid-May, with wanderers noted also in Franklin and Walla Walla Cos.

The species departs breeding grounds during mid-late Jul. Fall migrants are rarely encountered, with most records coming from areas heavily covered by observers in Aug. As with other rare species in the state, abundance data may be skewed by coverage of favorite birding localities, and not be completely representative of the actual breeding range.

Very rare vagrant in w. Washington with about 15 records between 14 May and 24 Nov. In B.C. there about 40 records, all but one coastal, in late May and Aug-early Nov (Campbell et al. 1997).

BBS data show that the N. American population is increasing. Peterjohn and Sauer (1995) estimated an increase of 2.6% per year in w. states (and see Cardiff and Dittman 2002), but Sharp (1992) found insufficient data to predict population status in Washington.

Remarks: Uncertainty over the population trend and concern over the limited habitat range led to monitored status in Washington. Limited distribution and available habitat are potential problems. Oak woodlands have only recently become a habitat of concern in Washington and protection is just beginning. Nesting success rates for Klickitat Co. birds, at the n. end of the species' range, were high: 86% of nests attempted successfully fledged young (JRS), higher than in s. populations. Birds are strong nest defenders and in two years of nest success research in Klickitat Co. no nests were stolen by starlings (JRS).

Noteworthy Records: East, *Fall, late*: 3 Sep 2000 at Two R. CP; 5 Sep 1975 at Richland; 7 Oct 1973 at Richland; 26

Sep 1999 at Royal City, Grant Co. *Outside normal range*: 16 Aug 1960 at Spokane (Weber and Larrison 1977); 6 Sept 1976 at Spokane Valley. **West**: 14 May 2000 at Spring Cr. Fish Hatchery, Skamania Co.; 24 May 1905 at Tacoma (Dawson and Bowles 1909); 26 May 1998 at Packwood; 27-30 May 1990 at Auburn; 1 Jul 1997 at Everett L., Skagit Co.; 18 Jul 1978 at Tacoma; 18 Jul 1978 at Flett Basin, Pierce Co. (Chappell 1979); 18 Jul 2001 at Lopez I.; 29 Jul 1979 at McKenna; 31 Aug 1975 at Seattle (Hunn 1982); 13 Sep 1979 at Orcas I.; 23 Sep 2000 at Tokeland; 9 Oct 1976 at Ocean Shores; 30 Oct 1994 at Seattle; 24 Nov 1956 at W. Seattle (Hunn 1982).

Jennifer R. Seavey

Tropical Kingbird *Tyrannus melancholicus*

Rare fall vagrant along outer coast. Very rare elsewhere in w. Washington.

Subspecies: Presumed to be *T. m. satrapa* (a.k.a. *T. m. occidentalis*).

The first of 25 records occurred in Nov 1916 on Destruction I. (Slipp 1942c). All subsequent records also were on the outer coast and the Olympic Pen., except for a bird that remained at Samish I., Skagit Co., 15-28 Nov 1992 and one at Stanwood 12-24 Nov 2001. Records span 1 Oct-4 Dec.

Washington records reflect an annual northward movement along the Pacific coast. Birds are mostly imms., presumably from populations in se. Arizona and nw. Mexico (Mlodinow 1998b). Numbers in Washington vary annually, with occurrences usually coinciding with those farther s. Populations are believed to be increasing in their normal range (Stouffer and Chesser 1998), and this may explain the apparent increase in vagrancy in the Pacific Northwest. Though Washington currently has no winter or inland records, there is at least one such record from Oregon (Mlodinow 1998b).

Remarks: Though main occurrence in Pacific and Grays Harbor Cos. appears logical, it may also reflect concentration of observers. Not all Washington records have eliminated the very similar Couch's Kingbird, *T. couchii,* from s. Texas and ne. Mexico. The two species can be separated only by in-hand measurements and voice. Couch's has occurred once in s. California (Mlodinow 1998b) and could potentially occur in Washington. Three published reports of Tropical Kingbirds were not accepted by the WBRC.

Records: 18 Nov 1916 at Destruction I. (Slipp 1942c); 26 Nov 1927 at Westport (Slipp 1942c); 17 Nov 1953 at Hoquiam (Burleigh 1954); 6-16 Nov 1976 at Ocean Shores; 1 Oct 1984 at Tokeland; 4 Oct 1984 at Aberdeen; 6 Oct 1984 at Port Angeles; 16 Nov 1986 at Tokeland; 10 Oct 1987 at La Push; 21-26 Oct1991 at Ocosta (WBRC); 30 Oct 1992 at Ruby Beach; 15-28 Nov 1992 at Samish Flats; 16 Oct 1995 at Ocean Shores; 2 Nov 1995 at Ocean Shores (WBRC); 2 Nov 1996 at Port Townsend; 20 Oct 1997 nr. S. Bend; 22 Oct 1997 nr. Sequim; 3 on 21-27 Oct 2000 at Ocean Shores, with 2 until 31 Nov, 1 until 4 Dec; 1 on 10 Nov 2000 at Tokeland; 1 on 28 Oct-6 Nov 2001 at Ocean Shores; 1 on 12-24 Nov 2001 at Stanwood; 1 on 12-13 Oct 2002 at Bay Center; 1 on 17and 20 Oct 2002 at Tokeland; 1 on 17 and 19 Nov 2002 at Samish Flats, 1 on 23, 25 Nov 2002 at Elma. Rejected by the WBRC: 27 Oct 1962 at White Swan, Yakima Co.; 14 Jul 1977 at Moxee (mistakenly reported as Maryhill SP, 14 Jun 1978); and 4 Nov 1993 at Tokeland (Tweit and Skriletz 1996).

Steven G. Mlodinow

Western Kingbird *Tyrannus verticalis*

Common summer resident and migrant in e. Uncommon spring migrant, local breeder, and very rare fall migrant in w. Washington.

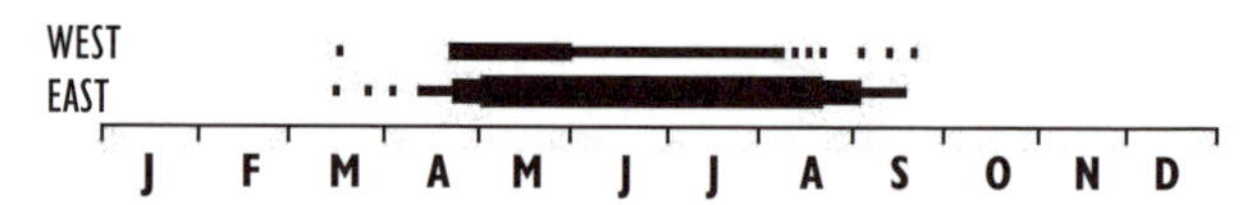

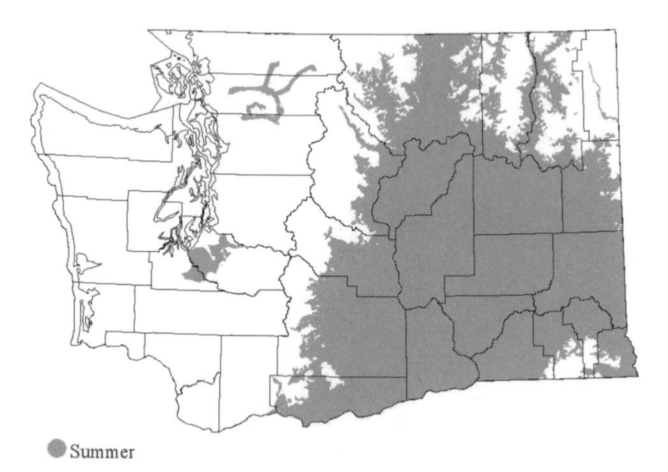

Habitat: Breeds in a wide variety of open habitats, especially steppe, agricultural areas in ponderosa pine zone and at ranch buildings and other structures. On the westside, migrants have been found in a variety of open areas, but nesting is local and limited to remnant prairies in Pierce Co. and agricultural areas along the Skagit R.

Occurrence: A characteristic open country species of summer in e. Washington, the Western Kingbird may be found locally close to and overlapping in foraging habitats with the Eastern Kingbird, but this species generally nests in more upland, drier areas in ponderosa pines and often next to power transformers on utility poles. Birds are widespread throughout the Columbia Basin and in suitable habitats in river valleys in surrounding counties in summer (Smith et al. 1997). Noted, at least in migration, occasionally at high elevation.

On the westside, birds are noted predominantly as a spring migrant, mostly from late Apr to late May, with an occasional non-breeder found in mid-Apr or early Jun. During an average spring, 15-20 are found, mostly in the Puget Trough and in Clark Co. Fall migrants are considerably rarer, however, averaging only about one per year, mostly during Aug. Breeders are quite scarce and local. In recent years known nesting areas are limited to the Ft. Lewis area and along the Skagit R. between Newhalem and Rockport (Wahl 1995, Smith et al. 1997) and Monroe. Nesting has also occurred previously nr. Packwood, Lewis Co. Occasional midsummer wanderers are also found away from known breeding locations.

Long-term population change in Washington uncertain. Rangewide BBS data appear to show an increase during the last 30 years (Gamble and Bergin 1996), though Sauer et al. (2001) indicated no change in Washington from 1966 to 2000. Smith et al. (1997) state that westside breeding numbers have declined, likely due to destruction of prairie habitat, though even 50 years earlier Jewett et al. (1953) considered the species rare to casual there. It is possible, however, that the clearing of forests—as in w. Washington river valleys—may have encouraged expansion (see Campbell et al. 1997).

Remarks: Four Oct reports of Western Kingbirds did not differentiate between Western and Tropical kingbirds: Tokeland, 9 Oct 1964; N. Jetty of Columbia R., 15 Oct 1966; Everett, 17 Oct 1913 (Jewett et al. 1953); Friday Harbor, 19 Oct 1975 (Lewis and Sharpe 1987) were not accepted by WBRC.

Noteworthy Records: East, *high count*: 350 on 3 Jun 2000 between Palouse Falls and Potholes Res. (SGM, DD). *Spring, early*: 14 Mar 1981 nr. Selah (Stepniewski 1999); 2 on 26 Mar 1995 at Joseph Cr., Asotin Co. *Fall, late*: 22 Sep 1904 at Cashmere (Jewett et al. 1953). **West**, *Spring, early*: 10 Mar 1919 at Everett (Jewett et al. 1953); 19 Mar 1979 at Lopez I. (Lewis and Sharpe 1987); 7 Apr 1908 at Everett (Jewett et al. 1953). *High elevation*: 1 in mid-Aug 2003 nr. Slate Pk. at 2400 m elevation (J. Duemmel p.c.). *High counts*: 12 on 22 May 1983 at Leadbetter Pt.; 7 on 22 May 2000 Point No Point. *Coast, Summer*: 24 Jun 1999 at Pt. Grenville; *Fall, late*: 3 Sep 1978 at Seattle; 11 Sep 1994 at Skagit WMA; 21 Sep 1982 at Leadbetter Pt.; 29 Sep 1976 at Anacortes.

Steven G. Mlodinow

Eastern Kingbird *Tyrannus tyrannus*

Common migrant and summer resident in e. Washington. Rare migrant and summer resident on westside.

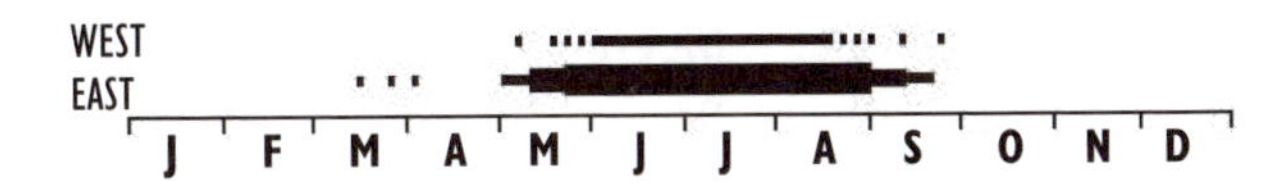

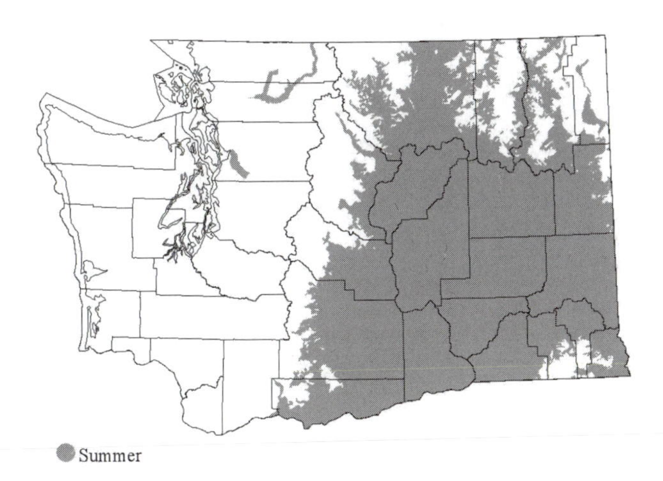

Habitat: Breeds in open areas in steppe and agricultural habitats with scattered large deciduous trees or groves of deciduous trees, particularly in moist locations in riparian areas, orchards, and golf courses. Forages also in adjacent habitats. Birds in w. Washington found almost exclusively in river valleys. Found in a wide variety of open habitats in migration.

Occurrence: The widespread Eastern Kingbird is typical of many areas of e. Washington in summer. Often sympatric with the more numerous Western Kingbird, the Eastern occupies somewhat different nesting habitat, utilizing deciduous trees in wetter, more riparian areas. The nesting range includes essentially all eastside counties at lower elevations (Smith et al. 1997). During spring migration, Eastern Kingbirds on the eastside are found mostly from mid-May into early or mid-Jun, and southbound birds are seen largely between early and late Aug.

On the westside, Eastern Kingbirds are rare migrants and local nesters in the Puget Trough and in Clark Co., and very rarely recorded from the Olympic Pen. and the outer coast. Birds breed regularly along the Snohomish R. from Everett to Snohomish and along the Skagit R. nr.

Marblemount and Corkindale and perhaps intermittently in Whatcom Co. at one or more scattered locations (Wahl 1995). Other recent breeding records include L. Joy, King Co., in 1978 and 1979 (Hunn 1982) and Ft. Lewis in 1993. Older records are from Nisqually in 1853; Dungeness in 1916; Seattle in 1893 and 1906; and nr. Lynden from 1930 to 1934 (Jewett et al. 1953). Numbers in the Snohomish R. valley appear to be increasing, with at least six pairs and three single birds there during Jun 2000 (SGM). Other migrants average about two per spring and one every other fall, and an occasional wandering bird or two in July defies classification. The peak for spring records is from 5 to 20 Jun.

BBS data indicate a significant decrease statewide of 3.4% per year from 1966 to 2000 (Sauer et al. 2001).

Noteworthy Records: East, *Fall, late*: 20 Sep 1904 at Cashmere (Jewett et al. 1953). *High count*: 125 on 25 Jul 1998 at Prosser. **West**, *Spring, early*: 8 May 1987 at Dungeness; 9 May 1998 at Sequim. *Fall, late*: 4 Sep 1999 at Steigerwald L.; 5 Sep 1978 at Arlington; 10 Sep 1965 at Seattle. *Outer coast*: 13 Sep 1974 and 28 Jun 1980 at Ocean Shores; 11 Jun 1980 at Leadbetter Pt.; 6-7 Jun 2001 and 11 Jun 2002 at Tatoosh I.

Steven G. Mlodinow

Scissor-tailed Flycatcher *Tyrannus forficatus*

Casual vagrant.

There are three Washington records of this wide-ranging species: an ad. (ph.) nr. Royal City, Grant Co., 4 Sep 1983 (Tweit and Paulson 1994); an ad. (ph.) at Desert WRA, 5 May 1985 (Tweit and Paulson 1994); and a specimen of a bird that landed on a boat 16 km nw. of Grays Harbor on 15 Jul 1990 and subsequently died. Additionally, in 2003, reported at George on 15 May and nr. Rockport on 31 May.

Washington occurrences, in May, Jul, and Sep, are similar to seasonal occurrence in Oregon, which has about 12 records, all between 7 May and 14 Nov.

Steven G. Mlodinow

Fork-tailed Flycatcher *Tyrannus savana*

Casual vagrant.

This C. and S. American species has been recorded once in Washington: one (ph.) at Chinook on 12-13 Sep 1995. Fork-tailed Flycatchers have been recorded over 100 times in the U.S. and Canada, mostly during Sep and Oct. Most of these, including the Washington bird, appear to be the migratory S. American race, *T. s. savana*, rather than the sedentary C. American race, *monachus* (McCaskie and Patten 1994). Individuals that wander to N. America may winter in the tropics, and then migrate n. instead of s. for their breeding season, arriving during fall (McCaskie and Patten 1994). There are only two other records w. of the Rocky Mts.: one in Blaine Co., Idaho, 25 Aug-7 Sep 1991 (Trost 1991), and one in Sonoma Co., California, 4-8 Sep 1992.

Steven G. Mlodinow

Loggerhead Shrike *Lanius ludovicianus*

Fairly common local summer resident in e., rare in winter. Very rare to rare migrant and winter visitor w.

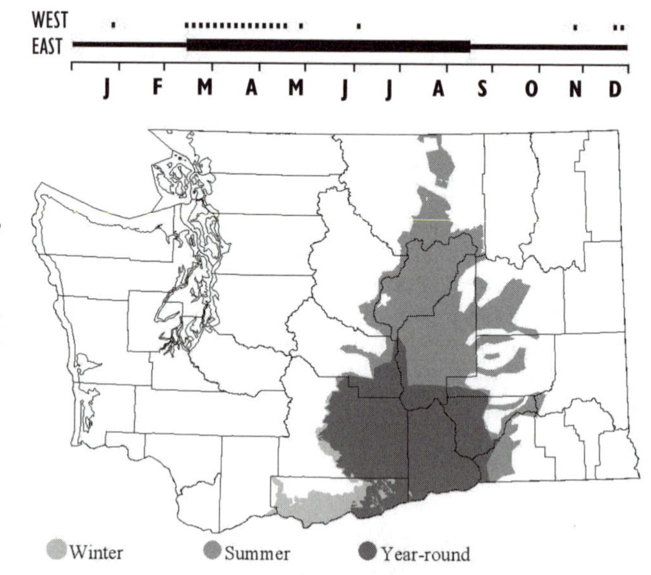

Subspecies: *L. l. gambelli*.

Habitat: Relatively undisturbed Columbia Basin shrub-steppe zone in open sagebrush community areas with patches of grassy areas (Smith et al. 1997), often in ravines or with scattered trees or hedgerows for nesting.

Occurrence: An indicator of the shrub-steppe ecosystem (Stepniewski 1999), and formerly more or less taken for granted in its primary range in e. Washington (Jewett et al. 1953), now a species of concern over much of its range. Nesting success is greater in greasewood shrubs than in big sagebrush, and winter survival may depend on critical habitat features and relatively mild weather (see Stepniewski 1999). Preservation of remaining habitat

Loggerhead Shrike (G. Scott Mills)

recommended as essential for maintenance of the species population (WDFW 1997d).

Washington summer distribution is mainly in low elevations along the Columbia R., Columbia Basin, and areas in Okanogan and Klickitat Cos. Status in se. is uncertain: first nesting record in 1976 in Asotin Co., where it was formerly a rare breeder (Larrison and Sonnenberg 1968, Smith et al. 1997). In 1983 numbers at Columbia NWR reported to have declined over 20-30 years. BBS data indicate a significant population decline in Washington from 1966 to 2000 (Sauer et al. 2001), and there was a decline in the Northwest (Paulson 1992b) and the w. U.S. (DeSante and George 1994, Smith et al. 1997, WDFW 1997d). Conversion of shrub-steppe habitats to agriculture is a likely cause (Smith et al. 1997).

Irregular in winter, with CBC occurrence most frequent at Tri-Cities and Toppenish. It is rare then in the Columbia Basin, very rare elsewhere on the eastside. Data indicate birds apparently were most numerous in 1970s, less so in 1990s. CBC and other winter reports were spread between Whatcom, Clallam, Grays Harbor, Clark, Klickitat, Okanogan, Spokane, and Walla Walla Cos., though some of these may have been Northern Shrikes.

It is very rare on the westside from late Oct through May, with most records from from mid-Mar to early May. There are late spring westside records (e.g., Jewett et al. 1953, Hunn 1982), but another Jun report (Lewis and Sharpe 1987) did not satisfactorily distinguish this from the Northern Shrike. An apparent increase in reports w. of the Cascades may be due to increase in numbers and effort of field observers.

Remarks: Classed SC, FSC, PHS, Blue Listed by NAS. Many authors point out the need for careful identification of wintering shrikes in Washington: Northern Shrikes are much more common. Summer reports may be more likely, with a bird on 4 Jun 1976 at Seattle "probably" this species (Hunn 1982), while one on 24 Jun 1982 on San Juan I. was considered hypothetical (Lewis and Sharpe 1987).

Noteworthy Records: West: 26 Jan 1942 at Dungeness (Jewett et al. 1953); 25 May 1975 at Seattle (Hunn 1982); 3 Jul 1975 at Redmond (Hunn 1982); 24 Nov 1983 at Tacoma (Jewett et al. 1953). *CBC high counts*: 11 in 1974, 7 in 1969 and 1971 at Tri-Cities; 6 in 1979 at Chewelah.

Terence R. Wahl

Northern Shrike *Lanius excubitor*

Widespread, uncommon to fairly common winter resident in lower-elevation, non-forested regions statewide.

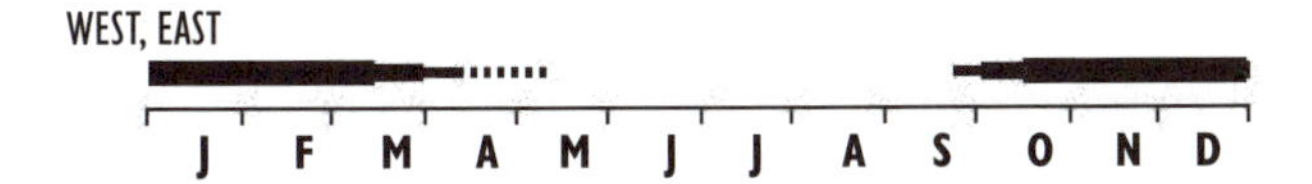

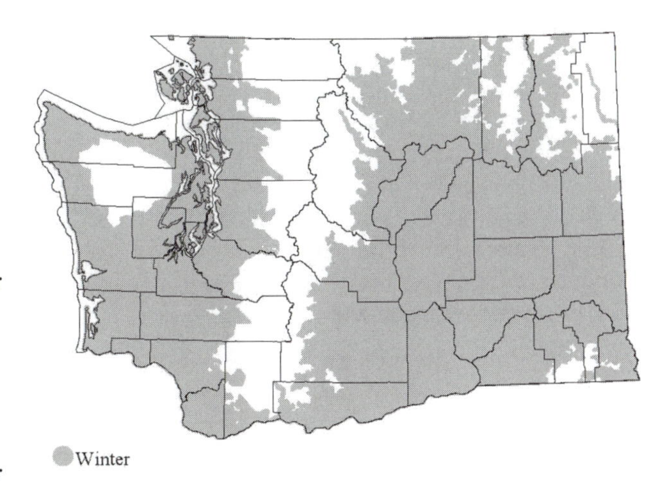

Subspecies: *L. e. invictus.*

Habitat: Primarily open habitats such as shrub-steppe, grasslands, agricultural areas, clear-cuts and estuaries; both rural and suburban areas, attracted to habitats that provide suitable prey (small mammals and small birds), hunting perches, and sites used for caching prey (Atkinson 1993).

Occurrence: This shrike has a circumpolar distribution and breeds in open or lightly forested habitats as well as taiga and bogs in the northernmost Canadian provinces and Alaska (AOU 1998). It occurs throughout the Columbia Basin and on the westside particularly at large estuaries and associated agricultural areas or open habitats such as Nisqually R. delta, Skagit Bay, Lummi Bay, and Willapa Bay (Jewett et al. 1953, Buchanan in press). It is also noted regularly in cities (e.g. Spokane, Bellingham, and Olympia.

Birds typically arrive during the first week of Oct, although there are several records from late Sep. Most birds leave the area by mid-Apr, but in some years birds appear to depart before the end of Mar.

This is a solitary species, but multiple birds will use an area when resources are plentiful. An analysis of CBC data indicated that the Northern Shrike is occasionally found in sizable numbers in a number of CBC circles with high counts of 420 birds. Noteworthy were 54 birds at Padilla Bay in 1996. Birds are generally more abundant at CBC sites in e. Washington (2.12 shrikes per 100 party miles) than at rural CBC sites in w. Washington (1.11 per 100 party miles); they are less common at suburban/urban sites such as Kent, Tacoma, and Seattle (0.52 per 100 party miles; Buchanan in press).

The abundance of Northern Shrikes varies substantially both interannually and within CBC sites. This variation does not appear to be cyclic when one examines count data from individual CBC sites. In fact, the abundance of shrikes at sites within a region appears to vary asynchronously. There is a weak indication of a pattern or cycle, however, when the CBC sites from e. Washington are considered as a group (Buchanan in press).

Recent analysis of CBC data indicates that wintering populations in both e. and w. Washington have not changed over the past two decades (Buchanan in press). The factors that dictate the annual abundance of this shrike remain unknown. Long-term CBC data indicate significant declines ($P < 0.01$) on the westside and ($P < 0.05$) on the eastside.

Noteworthy Records: West, *Spring, late*: 14 Apr 1984 at Seattle; 14 Apr 1998 at Ridgefield NWR; 16 Apr 1978 at Whidbey I.; 16 Apr 1999 at Kent; 17-19 Apr 1986 at Seattle; 19 Apr 1978 nr. Olympia; 20 May 1978 at Samish Flats; 23 Apr 1999 at Weir Prairie; 28 Apr 1986 at Kelso. **East**, *Spring late*: 12 Apr 1977 at Spokane; 12 Apr 1982 at Hyak; 25 May 1998 at Lyons Ferry SP. *Fall, early*: 22 Sep 1990 at Seattle. *Note*: A record for 23 Jul 1968 at McNary NWR is questionable.

Joseph B. Buchanan

White-eyed Vireo *Vireo griseus*

Casual vagrant.

A species breeding throughout se. N. America, the Caribbean and e. Mexico, with just one record of this very rare w. coast vagrant: a single-person sighting of a singing male on Vashon I. on 11 Jul 1981 (Tweit and Paulson 1994). As of 1998, there was just one other record n. of c. California, one at Point St. George, Del Norte Co., California, on 27 Jun 1998 (Erickson and Hamilton 2001).

Bill Tweit

Yellow-throated Vireo *Vireo flavifrons*

Casual vagrant.

A species that breeds throughout much of e. N. America and winters from s. Mexico to n. S. America: recorded just once in Washington. One (ph.) at Spencer I., Snohomish Co., 26-28 Oct 1995 (Aanerud and Mattocks 1997) was the first record for the Pacific Northwest. The scarcity of records is surprising, as there have been 69 accepted records in California through 1998 (Erickson and Hamilton 2001); approximately 70% have been spring records.

Bill Tweit

Cassin's Vireo *Vireo cassinii*

Fairly common summer resident and uncommon migrant.

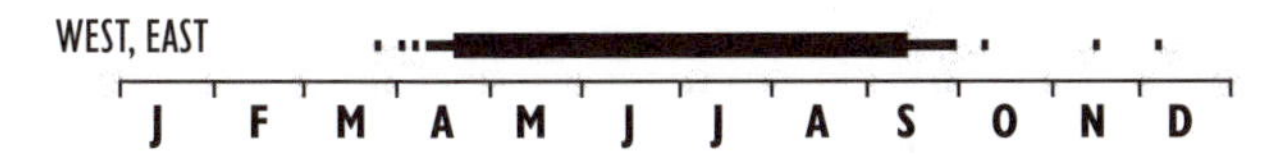

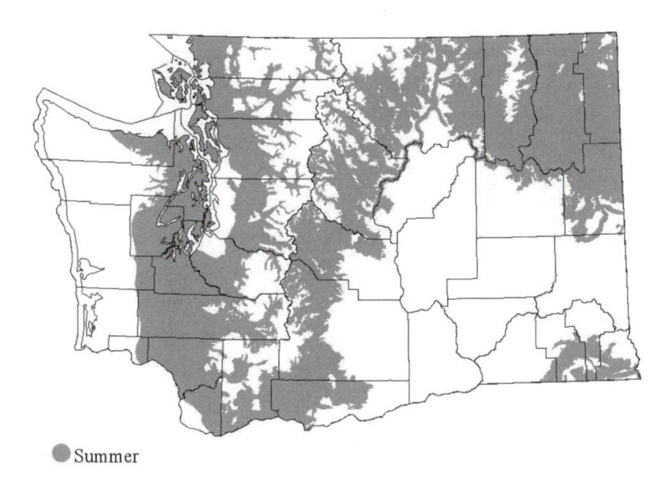

Summer

Habitat: Drier open coniferous or mixed forests. Migrants occur in a broad range of wooded habitats.

Occurrence: Cassin's Vireo is encountered more often during summer than in migration and is more common on the eastside than the westside. The species tends to be common in many eastside dry forests on the e. slope of the Cascades, the ne., and Blue Mts. (Smith et al. 1997). In w. Washington, they are dispersed and rather uncommon, especially in the n. counties, and they are absent as breeders from most of the Olympic Pen. and the outer coast. They are more common on the ne. dry rain-shadow side of the Olympics, drier woodlands of the San Juan Is., and in prairies in Thurston and Pierce Cos., where the species can be numerous.

Migrants seen relatively infrequently, perhaps passing mostly through the mountains, especially during fall. Spring migration occurs mainly between late Apr and late May, though occasional individuals appear weeks earlier and often a few northbound individuals are still found into early May. Fall migration is largely between mid-Aug and early Sep. Migrants are more numerous on the eastside than on the w. and are quite scarce along the outer coast.

Washington BBS data indicate an increase of 2.7 % per year between 1966 and 2000 (Sauer et al. 2001), while B.C. BBS data show an increase of 5% per year between 1968 and 1993 in the interior and no change on the coast (Campbell et al. 1997).

Remarks: Formerly Solitary Vireo.

Noteworthy Records: *High counts*: 35 on 39 Jun 2001 at Coppei Cr. (SGM). *Spring, early*: 22 Mar 1910 in Whatcom Co. (Wahl 1995); 29 Mar 1986 at Bellingham; 30 Mar 1995 at Seattle. *Fall, late*: 3 Oct 1989 at Walla Walla; 4 Oct 1999 at Doug's Beach, Klickitat Co.; 13 Oct 2001 at Rooks Park, Walla Walla Co.; 12 Nov 1995 at Spencer I.; 4 Dec 1997 at Port Orchard.

Steven G. Mlodinow

Blue-headed Vireo *Vireo solitarius*

Casual vagrant.

The Blue-headed Vireo was, until recently, considered conspecific with Plumbeous and Cassin's vireo under Solitary Vireo. Plumage, morphological, vocal, and genetic differences led to these species being split in 1998 (AOU 1998). The Blue-headed Vireo breeds from e. N. America w., through Canada, to ne. B.C.

Washington has two sight records: one on 8 Sep 1995 at Seattle and one on 28 Aug 2000 at Palouse Falls SP. Its exact status as a vagrant in w. N. America is still uncertain but a few occur each fall in California. Oregon has seven records as of summer 2000, split about evenly between eastside and westside, and all but one from 9 Sep and 5 Oct (Johnson 1998).

Remarks: For a discussion of Solitary Vireo identification, see Heindel (1996).

Steven G. Mlodinow

Hutton's Vireo *Vireo huttoni*

Uncommon year-round resident in w. Washington lowlands. Very rare breeder in e. Washington.

WEST
EAST
J F M A M J J A S O N D

Year-round

Subspecies: *V. h. huttoni.*

Habitat: Primarily forests and edges below 450 m elevation. Most often in hardwoods or second-growth conifer forests with a hardwood component, locally with garry oak and Pacific madrone. Absent from old-growth rain forests of the Olympic Pen. (Smith et al. 1997).

Occurrence: Usually inconspicuous but vocally obvious in season, Hutton's Vireo resides throughout the w. lowlands in suitable habitats. Jewett et al. (1953) described it as rare, and the species has probably benefitted from logging practices that have resulted in younger forests with a deciduous component (Smith et al. 1997). BBA surveys confirmed or showed probable breeding in lowlands of all westside counties except in the Gorge area of Skamania Co. (Smith et al. 1997, C. Chappell). Fall dispersal is indicated by reports out of normal habitat/range in Sep-Oct (SGM, TRW). There are three eastside records and status there is uncertain: birds reported there are at the periphery of the regular range (see Campbell et al. 1997).

Management practices that benefit Hutton's Vireo include maintaining a deciduous tree component in the subcanopy and understory and retaining patches of forest rather than dispersed trees when harvesting (Altman 1999). Manuwal and Pearson (1997) found this one of the more common species in private, managed forests of the s. Cascades in forests from pre-canopy stands (12-20 years) to harvest-age stands (50-70 years). Occurrence was significantly correlated with indices of vegetation complexity in a forest stand and negatively correlated with clear-cuts. In NF lands in the s. Cascades, found only in young forests (55-80 years) and not in older forests (Manuwal 1991, and see Carey et al. 1991).

BBS results in Washington from 1966 to 2000 indicate an increasing trend of 4.8% annually (Sauer et al. 2001). The number of birds per route in the S. Pacific region of Washington (1.1 per route) is indicative of low numbers present, though abundance is probably underestimated by the BBS because the species' song is most obvious weeks before surveys in Jun. Numbers of birds per party hour from long-term CBCs varied, but analysis of data from 1970 to 2000 indicated a significant increase (P <0.01).

Noteworthy Records: East: pair on 26 May 1990 at Wenas Cr.; 5 Sep 1998 at Umatilla NWR; 17 Jun 2000 at Husum, Klickitat Co.

John F. Grettenberger

Warbling Vireo *Vireo gilvus*

Common summer resident in deciduous forests throughout. Uncommon migrant in Columbia Basin.

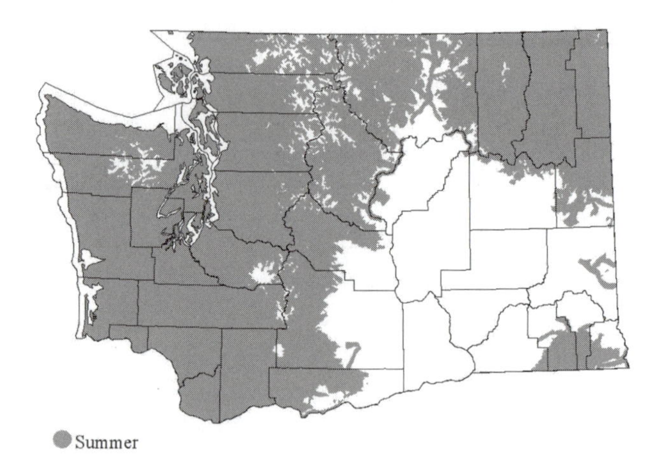

Subspecies: *V. g. swainsonii* breeds in w. Washington and ne. Cascades, and *leucopolius* in e. Washington.

Habitat: Common in deciduous forests throughout the state, but present in a variety of settings such as avalanche chutes, riparian zones, forested wetlands, and hardwoods patches in conifer forests, from lowlands to subalpine elevations (Smith et al. 1997, Stepniewski 1999).

Occurrence: The most abundant vireo species in Washington, the Warbling Vireo is typically not found in older coniferous forest (Carey et al. 1991, Manuwal 1991, Huff and Brown 1998), but it is present during stand regeneration in w. and ne. Washington (Carey et al. 1991, O'Connell et al. 1997). It is more abundant in wide riparian buffers after harvest than narrow buffers or preharvest riparian zones, and very local in managed coniferous forests in sw. Washington (Manuwal and Pearson 1997).

First northbound birds may be noted in late Apr, but typically occur in early May and the species is quite common by mid-May, with northbound migrants often noted into early Jun. Southbound birds may be noted as early as late Jul, but numbers peak in mid-late Aug to early to mid-Sep. Few are noted in late Sep.

BBS data indicate that the species has been increasing in Washington at an annual rate of 3.5% (P <0.01) from 1966 to 2000, with birds most abundant in the Cascade Mts. survey area (Sauer et al. 2001).

Remarks: This "modest species" (Jewett et al. 1953) is assumed to have benefitted from forest practices that have dramatically increased the amount of deciduous habitat (Smith et al.1997), but other forest management practices, such as herbicide spraying, which reduce deciduous regrowth, are detrimental to the species.

Noteworthy Records: *Spring, early*: 7 Apr 1991 at Tacoma; 10 Apr 1972 at Bellingham (N. Lavers): 19 Apr 1998 at Pysht; 24 Apr 1997 at Richland. *Fall, late*: 1 Oct 1995 at Crow Butte; 5 Oct 1972 at Spokane; 5 Oct 1997 at Everett; 11 Oct 1981 at Leadbetter Pt.; 19 Oct 1997 at Ocean Shores.

John F. Grettenberger

Philadelphia Vireo *Vireo philadephicus*

Casual vagrant.

A vireo breeding throughout much of n. N. America w. to e. B.C. A single-person sight record of one at Summer Falls SP, Grant Co., 25 Sep 1991 (Tweit and Paulson 1994) is the only accepted Washington record; three earlier reports were inadequately documented and two of those were from unlikely dates in Apr and Aug. The species' scarcity is surprising, because of the w. extent of its breeding range and relative frequency as a vagrant in California (116 records through 1998; Erickson and Hamilton 2001).

Bill Tweit

Red-eyed Vireo *Vireo olivaceus*

Locally fairly common to common summer resident and rare to uncommon migrant.

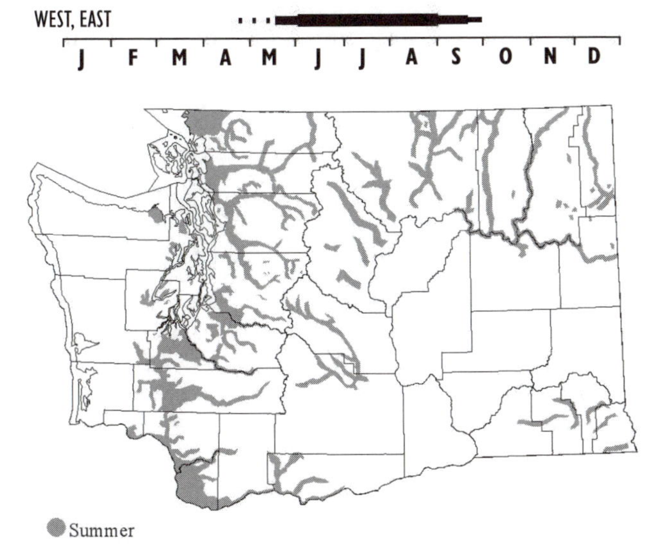

Subspecies: *V. o. caniviridis* breeds only in the Pacific Northwest. Due to identification problems, separation in the field from other subspecies is unlikely.

Habitat: Riparian areas, closely associated with black cottonwood (Smith et al. (1997) and also with bigleaf maple in w. Washington. Wide variety of deciduous habitats in migration.

Occurrence: Red-eyed Vireos apparently were not as numerous in Washington historically: Jewett et al (1953) described the species as an "uncommon and local migrant and summer resident." On the westside, it was not reported in Whatcom Co., for example, until 1913 (Wahl 1995). Gabrielson and Jewett (1940) noted an increase around Portland, Oregon, early in the 20th century. Jewett et al. (1953) state the species was probably previously overlooked and note possibly confusing similarity of vocalizations with that of *V. cassinii*.

Locally common, the Red-eyed Vireo has a somewhat limited breeding distribution in Washington. Highest breeding densities are in ne. counties along the Sanpoil, Kettle, Columbia, Colville, and Pend Oreille Rs. (Smith et al. 1997). In w. Washington, the Nooksack, Skagit, Stillaguamish, Skykomish, and Snoqualmie R. valleys and Fort Lewis have breeding populations and concentrations are found nr. the Columbia R. in Clark, Skamania, and Klickitat Cos. (Smith et al. 1997). There are few records at any season from w. of the Puget Trough (e.g., the species "probably" is a very rare migrant in the San Juan Is.; Lewis and Sharpe 1987). Nesting was noted on the Olympic Pen. nr. Sequim from 1999 to 2001.

Migrants are casual on the outer coast, rare elsewhere on the westside, and uncommon to rare on the eastside. This relative scarcity is probably due to this species' migration route: most n. to s. movement of Red-eyed Vireos takes place e. of 105° W and Washington breeders probably migrate e. or ne. before moving s. to the neotropics, following a reverse route in spring. Spring migration seems to occur mostly during late May and early Jun while fall passage appears to be largely from late Aug to mid-Sep. A dead bird found at 1800 m elevation in Whatcom Co. in Sep suggests migration in the mountains (Wahl 1995). Occasional wanderers are seen at non-breeding locations throughout the summer.

Washington BBS data indicate a decrease of 2.3% per year from 1966 to 2000 (Sauer et al. 2001), and B.C. BBS data from s. interior mountains indicated a 2% decrease per year from 1968 to 1993, and no detectible change on coastal routes (Campbell et al. 1997).

Noteworthy Records: West, *high count*: 20 on 17 Jun 2000 nr. Darrington (SGM). *Spring, early*: 7 May 1985 at Deer Harbor (Lewis and Sharpe 1987); 9 May 1976 at Seattle. *Fall, late*: 25 Sep 1999 at Mt. Pleasant, Skamania Co.; May 1984 at Cowlitz Bay, San Juan Is., 7 May 1985 at Deer Harbor, 16 Jun 1986 at San Juan I. (Lewis and Sharpe 1987). *Outer coast*: 7 Jun 1991 at Quinault.

Steven G. Mlodinow

Gray Jay *Perisoreus canadensis*

Fairly common to common permanent resident in mountains statewide; locally uncommon to rare permanent resident of westside lowlands. Rare winter visitor elsewhere in forested zones throughout the state.

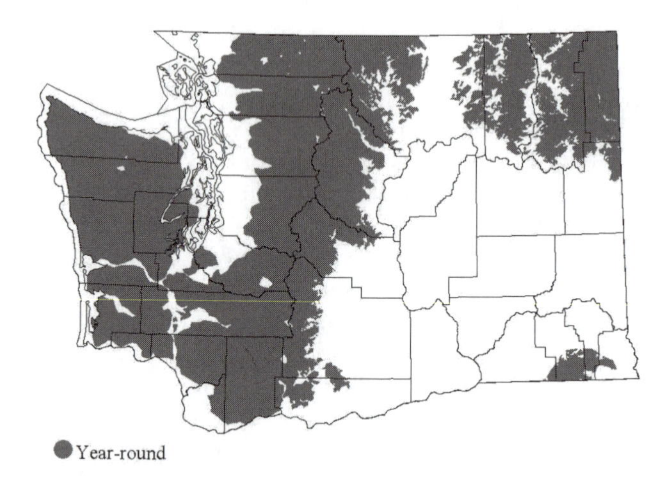

Subspecies: *P. c. obscurus* in sw. Washington and Olympic Pen., probably Puget lowlands; Cascade Gray Jay, *griseus*, in the Cascades; Idaho Gray Jay, *bicolor*, in the Okanogan Highlands and Blue Mts. (AOU 1957).

Habitat: Generally found in conifer forests above about 600-900 m elevation on the westside and about 1200 m on the eastside and interior zones with snowy winters and relatively closed canopy structure. Lower numbers in more open subalpine parkland habitat, nr. upper tree linetree line (Manuwal et al. 1987), in w. low-elevation older forests, and in eastside middle elevation, drier and warmer conifer forests (Dumas 1950, Smith et al. 1997).

Occurrence: The Gray Jay is a permanent resident of snowy forested mountainous landscapes throughout the state. Westside lowland range and abundance (Jewett et al. 1953) has probably been much reduced due to land clearing, development, and logging, and some of the species' historical sites are in areas where it does not now occur regularly (e.g. Seattle area).

Distribution is most continuous in large core areas of extensive montane forest typical of mountain landscapes in national parks and national forests. At low elevations on the westside, the species has a patchy breeding distribution, is relatively uncommon, and is restricted to middle to late seral forests. Small, scattered numbers are present in the Willapa Hills and Black Hills, coastal plains and hills on the w. side of the Olympic Pen., the Kitsap Pen., Fort Lewis, and the c. and n. Cascade foothills (Sharpe 1993, Smith et al. 1997).

Found primarily in conifer-dominated forests or parklands and avoids stands less than about 50 years in age. Also frequently encountered in mixed conifer-broadleaf riparian forests in major w. Olympic Pen. river valleys. It appears to be associated with older forests where it may be sensitive to forestry practices that favor early successional stages over mature or old-growth forest (Manuwal and Huff 1987, Huff et al. 1991, McGarigal and McComb 1995, Brooks 1997).

Though the species is not a migrant *per se*, there appears to be some downslope movement in fall and winter, both in Washington (e.g., Dawson and Bowles 1909) and in adjacent B.C. (Campbell et al. 1997), though many birds remain at high elevations in the winter.

BBS data suggest a population decline from 1966 to 2000 (Sauer et al. 2001). CBC data are limited and indicate no trend. On CBCs through 1999, 10 records were from eastside locations, four were from the westside lowlands, and 48 were from lowlands peripheral to populations in the Olympics and sw. counties, including Sequim-Dungeness, Olympia, and Grays Harbor. Eastside CBC records were from Chewelah, Spokane, and Ellensburg. This may suggest more regular dispersal of westside populations or merely a biased distribution of counts.

Remarks: Information desirable on habitat and area requirements for lowland westside residents, possible modification of forest practices to accommodate those, and the subspecific status of Puget Lowland birds.

Noteworthy Records: *Out of regular range, by county, Whatcom*: 1 on 1 May 1978 on Lummi Pen. (Wahl 1995); 2 on 21 Oct 1978 (Wahl 1995) and 2+ in early Jun 1992 at Chuckanut Mt. (CBCh); *San Juan*: 1 on 1 Feb 1975 on Cady Mt. (Lewis and Sharpe 1987); *King*: 1 on 24 Apr and 22 May 1976 at Seattle; pair on 14 Jul 1986 in Enumclaw (fide T. Bock p.c.); 1 on 3 Feb 1989; 2 on 15 Feb 1991 in Issaquah (R. Conway p.c.); 1 on 10 Feb 1996 in Renton (R. Conway p.c.); *Kitsap*: 2 on 21 Feb 2004 at Anderson L. SP (D. Paulson p.c.); *Pierce*: 31 Oct 1986 at Spanaway (fide T. Bock p.c.); 29 Aug 1976 at Fort Lewis (fide T. Bock p.c.); *Walla Walla*: 1 on 13 Feb 1997 at Walla Walla. *CBC, high counts*: 30 in 1964 at Mt. Rainier; 21 in 1982 and 1986, 23 in 1993 at Sequim-Dungeness; 14 in 1973 at Grays Harbor.

Christopher B. Chappell

Steller's Jay *Cyanocitta stelleri*

Fairly common to locally common resident in forested and adjacent habitats and urban areas statewide, some altitudinal and latitudinal movements or migration. Local in the San Juan Is.

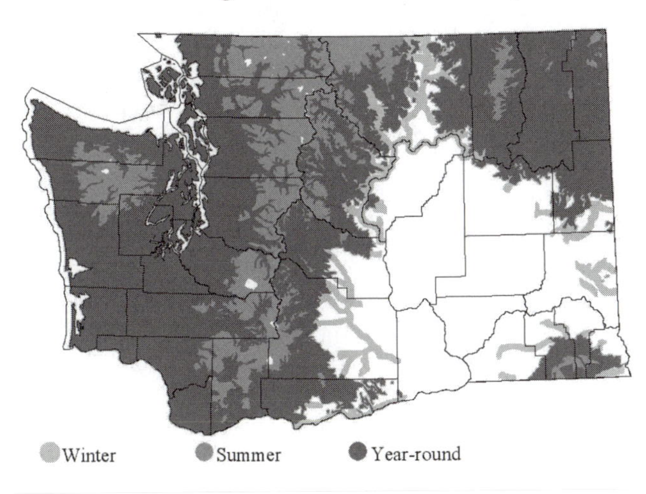

Subspecies: *C. s. stelleri* in w. Washington, *annectens* e.

Habitat: Westside Douglas-fir and hemlock forest, in edge habitats, clearings, forested parks, and urban neighborhoods. Can be in mixed forests, residential areas, orchards, and agricultural areas in forested landscapes (Smith et al. 1997).

Occurrence: Breeds to tree line at 900-1500 m, occurring peripherally in mountain hemlock, subalpine fir and alpine/parkland zones (Smith et al. 1997), where it incurs at high mountain passes into habitats used by the Gray Jay. Distribution apparently has extended to higher elevations with occurrence of people there (Smith et al. 1999). Steller's Jay is widespread on the westside, though it is uncommon locally in the San Juan Is., perhaps due to "a mistrust of crossing open water" (Lewis and Sharpe 1987), with most on Orcas I.

Noticeable movements take place in the e. between lower elevations to 2000 m in late summer, and downslope in late Aug and early Sep where it is found in fir and yellow pines above steppe zones (Jewett et al. 1953). In winter birds also occur in deciduous woodlands with diverse flora of shrubs and garry oaks (Stepniewski 1999).

Very rare from fall to spring away from coniferous woodlands and nearby towns; irregular, small incursions occur in the Columbia Basin, 80 km from coniferous forest, at Frenchman Hills Wasteway w. of Potholes Res., and others at Touchet, Wallula, the mouth of the Snake R., Clarkston, and Wapato in fall and winter.

Table 32. Steller's Jay: CBC average counts per 100 party hours by decade.

	1960s	*1970s*	*1980s*	*1990s*
West				
Bellingham		14.7	26.6	53.6
Cowlitz-Columbia			67.4	71.7
Edmonds			49.4	36.0
E. Lk. Washington			53.9	75.6
Grays Harbor		45.1	49.6	53.8
Kent-Auburn			40.4	82.4
Kitsap Co.		41.6	65.2	74.9
Leadbetter Pt		96.7	52.1	109.0
Olympia			61.5	104.1
Padilla Bay			6.1	15.9
Port Townsend			53.4	69.6
San Juan Is.			1.9	11.3
Seattle	57.2	56.5	25.2	48.2
Sequim-Dungeness			50.5	79.6
Tacoma		84.0	46.6	78.7
East				
Ellensburg			31.7	50.7
Spokane	1.5	4.3	1.3	0.5
Toppenish			2.6	2.5
Walla Walla		37.5	43.2	61.4
Wenatchee	21.0	47.1	44.3	64.5
Yakima Valley		0.0	4.5	0.3

Though birds are present year-round in breeding range, migration is often apparent on the westside from early Sep into early Oct, especially on the outer coast, with hundreds observed several times at Ocean Shores. Numbers crossing Saanich Inlet on Vancouver I., B.C., and in Seattle in early fall were other movements. Smaller flocks and concentrations in spring suggest similar, reverse movements.

Attraction to filbert orchards and residential trees and bird feeders widely noted. Birds reportedly eat stored grain, dig up vegetables in truck gardens, and victimize other nesting birds (Jewett et al. 1953).

BBS data showed a significant increase in Washington from 1966 to 2000 (Sauer et al. 2001). CBCs also suggested long-term winter increases at several locations (Table 32).

Remarks: Hybridization between Steller's and Western Scrub Jay was documented in 1999-2002 in Tacoma (S. Agnew p.c.).

Noteworthy Records: West, *Fall*: 50-100 per hr on 11 and 23 Sep 1979, 52 on 8 Sep 1991 moving s. in 12 min, 100+ on 10 Sep and 80 on 15 Sep 1992 moving through Ocean Shores area; 85 on 7 Sep 1986 moving s. along

Seattle shoreline; 90 on 10 Sep 1998 at Discovery Park, Seattle. **East**, *Fall*: noted in Nov 1990 at Touchet and Wallula, 1 on 28 Sep at Yakima Training Center, singles on 7 Oct and 8 Nov at College Place, and 5 on 27 Oct 1996 at Touchet. *Winter*: 1 on 20 Feb 1981 at Frenchman Hills Wasteway, 80 km from coniferous forest; 3 in winter 1990-91 at Snake R mouth, singles in Clarkston and Wapato. *CBC high counts*: 386 in 1996 at Olympia, 199 in 1995 at Tacoma, 181 in 1998 at Seattle; 176 in 1996 at Everett; 176 in 1995 at Sequim-Dungeness; 170 in 2000 at Bellingham; 138 in 1996 at Camas Prairie-Trout L.; 135 in 1996 at Kent-Auburn; 125 in 1975 at Grays Harbor; 100 in 1998 at Port Gamble.

Terence R. Wahl

Blue Jay *Cyanocitta cristata*

Rare winter visitor.

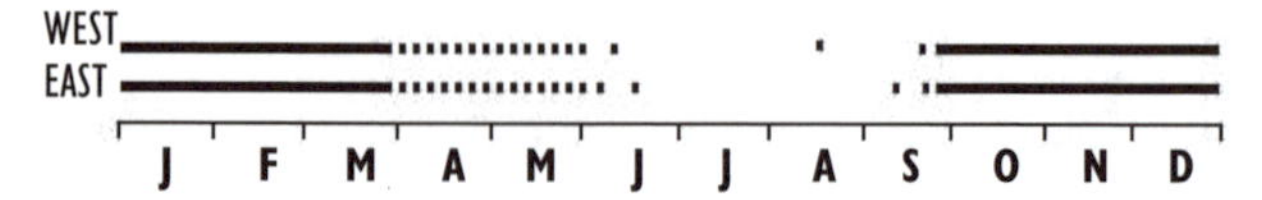

Subspecies: Presumably *C. c. bromia*.

Habitat: Deciduous woodlands, suburban areas.

Occurrence: The Blue Jay was first recorded in Washington in 1951, when specimens were secured in Jan and Oct at Pullman (Fitzner and Woodley 1976). The next record was a bird at Turnbull NWR in Sep. Beginning with a bird wintering along the L. Spokane R. during the winter of 1970-71, Blue Jays have appeared annually, with a total of 226 records through 1999. Two major incursions have been noted: the winter of 1976-77 with 51 records (Weber 1977) and the winter of 1994-95 with 32 records. Five other minor incursions (10 or more records per winter) have been noted: 1977-78 (12), 1990-91 (13), 1991-92 (13), 1996-97 (11), and 1997-98 (10). In other winters, Blue Jay records have averaged 3.7 records per year.

Possible breeding has been observed on two occasions after incursions: from 10 to 18 May 1991 a pair 10 km n. of Waitsburg, Walla Walla Co., described as "silent, secretive, carrying twigs"; and one was seen carrying nesting material in Clarkston Heights on 3 Apr 1995, but it disappeared after 13 Apr. These attempts are not unexpected, as successful breeding was documented in e. Oregon following the 1976-77 incursion (Roberson 1980).

Noting increased wintering numbers in the n. part of the range 1962-71, Bock and Lepthien (1976) theorized that increased winter feeding led to both increased populations and increased dispersion from their n. range, leading to increasing numbers in the Pacific Northwest. Smith (1979) and Weber (1977) concluded that, since Blue Jay migration patterns are n.-south diurnal movements, increased numbers in the region were from increasing populations in w. Canada, and that many followed the Columbia R. s. The geographic distribution of records supports this, as the lower Columbia Basin and the e. third of the state (e. of 120°) account for 65% of the records. Four of the most populous e. counties (Spokane, Whitman, Walla Walla, and Benton) account for 54% of state records. Westside counties only account for 26% of the total records, with almost all of those in the Puget Trough, and only four on the outer coast (2% of the total). A cluster of four recent records from late May (in 1993, 1996, 1999, 2000) from the Point No Point area, often with movements of Steller's Jays, provide an indication of a return movement.

Noteworthy Records: West, *Fall, early*: 15 Aug 1992 Freshwater Bay, Clallam Co.; 20 Sep 1998 Seattle. *Spring, late*: 9 Jun 1996 at Conway. **East**, *Fall early*: 10 Sep 1973 nr. Spokane, 22 Sep 1977 Kamiak Butte. *Spring, late*: 3 Jun 1979 Spokane R. Lincoln Co.; 15 Jun 1985 upper Wenatchee valley, Chelan Co.

Bill Tweit

Western Scrub-Jay *Aphelocoma californica*

Common resident in the s. Puget Trough, lower Columbia R. upstream through the Gorge and Klickitat valley. Uncommon resident to rare visitant n. and w. of Olympia. Very rare visitant in the interior and on the outer coast.

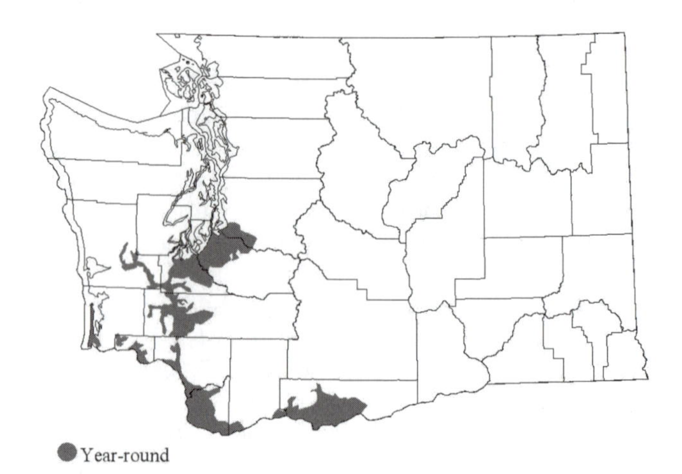

Subspecies: Two major groups recognized: a coastal *A. c. californica* group and an interior *woodhouseii* group (Pyle 1997, Dunn and Garrett 2001; see Curry et al. 2002). All but one Washington reports refer to *californica*.

Habitat: Deciduous woodlands, especially oaks, and suburban areas.

Occurrence: Scrub-jays have exhibited one of the most dramatic range expansions of any Washington species in the last several decades. Their increase has not been exclusive to Washington, as BBS data indicate a highly significant 1% per year increase in California and a somewhat significant 1.1% per year increase in Oregon from 1966 to 2000 (Sauer et al. 2001). Gilligan et al (1994) describe recent population increases in the Willamette Valley and range expansion towards the coast.

Jewett et al. (1953) described them as rare permanent residents in extreme sw. Washington, and speculated they may have been more common formerly. They cited only three records from the 1900s, and did not appear to have observed this species themselves. Similarly, Gabrielson and Jewett (1940) indicated that the species reached the northernmost limit of its range along the Columbia R. nr. Portland, and cited a thriving colony on Sauvies I. from which occasional individuals straggled to Washington. They also noted it occurred only in scattered pairs elsewhere in the n. Willamette Valley. It is unlikely that they understated the range and abundance of such a conspicuous species. Ample fieldwork had apparently been conducted in sw. Washington: Jewett et al. (1953) cite numerous distributional records of other species from Clark Co., and describe distributional surveys and observations from Clark, Skamania, and Klickitat Cos. By the mid-1970s, the expansion of scrub-jay range had been relatively minor; birds were found frequently at Ridgefield (BT), but reports elsewhere were infrequent.

Over the next quarter century, the range rapidly expanded in the Puget Trough to Olympia, e. along the Columbia R. into Skamania and Klickitat Cos. (W. Cady p.c.) and w. along the Columbia R. to Wahkiakum Co. (Smith et al 1997). The rate of expansion appears to be steady or increasing from the late 1970s on. Olympia CBC counts (Table 33), where the species was first noted in 1981 and first recorded on a CBC in 1989, show a rapid increase. Additionally, the limited number of BBS samples show a large increase of 11.7% per year from 1980 to 2000; they were unrecorded prior to 1980.

Table 33. Western Scrub-Jays on Olympia CBCs.

Year	*Number*	*Birds/100 party hours*
1989	2	1.7
1990	1	0.5
1991	5	2.2
1992	5	3.3
1993	4	2.5
1994	1	0.6
1995	11	6.6
1996	26	11.6
1997	39	16.3
1998	29	12.7
1999	25	10.3
2000	38	16.6

First confirmed breeding records from the Puget Sound and lower Chehalis Valley regions were obtained in the 1990s (Table 34), partially coincident with BBA work. A Steller'sXScrub-jay pair produced hybrid young in Spanaway in 1999 and 2000.

Away from Klickitat Co., interior records are scarce, with just 11 reports through 2002. Four are from Yakima Co., adjacent to Klickitat. The remainder are from the e. slope of the Cascades, with two e. of Snoqualmie Pass, or from the lower Columbia Basin. All but one appear to be of the coastal group; an apparent interior individual was at Asotin in Feb-Mar 2002. Records from the outer coast are also very scarce; there are only four reports through 2002.

Bill Tweit

Table 34. First scrub-jay confirmed breeding records for Puget Sound and Chehalis Valley locations.

Location	*County*	*Date*	*Reference*	*Note*
Olympia	Thurston	summer 1990	AB 44:1179	3 pairs bred
Elma	Grays Harbor	6 Mar 1994	FN 48:336	family group
Enumclaw	King	summer 1994	FN 48:983	first breeding record
Brady	Grays Harbor	11 Jun 1998	WOSN 58:10	nesting materials
Spanaway	Pierce	summer 1999	NAB 53:427, WOSN 64:8	2 ad. feeding young

Pinyon Jay *Gymnorhinus cyanocephalus*

Casual vagrant.

The Pinyon Jay, a resident in the Great Basin, formerly may have occurred on an irregular basis.

There are no recent records. Bendire found them numerous at Fort Simcoe, Yakima Co., in Jun 1881 (Tweit and Skriletz 1996) and two specimens were collected on 22 Apr 1967 from one of several small flocks present for about a week nr. Goldendale (Mattocks et al. 1976). Two published reports from the 1940s were not accepted by the WBRC because they lacked any corroborating details (Tweit and Skriletz 1996).

Bill Tweit

Clark's Nutcracker *Nucifraga columbiana*

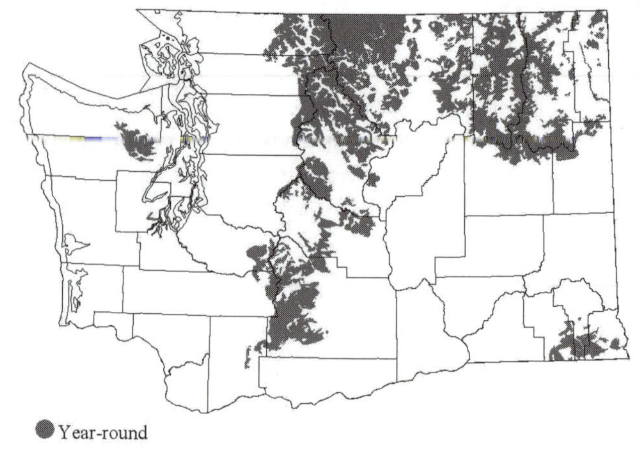

Fairly common resident, occasionally common migrant in coniferous forests, most numerous on the e. slope of Cascades, the Okanogan and ne. forests. Uncommon in e. Blue Mts. Erratic migrant and winter visitor away from coniferous forests, in deserts and urban areas. Very rare on westside and in Columbia Basin.

Habitat: Primarily in relatively open yellow pine habitat associated with cliffs at lower elevations and especially with whitebark pine parklands at higher elevations, also grand fir zone in dry e. Blue Mts. (Smith et al. 1997).

Clark's Nutcracker (G. Scott Mills)

Occurrence: A familiar summer resident in the mountains, occurs erratically in large numbers in downslope habitats, in non-forested lowlands and urban areas. Breeds in the Olympics, the w. slope of the N. Cascades, and areas like Mt. Rainier NP. Much more numerous from the Cascade crest e., downslope in forested mountains and in the Okanogan highlands. Very infrequent in sw. in Grays Harbor and Pacific Cos. at any season.

Like other high-latitude or high-elevation breeders, affected by variations in food supplies for both breeding success and winter survival and resultant irruptions are sometimes noteworthy (see Tomback 1998). In e. Washington, non-systematic reports indicated birds were noticeably abundant in winters of 1970-71, 1980-81, 1988-89, and fall 1996 at n. locations from Chelan Co. e. to Spokane. Reports of individuals or small numbers out of normal range occurred in 1965-66, 1967, 1973, 1975, and 1996-97. There were about 13 westside reports of single birds or small numbers at low elevations between 1966 and 1999, including summer occurrences in 1979, 1997, 1998, 1999.

Population trends in Washington, as over N. America (see Tomback 1998), appear uncertain but BBS data, based on minimal sampling, suggest a decline in the state (Sauer et al. 2001).

Remarks: Dispersal range of large numbers reported in fall irruptions would be of interest.

Noteworthy Records: West, *Summer*: 1 on 7 Jun 1979 at Tokeland; 1 on 25 July 1997 at Orcas I.; 1 on 26 Jul 1998 at Westport; 10 on 5 Jul 1999 on Sumas Mt. *Fall*: 1 on 22 Oct 1977 at Ocean Shores; singles at

Bellingham, Sequim-Dungeness and Seattle between 1959 and 1999 from Sep to early Oct. *Winter*: 3 in Jan 1973 nr. Ilwaco. **East**, *Summer*: 4 on 4 Jul 1966 at Turnbull NWR, 11 in summer 1971 suspected of breeding at Tower Mt. Fall: 190 passing Diamond Head, Chelan, 10-25 Sep 1996; hundreds along highways in Chelan and Okanogan Cos. in late Sep to mid-Oct 2000 (D. McNeely p.c.). *Winter*: flocks of up to 100 Sep 1970-Mar 1971 in Spokane; 75 on 15 Dec 1980 e. of Twisp. *CBC High counts*: 51 in 1999 at Chelan; 270 in 1970, 41 in 1972, 22 in 1981 and 63 in 1986 at Spokane; 21 in 1989 and 53 in 1997 at Twisp; 38 in 1976 at Wenatchee.

Terence R. Wahl

Black-billed Magpie *Pica hudsonia*

Common year-round resident at lower elevations in e. Washington. Rare, irregular visitor, mostly in fall and winter, in w.; most frequent in Whatcom and Skagit Cos.

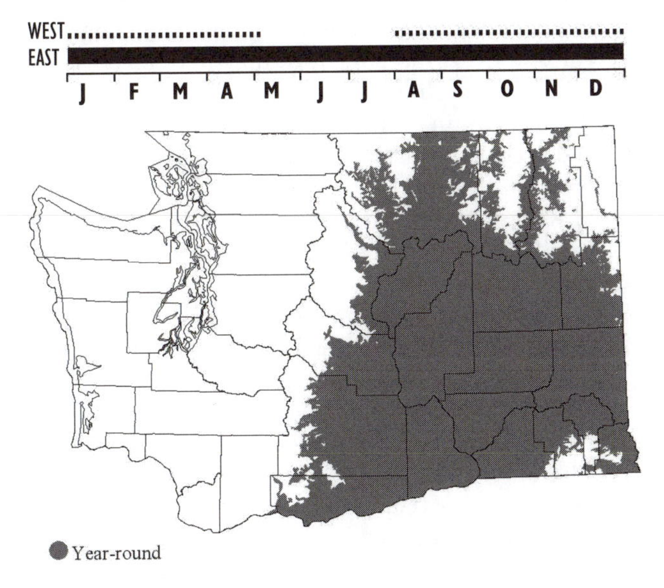

Subspecies: *P. hudsonia*.

Habitat: Shrub-steppe, agricultural, and other open habitats with scattered deciduous trees and shrubs. Favors thick cover in riparian habitat and may nest in loose colonies in scrub-filled draws (Cannings 1987). Prefers thickets and riparian areas for roosting, and varied open and human habitats for foraging (Campbell et al. 1997).

Occurrence: Magpies breed throughout many eastside lowland habitats below most forested regions and forage widely in agricultural and urban areas (Smith et al. 1997). Some move upslope into forested areas following breeding (Smith et al. 1997) to forage on conifer seeds (Stepniewski 1999), and to lower elevations in winter (Weber and Larrison 1977),

Table 35. Black-billed Magpie - CBC average counts per 100 party hours by decade.

	1960s	*1970s*	*1980s*	*1990s*
Ellensburg			288.1	532.9
Spokane	164.9	177.8	189.3	182.7
Toppenish			685.6	904.2
Tri-Cities		481.3	512.7	556.1
Walla Walla		427.5	385.3	495.7
Wenatchee	159.7	110.1	91.0	136.0
Yakima Valley		486.6	581.0	1351.7

when they aggregate at foraging sites like feedlots, garbage dumps, and grain elevators and roost communally in dense thickets. Birds "are wary of high-density development" in urban areas where crows are becoming more common (Smith et al. 1997).

Jewett et al. (1953) stated that birds make a "pronounced partial westward migration, wintering commonly" in w. Washington, and Dawson and Bowles (1909) stated it "migrates regularly but sparingly thru mountain passes to West-side at close of breeding season." It is apparent that the situation has changed. Even with many more observers afield and reporting findings regularly, numbers of westside reports declined in the 1980s and 1990s. The species has only been recorded three times on westside CBCs, the last one in 1988. Whether this decrease may be due to extensive habitat change or other factors is unknown.

Statewide, populations appear relatively stable. BBS data for Washington suggest a non-significant increase from 1966 to 2000 (Sauer et al. 2001). Winter numbers appeared to increase overall on CBCs in e. Washington (Table 35). All counts with regular magpie occurrence increased over time, with numbers in Ellensburg, Tri-Cities, and Yakima Valley significantly higher.

Remarks: Magpies were historically persecuted as vermin with a 10¢ bounty, but shooting and incidental mortalities from predator control probably are lower today. Crows and introduced fox squirrels impact egg survival at some locations in Idaho and elsewhere. Urban sprawl can eliminate habitat except where riparian vegetation is maintained (Trost 2000). Considered a superspecies with Yellow-billed Magpie (AOU 1998).

Noteworthy Records: West, *Spring*: 50+ on 4 Mar 1905 at Glacier, Whatcom Co. (Dawson and Bowles 1909). *Winter*: 6 in 1975-76 in w. Whatcom and Skagit Cos.; 7-10 in 1979-80 along Cowlitz R. nr. Morton.

Derek W. Stinson

American Crow *Corvus brachyrhynchos*
Northwestern Crow *Corvus caurinus*

Common to abundant resident statewide, fairly common to common in open, dry eastside areas, scarce and local at high elevations. Local, rapidly spreading, in Columbia Basin.

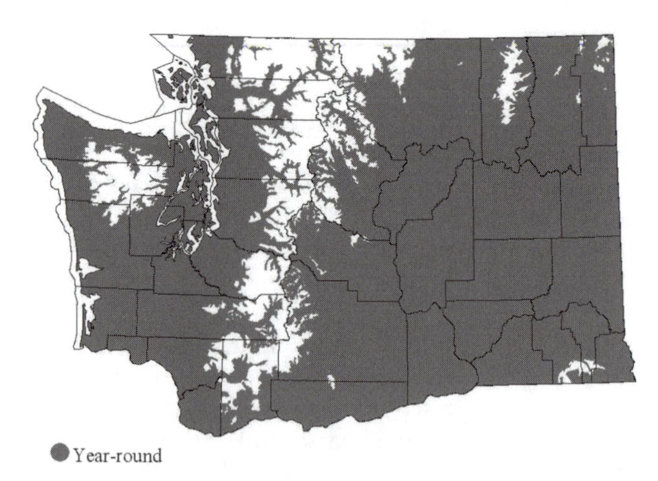

● Year-round

Subspecies: Described as distinct species (Baird 1858 in Verbeek and Butler 1999, AOU 1910, 1957, 1983, 1998, Jewett et al. 1953) or subspecies: American Crow, *C. brachyrhynchos hesperis,* and Northwestern Crow, *C. brachyrhynchos caurinus* (Dawson and Bowles 1909, AOU 1931, Mattocks et al. 1976, Smith et al. 1997, and see Verbeek and Butler 1999, Verbeek and Caffrey 2002).

Habitat: Nests widely in trees in urban areas, farms, parks, thickets, clearings in second-growth timber areas. Forages widely, opportunistically also at campgrounds, dumps, seabird colonies, beaches, westside intertidal areas.

Occurrence: Many w. Washington observers have not attempted to differentiate between the two species/forms, and reports and most data are "lumped" (see Smith et al. 1997). Historically the American Crow was described generally as occurring in e. Washington and upland westside areas and the Northwestern Crow as a saltwater-edge resident coastally and along inland marine waters. Georgia Strait, in Washington and B.C., is considered to be the "center of abundance" of Northwestern Crow (Campbell et al. 1997 in Verbeek and Butler 1999).

Crows are local in e. Washington and expanding into cities and settled areas, especially in the vicinity of irrigated farmlands They began nesting nr. Othello, where they were infrequent previously, in 1991 or 1992 (R. Hill p.c.). E. of Othello, in dryland agricultural areas generally dominated by magpies and ravens, crows are absent except in ponderosa pine areas and developed areas such as the Tri-Cities and Wallula, where they are common. Along the Cascade crest, crows occur at ski areas, towns, and nearby managed forests. In lowland forested areas, they are present in similar areas and in farmlands. In contiguous lowland conifer forest the crow competes with or is replaced by the raven (Smith et al. 1997).

It is probable that logging and development have contributed to expansion of historical ranges of both forms and result in overlap and intergradation. Verbeek and Butler (1999) suggested that the clearing of the coastal forests and rivers led to expansion of Northwestern Crows up the rivers and away from coasts. Adapted to Native American settlements, this form does well in modern urban areas. Campbell et al. (1997) reported Northwestern Crows from sea level to 1700 m elevation and inland as much as 120 km along major rivers in B.C. and Jewett et al. (1953) reported them up Columbia R. as far as Vancouver, Washington. The American form may have spread across the mountains, following highways, powerline rights-of-way, and adjacent agricultural development. Both may have increased also due to restrictions on shooting, increase in food supply, and elimination of predators (Wahl 1995).

Migration is uncertain (Jewett et al. 1953). Recent authors (e.g., Verbeek and Butler 1999) noted winter concentrations of "hundreds or thousands," likely indicating local movements in response to food availability at farms, processing plants, or garbage dumps. The Northwestern Crow is nonmigratory, moving locally (Gabrielson and Lincoln 1959, Campbell et al. 1997). Winter range of that form is almost identical to its breeding range except that birds move from the coast to more protected areas and tend to forage nr. human settlements rather than on coasts due to high daytime tides (Campbell et al. 1997, Verbeek and Butler 1999).

BBS data from 1966 to 2000 separate the two "species" but trends are inconclusive (Sauer et al. 2001). Long-term CBC data are variable by location but appear to indicate local increases. These interpretations are similar to findings in lower mainland B.C. by Verbeek and Butler (1999).

Remarks: Desirable studies include the taxonomic relationship between Northwestern and American crows in areas of integration, especially information on morphology and vocalization within breeding pairs and distributions in sw. Washington (Verbeek and Butler 1999).

Noteworthy Records: East, *Summer*: flocks of up to 200 in 1969 at McNary NWR; >200 in mid-Jul 1985 over Walla Walla. **West,** *Winter*: estimated 25,000 on 19 Feb 2001 over Pilchuck R. nr. Snohomish (S. Kostka p.c.). *CBC high counts*: 9931 in 1994 at Seattle; 5280 in 1998 at Tacoma; 4096 in 1985 at Kent-Auburn; 3437 in 1974 at Bellingham; 3413 in 1989 at Olympia; 3064 in 1992 at Grays Harbor; 2021 in 1985 at Kitsap Co.

Ann M. Eissinger

Common Raven *Corvus corax*

Fairly common resident in w., common resident in e. Washington.

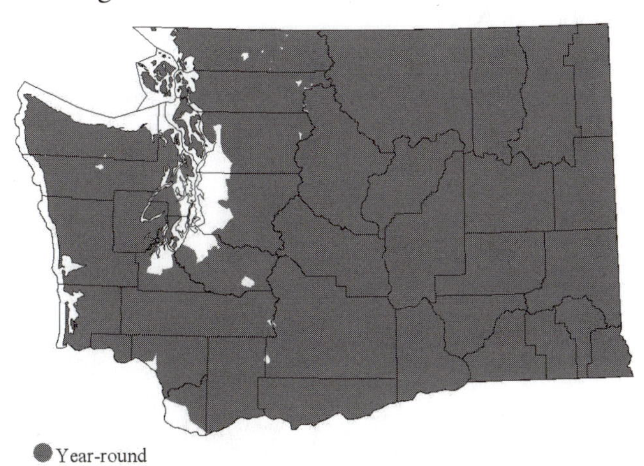

Subspecies: *C. c. sinuatus.*

Habitat: Mountains, coniferous forests, cliffs, sagebrush deserts. Forages widely, including fields, highway shoulders, beaches. Occasional in urban areas.

Occurrence: The raven has long been considered one of the most widespread species in the state (Booth 1942, Kitchin 1949, Wahl and Paulson 1991, Weber and Larrison 1977, Smith et al. 1997). It is less common in developed areas and more abundant at higher elevations, less conspicuous than and somewhat exclusive of crows in distribution.

Generally more abundant in mountainous regions, below snowline (Larrison 1947a, Kitchin 1949, Ehrlich et al. 1988), "fairly common" resident in uncultivated, open areas and cliffs along the Columbia R. in e. Washington (Booth 1942, Weber and Larrison 1977, Knight et al. 1982). Some birds wander to higher elevations in summer and fall (Stepniewski 1999).

Kitchin (1949) and Dawson and Bowles (1909) described birds using coastal beaches at low tides and carrying food back to inland nest sites, and ravens are notorious for feeding on road- and railroad-kills (e.g. Stepniewski 1999). Infrequently noted in large flocks, ravens sometimes roost in flocks of hundreds or more. Birds are generally excluded from major urban areas, probably due to harassment, crows, and lack of nest sites (Smith et al. 1997), though confirmed nesting in urban King Co. (Hunn 1982).

Table 36. Common Raven: CBC birds per 100 party hours by decade.

	1960s	*1970s*	*1980s*	*1990s*
West				
Bellingham		4.0	9.8	14.1
Grays Harbor		0.8	4.2	7.5
Kitsap Co.		0.0	3.4	4.1
Leadbetter Pt		2.3	8.2	26.6
Padilla Bay		12.6	33.1	41.9
Port Townsend		0.6	6.0	13.3
San Juan Is.		15.2	26.3	15.8
Sequim-Dungeness		14.5	47.0	80.3
East				
Spokane	0.3	12.9	25.5	48.6
Toppenish			82.6	97.8
Tri-Cities		0.6	0.9	7.5
Walla Walla		0.5	12.8	45.8
Wenatchee		3.7	14.5	53.5
Yakima Valley		1.8	7.0	47.5

Population decreases in other parts of N. America (Boarman and Heinrich 1999) are not evident in Washington. In fact, present widespread distribution (see Smith et al. 1997) shows an obvious change in status from when Dawson and Bowles (1909) described local distribution as nearly confined to the Olympic Pen. In 1989 observers reported a pattern of ravens becoming common in n. Puget Sound and uncommon in s. Puget Sound. CBC data indicate a significant increase ($P < 0.01$) in winter statewide (Table 36; Wahl 1995). BBS data also suggest an increase statewide (Sauer et al. 2001).

Remarks: Two subspecies, *sinuatus* and *principalis*, were described, the former in e. Washington and the latter on the westside (Dawson and Bowles 1909, Jewett et al. 1953), while Kitchin (1949) verified, through specimens, the occurrence of only *sinuatus*. The species is listed as a gamebird in B.C. (Campbell et al. 1997). The species has long been associated with negative impacts on other animals including endangered species (Boarman and Heinrich 1999).

Noteworthy Records: *High counts*, **East**, *Spring*: 50 on 20 May 1973 at Yakama Reservation. *Winter*: 35 on 22 Feb 1967 at Thomas L., nr. Tiger. **West**, *Fall*: 41 on 10 Nov 1973 at Samish I. *CBC high counts*: 499 in 1991 at Moses L.; 241 in 1997 at Twisp; 239 in 1998 at Sequim-Dungeness; 173 in 1995 at Camas Prairie-Trout L.; 169 in 1998 at Columbia Hills-Klickitat Valley; 167 in 1997 at Ellensburg; 162 in 1993 at Toppenish; 127 in 1996 at Spokane; 122 in 1992 at Padilla Bay; 115 in 1985 at Chewelah; 111 in 1993 at San Juan Is.

Jennifer R. Seavey

Sky Lark *Alauda arvensis*

Formerly very local common resident on San Juan I., apparently extirpated. Very rare winter visitor.

Subspecies: Nominate, *arvensis,* introduced. Siberian, *pekinensis,* recorded as far s. as California.

Habitats: Short-grass fields, including pastures, golf courses, and airfields. Nested in Washington amid rabbit warrens on grassy slopes.

Occurrence: Introduced into a number of areas across N. America (see Campbell et al. 1997), established successfully only on s. Vancouver I., B.C., following introduction in 1903. By 1960 birds had reached San Juan I. (Bruce 1961) and later nested successfully at the American Camp area at the s. end of the island (Wahl and Wilson 1971, Weisbrod and Stevens 1974). Birds concentrated there, in a sloping, grassy area not only apparently suitable to the birds but, with scenic views of the Olympic Mts. and the Strait of Juan de Fuca and essentially separated from noise and development, also esthetically pleasing to humans (Lewis and Sharpe 1987). Inclusion within San Juan National Historical Park assured preservation of habitat of grasses, thistles, and ferns, shared with and affected by a huge population of introduced European rabbits (Weisbrod and Stevens 1974). Between the early 1970s and 1988, 12-60 pairs were estimated to be present in the American Camp area.

Populations on Vancouver I. increased into the 1960s, with an estimated 1000 birds present in 1965. Numbers began declining in the late 1960s, attributed in part to severe winters and habitat loss. Though a few more birds may reside at a few unknown locations, just 64 singing birds were located in 1993, in four restricted areas nr. Victoria and around the airport at Sidney, B.C. (Campbell et al. 1997). Though habitat was protected, numbers on San Juan began to decline and by 1998 none were reported with certainty. This was attributed to the introduction of foxes. The birds' situation on San Juan I., with protected, suitable ground-nesting habitat, was probably unique in Washington and, unfortunately, also vulnerable to such threats. Two birds nr. Sequim in 1998-99 may have been vagrant *pekinensis.*

Remarks: The only possibility for re-establishment of a breeding population would appear to be the elimination of predators on San Juan I. Formerly called the Eurasian Skylark (AOU 1995).

Noteworthy Records: *Winter:* 2 on 23 Dec 1998-1 Jan 1999 nr. Sequim. *CBC:* average of 3.5 birds recorded on 11 of 16 counts between 1982 and 1997 on San Juan I., with a maximum of 12 in 1983.

Terence R. Wahl

Horned Lark *Eremophila alpestris*

Locally common summer and widespread common winter resident in e. Washington. Local breeder in high Cascades and Olympics; local, very rare breeder in w. lowlands. Locally common to abundant migrant and winter visitor in e., uncommon migrant and local in winter w.

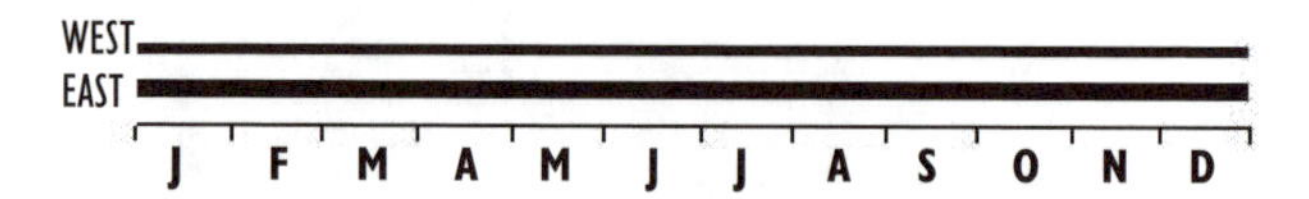

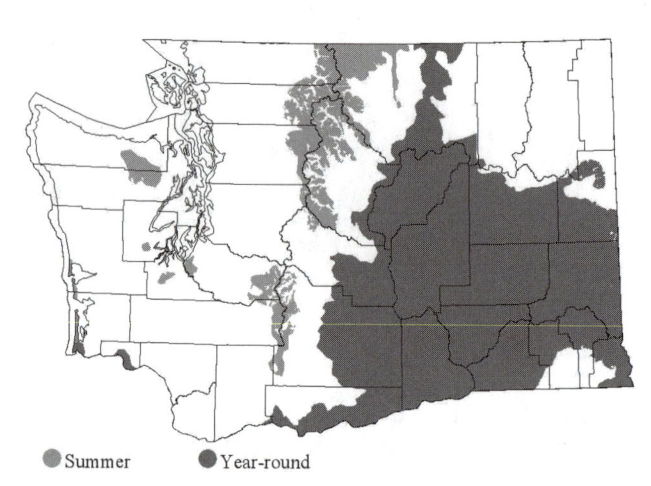

Subspecies: Dusky Horned Lark, *E. a. merrilli,* breeds in steppe and agricultural areas below ponderosa pine zone in e. Pallid, *alpina,* at high elevations in Cascades and ne. Olympics; Streaked, *strigata,* in remnant prairies s. of Puget Trough and on outer coastal beaches (Smith et al. 1997). "Arctic," *arcticola,* winters e.

Habitat: Open, short-grass habitats, plantings of crops like wheatgrass (Littlefield 1990), open shrub-steppe sagebrush, at least formerly at airport edges, ocean beaches. Fields and irrigated farmlands seasonally (Stepniewski 1999).

Occurrence: Description of the status of the Horned Lark is complicated by the number of subspecies occurring in Washington. While it remains common to seasonally abundant e. of the Cascades it has essentially disappeared from much of the w. lowlands as it has in sw. B.C. (Campbell et al. 1997).

Variations in nesting success e. of the Cascades likely relate to precipitation, with fewer birds present when vegetation is high following winters of heavy rainfall (Stepniewski 1999). Wheat fields are important habitat in winter, and heavy snows force large numbers of birds to gravel on highway shoulders (Stepniewski 1999). A species of open, short-grass habitats that has benefitted from extensive agricultural plantings e. of the Cascades (Littlefield 1990), it is widespread in Columbia Basin and adjacent open zones at low elevations, apparently absent from Methow Valley, sparse in Okanogan Valley (Smith et al. 1997).

Birds breed at high elevation in volcanic rainshadows of Glacier Peak, Mt. Rainier, Goat Rocks, Mt. Adams (Smith et al. 1997), possibly at Mt. Baker. Changes in status of this mountain subspecies are unknown. Status at Mt. St. Helens following the volcanic eruption in 1980 is uncertain.

In most w. lowlands, urbanization, conversion of prairies to agriculture, fire suppression, and vegetation changes probably caused extirpation of *strigata* (Smith et al. 1997). Birds similarly decreased in lower mainland B.C. with only a very few nesting sites known (Campbell et al. 1997). Reports indicated a decline in w. Washington lowland populations in the early 1980s, with only small numbers now noted during migration. Along with birds in the Ft. Lewis area, fields and prairies and along coastal dunes at Ocean Shores, birds in spring and summer 1998 and 1999 at airports at Shelton and Olympia suggested nesting at the n. edge of prairies s. of Puget Sound.

Birds are variable by season, and more widespread in winter. Westside migration is apparent from late Jul to Nov, peaking from late Aug to late Sep. Birds are most common on the outer coast (SGM), but have declined elsewhere. Spring migration is from late Feb to early Mar into mid-Apr. Except for 15 at Padilla Bay in 1996, the species has essentially disappeared from westside CBCs with the next-to-last record in the Puget Trough in 1983, and very small numbers coastally at Grays Harbor and Leadbetter Pt. as late as 1996.

BBS data indicate a decrease in w. U.S. over 1966-91 (DeSante and George 1994). Changes in interior and n. B.C. populations were not evident (Campbell et al.1997), though numbers decreased in e. Washington ($P<0.01$) from 1966 to 2000 (Sauer et al. 2001). No statistically significant changes are shown in eastside winter CBC abundance.

Remarks: *Strigata* classed as SC, FC, PHS. Breeding status at Mt. Baker and Mt. St. Helens should be determined.

Noteworthy Records: West, *Spring-Summer*: nesting nr. Cathlamet in 1990; nr. Port Angeles in 1994. *Fall*: 1 on 13 Sep 1970 64 km off Grays Harbor. *Winter*: 50 on 3 Feb 2000 at Steigerwald L. NWR and 28 on 1 Feb 2001 at Midway Beach (S. Mlodinow). **East**, *high count*: 5000 on 28 Jan 1979 nr. Moxee; 5000-10,000 on 19-20 Feb 1989 following heavy snowfall nr. Toppenish NWR (Stepniewski 1999); 13,000 on 20 Feb 1993 in e. Yakima Co. *CBC high counts*, **East**: 2800 in 1970 at Yakima Valley, 2611 in 1990 and 1324 in 1992 at Two Rivers, 1776 at Moses L. in 1992, 1100 in 1963 at Walla Walla; **West**: 47 in 1970 at Bellingham, 21 in 1983 at Sequim-Dungeness, 12 in 1976 at Grays Harbor. Not reported from 1962 to 1998 at Olympia or Tacoma, nr. remnant breeding populations.

Terence R. Wahl

Purple Martin *Progne subis*

Rare to uncommon migrant and local breeder in w. Washington. Very rare e. of the Cascades.

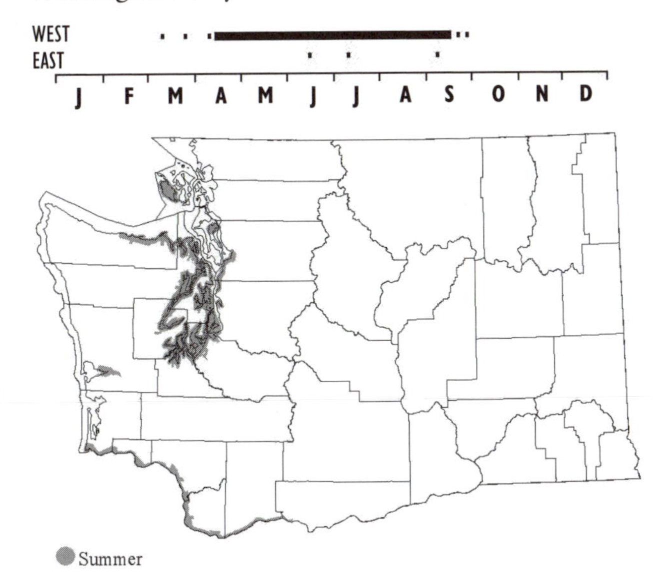

Subspecies: *P. s. arboricola.*

Habitat: Nests primarily in artificial cavities (nestboxes and gourds) specifically put up for martins on marine pilings, and incidentally in and on various human-made structures over or nr. interior marine waters, less commonly over lakes and marshes, rarely upland or in snag cavities. Forages nr. nestsite, usually higher than other swallows (Brown 1997).

Occurrence: Rarely noted in the mid-1800s (Suckley and Cooper 1860), martin populations increased following European settlement, apparently from increased incidental nestsite availability in human-made structures (Lund 1978, Milner 1988). By the early 20th century, martins were considered fairly common migrants and summer residents in w. Washington (Dawson and Bowles 1909, Edson 1919, Jewett et al. 1953), many nesting in cavities in marine pilings. A huge pre-migration roost of 7000-12,500 occurred in Seattle during the 1940s

(Higman 1944, Larrison 1945). Martins then declined as starlings became established (Smith et al. 1997) and possibly due also to competition from House Sparrows for nestsites (Jewett et al 1953, Campbell et al. 1997), cavity loss from clear-cutting without snag retention (USFWS 1985), nest parasites, and harsh weather (Milner 1988). Virtually all references have focused on availability of nest cavities as the primary factor affecting distribution and abundance (Williams 1998).

With remnant Puget Trough colonies in decline in the 1970s, nestbox installations began in an attempt to rebuild the population (Davis 1995). In 1980, 40 pairs were noted in nestboxes and on buildings. In 1981 WDFW recommended that the martin be designated a state threatened species and WDFW began monitoring. By the mid-1980s, 35-40 pairs nested in the Olympia-Shelton area. Twenty or more pairs nested at Ft. Lewis in martin houses and wood-duck boxes and isolated pairs nested on San Juan and Bainbridge Is. By the late 1980s numbers had stabilized and increased at some nestbox colony sites. These increases contrasted with nr. disappearance of the species in nw. Washington (e.g., Wahl 1995). Martins did well on Ft. Lewis, the n. extent of sizable martin populations in 1989, breeding in houses and snags. In the late 1990s moderately large colonies nested at Woodard Bay, Hylebos waterway, Vashon I., Seabeck marina, and L. Nahwatzel (M. Tirhi, WDFW, Western Purple Martin Working Group, unpubl. 1999). Ridgefield and Steigerwald NWRs reported 60 nests in 1999. By 2002, over 40 pairs nested in nestboxes at Everett, Tulalip, and Camano I. and in 2003 small numbers occupied nestboxes in Skagit and Whatcom Cos. (SK). Birds now breed locally n. to San Juan I., throughout the Puget Trough n. to Bellingham, w. to Hoquiam at Grays Harbor, s. to Washougal (WDFW) along the lower Columbia R. e. to Bingen (S. Johnston p.c.).

Arrival begins in Apr, subads. arrive in May-Jul. Birds depart in Aug-Sep. Large premigratory gatherings or roosts are currently unknown regionally. The species is very rare e. of the Cascades (Woodruff 1995, WDFW).

Population statewide is estimated at 600 nesting pairs with approximately 400 in the Puget Trough (WDFW, M. Tirhi p.c.), and 200 on the Columbia R. (D. Fouts p.c.). The recent Washington nestbox increase parallels events in Oregon (Horvath 1999) and B.C. (Copely et al. 1999). Though BBS data do not indicate a long-term trend in Washington, there was an apparent increase of 1% nationwide from 1966 to 1999 (Sauer et al. 2000).

Remarks: Also called Western Purple Martin, West Coast Purple Martin. Starlings are a moderate threat to martins at managed sites and capable of seriously reducing populations at unmanaged colonies (Brown 1977, 1981). During the decline in the Northwest virtually all martins nested in unmanaged colonies. Designated SC, PHS.

Noteworthy Records: West, *Spring, early*: 10 Mar 1996 at Ridgefield NWR. *Summer, late nest*: 12 Sep 1982 at Dugualla Bay. *Fall, late*: 20 on 22 Sep 1974 at Olympia. **East**: 1 on 14 Jun 1979 at Tucannon R., Columbia Co.; 1 on 4 Sep 1990 at Richland; 1 on 16 Aug 2001 at Wenatchee.

Stanley Kostka and Kelly McAllister

Tree Swallow *Tachycineta bicolor*

Locally common summer resident nr. water in w. and e. lowlands, very rare in winter w.

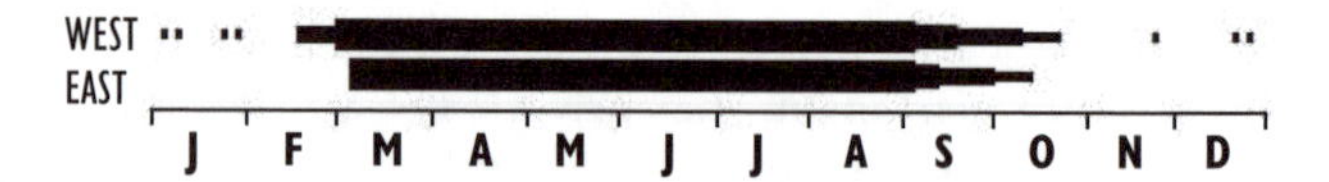

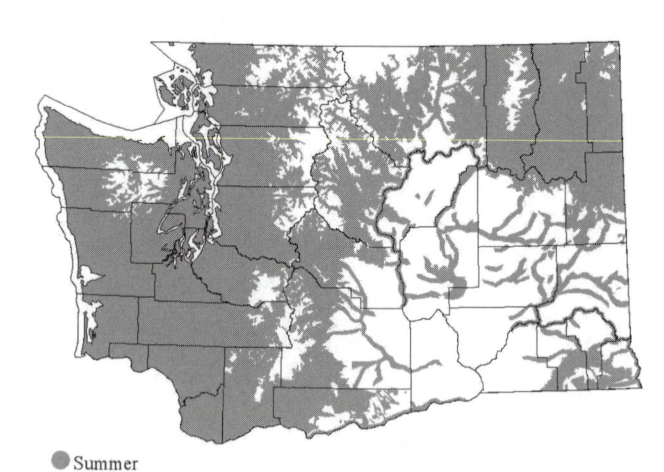

Summer

Habitat: Occurs at edges of forested habitats associated with marshes, beaver ponds, lakes, and rivers, nests in tree cavities especially over water, holes in buildings, and readily utilizes nestboxes. Forages widely, often over open water.

Occurrence: Common at low elevations w. and e. of the Cascades, more locally in foothills and at mid-elevations in mountains w., this is the earliest-arriving swallow in spring in the w. though it is later than the Violet-green Swallow in the e. Though recorded in Jan, most arrive in Feb in the w. and in Mar in the e. Like other swallows, first arrivals in spring concentrate very locally at lowland lakes with early insect hatches (e.g. Wahl 1995) and birds can utilize vegetation briefly during bad weather (Robertson et al. 1992). Found foraging more often nr. water during summer residence than the Violet-green Swallow.

Interannual variations in spring arrival and nesting success attributed to weather conditions during migration and particularly when nestlings are being fed. Fall migration is gradual, with a slow decline in numbers (Campbell et al. 1997).

In w. Washington, drainage of swamps negatively affected numbers, and conversion of forests was said to reduce numbers of this species (Jewett et al. 1953). Some changes (e.g., forests to open-country agriculture) may have had the opposite effect. Availability of nest sites controls numbers in a locale (Godfrey 1986). Nestboxes provided 69% of sites for sample in B.C., with preference for smaller holes than boxes used by starlings and bluebirds (Campbell et al. 1997). Larrison and Sonnenberg (1968) stated range and population in e. had increased, presumably due to creation of reservoirs, and irrigated areas.

Changes in historical range and abundance with those in 2000 are uncertain. E. of the Cascades habitat changes appear to have been favorable to this species, but an overall change in the w. lowlands from 1950 to 2000 is unknown. BBS data indicated a continent-wide increase from 1966 to 1991 (Peterjohn and Sauer 1993), but a decline from 1966 to 2000 is suggested statewide (Sauer et al. 2001).

Noteworthy Records: West, *Fall-Winter*: 14 Nov 1981 at Bellingham (Wahl 1995); 3 on 27 Dec 1981 at Ridgefield; 23 Jan 1986 at Ridgefield; 31 Jan 1988 at American L.; 31 Jan 1990 in s. Clark Co.; 2 on 19 Dec 1988 at Tacoma; 1 on 19 Dec 1995 at Ridgefield NWR with 11 there on 29 Jan 1996. 1 on 3 Jan 1992 at Seattle; 1 on 6 Jan 1994 at Nisqually; 1 on 6 Jan and 3 on 9 Jan 1999 at Elma; 1 on 27 Dec 1999 at American L. *CBC*: 1 in 1988 at Sequim-Dungeness, 1 in 1997 at Skagit Bay; CW in 1978 at Tacoma.

Terence R. Wahl

Violet-green Swallow *Tachycineta thalassina*

Widespread, common summer resident in forest zones to moderate elevations, in adjacent urban and agricultural habitats; more localized in urban areas in Columbia Basin.

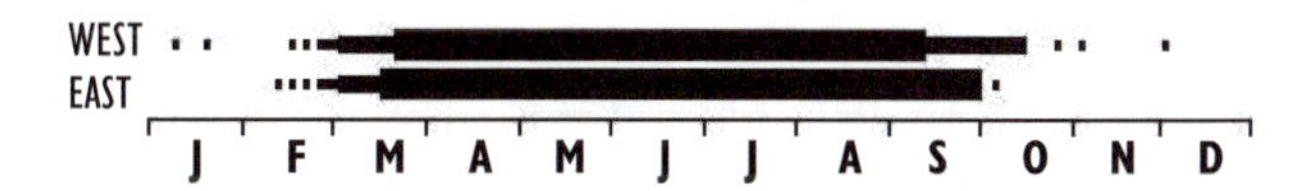

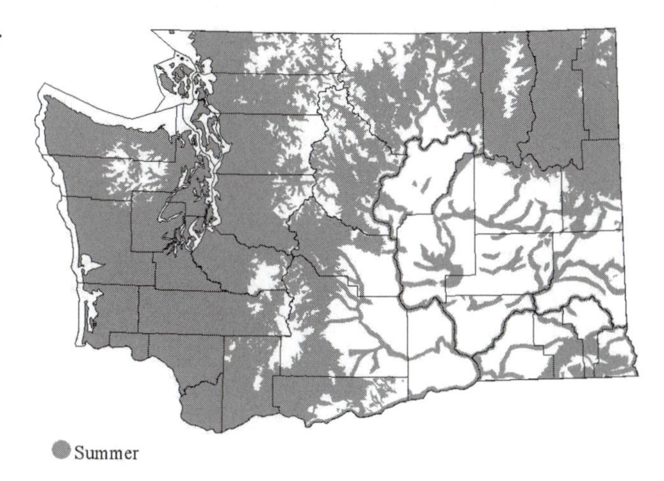

Subspecies: *T. t. lepida.*

Habitat: Nests in holes in trees, rock crevices, buildings, light standards, and readily in nestboxes, principally about settlements and prairies w.; in the coulee country, bunchgrass, and wheat-growing areas and especially in forested regions.

Occurrence: Likely to take advantage of nesting opportunities in urban and settlement areas. This species is more widespread than the Tree Swallow. It often nests in openings in forested areas, cliffs, away from water and in the mountains (at 1200-1500 m; see Smith et al. 1997).

Arrival dates and nesting success vary with weather conditions during migration and locally when food supplies and foraging are restricted due to bad weather. Like the Tree Swallow, early spring arrivals concentrate at lowland lakes. Though broadly described by regional authors as occurring in forested zones, birds may utilize forest edges for nesting but are primarily associated with open habitats within or adjacent forested areas (e.g., Smith et al. 1997).

Differences between departure of local breeders after young fledge and later passage of fall migrants noted (Campbell et al. 1997, Stepniewski 1999). The species often gathers conspicuously in single-species flocks in Aug-Sep before fall migration. Flocks are often seen during diurnal fall migration, sometimes with Vaux's Swifts (Wahl 1995). Birds are noted migrating at high elevations: 1500 m in the N. Cascades and 2700 m on Mt. Rainier (Jewett et al. 1953).

Like the Barn Swallow, the species may have benefitted from and maintained numbers through provision of nesting and foraging opportunities by humans (Campbell et al. 1997). BBS data indicate a significant increase in numbers in Washington from 1966 to 2000 (Sauer et al. 2001).

Remarks: Replacement of this species by more aggressive Tree Swallow in e. Washington (Larrison and Sonnenberg 1968) was not addressed by later authors. Fidelity of a migratory species to a nesting site shown by record of a female banded on Vancouver I. in 1983 and recaptured annually through 1989 (Campbell et al. 1997).

Noteworthy Records: *Winter*: 2 on 8 Nov 1974 at Ocean Shores; 1 on 15 Jan 1985 at L. Steilacoom; 4 Nov 1989 at Kent; 1 on 1 Dec 1998 at Kent; 3 on 6 Jan 1999 at Elma; 3 on 18 Jan 2003 at Ridgefield; 5 on 21 Jan 2003 nr. LaConner. *High counts,* **West**, *Spring*: 500 on 29 Apr 1997 at Ebey I., Snohomish Co. *Fall*: 1000 on 8 Sep 1996 at Admiralty Cove, Island Co. (SGM); 850 on 1 Oct 2000 at Tumwater; 600 on 31 Aug 2000 at Ridgefield; 250 on 1 Sep 1999 at Skagit flats; 200 on 3 Sep 2000 at Fir I. **East**, *Spring*: 100 on 18 Mar 2000 at Vantage; 500 on 15 Mar 2001 at Cow L., Adams Co. *Fall*: 200 on 2 Oct 1971 at Wenas L.; 2000 on 31 Aug 1987 nr. Winthrop; 800 on 26 Sep 1999 at Wanapum Dam.

Terence R. Wahl

Northern Rough-winged Swallow

Stelgidopteryx serripennis

Fairly common, locally common summer resident at low to moderate elevations statewide.

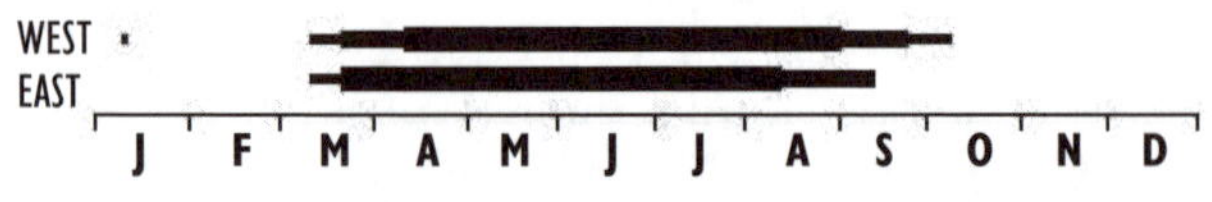

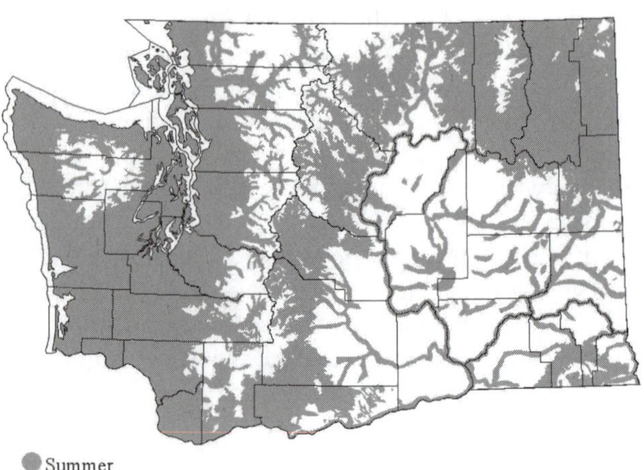

Subspecies: *S. s. serripennis.*

Habitat: Nests mainly along rivers, flood-plain streams, and coastal areas. Occurrence in upstream areas limited by lack of soft sand sediments (Smith et al. 1997). Requires holes in sand sediments for nesting and these limit numbers and nesting distribution to banks cut by stream flows, road cuts, inactive storage piles of road materials, and unused kingfisher holes (Campbell et al. 1997). Forages over water, but also over open fields and shrub-steppe habitats away from water.

Occurrence: Relatively inconspicuous and usually found locally only in small numbers, the bank-nesting Rough-winged is usually found with other swallow species. E. of the Cascades nesting birds are often almost hidden in large Bank Swallow colonies.

Nests as solitary pairs or few pairs ("loosely colonial" [Campbell et al. 1997]), essentially along streams statewide. First noted in spring with larger numbers of other swallows over water bodies with early insect hatches. Like other single-brooded swallows, fall migration not obvious but appears to start after young fledge, with numbers gradually diminishing.

In addition to cuts and bluffs resulting from highway construction, populations may have benefitted from erosion of stream banks caused by cattle (Stepniewski 1999). Along the coast, nests reportedly associated with sites used by Pigeon Guillemots (Jewett et al. 1953). Nesting habitat increased e. of the Cascades due to increased flooding of rivers (Smith et al. 1997) and, in B.C., 54% of nests were in human-influenced habitats (Campbell et al. 1997).

As in the case of the Bank Swallow, populations on the eastside may have peaked in the latter half of the 20th century during periods of highway expansion, dam and reservoir construction. BBS data suggest a decline from 1966 to 2000 in Washington (Sauer et al. 2001).

Noteworthy Records: West, *Winter*: 1 on 24 Dec 1994 at Point No Point, 1 on 10 Jan 2003 at Edmonds. **East**, *Spring*: 1000 on 20 Apr 1999 at Columbia NWR would be an exceptionally large concentration, presumably of migrants passing through, though Bank Swallows not mentioned at the same time. *Fall*: hundreds on 18 Jul 1975 "suggested a pre-migration gathering." *CBC*: 1 in 1981 at Yakima Valley.

Terence R. Wahl

Bank Swallow *Riparia riparia*

Locally common to abundant migrant and summer visitor at low elevations in e. Washington. Rare migrant w.

Subspecies: *R. r. riparia.*

Habitat: Nests in banks, using suitable soil layers while avoiding others, and, as cut banks degrade, birds often appear to seek newer habitat. Forages widely over water and over desert areas in e. Washington.

Occurrence: "Never conspicuous" (Jewett et al. 1953), the colonially nesting Bank Swallow has obviously changed in abundance since the 1950s. Perhaps more than in any other area species, the provision of nesting habitat in cut banks due to river erosion, highway and railroad construction appears to have been responsible (see Campbell et al. 1997, Stepniewski 1999).

Widely distributed over e. Washington along streams, though variable interannually. Breeds along Columbia, Snake, Okanogan, Touchet and Yakima Rs., at locations along smaller streams, and occasionally away from water at sites like soil storage piles up to at least 760 m elevation (TRW and see Campbell et al. 1997). Bowles and Hurley (in Jewett et al. 1953) stated that colonies are always small, and reports given by Jewett et al. supported that. By the 1960s, colonies of several thousand birds were known. Though some of those large colonies have been abandoned or greatly reduced in size in some areas (e.g., at confluence of Okanogan and Columbia Rs. [TRW]), overall populations may be sustained through establishment of new colonies like one of 10,000 burrows nr. Locke I. in Franklin Co. (A. Stepniewski in Smith et al. 1997).

In w. Washington, a nest was reported previous to 1900 on an island nr. Cape Disappointment at the mouth of the Columbia R., and nesting was reported on 28 Apr 1918 nr. Port Angeles (Jewett et al. 1953). Birds are now fairly common locally in summer along the Skagit and Skykomish Rs., and small colonies are present nr. Woodland, Toutle, and Auburn (SGM). In sw. B.C., small numbers nested in the 1960s and 1990 (Campbell et al. 1997). Origin of fall migrants in w. Washington is uncertain: birds nest as far n. as w. Alaska. The destination of a bird seen on 11 May 1980 at sea off Grays Harbor was unknown.

Early departure following nesting indicated by an estimated 1000 birds per hour flying s. on 15 Jul 1995 at the Walla Walla R. delta (Stepniewski 1999). Increased numbers are apparent in the Columbia Basin in Aug and Sep (Stepniewski 1999, SGM).

BBS data suggest a population increase in Washington from 1966 to 2000 (Sauer et al. 2001).

Remarks: In the late 1990s westside spring-fall reports increased noticeably, possibly indicating greater skill of observers. As noted by Campbell et al. (1997), some reports of sightings and nesting, particularly in w. Washington, require confirmation that the birds were not Rough-winged Swallows.

Noteworthy Records: West, *Summer*: nesting reported in 1993 along Skykomish R. nr. Monroe, in 1995 nr. Snohomish; up to 5 in summer 1997 nr. Marblemount; 4 reports in Jun-early Jul 1999 along Skagit/Sauk Rs. suggested nesting. *Fall*: 1 or 2 from 17 Aug-15 Sep at Kent; 1 on 25 Sep at Ocean Shores and 1 on 2 Oct 1977 nr. Everett. *High count*: 100 on 31 Aug 2000 at Ridgefield. **East**, *Summer*: 1500 birds banded/year at colonies nr. Brewster in late-1960s; 100 nest holes in summer 1978 at McNary NWR; large colony nr. Okanogan R. mouth and 300-400 birds along the Snake R in Franklin Co. in 1979; 200 pairs on 12 Jun 1981 at mouth of Okanogan R; 1000 on 31 Jul 1993 at Quincy; 300 on 28 Jun 1997 at Walla Walla R. delta; 1000 on 22 Jul 1998 at Umatilla NWR; 2000-4000 on 21 Jul 1999 at Columbia NWR; 2000 on 8 Jul 2000 at Walla Walla. *Fall*: 500 on 17 Jul nr. Spokane, 300-500 in Asotin Co. in summer 1966; 3000+ on 17 Aug 1975 nr. Spokane; 1 on 13 Oct 1995 at Walla Walla; 2000 on 21 Jul 1999 at Columbia NWR. *Winter*: 1 on 25 Jan 1992 at Ringold Fish Hatchery, Grant Co.

Terence R. Wahl

Cliff Swallow *Pterochelidon pyrrhonota*

Widespread, common to abundant migrant and summer resident e. of the Cascades, common in w. Very rare in winter.

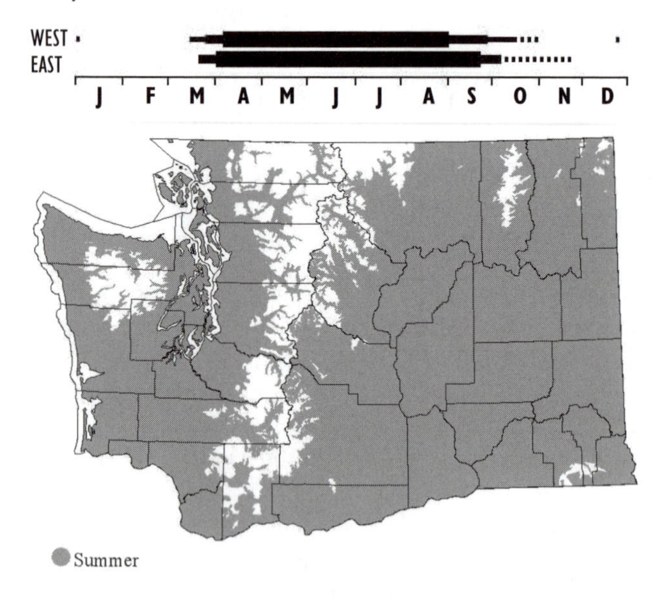

Subspecies: *P. p. pyrrhonota* breeds w., *hypopolia* e. of Cascades.

Habitat: Nests on cliff faces, buildings, and other construction and under bridges. Forages widely, often over water.

Occurrence: Like most other swallows, Cliff Swallow numbers increased with the expansion of human-provided habitats. This is shown by comparisons of reports dating back to 1860 (Jewett et al. 1953) with numerous reports and documented increases since 1950.

Following spring arrival, birds are widespread but concentrated at bodies of water with early insect hatches, as at gatherings of thousands on 30 May 1971 at Medical L. Colonies then locate in natural situations on big cliffs along rivers and coulees. Widespread reports, for example, include thousands nesting at Priest Rapids, Grand Coulee, and L. Lenore (Stepniewski 1999), large colonies occupying cliffs in the channeled scablands in se. counties (Larrison and Sonnenberg 1968), and, in the w., occasional nests on cliffs on the w. side of San Juan Is. (Lewis and Sharpe 1987). By far most birds reside in the lowlands, but some may occur to 900 m elevation (Smith et al. 1997). The Cliff Swallows' unique bottle-shaped mud nests, placed on suitable, usually vertically faced, nesting habitats, require nearby sources of suitable mud.

As in B.C., where 80% of breeding sites were in human-influenced situations (Campbell et al. 1997), colonies are found statewide at nesting opportunities at dams, under bridges and, sometimes unappreciated, on a wide range of buildings. W. of the Cascades, the majority of nests are probably associated with structures. In addition to provision of nest sites, reservoirs, fields, and urban areas add foraging opportunities in addition to natural open areas and wetlands (see Stepniewski 1999).

Dates of migration, reproductive success, and mortality rates vary greatly due to food supplies and weather. Reproductive failure due to nest parasites especially noted for this species (see Campbell et al. 1997). In 1980, like other swallows, reportedly suffered noticeable mortality locally due to effects of volcanic eruption at Mt. St. Helens on food supplies.

As with other most swallows, numbers noticeably decrease following fledging of young birds in Jul and Aug. On the eastside, sizable flocks occur again in late Aug and Sep with migration of populations from Canada (Stepniewski 1999), though westside numbers decrease abruptly in early Sep (SGM).

BBS data indicated that populations increased in w. N. America from 1966 to 1991 (DeSante and George 1994; see Brown and Brown 1995), though no change was indicated in Washington from 1966 to 2000 (Sauer et al. 2001).

Remarks: Formerly in genus *Hirundo*. Browning (1992) suggests that only one subspecies (*pyrrhonota*) occurs here. A hybrid CliffXBarn Swallow was reported on 29 Aug 1995 on Whidbey I.

Noteworthy Records: West, *Winter*: 1 on 29 Dec 1973 at Bellingham CBC. 1 on 1 Jan 1992 at Seattle. *High counts*: 500 on 19 May 2000 at Stanwood STP. **East**: "thousands" on 30 May 1971 at Medical L. and on 13 May 1972 at Pasco; 500 in May 1971 at Clarkston.

Terence R. Wahl

Barn Swallow *Hirundo rustica*

Common summer resident, common to locally abundant migrant throughout. Very rare to rare in winter in lowlands w. of the Cascades.

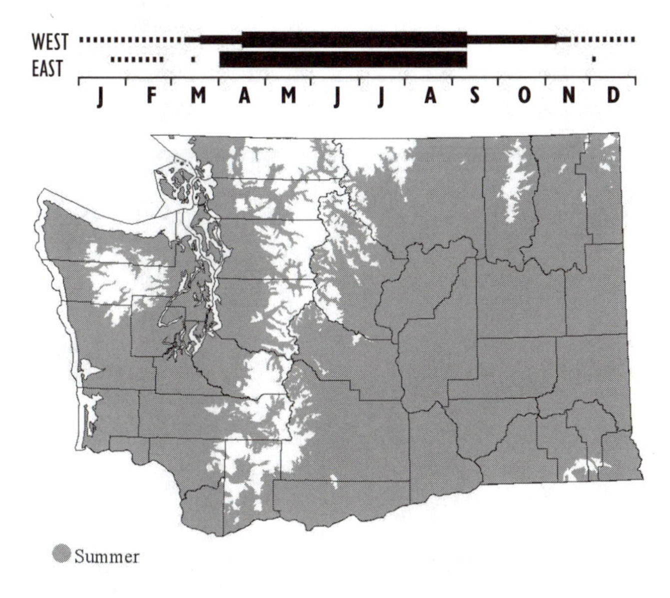

Subspecies: *H. r. erythrogaster.*

Habitat: Widespread in agricultural and urban areas, with nest-site availability determining local distribution. Barns, garages, porch overhangs, culverts, and other structures most often used, with the placement of nests on objects like light fixtures showing the adaptability of the species.

Occurrence: Early descriptions (see Jewett et al. 1953) indicated that this is another swallow benefitting from human association through provision of nesting locations on and in many structures. Building nests on ferries crossing Washington and B.C. waters (TRW, Campbell et al. 1997) would seem to indicate both adaptability and determination.

An apparent expansion into high elevations (e.g., Wahl 1995) where ski resort buildings provide nesting opportunities nr. foraging areas over ski slopes (Smith et al. 1997) noted in recent decades. Few reported to breed in natural habitats, such as locations in the San Juan Is. (Lewis and Sharpe 1987), and at cliffs or caves on the outer coast (TRW). Not colonial like the Cliff Swallow, nesting birds may aggregate under favorable conditions. Double-brooded with a consequently protracted nesting season, with young being fed as late as 4 Sep (Stepniewski 1999), and at Longmire, Mt. Rainier, on 15 Sep (in Smith et al. 1997).

Concentrations of large numbers noted prior to and during fall migration at roosts and along rivers and coastlines (e.g. Stepniewski 1999). Like other species migrating long distances, has been recorded on ships at sea: one was seen offshore from Grays Harbor on 11 May. Recorded more often in winter than other swallows. Numbers varied but reports increased noticeably: two birds were noted in Washington in Dec 2000, 62 were found in Jan and 29 in February 2001, compared with just 26 total during the previous eight winters. In winter 2002-03 about three times that number of birds occurred, including a number e. of the Cascades, where they were unrecorded a year earlier. This was part of a w. coast occurrence pattern, perhaps linked with anomalous weather. Birds have been noted in winter at sewage-treatment ponds that do not freeze and are near likely roosts in barns (e.g., Wahl 1995).

Populations in w. N. America increased over past 100 years (Paulson 1992b, DeSante and George 1994), but BBS surveys indicated a decreasing trend over the recent period from 1966 to 1991 (DeSante and George 1994) and from 1966 to 2000 in Washington ($P < 0.01$; Sauer et al. 2001), similar to trends in n. N. America and Europe (see Campbell et al. 1997).

Remarks: Associated with livestock on farms; removal of animals reduces swallow populations (Tate 1986), and increasing use of metal farm buildings and changes in farm practices also may reduce numbers (Campbell et al. 1997).

Noteworthy Records: West, *Summer*: pair nesting at Mt. Baker ski area on 28 Jul 1974 with 3 m of snow on the ground. *High counts*: 2000 on 3 Sept 1994 at Everett (SGM); 1800 on 15 Aug 1995 at Everett (SGM); 2800 on 31 Aug 2000 at Ridgefield. **East**, *Fall*: ca. 1000 on 21 Sep 1968 at Medical L.; >6700 on 24 Sep 1972 at Medical L.; 5000 on 2 Oct 1983 nr. Potholes Res.; 5500 on 11 Sep 1994 at Wahluke WRA. *CBC*: 2 in 1987, 1 in 1989 at Bellingham, 1 in 1991 at Leadbetter Pt., 1 in 1962, and 1 each during CP in 1974 and 1975 at Seattle.

Terence R. Wahl

Black-capped Chickadee *Poecile atricapilla*

Common year-round resident in w. Washington, rare in the San Juan Is. Common in wetter areas of e. Washington, uncommon and localized in the Columbia Basin.

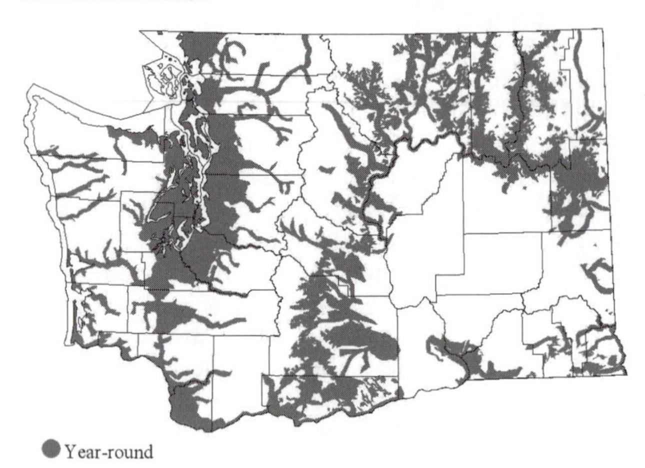

Subspecies: *P. a. occidentalis* in w., *fortuitous* in restricted areas on the eastside.

Habitat: Deciduous or mixed deciduous-coniferous forests, riparian woodland, thickets, shrubbery, and residential places. Uses nest cavities by enlarging small holes created by woodpeckers, natural holes in trees, nestboxes.

Occurrence: Common, locally numerous resident in lowlands (<900 m) in deciduous or mixed deciduous-coniferous woodlands w. of the Cascades. It is uncommon throughout most of the Olympic Pen. even in natural riparian areas, though found in towns (Sharpe 1993, in Smith et al. 1997), and it is rare in the San Juan Is. (Lewis and Sharpe 1987).

Found mainly in the wetter areas of e. Washington, the species is localized in the c. Columbia Basin even in apparently suitable habitat (Smith et al. 1997). Migration is unknown, though downslope movements and wanderings are implied by out-of-habitat occurrences. Birds have apparently increased locally and are now present in areas such as Richland (SGM) and they are commented upon by observers during irruptive appearances in Wenatchee and Walla Walla. Severe weather during late fall and winter months often precedes movements into cities in c. Washington.

The common chickadee of successional forests, replacing the Chestnut-backed Chickadee in areas where clear-cuts regenerate as deciduous forests and presumably benefitting locally from other habitat alterations including urbanization. BBS data reveal a significant population increase in N. America between 1966 and 1989 (Droege and Sauer 1990), while population appeared stable in Washington

Black-capped Chickadee (Shawneen Finnegan)

from 1966 to 2000 (Sauer et al. 2001). CBC data suggest an increase, though more intensive censusing at bird feeders may have biased some local numbers upward.

Remarks: Formerly in genus *Parus.*

Noteworthy Records: *San Juan Is.*: 1 on 5 Mar 1998 at American Camp, San Juan I.; 4 additional records from Lopez and Orcas Is. are given as unconfirmed by Lewis and Sharpe (1987).

Devorah A. N. Bennu

Mountain Chickadee *Poecile gambeli*

Common resident of dry coniferous forests throughout e. Washington and local in dry high-elevation mixed forests just w. of the Cascade crest and on the slopes of Mt. Rainier. Irregular downslope movements in winter.

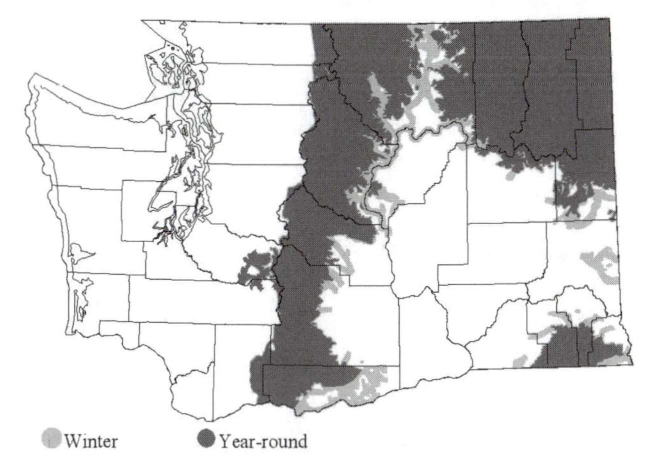

Subspecies: *P. g. grinnelli* in ne. and Blue Mts., *abbreviatus* in the Cascades (see McCallum et al. 1999).

Habitat: Dry montane coniferous forests, primarily pine, spruce-fir above 1800 m, locally in pinyon-juniper. Mixed forests, riparian woodlands, and suburban areas during non-breeding season, primarily in e. Washington (Smith et al. 1997) and infrequently at elevations below 900 m.

Occurrence: Common throughout dry coniferous forests of e. Washington; local and uncommon in high-elevation drier forests w. of the Cascade crest. Downslope movements are regular in e. Washington, where birds are rare but annual away from breeding range.

Irruptions into lower elevations on the westside are not annual. These are probably related to food scarcity in the breeding range, though birds involved do not necessarily originate in Washington. Lowland records are mostly from mid-Sep to early Apr, peaking from late Sep to late Dec. Irruptions were noted during winters of 1963-64, 1972-73, 1986-87, and 1998-99. A summer record from Redmond in 2001 was very unusual. Status on the Olympic Pen. and s. outer coast is not well understood.

BBS data suggest an increase from 1966 to 2000 (Sauer et al. 2001). CBC data are insufficient to determine trends; there are few long-term locations where the species occurs regularly.

Remarks: Formerly in genus *Parus*.

Noteworthy Records: West, *Spring*: 2 on 9 Apr 1974 n. of Forks; 3 on 14 Apr 1974 on Hurricane Ridge Rd. in Olympic NP; a few at feeders nr. Vancouver until late Apr-early May 1987. *Fall*: 68+ in Oct-Nov 1986 at lowland sites. *Winter*: 1 in 1973-74 at Aberdeen. 2 in 1987-88 at Grays Harbor. High count: 25 on 8 Nov 1986 in Discovery Park, Seattle. **East**, *high count*: 140 on 7 Mar 1998 in c. Okanogan Co.

Devorah A. N. Bennu

Chestnut-backed Chickadee *Poecile rufescens*

Common year-round resident throughout w. Washington and locally common in wet closed coniferous forests in higher elevations of e. Washington.

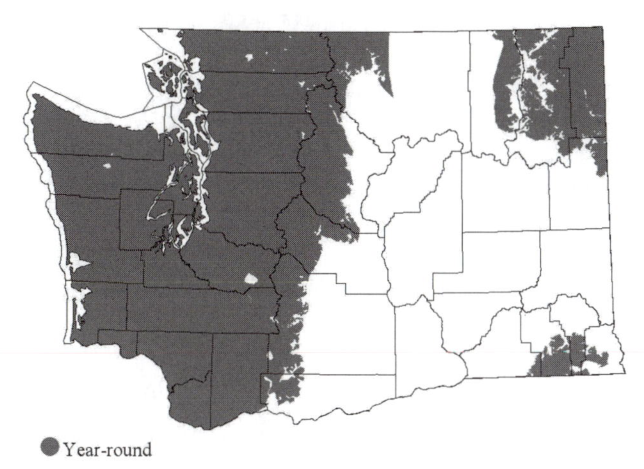

Subspecies: *P. r. rufescens.*

Habitat: Dense mature coniferous forests of fir, spruce, pine, cedar, tamarack, and hemlock (Bent 1964, Terres 1980). Less common in oak or mixed pine-oak woodlands, and found locally in riparian woodlands (AOU 1998). Moist dense coniferous forests in higher elevations of e. Washington (Smith et al. 1997).

Occurrence: The common chickadee of humid fog-shrouded coastal mountains and wet coniferous forests also occurs in conifers in city parks and gardens. Widespread in w. Washington, it is the most abundant chickadee outside the Puget Trough (Smith et al. 1997). Dispersal away from normal breeding locations is evident in w. Washington from late Sep into mid-Nov and birds often winter locally at sites with bird feeders attracting Black-capped Chickadees.

The species is local and uncommon in e. Washington in wetter forest zones and locally in damp areas within drier zones, such as ravines and north-facing slopes, in the e. Cascades, Ferry to

Pend Oreille Cos. and the Blue Mts. (Smith et al. 1997). Typically breeds from Apr through Jun after excavating or moving into a nest cavity in a dead tree (Bent 1964).

Steadily losing habitat to the Black-capped Chickadee when clear-cuts regenerate as deciduous forests in Washington State (Smith et al. 1997). CBC data indicate no certain trend in w. Washington, and very low numbers occur at eastside locations.

Remarks: Formerly in genus *Parus.*

Noteworthy Records: *CBC high counts by location*: 632 in 1999 at Olympia; 582 in 1997 at Padilla Bay; 515 in 1983 at San Juan Is.; 472 in 1986 at Bellingham; 322 in 1999 at Tacoma; 316 in 1991 at Grays Harbor; 300 in 1988 at Sequim-Dungeness; 300 in 1986 at Kent-Auburn; 261 in 1993 at Oak Harbor; 236 in 1987 at Skagit Bay; 233 in 1992 at Seattle; 218 in 1988 at Kitsap Co.

Devorah A. N. Bennu

Boreal Chickadee *Poecile hudsonica*

Locally fairly common resident of high-elevation forests of northernmost e. Washington; uncommon in n. Cascades.

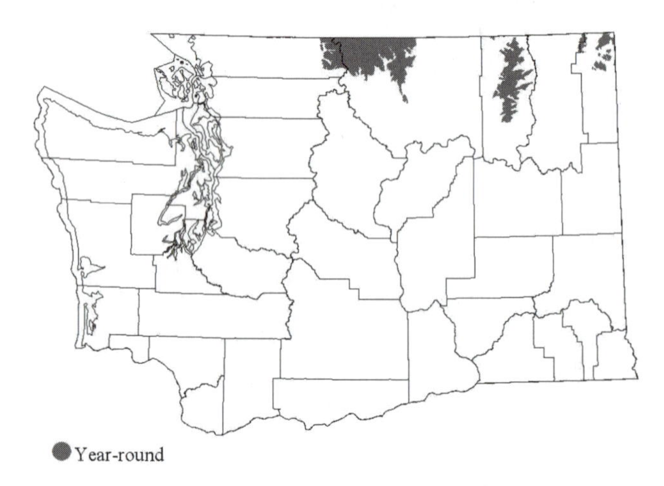

Subspecies: *P. h. cascadensis* in the Cascades; *columbianus* in the Kettle Range and Selkirk Mts.

Habitat: Damp, cool coniferous subalpine and alpine forests, closely associated with high-elevation Engelmann spruce (Smith et al. 1997).

Occurrence: The n.c. highlands of Washington represent one of the southern-most limits for the species' occurrence in the w. U.S. In Washington, the Boreal Chickadee's range overlaps with those of Mountain and Chestnut-backed chickadees but these three species differ in the type of conifers they favor (Smith et al. 1997; see Ficken et al. 1996). Little is known about population size or stability, winter movements, or flock sizes in the state. Though periodic irruptions into lowlands are described for the species' range in e. N. America (Ficken et al. 1996), there are no records of such movements here.

Two nests found in Okanogan Co. on 22 and 23 Jun 1991 appear to be the first in Washington. There are reports of probable nesting for this species from Whatcom, Okanagan, Ferry, and Pend Oreille Cos. (Smith et al. 1997), including one pair carrying food on the w. fork of the Pasayten R. in the N. Cascades in early Jun 1980.

Noteworthy Records: West, *Fall*: 1 on 15 Aug 1982 along Ptarmigan Ridge Trail on Mt. Baker; 10 there in early Sep 1985 were at the species' sw.-most limit. **East**, *high counts, Summer-Fall*: ca. 25 on 20 Aug and at least 8 on 6 Sep 1965 at Salmo Pass; ca. 20 in early Jun 1980 on the W. Fork of Pasayten R.; 22 on 2-3 Aug 1981 at Tiffany Mt, 36 km w. of Tonasket; 17 on 27 Jul 1998 at Freezeout Ridge, Okanogan Co.

Devorah A. N. Bennu

Bushtit *Psaltriparus minimus*

Fairly common resident in w. and locally fairly common e. of the Cascades.

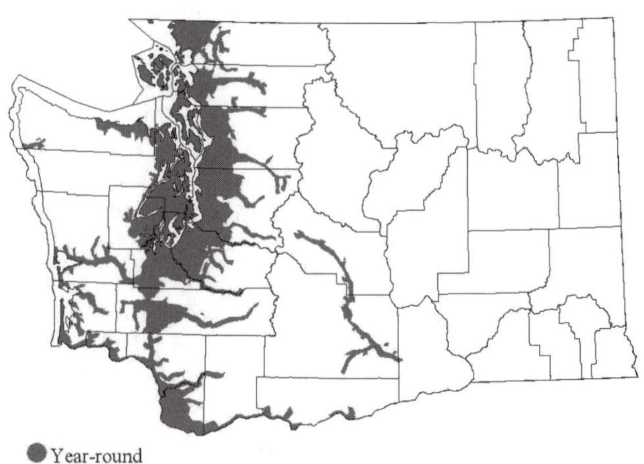

Subspecies: *P. m. minimus*; with *plumbeous* reported in Columbia Basin.

Habitat: Shrubs, hardwoods, and mixed urban habitats; dense riparian vegetation in e. Washington (Smith et al. 1997).

Occurrence: Bushtits have spread geographically over past decades, likely due to urbanization, changes in land use and vegetation (Opperman 1992). Widespread in the Puget Trough and westside

Table 37. Bushtit - CBC average counts per 100 party hours by decade.

	1960s	*1970s*	*1980s*	*1990s*
West				
Bellingham		127.3	175.7	85.3
Cowlitz-Columbia			125.1	50.9
Edmonds			580.9	220.4
E. Lk. Washington			204.3	138.6
Grays Harbor		78.5	255.7	123.0
Kent-Auburn			335.0	231.6
Kitsap Co.		111.4	87.9	119.3
Leadbetter Pt		232.4	49.2	22.5
Olympia			107.4	55.3
Padilla Bay			70.7	42.2
Port Townsend			74.9	200.8
San Juan Is.			31.0	21.4
Seattle	246.9	394.6	724.3	588.0
Sequim-Dungeness			111.3	116.9
Tacoma		329.2	407.7	214.6
East				
Toppenish			9.8	46.0

lowlands w. to Grays Harbor and Willapa Bay and s. to the Columbia R., the species has spread n. in B.C. (Cannings et al. 1987) and across the Cascades in Washington. It expanded its known 1950s distribution in Yakima Co. (see Jewett et al. 1953) and also into Kittitas Co. (Opperman 1992).

Local eastside expansion likely due to habitat creation in westside valleys along powerlines, highways, and railroad rights-of-way across mountain passes (see Opperman 1992). Occurrence along the Skagit R. in 1980 as far as Newhalem (BBS), nr. Skykomish, and middle fork of the Snoqualmie R. (Smith et al. 1997) support this, showing that birds move well up westside river valleys on occasion. Crossing higher mountain passes (1200-1500 m) in the n. Cascades may be much less likely than the lower route (900 m) over Snoqualmie Pass. In addition to the population nr. Toppenish known since 1947 (Jewett et al. 1953), occurrence documented in Kittitas Co. at several locations beginning in 1989 (Smith et al. 1997). In addition to establishment and expansion of a Klickitat population, small numbers occurred on Walla Walla CBC in 1976 (and see Weber and Larrison 1977) and were recorded also in Garfield Co. (Smith et al. 1997) and in summer 2002 at the Potholes Res.

Distribution and trends e. of the Cascades in Oregon appear similar (Gilligan et al. 1994). In B.C., Bushtits were unknown on Vancouver I. prior to 1940s, were common at Victoria by 1978. Ten in Queen Charlotte Is. on 10 May 1984 were possibly the first there. One at Kelowna, B.C., in winter 1995-96 was second Okanagan Valley record.

Weather apparently a major factor in winter survival and large variations in populations are reported. Record numbers reported on CBCs in the 1980s but no statewide trend noted on established CBCs from 1960s through 1998 (Table 37).

Remarks: Subspecies in Washington uncertain (see Opperman 1992), with *plumbeus* identified in the Columbia Basin (D. R. Paulson p.c.). Birds first reported from the San Juan Is. in 1938 (Lewis and Sharpe 1987), and in Queen Charlotte Is., B.C., in 1984.

Noteworthy Records: West, *Summer*: on BBS in 1980, 4 birds sw. of Newhalem and 2 ne. of Marblemount on Skagit R. **East**, *Summer*: breeding at Rock Cr., Klickitat Co., in 1996 and at Alder Cr. in spring 1998, noted as at "east edge of Washington range"; on 3 Jun 2002 at n. end of Potholes Res. (BT, R. Friesz p.c.) and 20 there on 20 Oct 2002 (G. Shugart, *fide* D. Paulson).

Terence R. Wahl

Red-breasted Nuthatch *Sitta canadensis*

Fairly common resident in forests statewide; some downslope winter movements and occasional winter irruptions from the n.

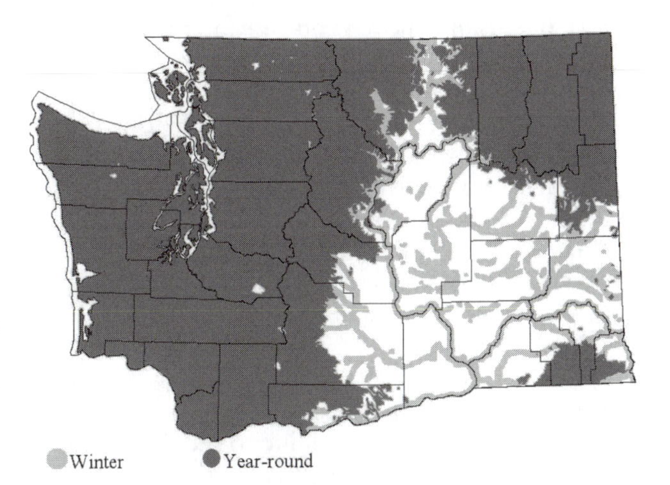

Habitat: Mature and old-growth stages of coniferous forests and mixed conifer-hardwood forests (Brown 1985), secondarily in deciduous and residential habitats.

Occurrence: The Red-breasted Nuthatch occurs statewide in forested areas, including the San Juan Is. and outlying Kamiak Butte. Apparently less numerous, with breeding not confirmed, on the w. side of the Olympic Pen. (Smith et al. 1997). Birds wander widely in winter and occur in many habitats including unforested areas like the Columbia Basin, where the species is a fairly common migrant and has nested in towns such as Othello (B. Flores p.c.).

On an industrial forest in Lewis Co., birds were numerous and showed highest occurrence in commercially mature conifers compared with recent clear-cuts, pole and sapling conifers, or hardwoods and non-forested areas (Bosakowski 1997). In the s. Cascades, breeding season detection rates were similar in old-growth, mature, and young stands, but winter detection rates were significantly higher in old-growth stands, probably due to the abundance of western hemlock cone crops (Manuwal and Huff 1987). In Oregon, breeding densities were higher in areas of large snags (Mannan and Meslow 1984), in stands from 75 to 200+ years (Mannan et al. 1980), though breeding abundance in old-growth forest decreased with forest age (Carey et al. 1991). During breeding and winter, abundance was higher in Douglas-fir stands that were thinned than in unthinned stands (Hagar et al. 1996).

As in the case of other n. forest birds, irregular winter irruptions occur, and cone crop size in Washington and Oregon was found to vary inversely with periodic winter irruptions into California (Widrlechner and Dragula 1984). Field reports suggested above-normal numbers in Spokane in 1962-63 and 1989-90 and in Bellingham in 1966-67. Migrations are poorly described in Washington. There are upslope movements in fall (Stepniewski 1999), downslope movements in winter (Godfrey 1986, Campbell et al. 1997, Stepniewski 1999), but migration patterns are little described. Four at-sea records, to at least 56 km offshore, in Sep off Grays Harbor (TRW) were apparently of lost migrants.

N. America-wide BBS data show a significant population increase from 1966 to 1991 (Peterjohn and Sauer 1993), though not in Washington from 1966 to 2000 (Sauer et al. 2001). CBC data in Washington indicate a significant westside increase ($P < 0.01$).

Remarks: Like other snag-dependent species, the species has probably suffered due to the shortage of snags, cull trees, and residual trees left in clear-cut units over the past 40-50 years (Neitro et al. 1985). Birds will use artificially created snags, created by chainsaw-topping of live trees (Chambers et al. 1997) and probably benefit from snags and green recruitment trees now required by law to be left in clear-cuts.

Noteworthy Records: *High counts, Spring*: 100 on 7 Mar 1998 in Okanogan. *Summer*: 30 on 23 Jul 1973 nr. Peola.

Thomas Bosakowski

White-breasted Nuthatch *Sitta carolinensis*

Fairly common resident in e. Washington. Local, declining resident in Clark Co., very rare vagrant elsewhere in w.

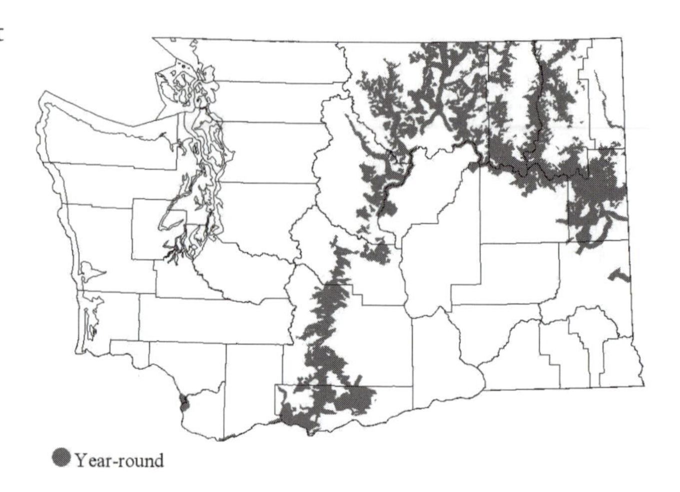

Subspecies: *S. c. aculeata* in w., *tenuissima* in e.

Habitat: Primarily open-canopy stands dominated by ponderosa pine or Oregon white oak (Dumas 1950, Chappell and Williamson 1984, Smith et al. 1997), some in urban areas or more closed-canopy forests with an abundance of either of those species. On the eastside, breeds mostly in the ponderosa pine and Douglas-fir zones, and to a lesser degree in the grand fir zone (Smith et al. 1997), on the westside in oak or oak-Douglas-fir woodlands.

Occurrence: A widespread permanent resident along the e. slope of the Cascades, ne. counties, Spokane Co., Kamiak Butte, and the Blue Mts. (Smith et al. 1997). There is some post-breeding dispersal, mostly from Aug through Dec (Jewett et al. 1953, Larrison and Sonnenberg 1968), likely late summer upslope movement to nr. upper tree line (Jewett et al. 1953) and downslope movements occur (see Stepniewski 1999). Occasional eastside fall or winter irruptions may occur, with 129 birds in 1989 on the Spokane CBC (see Pravosudov and Grubb 1993).

There is one known westside population, centered at Ridgefield, with possible breeding nr. Washougal, and birds rare in Skamania and Cowlitz Cos. (SGM). The s. Puget Trough population, centered in Lakewood, has apparently disappeared. Puget Sound and Willamette Valley populations may have been bridged by smaller numbers of birds from Woodland to Centralia, as suggested by recent records from nr. Kalama and Cowlitz prairie. The species was probably never common n. of the Tacoma area due to limited oak habitats (see Rathbun 1902). There are no records in the San Juan Is. (Lewis and Sharpe 1987) and birds are rare in sw. B.C. (Campbell et al. 1997).

Formerly common in the oak/prairie region of s. Puget Sound (Cooper 1860, Bowles 1929, Jewett et al. 1953), birds declined significantly in the Tacoma area in the early 1900s (Bowles 1929) and generally in w. Washington (Kitchin 1949). Another decline—from nine occupied breeding sites to none—occurred in the early 1980s in Pierce and Thurston Cos. (Chappell and Williamson 1984, Smith et al. 1997). The last noted occupied breeding site in Pierce Co. was in 1998. Up to seven birds were recorded annually on the Tacoma CBC until 1991; none were found subsequently. The species is also declining in the Willamette Valley of Oregon as evidenced by both CBC and BBS data (Sauer et al. 1999b). Records away from s. Puget Sound and the Clark Co. area may pertain to wandering birds from the eastside (SGM).

Declines possibly resulted from loss of oak woodland, scarcity of nesting cavities, competition for nest sites with starlings, inadequate mast for overwintering, genetic drift associated with small isolated populations, unknown disease, or natural disturbance. Changes in oak habitats due to fire suppression and other anthropogenic factors probably affected the population. Remaining oak habitat is very patchy, dense and closed-canopy in structure, has smaller trees, and is mostly in an urban or agricultural matrix (Chappell and Crawford 1997, Hanna and Dunn 1997). Bowles (1929) suggested that the Pierce Co. decline was related to the loss of large oaks (see Hagar and Stern 1997)—birds are now rare there (CBCh). Younger oaks have few natural cavities for nesting sites. Recent nests found in Pierce Co. are in artificial boxes.

The widespread eastside subspecies has evidently increased in numbers, at least locally, since the 1960s. Numbers reportedly increased in the Spokane area during the late 1960s. Weber and Larrison (1977) noted an increase in records from se. Washington beginning in 1973, as in interior B.C. since the 1970s (Campbell et al. 1997). BBS data indicate no overall trend statewide, but a decline for the Cascade Mts. region (Sauer et al. 1999a).

Remarks: The westside subspecies is SC, FSC, PHS. Narrow habitat requirements suggest the eastside subspecies should be carefully monitored.

Noteworthy Records: West, *outside current breeding range*: resident nr. Kalama 1985-95 (D. Caldwell p.c.). *Spring*: 1 on 25 Apr 1998 Hurricane Ridge (*fide* B. Norton p.c.); 1 on 9 May 1998 at Sequim (J. and P. Fletcher p.c.); 1 on 14 May 1998 at Sequim (E. Kridler p.c.). *Fall*: 1 at Westport, on 22 Oct 1928; 1 on 16 Sep 1965 at Bellingham; 1 on 11 Aug 1969 at Seattle; 1 on 21 Sep 1969 at Bellingham; 1 on 14 Sept 1974 and on 14 Sep 1975 at Olympia; 1 on 10 Oct 1982 at Bellingham; 1 on 20 Sep 1990 at Pt. Roberts; 1 on 5 Sep 1995 at Belfair. *Winter*: 4 in 1935, 1 in 1938 and in 1939, 1 in 1952, 1 in 1969 at Seattle CBC; 1 on 31 Dec 1972 at Bellingham; 1 on 18 Mar 1979 at Pt. Roberts; 1 on 6 Mar 1997 at Carnation; 1 in 1983 and in 1984 Kitsap Co. CBC; 1 on 20 Dec 1994 at Bellingham; 1 on 21 Dec 1995 at Cowlitz Prairie, 6 km ne. of Toledo (S. Johnston p.c.).

Christopher B. Chappell

Pygmy Nuthatch *Sitta pygmaea*

Fairly common to uncommon resident in dry, open ponderosa pine forest in ne. counties and along e. slopes of the Cascades, local in some areas. Very rare vagrant w.

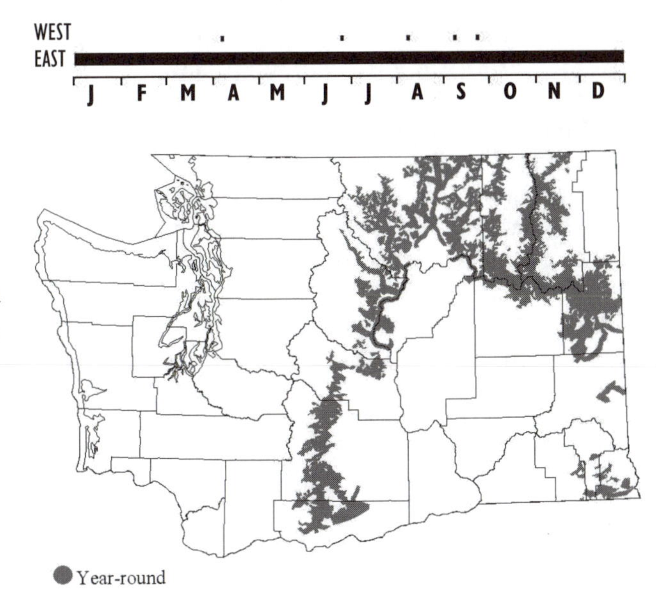

Subspecies: *S. p. melanotis.*

Habitat: Restricted almost completely to ponderosa pine forests, rarely in adjacent interior Douglas-fir stands (Smith et al. 1997).

Occurrence: One of the most habitat-specific birds in the state. Its association with yellow pine (ponderosa pine) forests described by Jewett et al. (1953) was shown again in subsequent decades by BBS and BBA surveys in summer (e.g., Smith et al. 1997) and by CBCs.

Primary distribution along the e. slope of the Cascades from Okanogan to Klickitat Cos. and from Okanogan e. to Stevens and Spokane Cos. to 1200-1500 m elevation (Jewett et al. 1953). Noted in Whitman Co. and locally in Garfield Co. in the Blue Mts. (Smith et al. 1997). Fall and winter movements remain, as stated by Jewett et al. (1953), little known, but downslope shifts occur in fall through spring possibly due to pine-seed shortages (Stepniewski 1999). Individuals or small groups recorded away from normal habitats from Yakima, Okanogan, and Chelan Cos. The few westside records, from Tatoosh I. to Leadbetter Pt., show the vagrancy possibilities of a small passerine assumed to be resident and closely associated with one forest type.

Birds are local, however, in some areas dominated by ponderosa pine, as in the Blue Mts., where extensive BBA field surveys recorded very limited occurrence (Smith et al. 1997) and on the Yakama Indian Reservation where RHL had just three records, including one nest, during eight years of fieldwork and suggested that forest understories there may have been too dense during the 1990s for this species.

Spokane CBCs indicate much greater winter abundance there than along the e. slope of the Cascades. From 1959 to 1999, up to 587 birds occurred on all 41 Spokane counts. Cheney CBCs, 1960-64, also reported numbers. In contrast, Bridgeport, Colville, Chewelah, Twisp, Ellensburg, Camas Prairie, Columbia Hills-Klickitat Valley, and Walla Walla CBCs had only 1-12 birds on just 15 counts. Numbers recorded on Spokane CBCs showed no statistically significant change over time. BBS data judged inadequate in routes to determine trends (Kingery and Ghalambor 2001).

Noteworthy Records: West: 6 on 6-8 Aug 1969 at Chuckanut Mt. nr. Bellingham; 1 on 9 Sep 1973 at Ocean Shores; 3 on 10 Apr 1974 nr. LaPush; 1on 25 Sep 1977 at Leadbetter Pt. in lodgepole pines; 1 on 26 Jun 1999 nr. Mt. Hardy, Skagit Co.

Rosemary H. Leach

Brown Creeper *Certhia americana*

Fairly common permanent resident in forests statewide; rare to uncommon migrant and winter visitor elsewhere.

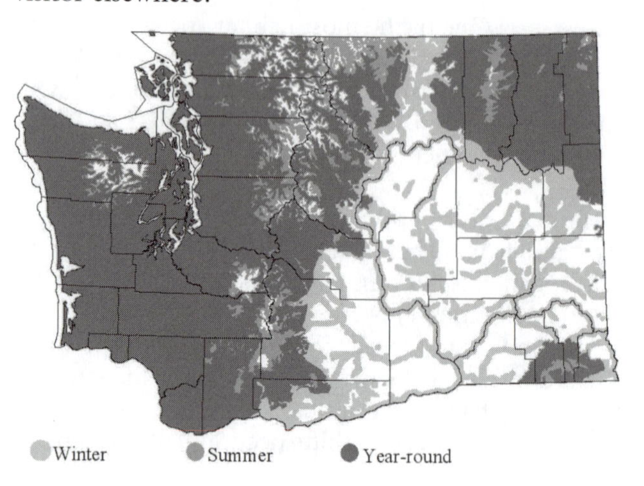

Subspecies: *C. s. montana* in e.; *occidentalis* on westside.

Habitat: Primary habitat typically mature and old-growth stages of coniferous forests and mixed conifer-hardwood forests (Brown 1985).

Occurrence: The creeper is a resident woodland specialist closely associated with moderate-age or mature forests in w., and at medium to high elevations e., often nr. riparian areas, though it is absent from bottomland forests of large ancient trees on the Olympic Pen. (Smith et al. 1997). Highest density of breeding birds is in old-growth forests in B.C. (Campbell et al. 1997). Noted foraging to 2000 m elevation in e. Washington (Jewett et al. 1953).

Though the species' song is heard in coniferous and mixed forests in nesting season, birds are noted by most observers in winter when they appear in city parks and residential areas. A species largely forced uphill, like other native-forest inhabitants, by development in the lowlands, the creeper breeds at low elevations where habitat remains.

Migration is uncertain. Some downslope post-breeding movement, if not true migration, is shown by lowland, non-breeding habitat records (and see Hejl et al. 2002b). It is a rare to uncommon migrant and winter visitor in the Columbia Basin (e.g., Jewett et al. 1953, Stepniewski 1999) from Aug to May (SGM).

In the s. Washington Cascades, detection rates in old-growth and mature stands were higher than in young stands during winter and breeding-season surveys (see Manuwal and Huff 1987, Carey et al. 1991). On a private industrial forest in Lewis Co., the species showed a significant positive correlation only with commercially mature conifers (>45 years old) and no younger forest stages (Bosakowski 1997). Year-round, creeper abundance in Oregon was significantly higher in Douglas-fir stands that were thinned than in un-thinned stands (Hagar et al. 1996), though the reason for this is not clear.

Population trends based on BBS data (see Hejl et al. 2002) and CBC data are uncertain.

Remarks: Has probably suffered from a shortage of snags, cull trees, and residual trees left in clear-cut units over the past 40-50 years (Neitro et al. 1985). Creepers are probably benefitting from the increasing numbers of snags and green recruitment trees now required by law to be left in clear-cut units. There is no evidence yet that creepers will nest in artificially created snags.

Thomas Bosakowski

Rock Wren *Salpinctes obsoletus*

Fairly common to common summer resident and rare winter resident in e. Washington. Very rare in w. Washington.

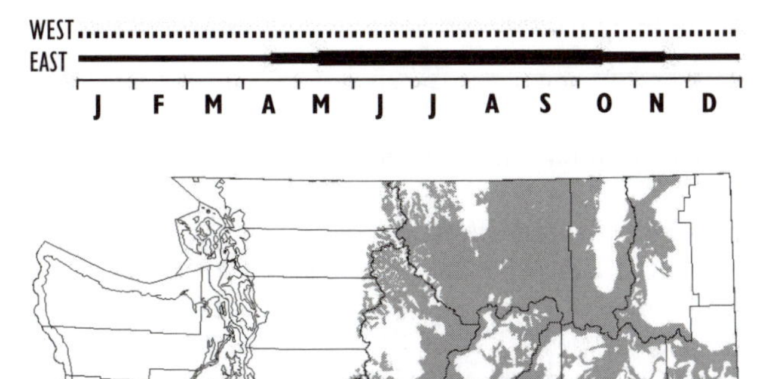

Subspecies: *S. o. obsoletus.*

Habitat: Rocky, arid habitats in shrub-steppe zones in dry forest zones of e. Washington, including canyons, rocky outcroppings, coulees; found to a lesser extent in clear-cuts and talus slopes (Smith et al. 1997) and around human-made concrete or stone structures.

Occurrence: Most numerous in the Columbia Basin but also found in talus slopes and clear-cuts at higher elevations in the Cascades and Blue Mts. (Smith et al. 1997). Most birds leave the state before winter, but some linger at lower elevations, especially along the Columbia R. Very rare in w. Washington, where it has occurred in all seasons. Most recent summer records have come from drier habitats nr. the Cascade crest, with singing birds reported from e. Whatcom Co. s. to Mt. St. Helens (e.g., Hunn 1982, Wahl 1995). The species may have bred formerly on Mt. Dallas on San Juan I. (Lewis and Sharpe 1987), but there have been no summer records from the San Juans since 1986. There are two reports during breeding season in the Olympics. Most migration and winter records are from the lowlands, with one on Protection I. and one at Grays Harbor.

Birds are most abundant from Apr to Oct but occasionally winter in both e. and w. Washington: some 24 CBC records from 1967 to 1999 included birds at Chelan, Columbia Hills-Klickitat Co., Grand Coulee, Spokane, Tri-Cities, Toppenish, Two Rivers, Wenatchee, and Yakima Valley, with three westside reports of single birds at Padilla Bay, Sequim-Dungeness, and Seattle. Larger numbers on CBCs may reflect mild weather prior to the count (Stepniewski 1999).

BBS data indicate a significant statewide decline ($P < 0.05$) from 1966 to 2000 (Sauer et al. 2001).

Noteworthy Records: West, *Spring*: 2 on 15 Apr 1978 on sw. San Juan I. were gone by 15 May; 1 on 28 May 1982 at Moran SP, Orcas I.; 3 on 18 Apr 1990 in Seattle; 2 on 27 May 1996 at Hurricane Ridge, Olympic NP. *Summer*: 1 on 5 Aug 1973 on San Juan I.; 1 in Jun 1978 in Gifford Pinchot NF, 30 m elev.; possible breeding in 1982 in King and Whatcom Cos.; pair in 1983 at nest site nr. Naches Pass, King Co.; 1 on 7 Jul 1984 on San Juan I.; 1 in summer 1984 and pair in 1992 feeding 2 fledglings at Mt. St. Helens; birds in 1985 nr. Mt. Rainier NP and on Mt. Si; 1 on 1 Jun singing nr. Naches Pass; 1 on 12 Jun on Hurricane Ridge, and 1 on 22 Jun 1986 atop Snoqualmie Mt., King Co.; birds in Jun 1988 on Goat I. in Skagit Bay, se. of Greenwater, Pierce Co.; singing on 4 Jul 1995 on Kelly Butte, King Co.; 1 in 1999 carrying food at Silver Star Mt., Clark Co.; 2 on 3 Jul 1999 carrying food at Steigerwald L. NWR; 3 on 30 May 2001 at Pyramid Rk., Clark Co.; 1 on 16 Jun 2001 at 1800 m elev. on Sleeping Beauty, Skamania Co. *Fall*: 1 on 29 Sep 1984 on Stuart I., San Juan Co.; 1 on 29 Oct 1984, 1 28-30 Sep 1998 at Seattle; 1 on 4 Oct 1998 at Kirkland; 1 on 12 Sep 1999 at Washougal; 1 on 12 Sep 1999 at Steigerwald L.; 1 on 28 Oct 2001 at Ocean Shores. Winter: 1 on 17 Dec 1983 on Protection I.; 1 on 21 Feb 1990 at Seattle. **East**: ca. 15 in summer 1988 entering holes in soft-cut banks nw. of Lowden in wheat fields area with no sign of rocky outcroppings. *Winter*: 8 on 2 Jan 2003 nr. Vantage.

Devorah A. N. Bennu

Canyon Wren *Catherpes mexicanus*

Locally fairly common year-round resident in in deep, rocky canyons in steppes of e. Washington and at low elevations along the Cascade e. slope. Very rare w. of the Cascades.

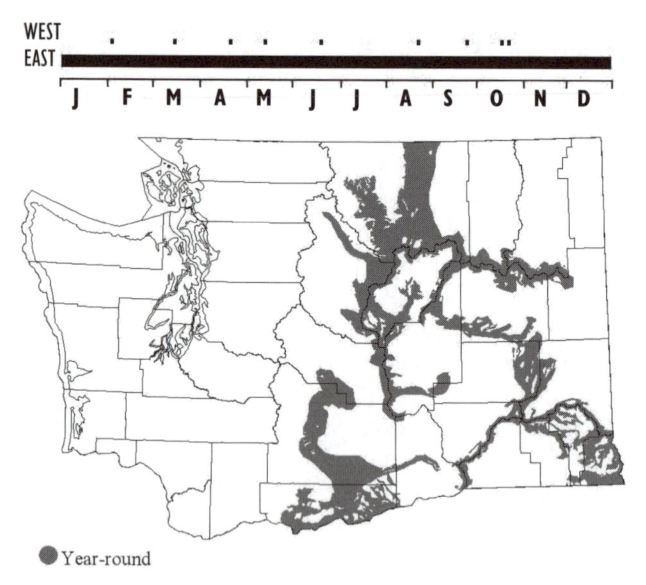

Year-round

Subspecies: *C. m. conspersus.*

Habitat: Steep cliffs, canyons, rocky outcroppings, and talus slopes adjacent to water (Smith et al. 1997), rarely in stone buildings and other human-built structures (Reilly and Pettingill 1968).

Occurrence: The hardy Canyon Wren is often noted on sheer, dry desert cliff faces and rocks. It is resident in e. Washington in deep rocky canyons, basaltic rims, and coulees of major rivers and their tributaries in steppe and desert zones, and also in montane regions on the e. slopes of the Cascades at low elevations (Smith et al. 1997).

A very rare wanderer to w. Washington, with fewer than 10 records, most from high elevations on the w. slope of the Cascades, including summer records in Skamania Co. (C. Chappell), at Carbon Glacier on Mt. Rainier, at Mt. Index, and a truly offbeat occurrence in Oct at Fort Canby on Cape Disappointment (Sharpe 1993).

Though generally non-migratory, populations at higher elevation may move downslope in severe winters (Stepniewski 1999). Occasional fall and winter sightings from n. and w. Washington suggest that birds may wander more widely than previously documented. Severe winters result in large population declines and occasional local extirpation. Last observed local seasonal extirpation in B.C. (Campbell et al. 1997) and in Washington was in 1969 where the species was only reported in the Clarkston area during the following summer.

BBS data inadequate to determine trends (Sauer et al. 2001); winter occurrence on CBCs is inconsistent except at Spokane where no real trend is apparent. As Campbell et al. (1997) point out, no new suitable habitat is being created: overall population size may be limited by this.

Remarks: Subspecies in Washington described as *griseus* by Jewett et al. (1953), while the interior B. C. race is *conspersus* (Campbell et al. 1997).

Noteworthy Records: West, *Spring*: 1 on 23 Apr 1988 at Carbon Glacier on n. side of Mt. Rainier. *Summer*: pair on 16 May 1992 at Carbon Glacier; 1 on 21 Jun 1993 on Mt. Index. *Fall*: 1 on 10-15 Oct 1981 at N. Head, Cape Disappointment; 1 on 20-21 Aug 1992 at Beacon Rk. SP; 1 on 26-27 Sep 1998 at Sunrise, Mt. Rainier. *Winter*: 1 on 6 Feb 1992 at Mt. Rainier. **East**, *high counts*: 6 on 9 Jan 1999 at Little Goose Dam; 14 on 29 Oct 1998 at Umtanum Canyon; 21 on 13 Nov 1998 between Heller Bar and Asotin.

Devorah A. N. Bennu

Bewick's Wren *Thryomanes bewickii*

Widespread, common resident in w. Washington lowlands. Locally fairly common to common in e. Washington, mostly along major rivers, expanding its range.

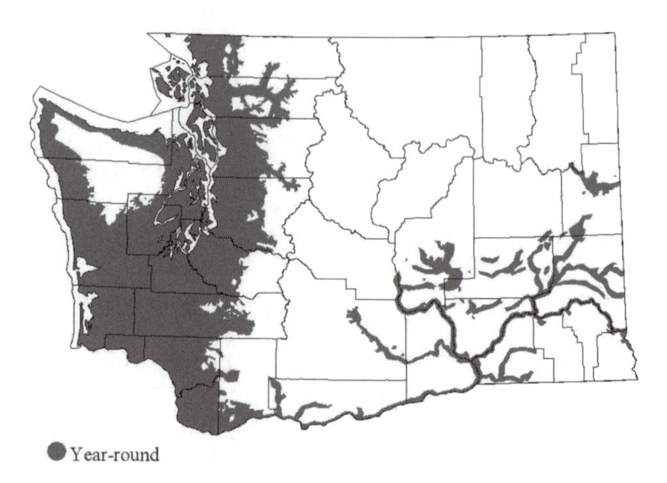

Year-round

Subspecies: *T. b. calophonus.*

Habitat: Farmland, urban, and suburban lowlands with brushy areas, thickets, scrub and forest edges, open and riparian woodland, chaparral, human-provided power line cuts, road edges, brushy parks, residential gardens. Forages widely, including at bird feeders, drift-logs on beaches.

Occurrence: Familiar resident in the Puget Trough, Olympic Pen., outer coast, and in westside river valleys mostly at low elevations to 100 m, with some at higher elevations in suitable habitat; conversion of Puget Trough habitats to urban areas and clear-cuts has expanded habitat (Smith et al. 1997).

Though overall statewide populations are heavily concentrated w. of the Cascades, there has been dramatic growth and spread on the eastside. In the 1950s Jewett et al. (1953) stated Bewick's Wrens were found only in a very limited area within the riparian belt along the Yakima R. on the Yakama Indian Reservation. During the early and mid-1970s, birds spread eastward along the Columbia R. and were common in the Tri-Cities and Walla Walla Co. by the mid-1980s. By 1990 the species was established in small numbers in Spokane Co. Currently, Bewick's Wrens are common in riparian areas along the Columbia R. to White Bluffs and along the Walla Walla R. to Walla Walla. The species is locally fairly common along the Yakima R. from the Tri-Cities to Yakima, and along the Snake R. to Asotin and along the Little Spokane R. Numerous small populations also exist in towns and other wooded habitats in the s. and c. Columbia Basin n. to Spokane and Potholes Res. CBC overall high count totals, generally indicative of winter status, included 129 at Two Rivers in 1989, comparable to some local high westside counts.

Resident in most of w. Washington. It may show latitudinal and elevational movements, particularly on the eastside during severe winters when the species is susceptible to localized extinctions, as in winter 1969 when birds were seen only nr. Yakima in spring and summer 1969, and where they remained scarce for several years afterwards.

Population trends are uncertain. BBS data indicate declines in parts of the U.S. for some time, but data suggest an increase from 1966 to 2000 in Washington (Sauer et al. 2001). BBS data in lower mainland B.C. indicated no change from 1969 to 1993 (Campbell et al. 1997). Washington CBC data show mixed trends with some declines on longest-term counts.

Remarks: Formerly *Troglodytes bewickii*. Jewett et al. (1953) described three subspecies for Washington.

Noteworthy Records: *High counts*, **West**: 31 on 5 Oct 2002 at Ridgefield; 25 on 23 Feb 2003 at Spencer I. (SGM). **East**: 18 on 14 Sep 1998 at Yakima R. delta; 16 on 2 Sep 2002 at Wahluke WMA. *CBC*: 249 in 1997 at Seattle; 148 in 1972 at Bellingham; 148 in 1975 at Grays Harbor; 135 in 1986 at Kent-Auburn; 130 in 1996 at Tacoma; 129 in 1989 at Two Rivers; 124 in 1999 at Olympia; 100 in 1995 at Everett; 80 in 1995 at Padilla Bay; 71 in 1989 at Skagit Bay; 57 in 1997 at Sequim-Dungeness; 55 in 1999 at Kitsap Co.; 51 in 1999 at Vashon I.

Devorah A. N. Bennu

House Wren *Troglodytes aedon*

Common, widespread summer resident in low-elevation dry brushy areas in e. Washington. Rare to locally uncommon in w. Washington except locally common in the San Juan Is. Very rare in winter.

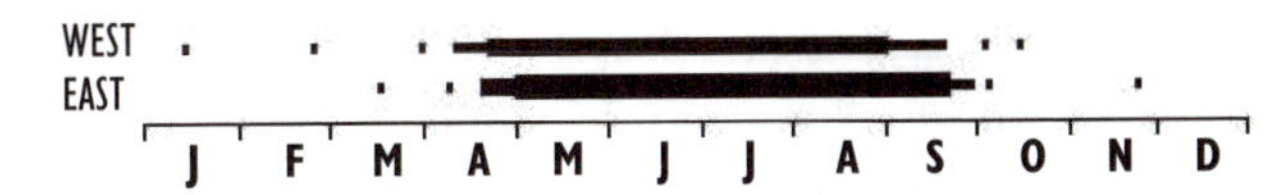

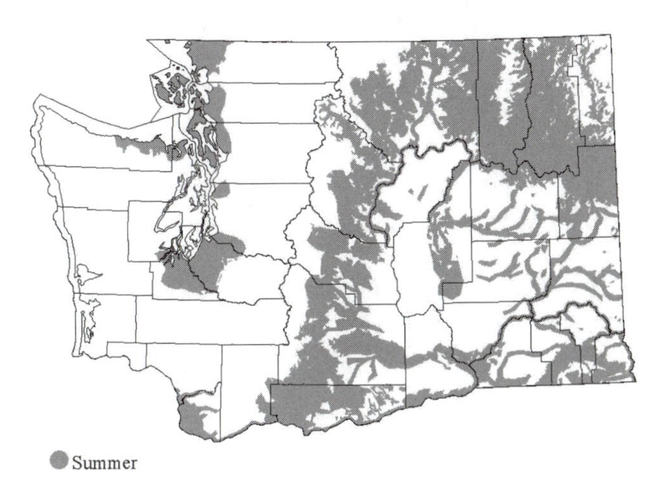

Subspecies: *T. a. parkmannii.*

Habitat: Low-elevation brushy areas, including towns, orchards, and shrub-steppe. Also found at higher elevations within stands of ponderosa pine, oak, and Douglas-fir (Smith et al. 1997).

Occurrence: The House Wren is widespread over much of e. Washington, though abundance is low in hottest parts of the Columbia Basin and reportedly in areas of the se. corner of the state (Smith et al. 1997).

On the westside, House Wrens were formerly more numerous than they are today. Edson (1919) considered the species a frequent summer resident in Whatcom Co. and Rathbun (1911) also considered it a common migrant and summer resident in King Co. Jewett et al. (1953) gave status similarly. Numbers have declined sharply since then (e.g., only 15 records from Whatcom Co. 1965-94 [Wahl 1995]). Currently a rare summer resident and migrant through most of w. Washington. It is common in dry woodland and prairie areas of the San Juan Is., and locally uncommon to fairly common at c. Whidbey I., the ne. Olympic Pen., the s. Puget Sound region, and Clark Co. Birds typically arrive in w. Washington during late Apr and are rare after late Aug, with almost no records after mid-Sep.

This adaptable species has nested in cavities around human homes or in bird houses, and places like abandoned paper hornet nests, old hats, teapots, and animal skulls (Terres 1991) and this has likely facilitated a range expansion. In e. Washington, House Wrens compete successfully for bluebird nestboxes.

BBS data indicate populations increased in Washington by 8.9% annually from 1966 to 2000 (Sauer et al. 2001).

Noteworthy Records: West, *Spring*: 23 Feb 1999 at Skagit WMA. *High count*: 37 at Ridgefield NWR on 9 Jul 1998. *Winter*: 1 on 12 Jan 2003 in Mason Co. **East**, *Spring*: 15 Mar 1996 at Burbank, Walla Walla Co.; 1 on 8 Apr 1997 at Maryhill. Fall: 25 Nov 1962 at L. Lenore. 1 on 2 Oct 1999 at Wahluke slope WMA.

Devorah A. N. Bennu

Winter Wren *Troglodytes troglodytes*

Widespread, uncommon to fairly common resident in coniferous forests statewide. Some move downslope into lowlands for the winter. Rare in other habitats.

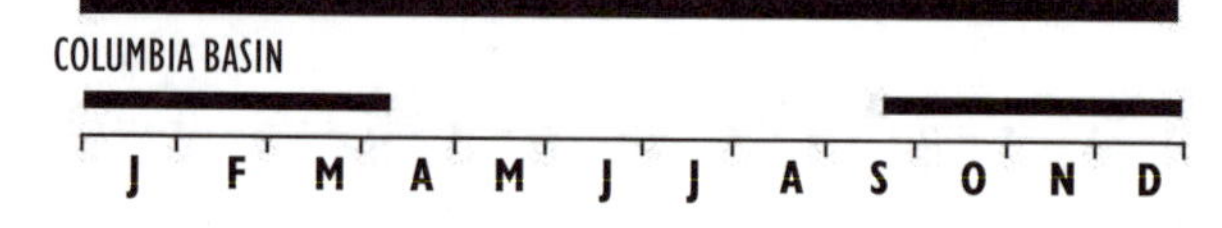

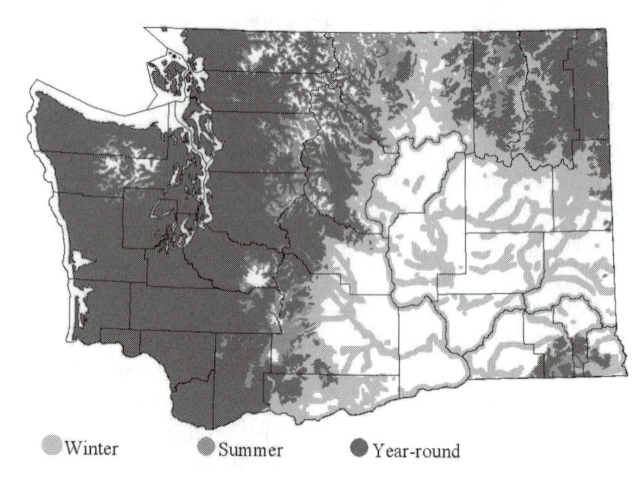

Subspecies: *T. t. pacificus* in w., *salebrosus* on eastside.

Habitat: Coniferous forest, particularly spruce and fir with dense understory nr. water, open areas with low cover along rocky coasts, cliffs, islands, or high mountain regions, including moors and steppes. Deciduous forest and woodland with understory, thickets, hedgerows, and brushy fields in migration and winter.

Occurrence: A vocally impressive, small brown bird found statewide in suitable habitats (Smith et al. 1997). For a common, well-known species, its seasonal movements are relatively poorly understood. Birds migrate from n. coastal and interior B.C. but magnitude and extent of migration, as well as downslope movements, in sw. areas are uncertain (Campbell et al. 1997). Similarly, numerous reports of birds wintering in urban backyards in Washington indicate migration or local downslope or other movements (e.g., Jewett et al. 1953, Wahl 1995, Stepniewski 1999).

On the coast, Winter Wrens are common and widespread, preferring dense and dark woods where mosses are dominant and where the forest floor is covered with litter and fallen trees in various stages of decomposition (Gabrielson and Jewett 1970). Also found on treeless islands in dense brush. In the drier interior, they are less common and more dispersed, usually avoiding dry ponderosa pine forests and often found on north-facing slopes and around streams and ponds, preferring cool, moist ravines with thick underbrush in dense woods at higher elevations (Smith et al. 1997). Wrens are also found at edges of human-made corridors such as clear-cuts, clearings for power lines, in park or garden shrubbery, and along golf courses.

Severe winters may cause sharp population reductions or local extirpations (e.g., Larrison and Sonnenberg 1968, Hejl et al. 2002a), though BBS data indicate a stable population from 1966 to 2000 (Sauer et al. 2001). Counts on westside CBCs may reflect improved observer coverage beginning in the 1970s and less severe winter weather mortality in the 1990s in particular.

Remarks: Known by many common names over its huge range including, in the Old World, simply the "Wren"(AOU 1983). Forest fragmentation and management a concern for maintenance of populations (Altman 1999 in Hejl et al. 2002).

Noteworthy Records: West, *Spring*: 20 pairs counted 5-11 May 1973 on Destruction I. *Summer*: birds on territory at 1800 m on Mt. Rainier during 3rd week of Jun 1991, with snow completely covering the ground. *Fall*: 1968: 1 on 7 Sep 1968 at the seaward extremity of the n. jetty of the Columbia R. **East**, *Columbia Basin, Summer*: 1 on 29 Jun 2001 at Bateman I. *Fall*: 1 on 14 Oct 1967 in sagebrush nr. Banks L.

Devorah A. N. Bennu

Marsh Wren *Cistothorus palustris*

Common summer resident and migrant and fairly common winter resident in w. Washington. Common summer resident and migrant and rare to uncommon winter resident in e.

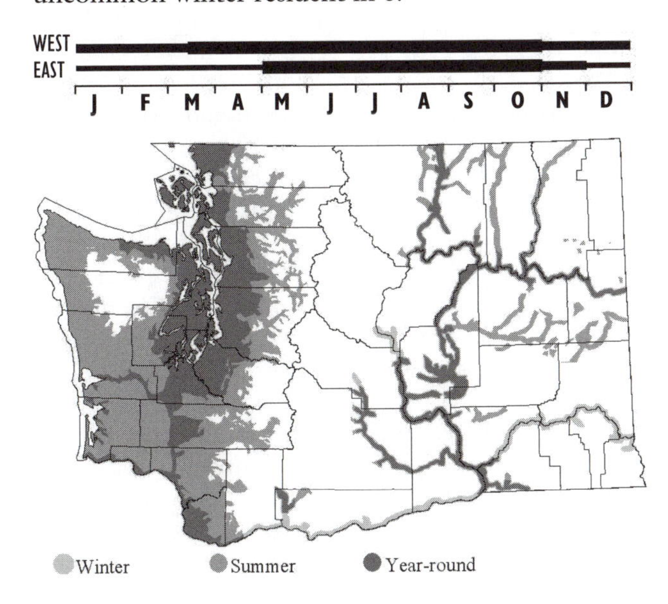

Subspecies: *C. p. browningi* in w. Washington, *paludicola* in sw., *pulverius* in e. Washington.

Habitat: Freshwater and brackish marshes with emergent vegetation, including cattail, tule, bulrush, and reeds. Noted in upland habitats in migration.

Occurrence: Marsh Wrens are widespread in w. Washington lowlands, utilizing habitats from extensive marshes to roadside ditches. E. of the Cascades, they are found in Grant and w. Adams Cos., in sloughs and marshes throughout the Yakima R. valley, in Grand Coulee, and in marshy areas along the Columbia R. (Smith et al. 1997). BBA surveys recorded few if any occurrences in Chelan Co., ne. and se. counties, and status there is uncertain.

Table 38. Marsh Wren - CBC average counts per 100 party hours by decade.

	1960s	*1970s*	*1980s*	*1990s*
West				
Bellingham	1.26	7.97	6.78	4.37
E. Lk Washington			8.90	10.83
Edmonds			11.55	20.29
Grays Harbor		62.00	46.50	23.91
Kent-Auburn			20.08	41.67
Leadbetter Point		22.07	12.45	13.15
Olympia	12.50	7.88	17.76	27.85
Padilla Bay		6.07	3.46	8.28
Port Townsend		2.44	3.61	7.55
Sequim-Dungeness		4.15	34.21	29.15
Seattle	7.08	4.98	11.69	8.37
East				
Spokane	0.57	0.41	0.96	2.65
Tacoma	1.79	4.52	8.22	12.86
Tri-Cities	10.07	7.18	5.49	13.99
Toppenish			17.05	18.01

The species is characteristic of low-elevation wetlands with emergent vegetation. Coastal birds occur also in brackish marshes and some freshwater breeders apparently winter on or nr. the coast in mild winters. During non-breeding seasons, birds occur at low elevations from sea level to about 1040 m (Campbell et al. 1997). Winter occurrence is primarily in Puget Trough (e.g., Skagit Bay and Olympia) and sw. lowlands where locally common, as in Grays Harbor. On the eastside, more numerous in winter in the lower Columbia Basin, where winter climate is relatively moderate. Increasingly rare at higher elevations and latitudes.

Severe winter weather can result in localized extirpations, and mild winters likely increase populations. Over the long term, Campbell et al. (1997) noted population increases in B.C., and Washington BBS data show a significant increase ($P < 0.01$) from 1966 through 2000 (Sauer et al. 2001). CBC data indicate a significant eastside increase ($P < 0.05$; Table 38).

Marsh Wren (Shawneen Finnegan)

Remarks: Numbers have probably declined locally due to loss of freshwater wetlands.

Noteworthy Records: *High counts,* **West**: 79 on 25 May 1997, 55 on 28 Jun 1997, 55 on 7 Jun 1998, 56 on 23 Feb 2003 at Spencer I. (SGM). **East**: 200 in spring 1969 at Turnbull NWR. *CBC*: 252 in 1977 at Grays Harbor; 126 in 1989 at Skagit Bay; 101 in 1997 at Olympia; 88 in 1995 at Kent-Auburn; 74 in 1989 at Sequim-Dungeness; 71 in 1997 at Everett; 34 in 1979 at Bellingham; 30 in 1999 at Toppenish; 28 in 1998 at Edmonds; 28 in 1984 at Tacoma; 26 in 1980 at Seattle; 23 in 1995 at E. Lk. Washington; 22 in 1999 at Wahkiakum; 22 in 1991 at Moses L.; 21 in 1997 at Tri-Cities.

Devorah A. N. Bennu

American Dipper *Cinclus mexicanus*

Fairly common resident along clear, fast-moving streams and rivers in mountainous areas and locally nr. sea level.

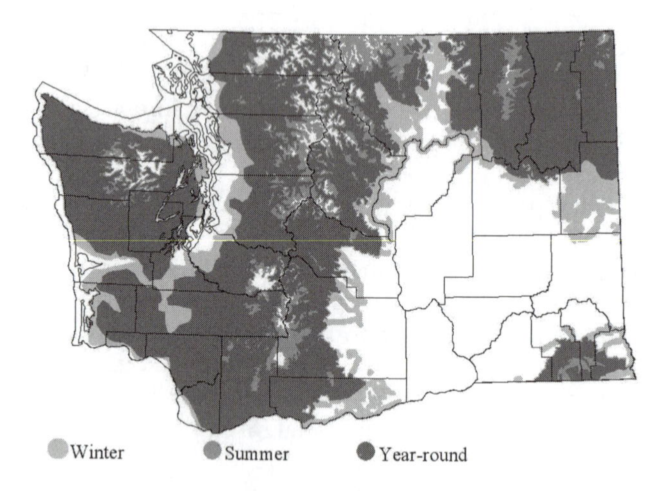

Subspecies: *C. m. unicolor.*

Habitat: Fast-moving, clear, and unpolluted streams and small rivers with conifers on both sides of the stream are optimal habitat, often in riparian and aquatic habitats in mountainous areas. Side streams and small rivers (usually <20m wide) are commonly used (Bakus 1959, Price and Bock 1983), nest sites over or directly adjacent to the stream channel, inaccessible to predators, and above flood levels.

Occurrence: Widespread from the coast to the Cascades, the Okanogan Highlands, ne. Washington and the Blue Mts. May breed on Orcas I. in the San Juans (Lewis and Sharpe 1987). Most found in mid-elevation habitats, from nr. sea level to 1500 m in the Olympics and 1675 m in the Cascades (Smith et al. 1997). Post-breeders wander to tree line, and there is downslope movement in winter (Smith et al. 1997, Stepniewski 1999). Local distribution limited by nest sites, with human-made structures and artificial nests possibly increasing breeding pairs along stream reaches (Hawthorne 1979, Kingery 1996, Loegering 1997). Nest also on support beams under bridges, in sluiceways and spillways, culverts, pipes, and buildings (Kingery 1996). In addition to streams, dippers have also been observed foraging on lakes, water-treatment ponds, and at fish-hatchery ponds (Jewett et al. 1953, Wahl 1995, Kingery 1996).

Winter freezes may force dippers to lower elevations (Jewett et al. 1953, Price and Bock 1983) and, at least occasionally, into marine environments (Hayward and Thorsen 1980). Recorded on many CBCs across the state, from counts within breeding areas in the mountains (e.g., N. Cascades, Sequim-Dungeness, Spokane, Twisp) to downslope areas adjacent to breeding habitats, in lowland breeding sites and areas like Leadbetter Pt. and Grand Coulee. Considered very rare in the Columbia Basin.

Remarks: An indicator of healthy streams and riparian ecosystems. Pollution, siltation, timber harvest, dams, and annual flooding events can negatively affect populations by reducing habitat quality or eliminating habitat. Dippers have few predators and appear to have little impact as a predator of fish species valued by humans. Dippers have received little if any formal study in Washington. Forest practices that protect or enhance riparian corridors would likely benefit populations. Nestbox placement or nest-site creation can enhance breeding habitats and increase local productivity.

Noteworthy Records: *CBC high counts*: 48 in 1993, 37 in 1995 and 33 in 1998 at N. Cascades; 15 in 1989 and 1998 at Twisp; 16 in 1993, 15 in 1977 and 14 in 1999 at Spokane; 12 in 1994 at Sequim-Dungeness; 12 in 1985 at Wenatchee.

Jeffrey C. Lewis

Golden-crowned Kinglet *Regulus satrapa*

Common summer resident in coniferous forests statewide. High-elevation populations move downslope in winter.

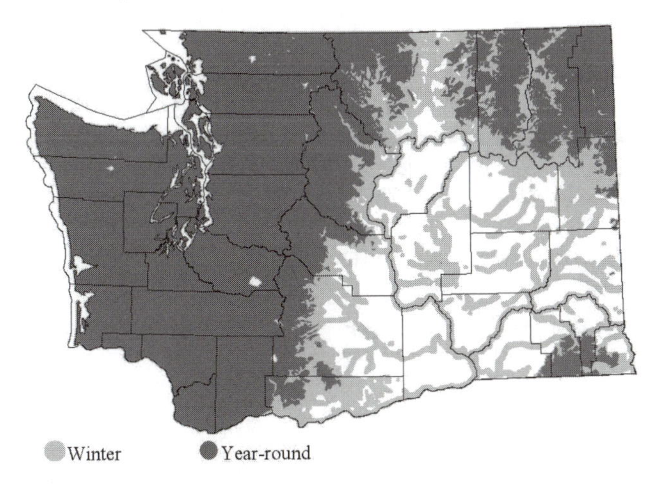

Subspecies: *R. s. olivaceous* in w. Washington, *amoenus* in e.

Habitat: Moist, closed-canopy coniferous forests. Less common especially in higher, dry forests above the ponderosa pine zone in e. Washington. Migrants on occasion in almost all habitats, including steppe zones in e.

Occurrence: A common, widespread forest resident, especially in w. Washington, may be overlooked in summer because of occurrence in tall conifer habitat. Found in westside forests, on the e. slope of the Cascades, in ne. counties, and the Blue Mts. It is less common than the Ruby-crowned Kinglet in e. Washington, especially in higher, dry forests above the ponderosa pine zone, including the Blue Mts.(Smith et al. 1997). Birds are uncommon in cut-over areas (Sharpe 1993, in Smith et al. 1997) and frequently forage on the ground in winter (e.g., Lewis and Sharpe 1987, TRW), which is unusual for a species that spends almost all its time in trees.

Abundances change seasonally, but overall patterns are uncertain. In winter Golden-crowns outnumber Ruby-crowned Kinglets in areas with coniferous trees. Birds move downslope in winter in e. Washington (Stepniewski 1997) as in interior B.C. (Campbell et al. 1997), and presumably in w. Washington, but effects of migration in or out of the state on total populations are unknown. Birds appear in non-breeding areas during early and mid-Sep. Reduced numbers in spring are apparent by mid-May with migration and as pairs of local residents disperse (Campbell et al. 1997).

In winter, CBC birds per party hour indicate about five times as many birds occur on counts w. of the Cascades than e. Apparently great declines in abundances are due to effects severe winter weather, as following 1955 (e.g., Larrison and Sonnenberg 1968, and see Campbell et al. 1997). BBS data indicate a statewide non-significant decline of 3.0% per year from 1966 to 2000 (Sauer et al. 2001). CBC counts suggested a significant long-term westside increase (P <0.01).

Remarks: Jewett et al. (1953) list the Eastern Golden-crowned Kinglet, ***R. s. satrapa***, as a casual migrant (see Campbell et al. 1997). It is not known whether population declines might be due to forest harvests or winter populations have increased with maturity of habitats in urban and suburban areas.

Noteworthy Records: East, *Fall, high counts*: 125-150 on 22 Oct 1989 nr. White Pass; 625 on 29 Sep 1996 along upper Cold Cr., Yakima Co. (Stepniewski 1999). *CBC high counts,* **West**: 1822 in 1989 at Olympia, 1538 in 1988 at Padilla Bay, 1242 in 1989 at Seattle, 1220 in 1995 at Tacoma, 1217 in 1991 at Sequim-Dungeness, 1127 in 1986 at Bellingham; **East**: 380 in 1980 at Spokane, 323 in 1974 at Walla Walla, 151 in 1992 at Yakima Valley, 99 in 1980 at Chewelah, 80 in 1988 at Ellensburg, 60 in 1993 at Chelan, 56 in 1974 at Tri-Cities, 42 in 1990 at Toppenish, 23 in 1990 at Wenatchee, 13 in 1993 at Twisp, 5 in 1989 at Moses L.

Terence R. Wahl

Ruby-crowned Kinglet *Regulus calendula*

Common summer resident in coniferous forest e. of the Cascade crest, local in Olympics. Common winter resident w., locally uncommon to fairly common e.

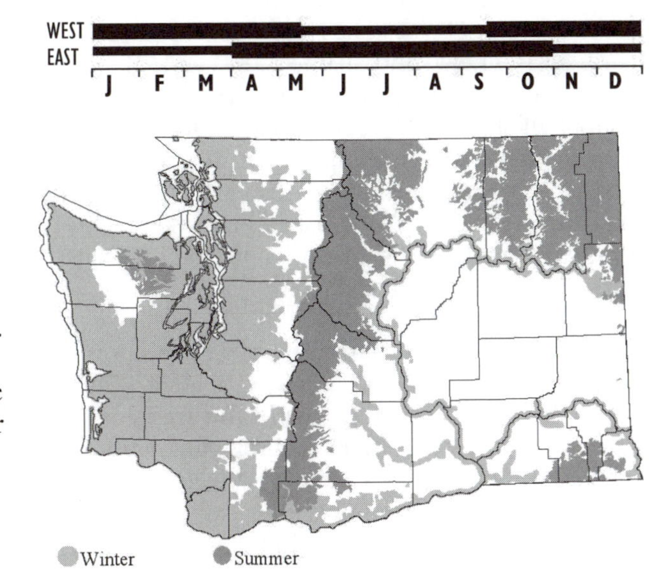

Subspecies: *R. c. cineraceous.*

Habitat: E. of the Cascade crest, preferred nesting habitats in small forest openings nr. water above dry interior Douglas-fir zone, with fewer below that. Nests in ne. Olympics at high elevations in dry coniferous forest above mountain hemlock zone, and w. of Cascade crest at very high elevations (Smith et al. 1997). In winter in lowlands, often in brush and gardens, and less frequently in conifers than the Golden-crowned Kinglet.

Occurrence: This small insectivore's seasonal pattern is quite different on e. and w. sides of the Cascades in Washington and illustrates the scale of migration. Birds are widespread in summer e. of the Cascades at high elevations ne. to Pend Oreille Co. and in the Blue Mts. Virtually absent in summer in w. Washington, though discovered nesting in 1986 in ne. Olympics at high elevations in dry coniferous forest above mountain hemlock zone, and subsequently w. of Cascade crest at very high elevations (Smith et al. 1997).

Moves to lower altitudes and latitudes to avoid winter cold. In B.C. largest numbers switch from c. and sub-boreal interior forests in summer to lowlands of sw. B.C. during winter (Campbell et al. 1997). Peak migrations are from mid-late Sep to Oct and mid-Apr to early May statewide. Highly variable in abundance, with eastside numbers apparently dependent on severity of weather. In Yakima Co. many winter in pear and apple orchards with high insect abundance (Stepniewski 1999). Birds winter in lower-elevation mixed habitats, with small numbers in many locations e. to Spokane and Clarkston. Highest eastside numbers are in relatively mild lower Columbia Basin and s. Klickitat Co.

Widespread numbers winter in w. lowlands, often in urban areas. Crosses open water to winter in the San Juan Is. (Lewis and Sharpe 1987) and, more dramatically, in fall to the Queen Charlotte Is., B.C. (Campbell et al. 1997). Seasonal abundance changes shown by CBCs to be about 5 times larger w. of the mountains than in the e. CBC data indicate increases w. ($P < 0.01$) and e. ($P < 0.05$). Stepniewski (1999) suggested that orchards in e. Washington may have increased winter occurrence.

Remarks: Subspecies occurrence not agreed upon (see Jewett et al. 1953, Campbell et al. 1997, Smith et al. 1997). Determination of possible westside breeding at lower elevations requires elimination of confusion with Hutton's Vireos. Ruby-crowns often sing loud, recognizable songs in lowlands late in the spring before or during migration (see Smith et al. 1997).

Noteworthy Records: East, *Summer*: nested at Pullman (Smith et al. 1997). *Fall*: 625 migrating on 29 Sep 1996 nr. Yakima (Stepniewski 1999). **West**, *Spring*: "several thousand" on 15 Apr 1984 on Tatoosh I. during migration. *Summer*: 1 on 21 Jun 1977 nr. McCleary; 1 on 17 Jul 1981 at Ocean Shores. *Breeding*: 2 singing along upper Royal Cr., ne. Olympics on 19-20 Jul 1986; 1 at 1300 m in Cat Basin, ONP, on 7 Aug 1989; birds at Dose Meadows and Seven Lks. Basin in Olympics in summer 1992. *CBC high counts*, **West**: 562 in 1996 at Tacoma, 525 in 1988 at Olympia, 513 at Seattle in 1997. **East**: 66 in 1986 at Yakima Valley, 47 in 1997 at Lyle, 46 in 1988 at Toppenish.

Terence R. Wahl

Blue-gray Gnatcatcher *Polioptila caerulea*

Very rare visitor.

The Blue-gray Gnatcatcher has been recorded in Washington seven times. Records are primarily from late fall or winter, and all but two from lowland areas of the westside (Tweit and Skriletz 1996). The first record was of a bird (ph.) at Whidbey I., 10-16 Nov 1978. Subsequent records included one at Ocosta, Grays Harbor Co., 21 Feb 1983; one at Seattle 6 Dec 1986-30 Jan 1987; one at Richland, 4 Nov 1997; one at Marymoor Park, King Co.,21-25 Oct 1999; one at Point No Point, 10-11 Nov 1999, one at Hood Park, Walla Walla Co., 23 Sep 2000; male (ph.) built nest, defended territory on 25-30 May 2002 at Hardy Canyon. Subspecies presumably *P. c. amoenissima*.

Bill Tweit

Western Bluebird *Sialia mexicana*

Locally fairly common and widely distributed summer resident in e. Washington and c. and sw. Washington except for high elevations, dense forests, and the Columbia Basin. Rare in winter.

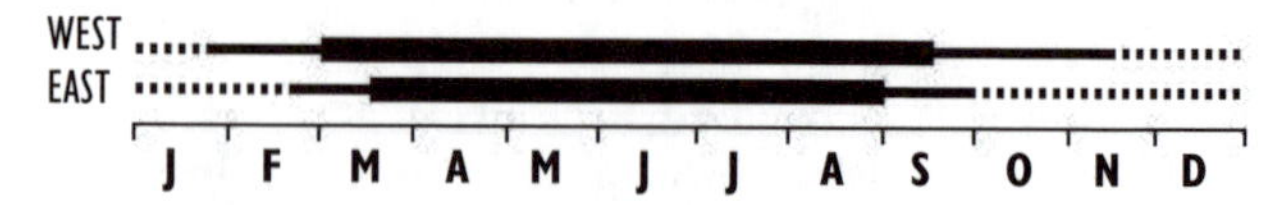

Subspecies: *S. m. occidentalis* (Guinan et al. 2000).

Habitat: In e. Washington most often open forest, burned forest, or forest edge (Herlugson 1978, Smith et al. 1997), also in agricultural areas, shrub-steppe, and grasslands (Smith et al. 1997). In w. Washington found in native prairie, burned forest, and clear-cut forest (Smith et al. 1997), and not in subalpine areas (Manuwal et al. 1987).

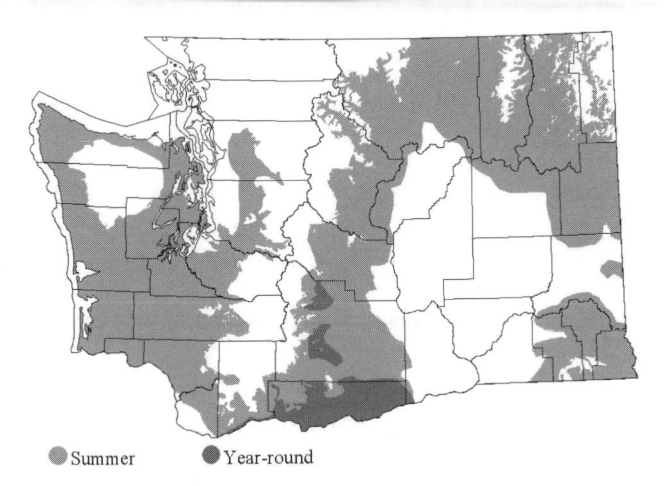

Occurrence: Nests between forest and steppe zones in e. Washington, in counties adjacent to the Columbia R., in the ne. and the Blue Mts. (Smith et al. 1997). Its known or suspected westside range includes Pierce and Thurston Co. prairies, isolated areas nr. Port Townsend, the mouth of the Naselle R., in Grays Harbor, Pacific, Lewis, Clallam, and Skamania Cos., and in forest clearings in Snohomish, King, Pierce, Mason, and Thurston Cos. (Zarnowitz and Manuwal 1985, Bosakowski 1997, Smith et al. 1997, JBB, WDFW database, SGM). Breeding birds are rare away from known breeding areas where they sometimes arrive in late Jan-Feb on the westside and late Feb on the eastside.

The Western Bluebird's status changed greatly in the 20th century. Once relatively common w. of the Cascades (Jewett et al. 1953, Herlugson 1978), its numbers were much diminished by about mid-century. In the San Juan Is., for example, where it was considered common in the 1930s, it was last recorded breeding in 1964 (Lewis and Sharpe 1987), and it has not been recorded in Whatcom Co., where it formerly nested, since 1975 (Wahl 1995). The decline appears to have resulted from habitat changes (e.g.. reduction in burned forest, loss of snags, conversion of oak and prairie habitats) before the arrival of the European Starling (Rogers 2000). This decline parallels that in sw. B.C. (Campbell et al. 1997).

Nestbox programs in both e. and w. Washington have resulted in substantial local population increases since the 1970s. A nestbox trail in the Spokane area produced an average of 173 fledglings between 1976 and 1981. The number of breeding pairs at Fort Lewis increased from four in 1981 to 215 in 1988, with 700+ fledglings being produced in 1989 (G. Walter p.c.).

Despite recent increases in abundance and promising levels of productivity, rangewide BBS trends indicate a significant decline in abundance between 1980 and 1998 (Sauer et al. 2000). Trends from regional BBS routes are mixed, with routes in the Columbia Plateau showing a non-significant increase, and routes in the Cascades showing a decline. There was a statistically significant decline of -3.6% in w. Washington between 1966 and 1998 (Sauer et al. 2000), and a statewide decline of -2.7% from 1966 to 2000 (Sauer et al. 2001).

Rare in winter: recorded 31 times at 10 CBC locations since 1960. Noted mostly at Walla Walla (10 counts 1973-98), Spokane (7 times 1966-87), and Tacoma (5 times 1991-97). Twenty-five CBC records are from e. Washington, and only Tacoma and Port Gamble had westside records.

Remarks: Because of successful response to nestbox programs it appears that a lack of nesting structures is the most limiting factor on populations. Continued work on nestbox programs, adequate retention of snags in clear-cuts and burned areas, and protection of prairies and other important habitats from conversion will be important conservation measures. Classed PHS.

Noteworthy Records: West, *Winter, outside Pierce and Thurston Cos.*: 11 on 18 Dec 1999 at Squamish Harbor, Jefferson Co.; 2 on 13 Jan 1989 in Cowlitz Co.; 10 on 2 Feb 1991 nr. Sequim; 6 on 11 Feb 1999 in Clallam Co.; 4 on 16 Feb 1980 at Centralia (WDFW); 5 on 29 Feb 1980 nr. Morton (WDFW); 10 on 3 Mar 1997 at PeEll. **East**, *High counts, Fall*: 50 on 26 Sep 1964 at Spokane; 10 on 20 Oct 1996 at Cold Cr. Divide, Yakima Co. (Stepniewski 1999). *Winter*: 20 on 11 Nov 1989 at Whitman Co. CBC high counts: 24 in 1998 at Lyle, 24 in 1999 at Lyle, 19 in 1980 at Spokane, 17 in 1974 at Spokane, 13 in 1995 at Camas Pairie-Trout L., 11 in 1999 at Port Gamble, 11 in 1989 at Walla Walla.

Joseph B. Buchanan

Mountain Bluebird *Sialia currucoides.*

Fairly common breeder and rare winter visitor in e. Washington from lower-elevation open habitats to alpine meadows and parklands. Very local breeder in w.. Widespread migrant.

Habitat: Burns, agricultural and grazed areas, clear-cuts, parklands, meadows, shrub-steppe, and forest edges from lower elevations to alpine zones serve as breeding habitats. Nests in cavities in snags (e.g., abandoned woodpecker nests), crevices in rocks and cliffs, buildings, and other human-made structures (Jewett et al. 1953). Readily use nestboxes placed in these open habitats, and nestbox placement has made available open habitats that lacked nesting structures.

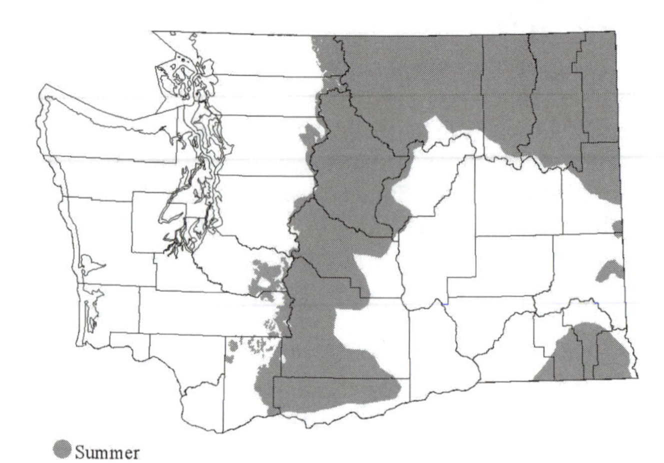

Occurrence: Breeds mainly on the e. slope of the Cascades, ne. Washington, and the Blue Mts. (Herlugson 1978, Smith et al. 1997). Breeding range also includes alpine habitats at Mt. Rainier and Mt. St. Helens in w. Washington. Though a common breeder in the Olympic Mts. in the 1940s (Jewett et al. 1953, Sharpe 1993, as cited by Smith, et al. 1997), absent in the 1990s (Smith et al. 1997). Placement of nestboxes in Klickitat Co. has expanded the breeding range (Herlugson 1978).

Migration poorly documented. Birds arrive on the eastside as early as late Feb. The species is a rare spring migrant in westside lowlands from late Feb to early May, with most records from Puget Sound lowlands and noted w. to the outer coast. It is very rare during the rest of the year, with most summer records from alpine areas high in the Cascades. The only westside lowland summer record is of a bird nr. Bellingham in Jun 1999. Fall records w. of the Cascade crest are few, but large numbers are occasional along the crest, with 60 at Mt. Rainier on 17 Aug 1994.

Generally winters s. of Washington, but observations of birds in winter suggest that a small number may be year-round or winter residents in e. Washington. CBCs from 1959 to 1999 reported birds only nine times, from Bellingham, Padilla Bay, San Juan Is., Sequim-Dungness, and Port Townsend in the w., and Bridgeport and Walla Walla e. of the Cascades.

BBS data indicate an increasing trend in Washington from 1980 to 2000 (trend estimate 4.9%; Sauer et al. 2001). Decreases in numbers, however, were widely noted before that period (e.g., Weber and Larrison 1977). This was less likely due to competition with starlings for nest sites than was the case with the Western Bluebird and was suggested to have been due to climatic factors (see Herlugson 1978).

Remarks: Efforts to create nesting habitat along bluebird nestbox trails seem to have offset any reductions in habitat caused by fire suppression and management for late-successional forests on some state and federal lands. Known to hybridize locally with Eastern and Western bluebirds.

Noteworthy Records: West, *Summer*: several in Jun 1963 on Hurricane Ridge (TRW) may be last report from the Olympics. 1 on 18 Jun 1999 nr. Bellingham. *Outer coast, Spring*: 1 on 9 Apr 1977 at Ocean Shores; 2 on 24 Apr 1997 at Leadbetter Pt. *Winter*: 1 in Jan-Feb 1967 and 1 on 18 Jan 1970 at Lummi Flats; 1 on 4 Jan 1971, and 3 reports in winter 1983-84 nr. Bellingham; 1 in Dec 1975 and 2 in winter 1977-78 at Samish Flats; 1 on 19 Dec 1981 nr. Port Townsend; 2 in winter 1983-84 on San Juan I.; 1 on 29 Dec 1987 at Graysmarsh, Clallam Co.; 1 on 12-26 Dec 1998 nr. Sequim; 1 on 15 Dec 2001 at Ebey's Prairie; 1 on 17 Feb 2002 nr. Burlington. *High counts*: 20-30 on 30 Sep 1971 at Ross L.; 12 on 9 Apr 1975 at Samish Flats. **East**, *Winter*: 1 on 16 Dec 1962 at Banks L.; 1 on 7 Feb 1967 at Spokane; 2 on 7-9 Jan 1980 nr. Clarkston; 1 on 12 Jan 1985 nr. Rogersburg; 1 on 16 Jan 1988 at Grand Coulee; 4 on 24 Jan 2000 at Crab Cr. *High counts*: 75-100 on 24 Sep 1972 nr. Spokane; 150 on 21 Aug 1972, 100 on 2 Aug 1973, and 150 on 6 Oct 1974 at Peola; 25 in fall 1981 at Wenas Cr.; 85 on 6 Mar 1984 at Bickleton; 41 on 21 May 1998 at Peola; 30 on 18 Oct at Ephrata; 175 on 20 Aug 1999 at N. Riggs Canyon, Kittitas Co.

Jeffrey C. Lewis

Townsend's Solitaire *Myadestes townsendi*

Fairly common migrant and summer resident and migrant, and uncommon winter resident in e. Washington. Locally fairly common summer resident and migrant in westside mountains, otherwise rare migrant and winter resident in w.

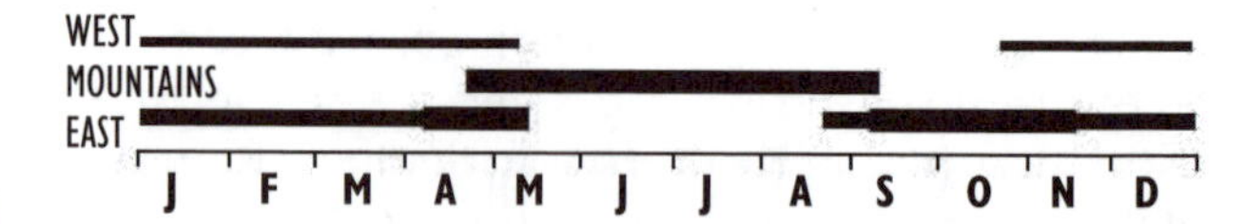

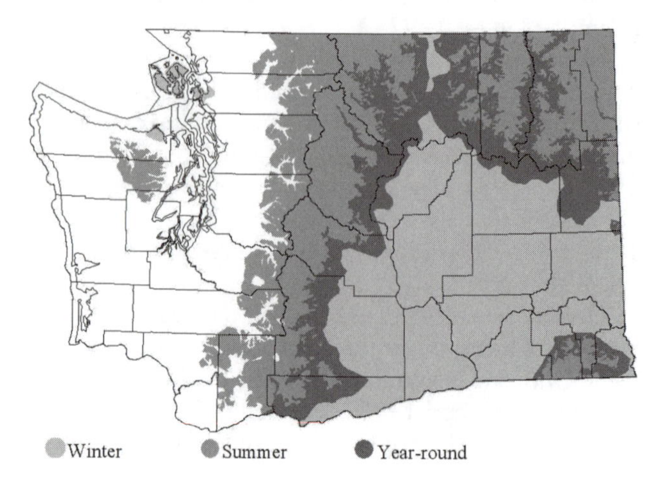

Subspecies: *M. t. townsendi.*

Habitat: Breeds in relatively drier, open coniferous forests with a combination of steep banks for nest placement and open areas with some tall trees for fly catching (Smith et al. 1997). During migration also in lowland riparian areas, towns, and suburbs. In winter, almost exclusively in areas with small fruit, particularly juniper or piñon-juniper woodlands.

Occurrence: A characteristic summer bird of eastside montane forests, but with a widespread and complex statewide distribution. Solitaires breed predominantly in montane woodlands, being widespread and fairly common in e. Washington, though they are uncommon in the w. Blue Mts. (Smith et al. 1997). On the westside, this species is a local but fairly common breeder above 1100 m in the Cascades and the ne. Olympics, particularly at clear-cuts, burns, and high ridges (Smith et al. 1997).

Fall migration begins in late Aug or early Sep and continues at least through Oct. Many birds move s. through the Cascades, but a fair number of fall migrants are also found in e. Washington lowlands, where this species is fairly common in riparian areas, especially during Sept. Fall migrants are less than annual on the westside and generally occur during Nov.

Downslope wandering is evident (Stepniewski 1999), with solitaires wintering in small numbers where berries are available in the foothills and towns of e. Washington. Typically only a small number are reported wintering in similar locations on the westside and CBC data indeed indicate much greater numbers in winter e. of the Cascades.

Spring migration mostly from early Apr to early May. In contrast to fall, early northbound migrants avoid high montane areas and birds are fairly common in eastside lowlands. In lowland w. Washington, the status is more variable, with birds almost fairly common during Apr some years and virtually absent other years. This is likely related to snow cover and food supplies and possibly temperature in the mountains.

BBS data indicates a statewide increase ($P < 0.05$) from 1966 to 2000 (Sauer et al. 2001). Interdecadal variations in occurrence are apparent in CBC data, indicating a significant westside decline ($P < 0.05$).

Noteworthy Records: East, *high counts*: 76 on 8 Sep 1996 and 60 on 19 Sep 1996 at Cold Cr., Yakima Co. (Stepniewski 1999); 50 on 17-18 Sep 1983 at Grand Coulee.

Steven G. Mlodinow

Veery *Catharus fuscescens*

Fairly common in lower and mid-elevation areas in much of e. Washington, absent from the Columbia Basin. Local, rare in n. Puget Trough river valleys.

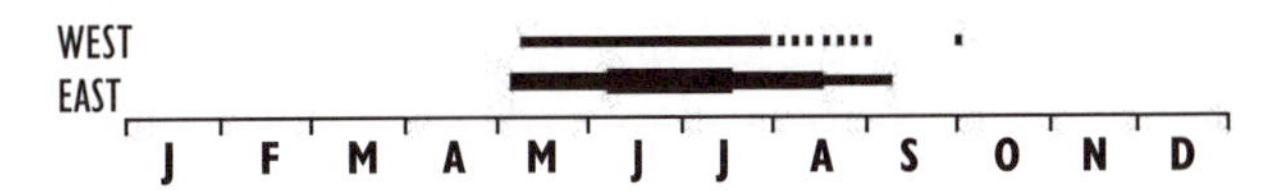

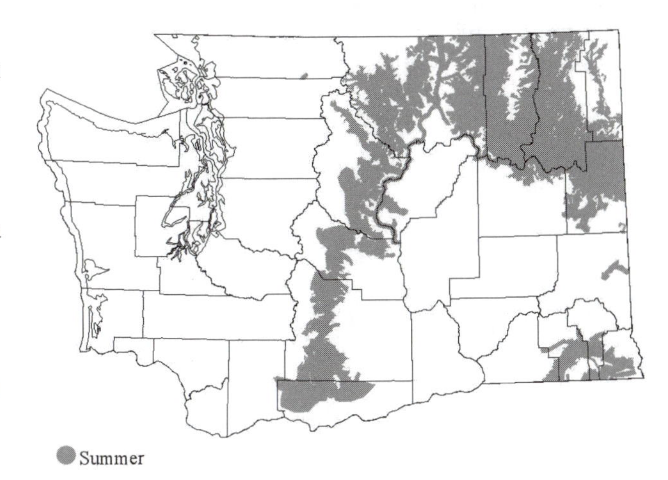

Subspecies: *C. f. salicicola.*

Habitat: Shrubby, moist deciduous habitats, particularly in riparian zones in Washington (Moskoff 1995).

Occurrence: The Veery is a summer resident in lower and mid-elevation areas throughout much of e. Washington (Smith et al. 1997), except for the Columbia Basin, where it is apparently absent as a breeder. It is locally common on the e. slope of the Cascades (e.g., at Teanaway: SGM). Records of migrants away from breeding areas are rare.

In w. Washington, a small breeding population is established in Whatcom and Skagit Cos., where up to several pairs have been detected in recent years (Wahl 1995, BBS). The species has also been recorded elsewhere in w. Washington, with three records in King Co. (Hunn 1982) and others in Snohomish, Skagit, Clallam, Pierce, and Pacific Cos. Birds assumed to be non-breeders have been recorded in spring-summer months on s. Vancouver I. and in lower mainland B.C. (Campbell et al. 1997).

This is a late migrant, typically arriving in mid-May or later. Infrequently seen and detected most often by its distinctive vocalizations. It becomes quiet, and therefore rather inconspicuous, after breeding and its fall departure is not often documented (Stepniewski 1999). Like the catbird, the Veery probably migrates e. before turning s. in fall.

The Veery responds negatively to heavy cattle grazing in its habitats (Mosconi and Hutto 1982), and similarly to local horse-riding activities in Spokane (J. Acton p.c.). BBS data indicate that the Veery population in e. Washington has been stable since 1966 (Sauer et al. 2001) although populations in e. N. America have declined (Sauer et al. 1999b).

Remarks: Phillips (1991) recognized this and two other subspecies, but agreed that only *C. f. salicicola* (synonymous with *subpallidus* as per Burleigh and Duvall 1959) occurred in Washington.

Noteworthy Records: West, *Spring*: 11 May 1985 at Carnation; 21 May 1978 e. of Redmond; 21 May 1978 e. of Duvall (Hunn 1982); 21 May 1984 at Sequim; 27 May 1980 se. of McKenna; 30 May 1981 at Colonial Cr. CG, Whatcom Co. (Wahl 1995). *Summer*: 1 Jun 1980 nr. Carnation (Hunn 1982); 5 Jun 1983 at Goodell CG along Skagit R. w. of Newhalem (Wahl 1995); pair on 20 Jun 1992, 2 pairs on 25 Jun 1994, and 5 pairs on 22-23 Jun 1995 about 1.5 km w. of Newhalem (Wahl 1995); 4 on 12 Jun and 4 Jul 1999 nr. Newhalem and 2 on 25 Jul 1999 at County Line Ponds, Skagit/Whatcom Cos.; 5 Jul 1980 at Deception Cr. trailhead e. of Skykomish (Hunn 1982); record in the "1990s" at Ross L. (Wahl 1995). *Fall*: 24 Aug 1980 at Leadbetter Pt.; first week in Oct 1966 at Tacoma. **East**, *Spring*: 12 May in se. Washington (Hudson and Yocom 1954); 14 May 1990 in Okanogan Co.; 18 May 2002 at Winddust SP, Franklin Co. *Fall*: 2 Sep 1941 in se. Washington (Lyman and Dumas 1951); 3 Sep 1999 at Trout L.; 22 Sep 2001 at Rattlesnake Ridge, Benton Co. *High count:* 36 on 10 Jun 2001 at Teanaway (SGM); 27 on 24 Jun 2000 at Teanaway.

Joseph B. Buchanan

Gray-cheeked Thrush *Catharus minimus*

Casual vagrant.

The single Washington record of this species was of a bird at Wallula on 6 Oct 1990. There are two Oregon records, both from late Sep. Inasmuch as Gray-cheeked Thrushes breed as far w. as Siberia and as close as nw. B.C., the species may occur in Washington more frequently than indicated by the single record. Sightings and confirmation of them are probably limited due to the species' skulking habits and identification problems. Subspecies presumably *minimus.*

Steven G. Mlodinow

Swainson's Thrush *Catharus ustulatus*

Common summer resident on westside and eastside. Common migrant on westside, uncommon migrant in eastside lowlands.

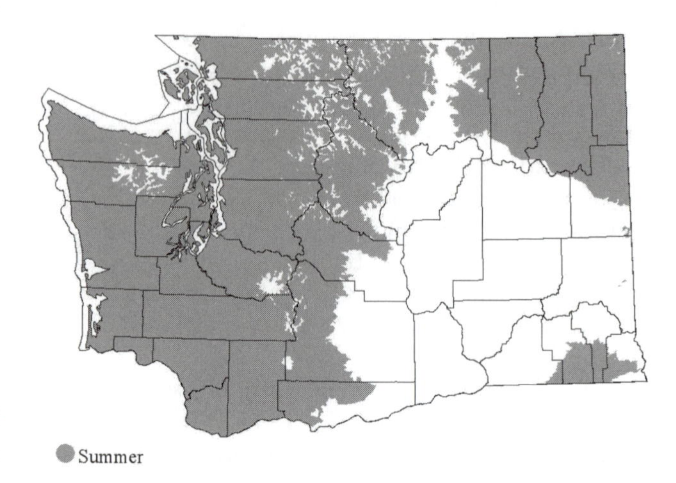

Subspecies: *C. u. ustulatus* breeds in w. Washington and Cascades e. slope; *swainsoni* in ne. and se. Washington.

Habitat: Breeds largely in moist wooded areas with substantial brushy undergrowth. In e. Washington, best areas are below the alpine/parkland zones, while in w. Washington, favored areas are below the mountain hemlock zone (Smith et al. 1997). Often a narrow strip of riparian land is sufficient on the westside. Migrants most often in similar habitat.

Occurrence: A widespread migrant and breeder, often heard calling during nocturnal migrations. Spring migration occurs mainly from mid-May to early Jun. Migrants are quite scarce in lowland e. Washington, especially during fall, being uncommon in lowland Yakima Co. (Stepniewski 1999) and rare in the Tri-Cities (Ennor 1991). Fall migration passes largely from mid-late Aug to mid-Sep.

Breeding birds are widespread in westside habitats, generally nesting at lower elevations than Hermit Thrushes. On the eastside this thrush is not as numerous and is limited to moister woods in the Cascades, ne. and Blue Mts. when breeding and it nests at generally higher elevations, to 1600 m (Smith et al. 1997), than on the westside. Though there is some overlap, Swainson's Thrush is distributed higher in elevation than the Veery and lower than the Hermit Thrush.

The populations and ranges have contracted over N. America (Mack and Yong 2000). In Washington, the species benefitted over decades following human population expansion and development, especially on the westside, from replacement of coniferous forests with deciduous growth (Smith et al. 1997). BBS data do not indicate change from 1966 to 2000 (Sauer et al. 2001).

Remarks: Authorities have recently agreed on two fairly distinct subspecies groups. The Russet-backed Thrush, *C. u. ustulatus,* breeds along the Pacific coast including w. Washington and the e. slope of the Cascades. The Olive-backed Thrush, *swainsoni,* breeds in ne. and se. Washington and the remainder of an extensive N. American range. Within the two broad taxonomic groupings, the number of subspecies described as from four (AOU 1957) to five (Lane and Jaramillo 2000a) to six (Pyle 1997). Differentiation between this finer classification is likely impossible in the field. For separation of Russet-backed versus Olive-backed Thrushes, see Lane and Jaramillo (2000b). Note that Smith et al. (1997) stated that the Olive-backed Thrush breeds mainly in w. Washington and the Russet-backed in e. Washington.

Noteworthy Records: *Spring, early*: 4 Apr 1999 at Skagit WMA; 14 Apr 1999 at Ocean Shores. *Fall, late*: 23 Oct 1973 at Spokane; 28 Oct 1998 at Post Office L., Clark C.; 29 Oct 1972 in Whatcom Co. (TRW); 1 Nov 1997 at Richland; 19 Nov 2000 at Medina, King Co. *High counts*: 130 on 28 Jun 1997 at Everett; 105 on 17 Jun nr. Darrington (SGM); 90 on 6 Jul 1994 at Spencer I.

Steven G. Mlodinow

Hermit Thrush *Catharus guttatus*

Common breeder in mountains. Uncommon migrant and winter resident in w. Washington. Fairly common migrant and rare winter resident in e. Washington.

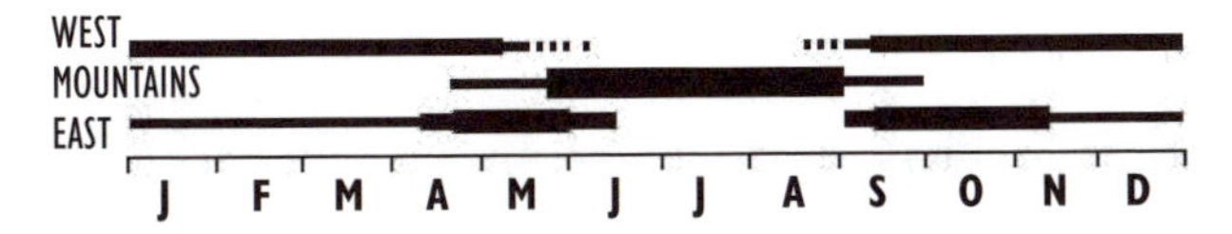

Habitat: Breeds in high-elevation coniferous forests. During migration and winter also occurs in low-elevation coniferous forests and in lowland deciduous riparian areas. Apparently an equally common breeder at higher elevations on both slopes of the Cascades, and breeds also in the ne. and Blue Mts. Reportedly bred, at least historically, to nr. sea level on the Olympic Pen. (Jewett et al. 1953), and in lower foothills (e.g., nr. Newhalem; Wahl 1995), but generally above 450 m (Smith et al. 1997).

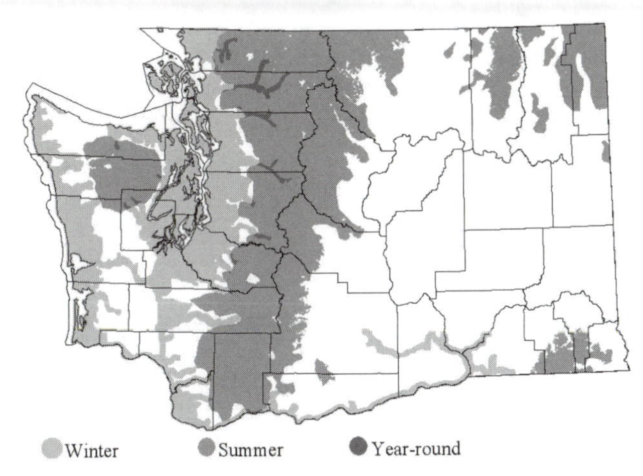

Typically breeds at higher elevations than Swainson's Thrush.

Occurrence: During migration Hermit Thrushes are distinctly more common on the eastside than the westside. Migration on the eastside is noted generally between mid-Apr and early Jun and again between mid Sep and early Nov. On the westside, migration dates are less certain due to overwintering birds, but appears to be mainly between mid-Apr and early May and from mid-Sep through mid-Oct.

Birds were recorded fairly consistently in small numbers on most westside CBCs. Overall, generally rare in winter on the eastside, uncommon in s. Columbia Basin and Klickitat Co., with eastside numbers apparently declining as winter progresses. Increasing numbers on CBCs at Tri-Cities and Two Rivers were apparent in the 1990s. Statewide, BBS data indicate a 1.1% per year increase from 1966 to 2000 (Sauer et al. 2001). CBC data, though variable, indicate no overall change in winter abundance.

Remarks: Authorities differ widely on number of subspecies (eight to 13) and their distribution (see AOU 1957, Pyle 1997, Lane and Jaramillo 2000a). Races fall into three major groupings, but authorities differ even on placement of races within these. Lane and Jaramillo (2000a) describe breeders in the Olympics and on both slopes of the Cascades as members of the Pacific Hermit Thrush group (Pyle 1997, Smith et al. 1997, but *contra* Lane and Jaramillo 2000a), migrants/wintering birds in w. Washington as mostly (possibly entirely) from the Pacific group, but eastside has more complex picture with both Pacific and Rocky Mt. individuals likely, and Eastern Hermit Thrushes possible as vagrants (see Lane and Jaramillo 2000a).

Noteworthy Records: *High counts, migration*: 55 on 19 Sep 1998 at Seattle; 22 on 1 Oct 1995 at ALE, Benton Co. *Winter*: 13 on 19 Dec 1998 nr. Lime Kiln SP., San Juan I.; 31 on 7 Dec 2002 at Wahluke Slope, Franklin Co.

Steven G. Mlodinow

Dusky Thrush *Turdus naumanni*

Casual vagrant.

A migratory thrush breeding in n. and e. Siberia, recorded rarely in Alaska. It has occurred once in B.C., at Langley, Jan-Apr 1993 (Campbell et al. 1997). There is one very unseasonal report of a bird at Mt. Vernon on 27 Jun 2002 (WOSN 85:14).

Terence R. Wahl

Redwing *Turdus iliacus*

Casual vagrant.

A Eurasian thrush recorded in winter about 10 times in North America and just once in the U.S. through 2000 (ABA 2002). The first Washington record occurred in Olympia on 21 Dec. 2004 (G. Revelas, BT) and the bird was subsequently noted (ph.) by many observers.

Steven G. Mlodinow

American Robin *Turdus migratorius*

Common permanent resident.

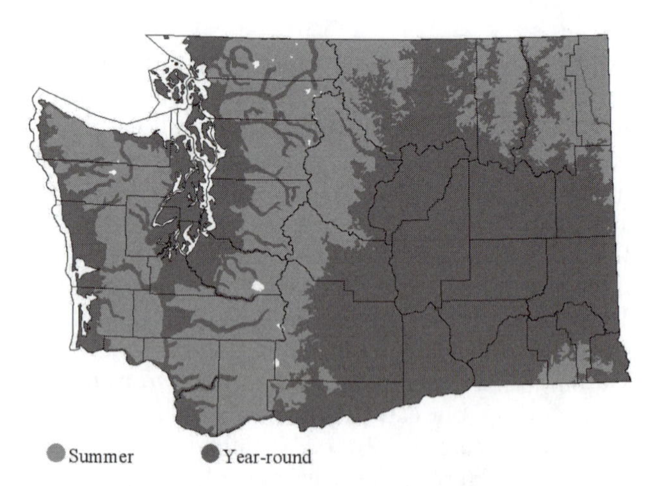

Subspecies: *T. m. caurinus* breeds in w. and *propinquus* in e. Washington (AOU 1957), and *migratorius* also winters (Sallabanks and James1999).

Habitat: Breeds virtually anywhere trees are present, absent only from dense rain forest, unvegetated alpine rocky areas, large dryland agricultural areas, and sagebrush stands. Forages widely, notably on lawns, pastures, and at agriculturally provided opportunities. During migration and winter found in a very broad range of habitats, sometimes far away from trees.

Occurrence: Familiar and ubiquitous, the robin is readily adaptable to humans and their habitat modifications. Over time, the species' population has benefitted from massive conversion of native forests to open habitats of many kinds and in effect replaced the Varied Thrush in many changed habitats in westside areas (e.g., Edson 1926 in Wahl 1995). It has also populated large areas of e. Washington where irrigation created suitable habitats from sagebrush desert.

A generalized status-summary statement does not indicate the large-scale changes in seasonal populations, locally and regionally, of individuals if not overall numbers. Despite being classed a permanent resident, significant n. to s. migrations and altitudinal movements take place. Peak migration appears to be during Sep and Oct in fall and during Feb and Mar in spring. In winter, flocks form and the species is noteworthy for winter roosts containing thousands of birds (e.g., Stepniewski 1999). Flocks wander widely, feeding on berries and fruits of many trees and shrubs in winter. Wintering birds are distinctly less numerous where snow cover is present.

Breeding population trends for N. America were stable or increasing from 1966 to 1979 (see Sallabanks and James 1999). In Washington, BBS data show a significant increase ($P<0.01$) of 1.4% per year from 1966 to 2000 (Sauer et al. 2001). CBC data appear to indicate no long-term trends statewide.

Remarks: Reliable field identification of races is unlikely.

Noteworthy Records: *High counts*: 200,000 on 15 Mar 2002 nr. Union Gap, Yakima Co.; 40,000 on 7 Feb 1988 at Yakima; 22,000 on 23 Jan 1975 at Yakima; 10,000 on 5 Mar 1999 at Ridgefield.

Steven G. Mlodinow

Varied Thrush *Ixoreus naevius*

Fairly common permanent resident in w. Washington. Fairly common summer resident/ migrant and uncommon winter resident in e. Washington.

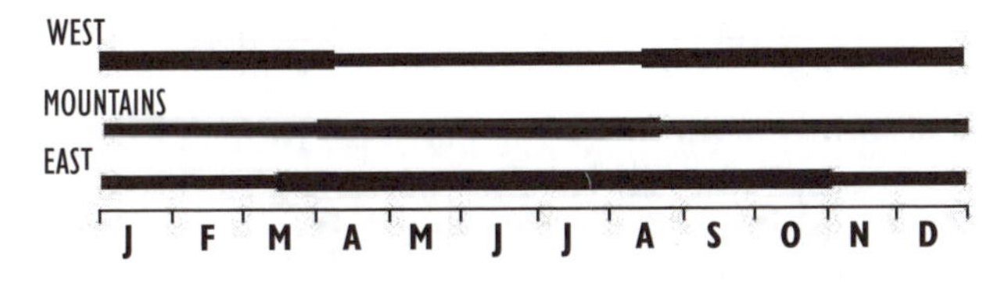

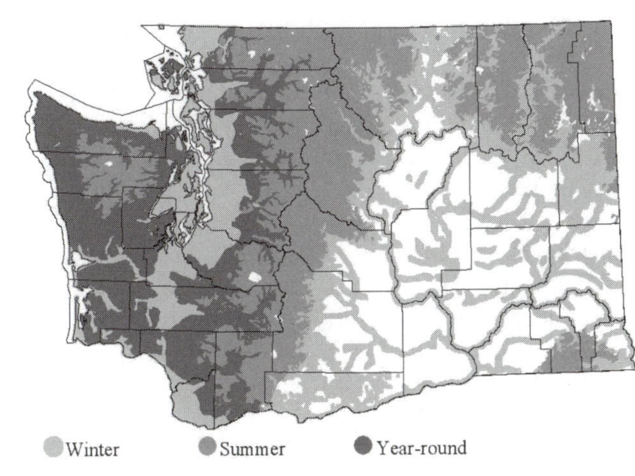

Subspecies: *I. n. naevius* breeds in w., *godfreii* and *meruloides* in e. Washington; *carlottae, naevius,* and *meruloides* occur in winter (George 2000).

Habitat: Breeds in coniferous woodlands throughout w. Washington. In e. Washington, nests mostly in subalpine forests, though also found in moist coniferous forests at lower elevations (Smith et al. 1997). During winter and migration found predominantly in mixed and coniferous woodlands, though broadleaf riparian areas are also used. Usually prefers dense vegetation, staying out of direct sunlight.

Occurrence: A characteristic bird of Washington's westside remnant rain forests, the Varied Thrush is often thought of today as a montane species. It formerly bred much more commonly down to sea level before logging and development eliminated most suitable habitat in the lowlands (Wahl 1995, Smith et al. 1997). Varied Thrushes are common from May into Aug in much of Washington's dense montane forests, with sizable numbers of migrants still possible in late Sep (SGM).

Migration is uncertain, but altitudinal movements into lowlands are apparent. Often the first few birds are seen in mid- or late Sep, but larger influxes typically do not occur until Nov or Dec. Sudden freezes or significant snowfall can cause a dramatic irruption into low-elevation woodlands and urban backyards at any time of fall and winter. Occasionally during such periods, birds can even be found in open fields hunting like robins, and high mortality during these episodes, particularly along roadsides, can result. Birds leave non-breeding areas during Apr and early May. E. of the Cascades, Varied Thrushes are more numerous during migration than during winter, with fall passage mostly occurring from late Sep through late Oct and spring passage occurring mostly from mid-Mar into Apr. A remarkably out-of-place bird was found in the Columbia Basin lowlands at Wilson Cr., Grant Co., on 21 Jul 1998.

Population trends are uncertain, perhaps due to cyclical occurrence. Inconsistent interpretations of BBA data results, with data showing an increase of 3.9% per year from 1966 to 1991 (Smith et al. 1997), and a statewide increase from 1966 to 2000 but a decrease from 1980 to 2000 (Sauer et al. 2001). CBC data show a mixed pattern with evident variability and no overall trend.

Remarks: From two to four weakly differentiated subspecies (see George 2000) that are unlikely to be reliably identified in the field.

Noteworthy Records: *High counts, Winter*: 500 on 23 Jan 1975 at Yakima; 500 on 10 Jan 1982 at Ocean Shores. *Fall*: 104 on 29 Sep 1996 at Cold Cr., Yakima Co. *CBC*: 873 in 1993 at

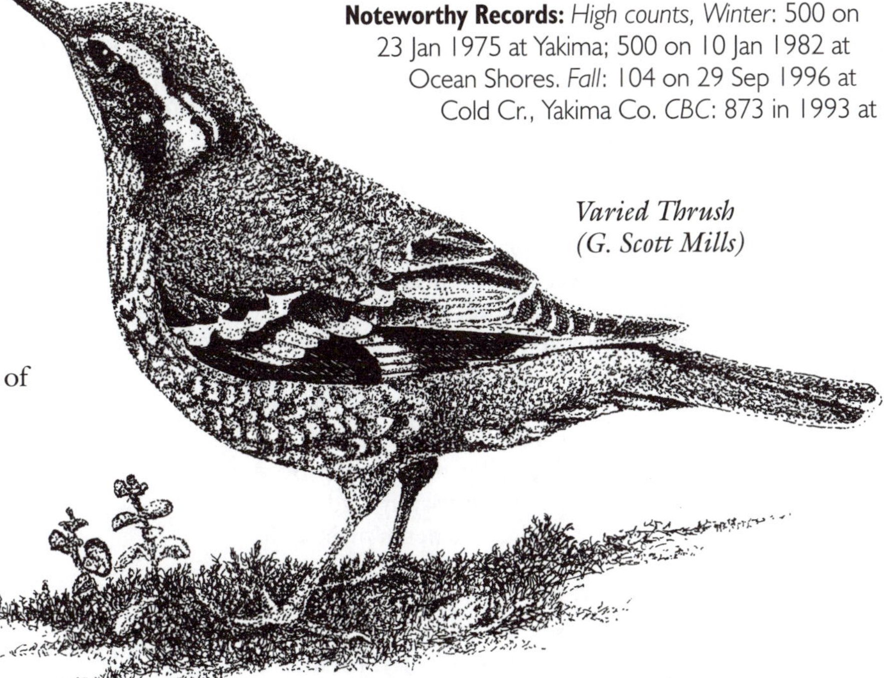

Varied Thrush (G. Scott Mills)

Quillayute R., 688 in 1996 at Padilla Bay, 602 in 1998 at San Juan Is., 560 in !987 at Wenatchee, 533 in 1983 at Bellingham.

Steven G. Mlodinow

Gray Catbird *Dumetella carolinensis*

Locally common summer resident in e. Washington, possibly occasional breeder in nw. Washington.

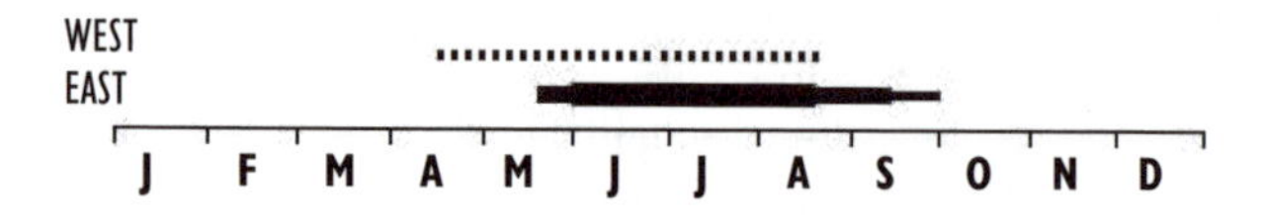

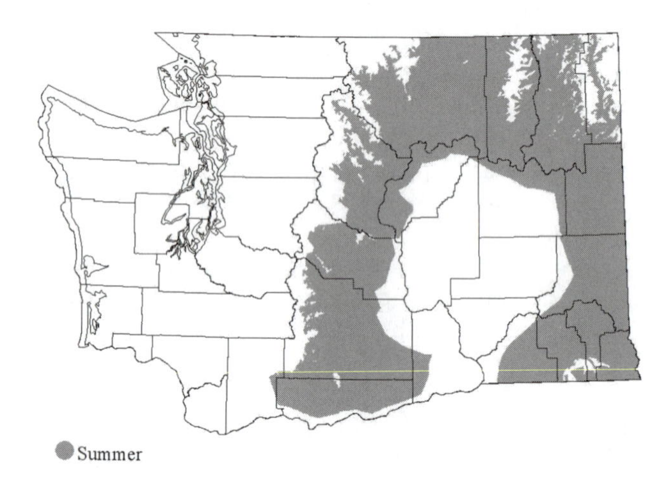

Summer

Habitat: Riparian habitats and low- to mid-elevation areas capable of supporting moist shrub growth, hedgerows, gardens, and orchards in e. Washington outside of the Columbia Basin (Smith et al. 1997).

Occurrence: A summer resident, breeding or presumed to breed throughout e. Washington at lower elevations (Smith et al. 1997), though apparently rare in much of the Columbia Basin where habitat is unsuitable. Its occurrence is generally within riparian areas, although in many suitable areas there are few records (e.g., Wing 1957). It is locally common at places such as Confluence SP in Wenatchee, Blackbird I. in Leavenworth, at Teanaway, and other locations. Suitable habitats are in substantial areas of Ferry, Pend Oreille, Spokane, Stevens, and e. Okanogan Cos. (Smith et al. 1997). Records in s. Walla Walla Co. (Dice 1918, Jewett et al. 1953) suggest wider distribution there, at least historically, than shown by Smith et al. (1997). Spring passage is mainly mid-May through early Jun, with fall movements mainly mid-Aug into early Sep. Migrants are uncommon in Columbia Basin riparian areas.

Recent fieldwork indicates more occurrence in the Columbia Basin than previously documented (SGM, B. Flores p.c., D. Beaudette p.c.), with records also from the Conboy NWR and the Rattlesnake R. in Klickitat Co. (JBB, BT), and it seems likely that the species may breed in suitable habitat associated with human habitation or crops due to water-management efforts.

A handful of records from Whatcom and Skagit Cos. suggest that the species may breed there sporadically (e.g., Jewett et al. 1953, Wahl 1995). Small numbers do breed regularly at Pitt Meadows, B.C. (Campbell et al. 1997). There are also three records from King Co., including two birds nr. Snoqualmie in Jul 1925 (Jewett et al. 1953).

Although population declines have been reported from the e. portion of its range, populations in w. N. America appear to be stable (Cimprich and Moore 1995). BBS data indicate a significant increase (P <0.05) in Washington from 1966 to 2000 (Sauer et al. 2001).

Remarks: No subspecies (AOU 1957). Jewett et al. (1953), Burleigh (1972), and others (see Cimprich and Moore 1995) believed there were enough differences between e. *A.carolinensis* and w. *A.ruficrissa* populations to warrant subspecific distinctions.

Noteworthy Records: West, *Spring*: 14 Apr 1974 at Port Angeles; 22 Apr 1976 n. of Bellingham (Wahl 1995). *Summer*: regular after 18 Jun 1976 at Butler Flats, Skagit Co.; pair in Jun 1986, possibly nesting at Thunder L., N. Cascades NP (Wahl 1995); 22 Jun 2002 at Sequim. **East**, *Spring*: 13 May 1951 at Potholes (Harris and Yocom 1952); 15 May in se. Washington (Hudson and Yocom 1954); 16 May 1990 at Wenas L. (Stepniewski 1999); 18 May 1974 along Union Flat Cr., Whitman Co. (Weber and Larrison 1977). *Fall, late*: 22 Sep 1963 at Spokane; 30 Sep 2001 at Lyons Ferry SP; 22 Sep 2002 at Lyons Ferry SP. CBC: 1 reported at Leadbetter Pt. in 1975. *High count*: 37 on 20 Aug. 1998 at Oroville.

Joseph B. Buchanan

Northern Mockingbird *Mimus polyglottos*

Rare visitant, very rare breeder.

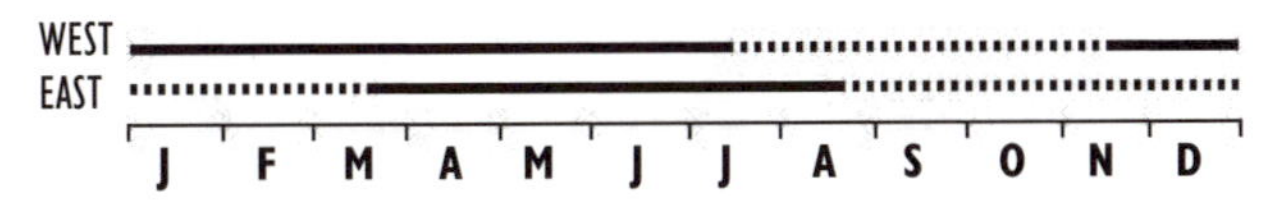

Subspecies: *M. p. leucopterus.*

Habitat: Scrubby habitat, including urbanized and rural areas.

Occurrence: Breeding across much of the s. U.S. n. to sw. Oregon, the mockingbird's occurrence in Washington is part of a recent continent-wide northward range expansion (Derrickson and Breitwisch 1992). Other than two hypothetical reports, one from Port Townsend in 1912 and the other from the Bellingham area in 1932 (Jewett et al 1953), mockingbirds were unrecorded in Washington until 17 Feb 1962 when a specimen was secured nr. Pullman (Moldenhauer and Bawdon 1962). Following that record, mockingbirds have occurred annually except in 1967, 1970, and 1971 and in apparently increasing numbers. Records averaged one per year in the 1960s, 2.5 per year in the 1970s and 4.5 per year in the 1980s and 1990s, and now total over 130 through 2000.

Records are evenly distributed between the eastside and the westside, but most of the records come from populous counties. King (16 records), Benton (15), and Yakima (12) Cos. combine for almost one-third of the total records.

Although there have been reports of singing males in potential breeding habitat since 1974, the first confirmed breeding record was not until the summer of 1990, when two ads. raised three young w. of Moses L. An ad. carrying food was seen at the same locale in 1991. The next confirmed nesting attempt was at Vernita, Grant Co., in the summer of 1995 when two ads. raised one young. Subsequent confirmed breeding records were at Richland in summer 1996 and Alderdale, Klickitat Co., in summer 2000.

Bill Tweit

Sage Thrasher *Oreoscoptes montanus*

Fairly common breeder in the shrub-steppe of e. Washington; very rare migrant w. of the Cascades.

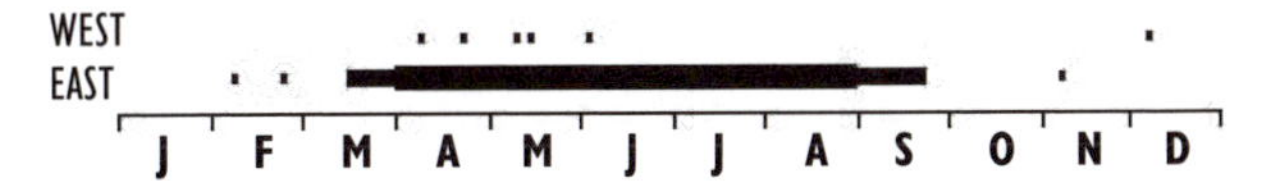

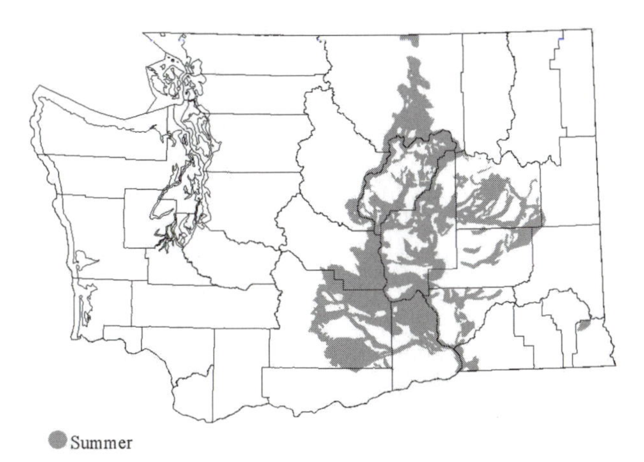

Habitat: Big sagebrush communities on deep soils and in shallow soil communities with a mix of big sagebrush and low-growing three-tip sagebrush. Also rarely in shrub-steppe stands with some bitterbrush (Stepniewski 1999) and locally in patches of native habitat that remain between agricultural fields and along roads.

Occurrence: An *Artemisia* obligate, the Sage Thrasher breeds locally in shrub-steppe habitats from extreme s.c. B.C. s. to e. California, n. Arizona, and New Mexico. Birds breed in Washington from the Okanogan Valley s. through Yakima Co. and as far e. as the scablands of e. Lincoln and locally in n. Asotin Cos. Birds rarely breed in the lower elevations of Benton Co., although they occur at higher elevations on adjacent ridges. Significant breeding sites in Washington include the Yakima Training Center, Yakama Indian Reservation, Quilomene WMA, Swanson Lks. WMA, and Moses Coulee. Spring migration is first noted in Mar and records after early Sep are noteworthy. There are about 15 records from w. Washington, mostly from mid-Apr to mid-May.

Jewett et al. (1953) described the Sage Thrasher as having an irregular distribution in the sagebrush habitats of e. Washington. The species may well be declining as its preferred habitat—big sagebrush/bunchgrass communities in good condition—becomes increasingly rare. Abundance declines with declining condition of the understory vegetation, making thrashers sensitive to changes brought about by livestock grazing and invasion by exotic plants (Vander Haegen et al. 2000). Unlike the Sage

Sparrow, which breeds primarily in large expanses of shrub-steppe, Sage Thrashers will establish territories and nest in small fragments of native habitat among agricultural fields (Vander Haegen et al. 2000; MVH). Birds can commonly be found in small patches of sagebrush among wheat fields in Douglas Co., where they nest alongside Brewer's and Vesper Sparrows. In c. Grant Co., Sage Thrashers can be found displaying and nesting in roadside Acircle corners—patches of native habitat that remain between irrigated crop circles. These fragments may be marginal breeding habitat, as research has found nesting success to be lower in fragmented shrub-steppe patches compared to large expanses of this habitat (WDFW unpubl.).

Based on limited field coverage, BBS data indicated a state increase from 1966 to 2000 (Sauer et al. 2001), but continued fragmentation and degradation of shrub-steppe pose threats to populations in Washington and throughout much of the Intermountain West. Classed SC, PHS.

Noteworthy Records: East, *Spring*: 19 Mar 1973 at Asotin Cr; 5 Feb1974 at Richland; 17 Mar 1996 at Yakima Training Center; *Fall*: 11 Dec 1975 at Tri-cities; 9 Nov 1978 at Davenport; Winter: 6 Feb 2002 along Crab Cr. *High counts*: ca. 50 on 11 Aug 1979 nr. Bridgeport; 30 on 12 Jul 1998 nr. Jameson L.; 30 on 1 Sep 1998 nr. Wanapum Dam. **West**: 21 Apr 1976 at Redmond; 1 May 1982 at Cape Flattery; 6 Dec 1986 at Tacoma; 10 May 1988 at Ocean Shores; 2 Jun 1988 at Tatoosh I.; 11 May 1989 at Seattle; 11 Apr 1993 at Willapa NWR; 20 Apr 1996 at Lummi Flats; 9 May 1996 at Ft. Lewis; 29 May 1999 in Skagit Co.; 24 Apr-2 May at Vancouver; 30 Apr 2000 at Steigerwald L.; 5 May 2000 at Granite Falls; 1 May 2001 at Bothell; 17 Apr 2002 at Marymoor Park, King Co., 24 Apr 2002 at Monroe.

W. Matthew Vander Haegen

Brown Thrasher *Toxostoma rufum*

Very rare vagrant.

Although this species breeds as nearby as s. Alberta, there are currently only five accepted Washington records: Nisqually NWR on 1 May 1994 (Tweit and Skriletz 1996); American Camp, San Juan I., on 26 Jun 1995; Coulee Cr., Spokane Co., on 15 Jan-19 Feb 1996; Tatoosh I., on 8-9 Sep 1999; and Wahluke Slope WMA on 10 Sep 1999. Another bird (record not reviewed by WBRC) was at Pasco 19-20 May 1963, and two published records were rejected by the WBRC: Skagit WMA, 14 Oct 1972 and Orcas I., 13 Oct 1988 (Tweit and Paulson 1994). Two subspecies, not separable in the field: state records presumably *longicauda*.

Steven G. Mlodinow

European Starling *Sturnus vulgaris*

Abundant in low- and mid-elevation areas nearly statewide.

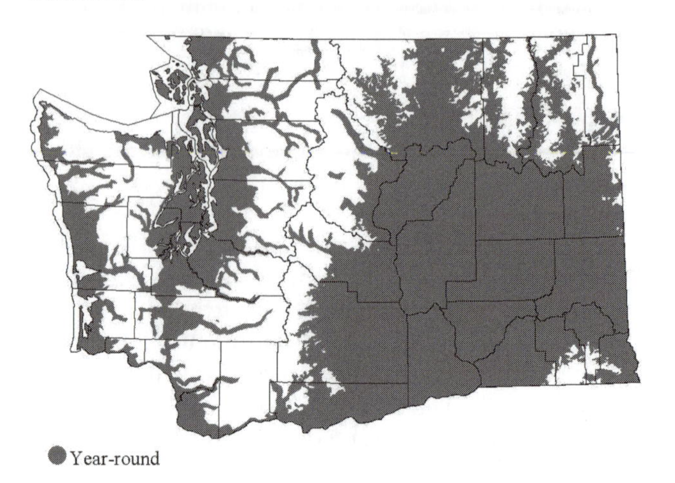

Subspecies: *S. v. vulgaris.*

Habitat: Absent only from forested areas and high elevations, found in nearly any open habitat including mud flats and beaches. Nests and roosts in essentially all urban settings. Often associated with human developments and activities.

Occurrence: The starling was introduced to New York City in 1890 and 1891 (Cabe 1993) and it is from these introductions that a huge, continent-wide population was established (Davis 1950, Kessel 1953). First introduced in the Pacific Northwest when 20 pairs were released in Portland and McMinnville, Oregon, in May 1889 (Jobanek 1993). The species fared well and nested in Portland until 1901 after which it no longer occurred until descendants from the New York introductions arrived several decades later (Jobanek 1993). Starlings were not known to have crossed the Columbia R. into Washington during this time.

Birds first recorded in Washington on 20 Mar 1943, when five were seen n. of Pullman (Wing 1943). Following additional sightings throughout the state over the next several years, breeding was first recorded in 1951 in Adams Co. (Hudson and Yocom 1954). Similar time lags between first occurrence and first breeding were also reported from Idaho, Oregon, and B.C. (e.g., Kessel 1953, Burleigh 1972, Jobanek 1993). Most records in early years were in fall and winter.

Starlings had been present in Washington for several years before they were recorded on a CBC. Few CBCs were established at that time and the species was not detected until 1959, when starlings were recorded at Moscow/Pullman (1070 birds), Seattle (681), and Spokane (116). Birds were then recorded at Cheney (one in 1960), Wenatchee (88 in 1960), Olympia (four in 1961), Orcas I. (25 in 1962), and Sequim (2400 in 1963). After this they were commonly encountered in all CBCs. CBC data from 30 sites between 1990 and 1999 showed highest counts at the Columbia R. estuary (mean = 16,677 birds), Bellingham (10,393), Kent/Auburn (7664), Moses L. (6368), and the Tri-Cities (6155). Four high counts exceeded 20,000 birds. Four counts in Portland, Oregon, between 1962 and 1966 reported at least one million birds (Jobanek 1993). During 1990-99 the starling was least abundant at N. Cascades (mean = five birds), Twisp (76), Kitsap Co. (578), Leadbetter Pt. (727), and Wenatchee (1362). Starlings may be under-represented on some CBCs because they tend to disperse away from nocturnal roosts to diurnal foraging areas outside count circle boundaries (Wahl 1995) and also because thorough counts of starlings, especially at night-time roosts, may not have been prioritized.

The starling has adapted well to conditions across N. America. Its continental population is thought to be somewhere between 140 and 200 million birds (Hygnstrom et al. 1994, Feare 1984). BBS data indicate that the population increased at an annual rate of 2.5 % between 1966 and 2000 (Sauer et al. 2001). Moreover, from 1994 to 1998 the USDA Animal and Plant Health Inspection Service (APHIS) killed over 1.5 million starlings in Washington, with unknown effect on the population (USDA 2000).

In recent years they have roosted in open-air lumber mills, open-doored warehouses and manufacturing facilities, under piers and bridges (Wahl 1995; R. Anderson p.c.). Starlings aggregate and roost in large flocks in midsummer, with larger flocks in winter. Enormous flocks have been documented at roost sites (e.g., one million birds under the Lewis and Clark bridge between Longview and Rainier, Oregon (R. Anderson p.c.).

Remarks: Starlings are aggressive cavity users and compete with other, native, species for these nest sites (see Cabe 1993). Starlings have been implicated in 382 bird strikes to aircraft in Washington 1990-98 (USDA 2000), and cause damage to property, agricultural crops, and livestock (Courtney et al. 1998).

Noteworthy Records: *Early colonization*, **West**, *Winter*: Dec 1945 at Fort Lawton, Seattle (Larrison 1947b); 3-10, 9-16 Nov 1947 at Woodland (Ransom 1948); 10 on 16 Jan 1949 between Bellevue and Redmond (Bennett and Eddy 1949); 2 on 29 Dec 1949 e. of L. Washington (Flahaut 1950); 6 on 2 Jan 1950 at Medina (Hagenstein 1950); 20-35 in Dec 1952-Jan 1953 on Eliza I. (Wick 1958). *Spring-Summer*: 14 Aug 1947 at Verlot Ranger Station (Larrison 1947c). *Nesting, first record*: unknown locality in 1955; 26 Jun 1959 at San Juan I. (Lewis and Sharpe 1987); 1962 in Des Moines (Hunn 1982). **East**, *Winter*: 2 on 3 Mar 1946 at Colton, Whitman Co. (Jewett 1946); 3 flocks on 29 Nov 1947 totaling ca. 120 nr. Almota, Whitman Co. (Hudson and Yocom 1954); flocks up to 500 nr. Walla Walla in Feb 1948 (Booth 1948); flocks >1000 in se. Washington in winter 1948-1949 (Booth 1949); 25 on 24 Nov 1950, 30 on 19 Dec 1950, and 15 on 9 Nov 1951 at the Potholes (Harris and Yocom 1952). *Spring-Summer*: 5 on 20 Mar 1943 n. of Pullman (Wing 1943); 6 km w. of Walla Walla on 28 Mar 1948 (Pope 1948). *Nesting, first record*: 26 Apr-8 Jun 1951 at Roxboro, Adams Co. (Hudson and King 1951, Hudson and Yocom 1954); 3 nests on 10 May nr. Clarkston (Braden 1953); 31 May 1953 at Pullman (Hudson and Yocom 1954); 15 May 1953 at Roxboro (Hudson and Yocom 1954). *CBC high counts since 1990*, **West**: 24,809 in 1992 at Kent/Auburn; 21,569 in 1993 at Columbia R. estuary; 21,232 in 1994 at Bellingham. **East**: 22,595 in 1995 at Moses L.; 19,099 in 1997 at Tri-Cities.

Joseph B. Buchanan

Siberian Accentor *Prunella montanella*

Casual vagrant.

There is one accepted Washington record: a bird (ph.) on Indian I., Jefferson Co., on 30 Oct 1983 (Tweit and Paulson 1994). A possible, likely, record was from Orcas I. on 10 Jan 1991 (see Tweit and Paulson 1994). Subspecies unknown, but likely *badea*.

There are only about 10 Alaskan records, mostly between mid-Sep and mid-Nov (ABA 1996). Other N. American records include one from Vancouver, B.C., 15 Dec 1993; two nr. Salmon Arm, e. B.C., 5 Mar to 10 Apr 1994; and one at Hailey, Idaho, 27 Dec 1996-Mar 1997. Like the bird reported from Orcas I., the Salmon Arm and Hailey birds were coming to suet feeders.

Steven G. Mlodinow

Yellow Wagtail *Motacilla flava*

Casual vagrant.

Recorded once in Washington: an ad. (ph.), likely *M. f. tschutschensis*, at Ocean Shores on 29 Jul 1992 (Tweit and Paulson 1994). An Old World species that breeds locally in in Alaska and the Yukon, it is very rare elsewhere in N. America: there are about 20 N. American records s. of the breeding range, with most between 29 Aug and 18 Oct, and peaking between 29 Aug and 20 Sep (Heindel 1999). The fact that the Washington bird was an ad., where most or all of the other records were of imms., may explain the exceptionally early date here (Heindel 1999).

Steven G. Mlodinow

White Wagtail *Motacilla alba*

Casual vagrant.

White Wagtails have been identified twice in Washington: an ad. male (ph.) at Crockett L. 14 Jan-5 May 1984; and an ad. (ph.) at Ocean Park on 26 Apr 1994 (Sibley and Howell 1998).

This species is scarce in Alaska, though it is distinctly more numerous than the Black-backed Wagtail there. S. of Alaska, there have been only 11 records in w. N. America, mostly between early Oct and early Mar (Sibley and Howell 1998). This pattern suggests that White Wagtails may be later fall migrants, earlier spring migrants, and more likely to winter than Black-backed Wagtails.

The status of these two wagtails in Washington is uncertain due to similar appearance and former conspecific status. Additionally, an apparent *M. a. leucopsis*, a black-backed form of White Wagtail, has been recorded in Oregon (see Morlan 1981, Howell 1990, and Sibley and Howell 1998). There are three records of unidentified Black-backed/White Wagtails from Washington: Seattle, 8-9 Nov 1981, at Keystone on 30 Apr 1990 (Tweit and Skriletz 1996), and 13 Dec 2002 in Benton Co. (NAB 57:250). The dates of these records may suggest that the Nov record was likely a White Wagtail, and the late Apr bird likely a Black-backed Wagtail.

Steven G. Mlodinow

Black-backed Wagtail *Motacilla lugens*

Casual vagrant.

Recorded three times in Washington: a sight record of a male at Azwell, Chelan Co., 19 May 1985; sight record of a female at Ocean Shores, 11 May 1986; and a male (ph.) at Point No Point, 5-7 May 1993.

Through summer 1997, there were only 15 records s. of Alaska, between late Jul to late May, with most between early Aug and late Sep and during May (Mlodinow and O'Brien 1996). There are reports of unidentified wagtails in Washington (see White Wagtail).

Steven G. Mlodinow

Red-throated Pipit *Anthus cervinus*

Casual vagrant.

The Red-throated Pipit, which breeds from n. Scandanavia to nw. Alaska, has been recorded once in Washington: a sight record at American Camp, San Juan I., 14-16 Sep 1979 (Tweit and Paulson 1994). The species is a regular fall migrant in California. From 1982 through 1994, over 160 birds were found in California, virtually all of them between 9 Sep and 11 Nov (Mlodinow and O'Brien 1996). Very few Red-throated Pipits have been seen between California and Alaska, with no records from Oregon and about six from B.C.. The B.C. sightings are from 4 Sep to 21 Nov except for one from mid-Jun (*Birders Journal* 6:161) and one from late Dec. Red-throated Pipits s. of Alaska have generally been with American Pipits very close to the ocean coast (Mlodinow and O'Brien 1996).

Steven G. Mlodinow

American Pipit *Anthus rubescens*

Fairly common breeder at high elevations in n. Cascades, Olympics and at volcanos in s. Cascades. Common to abundant migrant in lowlands; uncommon w., rare in winter e.

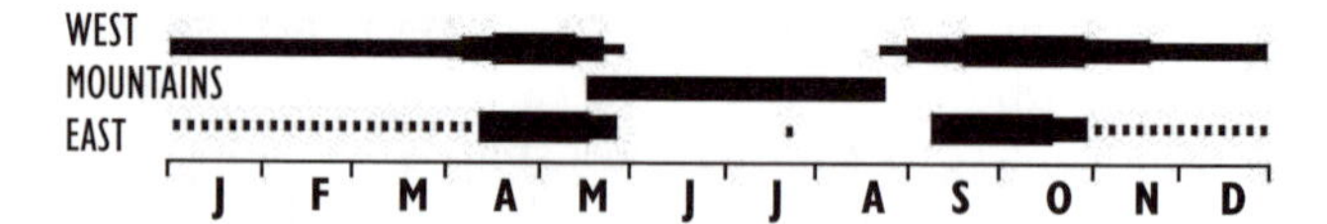

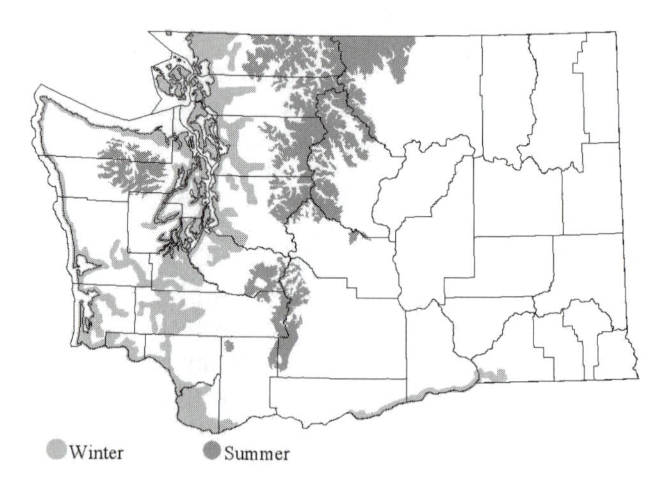

Subspecies: *A. r. pacificus.*

Habitat: In summer in high-elevation meadows, rocky areas, and seeps, as from snowfields, in alpine zones. In migration and winter in lowland fields, gravelly areas, river-banks, beaches, tidal flats.

Occurrence: One of the most widely distributed passerines in n. N. America. Its distinctive call-note, heard as it passes overhead, is often what calls the pipit to attention. Most apparent when migrating in spring and fall through the lowlands, when localized flocks of 500-2000 have been reported both e. and w. Migrates over extensive bodies of water as shown by occurrence in the Queen Charlotte Is. (Campbell et al. 1997) and at vessels off the Washington coast (Jewett et al. 1953).

Summer distribution probably incompletely known due to lack of coverage of high-altitude areas particularly in early summer. Relatively widespread in summer through the n. Cascades, apparently more restricted in the s. by lack of habitat except at volcanic peaks including Mt. Rainier and Mt. Adams, and wetter alpine regions of the Olympics (Smith et al. 1997). Breeding birds may be forced downslope by unseasonal snow storms (Stepniewski 1999). Nests nr. but in wetter situations than the Pallid Horned Lark in the mountains.

Irregular in occurrence and abundance in winter in w. lowlands, not recorded every year on CBCs anywhere. Most of the sparse, widespread winter records in the e. are of birds found along shorelines of Columbia and Snake Rs. Few CBC records e. of the Cascades in 1962-98: four occurrences at Two Rivers, one each at Columbia Hills-Klickitat Valley, Lyle, Tri-Cities, and Wenatchee. Statewide, no birds at all were recorded on CBCs in 1969,1972, 1973; highest numbers in 1981, 1983, 1984, 1988, 1989. CBC data from 1962 to 1992 indicated declines across the U.S. (Verbeek and Hendricks 1994).

Remarks: *A. r. alticola* also may occur (see Campbell et al. 1997, Smith et al. 1997). Referred to as Water Pipit prior to separation from that Eurasian species, *A. spinolettta* (AOU 1989). Status on Mt. St. Helens following eruption in 1980 uncertain (Smith et al. 1997).

Noteworthy Records: West, *Summer*: 1 on 18 Jun-4 Jul 1978 at Dungeness. *High counts*, **West**, *Spring*: 300 on 1 May 1999 at Stillaguamish R. delta (SGM). *Fall*: 500 on 21 Sep 1980 and 250 on 17 Sep 1982 at Ocean Shores. *Winter*: 100 on 24 Dec 2000 nr. Snohomish (SGM); 80 on 12 Dec 1998 at Samish Flats; 50 on 6 Dec 1998 at Silvana,. **East**, *Fall*: 2000 on 9 Oct 1971 nr. O'Sullivan Dam; at least 1000 on 21 Sep 1973 at West Medical L. and Reardan; 500 on 17-18 Sep 1983 nr. Banks L. *Winter*: 75 on 12 Dec 1998 at Dallesport. *CBC high counts*, **West**: 250 in 1981 at Tacoma, 116 in 1997 at Skagit Bay, 100 in 1996 and 1998 at Edmonds; **East**: 132 in 1989 at Two Rivers.

Terence R. Wahl

Bohemian Waxwing *Bombycilla garrulus*

Variable, uncommon to locally abundant winter visitor e. of the Cascades, very rare to uncommon w. Very rare breeder.

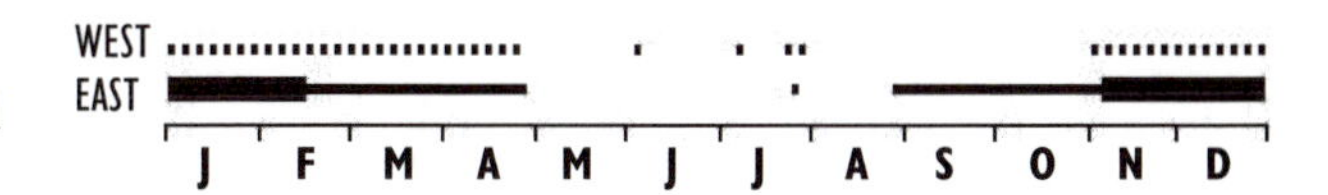

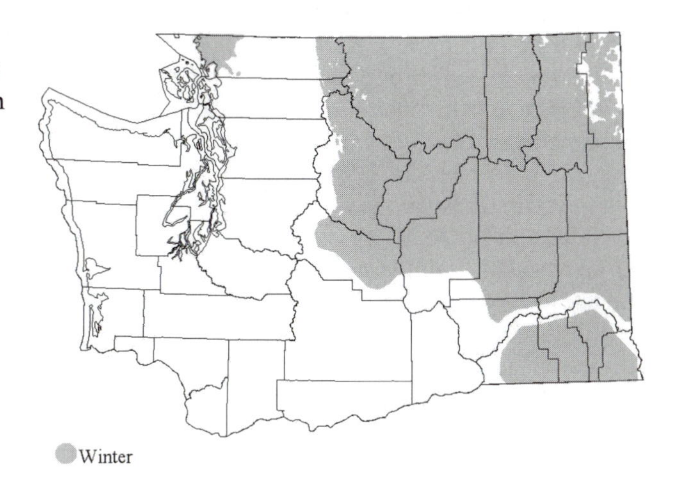

Subspecies: *B. g. pallidiceps.*

Habitat: Open and mixed woodlands in summer. In winter also in forest edges, parks and suburbs with shrubs and fruiting trees, especially mountain ash and apple orchards; uncommon in riparian woodlands and mixed coniferous-deciduous woodlands.

Occurrence: An attractive, irruptive species, variable but most regular in ne. counties Nov-Feb. Typically widespread, often in large foraging flocks—sometimes numbering in the thousands—that move erratically during peak years depending on local food crops (e.g., Jewett et al. 1953).

Regular on CBCs only at Spokane where the species was recorded in all but one year from 1959 to 1999. Birds are absent most winters in the w. Most frequent on Bellingham CBCs, though seen on only 11 counts from 1967 to 1999. Small numbers noted over time in the San Juan Is., at Dungeness, and s. to Chehalis and w. to Grays Harbor.

Bent (1950) and Jewett et al. (1953) noted Washington events in 1879-80, 1916-17, 1919-20 and 1931-32. CBCs showed irruptions in the e. in 1960-61 through 1963-64, 1966-67, 1970-71, 1975-76, 1978-79, 1984-85, and 1985-86, with flocks of up to 15,000 birds recorded at on Spokane in the 1960s. Low numbers apparent especially in 1965, 1968, 1993. Peaks w. of the Cascades were in 1966-67 and 1981-82. Winter numbers on long-

Table 39. Bohemian Waxwing: CBC average counts per 100 party hours by decade.

	1960s	*1970s*	*1980s*	*1990s*
Ellensburg			218.2	254.5
Spokane	4652.8	2590.4	1021.2	293.3
Walla Walla		585.6	681.0	142.8
Wenatchee	1900.0	1340.1	474.9	242.4
Yakima Valley		130.4	91.8	13.2

established Washington CBCs statewide have declined significantly (P <0.01; Table 39).

Very rare breeder: one confirmed and one probable record in Whatcom Co. in 1992, one probable from Harts Pass in 1979, and one in Pasayten Wilderness in 1987 (Smith et al. 1997). Post-breeding ads. and juvs. reported in the N. Cascades and high elevations of Okanogan and ne. counties in late summer and early fall, possibly indicating migration or, in some cases, failed breeding. Records of flocks associated with common junipers at high elevations in the Okanogan Highlands in mid-late October correspond with seasonal occurrence in s.c. B.C. (A. Stepniewski p.c.).

Noteworthy Records: West, *Summer*: 1 on 6 Jul 1976 at W. Dungeness; 1 on 16 Jun 1983 with Cedar Waxwings at Haida Pt., San Juan I. (Lewis and Sharpe 1987); nest sw. of Mt. Shuksan on 12 Jul 1992, and 12 ad., 4 juv. on 12 Jul 1992 suggested nesting at Pierce Camp, Whatcom Co. (Wahl 1995). *Winter*: flocks of up to 2000 in late Dec 1966-late Jan 1967 at Bellingham; 100 for ca. 3 weeks around Christmas 1968 at Bellingham; flocks of 25-250 during Jan-Feb 1969 nr. La Conner, at Seattle, and Chehalis. **East**, *Summer*: several uncertain nest records from Okanogan to Spokane Cos. (Jewett et al. 1953). *Fall*: Flocks of 25-35 as early as 6 Sep 1965 at Salmo Mt., flocks as early as 10 Oct 1968 at Spokane, 400-600 during week of 18 Oct 1971 in n. Okanogan Valley, 350 on 29 Oct 1988, 600 on 20 Oct 1991 and 10 on 18 Oct 1998 at ca. 2400 m elevation on Tiffany Mt. (A. Stepniewski p.c.). *Winter*: 4000 on 1 Jan 1971 at Spokane; thousands in Ellensburg, 300-400 at Walla Walla in Dec 1972, 10,000 in winter 1978-79 at Spokane. *CBC*: 450 in 1971 at Bellingham; 1300 in 1970 at Chelan; 1085 in 1985 at Chewelah; 1104 in 1978 at Ellensburg; 320 in 1995 at Grand Coulee; 240 in 1996 at Moses L.; 15,000 in 1961 and 1966, 8403 in 1975, and 7800 in 1963 at Spokane; 301 in 1979 at Tri-Cities; 701 at Twisp in 1995; 2271 in 1972 at Wenatchee; 1735 in 1985 at Walla Walla; 532 in 1975 at Yakima.

Terence R. Wahl and Devorah A. N. Bennu

Cedar Waxwing *Bombycilla cedrorum*

Uncommon to locally common resident in lowland woods, rare to uncommon winter visitor w. of the Cascades. Common summer resident, uncommon to fairly common winter resident in e. Washington.

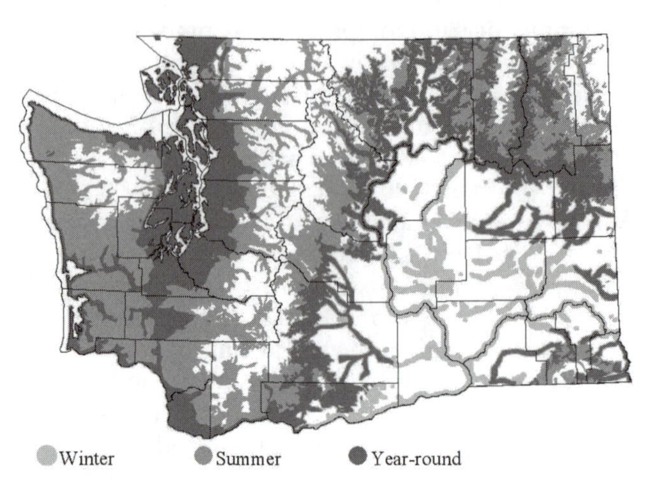

Subspecies: *B. c. larifuga.*

Habitat: Open woodland types, including both wet and dry, deciduous or coniferous forests, also forest edges, second growth, in parks, orchards, and suburbs in summer. Wooded or semi-open areas, fruiting trees and shrubs where berries are abundant during migration and winter.

Occurrence: Summer visitor throughout the lowlands of Washington, often arriving in mid-late May. Breeds commonly in lowland broadleaf forests, riparian areas, wetlands and bogs, and residential areas throughout Washington (Smith et al. 1997). Generally below 450 m on the Olympic Pen. (Sharpe 1993, in Smith et al., 1997), and below 900 m e. of Puget Sound. A few probable breeding records from above 1200 m. Tends to nest later in the season than other avian species. Late summer flocks often indicate start of fall migration period. Found in the mountains only during migration.

Often winters in small to medium sized flocks in c. and e. Washington. Wintering flocks can number up to 200 birds—a flock of 100 at the Tri-Cities on 22 Jan 1989 was judged noteworthy in field reports. Large flocks are notably smaller than irruptive numbers of Bohemian Waxwings.

In summer Cedar Waxwings feed primarily on insects including small moths, cankerworms, and various beetles and on outbreaks of insects like spruce budworms (Jewett et al. 1953). Often seen "hawking" flying insects over open water. Feed gregariously during much of the year on ripe berries and fleshy fruits including berries of cedar, mountain ash, hawthorn, *Pyracantha*, privet, mulberries, and may gorge on fermented fruits until they cannot fly. Found in orchards especially from mid-Jul into early Sep (Stepniewski 1999). Often feed alongside robins.

BBS data indicate no change from 1966 to 2000 (Sauer et al. 2001). CBC data show a significant decline ($P < 0.05$) w. of the Cascades and little overall change in the e.

Remarks: Subspecies not agreed upon and require further study (Witmer et al. 1997).

Noteworthy Records: *CBC high counts by location*: 1325 in 1982 at Tacoma; 1120 in 1983 at Wenatchee; 653 in 1959 at Seattle; 513 in 1997 at Tri-Cities; 399 in 1984 at Walla Walla; 356 in 1988 at E. Lk. Washington; 310 in 1997 at Ellensburg; 245 in 1995 at Spokane; 175 in 1998 at Sequim-Dungeness; 150 in 1992 at Toppenish; 128 in 1988 at Kitsap Co.; 120 in 1995 at Twisp; 120 in 1992 at Yakima Valley; 100 in 1970 at Chelan; 100 in 1993 at Moses L.

Devorah A. N. Bennu

Phainopepla *Phainopepla nitens*

Casual vagrant.

One record: a female in Seattle on 24 Sep 1994 (Tweit and Skriletz 1996). This desert sw. species has a great capacity to wander as evidenced by records from as far afield as Wisconsin, Massachusetts, and Rhode Island (AOU 1998). The seven Oregon records (through 2001) span 30 Aug-17 May. Subspecies, *lepida* or *nitens* (Pyle 1997), unknown. Several earlier Washington reports may have actually referred to poorly seen Steller's Jays.

Steven G. Mlodinow

Blue-winged Warbler *Vermivora pinus*

Casual vagrant.

One Washington record: a bird at Anacortes on 17 Sep 1990 (Tweit and Paulson 1994). The Blue-winged Warbler is one of the rarest eastern warblers found in w. N. America, with fewer than 40 records from w. of the Rocky Mts. (Dunn and Garrett 1997). There are only a few records from the Pacific Northwest, ranging from May to Aug.

Steven G. Mlodinow

Golden-winged Warbler *Vermivora chrysoptera*

Casual vagrant.

The only Washington record is of an imm. female banded (ph.) at Turnbull NWR on 20 Aug 1998 (Frobe 1999). The Golden-winged Warbler breeds no closer to Washington than sw. Manitoba or se. Saskatchewan, and it has been recorded only a few times in the Northwest, with one record from Idaho and two from Oregon (Dunn and Garrett 1997). The previous Northwest records were all from late spring, but California records are about evenly split between spring and fall, with the earliest fall records coming from mid-Aug (Dunn and Garrett 1997).

Steven G. Mlodinow

Tennessee Warbler *Vermivora peregrina*

Very rare spring and fall visitor.

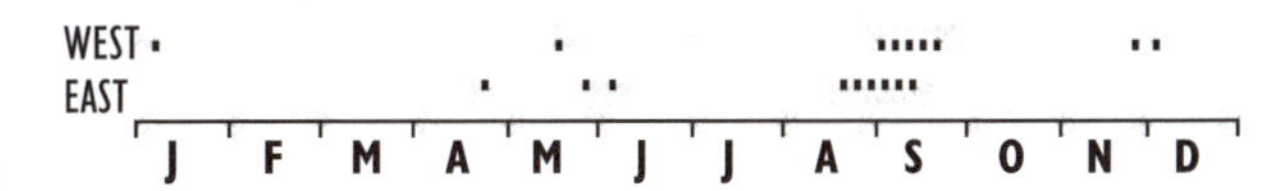

Habitat: Woodlands.

Occurrence: The Tennessee Warbler breeds across most of n. N. America, winters in C. and n. S. America (Rimmer and McFarland 1998). Although there have been over 2300 records in California through 1995 (Dunn and Garrett 1997), there are fewer than 20 reports for Washington. Through its fifth report, the WBRC had accepted only 13 records, with three other apparently valid reports. Eight of the nine westside records are fall season, while the seven eastside records are more evenly split between three spring and four fall. All but two of the westside records are from the Puget Trough; the spring and the winter record are from coastal areas. All of the interior records are from the e. third of the state, with the furthest w. interior record from Crow Butte SP, Benton Co., on 16 Aug 1998.

Noteworthy Records: *Spring, outer coast*: 20 May 1974 at Ruby Beach. *Winter*: 5 Dec 1993–5 Jan 1994 at Satsop.

Bill Tweit

Orange-crowned Warbler *Vermivora celata*

Common migrant and summer resident and uncommon winter resident in w. Washington. Common migrant, fairly common summer resident, and rare to uncommon winter resident in e.

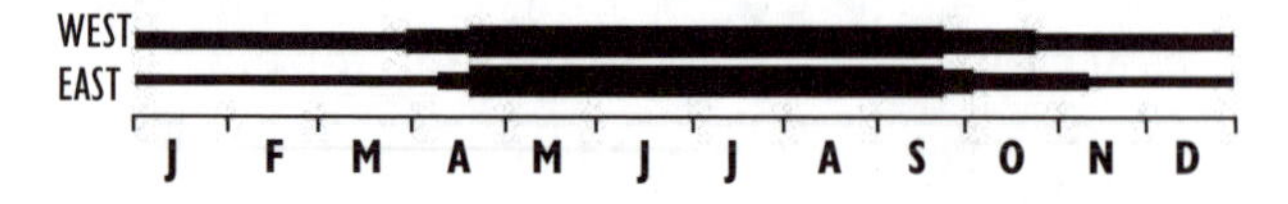

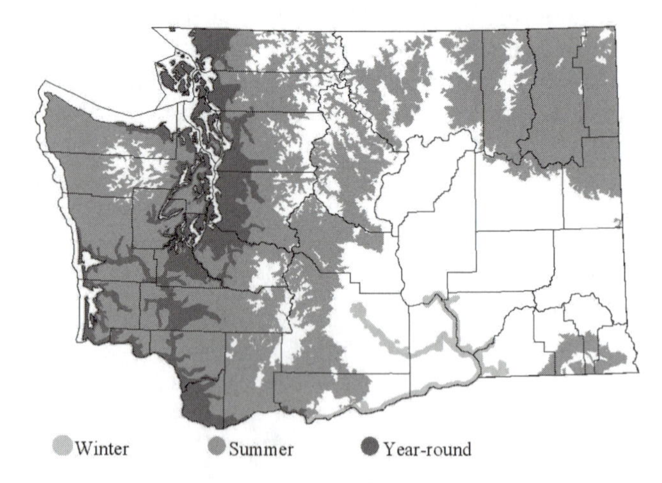

Subspecies: *V. c. lutescens* breeds in w. Washington, *orestera* breeds e.; *celata*, at least occasionally, a migrant and winter visitor (Jewett et al. 1953).

Habitat: Breeds in deciduous thickets, typically in or nr. deciduous or mixed woodlands. Sometimes breeds in suburban areas on westside. Migrants are found in broad range of habitats with moderate to dense undergrowth and are occasionally in more open circumstances, even weedy fields. Wintering birds usually found in moist areas with thick undergrowth.

Occurrence: One of the most common insectivores on the westside (the most abundant breeding warbler in the San Juan Is. [Lewis and Sharpe 1987]). Somewhat less common e. of the Cascades where its distribution was historically uncertain apparently due to confusion of migrant and breeding occurrences. Breeding in ne. Washington and the Blue Mts. was unknown to Jewett et al. (1953), though it was suspected that Orange-crowned Warblers might summer locally there. Status in e. Washington probably has not changed significantly over time (see Smith et al. 1997).

On the westside, first spring migrants are typically noted in late Mar or early Apr. End of spring migration is less clear, but is probably during mid-late May. First fall migrants appear in mid-Aug, and most are gone by late Sep, but a few stragglers continue to move through into early Nov. Eastside spring migrants first appear in mid-Apr and are largely through by mid-May, at least in lowlands. The first southbound birds can arrive as early as late Jul, but the bulk pass through from mid-Aug to late Sep, with a few stragglers appearing into early Nov. Wintering birds on both sides decrease from n. to s. From 1960 to 1999 birds were recorded 154 times on CBCs, with highest numbers (up to 11 birds at Everett in 1995) from 1989 to 1999. Westside records were relatively evenly spread but eastside numbers were highest in s. areas including Tri-Cities, Toppenish, and Yakima Valley. Numbers apparently decline throughout winter.

Though logging would appear to have increased habitat for the species over time, BBS data indicate that, statewide, breeding numbers declined significantly (P <0.01; 4.5% per year) from 1966 to 2000 (Sauer et al. 2001). CBC data indicate a significant westside increase (P <0.05).

Noteworthy Records: *High counts*: 100 on 2 May 1991 at Naches; 92 on 26 Aug 1998 at Mountlake Fill; 53 on 14 Sep 1999 at Skagit WMA (SGM). *Winter*: 9 on 19 Dec 1999 nr. Toppenish; 7 on 31 Dec 1995 at Everett (SGM).

Steven G. Mlodinow

Nashville Warbler *Vermivora ruficapilla*

Common breeder and migrant in e. Washington. Uncommon spring migrant, local breeder, and rare fall migrant in w. Casual in winter.

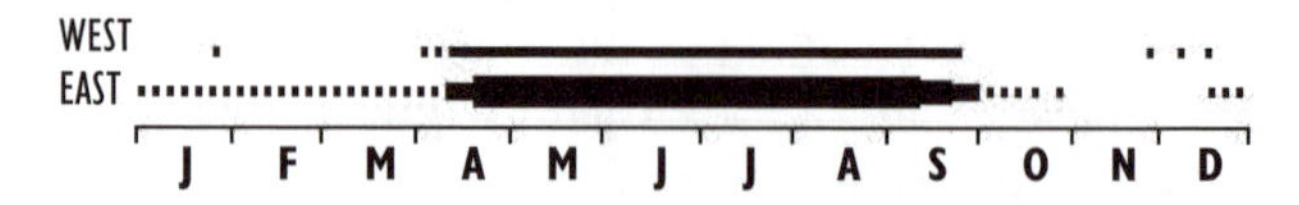

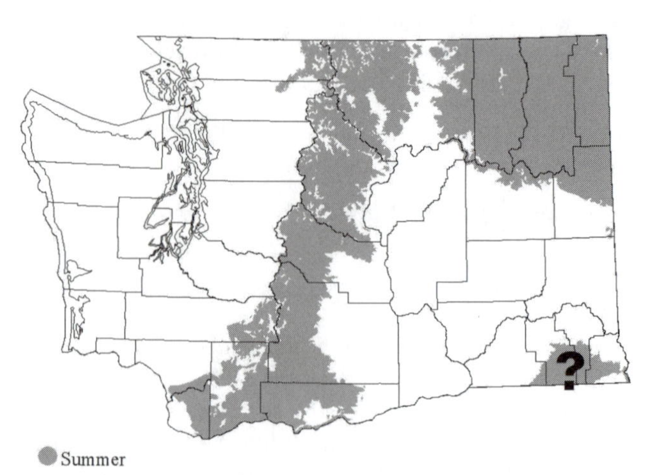

Subspecies: *V. r. ridgwayi*; *ruficapilla* possible.

Habitat: Predominantly oak woodlands and shrubby/hardwood microhabitats of drier coniferous woodlands (Smith et al. 1997). During migration found in a wider variety of habitats but still favors shrubby spots such as willow thickets and Russian olive patches in shrub-steppe.

Occurrence: A common breeder on the e. slope of the Cascades and in the ne., in all forested zones below alpine/parkland, with most in lower-elevation dry forests. Peak breeding densities are found in the garry oak woodlands of s. Klickitat Co. Birds may also breed in the Blue Mts., but BBA surveys recorded very few (Smith et al. 1997). Nashville Warblers are common migrants across e. Washington.

In w. Washington, breeding birds are localized, uncommon to common at scattered areas along the upper reaches of river valleys, particularly in Skamania, Whatcom, and Skagit Cos. Formerly, they were common breeders in the garry oak woodlands of Pierce and Thurston Cos. (Jewett et al. 1953) but have not been reported in recent decades. In migration, they are found mostly along the w. slope of the Cascades, and in the Puget Trough, where they are uncommon spring and rare fall migrants. W. of the Puget Trough they are rare to very rare either in spring or fall and there are few records from the San Juan Is. (Lewis and Sharpe 1987).

Spring migration is mostly from mid-Apr to mid-May, with a distinct peak during late Apr-early May on the westside. Fall movement can begin as early as late Jul but peaks from mid-Aug to early Sep on the eastside and is a week or so later w. of the Cascades.

Regenerating forests provide additional nesting habitat (see Williams 1996) for the species. BBS data indicate an increase in Washington from 1966 to 2000 (Sauer et al. 2001).

Remarks: The two subspecies may be separable in the field (see Dunn and Garrett 1997).

Noteworthy Records: *High count,* **East**: 20 on 20 Aug 1995 at White Bluffs, Grant Co. **West**, *outer coast*: 4 on 5 May 1916 at Destruction I. (Jewett et al. 1953). *Spring, early*: 5 Apr 1916 at Kiona, Benton Co. (Jewett et al. 1953); 9 Apr 1964 at Seattle; 9 Apr 1995 at College Place. *Summer, Olympic Pen.*: 16 Jul 1976 at Dungeness. *Late Fall/ Winter*: 28 Nov-6 Dec 1974 at Vashon I.; 5 Dec 1984 at Ocean Shores; 7 Dec 1974 at Seattle; 19 Dec 1993 at Seattle; 23 Dec 1976-12 Apr 1977 at Pullman (Weber and Larrison 1977); 24 Jan 1975 at Seattle.

Steven G. Mlodinow

Northern Parula *Parula americana*

Very rare vagrant.

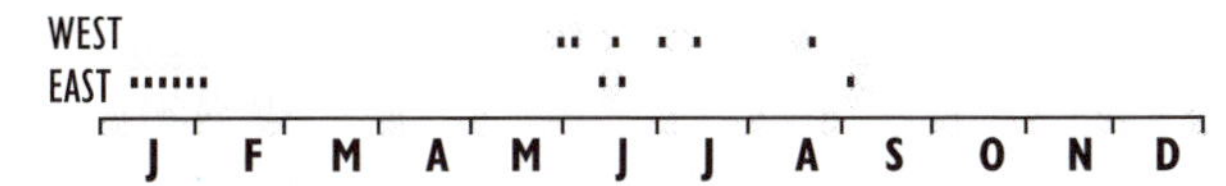

Recorded nine times in Washington: 10 Jan-3 Feb 1975 at Richland (Tweit and Paulson 1994); ad. male on 13 Jul 1979 at Clallam Bay; ad. male on 13 Jul 1979 at Humptulips; singing male on 21 Jun 1981 at Kamiak Butte; male on 18 Aug 1991 at Tokeland; singing male on 30 May 1992 at Seattle; singing male on 4 Jul 1992 at L. Quinault; 22 Jun 1995 at Granite Falls; 11 Jun 1996 at Hanford Site; 1 Sept 2000 at Vantage; singing male (ph.) on 2 Jun-3 Jul 2001 at L. Ozette. The two 1992 records coincided with a large influx into California that year (Tweit and Paulson 1994).

The Northern Parula breeds no closer to Washington than w. Minnesota, yet it is has been recorded over 800 times in California (Dunn and Garrett 1997), more than 35 times in Oregon (Gilligan et al. 1994), twice in B.C. , and once in Alaska (Dunn and Garrett 1997). Most of these records come from late spring and early summer. Notably, the midwinter record from Washington may be the northernmost for this species.

Steven G. Mlodinow

Yellow Warbler *Dendroica petechia*

Common migrant and summer resident, very rare in winter.

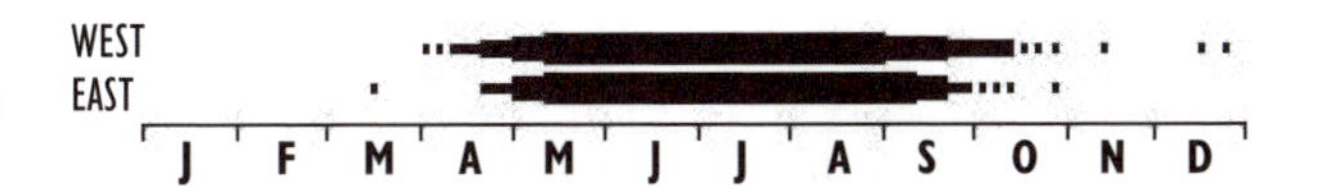

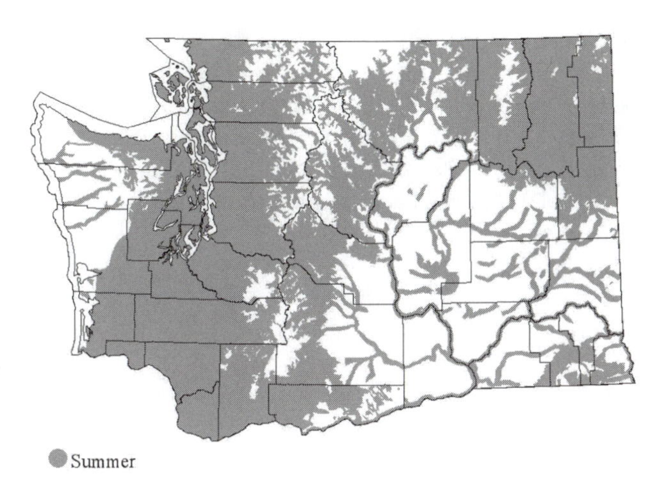

Summer

Subspecies: *D. p. morcomi* breeds e. of the Cascades and *brewsteri* w. (Dunn and Garrett 1997). *D. p. rubiginosa* known migrant.

Habitat: Breeds mainly in riparian areas containing willows, alders, and/or cottonwoods (Smith et al. 1997). During migration, also favors such areas but is found in a wider variety of deciduous woodlands.

Occurrence: With bright song and plumage, the Yellow Warbler characterizes summer in Washington lowlands. Spring migration peaks in mid-late May. It is then a widespread breeder, occurring in habitats with small wetlands in developed areas, but is less common in the high-rainfall w. Olympic Pen. (Smith et al. 1997). Though typically breeding at low elevations, birds are found to 1200 m in the Olympics (Smith et al. 1997) and they possibly breed to 1100 m in the Cascades (Jewett et al. 1953).

Fall migration peaks from mid- to late Aug, and apparently includes populations from n. nesting ranges. Birds were noted well off Grays Harbor from late Aug to late Sep during surveys from 1976 to 2000, with 11 records (including one exhausted apparent *rubiginosa*, which dropped to the surface and was eaten by a gull about 46 km off Westport on 22 Sep 1976 [TRW]).

Some human-provided habitats, such as in power-line rights-of-way, which are relatively protected from vegetative succession, may benefit species like this and stabilize local numbers. On the other hand, declines are attributed to cowbird parasitism (e.g. Stepniewski 1999). BBS data indicate a statewide 2.0 % annual decline in breeding birds from 1966 to 2000 (Sauer et al. 2001).

Remarks: Other n. races possibly occur (see Jewett et al. 1953, Smith et al. 1997, Lowther et al. 1999).

Noteworthy Records: *Spring, early*: 2 Apr 1995 at Nisqually NWR; 8 Apr 1984 at Dungeness Spit; 9 Apr 1974 at Sekiu; 9 Apr 1995 at Ridgefield NWR; *Fall, late*: 29 Oct 1972 at Walla Walla; 20 Nov 1990 at Everson (Wahl 1995); 23 Nov 1997 at Richland; 3 Dec 2000 at Neah Bay; 15-18 Dec 1974 at Hoquiam, 16 Dec 2002 nr. Sequim.

Steven G. Mlodinow

Chestnut-sided Warbler

Dendroica pensylvanica

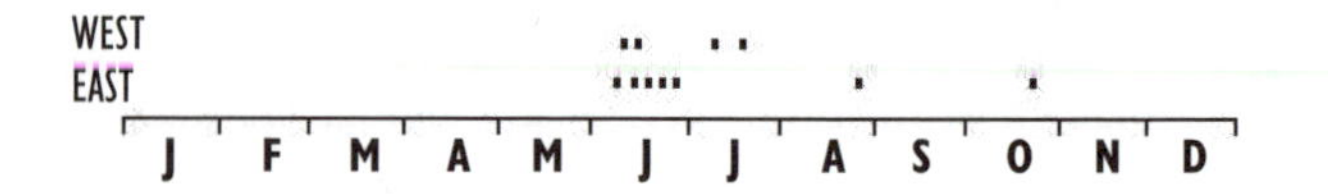

Very rare summer visitor.

Occurrence: Washington's first Chestnut-sided Warbler was collected in Grant Co. on 18 Jun 1960 (Marshall 1970). Since then, it has been found 14 more times, mostly between 11 Jun and 18 Jul. Ten records have been from the eastside and five from the w.

The species breeds regularly breeds as far w. as e. Alberta and is one of the more numerous "eastern" warblers on the w. coast. There are over 970 records from California (Dunn and Garrett 1997), more than 25 records from Oregon (Gilligan et al. 1994), and at least eight records from nr. Vancouver, B.C. (T. Plath p.c.).

Records: East: specimen on 18 Jun 1960 from nr. Othello (Marshall 1970); singing male on 19 Jun 1975 at L. Wenatchee; ad. male on 11 Jun 1977 at Palouse Falls SP (Climpson and Francik 1978); male on 17 Jun 1991 at Vantage (Tweit and Skriletz 1996); female on 27 Jun 1992 at Richland; male on 20 Jun 1995 at Hanford Site; ad. male on 23 Jun 1995 at Wapato; 1 on 27 Jun 1998 at Arid Lands Ecology Ctr.; 1 ph. on 20-24 Aug 1999 at Richland; 1 on 19-22 Oct 2000 at College Place. **West**: male on 13-14 Jun 1983 at Carnation (Tweit and Skriletz 1996); 1 on 18 Jul 1985 nr. Naches Pass; 1 on 10 Jul 1996 at Ridgefield NWR (I. McGregor to WBRC); male on 12 Jun at Rockport; 1 on 22-23 Jun 2002 at Graysmarsh.

Steven G. Mlodinow

Magnolia Warbler

Dendroica magnolia

Very rare spring and fall visitor.

An "eastern" Warbler that nests as close to Washington as c. B.C., with only 10 state records, half from the eastside and half from the w. Seven were during fall migration, mostly between 5 and 17 Sep. There are also one spring and two summer records.

Like a number of other species breeding ne. of Washington, recorded many more times s. of Washington: over 20 times in Oregon, mostly between 27 May and 14 Jun in spring and 19 Sep-

25 Oct during fall (Gilligan et al. 1994). There are more than 985 California records (Dunn and Garrett 1997).

Records: imm. on 17 Sep 1974 at Leadbetter Pt.; ad. on 21 Oct 1978 at N. Head, Long Beach Pen.; imm. on 7-8 Sep 1984 ph. at Olympia (Tweit and Paulson 1984); imm. on 13 Sep 1986 at Sullivan L.; imm. on 6-7 Sep 1987 at Vantage; imm. on 4 Oct 1988 at Protection I.; ad. male on 27 May 1994 at Ione (Tweit and Skriletz 1996); singing male (ph.), 15 Jun-4 Jul 1996 at Twisp; 1 on 10 Jul 1996 at Mt. Leona, Ferry Co.; ad. on 4 Sep 1999 at Skagit WMA (J. Flynn p.c., S. Terry p.c.). Not accepted by WBRC due to insufficient details: ad. on 22 Sp 1983 at n.jetty of the Columbia R. (Tweit and Paulson 1994).

Steven G. Mlodinow

Cape May Warbler *Dendroica tigrina*

Casual vagrant.

One Washington record: a bird with Yellow-rumped Warblers at Bellingham 21 Sep 1974. This spruce budworm specialist breeds as far w. as B.C. but is one of the rarer vagrants to the Pacific coast states. Through 2001, there were about 15 records from Oregon, mostly from May, Jun, Aug, and Sep. Vagrancy of this species to California seems closely tied to budworm populations on the Cape May's breeding grounds, and spruce budworm populations peaked during the early to mid-1970s (Patten and Burger 1998). A published report from Windust Park 10-11 Oct 1992 was rejected by the WBRC (Tweit and Paulson 1994).

Steven G. Mlodinow

Black-throated Blue Warbler *Dendroica caerulescens*

Very rare fall and winter vagrant.

Eight records: imm. male on 3 Nov at Ruby Beach; male on 8 Oct 1993 at Richland; male on 2 Nov 1994-5 Apr 1995 (ph.) at Mercer I.; male 5 Mar-13 Apr 1995 (ph.) at Olympia; 1 on 19 Sep 1998 at Bickleton; 1 on 20 Sep 1998 at Richland; male on 7-14 Dec 2001 (ph.) at Vancouver, male on 3 Nov 2002 at Brier, Snohomish Co. Another report from Spokane on 19 Feb 1992 was rejected by the WBRC (Tweit and Paulson 1994). Subspecies in Washington is presumably *caerulescens.*

In Oregon the species occurs annually, mainly from early Sep to mid-Nov. In California there are over 630 records, mostly from Oct to early Nov (Dunn and Garrett 1997).

Steven G. Mlodinow

Yellow-rumped Warbler *Dendroica coronata*

Common migrant and breeder throughout state. During winter, uncommon to fairly common on westside and rare to uncommon on eastside.

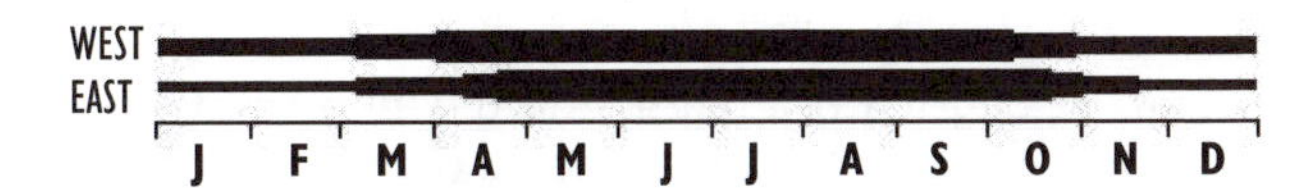

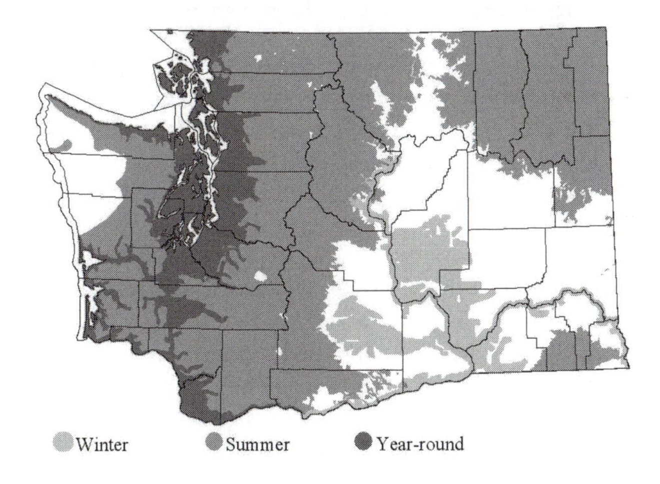

Subspecies: Audubon's Warbler, *D. c. auduboni,* is a migrant, winter visitor, and breeder; Myrtle Warbler, *coronota,* is a migrant and winter visitor.

Habitat: Breeds in coniferous and mixed woodlands, preferring drier forests overall and occurring only somewhat locally in wetter areas such as the Sitka spruce zone (Smith et al. 1997). During migration, found in a wide variety of woodlands, but during winter prefers deciduous woods and thickets.

Occurrence: Myrtle and Audubon's Warblers, once considered separate species, were found to hybridize freely in se. B.C. and sw. Alberta n. to se. Alaska and have been lumped together under Yellow-rumped Warbler (Barrowclough 1980, Dunn and Garrett 1997, AOU 1998). Intergrades, as well as both parental types, occur widely in Washington (Jewett et al. 1953).

Audubon's Warbler: Widespread, occurs across the state year-round. During summer, breeds locally in westside lowlands, more widely in the Cascades. It is apparently more numerous in e. Washington's drier coniferous forests in the Cascades, ne., and Blue Mts. During winter, abundance fairly well proportional to average temperature, with this taxa being most numerous in sw. areas and least numerous in the ne. Birds often resort to fruits and berries in winter (Jewett et al. 1953, Lewis and Sharpe 1987). Numbers appear to drop somewhat as the winter progresses, and typically more are present during warmer winters. Spring migration often begins in mid-late Mar, peaks from mid-Apr to early or mid-May, and ends in late May or early Jun. Fall movement starts in mid-late Aug, peaks from early to mid-Sep to early to mid-Oct, and then dwindles through the remainder of Oct into Nov.

Myrtle Warbler: Mostly a migrant and winter resident w. of the Cascades, though not rare on the eastside. Vagrant during summer. Noted by some authors (e.g., Lewis and Sharpe 1987, Wahl 1995) as less numerous than Audubon's during migration and winter, though reportedly almost as abundant at least locally in the Puget Trough. On the outer coast typically more common than Audubon's. On the eastside, Myrtle Warblers are uncommon to fairly common migrants and rare to uncommon during winter. Spring migration and fall migration dates similar to Audubon's except spring departure is a week or so earlier and fall arrival is a week or so later.

BBS data indicate a statewide increase of 1.7% per year from 1966 to 2000 (Sauer et al. 2001).

Largest CBC counts by far occurred at Grays Harbor (mainly Myrtle Warblers) and Tri-Cities (mainly Audubon's). CBC averages suggest no overall statewide trend in winter numbers.

Remarks: Field identification of birds within the Audubon's and Myrtle groups is not possible.

Noteworthy Records: Myrtle Warbler, *high count*: 180 on 17 Oct 1998 at Skagit WMA. *Summer*: 10 Jun 1976 at Spokane; 22 Jun 1995 at Seattle; 11 Jun 2002 at Vantage. **Audubon's Warbler**, *high count, Fall*: 1000 on 26 Sep 1998 at Bateman I. *Winter*: 105 on 21 Jan 1995 at Priest Rapids L. (Stepniewski 1999). **Yellow-rumped Warbler**, *CBC high counts*: 1248 in 1977, 1092 in 1981, 958 in 1985, 854 in 1991, 934 in 1993 at Grays Harbor; 540 in 1970, 632 in 1974, 404 in 1981 at Tri-Cities.

Steven G. Mlodinow

Black-throated Gray Warbler

Dendroica nigrescens

Common summer resident and migrant in w. lowlands, uncommon to common local summer resident on e. slope of Cascades in Klickitat, Yakima, and Kittitas Cos.

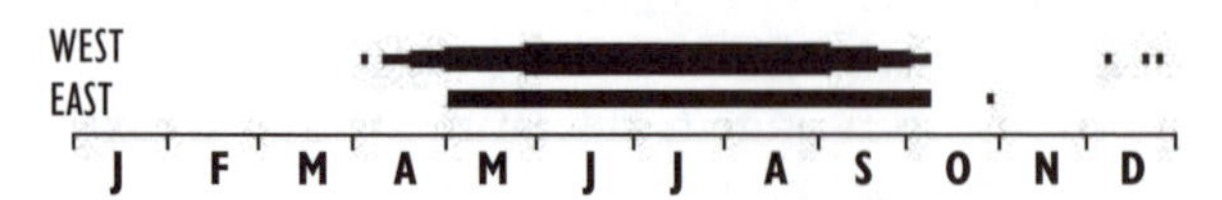

Subspecies: *D. n. nigrescens* breeds w. of the Cascade crest and *halseii* e. of the crest (Morrison 1990); birds in Kittitas Co. may be *nigrescens* (see Guzy and Lowther 1997).

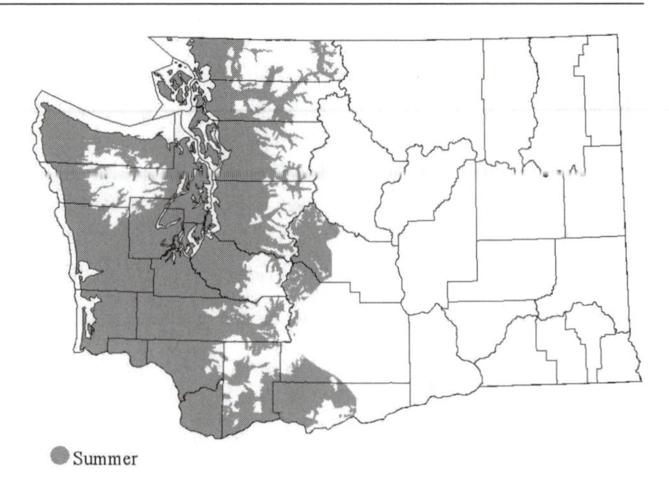

Habitat: Breeds in deciduous and evergreen broadleaf forests and woodlands, mixed broadleaf-conifer forests, young conifer forests, low-density urban areas, and westside riparian wetlands. Infrequent in mature or old-growth conifer forests (Manuwal 1991). Strongly associated with broadleaf trees and pole-size conifer stands (Morrison 1982, Gilbert and Allwine 1991, Bosakowski 1997), and streams within young conifer forests (Carey 1988).

Occurrence: One of the most common breeding warblers in westside lowlands and the oak woodlands of Klickitat Co. Its status appears unchanged since described by Rathbun (1902), Dawson and Bowles (1909), and Jewett et al. (1953). First migrants typically arrive in w. Washington about mid-Apr and the last depart around early Oct. Occasional stragglers stay longer into the fall, and very rarely winter. In e. Washington, migrants are very rare outside the breeding range. Unlike many other warblers, there appear to be no significant high montane fall movements, with most observations at low elevations.

Most breeders are w. of the Cascade crest, widespread and most common below about 450 to 600 m elevation (western hemlock and Sitka spruce zones), and uncommon to fairly common in the rainshadow of the Olympics, including the ne. peninsula, San Juan Is., Fidalgo and Whidbey Is., (Lewis and Sharpe 1987, Sharpe 1993, Smith et al. 1997). Birds are rare at higher elevations in the silver fir zone. E. of the crest, they are local and uncommon at middle elevations in w. Kittitas Co., and common in w. Klickitat Co. and far sw. Yakima Co. (Manuwal 1989, Smith et al. 1997, Stepniewski 1999). An isolated breeding population occurred at least formerly in the Juniper Dunes Wilderness Area in Franklin Co. (Weber and Larrison 1977).

Westside habitats increased since early 1900s timber harvests that expanded alder, maple, and pole-size young conifer stands. The species appears to be relatively tolerant of fragmentation and habitat modifications associated with urbanization or agriculture, and is often found as a summer resident in relatively small habitats in urban or agricultural habitats (Smith et al. 1997, CBCh). The open juniper woodland habitat at the Juniper Dunes Wilderness Area is unique in Washington but similar to habitat in se. Oregon.

Jewett et al. (1953) reported summer birds further e. than current known breeding areas: two sites in the Blue Mts. of se. Washington, two sites in Yakima Co., and at Leavenworth. Other than this possible contraction, populations appear to be relatively stable, with no significant regional trends in abundance on BBS data since the 1960s (Sauer et al. 2001).

Remarks: A breeding population on portions of the Yakama Indian Reservation is suspected (Stepniewski 1999). Status at the Juniper Dunes Wilderness Area is unknown. The only known Black-throated GrayXHermit warbler hybrid was collected nr. Trout L. on 16 Jun 1994 (Rohwer et al. 2000).

Noteworthy Records: West, *Spring*: 1 on 8 Apr 1997 at Tacoma. *Fall*: 31 Aug 1891 at head of the Cascade R. Skagit Co. (Jewett et al. 1953), 3 Oct 1919 at Tacoma (Bowles). *Winter*: 1 in 1977 on Grays Harbor CBC; 1 each in winters of 1988 and 1989 at Seattle (Sharpe 1993), 1 on 25 Dec 1997 at Seattle. **East**, *Spring*: 1 on 6 May 1973 nr. Ewartsville, Whitman Co. (Weber and Larrison 1977, Johnson), 1 on 18-19 May 1974 at Davenport. *Fall*: 27 Oct 1914 at Yakima (Ellis in Jewett et al. 1953), 6 Sep 1917 at Mt. Aix, Yakima Co. (Jewett et al. 1953), 1 on 22 Sep and 3 Oct 1975 at Walla Walla (Weber and Larrison 1977, Muse), 1 on 21 Sep 1991, 20 Aug 1993 and 21 Aug 1998 n. of Wapato (A. Stepniewski); 9 Sep 2002 at Kahlotus.

Christopher B. Chappell

Black-throated Green Warbler *Dendroica virens*

Casual vagrant.

One record: a singing male at Dishman SP, 2 Jul 1975. Another published record from Seattle, 12 Sep 1982, was rejected by WBRC (Tweit and Paulson 1994). Though this species breeds as far w. as B.C., it is one of the rarer vagrants to the Pacific coast states. There are fewer than 10 records from Oregon, mostly late May through late Jun.

Steven G. Mlodinow

Townsend's Warbler *Dendroica townsendi*

Fairly common to common summer resident in forests throughout most of Washington. Locally uncommon winter resident at low elevations in w., very rare e. Fairly common to common spring and fall migrant throughout.

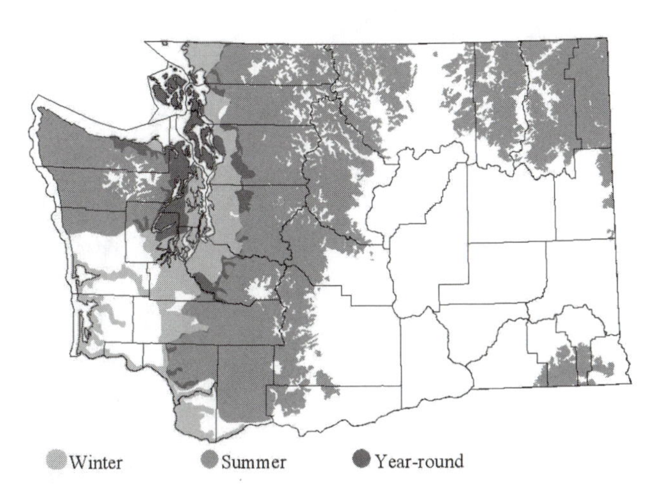

Habitat: Breeds in closed conifer-dominated forests, from nr. sea level to the upper limit of continuous forest. Favors Douglas-fir and the true firs as major stand components (Mannan and Meslow 1984, Pearson and Manuwal 2000), and avoids stands dominated by hemlocks or redcedar (Sharpe 1993, CBCh).

Occurrence: One of Washington's most common and widespread forest warblers, wintering relatively far n. for a warbler. Based on limited historical information, the breeding status of Townsend's Warbler does not appear to have changed much since the first half of the 20th century (Jewett et al. 1953).

Breeds throughout most of Washington, including the Olympic Pen., Kitsap Pen., Whidbey I., San Juan Is., the Cascades and foothills s. to Lewis, ne. Skamania, and nw. Klickitat Cos., the Blue Mts., the Okanogan Highlands, and forests nr. Spokane (Smith et al. 1997). It is largely absent as a breeder from most of sw. Washington and urbanized and agricultural areas of the Puget Trough, apparently depending upon fairly extensive conifer-dominated stands or a mosaic of conifer stands and clear-cuts. Breeding-season hybrids, or perhaps in a few cases pure Townsend's, are uncommon to rare in e. Thurston Co., the Black Hills, and parts of Clark, Cowlitz, and Skamania Cos. (Sharpe 1993, Rohwer and Wood 1998).

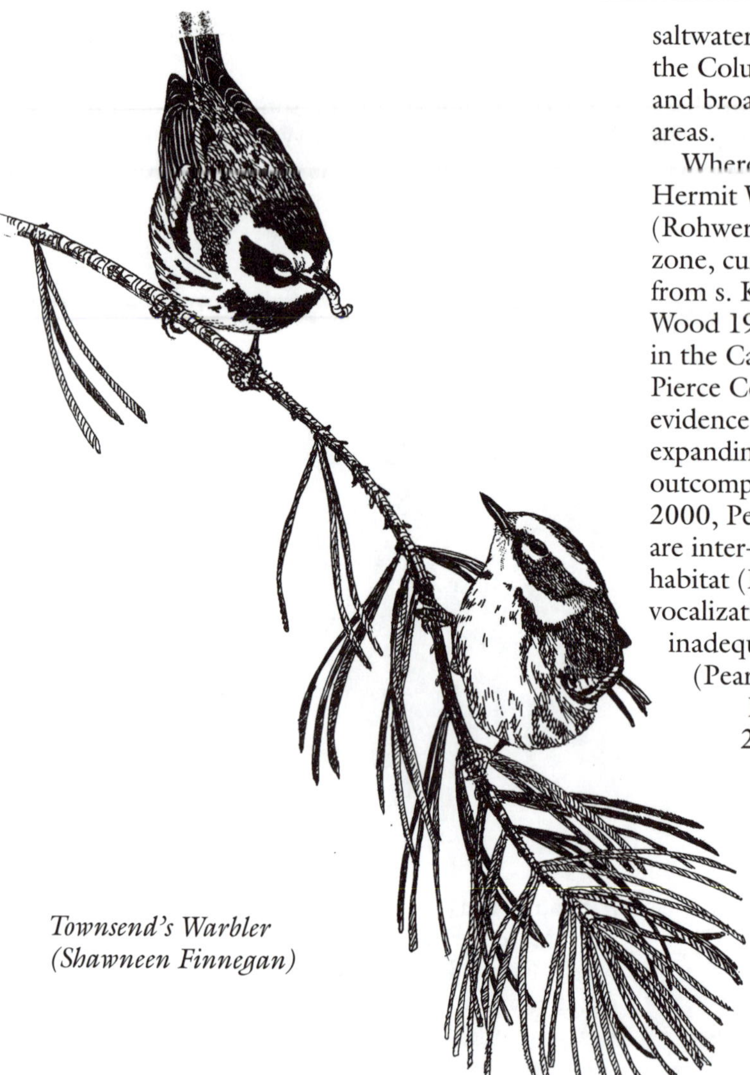

Townsend's Warbler
(Shawneen Finnegan)

The species is uncommon on the w. Olympic Pen., where most natural stands are dominated by western hemlock or western redcedar, and in the ponderosa pine zone on the eastside, where stands are generally open with little to no fir or Douglas-fir. In eastside forests, it is more numerous in old-growth than young managed stands, which correlates with the abundance of Douglas-fir and grand fir in the canopy (Mannan and Meslow 1984). On the westside, it appears to be more numerous in young than in old-growth forests (Manuwal 1991, Bryant et al. 1993).

Townsend's Warbler is a fairly common to common and widespread migrant in many wooded habitats, more common in the lowlands in spring and in the mountains in fall, where it moves upslope as far as the subalpine parkland after breeding. Westside numbers increase in mid-late Apr. On the eastside, typical arrival is from early May to early Jun and fall migration from mid-Aug to early Oct. Winter distribution is at very low elevation (mostly below 150 m), concentrated within a few miles of saltwater, mostly from Snohomish Co. s. and along the Columbia R. Habitats include conifer, mixed, and broadleaf deciduous forests, as well as urban areas.

Where distribution overlaps with that of the Hermit Warbler, the two species freely hybridize (Rohwer and Wood 1998) in a 100-125 km wide zone, currently the e. Olympic Pen., the Cascades from s. King Co. to the Mt. Hood area (Rohwer and Wood 1998), and the Kitsap Pen. (CBCh). Hybrids in the Cascades are most prevalent from Ashford, Pierce Co., sse. to Trout L., Klickitat Co. Recent evidence suggests Townsend's Warblers are expanding into the range of the Hermit Warbler and outcompeting it (Rohwer and Wood 1998, Pearson 2000, Pearson and Manuwal 2000). The two species are inter-specifically territorial, do not differ in habitat (Pearson and Manuwal 2000), and their vocalizations are extremely variable, overlapping, and inadequate to verify identification in Washington (Pearson 1997).

BBS data indicate no change from 1966 to 2000 in Washington (Sauer et al. 2001). Wintering birds recorded on westside CBCs have increased significantly ($P < 0.01$) since at least the 1970s (CBCh).

Remarks: Verification of apparent breeding range expansion and overlap with the Hermit Warbler is highly desirable.

Noteworthy Records: West, *high counts, Spring*: 30 on 13 May 1967 in Seattle; flock of 50 on 13 May 1984 on Bainbridge I. *CBC high counts*: 35 in 1973 at Leadbetter Pt.; 43 in 1992 at Tacoma; 43 in 1997 at Seattle. **East**, *Spring, late*: 1 on 15 Jun 1996 on Yakima Training Center (A. Stepniewski p.c.). *Fall, late*: 1 on 6 Nov 1997 at Wapato (A. Stepniewski p.c.); 1 on 18 Nov 1978 at Richland; 3 Dec 1998 at John Day Dam (S. Pink p.c.). Winter: 27 Feb 2003 at Wenatchee.

Christopher B. Chappell

Hermit Warbler *Dendroica occidentalis*

Locally common to uncommon summer resident and migrant in w. Washington in e. Jefferson, Kitsap, Pierce Cos., and in the s. Cascades from e.-c. King Co. to w. Klickitat Co. Very rare in winter.

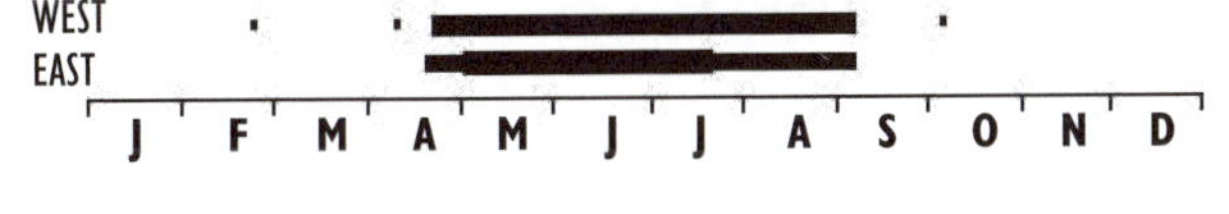

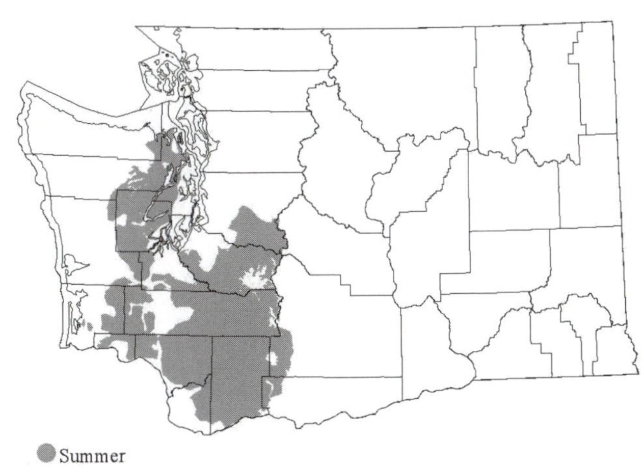

Summer

Habitat: Breeds in closed conifer-dominated forests, from nr. sea level to the westside mid-montane silver fir zone (Chappell and Ringer 1983) and high-montane eastside subalpine fir zone (Stepniewski 1999), rarely in the westside lower mountain hemlock zone (Manuwal et al. 1987, Sharpe 1993). Very local in eastside mixed-conifer forests.

Occurrence: The Hermit Warbler has a limited range in Washington and hybridizes prodigiously with Townsend's Warbler in a zone of overlap. Reports from around 1900 indicated possible breeding well n. and e. of the current breeding range, in the n. Cascades, ne. Washington, and the n. and w. Olympic Pen. (Dawson 1897, Jones 1900, Dawson and Bowles 1909, Rathbun 1916, Jewett et al. 1953).

First spring arrivals usually in late Apr (Pearson 1997), last fall departures typically in early Sep. Birds move upslope after breeding (Pearson 1997) and are uncommon in Aug and early Sep in subalpine parkland (Chappell and Ringer 1983). Migrants rarely detected outside breeding range, with spring records concentrated in mid-late May and fall records mostly from early Aug to early Sep.

Breeding in most forested areas of sw. Washington, the species favors Douglas-fir and true firs (Chappell and Ringer 1983, Pearson and Manuwal 2000). It is most abundant in young closed-canopy forests, less numerous in old-growth or very open stands (Carey et al. 1991, Manuwal 1991, Hansen et al. 1995, Bosakowski 1997). It is common w. of the Cascade crest and foothills from Mt. Rainier s., extending onto the eastside only on the e. slopes of Mt. Adams and in far w. Klickitat Co. N. of Mt. Rainier it is uncommon n. to c. King Co., and rare further n. It is fairly common to common in the e. Olympics above 150 m elevation (Sharpe 1993) and uncommon in lowlands and low hills of e. Jefferson, w. Kitsap, Mason, Thurston, sw. Pierce, Grays Harbor, Lewis, Cowlitz, and Clark Cos. In the Willapa Hills it appears to be rare to perhaps locally uncommon (Sharpe 1993, Smith et al. 1997), though Rohwer and Wood (1998) note that its song there is difficult to distinguish from that of the Black-throated Gray Warbler. It is absent from urbanized and intensely agricultural areas.

Hybridizes freely with Townsend's Warbler in a 100-125 km wide hybrid zone in the e. Olympic Pen., Kitsap Pen. (CBCh) , and Cascades from s. King Co. s. to the Mt. Hood area (Rohwer and Wood 1998). Hybrids most prevalent from Ashford, Pierce Co., se. to Trout L. Townsend's Warblers have apparently expanded into the range of the Hermit Warbler (Rohwer and Wood 1998, Pearson 2000, Pearson and Manuwal 2000). The two species are inter-specifically territorial, occupy the same habitats (Pearson and Manuwal 2000), have extremely variable vocalizations, and identification is confused in most of Washington (Chappell and Ringer 1983, Pearson 1997). Presumed vagrants are recorded as far n. as sw. B.C., including Vancouver I.

Populations appear relatively stable and, because the species is common in young Douglas-fir forests, there appear to be no major threats under current land-use patterns. BBS data show an increasing trend over the last 30 years in the Cascades from Washington to California, but no significant long-term trends elsewhere (Sauer et al. 1999a). Locally extirpated, probably by urbanization, from Spanaway and S. Tacoma where it was common around 1900, declined during the 1920s, and has not bred since 1974 (Bowles 1906b, 1929, Chappell and Ringer 1983).

Remarks: Populations are potentially vulnerable because of limited geographic range and habitat specialization (Andelman and Stock 1994, Pearson 1997). Documentation of abundance and distribution in peripheral areas of the species' range is needed to determine the spatial and temporal dynamics of the hybrid zone with Townsend's Warbler. Classed GL, NAS Watch List.

Noteworthy Records: West, *Spring*: 1 on 12 Apr 1954 at Seattle; 2 May 1964 at Spanaway; 1 on 8 May 1990 in Elwha R. Valley (F. Sharpe p.c.); 1 on 9 May 1970 at Bald Hill, Thurston Co.; 1 on 16 May 1981 at Bellingham (J.

Duemmel p.c.); 2 on 16 May 1982 at L. Crescent; 1 on 23 May 1982 nr. Sequim. *Summer*: 1 on 1 Jun 1975 at Tacoma; 1 on 2 Jun 1990 on Sumas Mt., Whatcom Co.; 17 Jun 1975 on Long Beach Pen.; 1 on 4 Jul 1981 at Bellingham. *Fall*: 1 on 16 Aug 1955 at Seattle; 1 on 5 Sep 1965 at Leadbetter Pt.; 1 on 23 Aug 1973 at Eld Inlet; 4 Aug 1990 at Naselle (A. Richards p.c.). Late: 1 on 5 Oct 1994 at Sequim; 10 on 6 Sep 1984 at Ashford, Pierce Co. *Winter*: 31 Jan 2001 at Sumner; 25 Feb 2003 at L. Sammamish. **East**, *Spring, out of range*: 21 May 1906 at Calispell L. (Dawson in Jewett et al. 1953); 1 on 26 May 1964, 1 on 26 May 1969 at Spokane; 1 on 29 May 1971 at Wenas Cr.; 1 on 18 May 1972 at Swakane WMA, Chelan Co. (Chappell and Ringer 1983); 1 on 1 May 1982 nr. Plain, Chelan Co.; 1 on 14-15 May 1977 at Ephrata; 1 on 9 May 1980 e. of White Pass; 1 on 28 May 1989 on Bald Mt. Kittitas Co. *Summer*: 1 collected nr. Stehekin (Dawson 1897); Cascade Pass (Jones 1900), Newport and Kalispell Range, Pend Oreille Co. (Dawson and Bowles 1909); 1 in Jul 1966 on Round Top Peak, Pend Oreille Co. (Larrison and Sonnenberg 1968); pair on 15 Jun 1978 at Swauk Pass (Chappell and Ringer 1983); 1 Jun 1992 at Windust SP; several 9 Jul 1995 along Gotchen Cr., Yakima Co., (Stepniewski 1999); 3 on 16 Jun 1997 nr. Manastash Ridge (D. Swayne p.c.); 17-24 Jun 1990 nr. Potato Hill, Yakima Co. *Fall*: 9 Aug 1896 at Pershall Cr, Chelan Co. (Dawson in Jewett et al. 1953); 1 on 19 Aug 1918 at Lyle (Jewett et al. 1953); 6 on 30 Aug 1954 on Mt. Spokane (LaFave 1955); 1 on 12 Sept 1972 at Wenatchee (Chappell and Ringer 1983). *Hybrid HermitXTownsend's*: 1 on 25 May 1986 Wicky Cr., Skamania Co.; 1 on 17 and 24 Jun 1990 at Potato Hill; 1 on 9 Jun 1990 at Cedar Falls, King Co.; 1 on 15 Jun 1991 at Raging R., King Co.; 1 in Jun 1994 at Bangor Subbase (CBCh); 1 on 21 Jun 1994 at Scenic Beach SP, Kitsap Co. (CBCh); 1 on 23 Jun 1994 at Camp Wesley Harris, Kitsap Co. (CBCh); 1 on 29 Jun 1994 nr. Camp Union, Kitsap Co. (CBCh).

Christopher B. Chappell

Blackburnian Warbler *Dendroica fusca*

Casual vagrant.

Like many "eastern" warblers, the Blackburnian breeds fairly far w., nesting as close as e.c. Alberta (Dunn and Garrett 1997). It also is regular on the w. coast during migration with more than 475 records from California, six records from Oregon (Gilligan et al 1994), and five from B.C. (Spitmann 1996). Most w. coast records have been during fall.

There are four sight records for Washington: an imm. male at Ocean Shores on 10 Sep 1979 (Tweit and Paulson 1994); a singing male at Richland on 31 May 1980 (Tweit and Paulson 1994); an imm. at Discovery Park, Seattle, on 4 Dec 1987 (Tweit and Paulson 1994); a male nr. Almira, Lincoln Co., on 25 Jun 1996. A report from Leadbetter Pt. on 4 Oct 1981 was rejected by the WBRC because documentation did not eliminate the possibility of Townsend's Warbler (Tweit and Paulson 1994).

Steven G. Mlodinow

Yellow-throated Warbler *Dendroica dominca*

Casual vagrant.

One Washington record of this se. U.S. species: a bird of the *albilora* subspecies was at Twisp from 8 Dec 2001 to 23 Jan 2002 and was presumed killed by a Sharp-shinned Hawk. There is one record in the Gulf Is., B.C., in Jan 1998 (Campbell et al. 2001) and three or four records in Oregon (Marshall et al. 2003).

Terence R. Wahl

Prairie Warbler *Dendroica discolor*

Casual vagrant.

One Washington record: a bird at Wallula on 20 Dec 1989 (Tweit and Paulson 1994). Through 2001, there were 11 Oregon records of this se. warbler, all but one coastal and most from mid-Sep to mid-Oct. There are two or three records for B.C., including a winter record from the Queen Charlotte Is. (Campbell et al. 2001). Subspecies presumably *discolor*.

Steven G. Mlodinow

Palm Warbler *Dendroica palmarum*

Rare fall and winter on the outer coast, very rare elsewhere.

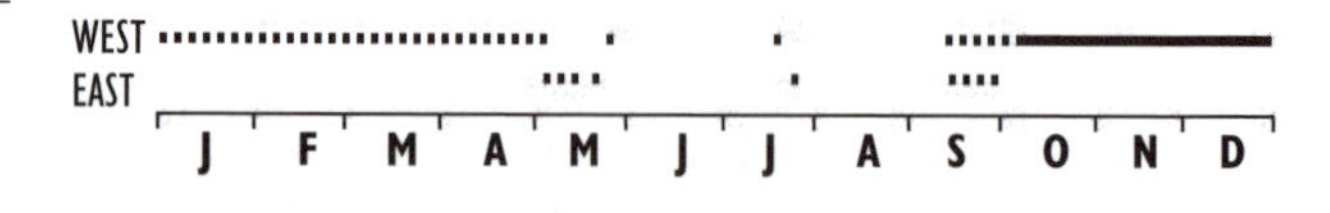

Subspecies: Apparently *D. p. palmarum*.

Occurrence: The most regularly occurring of "eastern" warblers in Washington, with approximately 130 records through Jan 2000, following the first record on 13 Dec 1964 at Wiser L., Whatcom Co. The majority (70%) are fall-winter records from the outer coast and another 14% are fall records from elsewhere in the westside lowlands. Since observers began regular coverage of the outer coast in the 1970s, the species has been recorded almost annually there in fall, averaging about three per year. None were reported in fall of 1973 and 1985, and high numbers were found in 1977 (11 birds), 1979 (10) and 1993 (16). Birds were exceptionally abundant in fall 1993 along the entire Pacific coast (Gilligan et al. 1994, Dunn and Garrett 1997).

Similar to the situation in both California (Roberson 1980) and Oregon (Gilligan et al. 1994), spring records and interior records are much less numerous than fall westside records. Westside spring records (late Mar-May) total only nine, five from the outer coast and four from the Puget Trough. Interior records are split between five spring and four fall records. There have also been three high-elevation records, including one of the two midsummer records: a singing male at Naches Pass.

Noteworthy Records: West, *Fall, coastal high counts*: 8 in 1979, 5 in 1975 at Ocean Shores. **East**: 26 Jun 1972 at Dishman Hills, Spokane Co.; 23 May 1976 at Goose Prairie, Yakima Co.; 14 Sep 1979 at Indian Canyon, Spokane Co.; 12 May 1984 at Davenport; fall 1984 at Lyons Ferry; 5 May 1996 at Yakima Training Center; 26 Sep 1996 at Davenport; 11 May 1997 at Toppenish; 18 Sep 1999 at Indian Canyon. Mountains: 1 Oct 1986 nr. Packwood; 17 Jun 1988 at Naches Pass; 25 Sep 1997 at Hurricane Ridge.

Bill Tweit

Bay-breasted Warbler *Dendroica castanea*

Casual vagrant.

A singing male nr. Granite Falls, Snohomish Co., on 27 Jun 2002 is the only record for Washington. Breeds in ne. B.C., with one fall record on Vancouver I. (Campbell et al. 2001). Recorded in late May, Jun, Aug, and Sep in e. interior Oregon (Marshall et al. 2003).

Terence R. Wahl

Blackpoll Warbler *Dendroica striata*

Very rare fall vagrant, casual in spring and summer.

Occurrence: An "eastern" warbler that breeds as far w. as w. Alaska and one of the most common "vagrant" warblers in California, with well over 3000 records (Dunn and Garrett 1997), and regular in Oregon. Blackpoll Warblers are rarely reported in Washington, with only 13 records. All but one have come from the eastside, and almost all were seen between 28 Aug and 20 Sep. Outliers include a late Jun bird on the outer coast and one during mid-May at Spokane.

Records: 1 on 2 Sep 1985 at Richland; 1 on 20 Sep 1986 at Davenport; imm. male 6-7 Sep 1987 collected at Vantage (Tweit and Paulson 1994); ad. male on 17 May 1991 at Spokane; imm. on 7 Sep 1991 at Davenport; 1 on 29 Aug-1 Sep at Burbank; 1 on 28 Aug 1993 at Moxee; 1 on 4-5 Sep 1994 at Saddle Mt., Grant Co.; 1 on 8 Sep 1994 at Vantage; ad. male 22-23 Jun 1997 ph. on Tatoosh I. (WBRC); 1on 25 Aug 1998 at Richland; 1 on 1 Sep 1998 at Wanapum Dam SP; 1 on 1 Sep 1998 at Wahluke WMA. Not accepted by WBRC due to insufficient details: 1 on 20 Sep 1976 at Ocean Shores (Tweit and Skriletz 1996); 1 on 7 Sep 1991 at Richland (Tweit and Paulson 1994); 1 on 5 Sep 1992 at Columbia Park, Benton Co. (Tweit and Paulson 1994). These reports do correspond with seasonal occurrence of the accepted records.

Steven G. Mlodinow

Black-and-White Warbler *Mniotilta varia*

Rare spring migrant. Very rare fall and winter vagrant. Casual in summer.

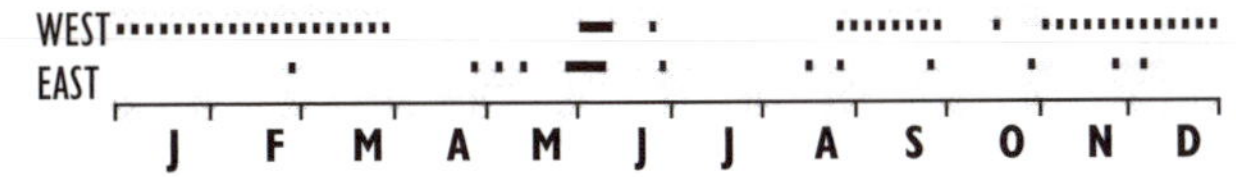

Washington's first Black-and-White Warbler was found nr. Pullman on 15 Aug 1948 (Jewett et al. 1953). The next one was reported on 21 Feb 1961 at Pasco. Since 1970 this species has been almost annual in Washington, perhaps due to increased observer numbers. There are now about 40 records. These have been split nearly evenly between w. and e. and have occurred throughout the calendar year. In e. Washington, most records have been during spring, with a peak in late May and early Jun. On the westside, records are almost evenly split between spring, fall, and winter, with one record from midsummer. By county: Whatcom (2 records), Skagit (2), Snohomish (1), King (5), Clallam (1), Kitsap (1), Cowlitz (1), Pacific (2), Clark (3), Pend Oreille (2), Douglas (1), Lincoln (1), Grant (3), Whitman (2), Yakima (1), Benton (4), Franklin (1), Columbia (1), Adams (1) and Asotin (5).

Noteworthy Records: *High counts*: 2 on 27 May 1973 nr. Clarkston; 2 on 5 Jun 1976 at Ridgefield NWR. *Midsummer*: 25 Jun 19881 nr. Cusick; 22 Jun 1998 at Rockport. *Winter*: 26 Nov-early Dec 1999 at Two Rivers CP, Benton Co.; 21 Feb 1961 at Pasco.

Steven G. Mlodinow

American Redstart *Setophaga ruticilla*

Uncommon migrant and breeder in hardwood forests in ne. Washington. Rare migrant and breeder in the ne. Cascades and Blue Mts. Locally rare migrant and very rare breeder in w.

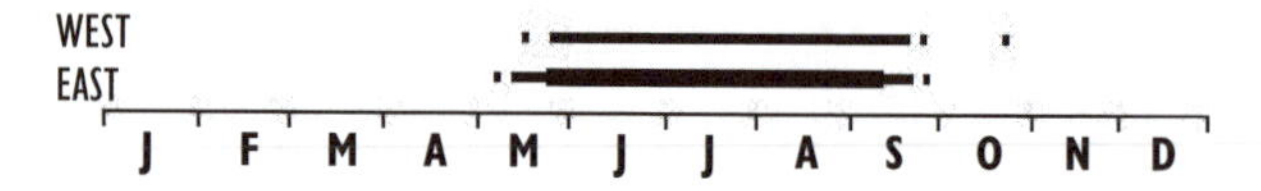

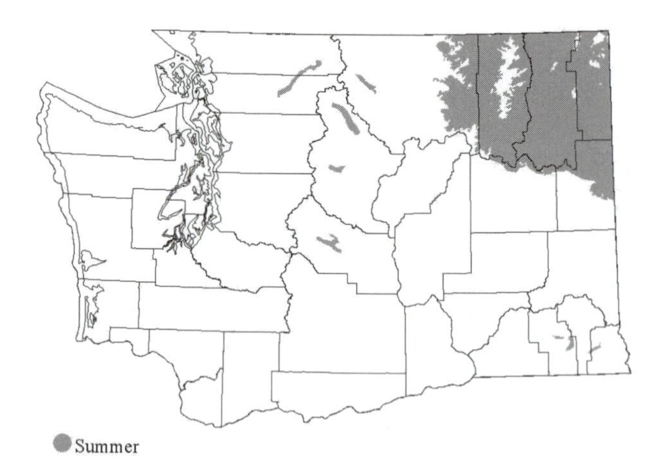

Summer

Habitat: Moist, deciduous woodlands usually nr. water. Especially low-elevation alder, willow, and dogwood woodlands bordering lakes or streams and birch and aspen stands at higher elevations. Found to an elevation of 1400 m in suitable habitat.

Occurrence: Jewett et al. (1953) considered the American Redstart an uncommon migrant and summer resident of the transition zone of e. Washington prior to 1950. Now an uncommon summer resident in the five ne. counties of Washington: Okanogan, Ferry, Stevens, Pend Oreille, and Spokane (Smith et al. 1997), a rare migrant and summer resident along the e. slope of Cascades and in the Blue Mts., and a rare migrant in the Columbia Basin.

Jewett et al. (1953) made no mention of occurrence in w. Washington. The first report was on 11 Jun 1972 at L. Ozette. Since 1978 redstarts have been recorded annually as rare migrants and were suspected to be locally very rare breeders when singing males were documented along the Skagit R., in Skagit and Whatcom Cos., from 1982 to 1986 and again in 1995-97. Breeding confirmed on 10 Aug 1996 when a pair of ads. was observed feeding four recently fledged young along the Skagit R. On 9 Jun 1997, an active nest with eggs was found there. Records nr. Darrington in May and Jun 1999 suggested breeding in Snohomish Co.

A late spring migrant, with first arrivals usually during the third or fourth weeks of May. Only three published records document arrival dates prior to 16 May. Fall migration begins in mid-Aug and continues into mid-Sep. Cannings et al. (1987) documented similar fall migration patterns for redstarts in the Okanagan Valley of B.C. There are only three published records after 21 Sep in Washington.

Over the past 30 years, continent-wide trend data from the BBS have documented slight declines (0.5 %/year) in redstart populations (Sauer et al. 1997). This decline is slightly accelerated in the w. U.S., s. B.C., and presumably in ne. Washington though BBS data are minimal (see Sauer et al. 2001).

Remarks: AOU (1957) designated the nw. breeding race as *tricolora*. Cannings et al. (1987) reported that redstarts are heavily parasitized by Brown-headed Cowbirds in s. B.C., where seven of 11 nests found contained one to three cowbird eggs.

Noteworthy Records: East, *Spring, early*: 1 May 1980 at Cle Elum; 3 May in se. Washington (Hudson and Yocum 1954). *Summer*: 29 Jun 1907 "breeding" at Spokane (SMNH PSM 16240); "3 collected" on 1 Jun 1993 at Hanson Cr., Stevens Co. (SMNH PSM 10967,10978,10970). *Fall, late*: 28 Sep 1985 at Washtucna; 10 Sep 1985 at Richland. **West**, *Spring, early*: 1 May 1976 at L. Terrell (Wahl 1995). *Summer*: 10 Aug 1996, 9 Jun 1997 at Skagit R. *Fall, late*: 25-26 Oct 1988 at Foster I., Seattle; 21 Oct 2000 at Seattle; 26 Sep 1978 at Samish I.; 21 Sep 1982 at Leadbetter Pt.; 19 Sep 1992 at Olympia. *Outer coast*: 11 Jun 1972 at L. Ozette; 11 Sep 1979 at Ocean Shores; 6 Jun 1981 at Dungeness Spit; 21 Sep 1982, 21 and 23 May 1983 at Leadbetter Pt.; 1 on 1 Jul 1998 at sea off n. coast landed on a vessel (BT).

Robert C. Kuntz II

Prothonotary Warbler *Protonotaria citrea*

Casual vagrant.

The Prothonotary Warbler has been recorded once in Washington: one (ph.) at Richland from 5 to late Sep 1970 (Mattocks et al. 1976). The species breeds no closer to Washington than s.c. Minnesota, and as with most other predominantly se. warblers, individuals wander to the Pacific coast relatively infrequently. California records number only around 160 (Dunn and Garrett 1997), and Oregon has about eight records, half of these between 19 Sep and 19 Oct. Classed GL.

Steven G. Mlodinow

Ovenbird *Seiurus aurocapilla*

Very rare visitor.

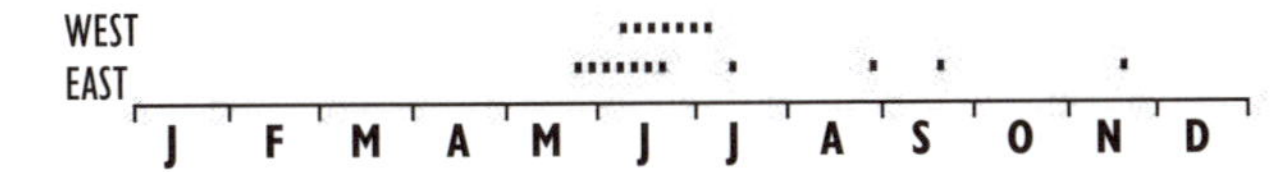

Subspecies: Unknown race in Washington.

Occurrence: The first Washington Ovenbird record is of a specimen collected at Spokane, 15 Nov 1956 (LaFave 1957). Of the subsequent 18 records, 12 have been from the eastside and all but three occurred between 24 May and 1 Jul.

The Ovenbird breeds nearest Washington in ne. B.C. and in w. Montana (Dunn and Garrett 1997). It is one of the more numerous "eastern" vagrants on the Pacific coast, with over 800 records from California and more than 35 from Oregon (Gilligan et al. 1994, Dunn and Garrett 1997).

Records: East: 1 collected on 15 Nov 1956 at Spokane (LaFave 1957); 1 ph. on 5 Jun 1972 at Richland 1972 (Tweit and Skriletz 1996); 1 taped singing on 16 Jun 1973 at Sullivan L.; 1 dead on 12 May 1977 at Pullman; 1 singing bird on 9 Jun 1979 at Teanaway (Tweit and Paulson 1994); 1 singing 12-20 Jul 1980 and 27 Aug 1980 at Hardy Canyon, Yakima Co.; 1 singing on 31 May 1988 at Richland; 11 Jun 1988 at Palouse Falls SP; 1 on 4-6 Jun 1991 nr. Richland; 1 on 18 Sep 1992 at Davenport; 1 on 24 May 1997 nr. Government Springs, Kittitas Co. (B. Senturia to WBRC); 1 on 24-27 May 1998 at Davenport; 1 singing on 16 Jun 2002 nr. Trout L. **West**: ad. male found dead on 26 Jun 1980 in W. Seattle (Tweit and Skriletz 1996); 1 singing on 17 Jun 1983 at Friday Harbor (Tweit and Skriletz 1996); 1 singing on 8 Jun 1992 on w. side of Ross L.; 1 singing 16 Jun-1 Jul 2000 at Packwood; 1 singing on 15 Jun 2001 at Vancouver; 1 specimen collected on 17 Jun 2001 at Seattle.

Steven G. Mlodinow

Northern Waterthrush *Seiurus noveboracensis*

Locally fairly common summer resident and breeder in forested swamps and wetlands in ne. counties. Very rare migrant elsewhere in e.; very rare migrant and winter visitor in w.

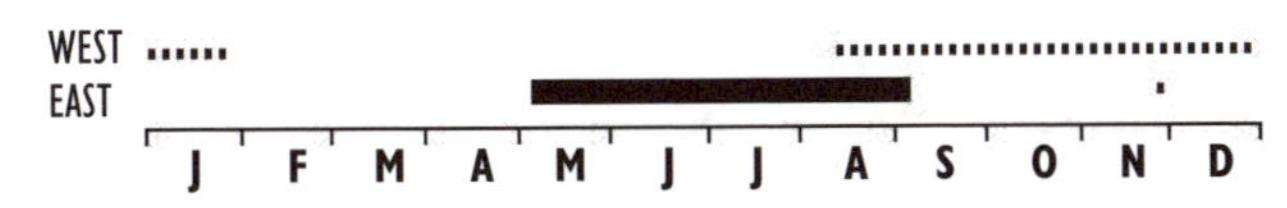

Subspecies: *S. n. notabilis.*

Habitat: Nr. water edge in bogs, swamps, and small streams.

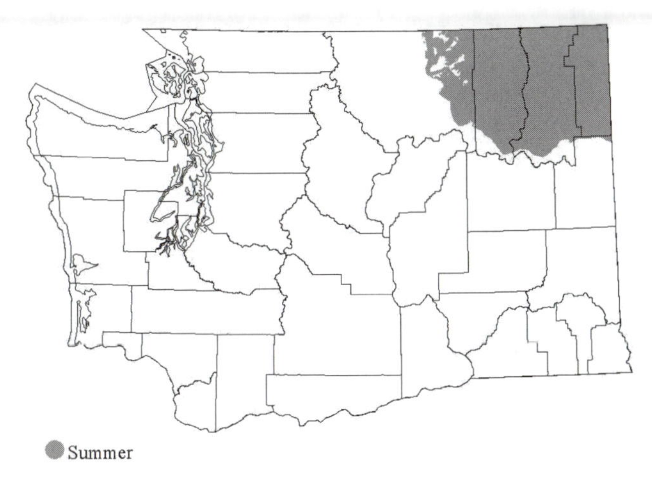

Occurrence: A change in this species' true status since it was listed as hypothetical by Jewett et al. (1953) is unlikely. Knowledge of its distribution may well be accounted for by the increase in fieldwork by dedicated amateurs beginning in the 1960s (see Smith et al. 1997) and expanded since by growing interest in birds through bird-watching and fieldwork like BBS and BBA surveys.

Birds breed widely in se. B.C. (Campbell et al. 1997), but in Oregon nest very locally in the sc. Cascades (Gilligan et al. 1994). In Washington, waterthrushes breed locally from e. Okanogan Co. to Pend Oreille and n. Spokane Cos. Records of singing birds elsewhere in e. Washington (e.g., Turnbull NWR) are attributed to migrants, but breeding is possible elsewhere, as at Wenas Cr. (Stepniewski 1999) and in the Blue Mts. (see Smith et al. 1997).

Increased field effort in appropriate habitats has resulted not only in fairly widespread records of migrants both e. and w. of the Cascades but also of apparently wintering birds. In w. Washington this species is a very rare fall migrant, with some birds lingering well into winter. Records are widely spread, mostly from areas with intensive coverage by field observers such as the e. Puget Trough from Whatcom s. to Thurston Cos.

Remarks: In Washington and B.C., records of actual nests are rare (Smith et al. 1997) relative to the number of reports of birds during the breeding season.

Noteworthy Records: *Outside breeding range*, **West**, *Fall*: 1 hypothetical in Jul 1976 at Cold Cr., San Juan I. (Lewis and Sharpe 1987); 1 on 28 Aug 1979 nw. of Slate Pk.; 1 on 26 Sep 1981 at Tennant L.; 1 on 19 Aug 1981 at Westport; 1 on 1 Oct 1983 at Ft. Canby SP; 1 in Oct 1986 at Diablo L. (Wahl 1995). 1 on 26 Aug 1986 at Bellingham; 1 on 8 Sep 1986 at Seattle; 1 on 17 Aug 1989 at Seattle; 1 on 9 Nov 1991 at Nisqually NWR; 1 on 18 Sept 1994 at Skagit WMA; 1 on 24-25 Nov 1996 at Skagit WMA; 1 on 12 Aug and 1 on 17 Oct-2 Nov 1998 at Skagit WMA; 1 on 30 Aug 1998 at Seattle. *Winter*: 1 on 17 Nov and 15 Dec 1968 at Seattle; 1 on 24 Jan 1974 at Skagit WMA; 1 on 8 Dec 1996-1 Jan 1997 at Skagit WMA; 1 on 25 Dec 1997 at Skagit WMA. CBC: 1 in 1986 at Skagit Bay. **East**, *Spring*: 1 on 25 May 1968 at Turnbull NWR; 1 on 17 May 1976 at Clarkston; 1 on 25 May 1986 at Potholes Res.; 1 on 25 May 1987 at Walla Walla. *Summer*: singing in early 1990s nr. Wenas CG (Stepniewski 1999); 1 singing on 10 Jun 2000 at Liberty; 1 singing on 10 Jun 1993 nr. Dayton (E. Hunn, in Smith et al. 1997). *Fall*: 1 probable on 22 Aug 1936 at Blue L., Grant Co. (Jewett et al. 1953); 30 Aug 2002 at Wenatchee.

Terence R. Wahl

Kentucky Warbler *Oporornis formosus*

Casual vagrant.

One record: singing male nr. Darrington 14 Jun 1992. The spring-summer of 1992 saw an unprecedented influx of se. warblers and vireos into the w. coast, especially California. This event was likely due, in part, to anomalous weather, but may well have also been influenced by range changes and population increases (Patten and Marantz 1996). Oregon has five records, all from May through Jul.

Steven G. Mlodinow

Mourning Warbler *Oporornis philadelphia*

Casual vagrant.

Washington's only record of the Mourning Warbler was of a bird at Lyon's Ferry SP on 26 May 2001. The species breeds as far w. as ne. B.C. (Dunn and Garrett 1997), but records w. of the Rocky Mts. are relatively scarce, and most of these are from fall, especially Sep. Spring-summer vagrants have occurred much less frequently in w. N. America though, coincident with the Washington record, a singing male was reported from Malheur NWR., Oregon, on the same day. W. coast spring records peak from 22 May to 7 Jun.

Steven G. Mlodinow

MacGillivray's Warbler *Oporornis tolmei*

Fairly common migrant and summer resident.

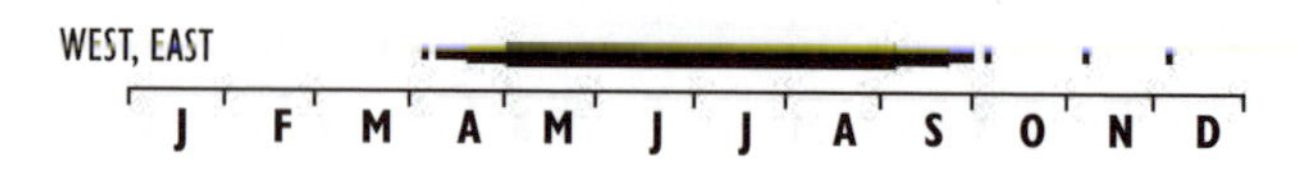

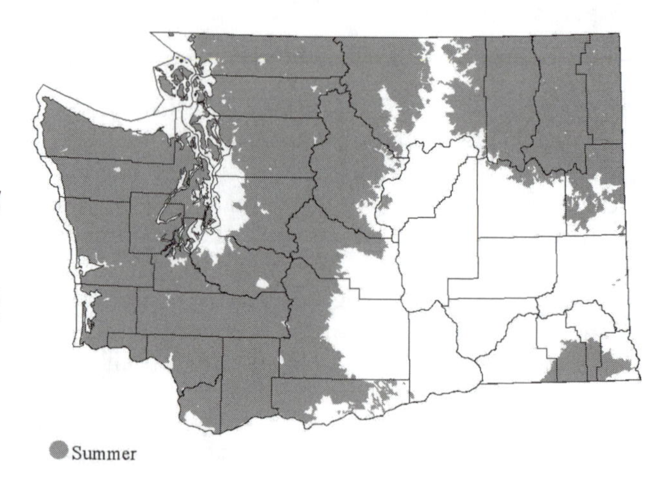

Summer

Subspecies: *O. t. tolmei* breeds; *monticola* is a potential migrant in e. Washington.

Habitat: For nesting, favors shrubby habitats within coniferous forests, including riparian zones, clear-cuts, avalanche chutes, and subalpine meadows (Smith et al. 1997). During migration, occurs in a wider variety of thickets and shrubs, typically in fairly moist areas with some trees nearby.

Occurrence: A brush-and-thicket warbler noted most often in its dense nesting habitat. On the westside, it is relatively uncommon away from breeding areas during migration, though it is uncommon to rare at any time on the outer coast. Spring passage is mostly from late Apr to early Jun with a peak during early to mid-May, tending to be slightly later on the eastside. Birds nest to 1200 m elevation (Jewett et al. 1953) and higher (Smith et al. 1997). Fall migration is largely from mid-Aug to mid-Sep but tends to be slightly earlier in fall on the eastside. Migrants can be found to relatively high elevations in the mountains in the fall (Jewett et al. 1953).

Shrubby, post-logging growth is attractive to this species (Smith et al. 1997). Clear-cut logging and fires increase habitat and abundance for a time, and then vegetative regrowth and succession result in local decreases and absence. For example, MacGillivray's Warblers were common breeders in the San Juan Is. during the late 1950s, but were virtually extirpated by the mid-1980s, a change that Lewis and Sharpe (1987) attributed to reforestation and cowbird parasitism. Though the overall status of MacGillivray's Warblers in Washington may appear not to have changed over time, BBS data indicate a statewide decline ($P < 0.05$) from 1966 to 2000 (Sauer et al. 2001).

Remarks: The two subspecies are not reliably separable in the field.

Noteworthy Records: *Spring, early*: 8 Apr 1997 at Olympia; 11 Apr 1983 at Bellevue. *Fall, late*: 8 Oct 1963 at Spokane; 12 Oct 1981 at Pullman (Johnson et al. 1984); 4 Nov-4 Dec 1994 at Everson (Wahl 1995).

Steven G. Mlodinow

Common Yellowthroat *Geothlypis trichas*

Common summer resident and migrant on westside. Fairly common to common resident and migrant e.

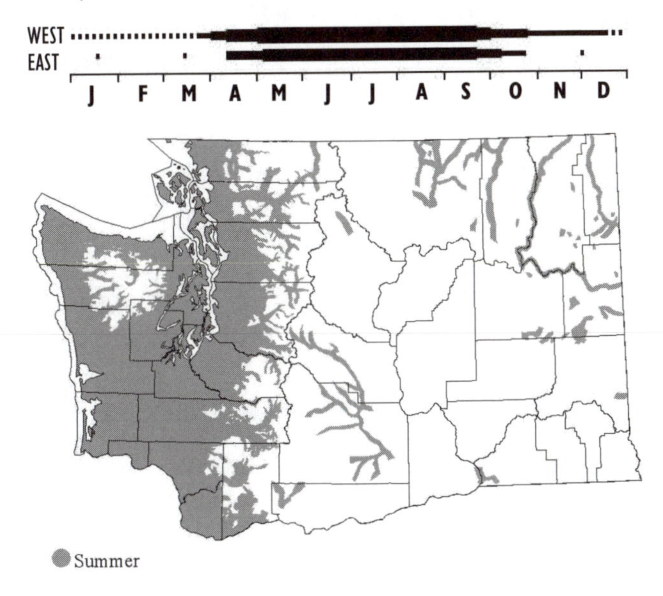

Subspecies: *G. t. arizela* breeds on the westside, *campicola* on the eastside (Jewett et al. 1953, Smith et al. 1997).

Habitat: Breeds in wet meadows, marshes, and even wetter parts of clear-cuts where emergent vegetation, especially cattails, is present. Often needs relatively little habitat, such as that provided by the marshy edges of a drainage ditch. During migration, occurs in a wider variety of habitats including drier brushy fields and areas of moist undergrowth well within woodlands, though still prefers moister and more open areas.

Occurrence: The only Washington warbler that favors marshes and wet meadows. Spring migrants begin to appear early, especially in the w., with the first few often noted in late Mar-early Apr well into May and, perhaps, early Jun. The yellowthroat is a widespread breeder on the westside, though rarely found at high elevations (Smith et al. 1997). It is relatively local in the e., especially during summer in se. Washington. Indeed, yellowthroats are somewhat uncommon there even as migrants. Formerly, this was considered a common nesting species in se. Washington (Jewett et al. 1953, Weber and Larrison 1977). For unknown reasons, the breeding range in e. Washington may be continuing to shrink, with ongoing declines noted in the c. Columbia Basin (Smith et al. 1997).

Southbound migrants begin to appear in mid-late Aug and are largely gone after early Oct. On the westside, a few often linger well into Nov or even early Dec—from 1960 to 2000 there were just 14 CBC records, on the westside, from Bellingham s. to Wahkiakum and w. to Grays Harbor. By late winter, this species is quite rare.

BBS data indicated a significant increase ($P < 0.01$) statewide from 1966 to 2000 Sauer et al. (2001).

Remarks: Races occurring in Washington are not agreed upon. Dunn and Garrett (1997) show *occidentalis* also nesting in e. Washington. N. races, *arizela*, *campicola*, and *yukonicola*, may also occur in migration, but distinction among these in the field is difficult, if not impossible (Dunn and Garrett 1997).

Noteworthy Records: West, *Summer, high count*: 41 on 27 Jun 1997 at Spencer I. *Fall, high count*: 120 on 25 Aug 1998 at Elma. *Winter*: 23 Feb 1983 at Nisqually NWR; 1 Jan-21 Feb 1984 at Seattle; 3 Jan 1987 at Kent; 25 Jan 1994 at Ebey Slough; 17 Jan 1998 at Seattle; 21 Jan 1999 at Skagit WMA; 19 Feb 2000 at Everett; 26 Feb 2000 at Seattle; 27 Jan 2001 nr. Snohomish; 10 Jan 2002 at Everett. **East**, *Spring, early*: 11 Mar 1995 at Bridgeport. *Winter*: 17 Jan 2000 at Lost I., Franklin Co.; 1 Dec 2001 at Wenatchee.

Steven G. Mlodinow

Hooded Warbler *Wilsonia citrina*

Casual vagrant.

The Hooded Warbler has been recorded three times in Washington: a male (ph.) at Discovery Park, Seattle, 31 Dec 1975-4 Apr 1976; a male at Kamiak Butte 15-21 Jun 1986; a male at Pullman 1-11 Dec 1989.

This se. warbler's breeding range is no closer to Washington than c. Iowa. There are, however, over 300 records from California (Dunn and Garrett 1997). There are at least 10 records from Oregon, three from s. B.C., and two records from Idaho scattered between late May and Jan.

Steven G. Mlodinow

Wilson's Warbler *Wilsonia pusilla*

Common summer resident/migrant and very rare winter visitor in w. Washington. Common migrant and fairly common summer resident in e. Washington.

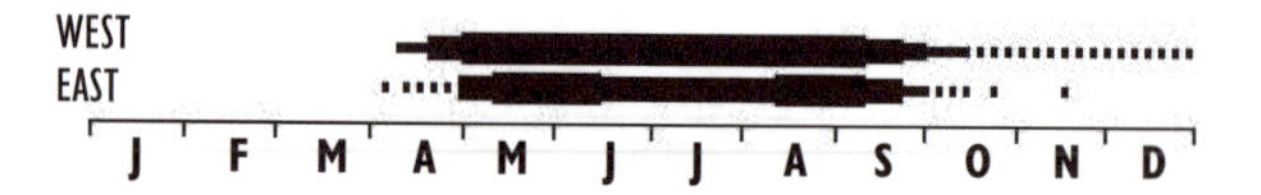

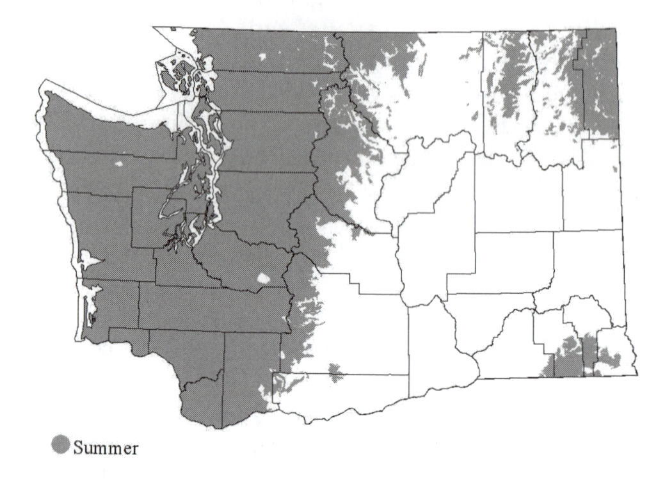

Subspecies: Three, probably not reliably separable in the field. *W. p. chryseola* breeds from Cascades w., *pileolata* to the e.; *pusilla* reported at least once during migration (Jewett et al. 1953, Dunn and Garrett 1997).

Habitat: Nests in dense undergrowth in moist microhabitats in w., in the moist woodlands of ne. Washington. During migration occurs in areas with deciduous trees and brush, including lowland riparian areas, subalpine areas with moist thickets, and towns, cities, and farmland where there are small clumps of trees with some undergrowth.

Occurrence: Wilson's Warbler is one of the most common migrant warblers throughout Washington and one of the most widespread breeders in forested areas on the westside. It is limited to middle- and high-elevation forests in the e. Cascades, and is very local in wet microhabitats with dense deciduous undergrowth in the Blue Mts. (Smith et al. 1997).

On the westside, spring migrants are evident into early Jun and southbound migrants typically appear in early to mid-Aug. On the eastside, apparent spring migrants are sometimes seen into mid-Jun. Records of singing migrants may lead to conclusions of wider breeding distribution than is real and require verification.

Westside winter records have increased dramatically over the last decade for reasons that are unclear, but possibly due to the increase in number of field observers or to milder winters in recent years.

Though the species is one of the first to repopulate logged-over areas, BBS data indicate a statewide decline of 0.7% per year from 1966 to 2000, similar to the downward trend noted elsewhere in the Pacific coast region (Ammon and Gilbert 1999, Sauer et al. 2001).

Noteworthy Records: West, *Spring, early*: 2 Apr 1997 at Tenino. *Winter*: 3 Dec 1994 at Samish Flats; 4 Dec 1992 at Seattle; 7 Dec 1997 at Seattle; 2 on 11 Dec 1976 at Bellevue; 17 Dec 1994 at Sequim; 21 Dec 1992 nr. Bayview, Skagit Co.; 25-26 Dec 1992 nr. Stanwood; 10 Feb 1982 at Juanita, King Co.; 1 Jan 2000 at Seattle; 16 Feb 2000 at Renton; 20 Feb 2000 at Blanchard, Skagit Co.; 26 Feb 2000 at Skagit WMA; 15 Dec 2001 at Lakewood, Pierce Co.; 19 Dec 2001 at Seattle; 2 Jan-3 Feb 2002 nr. Montesano; 14 Jan through Feb 2002 nr. Bayview. High count: 43 on 22 May 1999 at Woodland Park, Seattle. **East**, *Spring, early*: 6 Apr 1998 on Saddle Mt. Rd. *Fall, late*: 13 Oct 1981 at Pullman (Johnson et al. 1984); 15 Oct 1980 at Pullman (Johnson et al. 1984); 23 Oct 1994 nr. Touchet; 11 Nov 1999 at Richland; 16 Nov 1996 at Walla Walla; 17 Nov 1974 at Richland. *High counts*: 230 on 3 Sep 1989 in Adams Co.; 310 on 8 Sep 2001 at Washtucna.

Steven G. Mlodinow

Yellow-breasted Chat *Icteria virens*

Fairly common spring migrant and breeder and uncommon fall migrant in e. Washington. Very rare migrant and summer visitor on westside.

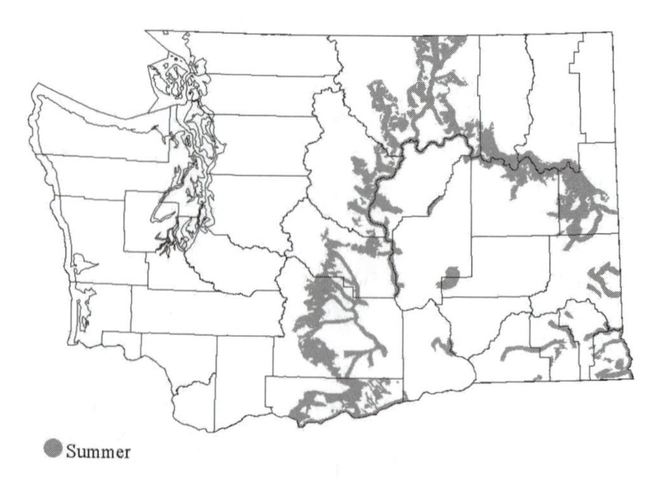

Subspecies: *I. v. auricollis.*

Habitat: Nests predominantly in shrubby riparian areas where shrub diversity is high, favoring red-osier dogwood, wild rose, chokecherry, bitter cherry, and willows (Stepniewski 1999). Migrants typically also noted in riparian habitats with dense cover.

Occurrence: A typical species of brushy riparian areas over much of e. Washington, and in shrubby areas throughout the Palouse region (Smith et al. 1997), the chat is conspicuous by its song, often even during the night.

Records away from breeding areas at the ecotone between eastside forest and steppe zones on the eastside are very scarce. No obvious changes in status on eastside, but on the westside this species was formerly a regular breeder locally. It was reportedly common in Clark Co. 50 years ago (Smith et al. 1997) and was an uncommon migrant and summer resident around L. Washington in the 1930s (Miller and Curtis 1940).

From 1970 through 2002 there have been 28 records from w. Washington. There were none in the 1960s, four in the 1970s, eight in the 1980s, 10 in the 1990s, and six between 2000 and 2002, likely due to increased observer effort. By county since 1970: Clark (6 records), Skamania (2), Cowlitz (1), Thurston (6), Pierce (4), King (4), Skagit (3), Whatcom (2). Most of these records are of singing males, many during summer, and breeding may well have occurred sporadically on the westside during the last 30 years. Particularly suggestive was a pair that spent the summer at Steigerwald L. NWR in 2002.

Though BBS data from 1966 to 2000 indicate a stable population in Washington (Sauer et al. 2001), chats have decreased rather dramatically over much of their range, presumably due to human developments in riparian habitats and, perhaps, cowbirds (Dunn and Garrett 1997).

Noteworthy Records: East, *Fall, late*: 12 Sep 1890 at Touchet (Jewett et al. 1953). **West**, *Spring, early*: 7 May 1935 at Vancouver (Jewett et al. 1935). *Summer*: 1 on 13 Jun and 20 Jun 1976 at Bacon Cr., Skagit Co. (BBS, TRW) and 1 on 16-17 Jun 2000 n. Rockport were the only summer records n. of King Co. Fall: ad. and imm. on 14 Aug 1978 at Olympia; 2 on 13 Sep 1980 at Pt. Roberts; early Sep 1984 at Deming Swamp, Whatcom Co. (Wahl 1995); 24 Sep 1953 at Bellevue.

Steven G. Mlodinow

Summer Tanager *Piranga rubra*

Casual vagrant.

There are two Washington records of this s. species. The first was of a bird ph. while spending much of winter 1997-98 at a feeder in Skagit Co. and the second is from Ridgefield on 26 May 2001. The first record is remarkable for being both on the westside and in winter. Oregon has 15-20 records, mostly from mid-May to mid-Jun in the se. part of the state. The six fall Oregon records were between 20 Sep and 25 Nov, and there is one winter record.

There are two races: *rubra* and *cooperi*. Washington and Oregon records have not been identified to subspecies. In California, most vagrant Summer Tanagers identified to subspecies have been of the e. *rubra*, even though *cooperi* breeds into e. California.

Steven G. Mlodinow

Western Tanager *Piranga ludoviciana*

Common migrant and summer resident. Very rare in winter.

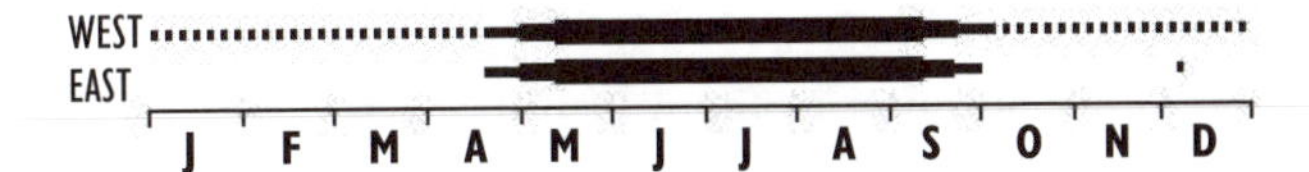

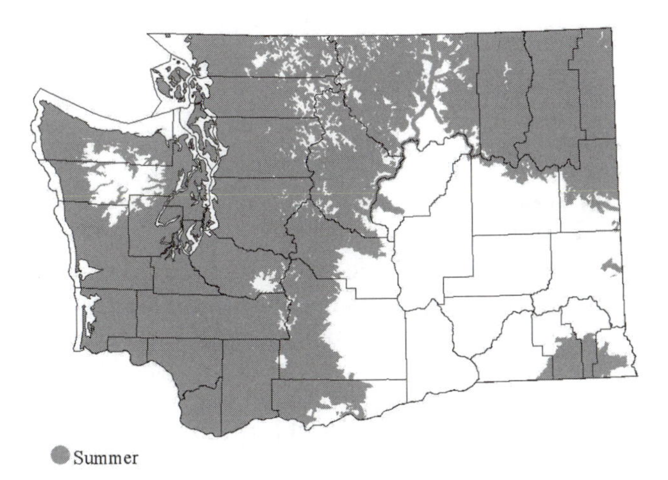

Habitat: Breeds in coniferous and mixed forests, especially favoring open coniferous woodlands or openings in coniferous forests (Smith et al. 1997). A wide variety of wooded habitats including suburban areas and broadleaf riparian woodlands in migration.

Occurrence: The most northerly of an essentially tropical- and subtropical-breeding family, the Western Tanager is one of Washington's common and more showy passerine migrants and summer residents.

Peak spring migration occurs during mid-late May, with a few migrants found into mid-Jun. During breeding season, occupies woodlands up to 900 m in the Cascades and the Olympics and to 1800 m in the Blue Mts. (Jewett et al. 1953, Sharpe 1993). Fall migration peaks from mid-Aug to early Sep, though apparently southbound birds have been found as early as late Jul. On rare occasions a straggler into winter, and at least once, successfully overwintered.

BBS data show an increase in Washington from 1966 to 2000 (P <0.01: Sauer et al. 2001), a decrease in Oregon, and little change rangewide from 1966 to 1996 (Hudon 1999).

Noteworthy Records: *Late Fall-early Spring*: 10 Dec 1999 at Mill Creek Rd., Walla Walla Co.; 25 Dec 1998-1 Jan 2000 at Federal Way; present at least through early Mar 1993 at Mercer I.; 9 Dec 1999 nr. Sequim; 4 Jan 2000 at Seattle; 2 Feb 2001 at Tacoma; 15-17 Feb 2001 at Tacoma; 30 Jan 2001 at Blaine; 20 Dec 2002-Feb 2003 at Seattle. *High counts*: 100 on 6 Sep at Mt. Pleasant, Skamania Co.; 100 on 7 May 2000 at Point No Point.

Steven G. Mlodinow

Green-tailed Towhee *Pipilo chlorurus*

Rare, local summer resident in Blue Mts. Casual winter visitor in w.

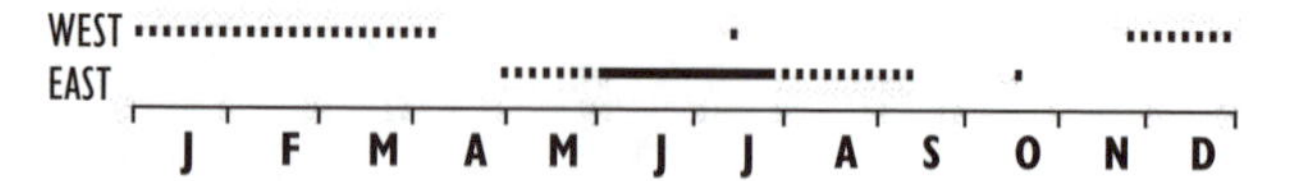

Habitat: Dry shrubby hillsides at moderate elevation.

Occurrence: Extremely local, found as a breeder almost exclusively in the Blue Mts. (Smith et al. 1997), with small colonies at Wenatchee Guard Station, at Lewis Pk., and nr. Dixie along Biscuit Ridge. Other reports from the Blue Mts. have come from Godman Springs, nr. Dayton, the Umatilla Ranger Station, and Coppei Cr. In spite of the small size and number of colonies, this species has maintained its foothold in Washington since at least 1885 (Johnson 1885), though there were only four records prior to 1953 (Jewett et al. 1953).

Because of the remote location of these nesting areas, arrival and departure dates are not well known. Most reports have been from early Jun to early Aug, but given dates in Oregon, arrivals may be during May and departure in Aug or Sep. There is only one eastside record away from the Blue Mts., at Hatton, Adams Co., on 16 May 2001. A very rare visitor w. of the Cascades, with just three certain records of overwintering birds.

Noteworthy Records: *Late*: 15 Oct 1961 nr. Dayton. **West**: 28 Nov 1965-24 Jan 1966 at Seattle; 7 Dec 1997-5 Apr 1998 at Skagit WMA; 16 Dec 2000-1 Jan 2001 at Ledgewood Beach, Island Co.; 17 Jul 1985 nr. Yellow Aster Butte, Whatcom Co. (Wahl 1995).

Steven G. Mlodinow

Spotted Towhee *Pipilo maculatus*

Common resident in w. Washington; common summer resident, migrant and fairly common winter resident in e.

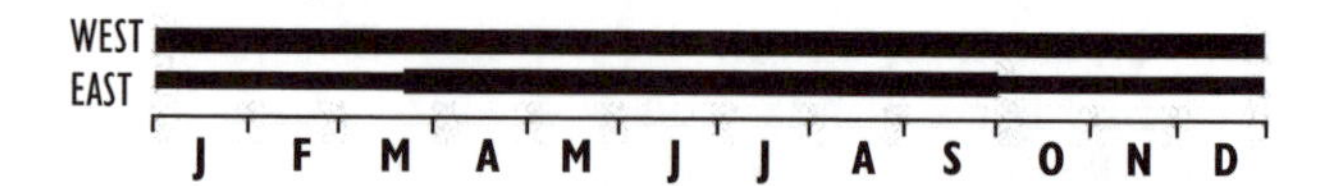

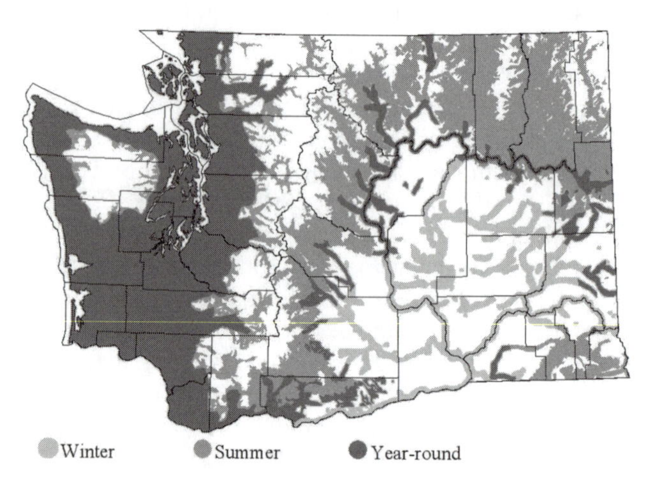

Subspecies: *P. m oregonus* in w., *curtatus* in e. Washington (Smith et al. 1997).

Habitat: Breeds in edges of lowland wooded areas with shrubby cover, including clear-cuts, suburbs, riparian zones, and wetland edges (Smith et al. 1997). In migration and winter, a wider variety of scrubby habitats including hedgerows and blackberry thickets, preferring some patchy woods nearby.

Occurrence: A widespread resident from low to mid elevations where forest practices have created brushy habitats, the Spotted Towhee, like some other species, is relatively unknown in some aspects of status in Washington. Birds were formerly rare along outer coast but have benefited from clear-cutting there and probably elsewhere on the westside along with other alterations of lowland habitats by humans (Smith et al. 1997).

The proportions and extent of migratory and resident populations are uncertain. Westside migration is hard to detect, though birds appear locally more numerous from Sep into May than during summer and some downslope movements are probable (e.g., Wahl 1995). On the eastside, spring passage is noted from mid-Mar to late May or early Jun. Southbound migrants, or at least post-breeding wanderers, are sometimes found as early as late Jun or early Jul, but the bulk of fall migration occurs from late Aug to late Sep, with a peak during early to mid-Sep. Wintering birds on the eastside are relatively uncommon in the n. and more common to the s.

BBS data indicate a significant increase of 3.5 % per year (P <0.01) in Washington from 1966 to 2000 (Sauer et al. 2001) and a 1.9% per year increase in B.C.) from 1968 to 1995 (Greenlaw 1996). CBC data also show a statewide increase (P <0.01) from 1960 to 1999.

Remarks: Occurrence and ranges of subspecies incompletely known. There are reliable reports of *curtatus*-type from the Puget Trough and one specimen from Leadbetter Pt., all mid-Sep to late Apr (SGM, D. Beaudette p.c., UWBM). This race may be a somewhat regular vagrant on the westside. The border zone between *oregonus* and *curtatus* in Klickitat Co. is not clearly defined. Jewett et al. (1953) reported a se. Washington specimen of *montanus* which is regular no closer than se. Idaho and often inseparable in hand (Greenlaw 1996). The more heavily spotted eastside forms (*curtatus*, *montanus*, and *arcticus*) are likely not reliably separable in the field, but the less spotted *oregonus* should usually be identifiable. Greenlaw (1996) shows *arcticus* breeding in ne. Washington. Formerly the Rufous-sided Towhee.

Noteworthy Records: *High counts*, **East**: 65 on 8 Sep 1996 at Cold Cr., Yakima Training Center (Stepniewski 1999); 45 on 14 Sep 1996 at Maryhill Museum (SGM, R. Rogers p.c.). **West**: 100 on 23 Nov 1996 at Spencer I.; 77 on 31 Dec 1995 at Smith I., Snohomish Co. (SGM).

Steven G. Mlodinow

American Tree Sparrow *Spizella arborea*

Irregular, uncommon to fairly common winter resident in e. Washington. Irregular, rare winter resident in w.

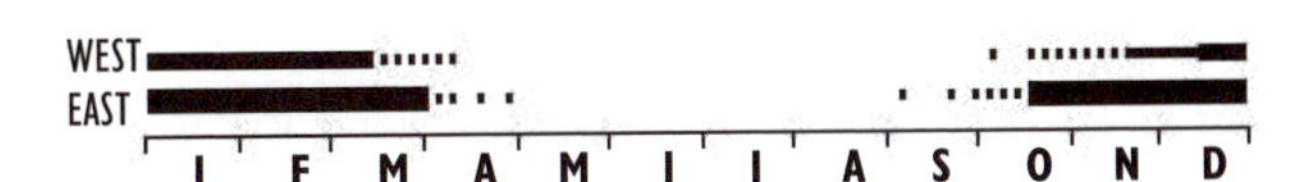

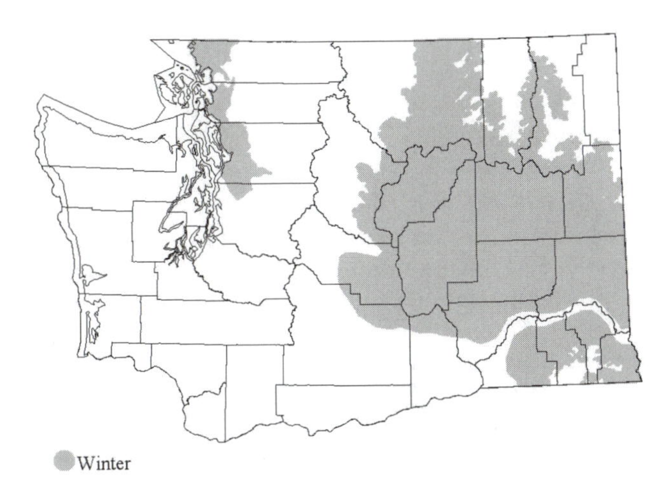

Subspecies: Presumably *S. a. ochracea* (see Rising 1996).

Habitat: Weedy, overgrown, and abandoned fields, gardens, often somewhat wet, typically with brush or a few scattered saplings nearby. Also in shrub-steppe, agricultural fields, and marsh edges.

Occurrence: Widespread, though irregular, winter resident through much of e. Washington, the American Tree Sparrow is one of the few sparrows not discouraged by snow cover. It is somewhat less numerous in relatively mild se. Washington, where it is currently uncommon and was considered very uncommon to rare by Weber and Larrison (1977). In w. Washington, the species is an uncommon winter resident, with most records from the Puget Trough. This species is rare on the outer coast and in southernmost w. Washington: the first record in Grays Harbor Co., an area intensively covered by field observers for years, was in October 2002. Peak years of occurrence on the eastside were winters of 1969-70, 1980-81, 1981-82, and 1990-91. Peak years on the westside were 1969-70, 1972-73, 1974-75, and 1998-99.

No obvious population changes over time have been noted. CBC data are limited but appear to indicate slightly higher numbers at eastside locations and a decline on the westside.

Noteworthy Records: East, *high counts,*: 100 on 29 Dec 1969 at Smyrna, Grant Co. West: 8 on 21-23 Jan 1999 at Skagit WMA. *Fall, early*: 7-14 Sep 1897 at Conconully (Jewett et al. 1953); mid-Sep 1897 at Wenatchee (Jewett et al. 1897); 18 Sep at Kamiak Butte (Weber and Larrison 1977); 26 Sep 1965 at Reardan; 2 on 27 Sep 1965 at Spokane. *Spring, late*: 17 Apr 1999 at Richland; 22 Apr (year unstated) in se. Washington (Hudson and Yocom 1954). **West**, *Fall, early*: 2 on 4 Oct 1971 nr. Bellingham. *Spring, late*: 5 Apr 1998 at Skagit WMA.

Steven G. Mlodinow

Chipping Sparrow *Spizella passerina*

Common summer resident and migrant in e. Washington. Locally fairly common to common breeder and uncommon to rare migrant w. Very rare in winter throughout.

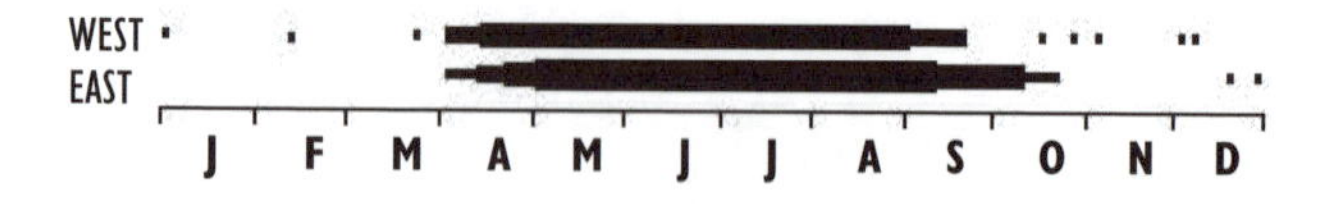

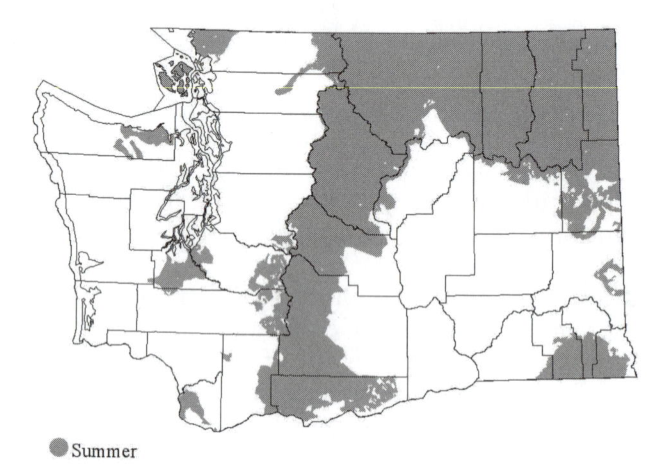

Subspecies: *S. p. arizonae.*

Habitat: Breeds in dry open forests, remnant prairie habitats, open farmlands, montane clear-cuts, and high alpine forests. During migration found in grassy or weedy areas, typically nr. trees or tall brush.

Occurrence: Chipping Sparrows are common to abundant in forests of e. Washington, common in the San Juan Is. and westside prairies nr. Sequim and Dungeness, and in Pierce, Thurston, and Clark Cos. They are also common in clear-cuts and high forests of the Olympics, and local and uncommon at high elevations in e. Whatcom, Skagit, and King Cos., on the sw. side of the Cascades crest, and at Mt. Rainier (Smith et al. 1997) and Mt. St. Helens (C. Chappell).

On the eastside, spring passage is mainly from late Apr to late May, and fall migration is mostly from mid-Aug to early Oct. In the w., migrants are scarce away from breeding areas, with most records coming from the Puget Trough. Westside spring passage is mostly from early Apr to mid-May, while southbound migrants are largely found from mid-Aug to mid-Sep. There is a scattering of westside midsummer records away from breeding areas, presumably failed breeders or unpaired individuals. Records from the outer coast are quite scarce. There were a number of published winter reports statewide, but many reports, particularly prior to the 1990s, have been questioned (e.g., Stepniewski 1999).

Numbers on the westside have apparently declined substantially since the early 1900s. Rathbun (1911) and Miller and Curtis (1940) both considered this species common around Seattle, but by the 1970s, this species was quite rare in King Co. ("endangered" [Hunn 1982]), somewhat decreased in the San Juans (Lewis and Sharpe 1987), and essentially gone from lowland Whatcom Co. (Wahl 1995). Loss of habitat and increasing cowbird numbers likely have both played a role. No significant changes have been reported on the eastside, though BBS data indicate a statewide decrease of 2.9% per year ($P < 0.05$) from 1966 to 2000 (Sauer et al. 2001).

Remarks: From one to 25 birds were reported 13 times on CBCs between 1967 and 1984, with six of these from the Yakima Valley (see Stepniewski 1999). Other reports, often questioned due to insufficient further documentation, include three in w. Washington. The last CBC report was of one at Spokane in 1998. C. and e. races might occur as vagrants, but are inseparable in field (Rising 1996).

Noteworthy Records: West, *high count*: 58 on 19 Apr 1966 at McChord AFB. *Spring, early*: 29 Mar 1986 at Ft. Lewis. *Late Fall-Winter*: 24-28 Nov 1998 nr. Bayview; 29 Nov 1998 at Sequim; 2 Dec 2001 at Lyle; 8 Dec 1998 at Port Angeles; 12 Dec 1990 at Washtucna; 21 Dec 1976 at Toppenish NWR; 27 Dec 1998 at Spokane; 10 Jan 1990 at Skagit Flats; Jan 1964 at Seattle; 15 Feb 1999 nr. Monroe. **East**, *Spring, early*: 7 Apr 1966 at Spokane.

Steven G. Mlodinow

Clay-colored Sparrow *Spizella pallida*

Rare, local summer resident and very rare migrant in e. Washington. Very rare in w. Washington.

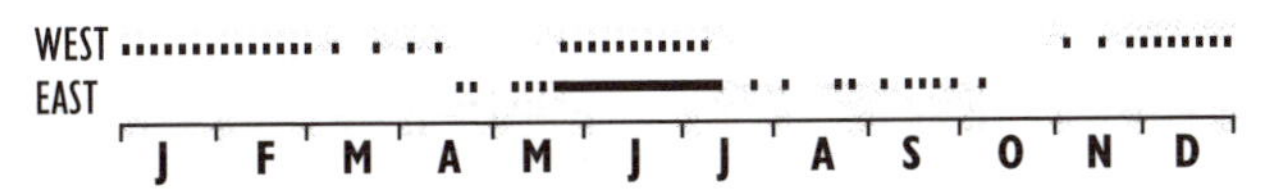

Habitat: Breeds in open grassland nr. Spokane, migrants and wintering birds found in weedy fields and blackberry thickets, usually in mixed flocks of sparrows where juncos and *Zonotrichia* are common. Mostly above sagebrush zone in moist swales in brushy patches with snowberry and rose in Okanogan Co. (Stepniewski 1999), remnant prairie habitat in Thurston Co.

Occurrence: The first record was of a bird collected in the Spokane Valley on 6 Jun 1960 (LaFave 1960b). The second was there on 7 May 1964. Since then, the species has been found somewhat irregularly there, having been recorded in at least 14 of the last 35 years, and annually since 1994. Breeding was likely there a number of times, but there are only a few instances of firm evidence. Most records were between mid-May and mid-Jul, and the peak number of singing males was four during Jun 1999.

Away from the Spokane area, there were only about 33 eastside records through 2002, almost half of which were from Okanogan Co. during Jun-Jul. Many of the Okanogan records involved singing males apparently on territory and the species may have bred there. A pair feeding young were found on the Kalispell Indian Reservation, Pend Oreille Co., during summer 2002. County records away from Spokane and Okanogan Cos.: Pend Oreille (3), Lincoln (2), Walla Walla (2), Douglas (1), Adams (2), Franklin (2), Chelan (1), Kittitas (2), Yakima (3), and Klickitat (1). Most of these were from Jun to Jul. Records of birds apparently migrating on the eastside are very few, from 21 Apr to 19 May and 10 Aug to 1 Oct.

There are only 16 westside records, mostly from late fall to early spring and these likely represent birds attempting to winter. There are also three records of singing males apparently on territory at Thurston Co. prairies during late spring and summer. There are no westside records of birds clearly on passage migration, though a number of such records exist from Vancouver, B.C. Numbers appear to be increasing in Washington, both as summer visitors and as vagrants, likely reflecting regional expansion (see Cannings et al. 1987).

Noteworthy Records: East, *Spring migrants*: 21 Apr 1979 at St. Andrews; 1 May 1998 at Madame Dorian SP; 10 May 1992 at Tonasket; 12-19 May 1998 at Davenport; 19 Apr 2002 at Ellensburg; 12 May 2001 at Tampico, Yakima Co. *Fall migrants*: 10 Aug 1974 nr. Ellensburg; 28 Aug 1994 at Washtucna; 22-24 Aug 2001 at Hooper, Adams Co.; 9 Sep 1998 at Windust, Franklin Co.; 11 Sep 1999 at Six Prong, Klickitat Co.; 16 Sep 1999 at Spokane; 19 Sep 1999 at Palouse Falls SP; 23-28 Sep 1994 at Spokane; 1 Oct 1999 at Iowa Beef. **West**: 29 May-10 Jun 1998 at Lacey; 30 Mar-20 Apr 1975 at Skagit flats; 13 Jun-8 Jul 1998 at Weir Prairie; 14 Jun-2 Jul 1995 at Olympia; 7-10 Nov 1999 at Seattle; 8 Nov 1976 at Skagit WMA; 11 Nov 2000 at Ebey I. (SM); 14 Nov 1999 nr. Bayview; 3 Dec 1988-28 Feb 1989 at Samish flats; 6-17 Dec 1998 at Leque I., Snohomish Co.; 2 Jan-3 Mar 1998 nr. Duvall; 3-9 Jan 1999 at Elma; 25 Jan-4 Feb 1992 at Kent; 14-30 Dec 2000 at Frenchman's Bar, Clark Co.; 24 Nov 2001-22 Jan 2002 at Woodland.

Steven G. Mlodinow

Brewer's Sparrow *Spizella breweri*

Common migrant and summer resident in e. Washington. Very rare in spring, casual in fall in w.

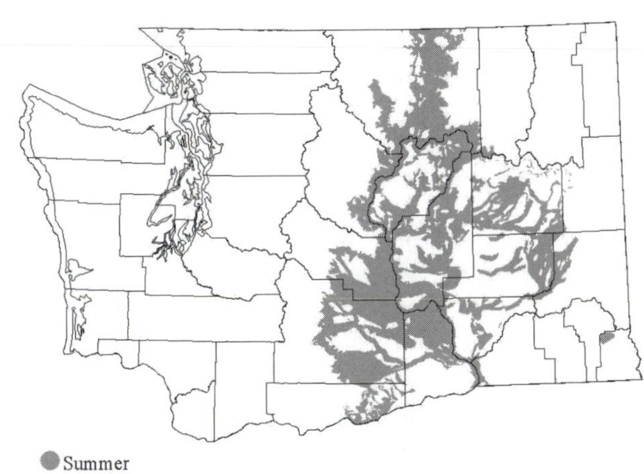

Subspecies: *S. b. breweri* breeds, *taverneri* at least casual in migration.

Habitat: Breeds in shrub-steppe habitats in higher and moister areas of Columbia Basin and Okanogan Valley, especially where big sagebrush or three-tip sage and cover of bluebunch wheatgrass predominate (Smith et al. 1997, Stepniewski 1999, LaFramboise and LaFramboise 1999). Post-breeding, disperses into a wider variety of shrub-steppe habitats, especially moister areas.

Occurrence: Remarkable for its prolonged, buzzing song from the sagebrush, Brewer's Sparrow is found mostly in appropriate habitat in Columbia Basin and

Okanogan Valley and is somewhat less common in Methow Valley and southernmost Columbia Basin. Formerly common breeder in grasslands e. to the Palouse in shrubby patches (Wing 1949), but conversion to agriculture has eliminated most appropriate habitat and Brewer's Sparrows are now local (Weber and Larrison 1977, Smith et al. 1997). Small scattered breeding populations still exist in these grasslands, including areas in Asotin Co., the Spokane valley, and Little Pend Oreille NWR. (Smith et al 1997).

Though it overlaps in breeding distribution and broadly in habitat with the Sage Sparrow, noted in Yakima Co. to generally occur at higher elevations (>900 m) in more grassy areas than that species (Stepniewski 1999). Timing of spring and fall passage is unclear. There are eight w. Washington records of Brewer's Sparrows: seven between 2 Apr and 5 Jun and one on 22 Sep. These were not identified to race.

The Timberline Sparrow, *taverneri,* breeds n. and e. of Washington (Semenchuk 1992, Rotenberry et al. 1999) and may well be a regular migrant here but, due to the difficulty of field identification, there are few certain records. Winter range is very poorly known but appears to stretch from s. California to New Mexico (Rising 1996).

BBS data from 1966 to 1996 show declines throughout the N. American survey area, especially from 1980 to 1996, and Washington data suggest a possible decline from 1966 to 2000 (Sauer et al. 2001). Surveys of wintering populations seem to be stable (Rotenberry et al. 1999).

Remarks: *S. b. taverneri* originally considered a separate species (Swarth and Brooks 1925), was later lumped with Brewer's Sparrow (AOU 1957), and then determined to be somewhat genetically divergent from *breweri* (see Rising 1996). The two are very similar—sight identifications may be questionable (see Pyle and Howell 1996, Rotenberry et al. 1999)—and even vocalizations may not be as distinct as commonly perceived (P. Sherrington). Specimens of *taverni* include two from Kittitas Co., 10 April 1932 and one nr. Moxee, 17 Apr 1932 (Munro 1935). Sight records include a singing bird at the Yakima Training Center, 17 May 1997 (Stepniewski 1999) and three there 9 May 1998. Classed GL, NAS Watch List.

Noteworthy Records: East, *Spring, early*: 22 Feb 1980 nr. Mondovi, Lincoln Co.; 19 Mar 1999 at Quincy. *Fall, late*: 7 Sep at Walla Walla R. delta; 7 Sep 1996 at Yakima Training Center (Stepniewski 1999); 14 Sep nr. Potholes Res. (Johnson and Murray 1976). **West**: 2 Apr 1995 at Seattle; 22 Apr 1995 at Ocean Park; 3 May 1995 at Seattle; 18 May 1997 in Pierce Co.; 23 May 1993 at Bellingham; 26-27 May 1973 at Samish I.; 4-5 Jun 1997 at Tatoosh I.; 22 Sep 1998 at Seattle. CBC: Reported at Port Townsend in 1978 and during CW at Walla Walla in 1981.

Steven G. Mlodinow

Vesper Sparrow *Pooecetes gramineus*

Common summer resident and migrant in e. Washington. Locally uncommon to rare summer resident, rare migrant, casual winter visitor in w.

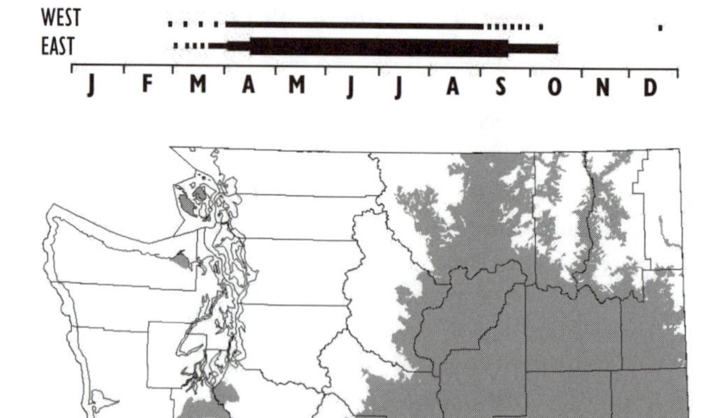

Subspecies: *P. g. affinis* in w. and *confinis* in e. Washington.

Habitat: Breeds in moister steppe habitats with lush cover of native grasses. also in other grasslands and agricultural areas (Smith et al. 1997). Found in a greater variety of open habitats during migration. Westside breeders found in remnant prairies. Migrants noted infrequently, in a variety of grassy habitats.

Occurrence: Widespread and common on the eastside. Jewett et al. (1953) considered this species to be relatively scarce in Yakima, Franklin, and Walla Walla Cos., but Stepniewski (1999) suggested that earlier observers probably were not familiar with the higher, lusher habitats favored by this species. Birds are more common in the Columbia Basin at higher elevations where native grasses are the ground cover (Stepniewski 1999), and the species replaces Brewer's and Sage sparrows where sagebrush is removed (Smith et al. 1997).

The westside population is in danger of extirpation. Birds were formerly much more widespread, with records listed by Jewett et al. (1953) along the Puget Trough n. to Skagit Co., and from Clallam and San Juan Cos. The breeding population now is limited almost entirely to remnant prairies in Thurston and Pierce Cos. and grasslands on San Juan I., though small numbers may still breed in e. Clallam Co. and nr. Shelton (Smith et al. 1997). This decline is likely due to habitat destruction from development, invasion of scotch broom, and encroachment by forests due to fire control (Smith et al. 1997).

Westside migrants are rare and mostly in the Puget Trough though a few have been found on the outer coast. Spring passage is mostly from early Apr to early May, and fall passage is mostly from early to late Sep. Spring records seem to outnumber those from fall. Summer records away from known breeding areas have become increasingly scarce, with only a couple since the 1970s. There are three westside winter records.

Overall, the population in Washington appears to be increasing, with BBS data showing a population increase of 4 % per year (P <0.01) from 1966 to 2000 (Sauer et al. 2001).

Remarks: The two races are not separable in the field and not always in hand (Rising 1996). *P.g. affinis* classed SC, FSC, PHS.

Noteworthy Records: East, *high counts, Fall*: 158 on 18 Aug 1996 at Greeley Pond, Yakima Training Center (Stepniewski 1999); 110 on 20 Aug 1998 at Horse Springs Coulee, Okanogan Co. *Spring*: 42 on 12 May 1999 at Horse Springs Coulee. *Spring, early*: 10 Mar 1921 at Yakima (Jewett et al. 1953). *Fall, late*: 10 Nov at the Potholes (Harris and Yocom 1952). **West**, *high count*: 20 on 18 Jul 1998 at Fort Lewis. *Spring, early*: 7 Mar 1983 at Fort Lewis. *Winter*: 26 Dec 1982 at Kent; 26 Feb-25 Mar 1998 at Leque I., Snohomish Co.; 28 Dec 1998 at Auburn.

Steven G. Mlodinow

Lark Sparrow *Chondestes grammacus*

Fairly common to locally common summer resident and migrant in e. Washington. Very rare vagrant in w. Casual in winter.

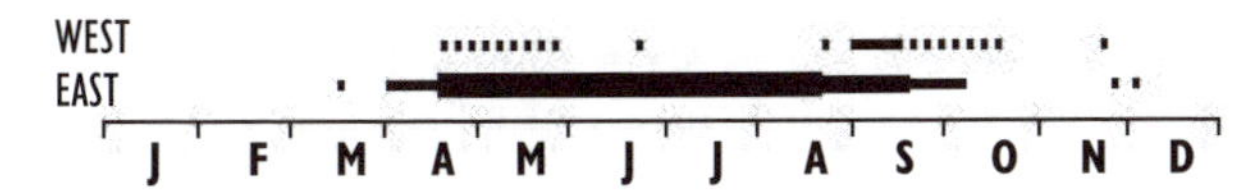

Subspecies: *C. g. strigatus.*

Habitat: Breeds mainly in lower-elevation open shrub-steppe habitats especially those with cheatgrass, also in agricultural areas and, sparsely, in dry pine woodland openings up to 1300 m (Smith et al. 1997, Stepniewski 1999). During post-

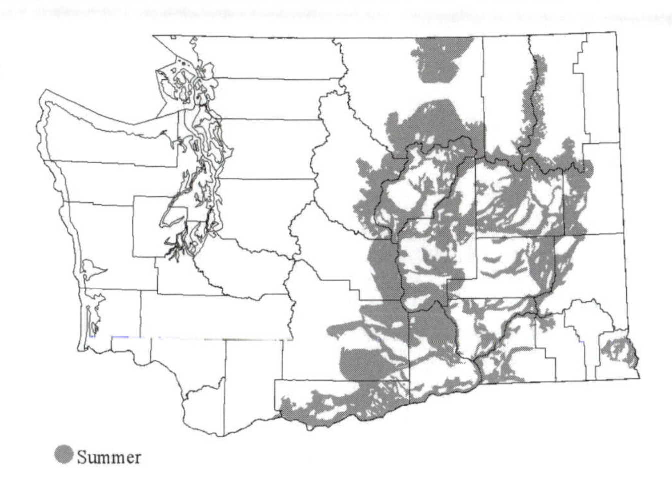

breeding dispersal and migration utilizes a wider range of weedy and grassy habitats.

Occurrence: An attractive, locally common sparrow of dry and open habitats throughout e. Washington. Spring migration occurs mostly from late Apr to mid-May. Post-breeding dispersal starts in mid-late Jul and blends into fall migration which lasts through Aug. In w. Washington, there are 20 records of vagrants, with 12 during fall, seven during spring, and one in summer. Spring dates span 26 Apr-28 May. Fall dates, mostly between 21 Aug and 4 Oct, peak in early Sep. Most westside records are from the outer coast, in contrast to most other "eastside vagrants." Strays to the westside were more numerous, without obvious reason, between 1978 and 1984, when 11 of the 19 records occurred. Westside records by county include Pacific (7), Grays Harbor (3), Clallam (2), San Juan (2), King (2), Skagit (3), and Kitsap (1).

Population is stable through most of w. N. America (Price et al. 1995), though BBS data in Washington suggested an increase from 1966 to 2000 Sauer et al. (2001).

Noteworthy Records: East, *Spring, early*: 27 Mar 1926 at Spokane (Jewett et al. 1953); 27 Mar 1926 at Moses L. (Jewett et al. 1953). *Fall, late*: 22 Nov 1977 at Walla Walla; 9 Dec 1980 at Walla Walla. **West**, *Spring*: 19 May 2002 at Concrete. *Summer*: 23 Jun 1965 at Seattle. *Fall, late*: 29 Nov 1980 at Fort Canby SP.

Steven G. Mlodinow

Black-throated Sparrow *Amphispiza bilineata*

Rare and local summer resident in e. Washington. Vagrant in w.

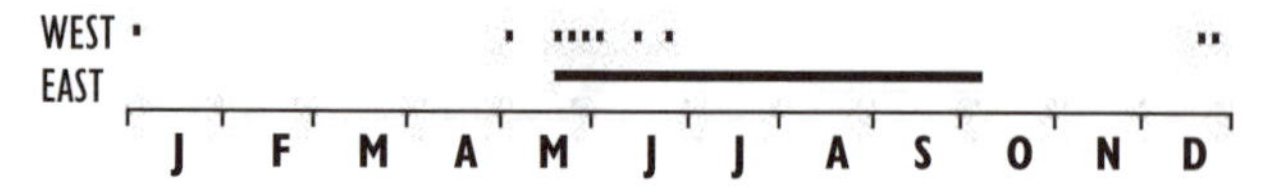

Subspecies: *A. b. deserticola.*

Habitat: Grasslands and shrublands with sparse vegetation in c. arid steppe zone (Smith et al. 1997). Often in sparse shrub-steppe dominated by small big sagebrush with a minor cover of spiny hopsage and stony ground underneath (Stepniewski 1999).

Occurrence: This species was first reported in Washington by Dawson (1908) at Brook L., Douglas Co., in May 1908, but not listed by Jewett et al. (1953). The first confirmed record was on 16 May 1976 of one (ph.) at Pt. Grenville. A Black-throated Sparrow was found the following spring nr. Walla Walla, on 10 May 1977.

Black-throated Sparrows subsequently have proved to be very local and somewhat irregular breeders in the c. Columbia Basin and irruptive late spring/early summer vagrants elsewhere in the state. The first breeding record was during summer 1987 when several territorial males and ads. carrying food were found s. of Vantage. Most subsequent nesting evidence comes from along the Columbia R. between Priest Rapids Dam and Vantage in Yakima, Kittitas, and Grant Cos. (Smith et al. 1997). Other probable nesting areas include the Rattlesnake Hills and the Yakima Training Center, nr. White Bluffs in Grant Co., nr. the Hanford Site (Smith et al. 1997, Stepniewski 1999) and Steptoe Butte in Whitman Co. Years with likely or confirmed breeding include 1987, 1988, 1989, 1990, 1993, 1994, and 1996-2001. It is not clear whether this species recently colonized the state, or if it has always been a scarce and irregular breeder here.

There are only 11 eastside records away from the Richland/Yakima/Vantage triangle. These span 29 Apr to 28 Jul and include two Jul records of multiple birds probably breeding. There are 19 westside records, 18 of which fall between 26 Apr and 23 Jun with a peak from mid-May to early Jun. The exceptional record comes from winter in Whatcom Co. Westside records by county are Grays Harbor (2), San Juan (1), Kitsap (2), Whatcom (1), Skagit (1), Snohomish (3), King (5), Pierce (1), Thurston (1), Clark (1), Skamania (1).

Occurrence varies interannually. Three years (1984, 1994, 1999) account for 17 of the 29 vagrant records and 1994 was also an exceptional year within the Washington breeding range. These irruptive years may be tied to drought within the species' core range in the s. U.S. (Hunn 1978).

Noteworthy Records: *High counts*: 12 on 19 Jun 1994 at Priest Rapids L. (Stepniewski 1999); 9 on 20 Jun-5 Jul 1994 nr. White Bluffs, Grant Co. **West**, *Spring, early*: 26-27 Apr 1993 at Monroe. *Spring, late*: 14 Jun 1981 at Redmond; 23 Jun 1984 at Swift Dam. *Winter*: 27 Dec 1987-9 Jan 1988 at Bellingham. **East**, *Spring, early*: 29 Apr 1996 at College Place. *Vagrants*: 1 on 10 May 1977 nr. Walla Walla; 1 on 27 May 1981 at Ephrata; 1 on 30 Jun-7 Jul 1984 at Monse; 3 on 30 Jun-28 Jul 1984 nr. Omak; 2 on 28 Jul 1984 at Steptoe Butte; 1 on 11 Jun 1988 nr. Wenatchee; 1 on 12-14 Jun 1989 at Burch Mt., Chelan Co.; 1 on 11 Jun 1994 at Bridgeport; 2 on 19 Jun 1994 at Rock Cr., Klickitat Co.; 29 Apr 1996 at College Place; 2 on 3-5 Jun 2000 exchanging food at Steptoe Butte.

Steven G. Mlodinow

Sage Sparrow *Amphispiza belli*

Uncommon migrant and summer resident in shrub-steppe of e. Washington; very rare migrant w. of the Cascades.

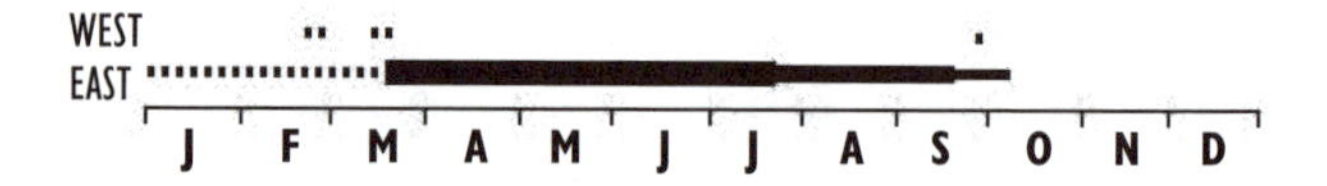

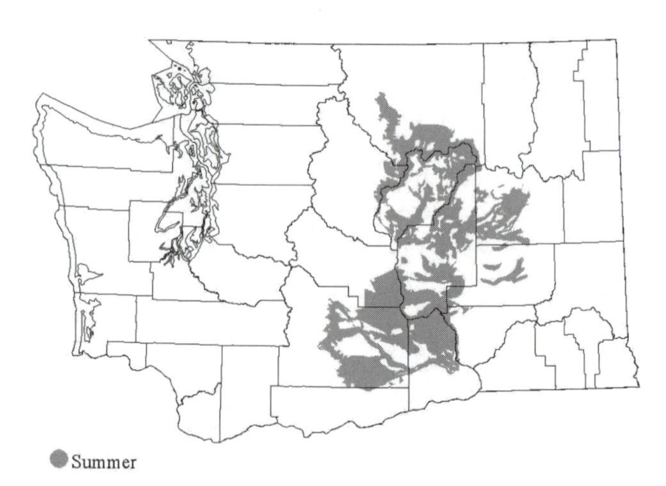

Subspecies: *A. b. nevadensis.*

Habitat: Big sagebrush; also in stands that include stiff sagebrush, rabbit brush, spiny hopsage, or bitterbrush as a component of the shrub layer.

Occurrence: The Sage Sparrow is a sagebrush obligate that breeds from n.c. Washington s. to Baja California and as far e. as c. Wyoming and w. Texas and winters in the sw. U.S. and n. Mexico. In Washington, it breeds from Douglas Co. s. to Benton Co., and as far e. as Lincoln Co.; it may also breed in the Methow and Okanogan valleys. Birds breed in unfragmented habitats, primarily in large expanses of sagebrush/bunchgrass shrub-steppe (Knick and Rotenberry 1995, Vander Haegen et al. 2000). Significant breeding areas in Washington include the Hanford Site, Yakama Indian Reservation, Yakima Training Center, Quilomene WMA, and Moses Coulee. Birds occur rarely in w. Washington, generally during spring migration coincident with their eastside arrival.

Smith et al. (1997) pointed out that BBA surveys recorded no birds in some apparently suitable habitat tracts, and also that over one-half of former Washington habitat is now cultivated. BBS data indicated a decline from 1966 to 2000 (Sauer et al. 2001). Population likely is declining as as the species' preferred habitat, large expanses of big sagebrush, disappears from the landscape due to conversion of shrub-steppe habitats to agriculture, fragmentation, and fire. Invasion of the understory by exotic annual grasses decreases the suitability of sagebrush stands (Dobler et al. 1996, Keany et al 1996, Fitzner 2000), probably by reducing the amount of open ground available to this ground-foraging omnivore.

Remarks: Keany et al. (1996) discussed the importance of perennial grass cover in habitat preference, though some optimum characteristics of that habitat component apparently are uncertain (A. Stepniewski p.c.). Classed SC, PHS, on NAS Watch List.

Noteworthy Records: East, *early*: late Feb at Hanford Site 1977; 31 Mar 1979 in Lincoln Co.; 16-17 Feb 1980 at Richland; 10 Feb 1980 at Toppenish NWR; 28 Feb 1981 at Dodson Rd. area, Grant Co.; 13 Feb 1983 at Richland; 7 Mar 1996 at Quilomene WRA; 17 Mar 1996 at Kennewick; 22 Feb 1997 at Richland; also Quilomene WRA. *Late*: 25 Dec at Pasco 1962. *Winter*: 4 wintered in 2001-02 at FEALE, Benton Co. *Out of range:* 28 May 2000 at Asotin. **West**: 16 Mar-3 Apr 1977 at Olympia; 17-19 Feb 1980 at Seattle; 29 Sep 1982 at Skamokawa; 14-15 Mar 1987 at Seattle; 13 Mar 1990 at Seattle.

W. Matthew Vander Haegen

Lark Bunting *Calamospiza melanocorys*

Very rare vagrant.

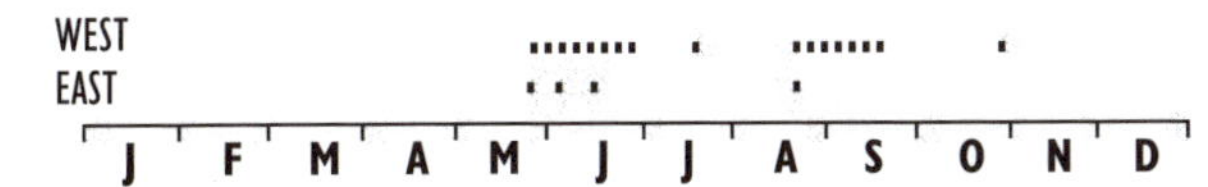

Occurrence: Lark Buntings breed throughout the c. U.S. and winter primarily in the sw. and Mexico. The first record in Washington occurred on 16 Jun 1967, when a male was found at Lummi Flats, Whatcom Co. (Mattocks et al 1976). Through 2002, there were 14 records, with over half since 1990. Nine are from coastal locations in either Jun or in fall, and five of those are from Clallam Co. Three are from the se. counties.

This semi-nomadic Great Plains breeder is also a very rare visitant to Oregon, with 16 records through 1994 (Gilligan et al. 1994) and 17 from B.C. (Campbell et al. 2001). Species on NAS Watch List.

Records: *Accepted by WBRC*: 1on 2 Sep 1973 at Cape Flattery; male on 21-23 Jun 1980 at Fort Flagler SP; male on 5 Jun 1981 at Sequim; 1 on 20 Aug 1988 nr. Ewan, Whitman Co.; 1 on 10 Sep 1991 at Westport; male on 12 Jun 1992 nr. Clarkston Heights, Asotin Co.; male on 3 Jun 1993 at Walla Walla; 1 on 19 Aug 1994 at Tatoosh I.; 1 on 31 Oct-1 Nov 1996 at Tokeland; 1st-yr male on 18 Jul 1998 at Fort Lewis. Other reports: male on 16 Jun 1967 at Marietta (Lummi Flats); 1 on 17-18 Sep 1986 on Tatoosh I.; 1 on 28 May 2000 at Ozette Trailhead, Clallam Co.; 24 May 2002 at Dodson Rd.

Bill Tweit

Savannah Sparrow *Passerculus sandwichensis*

Common migrant and summer resident, uncommon winter resident in w.; common migrant and summer resident, very rare winter resident e.

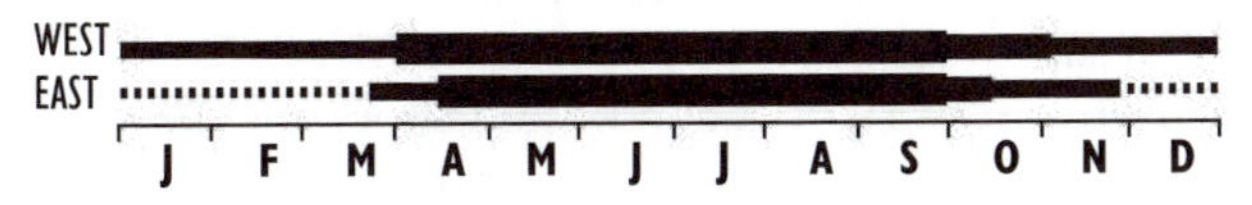

Subspecies: *P. s. brooksi* breeds in w., *nevadensis* in e. (Smith et al. 1997).

Habitat: Wide variety of grassy and weedy habitats, often disturbed areas. On the eastside apparently favors irrigated fields, and particularly fields of mint and alfalfa, but generally avoids pristine shrub-steppe (Smith et al. 1997). Often winters in grassy areas away from blackberry thickets and other dense cover and separately from other sparrow flocks.

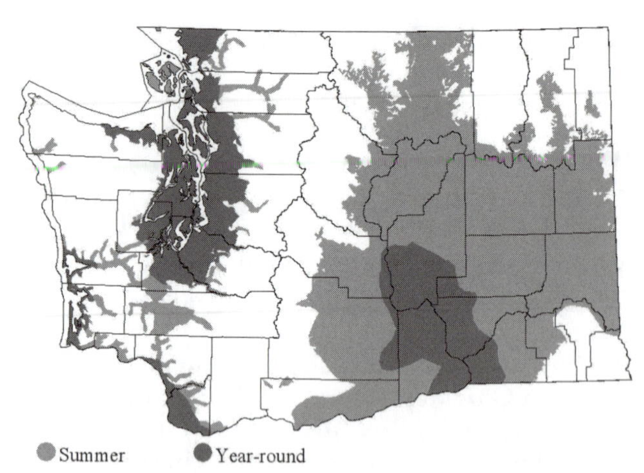

Occurrence: A common grassland bird statewide, breeding mainly at low altitudes but also found locally in subalpine meadows of n. Cascades such as those at Hart's Pass, Rainy Pass, and in Okanogan Co. (Smith et al. 1997). Migration is also predominantly through lowlands, but some movement occurs along the Cascades during fall (Jewett et al. 1953). Westside spring migration is mostly from early Apr to mid-May and fall passage is mostly from early to late Sep. On the eastside, spring migrants are found mostly between early Apr to mid-May and southbound migration peaks from late Aug to -late Sep. BBA surveys did not confirm breeding along the n. outer coast (Smith et al. 1997).

Typically found in grassy fields, where they consume large quantities of weed seeds (Godfrey 1986), including burned-over areas on the Yakima Training Center that furnish large areas of suitable habitat (Stepniewski 1999). Migrants are occasional at bird feeders (Wahl 1995). The effects on populations of mowing and harvesting fields during breeding season are unknown (Stepniewski 1999).

Wintering birds were apparently unknown early in the 20th century (Jewett et al. 1953). By the 1990s, the species was regular, albeit uncommon, on the westside, often in small flocks of up to 30 birds. Population increase may be due to creation of open grassy and weedy areas by human activity, but greater observer effort and other factors may be involved. Similar changes are apparent in the case of *Zonotrichia* and other sparrows. Savannah Sparrows may be less hardy than *Zonotrichia*, as perhaps indicated by their continued relative scarcity during winter in e. Washington.

BBS data show no summer population trends in Washington (Sauer et al. 2001). CBC data show no winter trend though increases are apparent at Sequim-Dungeness and, shorter term, at Skagit Bay and Toppenish.

Remarks: Subspecies fall into four main groups, only one of which occurs in Washington—the "typical" Savannah Sparrow (Rising 1996). Several subspecies within this group, not generally separable in the field, have been reported in Washington (e.g., *sandwichensis*, *crassus*, and *anthinus*; see Jewett et al. [1953], Smith et al. [1997]).

Noteworthy Records: *High counts*, **West**, *Spring*: 130 on 29 Apr 2001 at Fir I. *Fall*: 191 on 26 Sep 1998 at Fir I.; 150 on 15 Sep 1998 at Deer Lagoon; 150 on 15 Sep 1996 at Skagit WMA. *Winter*: 125 on 14 Dec 2000 at Shillapoo Bottoms, Clark Co.; 100 on 6 Dec 1998 at Leque I., Snohomish Co.

Steven G. Mlodinow

Grasshopper Sparrow

Ammodramus savannarum

Uncommon summer resident in e. Washington. Very rare in w.

WEST
EAST
J F M A M J J A S O N D

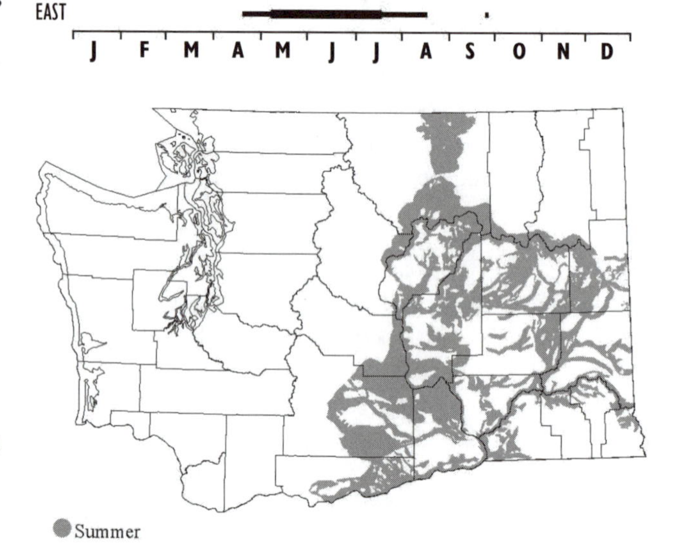

Subspecies: *A. s. perpallidus.*

Habitat: Bunchgrass, and steppe grasslands, especially those with wheatgrass, Idaho fescue and needle-and-thread; shrublands, shrub and tree savannahs; absent from areas of dense sage (Smith et al. 1997).

Occurrence: Status in Washington has been described by authors as difficult to determine, primarily due to the species' inconspicuous nature (even when singing) and apparent non-association with other sparrows. Records of certain migrants are consequently virtually unknown. Jewett et al. (1953) described status as occasional, not common. Particularly beginning in the 1960s, field observers in e. Washington began describing occurrence and habitats. Many factors affecting status, however, remain uncertain.

Nesting populations may be somewhat nomadic, responding to seasonal rainfall or grassland improved by fires (Smith et al. 1997, Stepniewski 1999), and as these occur somewhat sporadically, local populations are variable. At times, when habitat is suitable, can be locally fairly common to common. Smith et al. (1997) suggest that birds were formerly more numerous and widespread, with decline resulting from habitat destruction. Birds are widespread in suitable habitats in the Columbia Basin, Okanogan Valley, Yakima Valley, along the Snake R., and the Columbia R. w. into Klickitat Co. Absent from the Methow Valley (Smith et al. 1997). Common in Yakima Training Center big sagebrush habitat and may be locally fairly common elsewhere in Yakima Co. (Stepniewski 1999).

BBS analyses showed decreasing population trends in the U.S. (Peterjohn and Sauer 1993) and in the w. U.S., including Washington (DeSante and George 1994), though Sauer et al. (2001) indicate a stable population from 1966 to 2000.

Remarks: Overgrazing may make some habitats unsuitable (Smith et al. 1997). BBA records may understate distribution and density due to the species' high-pitched song and possible inability of some observers suffering from hearing loss to detect it (Smith et al. 1997). Three winter records from Lane Co. in Oregon (Gilligan et al. 1994) suggest the possibility of winter occurrence.

Noteworthy Records: *High counts*: 25 on 4 Jul 1964 nr. Newman L.; 20 on 5 Jun 1964 nr. Sprague. *Fall, late*: 27 Sep 1967 at Clarkston. West: 27 Jun-9 Jul 1998 at Weir Prairie, Thurston Co.

Steven G. Mlodinow

Le Conte's Sparrow *Ammodramus leconteii*

Casual vagrant.

Four records: the first was a specimen from Kennewick, 29 May 1964 (Aanerud and Mattocks 1997), followed by a bird from Willapa NWR, 15 Nov 1982, a singing male at Deep L., 18-29 Jun 1993; and a bird at Meadow Cr. CG, 2 Jun 1996. Le Conte's Sparrows likely occur more often, but their extremely secretive habits makes them very difficult to locate. There are two Oregon records, from Sep and Oct.

Steven G. Mlodinow

Nelson's Sharp-tailed Sparrow *Ammodramus nelsoni*

Casual vagrant.

For Washington there is one record, from Sullivan L. on 14 Sep 1986. Though regular in small numbers in California's saltmarshes during winter, inland records from Pacific coast states are very few (Patten and Radamaker 1991). There are no Oregon records, but there is an early Sep record from nr. Vancouver, B.C. (Campbell et al. 2001). There are two subspecies, *nelsoni* or *alterus*, which are not reliably separable in the field.

Steven G. Mlodinow

Fox Sparrow *Passerella iliaca*

Common migrant/ winter resident and local summer resident in w. Washington lowlands and coastal strip. Fairly common summer resident and migrant in Cascades, ne. mountains, Blue Mountains and e. woodlands. Uncommon migrant and rare winter resident in e. Washington steppe.

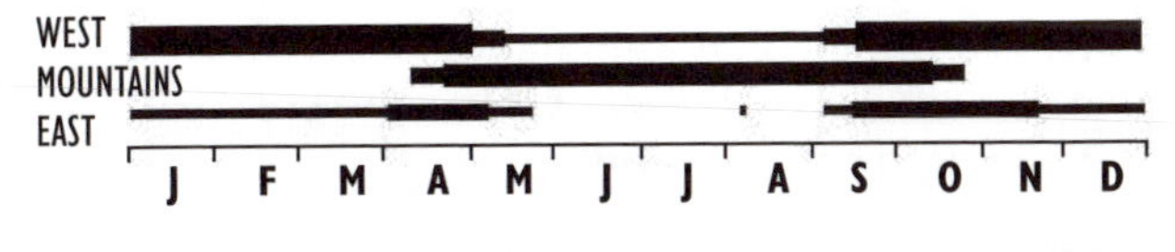

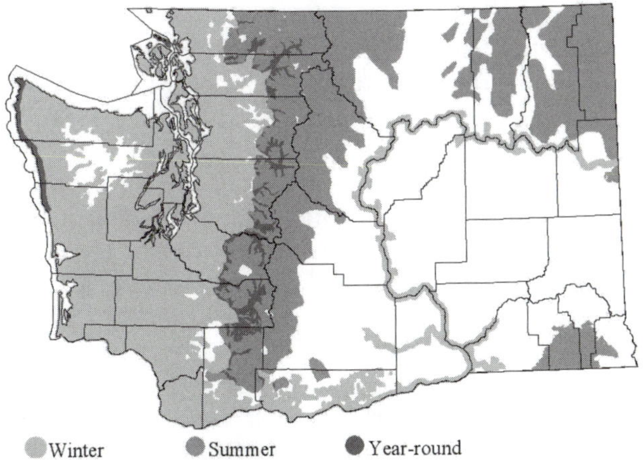

Subspecies: Three groups recorded in Washington: Sooty Fox Sparrow, *unalaschcensis*, Slate-colored, *schistacea*, and Red, *iliaca*.

Habitat: Breed very locally in shrubby portions of westside Puget Sound Douglas-fir habitat and woodland/prairie mosaic zones, coniferous forests in the Cascades and e. mountains (Smith et al. 1997); wooded areas with heavy undergrowth, dense cover away from woodlands in migration and winter.

Occurrence: *Sooty Fox Sparrow.* The common wintering race on the westside, probably also accounting for the great majority of wintering Fox Sparrows on the eastside, including Klickitat Co., where it is uncommon. Breeds locally in small numbers along the coast from Pt. Grenville to Cape Flattery, at the Sequim prairies and on Sucia and Yellow Is. in the San Juans (Smith et al. 1997). Migrants arrive and depart the outer coast about a week earlier than in the rest of w. Washington.

Slate-colored Fox Sparrow. Fairly common summer resident in the mountains (except for the Olympics) and in coniferous woodlands of ne. Washington. Status and distribution in Washington lowlands is unclear, but it probably represents a large percentage of eastside migrant Fox Sparrows. Formerly thought not to winter n. of California (Rising 1996), but there are now a number of winter records from se. and w. Washington. This subspecies is also a rare to very rare migrant in westside lowlands. A Slate-colored Fox Sparrow in the Columbia Basin at the Washington State University Tri-Cities campus on 10 Jul 1999 was unprecedented for midsummer (E. Rykiel).

Red Fox Sparrow. Vagrant, though the degree of rarity is unclear. Most records are undocumented, and rusty individuals of *altivagans* (a Sooty Fox Sparrow race) can lead to confusion.

BBS data indicate increased numbers in Washington from 1966 to 2000 (Sauer et al. 2001). CBC data show no statewide trend over two or three decades.

Remarks: The AOU (1957) recognized 18 races of Fox Sparrows, many of which were inseparable in the field. These races form four groups which are mostly separable in the field and may represent separate species (Zink 1986, 1994, Rising 1996, Zink and Kessen 1999).

Noteworthy Records: East, *Fall, early, away from breeding areas*: 6 Aug 1994 at Richland. **Slate-colored Fox Sparrow,** *late Fall-early Spring,* **West**: 28 Nov 1999 at Everett (K. Aanerud p.c.); 10 Jan 2002 at Everett; 12 Jan 1997 at Pt. Roberts (SGM, N. Ball p.c.); 26 Jan 2002 at Blynn, Clallam Co.; 14 Feb 1998 at Lummi Flats; 17 Feb 2002 at Edmonds. **East**: 7-8 Dec 2000 at Bateman I., Benton Co.; winter 2000-01 at N. Richland; 28 Dec 2001 - late Feb 2002 at Richland; Jan 2003 at Spokane (J. Acton p.c.). **Red Fox Sparrow, West**: 15 Feb 1929 at Renton and Seattle, 23 Apr 1922 at Seattle (Jewett et al. 1953). 31 Oct 1998 at Washougal; 13 Nov 1998 at Joyce; 12 Mar 2000 at Spencer I. (SGM); 11 Oct 2000 at Marymoor Park, King Co. (M. Hobbs p.c.). **East**: College Place, 9 Jan 1988 (K. Knittle p.c.); Maryhill, 22 Oct 1989 (D. Paulson p.c.); Richland, 25 Oct 1998; 8 May 1999 at Rock Cr., Yakima Co.; 25 Oct 1999 at Bateman I., Benton Co. (C. Corder p.c., J. Stevens p.c.).

Steven G. Mlodinow

Song Sparrow *Melospiza melodia*

Common permanent resident statewide.

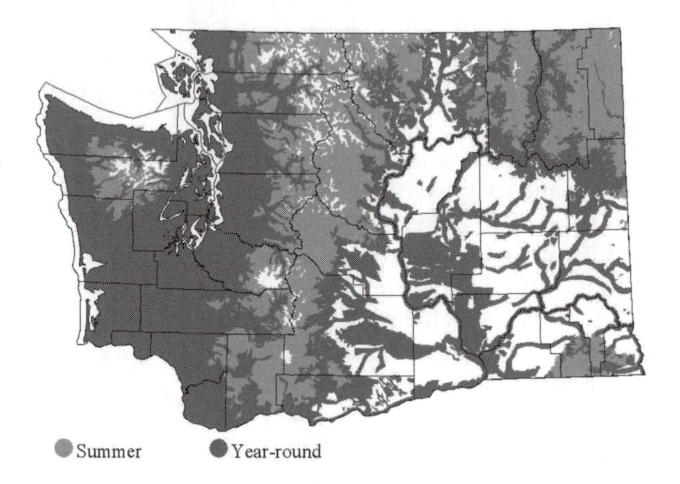

Subspecies: *M. s. merrilli* breeds in e., *fisherella* in s.c., and *morphna* in w. Washington (Smith et al. 1997, and see Arcese et al. 2002).

Habitat: Inhabits a wide variety of shrubby and brushy habitats, mostly below 1200 m, in cities, rural areas, and forest edges; generally absent from old-growth forest, subalpine forests, alpine parkland, and arid shrub-steppe (Smith et al. 1997). During winter, some downslope movements.

Occurrence: A relatively successful "little brown bird," essentially taken for granted by observers and consequently incompletely understood in population size and movements, the Song Sparrow is one of Washington's most widespread native landbirds, residing commonly in farmland hedgerows, riparian areas, open woods, wood edges, suburbs, clear-cuts, and other such open spots with good cover and readily coming to bird feeders. Birds occur to sea level where they forage at water's edge, and are noted on bare-rock breakwaters and ocean jetties (TRW). Many westside birds, at least, are year-round residents, defending territories through the winter. Some downslope migration for the winter is suggested (Wahl 1995, Stepniewski 1999) but this and effects of migration in or out of the state on population numbers are unquantified by systematic studies.

Over the long term, following human settlement and expansion, the species likely benefitted from conversion of forest to open land, especially on the westside (e.g., Smith et al. 1997). Recently, however, BBS data appear to indicate a decline (see Sauer et al. 2001), while CBC data on counts covering four decades indicate a significant eastside increase (P <0.01).

Remarks: The extent to which seasonal populations contain downslope migrants or birds from outside Washington is uncertain. Many of the 29 races occurring n. of Mexico are poorly differentiated and are inseparable in field and sometimes in hand (Rising 1996). Recorded regularly on almost all, if not all, BBS and CBC samples, and perhaps as frequently as any other species, the Song Sparrow is one of many species of regularly occurring passerines requiring study.

Noteworthy Records: West, *high counts, Summer*: 200 on 7 Jun 1998 at Spencer I.; *Fall*: 460 on 2 Nov 1997 at Ridgefield NWR; 400 on 23 Nov 1996 at Spencer I.; *Winter*: 300 on 25 Dec 1997 at Skagit WMA.

Steven G. Mlodinow

Song Sparrow (Shawneen Finnegan)

Lincoln's Sparrow *Melospiza lincolni*

Common breeder in mountains except Olympics. Common fall migrant and fairly common winter resident and spring migrant in w. Washington lowlands. Common migrant and rare in winter in e. lowlands.

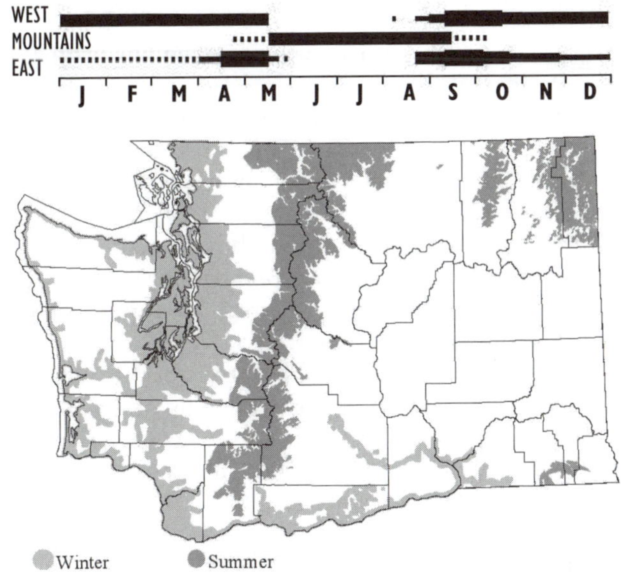

Subspecies: *M. l. lincolnii* breeds; *gracilis* occurs in migration and winter.

Habitat: Most breed above 900 m in subalpine bogs, swamps, marshes, and wet meadows, especially in areas with low willow cover and dense ground vegetation (Ammon 1995, Smith et al. 1997). Moist grassy/weedy areas with some nearby cover in migration and winter, occasionally in fairly wooded areas.

Occurrence: Like those of several other sparrow species, reports of wintering Lincoln's Sparrows have dramatically increased, especially on the westside. Jewett et al. (1953) considered this species only a possible winter visitor. Wintering birds were still noteworthy in 1968 when the winter AFN column stated "A number of Christmas Bird Count reports from the Region include Lincoln's Sparrows, but these records constitute such a departure from the established winter range of this species as to require substantial verification before acceptance." By the early to mid-1970s, a scattering of westside winter Lincoln's Sparrows was the norm, and by the 1990s, 10 or more in a winter's day were not unusual. On the eastside, the first winter Lincoln's Sparrow was recorded at Sunnyside Game Range, Yakima Co., 24 Jan 1970. Reported a number of times since, the species is still less than annual after December in e. Washington. Weather generally becomes more severe but observer effort concentrated on CBCs may play a part in a decrease in eastside reports after Dec.

In westside lowlands, fall migration is much more obvious than that of spring. Fall migrants mostly move through from early Sep to early Oct, though a push is sometimes seen in late Aug. Spring passage is mostly from early Apr to -early May. On the eastside, spring migration occurs mostly from mid-Apr to early May. Eastside fall passage is mostly from late Aug to early Oct, with a few birds often lingering as late as Dec.

The increase in wintering birds is probably due in large part to creation of weedy and brushy habitats by human activity, but greater observer effort may well play a role. A similar change has been noted in *Zonotrichia* and other sparrows, but Lincoln's Sparrows are less hardy than *Zonotrichia*, likely accounting for their continued scarcity during winter in e. Washington.

BBS data indicated a significant increase over N. America from 1966 to 1991 (Peterjohn and Sauer 1993). Based on relatively low numbers of birds, CBC data from long-term counts indicate a statewide increase in winter.

Remarks: Ammon (1995) updates subspecies occurrence in Washington.

Noteworthy Records: West, *Winter, high count*: 60 on 28 Sec 1996 at Ebey I., Snohomish Co. (SGM). *Fall, early*: 8 Aug 1971 at Lummi Flats.

Steven G. Mlodinow

Swamp Sparrow *Melospiza georgiana*

Rare in winter on the westside, very rare on the eastside.

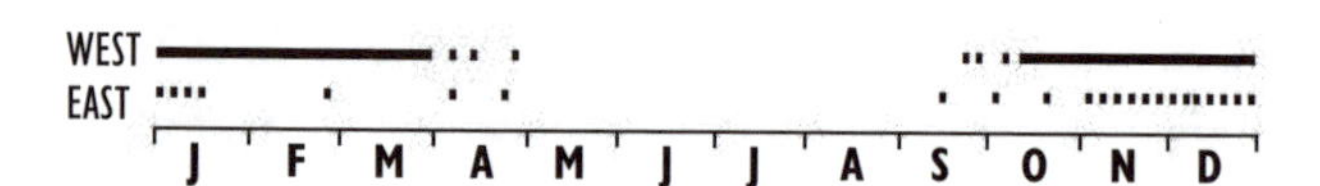

Subspecies: Presumably *M. g. ericrypta*.

Habitat: Freshwater marshes, blackberry thickets, damp, weedy fields.

Occurrence: The Swamp Sparrow is primarily a bird of e. and c. N. America, and was not recognized as a rare, but regular, winter visitant on the w. coast until the late 1960s and early 1970s (Roberson 1980, Gilligan et al 1994). The species was unrecorded in Washington until 1973 (Hunn 1973) and not found annually in Washington until the winter of 1986-87. Only five birds were found in the 1970s and 31 in the 1980s. Although they are now found annually, their numbers are highly variable between winters. Winter totals in the 1990s range from two to 38, averaging 12 per winter. Peak winters were 1993-94 (38 records), 1995-96 (23 records) and 2002-03 (26 birds).

Records are much more frequent on the westside than the interior: the westside accounted for 82% of the 158 records through winter 1999-2000. Reports are decidedly clumped, both on the westside and the interior. Highest numbers have been recorded in Skagit and Snohomish Cos.: 45 records there account for 40% of the westside records of known county origin. Other westside counties with relatively large totals include Clallam (16 records), King (14), and Clark (11). In the interior, 10 of 28 records are from Walla Walla Co. None of the other eastside counties have more than four records each.

Noteworthy Records: *Fall, early*: 14 Sep 86 at Sullivan L.; 20 Sep 1982 nr. Rockport; 25 Sep 1999 at Skagit WMA; 3 Oct 1993 at Ocosta. *Spring, late*: 26 Apr1995 at Ridgefield; 22 Apr 1997 at Maryhill SP; 11 Apr 2001 at Martha L., Grant Co. and at Monroe; 8 Apr 1972 at L. Sammamish; 7 Apr 1988 at Yakima; 4 Apr 1996 at Montlake Fill, Seattle. *High counts*: 6 on 7 Feb 2003 nr. Monroe; 5 mid-Dec 2002-early Jan 2003 nr. Elma. *Winter*, **East**: 3 Jan 1988 at Spokane, 4 Jan 1995 at McNary; mid-Jan 1988 at Walla Walla; 27 Jan 2001 at Bingen; 25 Feb 1995 at Whitcomb I., Benton Co.

Bill Tweit

White-throated Sparrow *Zonotrichia albicollis*

Uncommon migrant and winter resident. Very rare in summer.

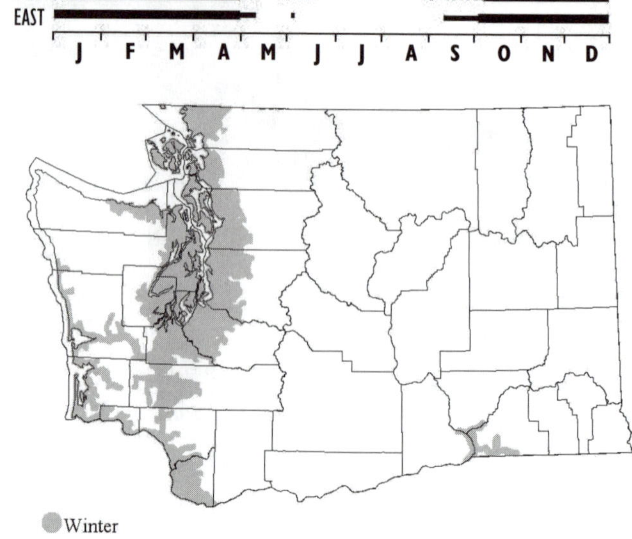

Habitat: Deciduous woodlands and edges with heavy undercover. Also found in more open areas with dense cover nearby.

Occurrence: A species much more frequently reported in recent decades, often noted with other *Zonotrichia* sparrows and possibly more likely with Golden-crowned than White-crowned sparrows (SGM). White-throated Sparrows used to be rarely reported: Jewett et al. (1953) listed only six state records. Reports have increased steadily since then and 30 or more in a year is no longer unusual. The reason for this change is not clear, but may be due in part to increased sparrow habitat resulting from human activities, more bird feeding, and increased field-observer attention to *Zonotrichia* flocks.

Generally more numerous on westside than the e., excepting perhaps in the lower Columbia Basin and Klickitat Co. where numbers may approach those of w. Washington. During winter, nearly absent on eastside n. of Spokane. Fall migration is apparent from early Oct into early or mid-Nov. Almost all Sep records are from e. Washington, implying that fall migration may be somewhat earlier there. Spring migrants rarely detected but northbound passage appears to occur from mid-Apr to early May.

BBS data showed a significant decrease from 1966 to 1991 over N. America (Peterjohn and Sauer 1993). CBC occurrence in Washington is limited, but indicate increases over time both on the westside and the eastside.

Noteworthy Records: *Fall, early*: 1 Sep 1998 at Spokane; 7 Sep 1981 at Leadbetter Pt. *Spring, late*: 21 Jun 1973 at Tide Pt., San Juan Co. (Lewis and Sharpe 1987); 8 Jun 1970 at Spokane. *High count*: 9 on 1 Jan 2002 nr. Monroe; 7 on 27 Nov 1999 at Post Office L., Clark Co.

Steven G. Mlodinow

Harris' Sparrow *Zonotrichia querula*

Uncommon migrant and winter visitor.

WEST, EAST

J F M A M J J A S O N D

Habitat: Weedy areas nr. dense brush, nr. deciduous wood edges.

Occurrence: Breeding in edge habitats between forest and tundra in n. Canada (Godfrey 1986), often found in Washington in more open habitat than the other *Zonotrichia* and more often with White-crowned than Golden-crowned sparrows.

Jewett et al. (1953) considered Harris' Sparrow a casual winter visitor and migrant, listing five records.

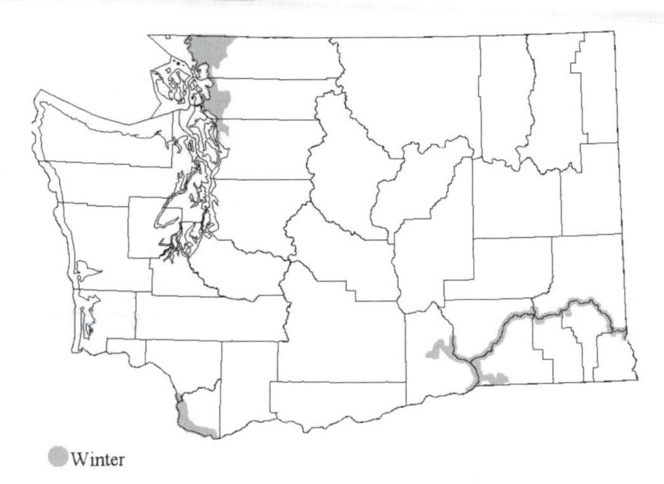

Over the next 30 years, reports increased similarly to those of the two regularly occurring wintering crowned sparrows, with an average of 10 to 15 per year in recent years. Over the 1980s and 1990s, reports of occurrence were similar w. and e. of the Cascades, and some 100 CBC records are similarly equally divided. Migration periods difficult to discern, but appear to be mostly from mid-Oct into early to mid-Nov and from mid-Apr to early May. Recorded irregularly statewide, but wintering birds are seldom reported from the ne. counties.

Reports peaked in the mid-1980s, and approached White-throated Sparrow reports in frequency. As with other *Zonotrichia*, the upswing in numbers was likely due, in part, to increased habitat availability to and much more intense observer effort. Harris' Sparrow reports declined in the 1990s from the 1980s, mirroring a decline in long-term CBC occurrences on the westside. Overall there have been no major population changes described (Norment and Shackleton 1993). Classed GL.

Noteworthy Records: *Fall, early*: 11 Sep 1971 at Spokane; 19 Sep 1999 at Lyons Ferry SP; 25 Sep 1975 at Spokane; 26 Sep 1999 at Wanapum SP. *Spring, late*: 26 May 1976 at Seattle; 2 Jun 1999 at Dot, Klickitat Co.; 7 Jun 1971 at L. Sammamish. *High count*: 4 on 23 Nov 1997-7 Jan 1998 at Madame Dorian SP. *CBC high counts*: 6 in 1997 at Two Rivers, 5 in 1986 at Tri-Cities, 4 in 1972 in Bellingham, 4 in 1975 at Walla Walla, 4 in 1985 at Mt. Vernon.

Steven G. Mlodinow

White-crowned Sparrow *Zonotrichia leucophrys*

Common year-round resident in w. Washington lowlands. Common in winter in eastside lowlands. Fairly common summer resident in Cascades. Uncommon summer resident in Pend Oreille Co. and Blue Mts.

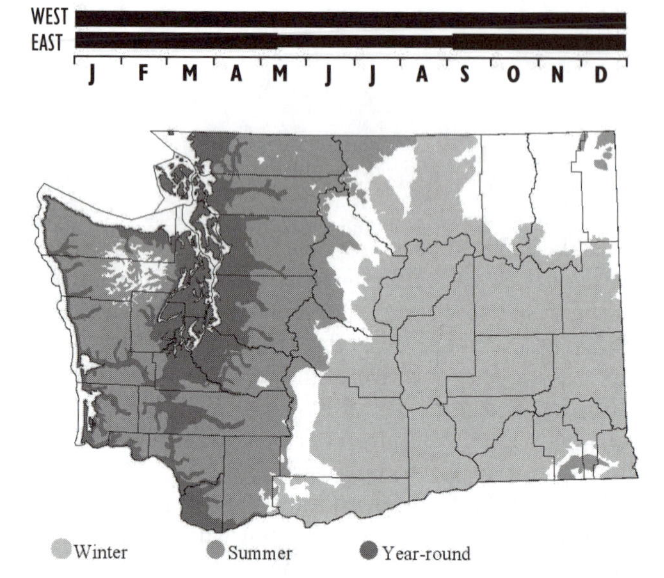

Subspecies: Puget Sound White-crowned Sparrow, *Z. l. pugetensis*, Gambell's, *gambelii*, and Mountain, *oriantha*, in Washington.

Habitat: Breeds from alpine and subalpine parkland (*oriantha*) to lowland in brushy areas in cities, suburbs, parks, clear-cuts, and early secondary growth (Smith et al. 1997). During winter, occupies a wide variety of edge habitats, brushy and weedy areas.

Occurrence: With one of the songs characterizing summer in w. Washington, White-crowned Sparrows have adapted to nest, as long as there are trees in planters or at edges, in otherwise bleak environments like shopping-center parking lots (Smith et al. 1997, TRW). Birds sometimes sing at night, later in the day and later in the nesting season than other species.

The species is present all year and overall seasonal abundances often appear similar. Westside winter populations include *pugetensis* and also *gambelli*. The former outnumbers the latter, though historically the reverse has been the situation at least locally, as in Whatcom Co. (Wahl 1995).

Puget Sound White-crowned Sparrow. The common breeding subspecies in w. Washington, and small numbers breed on the e. slope of the Cascades. Breeding habitat substantially increased by human development/logging over last century, with subsequent increases in breeding numbers and range. Local, fairly common during winter in w. lowlands, though sometimes outnumbered locally by *gambelii*. Spring migrants appear in late Mar and early Apr. Post-breeding movements begin in Jul and seem settled by Oct, with some birds probably leaving Washington entirely and others moving more locally. Non-breeding range in e. Washington poorly known, but one banded at College Place, 28 Jan 1929, was recaptured on 17 Jan 1930 (Jewett et al. 1953). Recent field observations in the Columbia Basin indicate this taxon is likely an uncommon, perhaps fairly common, migrant there from early Sep well into Oct and a rare winter visitor.

Gambel's White-crown. Common migrant and wintering race statewide. It is an uncommon breeder in w. Okanogan and nw. Chelan Cos., with scattered records s. to Naches Pass (Smith et al. 1997). Southbound migrants typically appear in early Sep and most non-breeders are gone after early May on the westside and mid-May on the e. Peak fall passage from mid-Sep to late Oct. Peak spring movement, often in very large flocks on the eastside, in late Apr-early May.

Mountain White-crown. Uncommon, local breeder in ne. Pend Oreille Co. and the Blue Mts., first recorded in Washington during 1970 at Salmo Pass (Smith et al. 1997). Migratory movements in Washington unknown.

The species' seasonal status has changed dramatically since it was described as "casual" in winter by Jewett et al. (1953). By the late 1960s large wintering populations were known (Lewis et al. 1968) and a general increase is apparent statewide. This upswing is likely due to increased winter habitat (weedy fields, blackberry brambles, etc.) brought about by human-induced changes. Winter populations were enhanced by eastside irrigation projects (Stepniewski 1999), and largest numbers in rural areas are likely at that season. CBC data indicate increases on the westside ($P < 0.05$) and the eastside. BBS data indicate no change from 1966 to 2000 (Sauer et al. 2001).

Remarks: Given the number of *pugetensis* found in e. Washington during fall migration, it is interesting that the fate of these birds remains unknown. There are relatively few eastside winter records, and birds are essentially unknown from fall or winter in e. Oregon. This situation is likely due to observer awareness of subspecies and poses a challenge for future field observers.

There are five subspecies, the three occurring in Washington plus *nuttalli* in coastal California and *leucophrys* in e. N. America. *Gambelli, leucophrys,* and *oriantha* interbreed where their ranges meet, and *nuttalli* and *pugetensis* interbreed extensively, but these two groupings are not known to interbreed with each other despite fairly extensive zones of sympatry (Chilton et al. 1995). Hybrid White-crownedXGolden-crowned sparrows have occurred several times in Washington: Jan 1966, in Apr in 1969, 1978, and 1988 in Bellingham (Wahl 1995); 1 May 1979 in Seattle; 5 May 1984 in Westport. A hybrid White-crownedXSong Sparrow was collected on San Juan I. in Jun 1959 (Lewis and Sharpe 1987).

Noteworthy Records: East, *high counts*: 2500 on 27 Feb 1999 at Big Flat HMU, Franklin Co.; 1200 on 28 Sep 1983 at Sunnyside Game Range, Yakima Co.; 1000 on 4 May 1963 nr. Banks L. Lowlands, Summer: 6 Jun 1965 at Meadow L., Spokane Co.; 15 Jun 1968 at Spokane.

Steven G. Mlodinow

Table 40. Golden-crowned Sparrow - CBC birds per 100 party-hours by decade.

	1960s	*1970s*	*1980s*	*1990s*
West				
Bellingham		26.3	42.8	62.2
Cowlitz-Columbia			80.5	122.4
Edmonds			19.5	61.8
E. Lk. Washington			3.7	43.9
Grays Harbor		89.3	118.9	160.9
Kent-Auburn			75.7	117.8
Kitsap Co.		9.7	16.1	31.5
Leadbetter Pt		53.1	23.9	50.8
Olympia			37.9	85.0
Padilla Bay			83.6	118.7
Port Townsend			45.1	55.4
San Juan Is.			61.3	93.7
Seattle	29.0	9.6	26.0	55.6
Sequim-Dungeness			135.6	128.3
Tacoma		32.2	43.6	93.2
East				
Ellensburg			7.0	1.8
Toppenish			10.8	59.6
Wenatchee	0.0	4.8	8.4	6.6
Yakima Valley		14.5	13.3	48.8

Golden-crowned Sparrow *Zonotrichia atricapilla*

Common migrant and winter resident in w. Washington. Uncommon to fairly common migrant and winter resident in e.; rare in easternmost counties. Very rare throughout in summer.

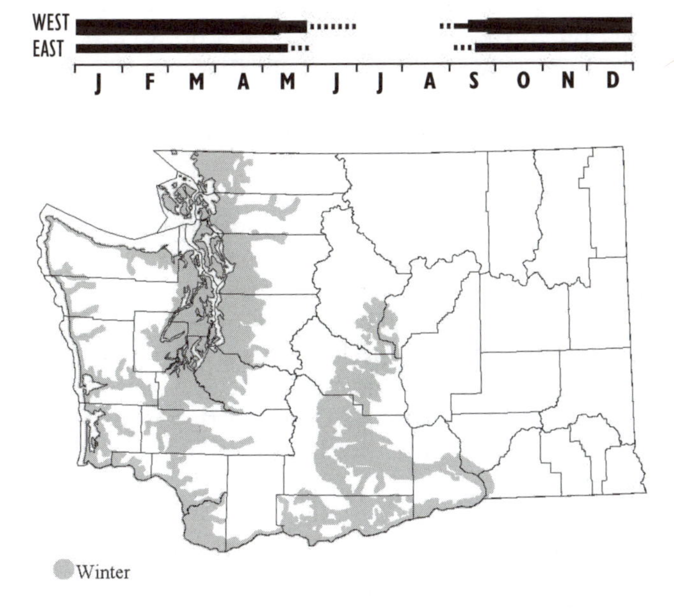

Habitat: Lowland brush and weedy areas in rural agricultural areas, parks, and habitats in urban habitats, including bird feeders.

Occurrence: Most Golden-crowned Sparrows winter in westside lowlands, though numbers are found on the eastside, primarily along the lower e. slope of the Cascades (e.g., Stepniewski 1999). Birds are fairly common through Klickitat and well into Yakima Cos., but are relatively uncommon e. of the Columbia R. Though there is much overlap, prefers generally more wooded areas than the White-crowned Sparrow.

In the lowlands, peak fall migration occurs from late Sep to early Nov, but southbound movement starts in earnest a week or so earlier in the Cascades. Spring migration peaks from mid-Apr to early May.

Golden-crowned Sparrows have bred at least once in Washington, at Harts Pass in Jul 1956 (Farner and Buss 1957), and may have also attempted breeding in the Seven Lakes Basin, Clallam Co., during Jun 1992. A bird singing on territory from late May to late Jun at Newhalem (Wahl 1995) may have been a late migrant. The species typically does not nest nr. the U. S. border in B.C., but there are breeding records from the s. coastal mountains and s. Vancouver I.

Jewett et al. (1953) stated that "a few birds winter regularly." Increases on the westside ($P < 0.01$) and eastside are indicated by long-term CBC data (Table 40; Wahl 1995). Increases presumably resulted from habitat created by human land-use changes (e.g. Wahl 1995) resulting in more winter habitat like weedy fields, brush, blackberry brambles, along with expanded availability of urban bird feeders. Other factors, including expanded breeding range and populations in Alaska and n. Canada are uncertain.

Remarks: Though there are some apparent local differences in habitat associations, birds are often noted with wintering White-crowned Sparrows which are attributed to more "weedy"situations (Bent 1968, Stepniewski 1999). Hybrids with White-crowned Sparrow have been noted on several occasions (e.g. Wahl 1995; see White-crowned Sparrow), especially during spring migration.

Noteworthy Records: *Summer*: 2 nesting pairs in Jul 1956 at Harts Pass (Farner and Buss 1957); singing 24 May-29 Jun 1975 at Newhalem (Wahl 1995); pair on 24 Jun 1992 at Seven Lks. Basin, Clallam Co.; 2 Jun 1998 at Two Rivers CP; 22 Jun 1998 at Sequim; 3 Jun 1999 at Vantage; 7 Jun 2002 at Hardy Canyon, Yakima Co. *High counts*, **West**: 220 on 25 Dec 1997 at Fir I. (SGM). **East**: 50 on 26 Nov 1999 nr. Maryhill (SGM, S. Finnegan p.c.).

Steven G. Mlodinow

Dark-eyed Junco *Junco hyemalis*

Common resident, migrant and winter visitor statewide.

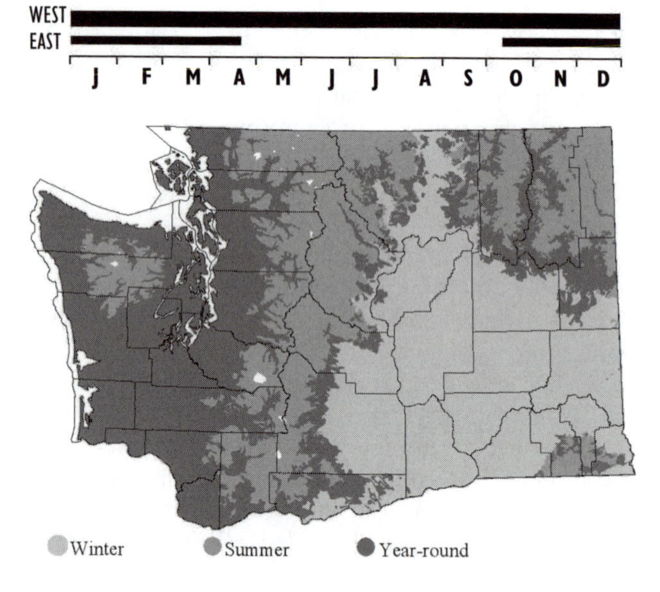

Subspecies: Oregon Junco, *J. h. oreganus*, and Slate-colored Junco, *hyemalis*.

Habitat: Deciduous and coniferous wooded habitats with a mixture of openings and dense herbaceous ground cover, most often open coniferous forest with brushy understory or openings (DeGraaf et al. 1991, Smith et al. 1997). During migration and winter, a wide range of habitats, especially in towns and suburbs, weedy fields and agricultural areas with nearby cover.

Occurrence: One of the most familiar species at winter bird feeders, the junco is a widespread summer resident in forested zones, where it readily adapts to changes, such as fires and clear-cuts, and also breeds in city parks with small forested stands. Historically the subject of species splits and subspecies realignments (e.g., Jewett et al. 1953, Rising 1996, Smith et al. 1997), but Washington races and status can be relatively simplified.

Oregon Junco. The common junco throughout the state. Most numerous in lowlands from fall to spring, likely due to some altitudinal migration as well as an influx from farther n. On the westside, movement into non-breeding areas typically first noted in mid-late Sep, and juncos have largely left these locations after mid-late Apr. Fall arrival and spring departure from non-breeding areas on the eastside is about a week earlier.

Slate-colored Junco. Migrant and winter visitor throughout, with peak numbers present from mid-Oct to mid-Apr. During this period, typically outnumbered by Oregon Juncos by about 200:1. Apparent intergrades between Oregon and Slate-colored juncos also occur between Oct and Apr. These intergrades are outnumbered by "pure" birds by about 4:1. Winter appearance of *hyemalis* during some major influxes of *oreganus* suggests relatively distant origin of some of the latter (Wahl 1995).

There have been few apparent changes in status over time. BBS data indicated a 0.2 % decrease per year from 1966 to 2000 statewide (Sauer et al. 2001). Long-term CBC data averages indicate a westside decrease ($P < 0.05$), while sizable increases in party hours (effort) probably resulted in more thorough censuses at bird feeders in urban areas and more time to check birds in brush in rural areas. Trends of Slate-colored Junco on CBCs are uncertain and confused by differences in identification criteria and numbers of juncos unidentified as to subspecies. Birds are recorded on CBCs statewide, with most frequent occurrence apparent e. of the Cascades, particularly at Spokane, which is closest to the breeding range of the subspecies.

Remarks: Both races in Washington have been further subdivided into forms not reliably separable in the field. There are 14 races classed in five somewhat phenotypically distinctive groups (see Rising 1996) or 15 subspecies (see Nolan Jr. et al. 2002).

Noteworthy Records: *Oregon Junco, high counts,* **Winter**: 2000 on 1969 CBC at L. Terrell; 600 on 26 Dec 1998 in Skagit Co.; 500 on 2 Jan 1999 at L. Terrell WMA (SGM, D. Duffy p.c.). Fall: 300 on 17 Oct 1998 at Fir I. (SGM). *Spring*: 300 on 4 Apr 1999 at Skagit WMA (SGM). *Slate-colored Junco, high count*: 4 on 2 Jan 1999 at L. Terrell WMA. *Fall, early*: 21 Sep 1998 at Skagit WMA (SGM); 22 Sep 1996 at Yakima Training Center (Stepniewski 1999); 22 Sep 2002 at Windust SP; 23 Sep 1965 at Spokane. *Spring, late*: 1 May 1975 at Seattle; 8 May 1975 at Nisqually NWR.

Steven G. Mlodinow

Lapland Longspur *Calcarius lapponicus*

Rare winter resident, and rare spring migrant in w. Washington. Casual in summer. Uncommon migrant and winter resident on eastside.

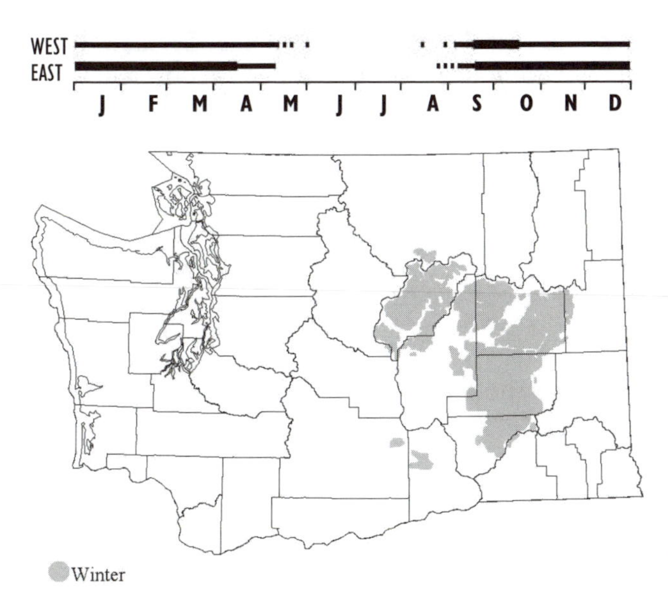

Subspecies: *C. l. alascensis.*

Habitat: Sparsely vegetated habitats: sandy coastal areas with beach grass, moist plowed agricultural fields and mudflat edges in w. Mostly agricultural fields, especially moist areas, often with Horned Larks, in the e.

Occurrence: Lapland Longspurs were considered "casual" by Jewett et al. (1953). Though this may not have been accurate for migrations, it might have reflected winter status. During recent decades most evident during fall migration, which occurs mainly between mid-Sep to mid-Oct. At this time, longspurs are most numerous on the outer coast, where they are fairly common. Large numbers occur during some fall migrations, with 1971, 1972, 1973, and 1997 particularly noteworthy. During winter, found mostly in e. Washington with flocks of Horned Larks. Spring passage less evident than fall migration, peaking in e. Washington between mid-Mar to early Apr, and in w. Washington during late Apr-early May. There is one w. Washington summer record. Birds have been recorded several times during Sep and Oct from alpine areas in the Cascades, implying a regular fall passage along the Cascade crest.

Based on the only quantitative data available there was an apparent decline in w. Washington. Longspurs occurred at nine w. Washington CBC locations between 1970 and 1999, with 199 of 282 birds found at two sites with historical occurrence—Bellingham and Grays Harbor. Some suitable habitat remained unaltered from 1960 to 2000, but wintering birds essentially disappeared from the Bellingham CBC in the 1970s and declined similarly at Grays Harbor, with a significant overall westside decrease ($P < 0.01$). Occurrence e. of the Cascades was too infrequent to suggest a trend.

Noteworthy Records: *High counts,* **West**, *outer coast*: 300 on 19 Oct 1971 at Leadbetter Pt.; 300 on 11 Oct 1997 at Ocean Shores; 200 on 3 Oct 1970 at Leadbetter Pt. **East**: 60 on 27 Dec 1973 at Pullman (Weber and Larrison 1977). **West,** *Winter*: 75 on 16 Jan-20 Feb 1993 at Lummi Flats (Wahl 1995). *Fall, early*: 15 Aug 1999 at Ocean Shores; 29 Aug 1998 at Calispell Peak; 29 Aug 1998 at Sunrise, Mt. Rainier; 30 Aug 1973 at Reardan; 30 Aug 1981 at Reardan. *Spring, late*: 22 May 1971 at Westport; 22 May 1971 at Mukkaw Bay; 22 May 1999 at Ocean Shores. 53). **East**, *Spring, late*: 23 Apr 1994 nr. St. Andrews. Summer: 1 Jul 1975 at Seattle (Hunn 1982).

Steven G. Mlodinow

Chestnut-collared Longspur *Calcarius ornatus*

Very rare visitor.

Breeding in the n. Great Plains region and wintering across much of the Southwest, this species occurs annually, primarily in fall-winter in California, but is much rarer further n. Oregon has approximately 10 records through 2001. Washington has five records to date, with three from spring-summer: one ad. male at Tokeland, 7 Jul 1974 (Harrington-Tweit 1979b), one ad. male (ph.) at Pt. Grenville, 26 Jun 1975 (Harrington-Tweit 1979), an ad. male along Soap L. Rd. between Brewster and Okanogan, Okanogan Co. on 27 May 1995 (Aanerud and Mattocks 1997), an ad. (ph.) at Seattle, 3-12 Dec 1995 (Aanerud and Mattocks 1997), an ad. male (ph.) at Hoquiam, 27-29 Nov 1999.

Bill Tweit

Rustic Bunting *Emberiza rustica*

Casual vagrant.

Three Washington records: an imm. (ph.) at Kent, 15 Dec 1986-22 Mar 1987, an ad. (ph.) at Kent, 11 Dec 1988-Apr 1989, one (ph.) at Leavenworth, 9-23 Jan 1999. The two Kent records may refer to the same bird, but the WBRC chose to treat these as separate records (Tweit and Skriletz 1996).

This Siberian species has been recorded nine other times in N. America s. of Alaska. These records stretch as far s. and e. as Kern Co., California, and span 26 Oct-30 Apr, with peak time from late Nov to late Feb. The e. Siberian *latifascia* is presumably the race occurring here.

Steven G. Mlodinow

Snow Bunting *Plectrophenax nivalis*

Uncommon and somewhat irregular migrant and winter visitor in e. Washington. Rare migrant and winter visitor w.

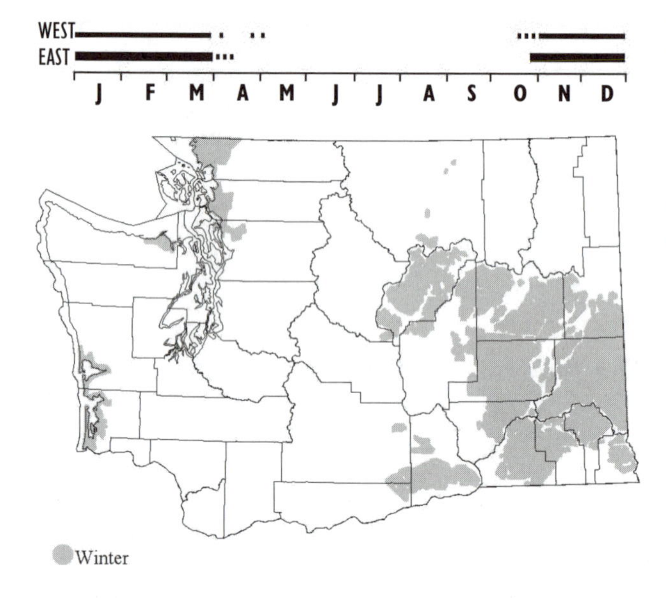

Subspecies: *P. n. nivalis.*

Habitat: Barren or sparsely vegetated fields, beaches, and mudflats.

Occurrence: E. of the Cascades, numbers appear to increase through fall and early winter and peak in Jan and Feb. They are reported most regularly in the n. Columbia Basin and appear to be in smaller numbers farther s. Largest concentrations have typically been reported in dryland wheat fields in Lincoln, n. Grant, and Douglas Cos. A few have also been reported during southbound passage in alpine areas

Table 41. Snow Bunting: CBC averages per 100 party hours by decade.

	1960s	1970s	1980s	1990s
West				
Bellingham		12.13	6.13	1.11
Grays Harbor		9.68	0.00	0.06
Leadbetter Pt		19.72	2.11	0.00
Sequim-Dungeness			4.54	2.51
East				
Spokane	1.43	0.66	0.00	0.00
Wenatchee	27.62	12.98	0.79	0.73

of Cascades. Numbers can vary greatly from year to year, presumably due to conditions farther n. Peak years in the e. do not necessarily correspond with peak years noted w. of the mountains. In w. Washington, peak numbers present from late Oct-early Nov into Dec and early Jan. Most reports are from the outer coast and the n. Puget Trough.

Winter numbers are represented on only a few CBCs statewide. Other reports are sporadic, with a number of locations reporting the species only once or twice over 20-plus years. Birds apparently occurred in greater numbers during the 1960s and early 1970s. They occurred with reasonable historic frequency on few counts and on these there was an obvious downward trend (Table 41), with a significant decrease ($P < 0.01$) w. of the Cascades.

Noteworthy Records: *High counts*, **East**: 15,000-20,000 on 19 Feb 1972 nr. Davenport; 7500 on 12 Feb 1978 nr. Davenport; 4000 on 19 Jan 1996 on Waterville Plateau; 3000 on 19 Jan 1996 at Mondovi, Lincoln Co.; 3000 on 9 Feb 1997 at Reardan; 2500 on 19 Dec 1993 nr. Davenport; 4000 in 2002-03 on Waterville Plateau. **West**: 225 on 1 Dec 1973 at Lummi Flats; 100 in late Dec 1972 to mid-Feb 1973 at Ocean Shores; 100 from 24 Nov 1966 to Jan 1967 at Sandy Pt. *Fall, early*: 10 Oct 1997 at Salt Cr. CP; 11 Oct 1987 at Ocean Shores; 12 Oct 1985 at Everett; 13 Oct 1983 at Ocean Shores. *Spring, late*: 21 Apr 1985 at Seattle; 21 Apr 1988 at Ocean Shores; 3 May 1975 at Westport, 18 and 23 May 2003 at Ocean Shores.

Steven G. Mlodinow

McKay's Bunting *Plectrophenax hypoboreus*

Casual vagrant.

Three Washington records: two birds (one ph.) at Ocean Shores, 16 Dec 1978-5 Mar 1979 (Aanerud and Mattocks 1997). One (ph.) at Ocean Shores in Jan-Feb 1988 (Aanerud and Mattocks 1997). One was at Lummi Flats on 27-28 Nov 1993 (Wahl 1995, Aanerud and Mattocks 1997). An apparent

McKay'sXSnow Bunting hybrid was at Midway Beach, 13 Feb 2001.

This species, which breeds on islands in the Bering Sea, is typically a short-distance migrant, but there are two other records away from Alaska: Clatsop Spit, Oregon, 23 Feb-9 Mar 1980 (Gilligan et al 1994) and Pacific Rim NP, B.C., 12 Feb 1980 (Campbell et al. 2001). Apparent hybrids with Snow Bunting are found regularly in Beringia, complicating identification (Sibley 2000). Classed GL, on NAS Watch List.

Steven G. Mlodinow

Rose-breasted Grosbeak

Pheucticus ludovicianus

Very rare late spring and summer visitant.

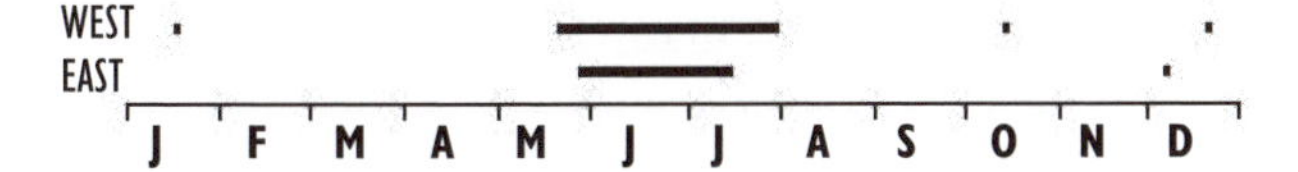

Habitat: Deciduous woodlands.

Occurrence: The first record of this e. N. American species was a specimen of an ad. male at Sprague on 2 May 1956 (Mattocks et al. 1976, Tweit and Skriletz 1996). The WBRC has accepted an additional 19 records through 2001 and another 10 reports with documentation await examination. Twenty-four of the 30 appeared between 22 May and 30 Jul. Virtually all of these were males. Records are evenly divided between the westside and the interior, and come primarily from populous areas apparently reflecting the distribution of observers.

As Roberson (1980) noted, this species is "unaccountably scarce" in Washington compared to its status in B.C., where it breeds in the ne. portion, and Oregon and California, where it is of annual and comparatively frequent occurrence. Its frequency in Washington is increasing; 24 of the 30 birds have appeared from 1990 on and they have been recorded annually since 1996.

Noteworthy Records: *Winter*: 13-16 Jan 1990 male at feeder at Chehalis; male at feeder 9-12 Dec 1998 at College Place was eaten by Sharp-shinned Hawk; 20 Dec 2001 nr. the Elwha R. mouth. *Fall*: ad. male on 3 Sep 1994 at Shaw I.; imm. male on 2 Oct 1992 at Spokane.

Bill Tweit

Black-headed Grosbeak

Pheucticus melanocephalus

Fairly common migrant and common summer resident in low-elevation forests and riparian corridors.

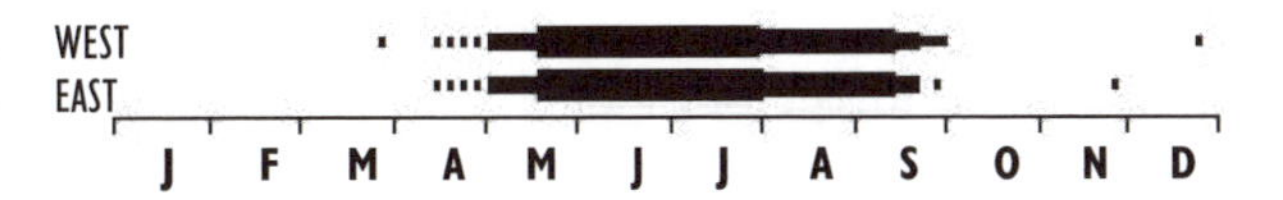

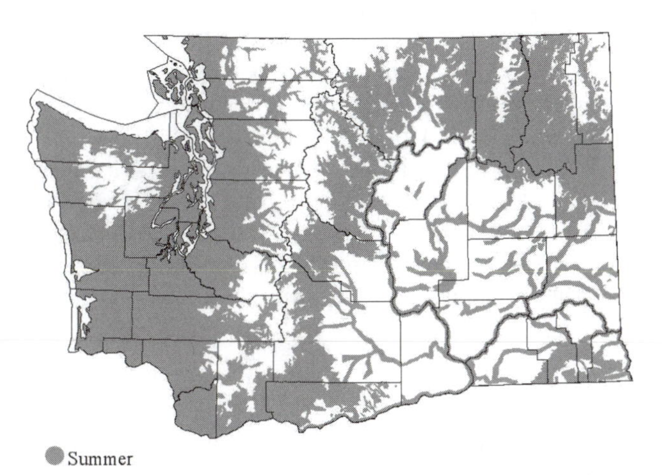

Subspecies: *P. m. melanocephalus* breeds in w. Washington, *maculatus* in e.

Black-headed Grosbeak (G. Scott Mills)

Habitat: Lower-elevation deciduous forests or hardwoods/shrubby habitats, including riparian zones, in coniferous forests.

Occurrence: The Black-headed Grosbeak historically has been described as uncommon, fairly common, and common in Washington by various authors (e.g., Jewett et al. 1953, Lewis and Sharpe 1987, Ennor 1991, Wahl 1995, Smith et al. 1997), with the sequence appearing to represent an increase in population size over time.

In e. Washington, birds are found in riparian forests in shrub-steppe (Smith et al. 1997), though local and not in all situations as in Yakima Co. (Stepniewski 1999). In managed forests in the s. Cascades, Black-headed Grosbeaks were found to be most abundant in open-canopied, early successional forests, which have a high percentage of deciduous vegetation. Abundance declined significantly in closed-canopy forests (Manuwal and Pearson 1997). In ne. Washington, the Black-headed Grosbeak was found to be most abundant in regenerating forests (O'Connell et al. 1997).

Increased numbers were apparent, at least locally, in the early 2000s in w. Wahington. This increase from historical abundance may have resulted from an increase in deciduous forest. BBS data from 1966 to 1996 show a statistically significant increase ($P < 0.01$) of 4.3% annually in Washington (Sauer et al. 2001).

Noteworthy Records: *Spring, early*: 28 Mar 1977 at Dungeness; 25 Apr 1964 at Drayton Harbor; 29 Apr 1998 at Sedro Woolley; 30 Apr 1999 at Trout L. *Fall, late*: 25 Sep 2001 at Mt. Pleasant, Klickitat Co.; 30 Sep 2001 at Cliffdell, Yakima Co.; Dec 1990 at Lummi Pen. (Wahl 1995); late Nov 1986 at College Place. *High count*: 64 on 30 Jun 2001 at Coppei Cr.

John F. Grettenberger

Lazuli Bunting *Passerina amoena*

Common migrant and summer resident in e. Washington. Uncommon and local in the Puget Trough, locally common in Clark and Skamania Cos., rare further w.

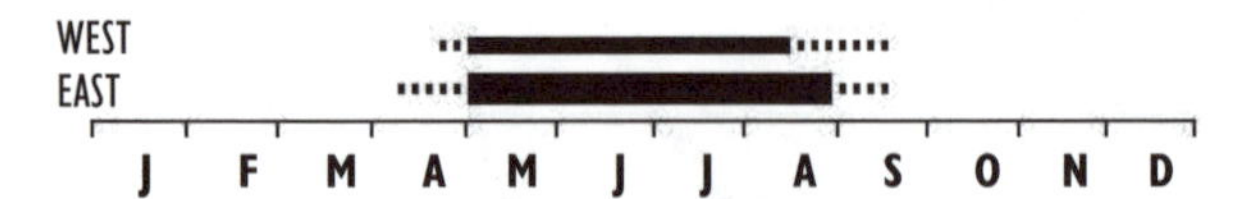

Habitat: Breeds in a variety of brushy habitats, especially brushy hillsides, riparian thickets, open woodlands, draws, and on canyon slopes from nr. sea level to over 1600 m. Will use recent post-fire habitats and, to a lesser extent, clear-cuts. Westside

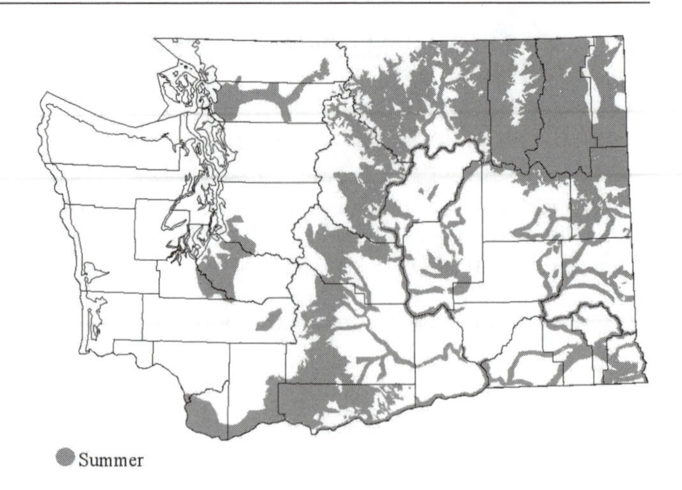

habitats include farmland edges, grasslands, hedges, and thickets, including power-line rights-of-way.

Occurrence: In 1950, considered a common migrant and summer resident in habitat in e. Washington, less commonly a migrant and breeder w. of the Cascades within the Puget Trough, and absent from the Olympic Pen. and outer coast (Jewett et al. 1953).

In addition to widespread distribution e. of the Cascades, BBA surveys showed most in the w. concentrated on the Fort Lewis prairie, in sw. King Co., around Vancouver and e. along the Columbia R. into Skamania Co., with local populations along the Skagit R. e. to Newhalem and the lower Cowlitz R. (Smith et al. 1997). Has been found locally along other westside river valleys such as the Snohomish. Recorded in Jun on Orcas I. (Lewis and Sharpe 1987), and on six occasions along the outer coast and on the Olympic Pen.

Migrants first found during early and mid-May, with only a few published records of arrivals before 1 May (*contra* Jewett et al. 1953). Similarly, Cannings et al. (1987) reported only two Apr arrivals in the Okanagan Valley of s. B.C. Peak numbers not present until mid-May.

Birds nest from nr. sea level to 1500 m (Jewett et al. 1953, Smith et al. 1997). It is generally believed that birds in the n. edge of their breeding range produce only one brood per season (Greene et al. 1996). Nests found from late May through late Jul, and later records may represent renesting. Parasitized (one of 11 nests) by Brown-headed Cowbirds in B.C. (Cannings et al. (1987).

By mid-Jul to mid-Aug, individuals begin drifting off territories and coalescing into small flocks during a premigratory fattening period. A flock or 40 to 50 along the Grande Ronde R. on 21 July 1966 is the largest concentration reported in Washington. Most fall migration in mid-Aug to early Sep, with only three published records later than 15 Sep in Washington: birds on 18 Sep along the Grande

Ronde R., on 19 Sep 1969 at Hart's Pass, and one on 31 Oct 1997 at Tokeland was the latest. Populations appear to be increasing in Washington. BBS data show a statewide annual population increase of 2.6% (P <0.05) from 1966 to 2000 (Sauer et al. 2001).

Noteworthy Records: East, *Spring, early*: 6 Apr 1913 at Prescott, Walla Walla Co. (Dice 1918); 25 Apr in se. Washington (Hudson and Yocum 1954); 28 Apr 1997 at Rock Cr., Klickitat Co.; 29 Apr 1934 at Wenatchee (UWBM 10376); 30 Apr 1997 at Vantage. *Fall, late:* 13 Sep at Grande Ronde R. (Weber and Larrison 1977). *High count*: 100 on 30 Jun 2001 at Coppei Cr. **West**, *Spring, early*: 22 Apr 1997 at Steigerwald L. NWR, 30 Apr 2000 at Steigerwald L. *San Juan Is.:* 28 Jun 1976 (Lewis and Sharpe 1987). *Outer coast*: 15 May 1982 at L. Ozette; 7 Sep 1995 at Ocean Shores; 31 Oct 1997 at Tokeland; 18 May 2000 at Dungeness; 22 May 2000 at Sequim; 7 Jul 2001 nr. Lebam, Pacific Co.

Robert C. Kuntz II

Indigo Bunting *Passerina cyanea*

Very rare spring and fall visitant.

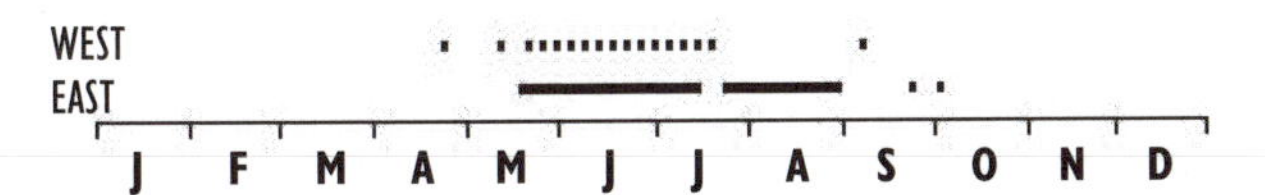

Habitat: Brushy edges, riparian areas, second-growth woodlands.

Occurrence: There have been 17 Washington records through spring 2002; 11 have been from the westside. Records are evenly divided between spring (late Apr–May) and summer (Jun– Jul), with only two in fall. Twelve records have been accepted by the WBRC, and four more are unreviewed. Except for the two fall records, all have been of ad. males, and eight of those have been birds that appeared at feeders. The predominance of ad. males at feeders suggests that many Indigo Buntings are likely undetected in Washington at present. The much greater numbers of records in California and Oregon indicates also suggests low frequency of detection: California averages about 55 records annually (Roberson 1980) and there are about 50 Oregon reports presently (NAB 56:351).

Records: West: 22 Apr 1992 at Diablo. Whatcom Co. (Wahl 1995); 10-11 May 2000 at Tacoma; 13-16 May 2003 nr. Rockport; 17 May 2001 at Washougal; 18 May 2000 at Nordland, Jefferson Co.; 19 May 1992 at San Juan I,; 1 Jun 1996 at Redmond; 7 Jun 2000 at Concrete; 10 Jun 2001 at Roy, Pierce Co.; 17 Jul 1984 at San Juan I.; 13 Sep 1988 at Seattle (Tweit and Skriletz 1996). **East**: 18 May 2002 at Yakima; 25-28 May 2002 at Potholes Res.; 29 May 2003 nr. Leavenworth; 25 Jun 1977 at Chewelah; 7-14 Jul 1973 at Pend Oreille SP.; 19 Jul-31 Aug 1999 nr. W. Richland; 23 Sep-2 Oct 1994 at Spokane (Aanerud and Mattocks 1997).

Bill Tweit

Painted Bunting *Passerina ciris*

Casual vagrant.

An ad. male in Seattle in Feb-Mar 2002 was the first reported in Washington. Breeds in the s. U.S. There are five records from e. Oregon, in Jun, Oct, Nov, and Dec (Marshall et al. 2003), and none in B.C. Classed GL.

Terence R. Wahl

Dickcissel *Spiza americana*

Casual vagrant.

A vagrant, with five records accepted in Washington: a male (ph.) at Beaver, Clallam Co., 4-16 Nov 1983; a male at Puget I., 23 Dec 1983-14 Apr 1984; a male (ph.) on Tatoosh I., 8-9 Oct 1987; a bird at Ocean Shores, 18-28 Feb 1996; one on Dodson Rd. nr. Frenchman Hills Rd., 11 Jun 1997 (Aanerud and Mattocks 2000). Another bird (reportedly ph.) was at Edison on 24 May 1995.

Oregon and B.C. have about 16 records combined, essentially falling into two periods. The majority of Dickcissels have been detected from late fall to early spring—likely birds that wintered or attempted to winter in the Northwest. Most of the remaining records are from the late May to mid-Jun passerine vagrant season.

Steven G. Mlodinow

Bobolink *Dolichonyx oryzivorus*

Locally common breeder in wet meadows and irrigated hayfields in ne. counties and along Toppenish Creek in Yakima Co. Rare away from breeding areas. Very rare migrant in w.

Habitat: Riparian wetlands, hayfields, and other open agricultural lands along rivers.

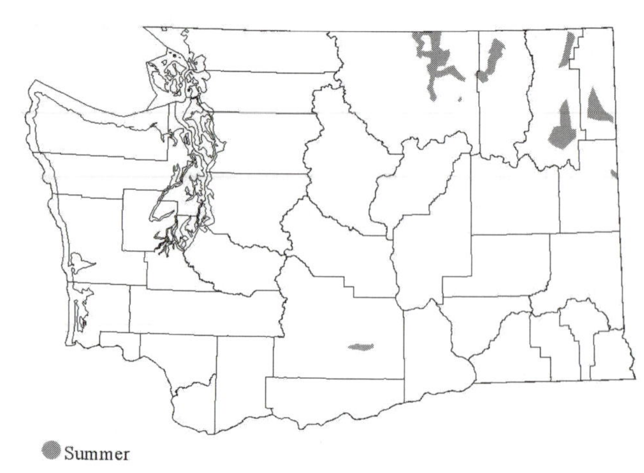

Occurrence: Washington's scattered Bobolink colonies are at the fringe of this highly migratory species' range in N. America. Limited knowledge of status in 1950 (Jewett et al. 1953) began to change in the 1960s, when field observers brought attention to the population at Cusick, Pend Oreille Co. Increasing field effort detailed several more locations. BBA surveys in the late 1980s-early 1990s showed birds nested in loose colonies in Washington at only 19 sites in Okanogan, Stevens, Ferry, Pend Oreille, Spokane, and Yakima Cos. (see Smith et al. 1997). Some of these (e.g., at Cusick, Curlew L., Aeneas Cr., and, marginally, Newman L.) are known to have existed for decades. The Yakima Co. site on Toppenish Cr. is the westernmost breeding area in the U.S., separated from the ne. Washington areas by 250 km, and appears to be relatively stable (Stepniewski 1999).

There are very few records of migrants on the eastside. Weber and Larrison (1977) described the species as a rare migrant and summer visitor in se. counties but there are few subsequent records. Bobolinks are very rare migrants on the westside, with most spring records from late May to early Jun and most fall records from late Aug to mid-Sep. A bird at Graysmarsh from 20 Dec 1999 to 8 Jan 2000 represents one of the few winter records from anywhere in N. America. From BBS data for the w. United States, DeSante and George (1994) noted a decreasing trend in populations from 1966 to 1991.

Remarks: Local populations may face problems due to mowing of hayfields and cattle grazing (Stepniewski 1999), though mowing is apparently often late enough in the season to allow successful reproduction (J. Acton p.c.). The unique Yakima Co. population is noted in Yakama tribal wildlife-management programs (Tracy Hames, Yakama Indian Nation Wildlife Resource Management Program). On NAS Watch List.

Noteworthy Records: West, *Spring-Summer*: 18 May 2002 at Cochreham I., Skagit Co.; 25 May 1979 at Seattle; 1 on 25 May 1982 at Ross L. (Wahl 1995); 26 May 1981 at Seattle; 1 on 28 May 1998 at Point No Point; 1 Jun 2001 at Seattle; 2-3 Jun 1980 at Seattle; 4-5 Jun 1977 at Tatoosh I.; 5 Jun 1989 at Tatoosh I.; 1 on 8 Jun 1974 at Cape Alava. *Fall*: 15 Aug 1982 at Seattle; 15 Aug 2000 at Point No Point; 3 and 14 Sep 1981 at Seattle; 12 Sep 1998 at Ocean Shores; 10 Oct 1983 at Seattle. *Winter*: 1 from 20 Dec 1999 to 8 Jan 2000 at Graysmarsh, Clallam Co. **East**, *Fall*: 20 Aug 1971 at Walla Walla; 27 Aug 1995 at Atkins L., Douglas Co.; 8 Sep 1974 at Spokane; 18 Sep 1976 at Calispell L., Pend Oreille Co.

Rosemary H. Leach

Red-winged Blackbird *Agelaius phoeniceus*

Common resident in lowlands statewide, summer resident in sub-alpine mountain valleys.

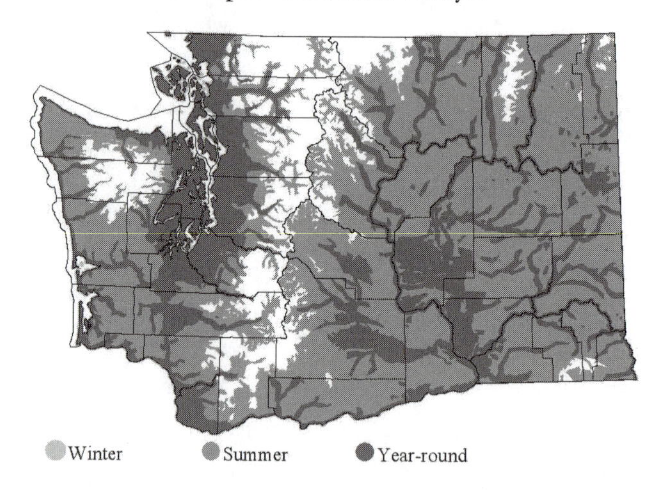

Subspecies: *A. p. nevadensis* breeds on the eastside and *caurinus* on the westside.

Habitat: Widespread, breeding in almost all habitats with suitable microhabitat of wetlands with emergent vegetation, especially cattails (Smith et al. 1997). During migration and winter, habitat types are expanded to agricultural fields, dairy farms, and feedlots.

Occurrence: Familar, ubiquitous, and found year-round virtually everywhere except dense forests and high mountains, this is one of the most abundant birds in N. America (Yasukawa and Searcy 1995). Though natural habitats were decreased by human actions, others were created, and this species adapted to very limited, localized, situations like roadside ditches (Smith et al. 1997) and to winter foraging in agricultural habitats. Resident statewide year round, some local movements occur, and birds retreat from higher elevations and concentrate in lowlands during winter and forage fields and feedlots (Stepniewski 1999). Regionally, birds are also distinctly less common in ne. Washington during winter than in summer.

Because of year-round occurrence and elevational movements, migration is often not obvious. Most northbound birds apparently arrive during Feb and Mar while most southbound birds likely leave during Oct and Nov (e.g., Jewett et al. 1953, Wahl 1995, Stepniewski 1999).

Winter populations and trends not apparent from CBC data, which often, irregularly, record sizable numbers of unidentified blackbirds. In addition, CBCs in rural areas, where largest numbers of blackbirds occur, have not been conducted for representative time-spans. BBS data indicated a significant breeding population increase statewide of 3.8% per year from 1966 to 1991 (Peterjohn 1991) and a statewide increase from 1966 to 2000 of 0.6 % per year (Sauer et al. 2001).

Remarks: Recognized races fall into two major groups generally separable in the field: Red-winged Blackbird (currently 26 races) and Bicolored Blackbird (three races: Jaramillo and Burke 1999; AOU designated 14 total races in 1957). Field identification down to subspecies level is otherwise dubious. Bicolored Blackbird, which nests mainly in California, has not been recorded in Washington. In many parts of the U.S. considered a "pest" species, damaging agricultural crops and subject to control programs.

Noteworthy Records: *High counts*: 5000 on 3 Aug 1963 at Cusick; 5000 on 25 Nov 1965 nr. Spangle, Spokane Co.; 5000 on 7 Jan 1995 nr. Conway (SGM).

Steven G. Mlodinow

Tricolored Blackbird *Agelaius tricolor*

Very local and rare resident in e. Washington. Casual in w. during fall and late winter. Status rapidly changing.

WEST
EAST
J F M A M J J A S O N D

Habitat: In Washington and Oregon nests almost exclusively in dense marshes of cattails or tules. Elsewhere utilizes wet brambles, nettles, thistles, and willow (Jaramillo and Burke 1999). There are almost no winter Washington records, but elsewhere typically in agricultural areas, especially where there is livestock and the vegetation is short, and often in mixed flocks with other blackbirds.

Occurrence: The status of the Tricolored Blackbird is one of rapid change. The species was first documented in Washington on 5 Jul 1998, when a colony of 30, perhaps 50, birds was located along Crab Cr., nr. the town of Wilson Cr. Subsequently, the species was found annually there, with sightings as early as 8 Apr. The colony is typically abandoned by early or mid-Jul. Peak counts were five during 1999, 25 in 2000, 28 in 2001, and 30-plus in 2002. During May-Jul, singles have been reported also from Othello, nr. Potholes Res., and nr. Alderdale, Klickitat Co. A flock of 30-40 birds in farmland nr. Texas L., Whitman Co., on 31 May 2002 lacked imms., was not nr. suitable nesting habitat, and likely represented failed breeders or a flock searching for a nesting site. Records from Aug to Apr are mostly from Othello and may represent birds from Wilson Cr. There are also two fall-winter records from Vancouver lowlands, perhaps from the nearby small breeding population in Portland, Oregon.

This species' occurrence in Washington is part of a wider range expansion in the Pacific Northwest. Historically, the species likely bred no farther n. than Klamath Co., Oregon, but over the last 30 or so years it has spread into n. Oregon, nesting at isolated colonies as far n. as Portland and in Wasco and Umatilla Cos. (Gilligan et al. 1994).

Noteworthy Records: West: 3 on 25-26 Nov 2000 and 1 on 13 Dec 2000 nr. Vancouver L.; 4 on 2-20 Feb 2000 and 2 on 3 Dec 2002 at Shillapoo Bottoms, Clark Co. **East**: 1 on 30 May 1999 at Othello; 3 on 4 Apr 2001 at Othello; 1 on 12 May 2001 at Alderdale, Klickitat Co.; 2 on 22 Mar-7 Apr 2002 at Othello; 30-40 on 31 May 2002 nr. Texas L., Whitman Co.; 1 on 20 Jul 2000 at Potholes Res.; 2 on 9 Sep 2002 at Othello; 8 on 9 Feb 2003 at Othello.

Steven G. Mlodinow

Western Meadowlark *Sturnella neglecta*

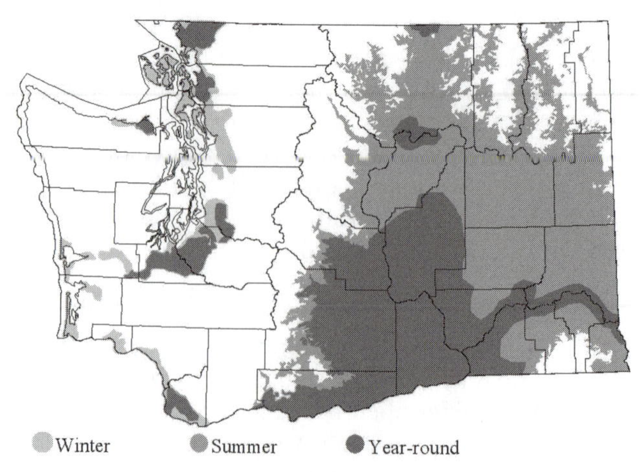

Common permanent resident in e. Washington. Fairly common migrant/winter resident and locally uncommon summer resident in w.

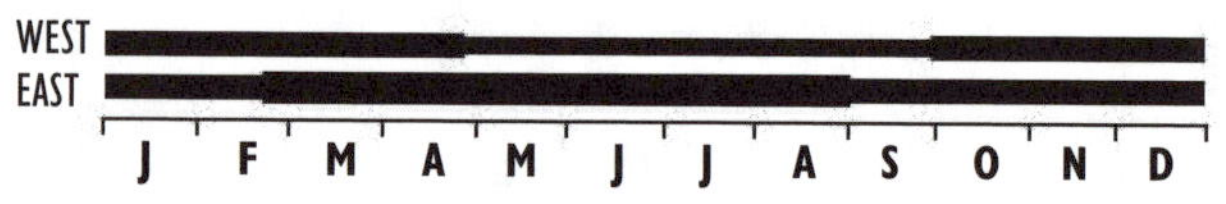

Subspecies: None (Jaramillo and Burke 1999), though others have considered nw. meadowlarks a separate race, *confluenta* (Jewett et al. 1953, AOU 1957; see Lanyon 1994).

Habitat: Breeds in grasslands, agricultural areas, and shrub-steppe (Smith et al. 1997). Winters in similar habitats, but often found in areas too marginal for breeding.

Occurrence: One of the most widespread and typical summer birds in much of the open country e. of the Cascades. In winter the Western Meadowlark withdraws from most of the n. Columbia Basin except locally as in snow-free agricultural areas and feedlots. Numbers decline through Sep and continue to do so, to some extent, throughout the winter, with an increase again beginning in Feb in s. parts of the Columbia Basin and in March in the n. (Stepniewski 1999). Wintering numbers are lower in the n. part of e. counties adjacent to the Canadian border. Breeding numbers may be increasing in ne. Washington due to conversion of wooded areas to farmland (Smith et al. 1997).

In w. Washington, meadowlarks were formerly a common breeder but development and intensive farming of grasslands have led to a sharp decline (Jewett et al. 1953, Smith et al. 1997). Now, breeding birds are quite local, restricted to a few remnant quality grasslands in the Puget Trough, in Clallam and Clark Cos., and rarely in Grays Harbor and Pacific Cos. Westside migrants typically start to arrive in mid-Sep, with main numbers appearing in early Oct. Wintering birds mostly exit by mid-late Apr, but a few remain into early to mid-May.

BBS data indicate a statewide decline from 1966 to 2000 (Sauer et al. 2001). CBC data show a significant decline (P <0.01) on the westside.

Meadowlark (G. Scott Mills)

Noteworthy Records: *High counts*, **East**: 400 on 8 Sep 1962 ne. Stratford, Grant Co. Winter: 225 on 2 Jan 1999 nr. Goldendale. **West**: 130 on 11 Dec 1997 at Ridgefield NWR; 100 on 14 Dec 1998 at Leque I., Snohomish Co.

Steven G. Mlodinow

Yellow-headed Blackbird

Xanthocephalus xanthocephalus

Common summer resident/migrant and rare winter resident in e. Washington. Uncommon spring migrant, very local breeder, rare fall migrant, and very rare winter resident in w.

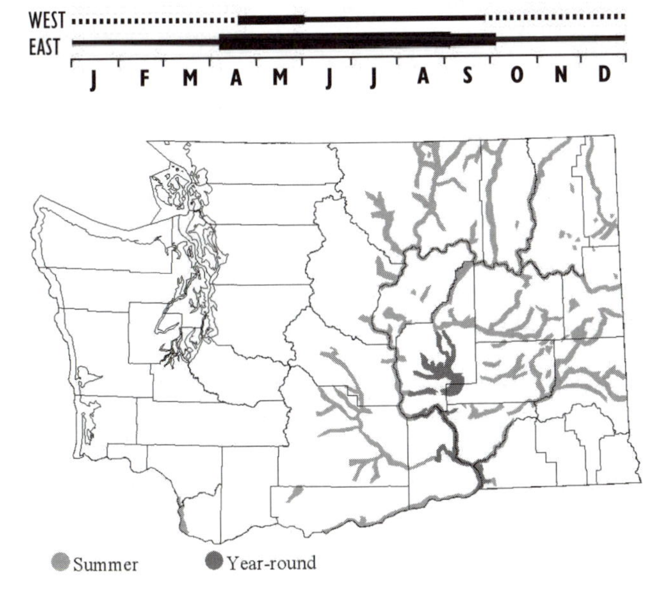

Habitat: Medium to large cattail, tule, and sedge marshes with open water. Found in migration and winter in a wide variety of wetlands, agricultural fields, feedlots, and dairy farms.

Occurrence: A conspicuous marsh bird with unmistakable vocalizations, the Yellow-headed Blackbird is a widespread breeder and migrant in e. Washington, occurring in lowland steppe habitats from the Columbia Basin and to 1100 m elevation in suitable habitats in forest zones. Favors more extensive marshes or those with deeper water than Red-winged Blackbirds (Smith et al. 1997).

Birds are relatively scarce in winter (e.g. Stepniewski 1999) except irregularly and very locally as at Moses L., where CBC numbers in the hundreds peaked in the mid-1990s. Most other winter records are concentrated in Tri-Cities, Yakima, and Walla Walla areas as shown by CBC numbers from those areas.

In w. Washington, most frequent—in low numbers—during spring migration with a peak from late Apr to late May. Established breeding colonies are few, with currently active sites only at Fort Lewis, Deer Lagoon, Ridgefield NWR, and nr. Wiser L., Whatcom Co. Isolated pairs breed occasionally elsewhere. Fall migrants are few, sporadic in most locations, with most between early Aug to mid-Sep. Records are almost entirely from the Puget Trough and Clark Co. and the species is very rare at any time on the outer coast and Olympic Pen.

BBS data indicate a decline in numbers in Washington from 1966 to 2000 (Sauer et al. 2001).

Noteworthy Records: West, *outer coast*: 26 Apr 1934 at Westport (Jewett et al. 1935); 17 Sep 1966 and 31 Aug 1975 at Westport; 3 on 23 Apr 1994, 2 on 30 Aug 1995 and 3 Sep 1997 at Ocean Shores; 14 Oct 1998 at Tokeland. *High counts*: 50 on 14 Aug 1968 in w. Skagit Co.; 30 on 11 May 1997 at Steigerwald L.; 30 on 27 Jul 2002 at Ridgefield NWR; 29 on 9 May 1970 at Skagit WMA; **East**, *high counts, Summer*: 1000 on 12 May 1918 sw. of Spokane (Jewett et al. 1953). *Winter*: 300+ on 15 Jan 1981 nr. Potholes Res.; 100 in Jan 1964 at Stratford; 100 on 2 Feb 2002 at Othello. CBC: 1500 in 1993 at Moses L.

Steven G. Mlodinow

Rusty Blackbird

Euphagus carolinus

Rare fall and winter visitant, particularly in the se. section.

WEST, EAST

J F M A M J J A S O N D

Subspecies: Presumably *E. c. carolinus.*

Occurrence: This species breeds throughout the boreal forest zone and winters in the se. U.S. The first Washington report, in 1927, was treated as hypothetical by Jewett et al. (1953) and rejected by the WBRC (Tweit and Skriletz 1996). The first unquestioned record was a bird (ph.) in the hand in Spokane on 26 Dec 1960 (Mattocks et al 1976). Through 2001 there were 59 records, with two-thirds of those occurring after 1989. Only two were recorded in the 1960s and they averaged under one per year in the 1970s and 1980s. From 1990 on, they have occurred annually and averaged 3.3 per year. This compares favorably with California, which averages fewer than 10 per year (Jaramillo and Burke 1999).

Records are concentrated in a few areas, with 56% of the total from the interior and 55% of those (18) from Walla Walla Co. The other interior counties have four or fewer records, most one or none. There are 27 westside records, 21 of those from the Puget Trough and only six from the outer coast and lower Columbia. Twelve of the westside records come from Snohomish, Skagit, and Whatcom Cos. Jaramillo and Burke (1999) speculate that this pattern of occurrence, more frequent in the interior than on the coast, is indicative of regular migration rather than vagrancy. There is little evidence, however, of spring northbound migration. Classed GL.

Noteworthy Records: *High counts*: 3 on 13 Oct 1973 at Turnbull NWR; 3 from 20 Dec 1987 to 9 Jan 1988 at Monroe. *Fall, early*: 24 Sep 1995 at Dungeness. *Spring-Summer*: 21 May 2000 at Ellensburg; 22-23 Jun 2001 at Tatoosh I. and a female was reported at Tatoosh I. on 24 Apr. 2003. *Winter*: 1 on 3 Dec 2002 at Shillapoo Bottoms, Clark Co.; 1 on 10-22 Jan 2003 nr. Silvana.

Bill Tweit

Brewer's Blackbird *Euphagus cyanocephalus*

Common in habitat year round statewide except uncommon in summer and absent in winter in ne. quadrant.

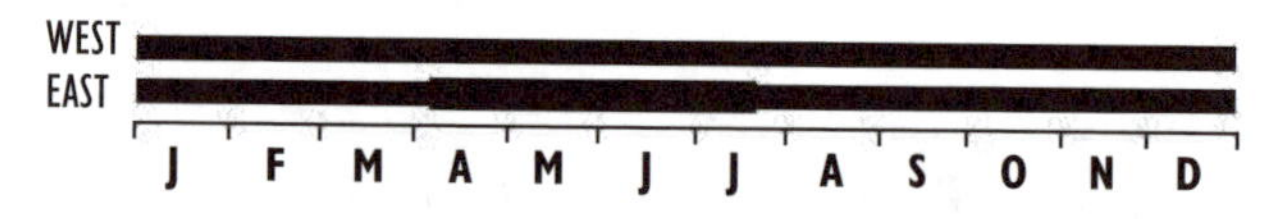

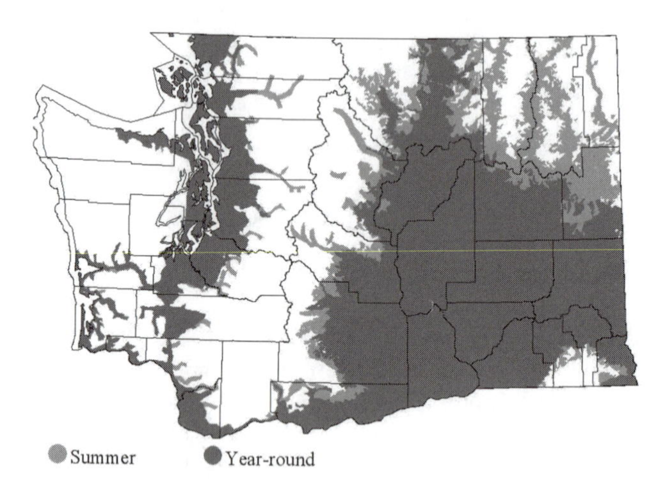

Habitat: Breeds mainly in agricultural areas and towns, generally avoiding larger municipalities. Also nests in shrub-steppe (typically nr. agricultural areas), grasslands, and clear-cuts. Winters mostly in agricultural areas, especially dairy farms and feedlots. During migration, occasionally seen in alpine areas.

Occurrence: In w. Washington, nesting has decreased in urban areas (e.g. Wahl 1995) but distribution is widespread in other habitats, including tree farms (Wahl 1995), from lowlands to suitable habitats, including ski areas and visitor centers, at higher elevations such as Snoqualmie Pass (Hunn 1982, Smith et al. 1997). Numbers increase and large flocks winter in lowland agricultural areas, with peak numbers from Oct to Mar. Whether these flocks comprise local populations or include birds from n. or inland nesting areas is uncertain.

E. of the Cascades, numbers peak from early Apr to late Jul (Stepniewski 1999), but a number linger through winter, especially in the s. Status in Ferry, Stevens, Pend Oreille Cos. incompletely known (J. Acton p.c.). Localized distribution (concentrated at feedlots in Yakima Co.) and westbound migrants noted in the mountains (Stepniewski 1999) and as high as 2300 m on Mt. Rainier indicate a likely westward shift for the winter. Banding recoveries show some Washington birds winter in California, though the extent of movements is unknown (Martin 2002).

BBS data indicate a population decline in the U.S. from 1966 to 1991 (DeSante and George 1994). Statewide population trends are uncertain (see Sauer et al. 2001). CBCs record highly variable numbers of Brewer's and also of unidentified blackbirds and this complicates interpretation of data. Though dense urbanization has eliminated some local populations, extensive conversion of steppe habitats to agriculture has increased food supply (Smith et al. 1997) and expansion of clear-cuts has increased that upland habitat.

Noteworthy Records: *High counts*: 4000 on 2 Dec 2001 at Othello (B. Flores p.c.); 3500 on 10 Nov 2002 nr. LaConner; 2000 on 27 Nov 1999 at Vancouver L. (SGM). CBC: 3020 in 1989 at Skagit Bay; 2926 in 1976 at Sequim-Dungeness; 2780 in 1993 at Moses L.; 2669 in 1974 at Bellingham; 2666 in 1989 at Toppenish.

Steve G. Mlodinow

Common Grackle *Quiscalus quiscula*

Very rare visitor.

Subspecies: Apparently *Q. q. versicolor.*

Occurrence: Although the first record was over 50 years ago, a specimen taken 22 Oct 1950 along the Idaho border (Jewett et al 1953), this species has remained very unusual in Washington. Grackles were found only once or twice per decade until 1995; they have been almost annual since then. As expected, 10 of the 13 records are from the eastside. This recent upsurge in reports coincides with strong westward range expansion in the last two decades (Jaramillo and Burke 1999).

Records: 22 Oct 1950 at Paradise Cr., Whitman Co.; 26-27 Jun 1965 at Seattle; 4 Dec 1974-18 Jan 1975 at Olympia; specimen on 3 Jul 1980 at Touchet; 1-6 Jan 1995 at College Place; 21 May 1996 at West Richland; 21 Jun 1997 at Tatoosh I.; 21 May 1998 at Grand Coulee; 12 May 2001 at Walla Walla; 8 Aug 2001 at Two Rivers Park, Franklin Co.; 9 Mar-30 Jun 2002 e. of Yakima; pair 3 Jun-12 Jul 2002 at Ephrata; 1 on 19 Oct and 13 Nov 2002-4 Jan 2003 at Kennewick. 1 reported on 9-27 Mar 2002 e. of Yakima.

Bill Tweit

Great-tailed Grackle *Quiscalus mexicanus*

Casual visitor.

Although its range has been rapidly expanding, the Great-tailed Grackle has been found only four times in Washington. First was an ad. male (ph.) at Union Gap, 25-26 May 1987; second an ad. male (ph.) at Stanwood, 2 Sep 2000-10 Mar 2001; third an ad. male at Othello on 15 Jul 2002; fourth a female 19-24 Oct 2002 at Kent.

In 1900, the Great-tailed Grackle's U.S. breeding range was limited to southernmost Texas (Ridgway 1902), but as of 2000, this species has been found breeding in 19 states and has been seen in four additional states and two provinces (Wehtje 2001). The first Pacific Northwest records came from Oregon in 1980 (Gilligan et al. 1994), and this species has been annual in se. Oregon since then, with breeding occurring in 1994 (Scheuring and Ivey 1995). Through 2000, however, there were only about 13 Oregon records from elsewhere in that state, almost all from May and Jun.

Subspecies occurring in the nw. is uncertain, but is likely *Q. m. monsoni* or *nelsoni* (Scheuring and Ivey 1995, Wehtje 2001).

Steven G. Mlodinow

Brown-headed Cowbird *Molothrus ater*

Common migrant and summer resident, fairly common winter resident in w. Washington. Common summer resident/migrant and uncommon winter resident in e.

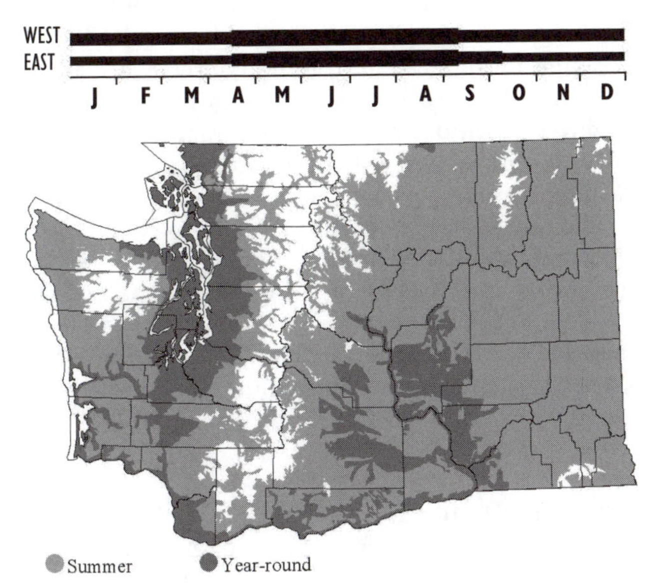

Subspecies: *M. a. artemisiae* in e. Washington. Westside birds may be *obscurus* (Jaramillo and Burke 1999).

Habitat: Most vegetation types below subalpine zones including clear-cuts and managed forests except large, mature, and intact forests (Smith et al. 1997); especially lowland riparian zones and hardwood secondary growth. During migration and winter, often in fields, feedlots, around dairies, and occasionally at bird feeders.

Occurrence: Notorious as a nest parasite, the Brown-headed Cowbird was not always part of the state's avifauna. Apparently, the species first arrived in e. Washington during the late 1800s as part of a range expansion from the Great Basin (Rothstein 1994). In w. Washington, there were a few records from the earlier part of the 20th century (Jewett et al. 1953), the earliest from Tacoma in the early 1900s (Bowles 1929). By the early 1950s, it was fairly common on the eastside, though rare on the westside (Jewett et al. 1953).

Status was relatively stable until the mid-late 1950s, when numbers arrived from w. California and w. Oregon (Rothstein 1994). Noted then n. to the San Juan Is. (Lewis and Sharpe 1987) and Whatcom Co. (Wahl 1995) with the first young noted in Seattle in Jul 1955. Numbers in w. Washington grew rapidly through the late 1950s and the 1960s and may still be continuing to increase, particularly in winter. Reasons for expansion include forest practices (Smith et al. 1997), creation of open or fragmented habitats, and availability of food at dairy farms and feedlots (Stepniewski 1999).

Spring migrants appear as early as mid-Mar. Summer birds typically do not associate with blackbirds, but migrants do so to some extent, particularly later in the fall. Fall movement begins in late Jul and gradually declines through Oct. On the eastside, wintering birds are more numerous towards the s., becoming rare during winter in northernmost counties. On the westside, winter blackbird flocks at dairies often include a few cowbirds.

Summer cowbird populations have increased over w. N. America including Washington (DeSante and George 1994), though BBS data from 1966 to 2000 (Sauer et al. 2001) do not indicate that and continent-wide documentation is limited and uncertain (Peterjohn and Sauer 1993). Winter numbers are relatively low and localized, with birds often mingling with blackbirds. Unidentified blackbird numbers, particularly on westside CBCs, probably include numbers of cowbirds.

Remarks: Subspecies generally not separable in the field except as nestlings: the gape is white in *artemisiae* and yellow in *obscurus* (Jaramillo and Burke 1999). Concerns expressed for impacts of cowbirds on populations of other passerines, including Willow Flycatcher and Yellow Warbler (e.g., Smith et al. 1997, Stepniewski 1999; see Paulson 1992b).

Noteworthy Records: *High counts, CBC*: 390 in 1992 and 258 in 1999 at Kent Auburn; 386 in 1990 at Padilla Bay; 333 in 1998 at Bellingham. *Other*: 250 on 5 Nov 1994 at Vancouver (SGM).

Steven G. Mlodinow

Orchard Oriole *Icterus spurius*

Casual vagrant.

One accepted Washington record: female or imm. (ph.) at Samish I., 15-27 Dec 1991. Previous reports, published in AB, were not documented. Oregon had fewer than 10 records, all from May to Jun and Sep to Jan. The one B.C. record is from late May (Campbell et al. 2001). Subspecies presumably was *I. s. spurius.*

Steven G. Mlodinow

Hooded Oriole *Icterus cucullatus*

Casual vagrant.

This sw. oriole has been spreading northward along the Pacific coast, establishing itself as a breeding species in nw. California as of 1981 (Harris 1996). It is now nearly annual in sw. Oregon, occurring mostly as a winter visitor and spring migrant (mid-Apr to early Jun). Through 2001, B.C. had seven records. There are four Washington records, all from late Apr through late Jul: a male (ph.) at Tokeland, 25 Apr 1992; a male (ph.) at Bellingham, 20-23 May 1996; a male at Joyce, 17-21 Jul 1999; a male (v.t.) at Bothell, 13-16 May 2001. Photographs appear to be of subspecies *nelsoni* from California.

Steven G. Mlodinow

Bullock's Oriole *Icterus bullocki*

Common migrant and summer resident in e. Washington. Uncommon migrant and summer resident in w. Very rare in winter.

Subspecies: Presumably *I. b. bullockii.*

Habitat: Lowland riparian areas, particularly where cottonwoods are present, orchards, suburban developments, agricultural areas with stands of deciduous trees, and other locations with similar vegetation structure.

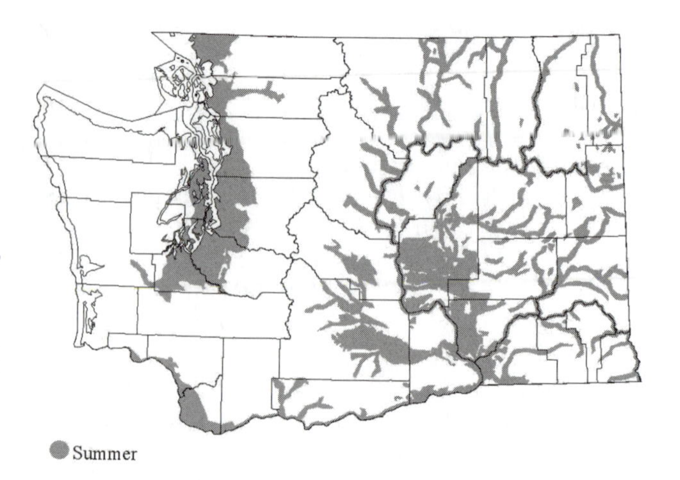

Occurrence: Widespread in habitat on the eastside (Smith et al. 1997), where this is one of the characteristic birds of lowland riparian habitats and may be seen perching on adjacent sagebrush on occasion. It is present locally in the Puget Trough where it appears more secretive, probably due to more dense westside vegetation, and often noted only by vocalizations.

Increase has been most evident on the westside where Bullock's Orioles were once considered casual (Jewett et al. 1953). Locally, for example, Edson (in Wahl 1995) did not mention occurrence in Whatcom Co., where birds are now regular, and Hunn (1982) describes increases in King Co. Irrigation and planting of orchards and windbreaks may well account for the increase on the eastside, whereas conversion of coniferous woodlands to deciduous is likely a factor in the increase in w. Washington. Birds are uncommon in the San Juan Is. (Lewis and Sharpe 1987) and very rare on the outer coast, seldom if ever breeding w. of Port Angeles, Elma in Grays Harbor Co., or Pe Ell in Pacific Co.

The population increase in Washington is indicated by BBS data, which showed a 2.5 % per year increase from 1966 to 2000 (Sauer et al. 2001).

Remarks: With the Baltimore Oriole, long considered a subspecies of the Northern Oriole, *I. galbula,* and recently recognized as distinct (see Rising and Williams 1999).

Noteworthy Records: East, *high counts*: 41 on 4 Jul 1998 at FEALE, Benton Co. (LaFramboise and LaFramboise 1999). *Late Fall-early Spring*: 17 Nov 1975 at Bellingham; 1 Dec 1998 at Yakima R. delta; 3-17 Dec 1995 at Ocean Shores; 15 Dec 1974 at Westport; 26 Dec 1973 at Bellingham; 28 Dec 1973 at Ferndale; late Dec 1983-Feb

1984 at Aberdeen; Jan-Feb 1993 at Mt. Vernon (Contreras 1997); 12 Apr 1978 at Tacoma.

Steven G. Mlodinow

Baltimore Oriole *Icterus galbula*

Casual vagrant.

The status of the Baltimore Oriole is somewhat uncertain because it was considered conspecific with Bullock's Oriole from 1973 to 1995. The species' breeding range has expanded westward during the last 50 years, with birds spreading into the Peace R. area of B.C., where it is now locally fairly common, as recently as the 1960s (Campbell et al. 2001). In Oregon, there are approximately 30 records, mostly from the eastside from mid-May to mid-Jun. There are three Washington records: an ad. male at Seattle, 5-8 Nov 1975; an ad. male nr. Cle Elum, 20 Jun 1987; and a bird at Echo Valley, Chelan Co., 31 May 1999.

Steven G. Mlodinow

Scott's Oriole *Icterus parisorum*

Casual vagrant.

There is one Washington record of Scott's Oriole: a male (ph.) at Chehalis, 11 Feb-13 Apr 1980. This sw. species is unrecorded in B.C. and has been recorded once in Oregon, at Fields, 4-8 Jun 1991 (Gilligan et al. 1994). Despite this relative scarcity of records from the Northwest, this species has some tendency to wander, as evidenced by records from such distant locales as Duluth, Minnesota, and w. Ontario (AOU 1998).

Steven G. Mlodinow

Brambling *Fringilla montifringilla*

Very rare winter visitor.

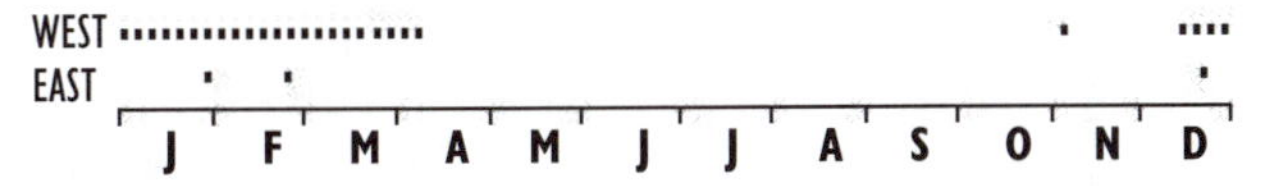

A famously irruptive and nomadic Eurasian species, with wintering flocks sometimes numbering in the millions (Clement et al. 1993). Individuals known to spend one winter in the British Isles and the next in Russia (Alerstam 1991). These wandering tendencies have undoubtedly led to this species being one of the more regular Old World passerines recorded in continental N. America. Through 1994, there were approximately 50 records from the contiguous U.S., mostly from w. states, ranging from late Oct through late Apr, with a peak from mid-Dec to late Feb (Mlodinow and O'Brien 1996). In addition to the invasion years above, the winter of 1983-84 saw an influx into w. N. America.

There are 14 Washington records, between 6 Nov and 12 Apr, with a peak from late Dec to late Feb. All but three records are from the westside, and eight of the 14 records occurred during three consecutive winters: 1990-91, 1991-92, and 1992-93.

Records: 1 (ph.) in winter 1968-69 at Aberdeen (not reviewed by WRBC); 6 Jan-22 Mar 1982 at Issaquah and L. Sammamish (ph.); 11-13 Jan 1984 at Tenino (ph.); 28 Dec 1988-1 Jan 1989 at Steilacoom; 6-10 Nov 1990 at Sedro Woolley (ph.); 14 Dec 1990-5 Jan 1991 at Port Angeles; 19 Jan-26 Feb 1991 at Elma (ph.); 15 Dec 1991-8 Feb 1992 at Westport (ph.); 30 Jan 1992 at Richland (ph.); 20-24 Feb 1992 at Walla Walla (ph.); 1-20 Jan 1993 at Lummi Flats (ph.); 9-12 Apr 1993 at Newhalem (ph.); 6 Dec 1995-10 Apr 1996 at Naselle (ph.); 23 Dec 2001 at Bridgeport.

Steven G. Mlodinow

Gray-crowned Rosy-finch *Leucosticte tephrocotis*

Uncommon, local breeder in rocky high-alpine areas of Cascades and Olympics; winters locally in the Columbia Basin, rarely in w. Washington.

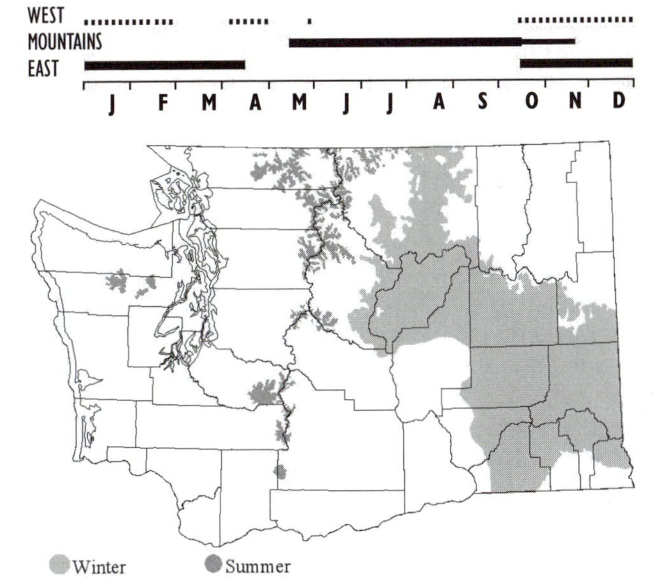

Subspecies: *L. t. littoralis* in Washington. *L. t. tephrocotis* breeds in s. B.C., ne. Idaho, and nw. Montana (AOU 1957, Burleigh 1972), *wallowa* breeds in the Wallowa Mts. of ne. Oregon (AOU

1957), though no breeding is described for the Blue Mts. (Smith et al. 1997). Jewett et al. (1953) reported wintering *tephrocotis* from extreme e. Washington.

Habitat: Nest on rocky terrain or on cliff faces in the alpine zone (Smith et al. 1997, AOU 1998). In winter, rocky level or sloping terrain, often with a sparse cover of grass (AOU 1998), rocky beaches, locally roosting in Cliff Swallow nests in the e. and opportunistic places like unoccupied, windowless buildings (Wahl 1995).

Occurrence: Breeding above tree line on high peaks in the Cascades from Whatcom and Okanogan Cos. s. to Yakima Co. and in the Olympics (Smith et al. 1997, Stepniewski 1999), rosy-finches migrate to lower elevations in winter, though this movement is erratic and inconsistent in destination. The species is well known for sporadic appearances across the eastside where, when present, it often occurs in flocks of hundreds, if not thousands. Between 1968 and 1999 there were 19 CBC totals of >100 birds and 17 totals of <100 in e. Washington. CBC results show great variability in wintering areas: 1300 were present at Wenatchee in 1970, 2100 were there in 1972, but the species was not recorded again until 1982 when 666 were counted. Abundance is likely related to weather conditions; birds may be forced out of higher elevations, or from further n., during severe winters (Burleigh 1972). Stepniewski (1999) describes winter distribution associated with availability of wheat blown from railroad cars and fields. Rosy-finches are early migrants and often depart their winter quarters in Mar; there are few e. lowland records later in the season. Use of swallow nests for roosts, noted along the Snake R. and the Columbia R. nr. Vantage (Jewett et al. 1953, Hudson and Yocom 1954, Burleigh 1972), may be an essential survival strategy.

Erratic, infrequent in westside lowlands. Most records are from Whatcom Co. or vicinity (Jewett et al. 1953, Wahl 1995), with occasional reports from areas such as King Co. (Hunn 1982) and Ocean Shores (Hoge and Hoge 1991). Wintering birds at Sandy Pt. and in Bellingham foraged in gravel beach habitats (Wahl 1995). Westside lowland records have declined substantially since the 1970s and early 1980s, in spite of a great increase in observer effort.

Noteworthy Records: East, *Fall*: 200 on 18 Oct 1998 at Chopaka Mt.; 1500 in 1984 nr. Hartline. *Winter, high counts*: 300 on 18 Nov 1972 at Kamiak Butte; 300 on 19 Nov 1972 at Blue L.; 1500+ on 22 Nov 1962 nr. Coulee City; 3000 on 3 Dec 1981 nr. Chelan; 1500 on 25 Dec 1964 nr. Hartline; 2 flocks of 200-300 in winter 1983-84 at Kahlotus. *CBC high counts*: 1246 in 1983 at Ellensburg; 310 in 1998 at Grand Coulee; 280 in 1991 at Twisp; 400 in 1969, 1300 in 1971, 2100 in 1973, and 666 in 1983 at Wenatchee; 400 in 1997 at Walla Walla. *Late Winter-early Spring*: 350 on 2 Mar 1963 in Columbia Basin; 250 on 2 Mar 1981 in Umtanum Cr. Canyon (Stepniewski 1999), 250 on 8 Mar 1997 at Asotin Cr.; 300+ on 23 Mar 1981 at Yakima Canyon (Stepniewski 1999); 700 on 11 Apr 1970 nr. Chelan; present on 13 Apr 1974 on cliffs above Park L., Grant Co.; 2500 on 16 Apr 2002 at Swakane Canyon, Chelan Co. **West**: 1 on 10 Apr 1984 on Mt. Constitution, Orcas I. (Lewis and Sharpe 1987); 10 on 13-16 Apr 1999 at Pt. Roberts; 1 on 30 Apr 1984 at Discovery Park, Seattle; 1 on 30 May 1984 on Mt. Erie, Fidalgo I. (Lewis and Sharpe 1987); 500 on 1 Oct 1998 at Mt. Rainier NP; 1000+ on 19 Oct 1972 at 1800 m on Mt. Baker; 500 on 23 Oct 1972 at Hart's Pass; 14 on 3 Nov 1980 on Silver Star Mt. Whatcom Co.: 2 on 12 Nov 1994 at Lummi Bay; 20 on 17-22 Nov 1975 at Bellingham; 60 on 18 Nov 1973 and 22 on 22 Nov 1975 at Post Pt., Bellingham; 3 on 8 Dec 1973 at Wiser L.; 60 in 1973 on Sehome Hill, Bellingham; 175 on 1973-74 CBC at Sandy Pt.; 3 on 3 Jan-10 Feb 1970 at Bellingham; 32 on 27 Jan 1973 at Chuckanut Mt. (Wahl 1995). Present in winter 1975-76 at Seattle, Ocean Shores, and in Skagit and Island Cos.; 30+ in winter 1977-78 at Samish Flats; 19 on 1973 CBC at Padilla Bay; 2 on 1983 CBC at E. Lk. Washington; present in winter 1992 at Skagit Bay; 21 Oct 2001 at Seattle; 2 May 2002 at Rosario Head, Skagit Co.

Joseph B. Buchanan

Pine Grosbeak *Pinicola enucleator*

Uncommon summer resident of mid- to higher-elevation forests. Rare in winter in eastside lowlands, very rare in westside lowlands.

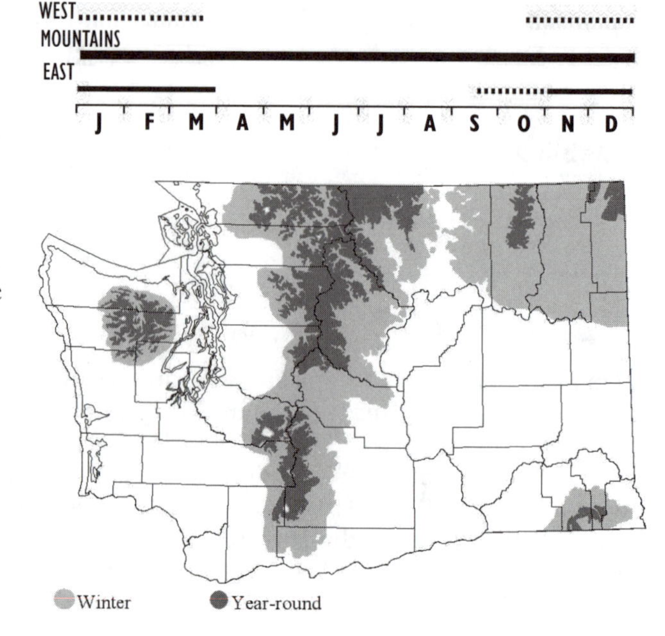

Subspecies: *P. e. montanus* breeds in Washington. Two subspecies that breed in Alaska, *alascensis* and *flammula*, have been recorded in winter in the Pacific Northwest (AOU 1957, Burleigh 1972).

Habitat: High-elevation coniferous forests, parkland, conifer-dominated wetlands. Also lowland forests, suburban, and urban plantings in winter.

Occurrence: A species of boreal forests, irruptive and variable in Washington like other n. finches. Jewett et al. (1953) described the breeding distribution as the n. Cascades. Recent information indicates that they breed s. as far as the White Pass area (Smith et al. 1997; they likely breed further s.—there are numerous records in the Oregon Cascades [Gilligan et al. 1994]). Birds breed in the Olympic Mts. and in the ne. part of the state (Smith et al. 1997). A pair was reported at Wickiup Springs in the Blue Mts. in 1946 where it was considered a rare breeder (Hudson and Yocom 1954, Weber and Larrison 1977).

Winter occurrence is irregular. Pine Grosbeaks occur most regularly in CBCs at Spokane, though found there in only 15 years between 1960 and 1999. Typically found in small flocks: data from 29 CBC count totals in e. Washington between 1980 and 1999 indicated a typical flock size of about six. High counts (>100 birds) occurred only at Spokane in 1960, 1963, 1973, and 1984, and at Wenatchee in 1965. Overall, numbers have decreased in recent years. In winter, birds are most often seen in association with berry-producing trees and shrubs (Cannings et al. 1987). In some years (e.g., 1984 and 1996) birds occur more widely than in other years when they are usually found in only one or a few localities.

Westside occurrence is even more infrequent. Birds are erratic in King Co. (Hunn 1982) and Lewis and Sharpe (1987) give just two records for the San Juan Is. CBC records are widespread but very infrequent, with eight records at Bellingham between 1969 and 1999 the most frequent occurrence. It is unclear whether movements to lower elevations are a response to weather conditions in the higher elevations or on a wide, irruptive scale due to scarcity of food resources.

Two aspects of distribution remain unclear. The breeding range has not been well documented: only 17 confirmed or probable breeding records were reported by Smith et al. (1997), and only three of these are in the modeled range beyond the Cascade Mts. And there is extremely limited information regarding the distribution of the subspecies occurring in winter. It is possible that the subspecies ranges overlap as they do in the Red Crossbill.

Noteworthy Records: *High counts*, **West**, *Winter*: 100 on 20 Dec 1969 and 50 on 11 Jan 1970 at Bellingham; (Wahl 1995); 12 on 28 Nov 1971 at Chuckanut Mt.; 12 on 28 Jan 1995 at Discovery Bay; 8 on 2 Feb 1964 on Whidbey I.; 8 on 18 Mar 1976 at Seattle. CBC: 75 in 1969 at Bellingham. **East**, *Winter*: 31 on 22 Dec 1998 at Trout L.; 20 on 5 Dec 1970 at Spokane; 12 on 3-27 Mar 1974 at Pullman (Weber and Larrison 1977); 9 as late as 25 Nov 1972 at Richland; 55 on 22 Dec 2001 nr. Winthrop; 51 on 26 Jan 2002 at Cle Elum. *CBC*: 216 in 1960, 270 in 1963, 212 in 1973 and 177 in 1984 at Spokane; 200 in 1965 at Wenatchee.

Joseph B. Buchanan

Purple Finch *Carpodacus purpureus*

Fairly common permanent resident in w. Washington, and locally in the e. Cascades and foothills. Rare to very rare farther e.

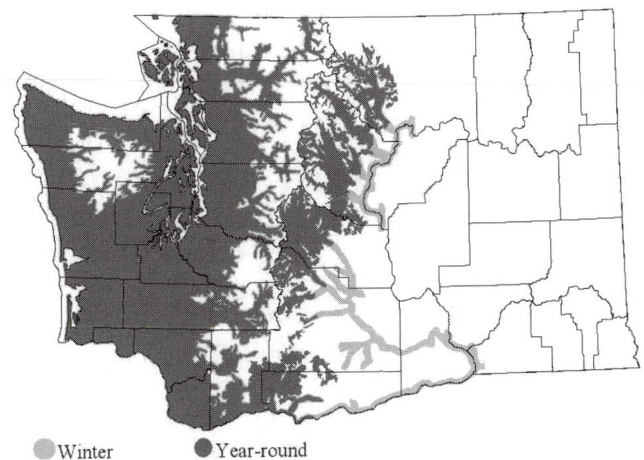

Subspecies: *C. p. californicus.*

Habitat: Wide range of habitats in proximity to forest edges in w. lowlands to silver fir zone and below grand fir zone downslope to steppe zone e. of the Cascade crest (see Smith et al. 1997).

Occurrence: The Purple Finch is a fairly common to common breeder in rural shrubby habitats and forested areas bordering lowland habitats on the westside. It is less common in forested areas at higher elevations. In e. Washington this species is somewhat local and mostly restricted to the Cascades and their foothills, most commonly from Kittitas Co. s. Records e. of there are relatively rare.

Table 42. Purple Finch - CBC average counts per 100 party hours by decade.

	1960s	1970s	1980s	1990s
West				
Bellingham		64.1	34.9	26.8
Cowlitz-Columbia			33.6	24.1
Edmonds			19.1	10.5
E. Lk. Washington			22.6	3.1
Grays Harbor		34.0	37.0	23.3
Kent-Auburn			27.7	10.8
Kitsap Co.		21.0	82.4	38.8
Leadbetter Pt		201.9	53.8	30.7
Olympia			34.1	40.2
Padilla Bay			66.3	24.2
Port Townsend			40.3	35.4
San Juan Is.			31.6	39.7
Seattle	63.0	31.2	19.6	3.1
Sequim-Dungeness			28.8	12.7
Tacoma		89.2	36.3	16.6
East				
Ellensburg			10.8	0.8
Toppenish			46.2	4.5
Wenatchee	0.0	18.3	0.9	0.4
Yakima Valley		3.9	1.3	2.6

The species has become less common in lowlands over past decades during which the related House Finch became established. The Purple Finch was widespread throughout w. Washington and eastside summer occurrence (reported in the late 1960s) was apparently in higher, wetter forest type than that of Cassin's Finch. Distribution in w. now in habitats essentially away from developed urban areas and at higher elevations than that of House Finch, which dominates at bird feeders.

More migratory than the House Finch and irruptive and more variable in numbers in winter range (Wootton 1996), with a downward trend in lowlands in summer range evident to many observers (Larrison and Sonnenberg 1968, Wahl 1995, Smith et al. 1997), though numbers in mountains may remain at levels prior to 1950s. Migration is little known and commented upon in Washington (see Stepniewski 1999), but qualitative reports suggest movements in e. in Apr-May and Aug-Oct. On the westside, birds move into non-breeding areas in late Aug-early Sep, reaching peak numbers by mid-Sep. Spring departure from those locations occurs mostly during early to mid-May.

BBS data from 1966 to 2000 indicate a slight decline statewide (Sauer et al. 2001). CBC data show declines in birds per party hour, with statistically significant (P <0.01) westside declines (see Table 42). During the same periods of years, House Finches increased on virtually all state CBCs.

Remarks: Population trends away from human-dominated lowland areas should be determined. Reports are lacking of severe winter mortality in Washington (see Wootton1996).

Noteworthy Records: *CBC high counts,* **West**: 248 in 1973 at Bellingham; 144 in 1975 at Grays Harbor; 100 in 1989 at Kent-Auburn; 346 in 1981 at Kitsap Co.; 140 in 1977 at Leadbetter Pt.; 175 in 1990 at Olympia; 420 in 1987 at Padilla Bay; 193 in 1989 at Skagit Bay; 60 in 1985 at Sequim-Dungeness; 120 in 1962 at Seattle; 196 in 1975 at Tacoma. **East**: 26 in 1981 at Ellensburg; 6 in 1992 at Tri-Cities; 104 in 1985 at Toppenish; 4 in 1988 and 1993 at Twisp; 64 in 1973 at Wenatchee; 1 in 1987 at Walla Walla; 20 in 1991 at Yakima Valley.

Terence R. Wahl

Cassin's Finch *Carpodacus cassinii*

Common breeder in open forests from upper Sonoran into Hudsonian zones in e. Local breeder w. of the s. Cascade crest, rare to the n. Variable, locally uncommon in lowlands in winter e., very rare w.

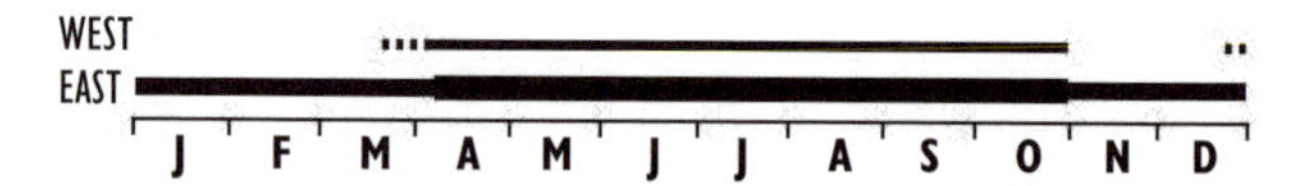

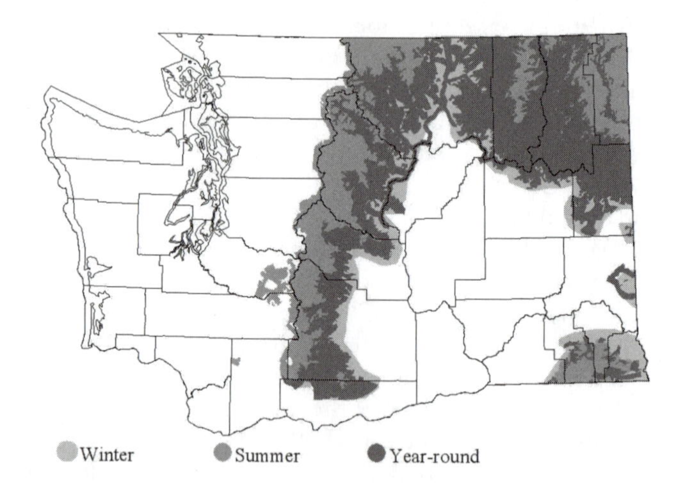

Habitat: Dry coniferous forsts, edges at mid- to high elevations. Winters in variety of habitats at lower elevations, including forest edge, urban areas.

Occurrence: Distribution is more completely known than it was 50 years ago (see Jewett et al. 1953). In summer found on e. Cascade slopes, in the Okanogan to Pend Oreille, and Lincoln and Spokane Cos., in the Blue Mts. to 2100 m (Jewett et al. 1953). More common in ponderosa pine zone (Stepniewski 1999). Anecdotal summer reports suggest absence locally in some years (e.g., "scarce or absent" except in the Okanogan in 1979).

W. of the Cascade crest recorded above Purple Finch habitat in subalpine fir zone usually above 1400 m in e. King Co. Noted nesting on s. sides of Mt. Rainier and Mt. St. Helens (Manuwal et al. 1987). In 1989 the location of a lower nest, at 1200 m, may have been due to creation of old-growth conifer clear-cut habitat in King Co.

More migratory than the other *Carpodacus* finches (Stepniewski 1999), with winter numbers quite variable. Spokane CBC numbers ranged from none in some years in the 1970s-80s to 1451 in 1988. Rare in winter at low elevations in se. Washington (Weber and Larrison 1977), unusual at Tri-Cities (first record in 1973). Relatively consistent at Spokane (15 of 19 CBCs, 1980-98), irregular elsewhere. Birds occur in lowlands w. of the Cascades in fall-spring, with first report in 1978. Subsequent records in fall, winter, spring records (e.g., Olympia, Sequim, Vashon I., Skagit Co.). Small numbers found breeding w. of the Cascade crest from 1976 on, locally nr. Packwood, Snoqualmie Pass, and in e. Skagit and Whatcom Cos.

Differences in summer distribution of Purple and Cassin's finches appear to be associated with dryness/wetness of forest types, with Purple Finches occurring e. of the crest in wetter habitats and Cassin's in dryer forests w. of the crest (see Smith et al. 1997). This species is reportedly "giving way" to the House Finch in winter in e. lowlands (Larrison and Sonnenberg 1968; see Smith et al. 1997). BBS data showed a decrease in numbers from 1966 to 2000 (Sauer et al. 2001). Local, low numbers have occurred on long-term CBCs.

Remarks: Described changes in occurrence may reflect in part distribution and also comparatively larger numbers of observers bird-watching and contributing to systematic field studies like CBCs, BBS and BBA surveys. Early records in King Co. are given with cautions by Hunn (1982).

Noteworthy Records: East: more than 100 on 21 Apr 1971 nr. Ahtanum; 100+ on 12 Apr 1998 in Wenas area (Stepniweski 1999). *CBC*: 1451 in 1988 and 159 in 1989 at Spokane. **West**: 1 on 22 Apr 1977 at Olympia; 1 on 20 Mar 1979 nr. Olympia; 1 netted from small flock on Vashon I. on 7 May 1984; 1 banded nr. Sequim 15 May 1987; 25-50 on Mt. Rainier in August 1919 (Jewett et al. 1953). 1 on 7-20 Nov 1996 nr. Burlington.

Terence R. Wahl

House Finch *Carpodacus mexicanus*

Common resident in urban residential areas, farmlands, and adjacent habitats statewide.

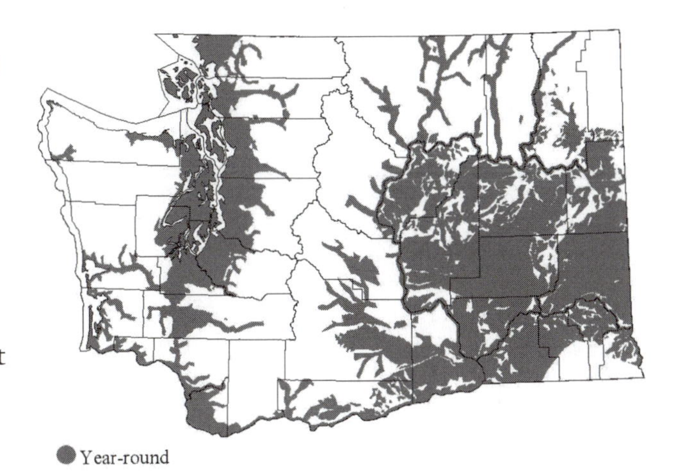

Subspecies: *C. m. frontalis.*

Habitat: Many habitats below coniferous forest zones statewide, ranging from heavily developed urban areas to agricultural and shrub areas. In steppe zone, habitats in agricultural areas require irrigation (Smith et al. 1997).

Occurrence: Dramatic, continent-wide expansion described and mapped in field guides, from pre-1940 (Hoffman 1927, Peterson 1941), through following decades (Peterson 1961, 1980, 1990) to the present (e.g., Gilligan et al. 1994, National Geographic Society 1999) and causes, ranging from introductions in e. U.S. to habitat changes and increases in bird feeders across N. America, discussed by many investigators (e.g., DeSante and George 1994).

One of the most conspicuous species in urban areas statewide today, but present only in s.c. Washington until the 1950s (Jewett et al. 1953). Then, as in other parts of the country, the population expanded both northward e. of the Cascades and into w. Washington. Reached the San Juan Is. by 1960 (Lewis and Sharpe 1987), Bellingham by 1961 (Wahl 1995), and continued expanding within habitats region-wide after that (see Smith et al. 1997).

Associates with bird feeders and distribution appears dependent on availability of feeders (Smith et al. 1997). In some areas, late-summer post-breeding local dispersal from residential areas made to beaches, fields, and treeless areas, and habitats with wild fruits and seed crops, returning to proximity to bird feeders in fall (Wahl 1995, Stepniewski 1999). Adaptable and bold, outcompetes and has supplanted Purple and Cassin's

Table 43. House Finch - CBC average counts per 100 party hours by decade.

	1960s	1970s	1980s	1990s
West				
Bellingham		327.6	356.8	340.5
Cowlitz-Columbia			95.8	108.9
Edmonds			289.6	329.9
E. Lk. Washington			58.9	279.1
Grays Harbor		304.7	227.3	217.5
Kent-Auburn			305.8	339.8
Kitsap Co.		64.2	123.1	115.8
Leadbetter Pt		259.2	123.6	125.7
Olympia			122.3	230.9
Padilla Bay			309.8	386.4
Port Townsend			259.0	187.8
San Juan Is.			68.4	115.1
Seattle	195.5	135.8	259.9	551.5
Sequim-Dungeness			335.2	385.4
Tacoma		260.1	265.8	411.0
East				
Ellensburg			597.6	1485.7
Spokane	521.1	502.1	839.2	1035.8
Toppenish			1891.8	1855.9
Tri-Cities		373.0	421.3	668.2
Walla Walla		110.1	422.9	501.8
Wenatchee	1192.8	313.5	804.9	1930.0
Yakima Valley		1010.1	1384.6	2081.1

finches in urban and agricultural areas (Wahl 1995, Smith et al. 1997). Nest location in shrubs may allow local replacement of House Sparrows if nest holes or boxes not available for that species.

BBS data from 1966 to 2000 indicate a significant increase (P <0.01) statewide (Sauer et al. 2001) and CBC counts (Table 43) also show significant increases (P <0.01 e.; P <0.05 w.).

Remarks: Considered a pest species in some agricultural areas in other states, feeds on cherries (Stepniewski 1999) and nips blossoms of various fruit trees (Wahl 1995). Essentially resident and non-migratory: explanations of "migrations" of very large numbers (1000+ reported at the mouth of the Columbia R. on 15 Nov 1992) desirable. Numbers declined from 300 to 180 million in e. U.S. after the winter of 1993-94, with numbers noted suffering from *mycoplasmal conjunctivitis* (Project Feeder Watch: Audubon 102:18), which has been noted in birds at feeders in Washington. Called "Linnet" in some past references.

Terence R. Wahl

Red Crossbill *Loxia curvirostra*

Irregular, uncommon to locally common statewide in conifer forests, rare in other habitats in migration and winter.

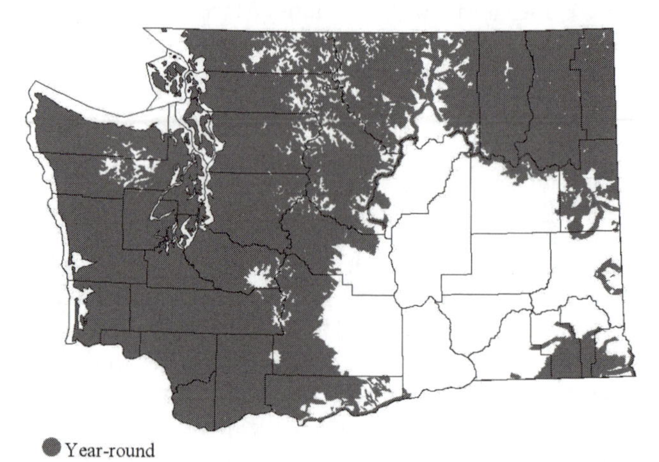

● Year-round

Subspecies: *L. c. sitkensis* and *benderei* in Washington (see below and Table 44).

Habitat: All forested regions. Rarely in migration and winter in unforested lowlands, attracted to conifers around human habitations.

Occurrence: The Red Crossbill is justly described as irruptive, nomadic, and irregular, and knowledge of its status is complicated further due to recent recognition of eight different types, possibly subspecies, of birds with distinctive call notes. Breeding has been documented or suspected in every area of the state except the Columbia Basin (Smith et al. 1997). Availability of cone crops is essential: crossbills breed where and when cone seed abundance is adequate (Smith et al. 1997); breeding behavior has been noted between Dec and Oct (Lewis and Sharpe 1987, Adkisson 1996). W. of the Cascade crest, lowland breeders are local, and notably consistently present in the San Juan Is. (e.g. Lewis and Sharpe 1987), Whidbey I., and the n. Olympic Pen.

In the breeding season in the s. Cascades, Manuwal (1991) found Red Crossbills to be most abundant in mesic old-growth sites compared to wet or drier old-growth sites and mature and young stands, with old-growth Douglas-fir forest as optimum habitat. During winter Red Crossbills were not detected in mid-seral stands in two of three years but were found abundantly in old-growth forest in all years at varying levels of abundance in response to changes in conifer seed production (Manuwal and Huff 1987, Huff et al. 1991). Abundance was strongly correlated with the basal area of western hemlock (r_s = 0.74, P <0.05; Huff et al. 1991).

Table 44. Distribution and preferred conifer species of Red Crossbill call types in Washington (from Groth 1993).

Call Type	*Geographic Range in Washington*	*Primary Conifer Species*
1	Olympic Pen., w. Cascades	western hemlock
2	E. side of state	Ponderosa pine, lodgepole pine
3	Olympic Pen., w. Cascades	western hemlock, Sitka spruce
4	Statewide	Douglas-fir (Sitka spruce?)
5	E. side of state	lodgepole pine, Englemann spruce
7	Likely occurs in e. side of state	Englemann spruce, lodgepole pine

Often in pairs or small numbers, but noted in flocks of >50 in winter. At lower elevations in winter crossbills are most regularly found in the Spokane area (present in 24 of 26 CBCs since 1974) with the state's six highest CBC totals since 1974 there. Widely recorded on the westside, with numbers noteworthy on CBCs at Grays Harbor and San Juan Is. It is often absent from CBC areas for extended periods (e.g., absent ≥5 consecutive years at long-term CBC such as Kitsap Co., Olympia, Walla Walla, Wenatchee). No trends are evident on BBS (Sauer et al. 2001) or on CBCs, though large variations between decades and locations occur.

There may be factors that influence patterns of occurrence in parts of the range, possibly influencing birds of a single call type only. Winter abundance at Walla Walla since 1974 was positively correlated with Spokane and Moscow/Pullman (Pearson correlation coefficients of 0.52 and 0.71, respectively, derived from CBC data). There were no correlations among other combinations of CBC sites in the state.

Remarks: Benkman (1993a) and Groth (1993) indicate the presence of eight distinctive types (subspecies) across N. America, at least five of which occur in Washington (see Adkisson 1996). These are differentiated by call types, morphology (mostly in bill size), and to a certain degree, by the conifer species with which they are most strongly associated (Table 44). Distinctive types are currently identified by call type only; subspecific nomenclature has not been applied (Groth 1993). Though research may demonstrate full species status in some of the types (Groth 1993), the general lack of reproductive isolation casts some doubt on this possibility (Knox 1992). Forest management practices may impact regional populations. The amount of conifer seed is generally higher in older forests than in comparatively younger ones (e.g., Fowells 1965). Consequently, reduction in forest harvest rotation length will result in reduced conifer seed production and may impact populations strongly associated with tree species managed on shorter rotations (Benkman 1993b).

Noteworthy Records: *CBC high counts,* **West**: 471 in 1992, 325 in 1996, 210 in 1993, 116 in 1994 and 1998 at Leadbetter Pt.; 102 in 1989 at Olympia. **East**: 901 in 1981, 884 in 1992, 806 in 1980, 762 in 1982, 651 in 1989, 538 in 1985 at Spokane.

Joeseph B. Buchanan and Steven M. Desimone

White-winged Crossbill *Loxia leucoptera*

Irregular visitant, primarily along the n. border.

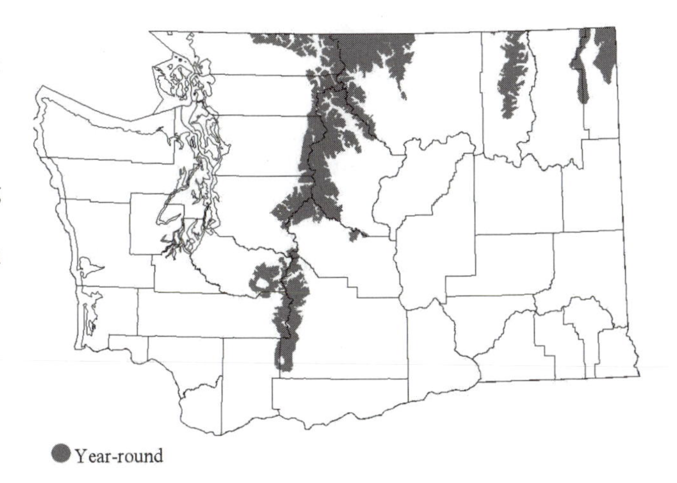

Subspecies: *L. l. leucoptera.*

Habitat: Coniferous forests and edges. Spruce and tamarack are preferred cones, with hemlock cones particularly preferred in ne. Washington (J. Acton p.c.).

Occurrence: Breeding across the boreal forest from Newfoundland to Alaska, this is an irregular, nomadic, winter wanderer. It sometimes stages very large irruptions, one coinciding with cone development in mid-late May and another with shedding of seeds in Oct and early Nov (Benkman 1992). Washington appears to lie on the edge of its regular range, as seasonal totals range from unreported to abundant; and there is nowhere in the state where birds can be found reliably. In flight years, they typically begin appearing in late summer or early fall (Table 45), between the movement times suggested by Benkman (1992).

Reports of irruptions appear to have increased from one per decade in the 1960-80 period, to three in the 1990s. This may be partly due to increased observer effort at high elevations in summer/fall.

Table 45. Recorded large irruptions of White-winged Crossbills into Washington.

Irruption Period	*Relative Abundance*	*Geographic Extent*	*Source*
Dec 1908-Apr 1909	Remarkable incursion	Puget Trough	Jewett et al 1953
Summer 1923	Predominating species	Mt. Baker	Jewett et al 1953
Sep-Oct 1965	Flocks of 100+	Ne. corner	AFN 20:75
Sep 1974	Commonest bird	N. Cascades, ne. corner	AB 29:92, 112
Jul-Sep 1978	Large numbers	Cascades, Olympics	AB 32:1203
Jul 1985-Mar 1986	Large numbers, most abundant bird at high elevations	Cascades, Olympics, Okanogan highlands, ne. corner	AB 39:941, 956; AB 40:145, 306, 324, 502, 517
Mar-Jun 1992	Moderate numbers, one report of 500	Cascades, Olympics	AB 46:472
Jul-Sep 1995	Best since 1992	N. Cascades, Okanogan highlands, ne. corner, Blue Mts.	FN 49:974
Jul 1997-Feb 1998	Very large numbers	Cascades, Olympics, Okanogan highlands, ne. corner	FN 51:1047; FN 52:118, 251
Aug-Sep 2001	Widely reported, smaller numbers	N. Cascades, Okanogan highlands, ne. corner	FN 56:100

There appears to have been only one large-scale winter irruption into the lowlands, in the winter of 1908-09 (Jewett et al. 1953). In recent years, lowland records appear to be associated with summer/fall movements in the mountains; they are never abundant and only infrequently linger. At higher elevations, they have been found in every portion of the state and are most commonly noted from the n. Cascades, the Okanogan highlands, and the Selkirks of ne. Washington. They are noted less frequently in the Olympics, the s. Cascades, and the Blue Mts., and there are a handful of records from the Willapa Hills. Records from low elevations are scattered across the state, including the outer coast, along the lower Columbia, and from most developed areas in e. Washington except the lowest-elevation portions of the Columbia Basin.

Although records of juvs. date back to 1919 (Jewett et al 1953), and there are numerous reports of singing and other forms of courtship activity (e.g., Smith et al 1997), there are no documented Washington breeding records to date.

Bill Tweit

Common Redpoll *Acanthis flammea*

Variably rare to common winter visitor e. of the Cascades, rare to uncommon w.

Subspecies: *A. f. flammea* (see Godfrey 1986, Monroe and Sibley 1993).

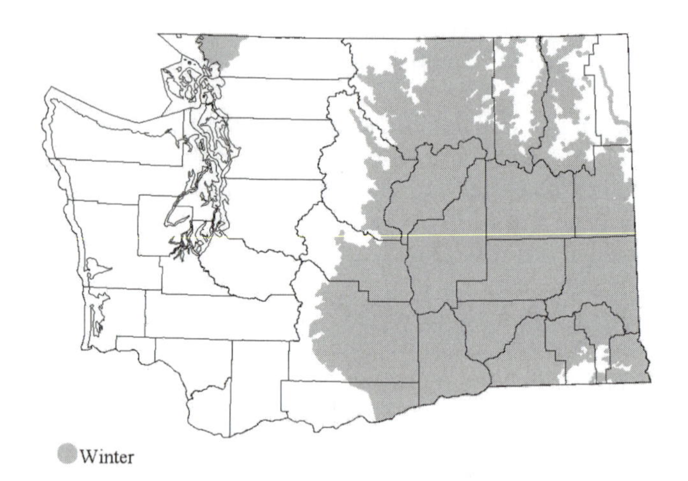

Habitat: Weedy open fields, catkin-bearing birch and alder trees, open woodlands, upper sandy beaches, occasional at bird feeders.

Occurrence: Breeding in subarctic and arctic Eurasia and in N. America in Alaska across n. Canada ; like other wintering "northern" finches, redpolls are irruptive and irregular anywhere in the state, presumably depending on reproductive success and food supplies to the n. Reports especially prior to 1980 or so likely reflected distribution of observers more than occurrence of the species.

Flocks or singles recorded from virtually all parts of the state, w. to Dungeness and Elma, e. to Clarkston and s. to Oregon. Much more numerous e. of the Cascades, especially across the n. from Wenatchee to the Okanogan Valley and to Spokane, than w. of the mountains. CBC counts, for example, ranged up to 870 at Spokane in 1965.

Large numbers not always recorded in same years in multiple locations within the region. Generally widespread numbers in recent peak years in

Table 46. Common Redpoll: CBC average counts per 100 party hours by decade.

	1960s	*1970s*	*1980s*	*1990s*
Ellensburg			10.84	0.20
Spokane	113.21	27.15	24.58	3.33
Tri-Cities		0.00	3.92	1.84
Toppenish			0.00	0.84
Wenatchee	22.65	27.72	6.89	0.88
Walla Walla		0.00	123.84	0.62
Yakima Valley		6.68	14.34	0.40

consistent e. locations include winters of 1960-61, 1965-66, 1969-70, 1971-72, 1975-76, 1977-78, 1981-82, 1984-85, 1985-86, 1999-2000, 2001-02. Numbers in w. much lower (the CBC high count was 18 at Bellingham in 1985) and very inconsistent, with no reports at all during some fall-through-spring seasons. Noticeable numbers in the w. in 1971-72, 1977-78, 1981-82, 1984-85, 2001-02 (and see Knox and Lowther 2000a). In 2001-02 over 800 birds were reported in w. Washington, with 90% of these n. of 47° N and 95% from n. of Seattle. Eastside CBC data (Table 46) indicate a decline in numbers in recent decades. Redpolls are often noted with siskins and goldfinches on the westside.

Noteworthy Records: West, *Winter*: flocks of at least 1000 in Pierce Co. in Feb 1917 (in Jewett et al. 1953); widespread in Jan-Feb 1982 with several at Leadbetter Pt., 30-50 at Skagit Flats, others in Seattle, Woodinville, Nisqually NWR; 35 on 24 Apr 1982 at Dungeness; 400+ on 19 Mar 2002 in Whatcom Co.; 150 on 1 Jan 2002 at Stanwood; 105 on 23 Dec 2001 at Camano I.; 100 on 9 Feb 2002 at Fir I. **East**: flocks of 150-250 in 1965-66 at Spokane ("biggest year ever"); 150 at Banks L. in fall-winter 1965-66; 250-300 on 11 Nov 1969 at Sullivan L.; 300 on 11 Mar 1972 nr. Tiger; 400 in winter 1984-85 at Pearrygin L.; 2000 in mid-Feb 2002 nr. Blue L., Okanogan Co.; 1300 on 24 Feb 2002 nr. Usk, Pend Oreille Co. CBC: 225 in 1960, 870 in 1965, 235 in 1981 at Spokane; 348 in 1981 at Chewelah; 194 in 1997 at Twisp; 660 in 1985 at Walla Walla; 660 in 1985 at Wenatchee.

Terence R. Wahl

Hoary Redpoll *Carduelis hornemanni*

Very rare winter and early spring vagrant.

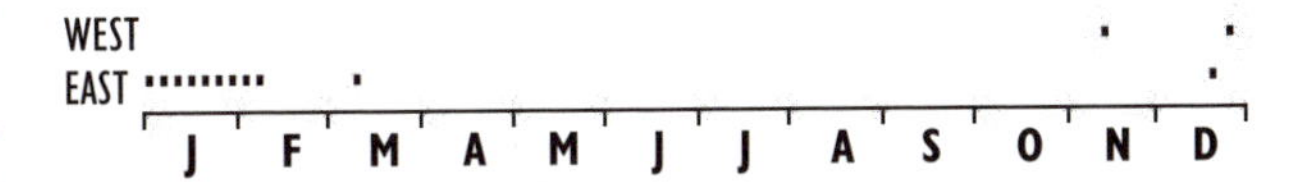

Subspecies: Most likely *C. h. exilpes* (Mlodinow and O'Brien 1996) with *hornemanni* possible (Clement et al. 1993).

Occurrence: An arctic species that wanders southward irregularly in small numbers, usually with Common Redpolls. The first Washington Hoary Redpoll was observed at Lummi Flats, Whatcom Co., on 28 Dec 1969. Fourteen of the subsequent 15 records have been for the eastside between 22 Dec and 16 Mar and are from Kittitas Co. n. Four winters—1981-82, 1984-85, 1999-2000 and 2001-02—account for 15 state records of this irruptive species. Knowledge of status and distribution s. of the species usual range is complicated by difficulties in separating this species and the Common Redpoll in the field. Clarification of status in Washington will likely become more clear as criteria for identification become more widely understood (see Czaplak 1995, Millington 1996).

Cannings et al. (1987) list 18 records of Hoary Redpolls from the Canadian Okanagan, mostly from mid-Dec to mid-Feb, with a specimen record on 9 Apr. There are also two Oregon records (Gilligan et al. 1994).

Remarks: See Knox and Lowther (2000b) for discussion of subspecies.

Reports: 1 on 28 Dec 1969 at Lummi Flats; ad. female on 30 Jan 1982 nr. Twisp (Aanerud and Mattocks 1997); ad. male on 31 Jan 1982 nr. Tonasket (Aanerud and Mattocks 1997); imm on 28 Dec 1984 nr. Winthrop; ad. on 9 Jan 1985 nr. Twisp (Aanerud and Mattocks 1997); 1 on 29 Jan 1998 nr. Curlew (R. Rowlett to WBRC); 1 on 22 Dec 1999 nr. Tonasket; 1 on 1 Feb 2000 nr. Winthrop; 1 (v.t.) on 14-16 Mar 2000 at Elk, Spokane Co.; 1 on 11 Nov 2001 at Lummi Flats; 1 (ph.) in early Jan 2002 at Cle Elum; 1 on 3 Jan 2002 at Ellensburg; 1 (ph.) 5-20 Jan 2002 at Electric City, Grant Co.; 1 on 19 Jan 2002 at Wenatchee; 1 on 27 Jan 2002 at Davenport; 1 on 26-27 Jan 2002 at Elk, Spokane Co.

Steven G. Mlodinow

Pine Siskin *Carduelis pinus*

Common breeder in coniferous forests statewide. Variable year to year, locally and statewide. Fairly common to abundant winter visitor.

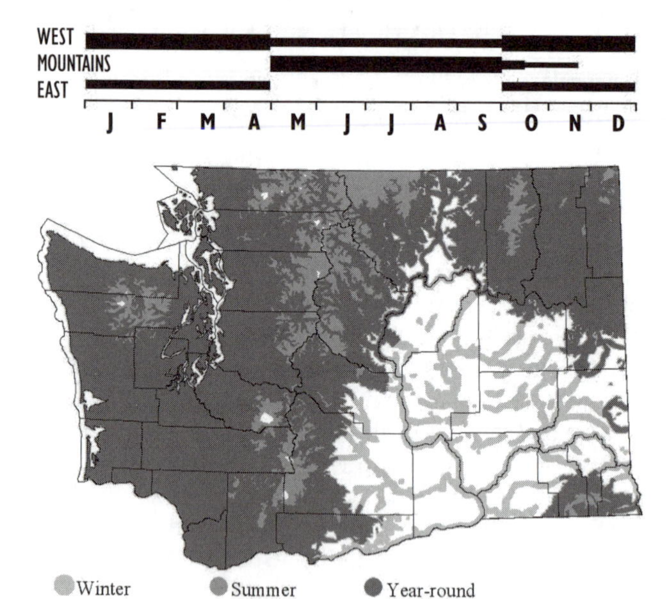

Subspecies: *C. p. pinus.*

Habitat: Coniferous and mixed woodlands in transition and Hudsonian zones, including trees in urban areas. Nests in conifers. Recorded feeding on wide range of conifers including hemlock and red cedar, deciduous buds, seeds of red alder, birch, dandelion, chickweed, and thistle. Readily attracted to bird feeders.

Occurrence: Variable in local and regional occurrence and abundance, one of a number of irruptive finch species dramatically affected by cone crops throughout its extensive range. Seasonal occurrence and nesting dates described universally by authors as variable, erratic, irruptive, or irregular in occurrence. A few individuals appear at non-breeding lowland locations as early as mid-Aug, but the main movement is typically in early to mid-Oct. Departure from wintering areas begins in Apr, with a few birds lingering well into May, sometimes into early Jun, even in the Columbia Basin.

Historical reports noted great abundances over time (e.g., fall 1913 and winter 1923-24; Jewett et al. 1953). In summer found in all conifer forests statewide, less commonly in mixed forests in w. lowlands (Smith et al. 1997). Migrations less well documented and usually noted and reported when flocks appear suddenly. Winter flocks may include goldfinches and, occasionally, redpolls. Like other finches, reported attracted to ash piles and salt licks.

CBC data provide general comparisons of winter occurrence in lowlands. Birds found per party hour on counts from 1973 to 1998 indicated over 10 times more siskins found on counts w. than e. of the Cascades. Abundance varied locally between years, between different regional counts during the same years, and peak numbers usually did not occur e. and w. during the same winters. In addition to a statewide peak in 1969, numbers from 1973 to 1998 peaked both e. and w. only in 1975 and 1995 and were widely noted also in qualitative field reports. Overall, CBC data suggest a westside decrease in recent decades. BBS data from 1966 to 2000 indicate a decreasing population trend ($P < 0.01$; Sauer et al. 2001; see DeSante and George 1994 and Dawson 1997).

On occasions when numbers in Washington remained low or unworthy of comment, field reports showed abundances nearby in B.C. (e.g., an estimated 1 million in winter 1978-79 in Douglas-fir stands) and in Idaho and Montana (e.g., winter 1985-86).

Remarks: One of most noticeable species affected by disease in winter, reportedly avian *Salmonella* contracted at bird feeders. Reported to have done extensive damage to vegetable gardens (Bent 1968).

Noteworthy Records: West, *Fall*: ca. 10,000 on 1 Aug 1996 at Hart's Pass. *CBC high counts by location*: 6049 in 1989 at Olympia; 5528 in 1975 at Tacoma; 4300 in 1969 at Seattle; 3723 in 1995 at Sequim-Dungness; 3476 in 1977 at Grays Harbor; 2996 in 1988 at Bellingham; 2793 in 1989 at Padilla Bay; 2209 in 1988 at Kent-Auburn; 2046 in 1988 at Skagit Bay.

Terence R. Wahl

Lesser Goldfinch *Carduelis psaltria*

Fairly common permanent resident in Klickitat Co. Very rare vagrant elsewhere in state.

Subspecies: *C. p. hesperophilus.*

Habitat: A wide variety of open habitats including orchards, towns, pastures, steppe, and open woodlands with garry oak essential (Smith et al. 1997).

Occurrence: Not reported for Washington by Jewett et al. (1953). AOU (1957) gives occurrence in Clark Co., apparently based on a report from Camas during mid-Aug 1951. The next report was on 15 Jun 1974, when four males and a female were found at Vancouver. In summer 1975 birds found breeding nr. Lyle. Last report from Clark Co. was in fall 1991, when 10 were at Ridgefield NWR, though the species still occurs nearby in Portland, Oregon. Distribution in Klickitat Co. is from Bingen to Rock Cr. and n. to Goldendale and the population appears to be increasing.

Birds were recorded in winter on Columbia Hills-Klickitat Valley and Lyle CBCs in the late 1990s. Winter flocks from 13 to over 100 at Lyle and Maryhill were found in the 1990s-2001.

Outside of Clark and Klickitat Cos., there are only nine records, mostly from late Jun through Oct, from Everett and Duvall e. and s. to Richland and Reardan.

Noteworthy Records: *High counts*: 90 on 24 Nov 2000 nr. Maryhill; 80 in Oct-Nov 1989 nr. Lyle; 15 on 7 Jan 2001 at Maryhill; 100+ on 31 Dec 2001 between Dallesport and Maryhill. *Records away from Klickitat/Clark Cos.*: 12 Aug 1988 at Reardan; Oct 1990 at Fort Simcoe (Stepniewski 1999); 3 Sep 1994 at Everett; 20 Jul 1996 at Everett (T. Peterson); 1 Jul 1997 at Duvall; 1-26 Mar 1999 at Richland; 1 Mar 2003 at Selah; 11 Apr 2002 at Quilomine WMA; 12 Apr 2003 w. of Yakima.

Steven G. Mlodinow

American Goldfinch *Carduelis tristis*

Common summer resident in open areas and edges in lowlands, uncommon to locally common in winter statewide.

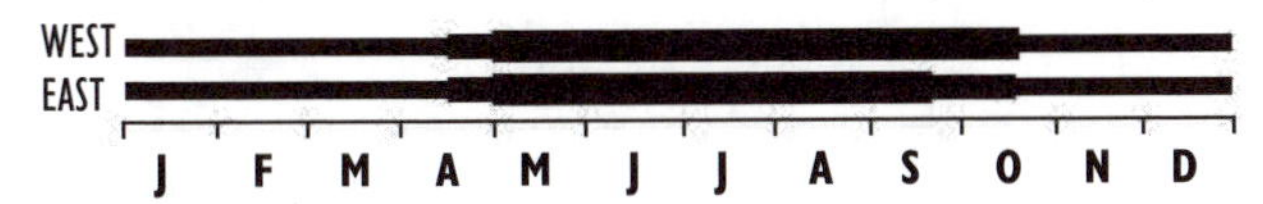

Subspecies: *C. t. jewettii* in w., and *pallidus* in e. Washington.

Habitat: Open, lowland agricultural, grasslands, clear-cuts, burns, and open urban habitats in transition zones below 300 m elevation w. and as high as 600 m in ne. (Smith et al. 1997).

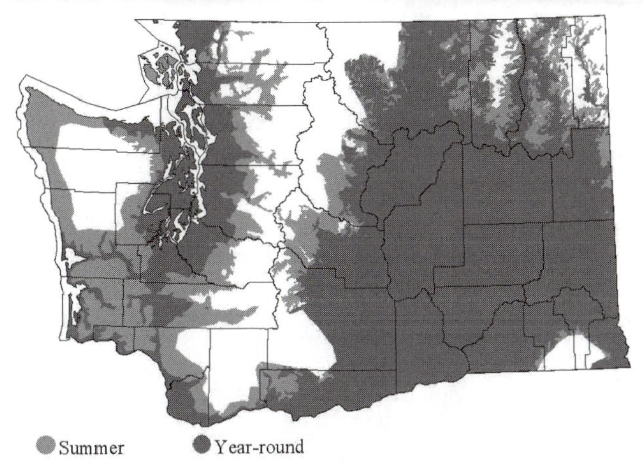

Occurrence: The conspicuous goldfinch is the official State Bird of Washington. It nests relatively late in the season, timed with abundance of seeds like thistles, dandelion, and other species in weedy fields, roadside brush; forages in winter in alders and other trees. Noted feeding on small caterpillars (Bent 1968). Attraction to bird feeders with niger and sunflower seeds brings ready contact with and recognition by humans. Much of summer population apparently migrates, but data on winter distribution of Washington breeders not described.

Distribution and numbers increased over past 100 years in w. N. America (DeSante and George 1994, Smith et al. 1997) with spread of agriculture with human development (see Edson 1919, 1926, Jewett et al. 1953, Wahl 1995) including recent industrial-forestry practice of poplar plantations (see Smith et al. 1997). Winter numbers appear higher e. of the Cascades than w., particularly in s. and e. areas. Peak CBC counts occurred at several of these areas in the 1990s (e.g., 1011 at Columbia Hills-Klickitat in 1998, 1400 at Two Rivers in 1997) and possibly reflected mild winter weather. Long-term increase in CBC numbers, significantly ($P < 0.01$) on the eastside.

Remarks: Noted in 1929 (Bent 1968) as a local problem species: feeding on seeds of cultivated sunflowers, cosmos, lettuce, and other plants.

Noteworthy Records: West, *Fall*: 300 on 22 Sep 1968 on *Salicornia* flats at Leadbetter Pt.

Terence R. Wahl

Evening Grosbeak *Coccothraustes vespertinus*

Variably fairly common to common summer resident statewide except Columbia Basin. Variably uncommon to common migrant and winter visitor statewide, rare to uncommon in Columbia Basin.

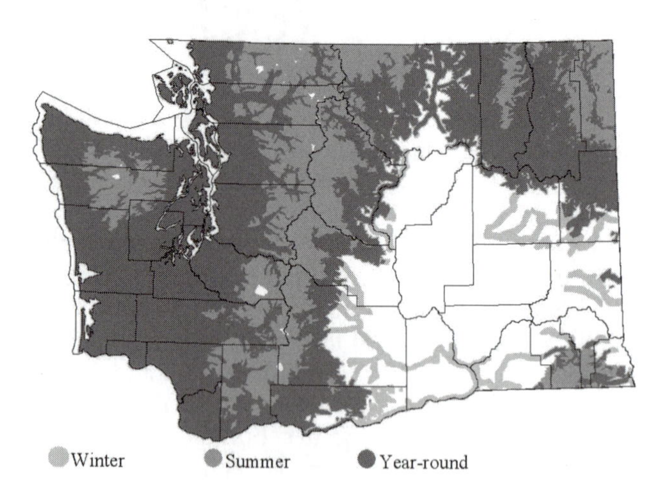

Subspecies: *C. v. brooksi* breeds in Washington.

Habitat: Summer habitat conifers and mixed woodlands, including second growth, occasionally in parks, with these and lower-elevation woodlands and urban parks, gardens, and bird feeders utilized in migrations and winter.

Occurrence: A conspicuous species famous for its inconsistency and noteworthy for its irruptive occurrence virtually everywhere within its wide range. Great variability, as in a number of n. finches and other species, often attributed to food supplies in breeding areas, migration routes, and wintering areas.

Breeds inconsistently in forested areas throughout Washington, including higher elevations in the Cascades, the Okanogan, and the Blue Mts. Less common in the Olympics (Smith et al. 1999). Migration most frequently reported in populated lowland areas, but often noticeable at high elevations in the mountains (Jewett et al. 1953, Stepniewski 1999). In Douglas Co. recorded at Badger Mt. (D. Stephens p.c.); few records for Grant, Adams, Franklin, Lincoln, Whitman Cos. (e.g., Smith et al. 1997; D. Stephens p.c.).

Variable presence affected by availability of natural foods: wild fruits including mountain ash and elder, winged seeds from a range of deciduous trees including maples, Pacific madrone seeds, conifer seeds, and infestations of insects such as spruce budworm (e.g., Smith et al. 1997, Stepniewski 1999, Campbell et al. 2001). Local occurrence determined by large-scale variations in availability of these foods. Large flocks may occur in some years in urban areas during spring migrations with none in several subsequent years. CBC numbers indicated widespread occurrence throughout the state, and anecdotal field observations of hundreds, occasionally thousands, of birds showed wide abundance in the Pacific Northwest in winters of 1973-74, 1974-75, 1975-76, spring and summer 1976, and w. of the Cascades in spring 1978 (an estimated 30,000 reported in Portland, Oregon), summer 1979, spring 1982, 1984, winter 1984-85, summer 1993, and fall 1998. CBC numbers (Table 47) indicate an overall decline statewide in recent decades. Apparent decline in breeding populations over time is statistically non-significant (see Smith et al. 1997, Sauer et al. 2001).

Table 47. Evening Grosbeak: CBC average counts per 100 party hours by decade.

	1960s	*1970s*	*1980s*	*1990s*
West				
Bellingham		66.9	44.0	37.3
Cowlitz-Columbia			111.1	47.8
Edmonds			94.0	7.1
E. Lk. Washington			28.8	15.6
Grays Harbor		2.3	30.6	0.9
Kent-Auburn			122.5	42.0
Kitsap Co.		6.0	21.8	11.1
Olympia			21.1	8.9
Padilla Bay			19.7	26.0
Seattle	11.1	22.4	14.4	10.8
Sequim-Dungeness			15.5	18.7
Tacoma		48.5	29.5	4.9
East				
Ellensburg			32.0	12.9
Spokane	52.5	37.5	201.6	9.5
Toppenish			7.5	2.0
Tri-Cities		57.7	56.7	37.4
Walla Walla		131.7	154.4	91.2
Wenatchee	22.7	21.2	7.8	3.1
Yakima Valley		40.1	2.5	6.2

Remarks: Universally described as erratic, variable, inconsistent, irruptive, irregular. An attractive species at bird feeders, probably a major consumer of commercially produced sunflower seeds. Subject to roadkills on occasion (e.g., several hundred killed on road shoulders during a spruce budworm outbreak in Jul 1980.) Formerly in the genus *Hesperiphona*.

Terence R. Wahl

House Sparrow *Passer domesticus*

Common resident statewide in urban areas and agricultural habitats.

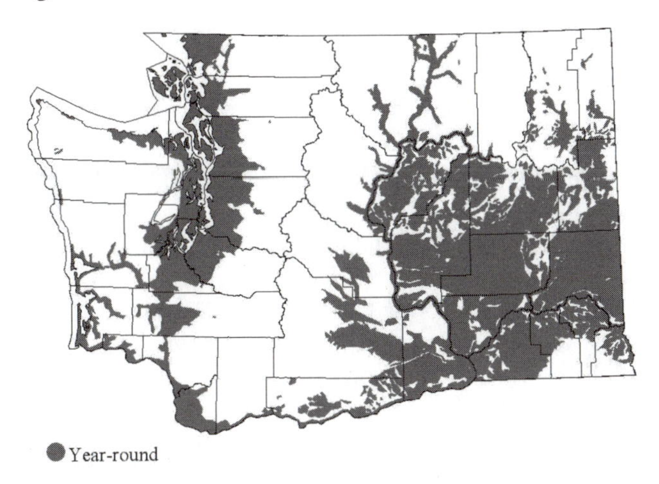

Habitat: Urban areas, agricultural structures, and feedlots. Utilizes holes in buildings and nest boxes in addition to cavities in suitable vegetation. Forages especially around habitations, at bird feeders, livestock-feeding and grain-handling areas.

Occurrence: The aggressive and adaptable House Sparrow numerically dominates many urban areas, quickly locating food sources, and outcompetes other species at bird feeders and nest holes. Some post-breeding and seasonal movements, probably local, are apparent (Stepniewski 1999). It is widespread throughout the state in cities and towns and locally so in many agricultural areas.

Recorded at Spokane in about 1895 and statewide soon after, with birds noted at LaPush in 1916 (Jewett et al. 1953, Smith et al. 1997). The source of the birds here, whether from the e. or California where they were noted in 1871 or 1872 at San Francisco (Johnston and Garrett 1994), is uncertain. Historical reports and subsequent data from many areas, including Washington, suggest that birds spread from urban areas (where they were associated with large numbers of horses) to farms and feedlots which, in e. Washington, had expanded with irrigation and large-scale farming (Jewett et al. 1953, Johnston and Garrett 1994, Smith et al. 1997). Apparent local population changes, perhaps cyclical, as in the Walla Walla Valley in the 1980s and 1990s (Smith et al. 1997) are unexplained.

CBC data from the 1960s through the 1990s indicate possible increases in several locations in w. Washington (e.g., Bellingham, Seattle, and Tacoma) and decreases in the e. (Tri-Cities, Wenatchee, Walla Walla, and Yakima Valley) but essentially no overall change over time. BBS data from 1966 to 2000 indicate a likely decline statewide. Counts from a larger region of w. N. America indicate stable populations over the past 20 years (Johnston and Garrett 1994).

Quite universally described as a pest species because of competition with other species for nest sites and food, consumption and possible contamination of grains; large numbers of House Sparrows at bird feeders waste food and also may attract accipiters and Merlins.

Remarks: Also called the English Sparrow. CBCs may have increased attention to and efficiency of counting these and other urban birds over time.

Terence R. Wahl

Introduced Species

A diverse array of introduced species have been recorded in Washington. Those that have clearly established self-sustaining populations are described in the species accounts. To define established, we employ the ABA Checklist Committee's definition (Dunn et al. 2002):

1. there is a more-or-less continuous wild population of potentially interacting individuals as opposed to a scattering of isolated individuals or pairs;

2. the population is large enough to survive a normal amount of mortality/ nesting failure;

3. sufficient numbers of offspring are being fledged annually to maintain or increase the population; and

4. the population is not directly dependent on human support.

Established Species

Feral species that currently or in the recent past meet these criteria in Washington are treated in the main section; they include: Chukar, Gray Partridge, Ring-necked Pheasant, Wild Turkey, Mountain Quail, California Quail, Northern Bobwhite, Rock Dove, Eurasian Collared Dove, Sky Lark, European Starling, and House Sparrow.

Accidental Introductions, Not Currently Established

Many other species have been reported, as a result of either accidental and purposeful introductions (Witmer and Lewis 2001). The majority of species—including waterfowl, raptors, psittids, and passerines—appear to result from accidental introductions. Two accidentally introduced species, Mute Swan and American Black Duck, have bred regularly and consequently appear on some previous checklists for Washington (e.g., Feltner et al 1994), but do not currently meet the criteria for established. Others, such as Mandarin Duck and Monk Parakeet, have bred occasionally, and have the potential to become established. Species in these categories are treated in additional detail in this section.

Mute Swan *Cygnus olor*

Widely introduced to N. America, has developed self-sustaining populations in various parts of the continent, including s. B.C. (Ciaranca et al.1997). In Washington, sporadic breeding has occurred at several sites around the Puget Trough and nr. Yakima. Not currently established in Washington, likely due at least in part to control activities by WDFW. Listed as a deleterious exotic wildlife species by WDFW in 1997, so the species cannot be kept in captivity or released without special authorization. The control activities and the listing are in response to reports that feral populations elsewhere in N. America compete with native waterfowl and impact aquatic vegetation (Kaufmann et al. 1996). The small number of winter records from coastal areas and the Puget Trough may represent dispersants from the small s. B.C. populations on Vancouver I., at Stanley Park in Vancouver, and at Pitt L. (W. Weber p.c.). Records elsewhere, including many summer records, are likely escapees from captivity.

Mandarin Duck *Aix galericulata*

Males in particular are frequently reported from lowland westside areas, averaging several per year, primarily in the fall, winter, and spring seasons. Birds have appeared for several consecutive winters at some locations. Has bred successfully at least once, in Redmond, when a pair produced three downy young in 2001, none of which survived to fledging (E. Woods p.c.). Fewer than five eastside reports to date. The presence of a well-established feral population in Great Britain (Ogilvie and Young 1998) provides evidence of their potential to establish in the maritime Pacific Northwest as well.

American Black Duck *Anas rubripes*

Status uncertain. There are records of likely vagrants, plus an apparently recently vanished population in Snohomish Co. established by birds escaped from captivity. That population was reported around 1970 and in 1972-73 birds were reported also from King Co. Numbers appeared to increase during the 1970s and perhaps into the 1980s and to decrease during the 1990s. None were found in 2000, though one apparently pure-bred bird was seen at L. Cassidy, Snohomish Co., in Feb 2001. Extensive interbreeding with Mallards probably played a major role in the decline, and a number of hybrids were still present in the Everett area as of 2001. Birds reported from Battle Ground in 1985 and Nisqually NWR. in 1994 were assumed to originate from the Snohomish Co. population.

Prior to the establishment of Snohomish Co. breeders, there were two undocumented Washington records: one shot nr. Port Ludlow on 30 Oct 1946 (Jewett et al. 1953) and one nr. Texas L., Whitman Co., on 17 Aug 1949 (Weber and Larrison 1977). The origins of these birds are unknown. During the 1970s, there were two introduced populations of Black Ducks in s. B.C., one on s. Vancouver I. and another nr. Vancouver (Campbell et al. 1990a), with the latter since extirpated (Toochin 1998). The species is also reported from other locations along the Fraser R. lowlands in s. B.C. (Campbell et al. 1990a). Black Ducks are kept in captivity in Washington and this, with vagrants or possible escapes from the Vancouver I. population, complicates determination of the origin of birds reported in Washington.

Monk Parakeet *Myiopsitta monachus*

Although nesting colonies have become established in many N. American cities, including Portland, Oregon (Gilligan et al 1994, Kaufmann et al. 1996, Moskoff 2003) there is no evidence that self-sustaining populations exist in Washington. In the mid-1980s, Monk Parakeets were reported from several Puget Sound locations, including Seattle, Auburn, and Tacoma (AB 37:331, 38:950 and 40:320), with indications of breeding at Seattle and Auburn. They have been unreported since then. The small Portland-area population was still extant in 2002 (Marshall et al 2003).

Reports of infrequently noted species that could conceivably establish feral populations in Washington are listed in Table 48; all of these species have been reported less than 10 times. On occasion, individuals from some of these accidental releases have apparently persisted in the wild for years, accounting for several reports, such as the Humboldt Penguin, Chilean Flamingo, and several of the waterfowl.

Additionally, there are numerous reports of likely escaped birds, many from the cage bird trade, whose chances of becoming established in Washington are remote.

Table 48. Introduced species infrequently noted.

Species	*Taxonomic Name*	*County*
Humboldt Penguin	*Sphensicus humboldti*	Grays Harbor
Chilean Flamingo	*Phoenicopterus chilensis*	Whatcom, Island
Greater Flamingo	*Phoenicopterus ruber*	Grays Harbor
White Stork	*Ciconia ciconia*	King
Black Swan	*Cygnus atratus*	King, Snohomish, Whatcom
Barnacle Goose	*Branta leucopsis*	King, Skagit, Walla Walla
Bar-headed Goose	*Anser indicus*	Skamania, Whatcom, Snohomish
Egyptian Goose	*Alpochen aegyptiacus*	King
Ruddy Shelduck	*Tadorna ferruginea*	Skamania
Black-bellied Whistling Duck	*Dendrocygna autumnalis*	Clark
Muscovy Duck	*Cairina moschata*	Whatcom, Snohomish
Cape Teal	*Anas capensis*	Snohomish
Red-crested Pochard	*Netta rufina*	Whatcom, Pierce
Impeyan Pheasant	*Lophophorus impeyanus*	Whatcom
Sulphur-crested Cockatoo	*Cacatua galerita*	Walla Walla
Cockatiel	*Nymphicus hollandicus*	Walla Walla, Snohomish
Budgerigar	*Melopsittacus undulatus*	Walla Walla, Snohomish
Green Parakeet	*Aratinga holochlora*	King
Mitred Conure	*Aratinga mitrata*	King
Peach-faced Lovebird	*Agapornis roseicollis*	Snohomish
Ringed Turtle-Dove	*Streptopelia risoria*	King, Snohomish
Yellow-crowned Bishop ??	*Euplectes afer*	King
Orange Bishop	*Euplectes franciscanus*	Skagit, Clark
Zebra Finch	*Poephilia guttata*	Walla Walla
Northern Cardinal	*Cardinalis cardinalis*	King

Table 49. Introduced species that have failed to establish.

Species	Taxonomic Name	Locations	Reference
Chilean Tinamou	*Nothoprocta perdicaria*	Clark Co., Lopez I.	WDFW 1979; Lewis and Sharpe 1987
Scaled Quail	*Callipepla squamata*	Benton, Yakima, and Grant Cos.	Jewett 1953; WDG 1979; Stepniewski 1999
Gambel's Quail	*Callipepla gambellii*	Priest Rapids area	WDFW 1979
Red-legged Partridge	*Alectoris rufa*	Yakima, Kittitas, Whatcom	WDFW 1979; Wahl 1995
Chinese Bamboo-Partridge	*Bambusicola thoracica*	Near Grandview, Yakima Co.; Whatcom Co	WDFW 1979; Wahl 1995
Kalij Pheasant	*Lophura leucomelanos*	Southwest Washington	WDFW 1979
Reeve's Pheasant	*Syrmaticus reevesii*	Many areas, Yakima Co	WDFW 1979
Japanese Green Pheasant	*Phasianus versicolor*	Cowlitz Valley, Lewis Co.	WDFW 1979
Peafowl	*Pavo cristatus*	Protection I., Jefferson, Maryhill, Klickitat Co.; L. Whatcom	Wahl 1995; Tweit

Purposeful Introductions, Not Currently Established

The purposeful introduction category in Washington is primarily limited to releases of gamebird species that have failed to establish sustaining populations (Table 49; WDFW 1979). Efforts in former centuries to introduce songbirds, such as those in the late 1800s by the Portland Bird Club, were generally unsuccessful and are sufficiently described in earlier publications (Jewett et al 1953; Witmer and Lewis 2001) that they do not merit further treatment here.

Rock Pigeon (G. Scott Mills)

Hypothetical List

Since Mattocks et al. (1976) provided the last comprehensive treatment of Washington's avifauna, several species have been reported as occurring in the state without sufficient evidence to conclusively document their presence. In many instances, these recent reports may be correct, but due to the difficult nature of field identification for some of these species, at present they are best treated as hypothetical. In some cases, additional information since the Mattocks et al. (1976) review adds to our understanding of some of the hypothetical records.

Of the 23 species classified as hypothetical by Mattocks et al., the occurrence of 13 are now well documented in Washington. These include Manx Shearwater, Snowy Egret, Iceland Gull, Common Black-headed Gull, Red-legged Kittiwake, Broad-tailed Hummingbird, Black Phoebe, Brown Thrasher, Blue-gray Gnatcatcher, Magnolia Warbler, Orchard Oriole, Black-throated Sparrow, and Hoary Redpoll. Three species of waterfowl on their list, Mute Swan, Barnacle Goose, and American Black Duck, are treated as introduced or escaped, but not self-sustaining (see Introduced Species). One species, Gray-headed Junco, is now treated as a subspecies of Dark-eyed Junco; the occurrence of this subspecies in Washington remains hypothetical. Five others (Terek Sandpiper, Passenger Pigeon, Crested Myna, Blue Grosbeak and Baird's Sparrow) are still hypothetical in Washington. Including the species discussed below, the current list of hypothetical species in Washington totals 11 species.

Juan Fernandez Petrel *Pterodroma externa*

A S. Hemisphere species occurring s. of the subarctic boundary in the n. Pacific Ocean (Wahl et al. 1989). One flew through a flock of birds at a chum at 45° 51'N 125° 00'W, about 56 km west of Grays Harbor on 14 Sep 1990. Due to the brief nature of the sighting, and the lack of experience of most of the observers, some of whom submitted conflicting details, this report is currently treated as hypothetical. The sole experienced observer (TRW) believes it was almost certainly this species. Recently, another was identified by experienced observers 80 km off Brookings, Oregon on 7 Jun 2002 (NAB 56: 476).

Black-vented Shearwater *Puffinus opisthomelas*

Jewett et al. (1953) suggested "more thorough investigation" off Washington would confirm this expected species. There are four specimens and seven sight records in B.C. n. to Victoria from 1891 to 1986 (Campbell et al. 1990a) and another reportedly sighted off Amphitrite Pt. in May 1999 (NAB 53: 318). These birds certainly passed through Washington waters. There is one accepted record for Oregon, a bird flying over the surf at Bandon, on 22 Nov 1992 (ORBC).

This species breeds in Mexico and, with two other small, "white-bellied" shearwaters (Townsend's, *P. auricularis*, and Newell's, *P. newelli*), breeding in the N. Pacific, could occur along the Washington coast. Sightings of such birds were not infrequent but identification remained uncertain and debated for many years (see Roberson 1980, Howell et al. 1994 and Roberson 1996). The situation was complicated in part because the species occurring in the Pacific Ocean were for many years classed as subspecies of the Manx Shearwater.

Crested Caracara *Polyborus plancus*

The first record was a bird shot at Westport on 21 Jun 1936 (Balmer 1936, UWBM #47862), after one had escaped from captivity near Portland, Oregon, during Jun 1936, suggesting the origin of the Westport bird (Jewett et al. 1953). The second was ph. at Ocean Shores on 13 Aug 1983, where it was observed eating small crabs (Hoge and Hoge 1991; photograph subsequently lost). The third record was a bird at Neah Bay from 4 Jan to 1 Feb 1998 (Anderson and Shiflett 1998). A caracara, likely this same individual, was found at a logging camp on Vancouver I., B.C., during May 1998, suggesting a 24 km cross-Strait movement.

While the original sighting was likely of a known escape, origins of the later records are less certain. Possession of the species has been illegal for several decades, trade in pet birds ceased in the U.S. by the early 1970s, and it has been illegal for many years, even with a falconry license, to possess a bird in Washington, Oregon, and California. A check subsequent to the 1998 Neah Bay sighting, revealed none had escaped from w. coast zoos. The last two records occurred in El Niño years. The 1998 record involved a winter bird that continued moving n., negotiating a wide water crossing into Canada. On the other hand, there are very few records from n. of the species' range in Texas and since the B.C. bird, like one of about 10 California birds, was begging food, it is likely at least some reports involved escapes and the species is considered hypothetical pending further information. Subspecies presumably *P.p. audubonii.*

Wood Sandpiper *Tringa glareola*

Three single-person sight reports have been considered and rejected by the WBRC (Tweit and Paulson 1994), primarily for lack of conclusive evidence substantiating this difficult field identification. Those reports include one at Tokeland on 9 Oct 1988 (AB 43: 159), one at Ocean Shores on 13 Oct 1989, and one at Dry Falls Dam on 4-5 Jul 1991.

Red-necked Stint *Calidris ruficollis*

A bird reported as an alternate-plumaged Red-necked Stint at Crockett L. on 18 Jul 1993 (AB 47:1144) corresponded with the peak period of late Jun-late Jul occurrence with a number of w. coast records of this species from B.C. s. to California. The report was not accepted by the WBRC (Tweit and Skriletz 1996), due to the lack of photograph, specimen, or corroborating observers.

Crested Auklet *Aethia cristatella*

Jewett et al. (1953) report a specimen from "Vance" (possibly in Grays Harbor Co.) found dead on the highway, reportedly 29 Jun 1937. Jewett and Gabrielson verified the identity of the specimen, and treated it as hypothetical since there was doubt regarding its origin. Recent summer records from California and Baja California (Garrett and Singer 1998) indicate that this record could have had a natural origin.

Passenger Pigeon *Ectopistes migratorius*

Jewett et al. (1953) considered the Passenger Pigeon to be "formerly occurring casually, at least in the n. and ne. parts of the state," based largely on the reports of J. G. Cooper. Apparently, a pair was collected at Spokane Falls, and others were observed in the Puget Sound (Cooper 1869, Cooper 1870). Unfortunately, no specimens or descriptions remain, and consequently Mattocks et al. (1976) dissented, treating their occurrence as hypothetical.

Documented records from B.C., however, suggest the likelihood of historical occurrence in Washington. A bird was collected in nearby Chilliwack, B.C., in 1859 (Duvall 1946) and skeletal elements of two Passenger Pigeons were found in an archaeological site in ne. B.C. (Campbell et al. 1990b). Their regular distribution extended w. through s. Alberta (Blockstein 2002); and the proximity of their range, considered with their great abundance and highly nomadic nature, provides some support for the conclusion by Jewett et al.

The other reports considered by Jewett et al. seem less likely. Rhoads' (1891) report of large numbers in the Pierce Co. prairies may well relate to Band-tailed Pigeons (Jewett et al. 1953), and Ridgway's (1916) reference to a bird collected in Puget Sound actually refers to the Chilliwack, B.C., specimen.

Crested Myna *Acridotheres cristatellus*

There have been several undocumented reports of this species (Mattocks et al. 1976), which have formed the basis for the presence of the species on several Washington checklists. The probable source for these would have been the introduced population at Vancouver, B.C., which apparently died out in early 2003.

References

Aanerud, K. R. 2002. Fifth report of the Washington Bird Records Committee. Wash. Birds 8: 1-18.

Aanerud, K., and P. W. Mattocks, Jr. 1997. Third report of the Washington Bird Records Committee. Wash. Birds 6:7-31.

Aanerud, K. R., and P. W. Mattocks, Jr. 2000. Fourth report of the Washington Bird Records Committee. Wash. Birds 7:7-24.

Aborn, D. A. 1994. Correlation between raptor and songbird numbers at a migratory stopover site. Wilson Bull. 106:150-54.

Ackerman, S. A. 1994. American White Pelicans nest successfully at Crescent Island, Washington. Wash. Birds 3:44-49.

Ackerman, S. A. 1997. Update: American White Pelican colony. WOSN 51:1, 6.

Adkisson, C. S. 1996. Red Crossbill (*Loxia curvirostra*). *In* A. Poole and F. Gill, eds. The birds of North America, No. 256. The Acad. of Nat. Sci., Philadelphia, and Amer. Ornithol. Union, Washington, D.C.

Agee. J. K. 1993. Fire ecology of Pacific Northwest forests. Island Press, Washington, D.C.

Ainley D. G. 1980. Geographic variation in Leach's Storm-Petrel. Auk 97:837-53.

Ainley, D. G., R. J. Boekelheide, S. H. Morrell, and C. S. Strong. 1990a. Pigeon Guillemot. Pp. 276-305 *in* Ainley, D. G., and R. J. Boekelheide, eds. Seabirds of the Farallon Islands. Stanford Univ. Press, CA. 450 pp.

Ainley, D. G., R. J. Boekelheide, S. H. Morrell, and C. S. Strong. 1990b. Cassin's Auklet. Pp. 307-38 *in* Ainley, D. G. and R. J. Boekelheide, eds. Seabirds of the Farallon Islands. Stanford Univ. Press, CA. 450 pp.

Ainley, D. G., H. R. Carter, D. W. Anderson, K. T. Briggs, M. C. Coulter, F. Cruz, J. B. Cruz, C. A. Valle, S. I. Fefer, S. A. Hatch, E. A Schreiber, R. W. Schreiber and N. G. Smith. 1988. Effects of the 1982-83 El Niño-Southern Oscillation on Pacific Ocean bird populations. Proc. Int. Ornithol. Congr. 19:1747-58.

Ainley, D. G., and B. Manolis. 1979. Occurrence and distribution of the Mottled Petrel. West. Birds 10:113-23.

Ainley, D. G., J. Norton, and W. J. Sydeman. 1995. Upper trophic level predators indicate interannual negative and positive anomalies in the California Current food web. Mar. Ecol. Prog. Ser. 118:69-80.

Ainley, D. G., L. B. Spear, S. G. Allen, and C. A. Ribic. 1996. Temporal and spatial patterns in the diet of the Common Murre in California waters. Condor 98:691-705.

Ainley, D. G., W. J. Sydeman, S. A. Hatch, and U. W. Wilson. 1994. Seabird population trends along the west coast of North America: causes and the extent of regional concordance. Pp. 119-33 *in* J. R. Jehl, Jr., and N. K. Johnson,eds. A century of avifaunal change in western North America. Studies in Avian Biology No. 15.

Ainley, D. G., R. L. Veit, S. G. Allen, L. B. Spear, and P. Pyle. 1995. Variations in marine bird communities of the California current, 1986-1994. CalCOFI Rep. 36:72-77.

Alcorn, G. D. 1941. The western Burrowing Owl in Grays Harbor County, Washington. Murrelet 22:57-58.

Alcorn, G. D. 1958. Nesting of the Caspian Tern in Grays Harbor, Washington. Murrelet 39:19.

Alcorn, G. 1959a. Puffins on the south Grays Harbor beaches. Murrelet 40:21.

Alcorn, G. D. 1959b. Another record of Lewis' Woodpecker in western Washington. Murrelet 40:10.

Alcorn, G. D. 1962. Checklist of the birds of the state of Washington. Dept. Biol., Univ. Puget Sound (Tacoma, WA.), Occas. Paper No. 17.

Alerstam, T. 1991. Ecological causes and consequences of bird orientation. *In* P. Berthold, ed. Orientation in Birds. Birkhauser, Basel, Switzerland.

Allen, H. A. 1991. Status and management of the Peregrine Falcon in Washington. *In* J. E. Pagel, ed. Symposium on Peregrine Falcons in the Pacific Northwest. USFS, Medford, OR. 119 pp.

Allen, J. A. 1980. The ecology and behavior of the Long-billed Curlew in southeastern Washington. Wildl. Monogr. 73:1-67.

Altman, B. 1999. Conservation strategy for landbirds in coniferous forests of western Oregon and Washington. Version 1.0. Amer. Bird Conserv., Boring, OR.

Altman, B. 2000a. Conservation strategy for landbirds of the east-slope of the Cascade Mountains in Oregon and Washington. Version 1.0. Amer. Bird Conserv., Boring, OR.

Altman, B. 2000b. Conservation strategy for landbirds in lowlands and valleys of western Oregon and Washington. Version 1.0. Amer. Bird Conserv., Boring, OR.

Altman, B. 2000c. Conservation strategy for landbirds in the northern Rocky Mountains of eastern Oregon and Washington. Version 1.0.

Altman, B., M. Hayes, S. Janes, and R. Forbes. 2001. Wildlife of westside grasslands and chaparral habitats. Pp. 261-91 *in* D. H. Johnson and T. A. O'Neil, eds. Wildlife-habitat relationships in Oregon and Washington. Oreg. State Univ. Press, Corvallis, OR.

Altman, B., and A. Holmes. 2000. Conservation strategy for landbirds in the Columbia Plateau of eastern Oregon and Washington. Version 1.0. American Bird Conservancy, Boring, OR, and Pt. Reyes Bird Obs., Stinson Beach, CA.

American Birding Association (ABA) 1996. ABA Checklist: Birds of the Continental United States and Canada, 5th ed. Colorado Springs, CO.

American Birding Association (ABA) 2002. ABA Checklist: Birds of the Continental United States and Canada, 6th ed. Colorado Springs, CO.

American Ornithologists' Union (AOU). 1910. Check-list of North American Birds. Amer. Ornithol. Union, 4th ed. New York.

American Ornithologists' Union (AOU). 1931. Check-list of North American Birds. Amer. Ornithol. Union, 5th ed. Lancaster, PA.

American Ornithologists' Union (AOU). 1957. Check-list of North American birds. Amer. Ornithol. Union. Allen Press. Lawrence, KS.

American Ornithologists' Union (AOU). 1973. Thirty-second supplement to the American Ornithologists' Union check-list of North American birds. Auk 90:411-19.

American Ornithologists' Union (AOU). 1983. Check-list of North American birds. Amer. Ornithol. Union, 6th ed. Lawrence, KS.

American Ornithologists' Union (AOU). 1985. Thirty-fifth supplement to the American Ornithologists' Union Check-list of North American birds. Auk 102:680-86.

American Ornithologists' Union (AOU). 1989. Thirty-seventh supplement to the American Ornithologists' Union Check-list of North American birds. Auk 106:532-38.

American Ornithologists' Union (AOU). 1993. Thirty-ninth supplement to the American Ornithologists' Union Check-list of North American birds. Auk 110:675-682.

American Ornithologists' Union (AOU). 1995. Fortieth supplement to the American Ornithologists' Union Check-list of North American birds. Auk 112:819-30.

American Ornithologists' Union (AOU). 1997. Forty-first supplement to the American Ornithologists' Union Check-list of North American birds. Auk 114:542-52.

American Ornithologists' Union (AOU). 1998. Check-list of North American birds. Seventh edition. American Ornithologists' Union, Washington, D.C.

American Ornithologists' Union (AOU). 2002. Forty-third supplement to the American Ornithologists' Union check-list of North American birds. Auk 119:897-906.

American Ornithologists' Union (AOU). 2003. Forty-fourth supplement to the American Ornithologists' Union check-list of North American birds. Auk 120: 923-31.

American Ornithologists' Union (AOU. 2004. Forty-fifth supplement to the American Ornithologists' Union check-list of North American birds. Auk 121: 985-95.

Ammon, E. M. 1995. Lincoln's Sparrow (*Melospiza lincolni*). *In* A. Poole and F. Gill, eds.The Birds of North America, No. 191. Acad. of Nat. Sci., Philadelphia, and Amer. Ornithol. Union, Washington, D.C.Ammon, E. M., and W. M. Gilbert. 1999. Wilson's Warbler (*Wilsonia pusilla*). *In* A. Poole and F. Gill, eds.The birds of North America, No. 478. Acad. of Nat. Sci., Philadelphia, and Amer. Ornithol. Union, Washington, D.C.

Andelman, S. J., and A. Stock. 1994. Management, research and monitoring priorities for the conservation of neotropical migratory landbirds that breed in Washington State. Wash. Nat. Heritage Prog., Wash. Dept. Nat. Resour., Olympia.

Anderson, C. M., and D. M. Batchelder. 1990. First confirmed nesting of the Black-shouldered Kite in Washington. West. Birds 21:37-38.

Anderson, C. M., and P. M. DeBruyn. 1979. Behavior and ecology of Peregrine Falcons wintering on the Skagit Flats, Washington. Unpub. Rep., Wash. Dept. of Game. 53 pp.

Anderson, C. M., P. M. DeBruyn, T. Ulm, and B Gaussoin.1980. Behavior and Ecology of Peregrine Falcons wintering on the Skagit Flats, Washington. Unpub. Rep., Wash. Dept. of Game. 54 pp.

Anderson, C.M., J. Fackler, D. Drummond, M. Finger, and A. Brewer. 1983. The spring hawk migration at Cape Flattery Peninsula, Clallam County, Washington, 1983. Unpubl. rep.

Anderson, C. M., D. G. Roseneau, B. J. Walton, and P. J. Bente. 1988. New evidence of a Peregrine migration on the west coast of North America. Pp. 507-516 *In* T. J. Cade, J. H. Enderson, C. G. Thelander, and C. M. White, eds. Peregrine Falcon populations:their management and recovery. The Peregrine Fund, Boise, ID.

Anderson, C. M., and J. T. Shiflett. 1998. Crested Caracara in Washington: A re-examination of the records. WOSNews 53:1, 7.

Anderson, D. A. 1988. Additional inland turnstone records. Oreg. Birds 14:389.

Anderson, D. A., and B. Bellin. 1988. Inland records of Ruddy and Black turnstones in Oregon. Oreg. Birds 14:47-50.

Anthony, R. G., E. D. Forsman, A. B. Franklin, D. R. Anderson, K. P. Burnham, G. C. White, C. J. Schwarz, J. Nichols, J. Hines, G. S. Olson, S. H. Ackers, S. Andrews, B. L. Biswell, P. C. Carlson, L. V. Diller, K. M. Dugger, K. E. Fehring, T. L. Fleming, R. P. Gerhardt, S. A. Gremel, R. J. Gutierrez, P. Happe, D. R. Herter, J. M. Higley, R. B. Horn, L. L. Irwin, P. J. Loschl, J. A. Reid, and S. G. Sovern. 2004. Status and trends in demography of Northern Spotted Owls, 1985-2003. Draft report to Interagency Reg. Monitoring Progr., Portland, OR.

Anthony, R. G., and F. B. Isaacs. 1989. Characteristics of bald eagle nest sites in Oregon. J. Wildl. Manage. 53(1):148-59.

Anthony, R. G., R. L. Knight, G. T. Allen, B. R. McClelland, and J. I. Hodges. 1982. Habitat use by nesting and roosting bald eagles in the Pacific Northwest. Trans. N. Amer. Wildl. Nat. Resour. Conf. 47:332-432.

Arcese, P., M. K. Sogge, A. B. Marr, and M. A. Patten. 2002. Song Sparrow (*Melospiza melodia*). *In* A.Poole and F.Gill, eds.The birds of North America, No. 704. Acad. of Nat. Sci., Philadelphia, and Amer. Ornithol. Union, Washington, D.C.

Armstrong, H. 1994. Cascade White-tailed Ptarmigan. Wash. Birder 2(3):1.

Arnold, K. A. 1994. Common Snipe. Pp. 117-25 in T. C. Tacha and C. E. Braun, eds. Migratory shore and upland game bird management in North America. Allen Press, Lawrence, KS.

Atkinson, E .C. 1993. Winter territories and night roosts of Northern Shrikes in Idaho. Condor 95:515-27.

Aubry, K. B., and C. M. Raley. 1996. Ecology of pileated woodpeckers in managed landscapes on the Olympic Peninsula. *In* Wildlife Ecology Team 1996 Annual Report: Ecology, Management, and Conservation of Sensitive Wildlife Species. USDA. For. Serv., Pacific Northwest Res. Station, Olympia, WA. pp 70-74.

Aubry, K. B., and C. M. Raley. 2002a. Selection of nest and roost trees by pileated woodpeckers in coastal forests of Washington. J. Wild. Manage. 66(2):392-406.

Aubry, K. B., and C. M. Raley. 2002b. The pileated woodpecker as a keystone habitat modifier in the Pacific Northwest. *In* W. F. Laudenslayer, Jr., P. J. Shea, B. Valentine, C. P. Weatherspoon, and T. E. Lisle, technical coordinators. Proceedings of the symposium on the ecology and management of dead wood in western forests. U.S. For. Serv. Gen. Tech. Rep. PSW-GTR-181.

Austin, J. E., C. M. Custer, and A. D. Afton. 1998. Lesser Scaup (*Aythya affinis*). *In* A. Poole and F. Gill, eds.The birds of North America, No. 338. Acad. of Nat. Sci., Philadelphia, and Amer. Ornithol. Union, Washington, D.C.

Austin, O. L., Jr. 1929. Labrador records of European birds. Auk 46:207-13.

Aversa, T. 1997. An observation of carrion feeding in Cooper's Hawk. Wash. Birds 6:32-33.

Aversa, T. 2000. Franklin's Gull. WOSNews 68:4.

Aversa, T. 2001. Nominate Rock Sandpiper at Ocean Shores, Washington. N. Amer. Birds 55:242-44.

Avery, M. L., P. F. Springer and N. S. Dailey. 1980. Avian mortality at man-made structures: an annotated bibliography. U.S. Fish and Wildl. Serv., Biol. Serv. Prog. National Power Plant Team, FWS/OBS-80/54. U.S. Fish and Wildl. Serv., Washington, D.C.

Bakus, G. J. 1959. Observations of the life history of the Dipper in Montana. Auk 76:190-207.

Balfour E., and C. J. Cadbury. 1979. Polygyny, spacing and sex ratio among Hen Harriers *Circus cyaneus* in Orkney, Scotland. Ornis. Scand. 10:133-41.

Ball, J. E., E. L. Bowhay, and C. F.Yocum. 1981. Ecology and management of the western Canada Goose in Washington. Wash. Department of Game Biol. Bull. No. 17. Olympia, WA.

Ballard, G., G. R. Guepel, N. Nur, and T. Gardali. 2003. Long-term declines and decadal patterns in population trends of songbirds in western North America, 1979-1999. Condor 105:737-55.

Balmer, A. 1936. Audubon's Caracara taken in Gray's Harbor County, Washington. Murrelet 17:54.

Baltosser, W. H., and S. M. Russell. 2000. Black-chinned Hummingbird (*Archilochus alexandri*). *In* A. Poole and F. Gill, eds.The birds of North America, No. 495. Acad. of Nat. Sci., Philadelphia, and Amer. Ornithol. Union, Washington, D.C.

Barrowclough, G. F. 1980. Genetic and phenotypic differentiation in a wood warbler (genus Dendroica) hybrid zone. Auk 97:655-68.

Bart, J., B. Collins, and R. I. G. Morrison. 2004. Estimating trends with a linear model: reply to Sauer et al. Condor 106:440-43.

Bartle, J. A., D. Hu, J.-C. Stahl, P. Pyle, T. R. Simons. and D. Woodby. 1993. Status and ecology of gadfly petrels in the temperate North Pacific. Pp. 101-111 *in* K. Vermeer, K., K. T. Briggs, K. H. Morgan, and D. Siegel-Causey, eds. The status, ecology and conservation of marine birds of the North Pacific. Can. Wildl. Serv. Spec. Publ., Ottawa.

Batey, K. M., H. H. Batey, and I. O. Buss. 1980. First Boreal Owl fledglings for Washington state. Murrelet 61:80.

Bechard, M. 1980. Factors affecting nest productivity of Swainson's Hawks (*Buteo swainsoni*) in southeastern Washington. Ph.D. diss., Wash. State Univ., Pullman.

Bechard, M.J ., and T. R. Swem. 2002. Rough-legged Hawk (*Buteo lagopus*). *In* A. Poole and F. Gill, eds.The birds of North America, No. 641. Acad. of Nat. Sci., Philadelphia, and Amer. Ornithol. Union, Washington, D.C.

Beebe, F. L. 1974. Field studies of the Falconiformes of British Columbia. Occasional Paper Series, No. 17, British Columbia Provincial Museum, Victoria, British Columbia, Canada.

Beer, J. 1944. Western Burrowing Owl in Clark County, Washington. Auk 61:652-53.

Behrstock, R. A. 1981. Prey-induced mortality of a Pied-billed Grebe. West. Birds 12:183-84.

Beissinger, S. B. 1995. Population trends of the Marbled Murrelet projected from demographic analyses. Pp. 385-94 *in* C. J. Ralph, G. L. Hunt, Jr., M. G. Raphael, and J. F. Piatt, eds. Ecology and conservation of the Marbled Murrelet. Gen. Tech. Rep. PSW-GTR-152. Albany, CA: Southwest Res. Station, For. Serv., USDA. 420 p.

Bell, D. A. 1996. Genetic differentiation, geographic variation and hybridization of gulls of the *Larus glaucenscens-occidentalis* complex. Condor 98:527-646.

Bellrose, F. C. 1976. Ducks, geese, and swans of North America. Stackpole Books, Harrisburg, PA.

Bemis, C., and J. D. Rising. 1999. Western Wood-Pewee (*Contopus sordidulus*). *In* A. Poole and F. Gill, eds. The birds of North America, No. 287. The Academy of Natural Sciences, Philadelphia, PA, and the American Ornithologists' Union, Washington, D.C.

Benkman, C. W. 1992. *(White-winged Crossbill)*. *In* A. Poole and F. Gill, eds. The birds of North America, No. 27. The Academy of Natural Sciences, Philadelphia, PA, and The American Ornithologists' Union, Washington, D.C.

Benkman, C. W. 1993a. Adaptation to single resources and the evolution of crossbill diversity. Ecological Monographs 63:305-25.

Benkman, C. W. 1993b. Logging, conifers, and the conservation of crossbills. Conserv. Biol. 7:473-79.

Bennett, H. S., and G. Eddy. 1949. European Starling in King County, Washington. Murrelet 30:18.

Bent, A. C. 1932. Life histories of North American gallinaceous birds. Dover Publ., New York.

Bent, A. C. 1939. Life histories of North American woodpeckers. Dover Publ., New York.

Bent, A. C. 1940. Life histories of North American cuckoos, goatsuckers, hummingbirds and their allies. U.S. Gov. Printing Off., Smithsonian Inst. U. S. Natl. Mus. Bul. 176.

Bent, A. C. 1950. Life histories of North American wagtails, shrikes, vireos and their allies. Dover Publ., New York.

Bent A. C. 1964. Life histories of North American jays, crows and titmice, Part 2. Dover Publ., New York.

Bent, A. C. 1968. Life histories of North American cardinals, grosbeaks, buntings, towhees, finches, sparrows, and allies. Part 1. Dover Publ., New York.

Bettesworth, J. 1991. Fall raptor migration at Slate Peak, Washington, during September 1991. Unpubl. Falcon Res. Group report, Hawk Watch International.

Bettinger, K. A., and R. Milner. 2000. Sandhill Crane (*Grus canadensis*). *In* E. M. Larsen and N. Nordstrom, eds. Management recommendations for Washington's priority species, Volume IV: Birds [Online]. Available http://www.wa.gov/wdfw/hab/phs/vol4/sndhlcrn.htm.

Bevis, K. R. 1994. Primary cavity excavators in grand fir forests of central Washington's east Cascades. Summary Rep. Master's Thesis, Central Wash. Univ., Ellensburg, WA.

Bildstein, K. L., and K. Meyer. 2000. Sharp-shinned Hawk (*Accipiter striatus*). *In* A. Poole and F. Gill, eds. The birds of North America, No. 482. Acad. of Nat. Sci., Philadelphia, and Amer. Ornithol. Union, Washington, D.C.

Binford, L. C., and D. B. Johnson. 1995. Range expansion of the Glaucous-winged Gull into interior United States and Canada. West. Birds 26:169-88.

Binford, L. C., and M. Perrone, Jr. 1971. First record of the Ruff in Washington State.

Calif. Birds 2:103-4.

Binford, L. C., and J. V. Remsen. 1974. Identification of the Yellow-billed Loon *(Gavia adamsii)*. West. Birds 5:111-26.

Bird, F. 1994. The colorful history and provocative future of Everett's terns. WOSNews 32:2-5.

Bird, F. 1995. Navy prepares to tern the tide - while volunteers struggle to 'build' a new Jetty Island home for Caspians and Arctics. WOSNews 35:5.

Blockstein, D. E. 2002. Passenger Pigeon (*Ectopistes migratorius*). *In* A. Poole and F. Gill, eds. The birds of North America, No. 611. Acad. of Nat. Sci., Philadelphia, and Amer. Ornithol. Union, Washington, D.C.

Bloxton, T. D. 2002. Prey abundance, space use, demography, and foraging habitat of Northern Goshawks in western Washington. M.S. thesis, Univ. of Wash., Seattle. 70 p.

Boal, C. W., and R. W. Mannan. 1999. Comparative breeding ecology of Cooper's Hawks in urban and exurban areas of southeastern Arizona. J. Wildl. Manage. 63(1):77-84.

Boarman, W. I., and B. Heinrich. 1999. Common Raven (*Corvus corax*). *In* A. Poole and F. Gill, eds. The birds of North America, No. 476. Acad. of Nat. Sci., Philadelphia, and Amer. Ornithol. Union, Washington, D.C.

Bock, C. E. 1970. The ecology and behavior of Lewis' Woodpecker (*Asyndesmus lewis*). Univ. Calif. Publ. Zool. 92:1-100.

Bock, C. E., and L. W. Lepthien. 1976. Changing winter distribution and abundance of the Blue Jay, 1962-1971. Amer. Midland Nat. 96: 232-36.

Boekelheide, B. 1998. A daring nomenclatural proposition - Olympic Gull. WOSNews 57:3.

Boekelheide, R. J., D. G. Ainley, S. H. Morell, H. R. Huber and T. J. Lewis. 1990. Common Murre. Pp.244-75 *in* D. G. Ainley and R.J. Boekelheide, eds. Seabirds of the Farallon Islands. Stanford Univ. Press, CA. 450 pp.

Boersma, P. D., and M. J. Groom. 1993. Conservation of storm-petrels in the North Pacific. Pp. 112-21 *in* K. Vermeer, K. T. Briggs, K. H. Morgan, D. Siegel-Causey, eds. The status, ecology and conservation of marine birds of the North Pacific. Can. Wildl. Serv. Spec. Publ., Ottawa.

Bolsinger, C. E., and K. L. Waddell. 1993. Area of old-growth forests in California, Oregon and Washington. USDA For. Serv., Portland, OR. Resour. Bull. PNW-RB-197. 26 pp.

Bolsinger, C. L., N. McKay , D. R. Gedney, and C. Alerich. 1997. Washington's public and private forests. USDA For. Serv., Portland, OR. Resour. Bull. PNW-RB-218. 144 pp.

Bond, S. I. 1971. Red Phalarope mortality in Southern California. Calif. Birds 2:97.

Booth, E. 1942. A field guide to the birds of eastern Washington, eastern Oregon, and Idaho. Walla Walla College, College Place, WA.

Booth, E. S. 1948. Starlings in Washington State. Condor 50:165.

Booth, E. S. 1949. Winter season: Palouse, northern Rocky Mountain region. Aud. Field Notes 3:177-78.

Booth, E. 1952. Ecological distribution of the birds of the Blue Mountains region of southeastern Washington and northeastern Oregon. Walla Walla Coll. Publ. Dept.of Biol. Sci.

Booth, E. S. 1957. Further records of the Red-backed Sandpiper in eastern Washington. Murrelet 38:31.

Bordage, D., and J. L. Savard. 1995. Black Scoter (*Melanitta nigra*). *In* A. Poole and F. Gill, eds. The birds of North America, No. 177. Acad. of Nat. Sci., Philadelphia, and Amer. Ornithol. Union, Washington, D.C.

Bosakowski, T. 1997. Breeding bird abundance and habitat relationships on a private industrial forest in the western Washington Cascades. Northwest Sci. 71:87-96, 244-53.

Bourne, W. R. P., and T. J. Dixon. 1975. Observations of seabirds 1970-1972. Sea Swallow 24:65-88.

Bowles, J. H. 1906a. Pacific Eider in Washington. Condor 8:57.

Bowles, J. H. 1906b. The Hermit Warbler in Washington. Condor 8: 40-42.

Bowles, J. H. 1918. The Limicolae of the state of Washington. Auk 35:326-33.

Bowles, J. H., 1921. Breeding dates for Washington birds. Murrelet 2:8-12.

Bowles, J. H. 1926. Summer record of the Burrowing Owl in Clallam County, Washington. Murrelet 7:39.

Bowles, J. H. 1929. Changes in Bird Population. Murrelet 10: 52-55.

Braden, G. 1953. Starlings nesting in Washington. Murrelet 34:47.

Brazil, M. A. 1991. The birds of Japan. Smithsonian Institution Press, Washington, D.C.

Brennan, L. A. 1999. Northern Bobwhite (*Colinus virginianus*). *In* A. Poole and F. Gill, eds. The birds of North America, No. 397. Acad. of Nat. Sci., Philadelphia, and Amer. Ornithol. Union, Washington, D.C.

Brennan, L. A., J. B. Buchanan, S. G. Herman, and T. M. Johnson. 1985. Interhabitat movements of wintering Dunlins in western Washington. Murrelet 66:11-16.

Briggs, K. T., W. B. Tyler, D. B. Lewis and D. R. Carlson. 1987. Bird communities at sea off California: 1975-1983. Studies in Avian Biol. No. 11, Cooper Ornithol. Soc.

Briggs, K. T., D. H. Varoujean, W. W. Williams, R. G. Ford, M. L. Bonnell and J. L. Casey. 1992. Chapter III: Seabirds of the Oregon and Washington OCS, 1989-1990 *in* J. J. Brueggeman, ed. Ebasco Env. and Ecol. Cons., Inc., Oreg. and Wash. marine mammal and seabird surveys. Final report. OCS Study 91-0093.

Brooks, A. C. 1918. Brief notes on the prevalence of certain birds in British Columbia. Can. Field-Nat. 31:139-41.

Brooks, J. P. 1997. Bird-habitat relationships at multiple spatial rResolutions in the Oregon Coast Range. M.S. thesis, Oregon St. Univ., Corvallis.

Brown, C. R. 1977. Purple Martins versus starlings and House Sparrows in nest site competition. Bull. Texas Ornithol. Soc. 10: 31-35.

Brown, C. R. 1981. The impact of starlings on Purple Martin populations in unmanaged colonies. Amer. Birds 35: 266-68.

Brown, C. R. 1997. Purple Martin (*Progne subis*). *In* A. Poole and F. Gill, eds. The birds of North America, No. 287. Acad. of Nat. Sci., Philadelphia, and Amer. Ornithol. Union, Washington, D.C.

Brown, C. R., and M. B. Brown. 1995. Cliff Swallow (*Hirundo pyrrhonota*). *In* A. Poole and F. Gill, eds. The birds of North America, No. 149. Acad. of Nat. Sci., Philadelphia, and Amer. Ornithol. Union, Washington, D.C.

Brown, D. E. 1924. The Burrowing Owl (*Speotyto cunicularia hypogaea*) in western Washington. Murrelet 5:8.

Brown, D. E. 1926. Birds observed at Moses Lake, Grant County, Washington. Murrelet 7:48-51.

Brown, D. E. 1934. Dottrel - a first record for the United States. Murrelet 15:79.

Brown, D. E. 1935. Dottrel in western Washington. Condor 37:82.

Brown, E. R., tech. ed. 1985. Management of wildlife and fish habitats in forests of western Oregon and Washington. USDA For. Serv. R6-F&WL-192, Pacific Northwest Reg., Portland, OR. 332 pp.

Brown, J. M. 1976. Western Burrowing Owl. Pp. 290 301 in Marine shoreline fauna of Washington: Volume II. Wash. Depts. of Game and Ecol.

Brown, P. W., and L. H. Fredrickson. 1997. White-winged Scoter (*Melanitta fusca*). *In* A. Poole and F. Gill, eds. The birds of North America, No. 274. Acad. of Nat. Sci., Philadelphia, and Amer. Ornithol. Union, Washington, D.C.

Brown, S., C. Hickey, B. Harrington, and R. Gill. 2001. United States shorebird conservation plan. Manomet Center for Conservation Sciences, Manomet, MA.

Browning, M. R. 1992. Geographic variation in Hirundo pyrrhonota (*Cliff Swallow*) from northern North America. West. Birds 23:21-29.

Bruce, A. M., R. J. Anderson, and G. T. Allen. 1982. Observation of golden eagles nesting in western Washington. Raptor Res. 16:132-34.

Bruce, J. A. 1961. First record of European Skylark on San Juan Island, Washington. Condor 63:418.

Bryan, T, and E. D. Forsman. 1987. Distribution, abundance, and habitat of Great Gray Owls in southcentral Oregon. Murrelet 68:45-49.

Bryant, A. A., J. P. L. Savard, and R. T. McLaughlin. 1993. Avian communities in old-growth and managed forests of western Vancouver Island, B.C. Can. Wild. Serv. Tech. Rep. Series, No. 16, Delta, B.C.

Buchanan, J. B. 1988a. North American Merlin populations: An analysis using Christmas Bird Count data. Amer. Birds 42:1178-80.

Buchanan, J. B. 1988b. Migration and winter populations of Greater Yellowlegs, *Tringa melanoleuca*, in western Washington. Can. Field-Nat. 102:611-16.

Buchanan, J. B. 1988c. The abundance and migration of shorebirds at two Puget Sound estuaries. West. Birds 19:69-78.

Buchanan, J. B. 1992. Winter abundance of shorebirds at coastal beaches of Washington. Wash. Birds 2:12-19.

Buchanan, J. B. 1997. Spotted owl management activities of the Washington Department of Fish and Wildlife: Final report for 1992-1995. U.S. Fish and Wildlife Service, Section 6, Project E-1, Segment 34. Olympia, WA.

Buchanan, J. B. 1999. Recent changes in the winter distribution and abundance of Rock Sandpipers in North America. West. Birds 30:193-99.

Buchanan, J. B. 2000. Shorebirds: plovers, oystercatchers, avocets and stilts, sandpipers, snipes and phalaropes. *In* J. Azerrad, E.M. Larsen, and N. Nordstrom, eds. Management recommendations for priority bird species. Wash. Dept. of Fish and Wildl., Olympia, WA.

Buchanan, J. B. In press. Distribution and abundance of wintering Northern Shrikes in Washington. Wash. Birds.

Buchanan, J. B., and J.R. Evenson. 1997. Abundance of shorebirds at Willapa Bay, Washington. West. Birds 28:158-68.

Buchanan, J. B., and J. D. Horn. 1992. Attempted predation of Rock Doves by Merlins. Wash. Birds 2:9-11.

Buchanan, J. B., L. L. Irwin, and E. L. McCutchen. 1993. Characteristics of spotted owl nest trees in the Wenatchee National Forest. J. Raptor Res. 27(1):1-7.

Buchanan, J. B., L. L. Irwin, and E. L. McCutchen. 1995. Within-stand nest site selection by spotted owls in the eastern Washington Cascades. J. Wildl. Manage. 59:301-10.

Buchanan, J. B., D. H. Johnson, E. L. Greda, G. A. Green, T. R. Wahl, and S. J. Jeffries. 2001. Wildlife of coastal and marine habitats. Pp. 389-422 in D. H. Johnson and T.A. O'Neil, eds. Wildlife-habitat relationships in Oregon and Washington. Oreg. State Univ. Press, Corvallis, OR.

Buehler, D. A. 2000. Bald Eagle (*Haliaeetus leucocephalus*). *In* A. Poole and F. Gill, eds. The birds of North America. No. 506. Acad. of Nat. Sci., Philadelphia, and Amer. Ornithol. Union, Washington, D.C.

Bull, E .L., and C. T. Collins 1993.Vaux's Swift (*Chaetura vauxi*). *In* A. Poole and F. Gill, eds. The birds of North America, No.77. Acad. of Nat. Sci., Philadelphia, and Amer. Ornithol. Union, Washington, D.C.

Bull, E. L., and H. D. Cooper. 1991. Vaux's swift nests in hollow trees. West. Birds 22:58-91.

Bull, E. L., and M. G. Henjum. 1990. Ecology of the Great Gray Owl. Gen. Tech. Rep. PNW-GTR-265. Portland, OR: U.S. Dept. Agric. For. Serv.

Bull, E. L., and J. E. Hohmann. 1993. The association between Vaux's swifts and old growth forests in northeastern Oregon. West. Birds 24:38-42.

Bull, E. L, and J. A. Jackson. 1995. Pileated Woodpecker (*Dryocopus pileatus*). *In* A. Poole and F. Gill, eds. The birds of North America, No. 148. Acad. of Nat. Sci., Philadelphia, and Amer. Ornithol. Union, Washington, D.C.

Burdick, A. W. 1944. Birds of the northern Cascade Mountains of Washington. Condor 46:238-42.

Burger, A. E. 1993. Possible El Niño effected detected in beached bird survey data in 1992. B.C. Field Ornithol. 3:10-11.

Burger, A. E. 1995. Marine distribution, abundance, and habitats of Marbled Murrelets in British Columbia. Pp. 295-312 *In* C. J. Ralph, G. L. Hunt, Jr., M. G. Raphael, and J. F. Piatt, eds. Ecology and conservation of the Marbled Murrelet. Gen. Tech. Rep. PSW-GTR-152. Albany, CA: Southwest Res. Sta., For. Serv., USDA. 420 p.

Burger, A. E., and D. M. Fry. 1993. Effects of oil pollution on seabirds in the northeast Pacific. Pp. 254-63 *in* K. Vermeer, K. T. Briggs, K. H. Morgan, and D. Siegel-Causey, eds.The status, ecology and conservation of marine birds of the North Pacific. Can. Wildl. Serv. Spec. Publ., Ottawa.

Burger, J. 1972. Dispersal and post-fledging survival of Franklin's Gulls. Bird Banding 43:267-75.

Burger, J., and M. Gochfeld. 1994. Franklin's Gull (*Larus pipixcan*). *In* A. Poole and F. Gill, eds. The Birds of North America, No. 116. The Academy of Natural Sciences, Philadelphia, and The American Ornithologists' Union, Washington, D.C.

Burleigh, T. D. 1954. Another record for the occurrence of the West Mexican Tropical Kingbird in the State of Washington. Murrelet 35:49.

Burleigh, T. D. 1972. Birds of Idaho. Caxton Printers. Caldwell, ID. 467 pp.

Burleigh, T. D., and A. J. Duvall. 1959. A new subspecies of Veery from the northwestern United States. Proc. Biol. Soc. of Wash. 72:33-35.

Burness, G. P., K. Lefevre, and C. T. Collins. 1999. Elegant Tern (*Sterna elegans*). *In* A. Poole and F. Gill, eds. The birds of North America, No. 404. Acad. of Nat. Sci., Philadelphia, and Amer. Ornithol. Union, Washington, D.C.

Butler, R. W. 1992. Great Blue Heron (*Ardea herodias*) *In* A. Poole, P. Stettenhiem, and F. Gill, eds. The birds of North America, No. 25. Acad. of Nat. Sci., Philadelphia, and Amer. Ornithol. Union, Washington, D.C.

Butler, R. W. 1996. Status of the subspecies of Great Blue Heron, *Ardea herodias fannini* in Canada. Can. Wildl. Serv., Pacific Res. Centre, Delta. B.C..

Byrd, G. V., and J. C. Williams. 1993a. Red-legged Kittiwake (*Rissa brevirostris*). *In* A. Poole and F. Gill, eds. The birds of North America, No. 60. Acad. of Nat. Sci., Philadelphia, and Amer. Ornithol. Union, Washington, D.C.

Byrd, G. V., and J. C. Williams. 1993b. Whiskered Auklet (*Aethia pygmaea*). *In* A. Poole and F. Gill, eds. The birds of North America, No. 76. Acad. of Nat. Sci., Philadelphia, and Amer. Ornithol. Union, Washington, D.C.

Byrd, G. V., E. C. Murphy, G. W. Kaiser, A. Y. Kondratyev, and Y. V. Shibaev. 1993. Status and ecology of offshore fish-feeding alcids (murres and puffins) in the North Pacific. Pp. 176-86 *in* K. Vermeer, K. T. Briggs, K. H. Morgan, and D. Siegel-Causey, eds. The status, ecology and conservation of marine birds of the North Pacific. Can. Wildl. Serv. Spec. Publ., Ottawa.

Cabe, P. R. 1993. European Starling (*Sturnus vulgaris*). *In* A. Poole and F. Gill, eds. The birds of North America, No. 48.Acad. of Nat. Sci., Philadelphia, and Amer. Ornithol. Union, Washington, D.C.

Cade, T. J. 1982. The falcons of the world. Cornell Univ. Press, Ithaca, NY.

Cade, T. J., J. H. Enderson, C. G. Thelander, and C. M. White. 1988. Peregrine Falcon populations:their management and recovery. The Peregrine Fund, Boise, ID.

Calkins, J. D., J. C. Hagelin, and D. F. Lott. 1999. California Quail (*Callipepla californica*). *In* A. Poole and F. Gill, eds. The birds of North America, No. 473. Acad. of Nat. Sci., Philadelphia, and Amer. Ornithol. Union, Washington, D.C.

Campbell, R. W. 1985. First record of the Eurasian Kestrel for Canada. Condor (87) 294.

Campbell, R. W., N. K. Dawe, I. McTaggert-Cowan, J. M. Cooper, G. W. Kaiser, and M. C. E. McNall. 1990a. The birds of British Columbia. Vol. 1. Nonpasserines: loons through waterfowl. Royal B.C. Mus., Victoria. 514 pp.

Campbell, R. W., N. K. Dawe, I. McTaggert-Cowan, J. M. Cooper, G. W. Kaiser and M. C. E. McNall. 1990b. The birds of British Columbia. Vol. 2. Nonpasserines: diurnal birds of prey through woodpeckers. Royal B.C. Mus.. Victoria. 636 pp.

Campbell, R. W., N. K. Dawe, I. McTaggert-Cowan, J. M. Cooper, G. W. Kaiser, M. C. E. McNall, and G. E. J. Smith. 1997. The birds of British Columbia. Vol. 3. Passerines: flycatchers through vireos. Can. Wildl. Serv./B.C. Wildl. Branch. UBC Press, Vancouver. 693 pp.

Campbell, R. W., N. K. Dawe, I. McTaggert-Cowan, J. M. Cooper, G. W. Kaiser, A. C. Stewart, and M. C. E. McNall. 2001. The birds of British Columbia. Vol. 4. Wood-warblers through Old World sparrows. Can. Wildl. Serv./B.C. Wildl. Branch and Resources Inventory Branch/Royal B. C. Museum. UBC Press, Vancouver. 741 pp.

Campbell, R. W., M. G. Shepard, B. A. MacDonald, and W. C. Weber. 1974. Vancouver birds in 1972. Vancouver Nat. Hist. Soc. Spec. Publ., Vancouver, B.C.

Canadian Wildlife Service, U.S. Fish and Wildlife Service and Instituto Nacional de Ecologica - SEMARNAP. 1998. North American waterfowl plan. 1998 update. Canadian Wildl. Serv., Hull, Quebec, Canada; U.S. Fish and Wildl. Serv., Arlington, VA; and Instituto Nacional de Ecologica - SEMARNAP, Mexico City, Mexico.

Canaris, A. G. 1950. Sight record of American Egret in eastern Washington. Murrelet 47:45.

Canning, D. J., and S. G. Herman. 1983. Gadwall breeding range expansion into western Washington. Murrelet 64: 27-31.

Cannings, R. A., R. J. Cannings, and S. G. Cannings. 1987. Birds of the Okanogan Valley, British Columbia. Royal B.C. Mus., Victoria. 420 pp.

Cannings, R. J. 1987. Gray Flycatcher: a new breeding bird for Canada. Amer. Birds 41:376-78.

Cannings, R. J. 1993. Northern Saw-whet Owl (*Aegolius acadicus*). *In* A. Poole and F. Gill, eds. The birds of North America, No. 42. Acad. of Nat. Sci., Philadelphia, and Amer. Ornithol. Union, Washington, D.C.

Cannings, R. J., and T. Angell. 2001. Western Screech-Owl (*Otus kennicottii*). *In* A. Poole and F. Gill, eds. The birds of North America, No. 597. Acad. of Nat. Sci., Philadelphia, and Amer. Ornithol. Union, Washington, D.C.

Cannings, R. J., and E. Hunn. Vocalizations of the Western Flycatcher *(Empidonax difficilis)* complex in the interior of the Pacific Northwest. Unpubl.

Cardiff, S. W., and D. L. Dittman. 2002. Ash-throated Flycatcher (*Myiarchus cinerascens*). *In* A. Poole and F. Gill, eds. The birds of North America, No. 664 . Acad. of Nat. Sci., Philadelphia, and Amer. Ornithol. Union, Washington, D.C.

Carey, A. B. 1988. The influence of small streams on the composition of upland bird communities. Pp. 153-61 *in* K. J. Raedeke, ed. Streamside management: Riparian wildlife and forestry interactions. Univ. of Wash., Inst. of For. Res. Contr. No. 59, Seattle, WA, USA.

Carey, A. B., M. M. Hardt, S. P. Horton, and B. L. Biswell. 1991. Spring bird communities in the Oregon Coast Range. Pp. 123-44 *in* L. F. Ruggiero, K. B. Aubry, A. B. Carey, and M. H. Huff, tech. coords. Wildlife and vegetation of unmanaged Douglas-fir forests. USDA For. Serv. Gen. Tech. Rep. PNW-GTR-285. 533 pp.

Carter, H. R., and K. J. Kuletz. 1995. Mortality of Marbled Murrelets due to oil pollution in North America. Pp. 261-70 *in* C. J. Ralph, G. L. Hunt, Jr., M. G. Raphael, and J. F. Piatt eds. Ecology and conservation of the Marbled Murrelet. Gen. Tech. Rep. PSW-GTR-152. Albany, CA: Southwest Res. Sta., For. Serv., USDA. 420 p.

Carter, H. R., and S. G. Sealy. 1986. Year-round use of coastal lakes by Marbled Murrelets. Condor 88:473-77.

Carter, M. F., W. C. Hunter, D. N. Pashley, and K. V. Rosenberg. 2000. Setting conservation priorities for landbirds in the United States: the Partners in Flight approach. Auk 117:541-48.

Cassidy, K. M. 1997a. Land cover of Washington State: Description and management. Volume 1 *in* K.M. Cassidy, C.E. Grue, M.R. Smith, and K.M. Dvornich, eds. Washington State Gap Analysis Project Final Report. Wash. Coop. Fish and Wildl. Res. Unit, Univ. Wash., Seattle, 270 p.

Cassidy, K. M. 1997b. Snowy Owl irruption into Washington and vicinity during the winter of 1996-1997. Wash. Birds 6:68-82.

Cassidy, K. M., M. R. Smith, C. E. Grue, K. M. Dvornich, J. E. Cassady, K. R. McAllister, and R. E. Johnson. 1997. Gap analysis of Washington State: An evaluation of the protection of biodiversity. Volume 5 *in* K.M. Cassidy, C.E. Grue, M.R. Smith, and K.M. Dvornich, eds. Washington State Gap Analysis - Final Report Wash. Coop. Fish and Wildl. Res. Unit, Univ. Wash., Seattle, 192 pp.

Cassidy, K. M., and C. E. Grue. 2000. The role of private and public lands in conservation of at-risk vertebrates in Washington state. Wildl. Soc. Bull. 28:1060-76.

Cassirer, E. F. ed. 1993. Status of Harlequin Ducks in North America. Harlequin Duck Working Group, Boise, ID.

Chadwick, D. H, and B. Littlehales. 1993. Birds of white waters. Nat. Geogr. 185:116-32.

Chambers, C. L., T. Carrigan, T. E. Sabin, J. Tappeinner, and W. C. McComb. 1997. Use of artificially created snags by cavity-nesting birds. West. J. Appl. Forestry 12:93-97.

Chappell, C. B. 1979. The birds of Lakewood and Steilacoom. Nongame Progr., Wash. Dept.of Game, Olympia, WA.

Chappell, C. B., and R. C. Crawford. 1997. Native vegetation of the South Puget Sound prairie landscape. Pp. 107-22 *in* P. Dunn and K. Ewing, eds. Ecology and conservation of the South Puget Sound prairie landscape. The Nature Conservancy, Seattle, WA.

Chappell, C. B., R. C. Crawford, C. Barrett, J. Kagan, D. H. Johnson, M. O'Mealy, G. A. Green, H. L. Ferguson, W. D. Edge, E. L. Greda, and T. A. O'Neil. 2001.Wildlife habitats: descriptions, status, trends, and system dynamics. Pp.s 22-114 *in* D.H. Johnson and T. A. O'Neil, dirs. Wildlife-habitat relationships in Oregon and Washington. Oreg.n State Univ. Press, Corvallis.

Chappell, C. B., and B. J. Ringer. 1983. Status of the Hermit Warbler in Washington. West. Birds 14: 185-96.

Chappell, C. B., and T. A. Williamson. 1984. First documented breeding of the White-breasted Nuthatch in western Washington. Murrelet 65:51-52.

Chilton, G., M. C. Baker, C. D. Berrentine and M. A. Cunningham. 1995. White-crowned Sparrow (*Zonotrichia leucophrys*). *In* A. Poole and F. Gill, eds. The birds of North America, No.183. Acad. of Nat. Sci., Philadelphia, and Amer. Ornithol. Union, Washington, D.C.

Christensen, G. C. 1996. Chukar (*Alectoris chukar*). *In* A. Poole and F. Gill, eds. The birds of North America. No. 258. Acad. of Nat. Sci., Philadelphia, and Amer. Ornithol. Union, Washington, D.C.

Ciaranca, M. A., C. C. Allin and G. S. Jones. 1997. Mute Swan (*Cygnus olor*). *In* A. Poole and F. Gill, eds. The birds of North America, No. 273. Acad. of Nat. Sci., Philadelphia, and Amer. Ornithol. Union, Washington, D.C.

Cicero, C., and N. K. Johnson. 1995. Speciation in sapsuckers (Sphyrapicus): III. Mitochondrial-DNA sequence divergence at the cytochrome-B locus. Auk 112:547-63.

Cimprich, D. A., and F. R. Moore. 1995. Gray Catbird (*Dumetella carolinensis*). *In* A. Poole and F. Gill, eds. The birds of North America, No. 167. Acad. of Nat. Sci., Philadelphia, and Amer. Ornithol. Union, Washington, D.C.

Clapp, R. B., P. A. Buckley, and F. G. Buckley. 1993. Conservation of temperate North Pacific terns. Pp. 154-63 *in* K. Vermeer, K. T. Briggs, K. H. Morgan, and D. Siegel-Causey, eds. The status, ecology and conservation of marine birds of the North Pacific. Can. Wildl. Serv. Spec. Publ., Ottawa.

Clark, W., and V. Clark. 1998. Spring raptor migration studies at Cape Flattery, Washington in 1990-1997. Unpubl. rep. 35 pp.

Clark, W., V. Clark, K. E. Wiersema, and L. Liu. 1998. Spring raptor migration studies at Cape Flattery, Washington in 1990-1997. Hawk Watch International, Inc., Salt Lake City, UT.

Clark, W. S. 1974. Second record of the Kestrel (*Falco tinnunculus*) for North America. Auk (91) 172.

Clark, W. S., and C. M. Anderson. 1984. First specimen record of the Broad-winged Hawk for Washington. Murrelet 65:93-94.

Clement, P., A. Harris, and J. Davis. 1993. Finches and sparrows. Princeton Univ. Press, NJ.

Climpson, J. T., and J. G. Francik. 1978. A Chestnut-sided Warbler in southeast Washington. Murrelet 59:102.

Clum, N. J., and T. J. Cade. 1994. Gyrfalcon (*Falco rusticolus*). *In* A. Poole and F. Gill, eds. The birds of North America, No. 114. Acad. of Nat. Sci., Philadelphia, and Amer. Ornithol. Union, Washington, D.C.

Cohen, A. N., H. D. Berry, C. E. Mills, D. Milne, K. Britton-Simmons, M. J. Wonham, D. L. Secord, J. A. Barkas, B. Bingham, B. E. Bookheim, J. E. Byers, J. W. Chapman, J. R. Cordell, B. Dumbauld, A. Fukuyama, L. H. Harris, A. J. Kohn, K. Li, T. F. Mumford, Jr., V. Radashevsky, A. T. Sewell, and K. Welch. 2001. Wash. State exotics expedition 2000: a rapid survey of exotic species in the shallow waters of Elliott Bay, Totten and Eld Inlets, and Willapa Bay. Wash. Dept. of Nat. Resour., Olympia, WA.

Cole, L. W. 2000. A first Shy Albatross, *Thalassarche cauta*, in California and a critical re-examination of Northern Hemisphere records of the former *Diomedea cauta* complex. N. Amer. Birds 54:124-35.

Collins, C. T. 1997. Hybridization of a Sandwich and Elegant tern in California. West. Birds 28:168-73.

Colwell, M. A., and J. R. Jehl, Jr. 1994. Wilson's Phalarope (*Phalaropus tricolor*). *In* A. Poole and F. Gill, eds. The birds of North America, No. 83. (Acad. of Nat. Sci., Philadelphia, and Amer. Ornithol. Union, Washington, D.C.

Connelly, J. W. Jr. 1978. Trends in Blue-winged and Cinnamon Teal populations of eastern Washington. Murrelet 59:2-6.

Connelly, J. W., M. W. Gratson and K. P. Reese. 1998. Sharp-tailed Grouse (*Tympanuchus phaisianellus*). *In* A. Poole and F. Gill, eds. The birds of North America, No. 354. Acad. of Nat. Sci., Philadelphia, and Amer. Ornithol. Union, Washington, D.C.

Conover, B. 1944. The North Pacific allies of the Purple Sandpiper. Field Mus. Nat. Hist., Zool. Ser. 29:169-79.

Conover, M. R. 1983. Recent changes in Ring-billed and California gull populations in the western United States. Wilson Bull. 95:362-83.

Conover, M. R., B. C. Thompson, R. E. Fitzner and D. E. Miller. 1979. Increasing populations of Ring-billed and California gulls in Washington state. West. Birds 10:31-36.

Contreras, Alan. 1997. Northwest birds in winter. Oreg. State Univ. Press, Corvallis. 264 pp.

Conway, C. J. 1995. Virginia Rail (*Rallus limicola*). *In* A. Poole and F. Gill, eds. The birds of North America, No. 173. Acad. of Nat. Sci., Philadelphia, and Amer. Ornithol. Union, Washington, D.C.

Cooper, J. G. 1860. Report upon the birds collected on the survey. Land birds. Chap. 1. U.S. Pacific R.R. Expl. & Surv. 12 (Book 2: Zoology) (3): 140-226.

Cooper, J. G. 1869. The fauna of the Montana Territory. II. Birds. Amer. Naturalist 3:73-84.

Cooper, J. G. 1870. Ornithology of California. I. Land birds. From manuscript and notes of J. G. Cooper, ed. S. F. Baird.

Copely, D., D. Fraser, and J. Cam Finlay. 1999. Purple Martins, *Progne subis*: A British Columbian success story. Can. Field-Nat. 113(2):226-29.

Cordell, H. K., and N. G. Herbert. 2002. Popularity of birding is still growing. Birding 34 (1):54-59.

Cory, C. B. 1888. The European Kestrel in Massachusetts. Auk 5:110.

Cottam, C. 1939. Food habits of North American diving ducks. U.S. Dept. of Agri. Tech. Bull. 643.

Courtney, S. P., N. I. Litchi, and Z. A. Bassett. 1998. A literature review on Starling (*Sturnus vulgaris*) biology and control methods. Unpubl. rep. to Weyerhaeuser Co., Tacoma, WA.

Cramp, S., D. J. Brooks, E. Dunn, R. Gillmor, P. A. D. Hollum, R. Hudson, E. M. Nicholson, M. A. Ogilvie, P. J. S. Olney, C. S. Roselaar, K. E. L. Simmons, K. H. Voous, D. I. M. Wallace, J. Wattel, and M.G. Wilson. 1985a. Handbook of the birds of Europe, the Middle East and North Africa: The birds of the Western Palearctic. Vol. 4. Terns to woodpeckers. Oxford Univ. Press, Oxford. 960 pp.

Cramp, S., and K. Simmons. 1980. Handbook of the birds of Europe, the Middle East and North Africa. Vol.2: Hawks to bustards. Oxford Univ. Press, NY.

Cramp, S., K. E. L. Simmons, I. J. Ferguson-Lees, R. Gillmor, P. A. D. Hollum, R. Hudson, E. M. Nicholson, M. A. Ogilvie, P. J. S. Olney, K. H. Voous, and J. Wattel. 1977. Handbook of the birds of Europe, the Middle East and North Africa: The birds of the Western Palearctic. Vol. 1. Ostrich to ducks. Oxford Univ. Press, Oxford. 722 pp.

Cramp, S., K. E. L. Simmons, D. J. Brooks, N. J. Collar, E. Dunn, R. Gillmor, P. A. D. Hollum, R. Hudson, E. M. Nicholson, M. A. Ogilvie, P. J. S. Olney, C. S. Roselaar, K. H. Voous, D. I. M. Wallace, J. Wattel, and M. G. Wilson. 1983. Handbook of the birds of Europe, the Middle East and North Africa: Tthe birds of the Western Palearctic. Vol. 3. Waders to gulls. Oxford Univ. Press, Oxford. 913 pp.

Crawford, J., and M. Pope. 1999. Mountain quail research: annual report. Game bird Research Progr, Oreg. State Univ.

Crocker-Bedford, D. C. 1990. Goshawk reproduction and forest management. Wildl. Soc. Bull. 18(3):262-69.

Csada, R. D., and R. M. Brigham. 1992. Common Poorwill (*Phalaenoptilus nuttallii*). *In* A. Poole, P. Stettenheim, and F. Gill, eds. The birds of North America, No. 32). Acad. of Nat. Sci., Philadelphia, and Amer. Ornithol. Union, Washington, D.C.

Cullinan, T. 2001. Important bird areas of Washington. Aud. Wash., Olympia,.

Curry, R. L., A. T. Peterson, and T. A. Langen. 2002. Western Scrub-Jay *(Aphelocoma californica*). *In* A. Poole and F. Gill, eds. The birds of North America, No. 712. Acad. of Nat. Sci., Philadelphia, and Amer. Ornithol. Union, Washington, D.C.

Cuthbert, F. J., and L. R. Wires. 1999. Caspian Tern (*Sterna caspia*). *In* A. Poole and F. Gill, eds. The birds of North America. No. 403. Acad. of Nat. Sci., Philadelphia, and Amer. Ornithol. Union, Washington, D.C.

Czaplak, D. 1995. Identifying Common and Hoary redpolls in winter. Birding 27:447-57.

Dahlquest, W. W. 1948. Mammals of Washington. Univ. Kans. Publ. Mus. Nat. Hist. No. 2. 444 pp.

Dark, S. J., R. J. Gutiérrez, and G. I. Gould, Jr. 1998. The Barred Owl (*Strix varia*) invasion in California. Auk 115:50-56.

Daubenmire, R. F. 1970. Steppe vegetation of Washington. Wash. State Univ. Agri. Experiment Sta. Tech. Bull. No. 62. 131 pp.

Davis, D. E. 1950. The growth of starling, *Sturnus vulgaris*, populations. Auk 67:460-65.

Davis, J. 1995. Purple Martins in Puget Sound, How to attract and maintain them. Unpubl. ms.

Davis, W. E., Jr., and J. A. Kushlan. 1994. Green Heron (*Butorides virescens*). *In* A. Poole and F. Gill, eds. The birds of North America, No. 129). Acad. of Nat. Sci., Philadelphia, and Amer. Ornithol. Union, Washington, D.C.

Davis, W. J. 1980. The Belted Kingfisher, *Megaceryle alcyon*: Its ecology and territoriality. M.S. Thesis, Univ. of Cincinatti. 100 pp.

Dawson, W. L. 1906. A new sport bird horizoning, with a Puget Sound example. Pacific Sportsmen 3:85-86.

Dawson, W. L. 1908. New and unpublished records from Washington. Auk 25:482-85.

Dawson, W. L., and J. H. Bowles. 1909. The birds of Washington. Occidental Publishing: Seattle, WA.

Dawson, W. R. 1997. Pine Siskin (*Carduelis pinus*). *In* A. Poole and F. Gill, eds. The birds of North America, No. 280. Acad. of Nat. Sci., Philadelphia, and Amer. Ornithol. Union, Washington, D.C.

Day, R. D., H. S. Wehle, and F. C. Coleman. 1985. Ingestion of plastic pollutants by marine birds. Pp. 344-86 *in* R. S. Shomura and H. O. Yoshida, eds. Proceedings of the workshop on the fate and impact of marine debris. U.S. Dept. Comm., NOAA Tech. Memo, NMFS, NOAA-TM-NMFS-SWFC-54.

Decker, F. R., and J. H. Bowles. 1926. The status of the ferruginous rough-leg in the state of Washington. Murrelet 7:54.

Decker, F. R., and J. H. Bowles. 1930. The Prairie Falcon in the state of Washington. Auk 47:25-31.

Decker, F. R., and J. H. Bowles. 1933. The spring of 1933 in Benton County, Washington. Murrelet 14:43-44.

DeGraaf, R. M., V. E. Scott, R. H. Hamre, L. Ernst, and S. H. Anderson. 1991. Forest and rangeland birds of the United States: Natural history and habitat use. Agriculture Handbook 688, U.S. Dept. of Agric., For. Serv., Washington, D.C.

Denny, M. 1995. Jaegers in southeastern Washington. Wash. Birder 3 (3):1-2, 7.

Derrickson, K. C., and R. Breitwisch. 1992. Northern Mockingbird (*Mimus polyglottos*). *In* A. Poole, P. Stettenheim, and F. Gill, eds. The birds of North America, No. 7. Acad. of Nat. Sci., Philadelphia, and Amer. Ornithol. Union, Washington, D.C.

DeSante, D. F., and T. L. George. 1994. Population trends in the landbirds of western North America. *In* J. R. Jehl, Jr., and N. K. Johnson, eds. A century of avifaunal change in western North America. Studies in Avian Biol. No. 15.

DesGranges, J. L. 1988. Biology of the Great Blue Heron and probable response of nesting birds to cottage development near their breeding colony on Boughton Island, Prince Edward Island. Can. Wildl. Serv., Environment Canada.

Desimone, S. M. 1997. Occupancy rates and habitat relationships of Northern Goshawks in historic nesting areas in Oregon. M.S. Thesis. Oregon St. Univ., Corvallis. 78 p.

DeStefano, S., S. K. Daw, S. M. Desimone, and E. C. Meslow. 1994. Density and productivity of Northern Goshawks: Implications for monitoring and management. Studies in Avian Biol. 16:88-91.

Devillers, P. 1972. The juvenal plumage of Kittlitz's Murrelet. California Birds 3:33-38.

Devillers, P. 1977. The skuas of the North American Pacific Coast. Auk 94:417-29.

Dice, L. R. 1918. The birds of Walla Walla and Columbia counties, southeastern Washington. Auk 35:40-51.

Divoky, G. 1992. Regional reports: Washington/ British Columbia. PSG Bull.19:17.

Dixon, R. D., and V. A. Saab. 2000. Black-backed Woodpecker (*Picoides arcticus*). *In* A. Poole and F. Gill, eds. The birds of North America, No. 509. . Acad. of Nat. Sci., Philadelphia, and Amer. Ornithol. Union, Washington, D.C.

Dobbs, R. C., T. E. Martin, and C. J. Conway. 1997. Williamson's Sapsucker (*Sphyrapicus thyroideus*). *In* A. Poole and F. Gill, eds. The birds of North America, No. 285. Acad. of Nat. Sci., Philadelphia, and Amer. Ornithol. Union, Washington, D.C.

Dobkin, D. S., J. A. Holmes, and B. A. Wilcox. 1986. Traditional nest-site use by White-throated Swifts. Condor 88:252-53.

Dobler, F. C. 1993. Wintering Peregrine Falcon (*Falco peregrinus*) habitat utilization near Sequim, Washington. Northwest Sci. 67:231-37.

Dobler, F. C., J. Eby, C. Perry, S. Richardson, and M. Vander Haegen. 1996. Status of Washington's shrub-steppe ecosystem: Extent, ownership, and wildlife/ vegetation relationships. Research Report. Wash. Dept. Fish and Wildl., Olympia.

Dowlan, S. 1996. A verified breeding record for Wilson's Phalarope at Baskett Slough National Wildlife Refuge, Polk County, Oregon. Oreg. Birds 22:74.

Droege, S., and J. R. Sauer. 1990. North American Breeding Bird Survey, annual summary, 1989. U.S. Fish and Wildl. Serv., Biol. Rep. 90(8).

Drost, C. A., and D. B. Lewis. 1995. Xantus' Murrelet (*Synthliboramphus hypoleucus*). *In* A. Poole and F. Gill, eds. The birds of North America, No. 164. Acad. of Nat. Sci., Philadelphia, and Amer. Ornithol. Union, Washington, D.C.

Drut, M. S., and J. B. Buchanan. 2000. U.S. National shorebird conservation plan: Northern Pacific coast working group regional management plan. U.S. Fish and Wildl. Serv., Portland, OR.

Dugger, B. D., K. M. Dugger, and L. H. Fredrickson. 1994. Hooded Merganser (*Lophodytes cucullatus*). *In* A. Poole and F. Gill, eds. The birds of North America, No. 98. Acad. of Nat. Sci., Philadelphia, and Amer. Ornithol. Union, Washington, D.C.

Dumas, P. C. 1950. Habitat distribution of breeding birds in southeastern Washington. Condor 52:232-37.

Duncan, J. R., and P. A. Duncan. 1998. Northern Hawk Owl (*Surnia ulula*). *In* A. Poole and F. Gill, eds. The birds of North America, No. 356. Acad. of Nat. Sci., Philadelphia, and Amer. Ornithol. Union, Washington, D.C.

Dunk, J. R. 1995. White-tailed Kite (*Elanus leucurus*). *In* A. Poole and F. Gill, eds. The birds of North America, No. 178. Acad. of Nat. Sci., Philadelphia, and Amer. Ornithol. Union, Washington, D.C.

Dunn, E. H., and D. J. Agro. 1995. Black Tern (*Chlidonias niger*). *In* A. Poole and F. Gill, eds. The birds of North America, No 147. Acad. of Nat. Sci., Philadelphia, and Amer. Ornithol. Union, Washington, D.C.

Dunn, J. L., D. L. Dittmann, K. L. Garrett, G. Lasley, M. B. Robbins, C. Sexton, S. Tingley, and T. Tobish. 2002. ABA Checklist: Birds of the Continental United States and Canada, 6th ed. Amer. Birding Assoc., Colorado Springs.

Dunn, J. L., and K. L. Garrett. 1997. A field guide to the warblers of North America. Houghton Mifflin Co., Boston, MA.

Dunn, J. L, and K. L. Garrett. 2001. Parapatry in Woodhouse' and California Scrub-Jays revisited. West. Birds 32: 186-87.

Duvall, A. J. 1946. An early record of the Passenger Pigeon for British Columbia. Auk 63:598.

Ebbesmeyer, C. C., D. R. Cayan, D. R. McLain, F. H. Nichols, D. H. Peterson, and K. T. Redmond. 1991. 1976 step in the Pacific climate: Forty environmental changes between 1968-1975 and 1977-1984. *In* J. L. Betancourt and V. L. Tharp, eds. Proc. of the Seventh Annual Pacific Climate (PACLIM) Workshop, April 1990. Calif. Dept. Water Res. Interagency Ecol. Studies Prog. Tech. Rep. 26.

Eddy, G. 1956. Western Willet (*Catoptrophorus semipalmatus inornatus* (Brewster) at Seattle. Murrelet 37:25.

Eddy, G. 1982. Glaucous-winged Gulls nesting on buildings in Seattle, Washington. Murrelet 63:27-29.

Eddy, G., and Mrs. Eddy. 1948. Turkey Vulture concentration. Murrelet 29:11.

Edgell, M. C. R. 1984. Trans-hemispheric movements of Holarctic Anatidae: the Eurasian Wigeon (*Anas penelope* L.). North Amer. J. Biogeography 11:27-39.

Edson, J. M. 1908. Birds of the Bellingham Bay region. Auk 25:425-39.

Edson, J. M. 1919. Birds of the Bellingham Bay region. State Normal School Exch. 2:1-8.

Edson, J. M. 1926. The birds and wild animals. Pp. 915-26 *in* L. R. Roth. History of Whatcom County. Pioneer Historical Publ. Co., Seattle, WA.

Edwards, J. L. 1935. The Lesser Black-backed Gull in New Jersey. Auk 52:85.

Ehrlich, P. R., D. S. Dobkin, and D. Wheye. 1988. The birder's handbook: A field guide to the natural history of North America's birds. Simon and Schuster Inc., New York.

Einarsen, A. S. 1965. Black Brant, sea goose of the Pacific coast. Univ. of Washington Press, Seattle. 142 pp.

Eissinger, A. E. 1994. Significant wildlife areas, Whatcom County, Washington. Whatcom Co. Planning and Development Serv., Bellingham. 270 pp.

Ellarson, R. S. 1956. A study of the oldsquaw duck on Lake Michigan. Ph.D. Thesis, Univ. of Wisconsin, Madison.

Ellis-Joseph, S., N. Hewston, and A. Green. 1992. Global waterfowl conservation assessment and management plan: First review draft. Report by the IUCN Captive Breeding Specialist Group and the Wildfowl and Wetlands Trust.

Ely, C. R., and A. X. Dzubin. 1994. Greater White-fronted Goose (*Anser albifrons*). *In* A. Poole and F. Gill, eds. The birds of North America, No. 131. Acad. of Nat. Sci., Philadelphia, and Amer. Ornithol. Union, Washington, D.C.

Emilio, S. G, and L. Griscom. 1930. The European Black-headed Gull (*Larus ridibundus*) in America. Auk 47:243.

England, A. S., M. J. Bechard, and C. S. Houston. 1997. Swainson's Hawk (*Buteo swainsoni*). *In* A. Poole, ed. The birds of North America, No. 265. The Birds of North America, Inc., Philadelphia, PA.

Engler, J. D., and E. D. Anderson. 1998a. The status of Greater Sandhill Cranes in Washington. Poster presented The Wildlife Society, Spokane, WA.

Engler, J. D., and E. D. Anderson. 1998b. Greater Sandhill Cranes at Conboy National Wildlife Refuge, final report 1997-field season.

Ennor, H. R. 1991. Birds of the Tri-Cities and vicinity. Lower Col. Basin Aud. Soc., Richland. 292 pp.

Enticott, J., and D. Tipling. 1997. Seabirds of the world. Stackpole Books. Mechanicsville, PA. 234 pp.

Erickson, R. A., and R. A. Hamilton. 2001. Report of the California Bird Records Committee: 1998 records. West. Birds 32: 13-49.

Evans, R. M., and F. L. Knopf. 1993. American White Pelican (*Pelecanus erythrorhynchos*). *In* A. Poole and F. Gill, eds. The birds of North America, No. 57. Acad. of Nat. Sci., Philadelphia, and Amer. Ornithol. Union, Washington, D.C.

Evenson, J. R, and J. B. Buchanan. 1997. Seasonal abundance of shorebirds at Puget Sound estuaries. Wash. Birds 6:34-62.

Everett, W. T., and R. L. Pitman. 1993. Status and conservation of shearwaters of the North Pacific. Pp. 93-100 *in* K. Vermeer, K. T. Briggs, K. H. Morgan, and D. Siegel-Causey, eds. The status, ecology and conservation of marine birds of the North Pacific. Can. Wildl. Serv. Spec. Publ., Ottawa.

Ewins, P. J. 1993. Pigeon Guillemot (*Cepphus columba*). *In* A. Poole and F. Gill, eds. The birds of North America, No. 49. Acad. of Nat. Sci., Philadelphia, and Amer. Ornithol. Union, Washington, D.C.

Ewins, P. J., H. R. Carter, and Y. V. Shibaev. 1993. The status, distribution and ecology of inshore fish-feeding alcids *(Cepphus* guillemots and *Brachyramphus* murrelets) in the North Pacific. Pp. 164-175 *in* K. Vermeer, K. T. Briggs, K. H. Morgan, and D. Siegel-Causey, eds. The status, ecology and conservation of marine birds of the North Pacific. Can. Wildl. Serv. Spec. Publ., Ottawa.

Ewins, P. J., and D. V. Weseloh. 1999. Little Gull (*Larus minutus*). *In* A. Poole and F. Gill, eds. The Birds of North America, No. 428. Academy of Natural Sciences, Philadelphia, and American Ornithologists' Union, Washington, D.C.

Falcon Res. Group. 1986. Newsletter no.1, Winter 1985-86.

Falcon Res. Group. 2001. Winter Bull. http://www.frg.org.

Farber, P. L. 2000. Finding order in nature: the naturalist tradition from Linnaeus to E.O. Wilson. Johns Hopkins Univ. Press, Baltimore, MD.

Farner, D. S., and I. O. Buss. 1957. Summer records of the Golden-crowned Sparrow in

Okanogan County, Washington. Condor 59:141.

Feare, C. J. 1984. The Starling. Oxford University Press, Oxford, U.K.

Feltner, T. B., K. Aanerud, E. S. Hunn, P. W. Mattocks, Jr., D. R. Paulson, J. Skriletz, R. A. Sundstrom, and B. Tweit. 1994. Check-list of Washington birds (Second Edition). Wash. Birds 3: 1-10.

Ferguson, H. L., K. Robinette, and K. Stenberg. 2001. Wildlife of urban habitats. Pp. 317-41 *in* D. H. Johnson and T. A. O'Neil, eds. Wildlife-habitat relationships in Oregon and Washington. Oreg. State Univ. Press, Corvallis, OR.

Ficken, M. S., M. A. McLaren, and J. P. Hailman. 1996. Boreal Chickadee *(Parus hudsonicus)*. *In* A. Poole and F. Gill, eds. The birds of North America, No. 254. Acad. of Nat. Sci., Philadelphia, and Amer. Ornithol. Union, Washington, D.C.

Finn, S. P. 1994. Northern Goshawk nest stand characteristics in Okanogan County, Washington. Unpubl. Rep., Wash. Dept. of Fish and Wildl., Ephrata. 21 p.

Finn, S. P. 2000. Multi-scale habitat influences on northern goshawk occupancy and reproduction on the Olympic Peninsula, Washington. M.S. thesis, Boise St. Univ., Boise, ID. 116 p.

Finn, S. P., T. D. Bloxton, J. M. Marzluff and D. E. Varland. 1999. Northern Goshawk occupancy, productivity, and habitat use in managed forests of western Washington: Third annual report. Unpubl., Raptor Research Center, Boise St. Univ., Boise ID. 48 p., append.

Finn, S. P., J. Marzluff, and D. E. Varland. 2002a. Effects of landscape and local habitat attributes on Northern Goshawk site occupancy in western Washington. Forest Sci. 48(2):1-10.

Finn, S. P., D. E. Varland, and J. Marzluff. 2002b. Does Northern Goshawk breeding occupancy vary with nest-stand characteristics on the Olympic Peninsula, Washington? J. Raptor Res. 36(4):265-79.

Fisher, J., N. Simon, and J. Vincent. 1969. Wildlife in danger. Viking Press, New York. 368 pp.

Fitch, H. S., F. Swenson, and D. F. Tillotson. 1946. Behavior and food habits of the Red-tailed Hawk. Condor 48: 205-57.

Fitzner, R. E. 2000. Reproductive ecology of the sage sparrow (*Amphispiza belli*) and its relationship to the characteristics of native and cheatgrass (*Bromus tectorum*) dominated habitats. M.S. Thesis, Wash. State Univ., Pullman.

Fitzner, R. E., D. Berry, L. L. Boyd, and C. A. Rieck. 1977. Nesting of ferruginous hawks (*Buteo regalis*) in Washington 1974-75. Condor 79:245-49.

Fitzner, R. E., and R. E. Woodley. 1976. Blue Jays wintering in the Columbia Basin of southeastern Washington. Murrelet 57: 64-66.

Flahaut, M. R. 1947. Red-billed Tropic-bird taken at Westport, Washington. Murrelet 28:6.

Flahaut, M. R. 1950. Winter season: Northern Pacific Coast region. Aud. Field Notes 4:216-218.

Flather, C. H., and H. K. Cordell. 1995. Outdoor recreation: historical and anticipated trends. Pp. 3-16 in R. L. Knight and K. J. Gutzwiller, eds. Wildlife and recreationists: Coexistence through management and research. Island Press, Washington, D.C.

Fleming, T. L. 1987. Northern Goshawk status and habitat associations in western Washington with special emphasis on the Olympic Peninsula. Unpubl. rep., USDA, For. Serv., Pacific Northwest Res. Sta., Olympia. 45 p.

Floberg, J., M. Goering, G. Wilhere, C. MacDonald, C. Chappell, C. Rumsey, Z. Ferdana, A. Holt, P. Skidmore, T. Horsman, E. Alverson, C. Tanner, M. Bryer, P. Iachetti, A. Harcombe, B. McDonald, T. Cook, M. Summers, and D. Rolph. 2004. Willamette Valley-Puget Trough-Georgia Basin ecoregional assessment, volume one: report. The Nature Conservancy.

Foerster, K. S., and C. T. Collins. 1990. Breeding distribution of the Black Swift in southern California. West. Birds 21:1-9.

Ford, R. G., D. H. Varoujean, D. R. Warrick, W. A. Williams, D. . Lewis, C. L. Hewitt, and J. L. Casey. 1991. Final report: Seabird mortality resulting from the Nestucca oil spill incident, winter 1988-89. Report for Washington Dept. of Wildlife. Ecol. Consulting, Inc. 77 pp.

Forest Ecosystem Management Assessment Team (FEMAT). 1993. Forest Ecosystem management: An ecological, economic, and social assessment. U.S. Govt. Printing Office 1993-793-071. Washington, D.C.

Forsman, E. D., and A. R. Geise. 1997. Nests of Northern Spotted Owls on the Olympic Peninsula, Washington. Wilson Bull. 109:28-41.

Forsman, E. D., E .C. Meslow, and H. M. Wight. 1984. Distribution and biology of the spotted owl in Oregon. Wildl. Monograph 87:1-64.

Forsman, E. D., S. G. Sovern, D. E . Seaman, K. J. Maurice, M. Taylor, and J. J. Zisa. 1996. Demography of the Northern Spotted Owl on the Olympic Peninsula and east slope of the Cascade Range, Washington. Studies in Avian Biol. 17:21-30.

Fowells, H. A. 1965. Silvics of forest trees of the United States. Agric. Handbook No. 271. U.S. Dept. Agr., Washington, D.C.

Fox, G. A. 1971a. Recent changes in reproductive success of the Pigeon Hawk. J. Wildl. Manage. 35:122-28.

Fox, G. A. 1971b. Organochlorines and mercury in Merlin eggs and protection of raptorial birds. Can. Field-Nat. 85:335-36.

Fox, G. A., and T. Donald. 1980. Organochlorine pollutants, nest defense behavior and reproductive success in Merlins. Condor 82:81-84.

Franklin, A. B. 1987. Breeding biology of the Great Gray Owl in southeastern Idaho and northwestern Wyoming. M.S. Thesis. Humboldt State Univ. Arcata, CA.

Franklin, J. F., and C. T. Dyrness. 1973. Natural vegetation of Oregon and Washington. USDA For. Serv. Tech. Rep. PNW-8. 417 pp. Repr. Oreg. State Univ. Press, Corvallis, 1988.

Franklin, J. F., and T. A. Spies. 1991. Composition, function, and structure of old-growth Douglas-fir forests. Pp.s 71-80 in L. F. Ruggiero, K. B. Aubry, A. B. Carey, and M. H. Huff, (tech. coord.. Wildlife and vegetation of unmanaged Douglas-fir forests. USDA For. Serv. Genl. Tech. Rep. PNW-GTR-285. Portland, OR.

Friedmann, H. 1950. Birds of North and Middle America. U.S. Nat. Mus. Bull. 50:1-793.

Friesz, R. 1979. Western Burrowing Owl. Unpubl. Wash. Dept. Game rept., Olympia. 25 pp.

Friesz, R., and H. Allen. 1981. Status classification for the nesting sites of the Ferruginous Hawk. Pittman-Robertson Proj. No. E-1, Wash. Dept. of Game, Ephrata.

Frobe, M. E. 1999. Banding bonanza - Golden-winged Warbler. WOSNews 58:1,6.

Fyfe, R. W., R. W. Risebrough, and W. Walker, II. 1976. Pollutant effects on the reproduction of the Prairie Falcons and Merlins of the Canadian prairies. Can. Field-Nat. 90:346-355.

Gabrielson, I. N., and S. G. Jewett. 1940. Birds of Oregon. Oreg. State Coll., Corvallis.

Gabrielson, I. N., and S. G. Jewett. 1970. Birds of the Pacific Northwest. Dover Publ., New York. 650 pp.

Gabrielson, I. N., and F. C. Lincoln. 1959. The birds of Alaska. The Stackpole Company, Harrisburg, PA., and Wildlife Manage. Inst., Washington, D.C.

Gaines, D. and S. A. Laymon. 1984. Decline, status and preservation of the yellow-billed Cuckoo in California. West. Birds 15: 49-80.

Gaines, W. L., and R. E. Fitzner. 1987. Winter diet of the Harlequin Duck at Sequim Bay, Puget Sound, WA., Northwest Sci. 61:213-15.

Gallo-Reynoso, J.-P. and A.-L. Figueroa-Carranza. 1996. The breeding colony of Laysan Albatrosses on Isla de Guadalupe, Mexico. West. Birds 27:70-76.

Galusha, J. G., B. Vorvick, M. R. Opp, and P. T. Vorvick. 1987. Nesting season censuses of seabirds on Protection Island, Washington. Murrelet 68:103-7.

Gamble, L. R., and T. M. Bergin. 1996. Western Kingbird (*Tyrannus verticalis*). *In* A. Poole and F. Gill, eds. The birds of North America, No. 227. Acad. of Nat. Sci., Philadelphia, and Amer. Ornithol. Union, Washington, D.C.

Garrett, K., and J. Dunn. 1981. Birds of Southern California: Status and distribution. Los Angeles Aud. Soc.

Garrett, K. L., and D. S. Singer. 1998. Report of the California Bird Records Committee: 1995 Records. West. Birds 29:133-56.

Gaston, A. J. 1992. The Ancient Murrelet: A natural history in the Queen Charlotte Islands. Academic Press, San Diego, CA. 249 pp.

Gaston, A. J. and I. L. Jones. 1998. The Auks. Oxford University Press, Oxford, U.K.

George, T. L. 2000. Varied Thrush (*Ixoreus naevius*). *In* A. Poole and F. Gill, eds. The birds of North America. No. 541. (Acad. of Nat. Sci., Philadelphia, and Amer. Ornithol. Union, Washington, D.C.

Gibbs, J. P., S. Melvin and F. A. Reid. 1992. American Bittern (*Botaurus lentiginosus*) *In* A. Poole and F. Gill, eds. The birds of North America, No. 18. Acad. of Nat. Sci., Philadelphia, and Amer. Ornithol. Union, Washington, D.C.

Gibbs, J. P., S. Woodward, M. L. Hunter, and A. E. Hutchinson. 1987. Determinants of Great Blue Heron colony distribution in coastal Maine. The Auk 104: 38-47.

Gibson, D. D. 1977. First North American nest and eggs of the Ruff. West. Birds 8:25-26.

Gibson, D. D., and B. Kessel. 1989. Geographic variation in the Marbled Godwit and description of an Alaska subspecies. Condor 91:436-43.

Gibson, D. and B. Kessel.1992. Seventy-four new avian taxa documented in Alaska 1976-1991. Condor 94:454-67.

Gibson, D. D., and B. Kessel. 1997. Inventory of the species and subspecies of Alaska birds. Western Birds 28:45-95.

Giese, A. R. 1999. Habitat selection by Northern Pygmy-owls on the Olympic Peninsula, Washington. M.S. thesis. Oreg. State Univ., Corvallis, OR.

Gilbert, F. F., and R. Allwine. 1991. Spring bird communities in the Oregon Cascade Range. Pp. 145-60 *in* L. F. Ruggiero, K. B. Aubry, A. B. Carey and M. H. Huff, tech. coors. Wildlife and vegetation of unmanaged Douglas-fir forests. U.S.D.A. For. Serv., Gen. Tech. Rep. PNW-GTR-285.

Gill, R. E., Jr., and C. M. Handel. 1981. Shorebirds of the eastern Bering Sea. Pp. 719-38 in D. W. Hood. and J. A. Calder, eds. The eastern Bering Sea shelf: Oceanography and resources (vol. 2). U.S. Dept. Commerce and U.S. Dept. Interior.

Gill, R. E., Jr. and C. M. Handel. 1990. The importance of subarctic intertidal habitats to shorebirds: A study of the central Yukon-Kuskokwim Delta, Alaska. Condor 92(3): 709-25.

Gill, R. E., Jr., and L. R. Mewaldt. 1979. Pacific coast Caspian Terns: Dynamics of an expanding population. Auk 100:369-81.

Gill, R E., and T. L. Tibbitts. 1999. Seasonal shorebird use of intertidal habitats in Cook Inlet, Alaska. U.S. Dept. Interior, U.S. Geological Survey, Report MMS 99-0012. Alaska Biological Sciences Center, Anchorage.

Gilligan, J., M. Smith, D. Rogers, and A. Contreras. 1994. Birds of Oregon: Status and distribution. Cinclus Publ., McMinnville, OR.

Godfrey, W. E. 1986. The birds of Canada, rev. ed. Nat. Mus. of Canada. Ottawa, Ontario. 595 pp.

Goodrich, L. J., S. C. Crocoll and S. E. Senner. 1996. Broad-winged Hawk (*Buteo platypterus*). *In* A. Poole and F. Gill, eds. The birds of North America, No. 218. Acad. of Nat. Sci., Philadelphia, and Amer. Ornithol. Union, Washington, D.C.

Gould, P. J., D. J. Forsell, and C. J. Lensink. 1982. Pelagic distribution and abundance of seabirds in the Gulf of Alaska and eastern Bering Sea. U.S. Fish and Wildl. Service. FWS/OBS-82/48.

Green, C. 1977. The exotic birds of New Providence. The Bahama Naturalist 2(2): 11-16.

Green, G. A., and R. G. Anthony. 1989. Nesting success and habitat relationships of Burrowing Owls in the Columbia Basin, Oregon. Condor 91:347-54.

Green, G. A., and R. G. Anthony. 1997. Ecological considerations for management of breeding Burrowing Owls in the Columbia Basin. J. Raptor Res. Rept. 9:117-21.

Greene, E., V. R. Muehter, and W. Davison. 1996. Lazuli Bunting (*Passerina amoena*). *In* A. Poole and F. Gill, eds. The Birds of North America, No. 277. Acad. of Nat. Sci., Philadelphia, PA, and Amer. Ornithol. Union, Washington D.C.

Greenlaw, J. S. 1996. Spotted Towhee (*Pipilo maculatus*). *In* A. Poole and F. Gill, eds. The birds of North America, No. 263. Acad. of Nat. Sci., Philadelphia, and Amer. Ornithol. Union, Washington, D.C.

Groth, J. G. 1993. Evolutionary differentiation in morphology, vocalizations, and allozymes among nomadic sibling species in the North American Red Crossbill (*Loxia curvirostra*) complex. Univ. Calif. Publ. in Zoology No. 127. Berkeley.

Grubb, T. G. 1980. An evaluation of Bald Eagle nesting in Western Washington. Pp. 87-103 *in* Knight et al., eds. Proc. Wash. Bald Eagle symposium. Wash. Dept. Wildl., Olympia.

Grubb, T. G., D. A. Manuwal, and C. M. Anderson. 1975. Nest distribution and productivity of Bald Eagles in Western Washington. Murrelet. 56(3):2-6.

Guberlet, J. E. 1930. Notes on relationships of parasitic flatworms to birds and mammals. Murrelet 11:15-17.

Guinan, J. A., P. A. Gowaty, and E. K. Eltzroth. 2000. Western Bluebird (*Sialia mexicana*). *In* A. Poole and F. Gill, eds. The birds of North America, No 510. Acad. of Nat. Sci., Philadelphia, and Amer. Ornithol. Union, Washington, D.C.

Gutierrez, R. J. ,and D. J. Delehanty. 1999. Mountain Quail (*Oreoryx pictus*). *In* A. Poole and F. Gill, eds. The birds of North America, No. 457. Cornell Lab. of Ornithol. and the Acad. of Nat. Sci.

Gutiérrez, R. J., E. D. Forsman, A. B. Franklin, and E. C. Meslow. 1996. History of demographic studies in the management of the Northern Spotted Owl. Studies in Avian Biol. 17:6-11.

Guzman, J. R., and M. T. Myres. 1983. The occurrence of shearwaters (*Puffinus* spp.) off the west coast of Canada. Can. J. Zool. 61:2064-77.

Guzy, M. J., and P. E. Lowther. 1997. Black-throated Gray Warbler (*Dendroica nigrescens*). *In* A. Poole and F. Gill, eds. The birds of North America, No. 319. Acad. of Nat. Sci., Philadelphia, and Amer. Ornithol. Union, Washington, D.C.

Habeck, J. R. 1994. Dynamics of forest communities used by Great Gray Owls. Chapter 15 *in* G.D. Hayward and J. Verner, tech. eds. Flammulated, boreal, and great gray owls in the United States: A technical conservation assessment. Gen. Tech. Rep. RM-253. Fort Collins, CO. USDA For. Serv.

Hagar, J. C., W. C. McComb, and W. H. Emmingham. 1996. Bird communities in commercially thinned and unthinned Douglas-fir stands of western Oregon. Wildl. Soc. Bull. 24:353-66.

Hagar, J. C., and M. A. Stern. 1997. Avifauna in oak woodland habitats of the Willamette Valley, Oregon 1994-1996. Unpubl. report, Oreg. Nat. Heritage Progr., Portland, OR.

Hagenstein, W. 1950. European Starling (*Sturnus vulgaris*) at Medina, King County, Washington. Murrelet 31:11.

Hall, W. A., and L. D. LaFave. 1958. Recent migration and collection records of shore birds in eastern Washington. Murrelet 39:21-22.

Hamas, M. J. 1974. Human incursion and nesting sites of the Belted Kingfisher. Auk 91:835-36.

Hamas, M. J. 1994. Belted Kingfisher (*Ceryle alcyon*). *In* A. Poole and F. Gill, eds. The birds of North America, No. 84. Acad. of Nat. Sci., Philadelphia, and Amer. Ornithol. Union, Washington, D.C.

Hamer, T. E., E. D. Forsman, A. D. Fuchs, and M. L. Walters. 1994. Hybridization between Barred and Spotted owls. Auk 111:487-92.

Hamer, T. E., and S. K. Nelson. 1995. Characteristics of Marbled Murrelet nest trees and nesting stands. Pp. 69-82 *in* C. J. Ralph, G. L. Hunt, Jr., M. G. Raphael, and J. F. Piatt, eds. Ecology and conservation of the Marbled Murrelet. Gen. Tech. Rep. PSW-GTR-152. Albany, CA: Southwest Res. Sta., For. Serv., USDA. 420 p.

Hamilton, R. A., and J. L Dunn. 2002. Red-naped and Red-breasted sapsuckers. West. Birds 33:128-30.

Hammerstrom, F., and M. Kopeny. 1981. Harrier nest-site vegetation. Raptor Res. 15:86-88.

Hancock, J., and J. Kulshan. 1984. The herons handbook. Harper and Row, New York.

Hand, K. D., L. L. Cadwell, and L. E. Eberhardt. 1994. Long-billed Curlews on the Yakima Training Center: Information for base realignment. PNL-9214. Pacific Northwest Lab., Richland, WA.

Hanna, G. D. 1920. New and interesting records of Pribilof Island birds. Condor 22:173-75.

Hanna, I., and P. Dunn. 1997. Restoration goals for Oregon white oak habitats in the south Puget Sound region. Pp. 231-45 *in* P. Dunn and K. Ewing, eds. Ecology and conservation of the South Puget Sound prairie landscape. Nature Conservancy, Seattle, WA.

Hansen, A. J., W. C. McComb, R. Vega, M. G. Raphael, and M. Hunter. 1995. Bird habitat relationships in nNatural and managed forests in the west Cascades of Oregon. Ecol. Applications 5(3): 555-69.

Hanson, E., D. Hays, L. Hicks, L. Young, and J. Buchanan. 1993. Spotted owl habitat in Washington: A report to the Washington State Forest Practices Board. Wash. Dept. Nat. Resour., Olympia.

Hanson, W. C. 1971. The 1966-67 Snowy Owl incursion in southeastern Washington and the Pacific Northwest. Condor 73: 114-16.

Harlow, D. L., and P. H. Bloom. 1989. Buteos and golden eagle. Pp. 102-10 *in* Proc. of the Western raptor management symposium and workshop. Natl. Wildl. Fed. Sci. and Tech.l Series No. 12.

Harrington, B. A. 2001. Red Knot (*Calidris canutus*). *In* A. Poole and F. Gill, eds. The birds of North America, No. 563. Acad. of Nat. Sci., Philadelphia, and Amer. Ornithol. Union, Washington, D.C.

Harrington-Tweit, W. 1979a. A seabird die off on the Washington coast in mid-winter 1976. West. Birds 10:49-56.

Harrington-Tweit, B. 1979b. First records of the Chestnut-collared Longspur in Washington. Murrelet 60: 107-8.

Harrington-Tweit, B. 1980. First records of the White-tailed Kite in Washington. West. Birds 11: 151-53.

Harris, A., L. Tucker, and K. Vinicombe. 1989. The MacMillan field guide to bird identification. MacMillan Press, Ltd., London, U.K.

Harris, S. W. 1951. Four unusual bird records for eastern Washington. Murrelet 32:43.

Harris, S. W. 1996. Northwestern California birds, 2nd ed. Humboldt State Univ., Arcata, CA.

Harris, S. W., and C. F. Yocom. 1952. Birds of the Lower Grand Coulee and Potholes area, Grant County, Washington. Murrelet 33:18-28.

Harrison, C. Conservation news. 2001. Pacific Seabirds 28:68.

Harrison, C. Conservation news. 2002. Pacific Seabirds 29:11-12.

Harrison, P. 1983. Seabirds: an indentification guide. Houghton Mifflin Co., Boston, MA. 448 pp.

Hasegawa, H., and A. R. DeGange. 1982. The Short-tailed Albatross *Diomedea albatrus*, on Torishima, 1979/80-1980-81. Jour. Yamashina Inst. Ornithol. 14:16-24.

Hatch, J. J,. and D. V. Weseloh. 1999. Double-crested Cormorant (*Phalacrocoraz auritus*). *In* A. Poole and F. Gill, eds. The birds of North America, No. 441. Acad. of Nat. Sci., Philadelphia, and Amer. Ornithol. Union, Washington, D.C.

Hatch, S. A. 1987. Did the 1982-83 El Niño - Southern Oscillation affect seabirds in Alaska? Wilson Bull. 99:468-74.

Hatch, S. A. 1993a. Ecology and population status of Northern Fulmars *Fulmarus glacialis* of the North Pacific. Pp. 82-92 *in* K. Vermeer, K. T. Briggs, K. H. Morgan, and D. Siegel-Causey, eds. The status, ecology and conservation of marine birds of the North Pacific. Can. Wildl. Serv. Spec. Publ., Ottawa.

Hatch, S. A. 1993b. Population trends of Alaskan seabirds. Pacific Seabird Group Bull. 20:3-12.

Hatch, S. A. 2003. Geographic variation in Pacific Northern Fulmars: are there two subspecies? Abstr. Pacific Seabirds 30:35.

Hatch, S. A., G. V. Byrd, D. B. Irons, and G. L. Hunt, Jr. 1993. Status and ecology of kittiwakes *(Rissa tridactyla* and *brevirostris)* in the North Pacific. Pp. 140-153 *in* K. Vermeer, K. T. Briggs, K. H. Morgan, and D. Siegel-Causey, eds. The status, ecology and conservation of marine birds of the North Pacific. Can. Wildl. Serv. Spec. Publ., Ottawa.

Hawaii Audubon Society. 1993. Hawaii's birds, 4th ed. Hawaii Aud. Soc., Honolulu.

Hawthorne, V. M. 1979. Use of nest boxes by Dippers on Sagehen Creek, California. West. Birds 10:215-16.

Hayes, G. E., and J. B. Buchanan. 2002. Washington State status report for the Peregrine Falcon. Wash. Dept. Fish and Wildl., Olympia. 77 pp.

Hayman, G. D., and J. Verner, tech eds. 1994 Flammulated, boreal and great gray owls in the United States: A technical conservation assessment. Gen, Tech. Rep. RM-253. Fort Collins, CO: U.S. Dept. of Agriculture, For. Serv., Rocky Mt. For. and Range Experiment Sta.. 214 pp.

Hayman, P., P. Marchant and R. Prater. 1986. Shorebirds: An identification guide to the waders of the world. Houghton Mifflin Co., Boston, MA.

Hays, D. W., and R. Milner. 1999. Peregrine falcon (*Falco peregrinus*). *In* E. M. Larsen and N. Nordstrom, eds. Management recommendations for Washington's priority species, volume IV: Birds [Online]. Available http://www.wa.gov/wdfw/hab/phs/vol4/peregrin.htm.

Hayward, G. D. 1994. Conservation status of great gray owls in the United States. Chapter 16 *in* G. D. Hayward and J. Verner, tech. eds. Flammulated, boreal, and great gray owls in the United States: A technical conservation assessment. Gen. Tech. Rep. RM-253. Fort Collins, CO. USDA For. Serv.

Hayward, G. D., and P. H. Hayward. 1993. Boreal Owl (*Aegolius funereus*). *In* A. Poole and F. Gill, eds. The birds of North America, No. 63. Acad. of Nat. Sci., Philadelphia, and Amer. Ornithol. Union, Washington, D.C.

Hayward, G. D. and J. Verner, tech. eds. 1994. Flammulated, boreal, and great gray owls in the United States: A technical conservation assessment. Gen. Tech. Rep. RM-253. Fort Collins, CO. USDA For. Serv.

Hayward, J. L., Jr., and A. C. Thoresen. 1980. Dippers in marine habitats in Washington. West. Birds 11:60.

Hazlitt, S. 2001. Where have all the Western Grebes gone? Birdwatch (publ. Bird Studies Canada), Winter 2001. 15:18-19.

Heindel, M. T. 1996. Field identification of the Solitary Vireo complex. Birding 28:458-71.

Johnson, J. 1998. Status of Blue-headed and Plumbeous vireos in Oregon. Oreg. Birds 24:74-76.

Heindel, M. 1999. The Yellow Wagtail in North America. Birders Jour. 8:182-93.

Heinl, S. C. 1997. New information on gulls in southeastern Alaska. West. Birds 28:19-29.

Hejl, S. J. 1994. Human-induced changes in bird populations in western North America during the past 100 years. Pp. 232-46 *in* J. R. Jehl, Jr., and N. K. Johnson, eds. A Century of avifaunal changes in western North America. Studies in Avian Biol. No. 15.

Hejl, S. J., J. A. Holmes, and D. E. Kroodsma. 2002a. Winter Wren (*Troglodytes troglodytes*). *In* A. Poole and F. Gill, eds. The birds of North America, No. 623). Acad. of Nat. Sci., Philadelphia, and Amer. Ornithol. Union, Washington, D.C.

Hejl, S. J., K. R. Newlon, M. E. McFadzen, J. S. Young, and C. K. Ghalambor. 2002b. Brown Creeper (*Certhia americana*). *In* A. Poole and F. Gill, eds. The birds of North America, No. 669. Acad. of Nat. Sci., Philadelphia, and Amer. Ornithol. Union, Washington, D.C.

Henny, C. J., and G. L. Brady. 1994. Partial migration and wintering localities of American Kestrels nesting in the Pacific Northwest. Northw. Natur. 75:37-43.

Henny, C. J., and J. E. Cornely. 1985. Recent Red-shouldered Hawk range expansion north into Oregon, including first specimen record. Murrelet 66:29-31.

Henny, C. J., R. A. Olson, and T. L. Fleming. 1985. Breeding chronology, molt, and measurements of accipiter hawks in northeastern Oregon. J. Field Ornithol. 56:97-112.

Herlugson, C. J. 1978. Comments on the status and distribution of Western and Mountain bluebirds in Washington. West. Birds 9:21-32.

Herman, S. G., and J. B. Bulger. 1981. The distribution and abundance of shorebirds during the 1981 spring migration at Grays Harbor, Washington. U.S. Army Corps of Engineers, Seattle District.

Herman, S. G., C. B. Chappell, G. Wallace, P. Martin, R. DelCarlo, J. R. Blair, and T. Williamson. 1983. The spring shorebird migration at Willapa Bay. Unpubl. ms.

Herter, D. R., and L. L. Hicks. 2000. Barred Owl and Spotted Owl populations and habitat in the central Cascade range of Washington. J. Raptor Res. 34:279-86.

Hickey, J. J., ed.) 1969. Peregrine Falcon populations: their biology and decline. Univ. Wisconsin Press, Madison.

Hicks, L. L., and D. R. Herter. 2003. Northern Spotted Owl research in the central Cascade Range, Washington. Plum Cr. Timber Co., Seattle, WA.

Higman, H. W. 1944. Autumn communal roosting of Purple Martins with the city limits of Seattle, Washington. Murrelet 25:43-44.

Hirsch, K. V. 1980. Winter ecology of sea ducks in the inland marine waters of Washington. M.S. Thesis, Univ. of Wash., Seattle.

Hodder, J. 2002. Regional reports, Oregon and Washington. Pacific Seabirds 29:115.

Hodder, J., and M. R. Graybill. 1985. Reproduction and survival of seabirds in Oregon during the 1982-1983 El Niño. Condor 87:535-41.

Hoffman, R. 1927. Birds of the Pacific states. Houghton Mifflin Co., Boston, MA. 353 pp.

Hoffman, S. W., W. R. DeRagon, and J. C. Bednarz. 1992. Patterns and trends in counts of migrant hawks in western North America, 1977-1991. Hawkwatch International, Albuquerque, NM.

Hoffman, W., W. P. Elliott, and J. M. Scott. 1975. The occurrence and status of the Horned Puffin in the western United States. West. Birds 6:87-94.

Hoffman, W., J. A. Wiens and J. M. Scott. 1978. Hubridiaztion between gulls *(Larus glaucescens and L. occidentalis*) in the Pacific Northwest. Auk 95:441.

Hoge, G., and W. Hoge. 1991. Birds of Ocean Shores. Black Hills Aud. Soc., Olympia. 50 pp.

Hohman, W. L., and R. T. Eberhardt. 1998. Ring-necked Duck (*Aythya collaris*). *In* A. Poole and F. Gill, eds. The birds of North America, No. 329. Acad. of Nat. Sci., Philadelphia, and Amer. Ornithol. Union, Washington, D.C.

Holt, D. W., and J. L. Petersen. 2000. Northern Pygmy-Owl (*Glaucidium gnoma*). *In* A. Poole and F. Gill, eds. The birds of North America. No. 494. Acad. of Nat. Sci., Philadelphia, and Amer. Ornithol. Union, Washington, D.C.

Horn, K. M., and D. B. Marshall. 1975. Status of Poorwill in Oregon and possible extension due to clearcut timber harvest methods. Murrelet 56:4-5.

Horvath, E. 1999. Distribution, abundance, and nest site characteristics of Purple Martins in Oregon. Prep. for ODFW Tech. Rep. #99-1-01.

Houston, C. S. 1997. Banding of *Asio* owls in south-central Saskatchewan. Pp. 237-42 *in* J. R. Duncan, D. H. Johnson, and T. H. Nicholls, eds. Biology and conservation of owls of the Northern Hemisphere: 2nd International Symposium, 1997 February 5-9, Winnipeg, Manitoba, Canada. General Technical Report NC-190, St Paul, Minnesota, U.S. Dept. Agric., For. Serv., N. Cent. Res. Sta.

Houston, C. S., D. G. Smith, and C. Rohner. 1998. Great Horned Owl (*Bubo virginianus*). *In* A. Poole and F. Gill, eds. The birds of North America, No. 372. Acad. of Nat. Sci., Philadelphia, and Amer. Ornithol. Union, Washington, D.C.

Howe, M. A., P. H. Geissler, and B. A. Harrington. 1989. Population trends of North American shorebirds based on the international shorebird survey. Biol. Conserv. 49:185-99.

Howell, S. N. G. 1990. The identification of White and Black-backed wagtails in alternate plumage. West. Birds 21:41-49.

Howell, S. N. G. 1994. Magnificent and Great frigatebirds in the eastern Pacific. Birding 26:400-415.

Howell, S. N. G., L. B. Spear, and P. Pyle. 1994. Identification of Manx-type shearwaters in the eastern Pacific. West. Birds 25:169-77.

Howell, S. N. G., and S. Webb. 1995. A guide to the birds of Mexico and northern Central America. Oxford Univ. Press, Oxford, U.K.

Howell, T. R. 1952. Natural history and differentiation in the Yellow-bellied Sapsucker. Condor 54: 237-82.

Hudon, J. 1999. Western Tanager (*Piranga ludoviciana*). *In* A. Poole and F. Gill, eds. The birds of North America, No. 432. Acad. of Nat. Sci., Philadelphia, and Amer. Ornithol. Union, Washington, D.C.

Hudson, G. E., and J. R. King. 1951. Nesting of the European Starling in Adams County, Washington. Murrelet 32:24.

Hudson, G. E., and C. F. Yocom. 1954. A distributional list of the birds of southeastern Washington. Res. Studies of the State Coll. of Wash. 22:1-56.

Huff, M., and M. Brown. 1998. Four years of bird point count monitoring in late-successional conifer forests and riparian areas from the Pacific Northwest National Forests, interim results. U.S. For. Serv. Partners in Flight Progr. 142 pp.

Huff, M. H., D. A. Manuwal, and J. A. Putera. 1991. Winter bird communities in the southern Washington Cascade Range. Pp. 207-18 *in* L. F. Ruggiero, K. B. Aubry, A. B. Carey and M. H. Huff, tech. coors. Wildlife and vegetation of unmanaged Douglas-fir forests. USDA. For. Serv. Gen. Tech. Rep. PNW-GTR-285.

Hunn, E. 1973. First record for the Swamp Sparrow in Washington State. West. Birds 4: 31-32.

Hunn, E. 1994. More White-tailed Ptarmigan records. Wash. Birder 2(4):2.

Hunn, E. S. 1978. Black-throated Sparrow vagrants in the Pacific Northwest. West. Birds 9:85-89.

Hunn, E. S. 1982. Birding in Seattle and King County. Seattle Audubon Society, Seattle, WA.

Hurley, J. B. 1921. Annotated list of Yakima County birds. Murrelet 2: 14-16.

Hutto, R. L. 1995. Composition of bird communities following stand-replacement fires in northern Rocky Mountain (U.S.A.) conifer forests. Cons. Biol. 9:1041-58.

Hygnstrom, S. E., R. M. Timm, and G. E. Larson, eds. 1994. Prevention and control of wildlife damage. Univ. Nebr. Coop. Ext. Serv., LincolnB.

Hyslop, C., and J. A. Kennedy, eds. 1996. Bird trends, No. 5. Canadian Wildl. Serv. Headquarters.

Irwin, L. L., T. L. Fleming, S. M. Speich, and J. B. Buchanan. 1991. Spotted Owl presence in managed forests of southwestern Washington. National Council of the Paper Industry for Air and Stream Improvement, Tech. Bull. 601. Corvallis, OR.

Islam, K. 2002. Heermann's Gull (*Larus heermanni*). *In* A. Poole and F. Gill, eds. The birds of North America, No. 643. Acad. of Nat. Sci., Philadelphia, and Amer. Ornithol. Union, Washington, D.C.

Isleib, M. E. 1979. Migratory shorebird populations on the Copper River Delta and eastern Prince William Sound, Alaska. Studies in Avian Biol. 2:125-30.

Iten, C., T. A. O'Neil, K. A. Bettinger, and D. H. Johnson. 2001. Extirpated species of Oregon and Washington. Pp. 452-73 in D. H. Johnson and T. A. O'Neil, man. dirs. Wildlife-habitat relationships in Oregon and Washington. . Oreg. State Univ. Press, Corvallis.

Ivey, G., and C. Baars. 1990. A second Semipalmated Plover nest in Oregon. Oreg. Birds 16:207-8.

Ivey, G. L., and C. P. Herziger (compilers). 2005. U.S. Intermountain West Waterbird Conservation Plan. Intermountain West Joint Venture, West Valley City, UT.

Jackson, J. A., H. R. Ouellet, and B. J. S. Jackson. 2002. Hairy Woodpecker (*Picoides villosus*). *In* A. Poole and F. Gill, eds. The birds of North America, No. 702. Acad. of Nat. Sci., Philadelphia, and Amer. Ornithol. Union, Washington, D.C.

Jaques, D. L., R. W. Lowe and D. Pitkin. 2003. Abstract. Island erosion in Willapa Bay, Washington, and effects on Brown Pelicans and other birds. Pac. Seabirds 30:40.

Jaramillo, A., and P. Burke. 1999. New World blackbirds: the icterids. Princeton Univ. Press, Princeton, NJ.

Jeffrey, R., and E. Bowhay. 1972. Washington Pacific Flyway Report. *In* J. Chattin, ed. Pac. Waterfowl Flyway Rep. 68.

Jehl, J. R. 1979. The autumn migration of Baird's Sandpiper. Studies in Avian Biol. 2:55-68.

Jehl, J. R., Jr. 1996. Mass mortality events of Eared Grebes in North America. J. Field Ornithol. 67: 471-76.

Jehl, J. R., Jr., and S. I. Bond. 1975. Morphological variation and species limites in murrelets of the genus Endomychura. Trans. San Diego Soc. Nat. Hist. 18 (2), 23 pp.

Jehl, J. R. Jr., J. Klima, and R. E. Harris. 2001. Short-billed Dowitcher (*Limnodromas griseus*). *In* A. Poole and F. Gill, eds. The birds of North America, No. 564. Acad. of Nat. Sci., Philadelphia, and Amer. Ornithol. Union, Washington, D.C.

Jewett, S. G. 1946. The Starling taken in the state of Washington. Condor 48:143.

Jewett, S. G., W. P. Taylor, W. T. Shaw, and J. W. Aldrich. 1953. Birds of Washington State. Univ. Wash. Press, Seattle.

Jobanek, G. 1993. The European Starling in Oregon. Oreg. Birds 19:93-96.

Johnsgard, P. A. 1954. Birds observed in the Potholes region during 1953-1954. Murrelet 35:25-31.

Johnsgard, P. A. 1975. Waterfowl of North America. Indiana Univ. Press, Bloomington, IN.

Johnsgard, P. A. 1990. Hawks, eagles, and falcons of North America: Biology and natural history. Smithsonian Inst. Press, Washington, D.C.

Johnson, D. H. 1993. Spotted Owls, Great Horned Owls, and forest fragmentation in the central Oregon Cascades. M.S. thesis, Oreg. State Univ., Corvallis.

Johnson, D. H., and T. A. O'Neil, man. dirs. 2001. Wildlife-habitat relationships in Oregon and Washington. Oreg. State Univ. Press, Corvallis.

Johnson, D. H., D. E. Timm, and P. F. Springer. 1979. Morphological characteristics of Canada geese in the Pacific Flyway. *In* R. L. Jarvis. and J. C. Bartonek, eds. Management and biology of Pacific Flyway geese: Proc.: 16 February 1979; 56-80; Oreg. State Univ. Press, Corvallis, OR. 346 pp.

Johnson, J. 1998. Field notes: western Oregon, summer 1997. Oreg. Birds 24:34-36.

Johnson, N. K. 1980. Character variation and evolution of sibling species in the *Empidonax difficilis-flavescens* complex. Univ. Calif. Press, Berkeley.

Johnson, N. K. 1994a. Pioneering and natural expansions of breeding distributions in western North American Birds. *In* J. R. Jehl, Jr., and N. K. Johnson, eds. A century of avifaunal change in western North America. Studies in Avian Biol. No. 15.

Johnson, N. K. 1994b. Old-school taxonomy versus modern biosystematics: species level decisions in *Stelgidopteryx* and *Empidonax*. Auk 11:773-80.

Johnson, N. K., and C. B. Johnson. 1985. Speciation in sapsuckers (*Sphyrapicus*): II. sympatry, hybridization, and mate preference in *S. ruber daggetti* and *S. nuchalis*. Auk 102:1-15.

Johnson, N. K., and J. A. Marten. 1988. Evolutionary genetics of flycatchers. II. Differentiation in the *Empidonax difficilis* complex. Auk 105:177-91.

Johnson, N. K., and R. M. Zink. 1983. Speciation in sapsuckers (*Sphyrapicus*): I. Genetic differentiation. Auk 100:871-84.

Johnson, O.W., and P. G. Connors. 1996. American Golden-Plover (*Pluvialis dominica*) and Pacific Golden-Plover (*Pluvialis fulva*). *In* A. Poole and F. Gill, eds. The birds of North America, No. 201-202. Acad. of Nat. Sci., Philadelphia, and Amer. Ornithol. Union, Washington, D.C.

Johnson, R. E., and G. E. Hudson. 1976. A Washington record of the Boreal Owl. Auk 93:195-96.

Johnson, R. E., and G. A. Murray. 1976. New bird records for southeastern Washington. Murrelet 57:54-57.

Johnson, R. E., J. D. Reichel, and C. J. Herlugson. 1984. Recent bird records from eastern Washington. Murrelet 65:60-61.

Johnston, R. F. 1992. Rock Dove (*Columba livia*). *In* A. Poole and F. Gill, eds. The birds of North America, No. 13). Acad. of Nat. Sci., Philadelphia, and Amer. Ornithol. Union, Washington, D.C.

Johnston, R. F., and K. L. Garrett. 1994. Population trends of introduced birds in western North America. Pp. 221-31 in J. R. Jehl, JJr., and N. K Johnson, eds. A century of avifaunal change in Western North America. Studies in Avian Biol. No. 15.

Jones, L. 1900. A summer reconnaissance in the west. Wilson Bull. 33: 10-38.

Kaiser, G. W. 1993. Seabird wreck in Boundary Bay. PSG Bull. 20:18-19.

Kaufman, K. 1996. Lives of North American birds. Houghton Mifflin, New York. 675 p.

Kauffman, J. B., M. Mahrt, L. A. Mahrt, and W. D. Edge. 2001. Wildlife of riparian habitats. Pp. 361-88 *in* D. H. Johnson and T. A. O'Neil, eds. Wildlife-habitat relationships in Oregon and Washington. Oreg. State Univ. Press, Corvallis.

Keany, J., J. Tims, M. Rector, and B. Hollen. 1996. Sage Sparrow and Sage Thrasher study on the Yakima Training Center and Expansion Area, Yakima, WA. Shapiro and Associates for the U.S. Army Ft. Lewis/Yakima Training Center, Yakima WA.

Kelly, E. G., E. D. Forsman, and R. G. Anthony. 2003. Are Barred Owls displacing Spotted Owls? Condor 105:45-53.

Kennedy, P. L. 1997. The Northern Goshawk (*Accipiter gentilis atricapillus*): Is there evidence of a population decline? J. Raptor Res. 31:95-106.

Keppie, D. M., and C. E. Braun. 2000. Band-tailed Pigeon (*Columba fasciata*). *In* A. Poole and F. Gill, eds. The birds of North America, No. 530. Acad. of Nat. Sci., Philadelphia, and Amer. Ornithol. Union, Washington, D.C.

Kessel, B. 1953. Distribution and migration of the European Starling in North America. Condor 55:49-67.

Kingery, H. E. 1996. American Dipper (*Cinclus mexicanus*). *In* A. Poole and F. Gill, eds.The birds of North America, No. 229. Acad. of Nat. Sci., Philadelphia, and Amer. Ornithol. Union, Washington, D.C.

Kingery, H. E., and C. K. Ghalambor. 2001. Pygmy Nuthatch (*Sitta pygmaea*). *In* A. Poole and F. Gill, eds. The birds of North America, No. 567. Acad. of Nat. Sci., Philadelphia, and Amer. Ornithol. Union, Washington, D.C.

Kitchin, E. A. 1923. The Prairie Falcon recorded in western Washington. Murrelet 4:13.

Kitchin, E. A. 1934. Distributional check-list of the birds of the state of Washington. Pac. Northwest Bird and Mammal Soc., Seattle, WA.

Kitchin, E. A. 1949. Birds of the Olympic Peninsula. Olympic Stationers, Port Angeles, WA.

Klima, J., and J. R. Jehl, Jr. 1998. Stilt Sandpiper (*Calidris himantopus*). *In* A. Poole and F. Gill, eds. The birds of North America, No. 341. Acad. of Nat. Sci., Philadelphia, and Amer. Ornithol. Union, Washington, D.C.

Knick, S. T., D. S. Dobkin, J. T. Rottenberry, M. A. Schroeder, W. M. Vander Haegen, and C. van Riper III. 2003. Teetering on the edge or too late? Conservation and research issues for avifauna of sagebrush habitats. Condor 105:611-34.

Knick, S. T., and J. T. Rotenberry. 1995. Landscape characteristics of fragmented shrubsteppe habitats and breeding passerine birds. Conserv. Biol. 9:1059-71.

Knight, R. L., and K. J. Gutzwiller, eds). 1995. Wildlife and recreationists: Coexistence through management and research. Island Press, Washington, D.C.

Knight, R. L., D. G. Smith, and A. Erickson. 1982. Nesting raptors along the Columbia River in north-central Washington. Murrelet 63:2-8.

Knox, A. G. 1992. Species and pseudospecies: The structure of crossbill populations. Biol. J. Linn. Soc. 47:325-35.

Knox, A. G., and P. E. Lowther. 2000a. Common Redpoll (*Carduelis flammea*). *In* A. Poole and F. Gill, eds. The birds of North America, No. 543. Acad. of Nat. Sci., Philadelphia, and Amer. Ornithol. Union, Washington, D.C.

Knox, A. G., and P. E. Lowther. 2000b. Hoary Redpoll (*Carduelis hornemanni*). *In* A. Poole and F. Gill, eds. The birds of North America, No. 544. Acad. of Nat. Sci., Philadelphia, and Amer. Ornithol. Union, Washington, D.C.

Knue, A. J. 2001. Pterodroma inexpectata! WOSNews 70:1.

Kobbe, W. H. 1900. The birds at Cape Disappointment, Washington. Auk17:349-58.

Kochert , M. N. 1989. Response of raptors to livestock grazing in the Western United States. Pp. 194-203 *in* Proceedings of the Western raptor management symposium and workshop. Natl. Wildl. Fed. Scientific and Tech. Series No. 12.

Kushlan, J. A., M. J. Steinkamp, K. C. Parsons, J. Capp, M. Acosta Cruz, M. Coulter, I. Davidson, L. Dickson, N. Edelson, R. Elliot, R.M. Erwin, S. Hatch, S. Kress, R. Milko, S. Miller, K. Mills, R. Paul, R. Phillips, J.E. Saliva, B. Sydeman, J. Trapp, J. Wheeler, and K. Wohl. 2002. Waterbird conservation for the Americas: The North American waterbird conservation plan, version 1. Waterbird Conserv. for the Americas, Washington, D.C.

LaFave, L. D. 1955. Another sight record of the Hermit Warbler for eastern Washington. Murrelet 36:11.

LaFave, L. D. 1957. The Ovenbird (*Seirurus aurocapillus*), a new bird for Washington. Murrelet 38:7.

LaFave, L. D. 1959. Additional migration and collection records of shore birds in eastern Washington. Murrelet 40:33.

LaFave, L. D. 1960a. The Hudsonian Godwit, a new bird for the state of Washington. Murrelet 41:16.

LaFave, L. D. 1960b. The Clay-colored Sparrow, a new bird for the state of Washington.
Murrelet 43:50.

LaFramboise, B., and N. LaFramboise. 1999. Birds of the Fitzner-Eberhardt Arid Lands Ecology Reserve. Nat. Conserv. of Washington.

Lanctot, R. B., and C. D. Laredo. 1994. Buff-breasted Sandpiper (*Tryngites subruficollis*). *In* A. Poole and F. Gill, eds. The birds of North America, No. 91. Acad. of Nat. Sci., Philadelphia, and Amer. Ornithol. Union, Washington, D.C.Lane, D., and A. Jaramillo. 2000a. Identification of *Hylocichla/Catharus* thrushes, part 1. Birding 32:120-35.

Lane, D., and A. Jaramillo. 2000b. Field identification of *Hylocichla/Catharus* thrushes, part 2. Birding 32:242-54.

Lane, R. C., and W. A. Taylor. 1997. Washington's wetland resources: Tacoma, Wash., U. S. Geological Survey, on-line at URL http://wwwdwatcm.wr.usgs.gov/reports/wetlands, accessed October 14, 1997, HTML format [based on R, C. Lane and W. A. Taylor. 1996. Washington's wetland resources. *In* J. D. Fretwell, J. S. Williams, and P. J. Redman, comps., Natl. water summary on wetland resources: US. Geol. Survey Water-Supply Pap. 2425, p. 393-97].

Lanyon, W. E. 1994. Western Meadowlark (*Sturnella neglecta*). *In* A. Poole and F. Gill, eds. The birds of North America, No. 104. Acad. of Nat. Sci., Philadelphia, and Amer. Ornithol. Union, Washington, D.C.

Larrison, E. J. 1945. Albino Purple Martin at Seattle martin roost. Murrelet 26:45-46.

Larrison, E. J. 1947a. Field guide to the birds of King County, Washington. Seattle Aud. Soc.

Larrison, E. J. 1947b. General notes. Murrelet 28:11-13.

Larrison, E. J. 1947c. Eastern Starling in Snohomish County, Washington. Murrelet 28:21.

Larrison, E. J. 1971. Sight record of the Gray Flycatcher in Washington. Murrelet 52:240.

Larrison, E. J. 1977. A sighting of the Broad-winged Hawk in Washington. Murrelet 58:18.

Larrison, E. J., and E. N. Franca. 1962. Field guide to the birds of Washington state. Seattle Aud. Soc..

Larrison, E. J., and K. G. Sonnenberg. 1968. Washington birds: their location and identification. Seattle Aud. Soc.

Larson, E. M., and S. A. Richardson. 1990. Some effects of a major oil spill on wintering shorebirds at Grays Harbor, Washington. Northwestern Naturalist 71:88-92.

Lavers, N. 1975a. Status of the Harlan's Hawk in Washington, and notes on its identification in the field. West. Birds 6:55-62.

Lavers, N. 1975b. The status of the Gray Flycatcher in Washington State. West. Birds 6:25-27.

Layman, S. A, and M. D. Halterman. 1987. Can the western subspecies of the Yellow-billed Cuckoo be saved from extinction? West. Birds 18: 19-25.

Leach, R. H. 1995. Confirmed sandhill crane nesting in Yakima County, Washington. Northwestern Naturalist 76:148.

Leal, D. A. 1999. A specimen record of Wedge-tailed Shearwater (*Puffinus pacificus*) for Oregon. Oreg. Birds 25:96.

Leder, J. E., and M. L. Walters. 1980. Nesting observations for the Barred Owl in western Washington. Murrelet 110-12.

Legg, K. 1956. A sea-cave nest of the Black Swift. Condor 58:183-87.

Leonard, D. L. 2001. Three-toed Woodpecker (*Picoides tridactylus*). *In* A. Poole and F. Gill, eds. The birds of North America, No. 588. Acad. of Nat. Sci., Philadelphia, and Amer. Ornithol. Union, Washington, D.C.

Lethaby, N., and J. Bangma. 1998. Identifying Black-tailed Gull in North America. Birding 30:470-83.

Lethaby, N., and J. Gilligan. 1992. Great Knot in Oregon. Amer. Birds 46:46-47.

Leu, M. 1995. The feeding ecology and the selection of nest shrubs and fledgling roost sites by Loggerhead Shrikes (*Lanius ludovicianus*) in the shrub-steppe habitat. M.S. thesis. Univ. of Wash., Seattle.

Lewis, A., M. Morton, and D. Farner. 1968. A second large wintering population of White-crowned Sparrows (*Zonotrichia leucophrys gambelii*) in Washington. Condor 70:280.

Lewis, M. G., and F. A. Sharpe. 1987. Birding in the San Juan Islands. Mountaineers, Seattle, WA.

Ligon, S. New Mexico birds and where to find them. Univ. of New Mexico Press, Albuquerque.

Littlefield, C. D. 1990. Birds of Malheur National Wildlife Refuge, Oregon. Oreg. State Univ. Press, Corvallis. 294 pp.

Littlefield, C. D., and G. L. Ivey. 2002. Washington State recovery plan for the Sandhill Crane. Wash. Dept. Fish and Wildl., Olympia,. 71 p.

Loegering, J. P. 1997. Abundance, habitat associations, and foraging ecology of American Dippers and other riparian-associated wildlife in the Oregon Coast Range. Ph.D. dissertation. Oreg. State Univ., Corvallis.

Lowe, R. 1993. El Niño hard on Oregon seabirds. PSG Bull. 20:62.

Lowe, R. 1996. Regional reports, Washington and Oregon. Pacific Seabirds 23:21.

Lowe, R. 1997a. Regional reports, Washington and Oregon. Pacific Seabirds 24:89-93.

Lowe, R. 1997b. Protecting roosting California Brown Pelicans at Willapa Bay. Pacific Seabirds 24:78.

Lowe, R. 1998. Regional reports, Washington and Oregon. Pacific Seabirds 25:84.

Lowther, P. E. 2000. Cordilleran Flycatcher (*Empidonax occidentalis*). *In* A. Poole and F. Gill, eds. The birds of North America, No. 556. Acad. of Nat. Sci., Philadelphia, and Amer. Ornithol. Union, Washington, D.C.

Lowther, P. E., C. Celada, N. K. Klein, C. C. Rimmer and D. A. Spector. 1999. Yellow Warbler (*Dendroica petechia*). *In* A. Poole and F. Gill, eds. The birds of North America, No. 454. Acad. of Nat. Sci., Philadelphia, and Amer. Ornithol. Union, Washington, D.C.

Lund, T. 1978. The Purple Martin in the Western United States. Part two: It's a question of holes. Oreg. Birds 4(2):1-9.

Lyman, S. H., and P. C. Dumas. 1951. Additional notes on unusual birds from southeastern Washington. Murrelet 32:23-24.

Lyons, D. E., D. D. Roby, and K. Collis. 2002. Assessing impacts of Caspian Tern predation on juvenile salmonids in the Columbia River estuary. Abstr. Pac. Seabirds 29:51.

Mack, D. E., and W. Yong. 2000. Swainson's Thrush (*Catharus ustulatus*). *In* A. Poole and F. Gill, eds. The birds of North America, No. 540. Acad. of Nat. Sci., Philadelphia, and Amer. Ornithol. Union, Washington, D.C.

MacRae, D. 1983. Pacific region *in* The Newsletter of the Hawk Migration Assoc. of N. Amer., Vol. 8 (2), Concord, MA.

MacRae, D. 1998. Turkey Vulture migration over the strait of Juan de Fuca between Canada and the United States. Vulture News, J. of the Vulture Study Group, No. 38 (2-9), Kimberley, South Africa.

MacRae, D. 1999. Olympic Vulture study. Unpubl.

MacWhirter, B., P. Austin-Smith, Jr., and D. Kroodsma. 2002. Sanderling (*Calidris alba*). *In* A. Poole and F. Gill, eds. The birds of North America, No. 653. Acad. of Nat. Sci., Philadelphia, and Amer. Ornithol. Union, Washington, D.C.

MacWhirter, R. B., and K. L. Bildstein. 1996. Northern Harrier (*Circus cyaneus*). *In* A. Poole and F. Gill, eds. The birds of North America, No. 210. Acad. of Nat. Sci., Philadelphia, and Amer. Ornithol. Union, Washington, D.C.

Madsen, S. J. 1985. Habitat use by cavity-nesting birds in the Okanogan National Forest, Washington. M.S. Thesis, Univ. of Wash., Seattle.

Mallory, M., and K. Metz. 1999. Common Merganser (*Mergus merganser*). *In* A. Poole and F. Gill, eds. The birds of North America, No. 442. Acad. of Nat. Sci., Philadelphia, and Amer. Ornithol. Union, Washington, D.C.

Mannan, R. W., and E. C. Meslow. 1984. Bird populations and vegetation characteristics in managed and old-growth forests, northeastern Oregon. J. Wildl. Manage. 48(4):1219-38.

Mannan, R. W., E. C. Meslow, and H. M. Wight. 1980. Use of snags by birds in Douglas-fir forests, western Oregon. J. Wildl. Manage. 44:787-97.

Manuwal, D. A. 1989. Birds of the Klickitat National Scenic River Area. Unpubl. report to Wash. Dept. of Fish and Wildl.

Manuwal, D. A. 1991. Spring bird communities in the southern Washington Cascade range. Pp. 161-74 in L. F. Ruggiero, K. B. Aubry, A. B. Carey, and M. H. Huff, tech. coord. Wildlife and vegetation of unmanaged Douglas-fir forests. USDA For. Serv., Genl. Tech. Rep. PNW-GTR-285.

Manuwal, D. A., H. Carter, R. W. Lowe, J. Takekawa, and U. Wilson. 1996. Oral presentation: Population trends in Murre populations from California, Oregon and Washington, 1979-1995. Pac. Seabirds 23 (42).

Manuwal, D. A., and M. H. Huff. 1987. Spring and winter bird populations in a Douglas-fir forest sere. J. Wildl. Manage. 51:586-95.

Manuwal, D. A., M. H. Huff, M. R. Bauer, C. B. Chappell ,and K. Hegstad. 1987. Summer birds of the upper subalpine zone of Mount Adams, Mount Rainier, and Mount St. Helens, Washington. Northwest Science 61: 82-92.

Manuwal, D. A., P. W. Mattocks, Jr., and K. O. Richter. 1979. First Arctic Tern colony in the contiguous western United States. Amer. Birds 33:144-45.

Manuwal, D. A., and S. Pearson. 1997. Bird populations in managed forests in the western Cascade Mountains, Washington. *In* Wildlife use of managed forests: A landscape perspective. Volume 2. Wash. Dept. Nat. Resour. Olympia.

Manuwal, D. A., and A. C. Thoreson. 1993. Cassin's Auklet (*Ptychoramphus aleuticus*). *In* A. Poole and F. Gill, eds. The birds of North America No. 50. Acad. of Nat. Sci., Philadelphia, and Amer. Ornithol. Union, Washington, D.C.

Marks, J. S. 1997. Is the Northern Saw-whet Owl (*Aegolius acadicus*) nomadic? P. 260 in J. R. Duncan, D. H. Johnson, and T. H. Nicholls, eds. Biology and conservation of owls of the Northern Hemisphere. USDA For. Serv. Genl. Tech. Rep. NC-190.

Marks, J. S, D. L. Evans, and D. W. Holt. 1994. Long-eared Owl (*Asio otus*). *In* A. Poole and F. Gill, eds. The birds of North America, No. 195. Acad. of Nat. Sci., Philadelphia, and Amer. Ornithol. Union, Washington, D.C.

Marshall, D. B. 1970. Chestnut-sided Warbler in Washington. Condor 72:246.

Marshall, D. B. 1997. Status of the white-headed woodpecker in Oregon and Washington. Aud. Soc. of Portland, Oregon.

Marshall, D. B., M. G. Hunter, and A. L. Contreras, eds. 2003. Birds of Oregon: A General Reference. Oreg. State Univ. Press, Corvallis. 768 pp.

Marshall, J. T. 1988. Birds lost from a giant sequoia forest during fifty years. Condor 90:359-72.

Martell, M. S., C. J. Henny, P. E. Nye and M. J. Solensky. 2001. Fall migration routes, timing and wintering sites of North American Ospreys as determined by satellite telemetry. Condor 103:715-24.

Marti, C. D. 1992. Barn Owl (*Tyto alba*). *In* A. Poole, P. Stettenheim, and F. Gill, eds. The birds of North America, No. 1 . Acad. of Nat. Sci., Philadelphia, and Amer. Ornithol. Union, Washington, D.C.

Martin, S. G. 2002. Brewer's Blackbird (*Euphagus cyanocephalus*). *In* A. Poole and F. Gill, eds. The birds of North America, No. 616. Acad. of Nat. Sci., Philadelphia, and Amer. Ornithol. Union, Washington, D.C.

Massey, B. W., D. W. Bradley, and J. L. Atwood. 1992. Demography of a California Least Tern colony including effects of the 1982-1983 El Niño. Condor 94:976-83.

Mattocks, P. 1999. Third summary report of the Washington Ornithological Society Bird Records Committee. WOSNews 61: 1, 4-5.

Mattocks, P. W., Jr., E. S. Hunn, and T. R. Wahl. 1976. A checklist of the birds of Washington State, with recent changes annotated. West. Birds 7:1-24.

Mazur, K. M., and P. C. James. 2000. Barred Owl (*Strix varia*). *In* A. Poole and F. Gill, eds. The birds of North America, No. 508. Acad. of Nat. Sci., Philadelphia, and Amer. Ornithol. Union, Washington, D.C.

McAllister, K. R. 1995. Washington State recovery program for the Upland Sandpiper. Washington Dept. of Fish and Wildlife, Olympia,.

McAllister, K. R., T. E. Owens, L. Leschner, and E. Cummins. 1986. Distribution and productivity of nesting bald eagles in Washington. The Murrelet 67(2):45-50.

McCabe, T. R. 1976. First record of a Magnificent Frigatebird in inland Pacific Northwest. Murrelet 57:43-44.

McCaffrey, B. J., C. M. Harwood, and J. R. Morgart. 1997. First nest of Caspian Terns *(Stern caspia)* for Alaska and the Bering Sea. Pac. Seabirds 24:71-73.

McCallum, D. A., R. Grundel, and D. L. Dahlsten. 1999. Mountain Chickadee (*Poecile gambeli*). *In* A. Poole and F. Gill, eds. The birds of North America, No. 453. Acad. of Nat. Sci., Philadelphia, and Amer. Ornithol. Union, Washington, D.C.

McCaskie, G., J. L. Dunn, C. Roberts, and D. A. Sibley. 1990. Notes on identifying Arctic and Pacific Loons in alternate plumage. Birding 22:70-73.

McCaskie, G., and M. A. Patten. 1994. Status of the Fork-tailed Flycatcher (*Tyrannus savana*) in the United States and Canada. West. Birds 25:113-27.

McCaskie, G., and M. San Miguel. 1999. Report of the California Bird Records Committee: 1996 records. West. Birds 30:57-85.

McDermond, D. K., and K. H. Morgan. 1993. Status and conservation of North Pacific albatrosses. Pp. 70-81 *in* K. Vermeer, K. T. Briggs, K. H. Morgan, K.H., and D. Siegel-Causey, eds. The status, ecology and conservation of marine birds of the North Pacific. Can. Wildl. Serv. Spec. Publ., Ottawa.

McDermott, F. 1998. Pacific Northwest. Hawk Migration Association of North America (HMANA). Hawk Migration Studies 25 (1):24-32.

McGahan, J. 1967. Quantified estimates of predation by a golden eagle population. J. Wildl. Manage. 31:496-501.

McGarigal, K., and W. C. McComb. 1995. Relationships between landscape structure and breeding birds in the Oregon Coast Range. Ecol. Monogr. 65: 235-60.

McGrath, M. T. 1997. Northern Goshawk habitat analysis in managed forest landscapes. M.S. Thesis, Oregon State Univ., Corvallis. 127 p.

McIntyre, J. W. 1990. Yellow-billed Loons and the Alaska oil spill. Abstr. Pac. Seabird Group Bull. 17:26.

McIntyre, J. W., and J.F. Barr. 1997. Common Loon (*Gavia immer*). *In* A. Poole and F. Gill, eds. The birds of North America, No. 313. Acad. of Nat. Sci., Philadelphia, and Amer. Ornithol. Union, Washington, D.C.

McKee, T., and P. Pyle. 2002. Plumage variation and hybridization in Black-footed and Laysan Albatrosses. N. Amer. Birds 56:131-38.

McLean, D. D. 1939. European Jack Snipe and Franklin's Gull in California. Condor 41:164.

McNaughton, M., K. Mills, M. Rauzon, W.J. Sydeman, L. Takano, H. Carter, M. Fry, T. Work, S. Newman, R. Helm, and T. Zimmerman. 2004. Draft regional seabird conservation plan, U.S. Fish and Wildlife Service, Pacific Region. U.S. Fish and Wildl. Serv., Portland, OR.

McNicholl, M. K., P. E. Lowther, and J. A. Hall. 2001. Forster's Tern (*Sterna forsteri*). *In* A. Poole and F. Gill, eds. The birds of North America, No. 595. Acad. of Nat. Sci., Philadelphia, and Amer. Ornithol. Union, Washington, D.C.

Meeth, P., and K. Meeth. 1988. A Shy Albatross off Somalia. Sea Swallow 37:66.

Melillo, J. M. 1999. Warm, warm on the range. Science 283:183-84.

Melvin, S. M., and J. P. Gibbs. 1996. Sora (*Porzana carolina*) *In* A. Poole and F. Gill, eds. The birds of North America, No.250. Acad. of Nat. Sci., Philadelphia, and Amer. Ornithol. Union, Washington, D.C.

Meyer, D. 1973. Observations of two rare sandpipers in eastern Washington. Murrelet 54:21-22.

Miller, R. C., and E. L. Curtis. 1940. Birds of the University of Washington campus. Murrelet 21:35-46.

Millington, R. 1996. Identification forum: Arctic Redpolls revisited. Birding World 9:65-69.

Mills, E. L. 1989. Eurasian Kestrel (*Falco tinnunculus*) in Nova Scotia. Nova Scotia Birds 30 (2):49-51.

Milner, R. L. 1988. Guidelines for establishing and maintaining a Purple Martin nest colony. Prep. for WDFW.

Mindell, D. P. 1983. Harlan's Hawk, *Buteo Jamaicensis harlani*, as a valid subspecies. Auk100:161-67.

Ministry of Environment, Lands and Parks. 1998. Inventory methods for nighthawk and poorwill standards for components of British Columbia's biodiversity No. 9. Ministry of Environment, Lands and Parks, Victoria, B.C., Canada.

Mirarchi, R. E., and T. S. Baskett. 1994. Mourning Dove (*Zenaida macroura*). *In* A. Poole and F. Gill, eds. The birds of North America, No. 117. Acad. of Nat. Sci., Philadelphia, and Amer. Ornithol. Union, Washington, D.C.

Mlodinow, S. 1984. Chicago area birds. Chicago, Illinois: Chicago Review Press.

Mlodinow, S. 1995. The Snohomish County Emperor Geese. WOSNews 36:1-6.

Mlodinow, S. G. 1997. The Long-billed Murrelet (*Brachyramphus perdix*) in North America. Birding 29:461-75.

Mlodinow, S. G. 1998a. The Magnificent Frigatebird in western North America. Amer. Birds 52:413-19.

Mlodinow, S. G. 1998b. The Tropical Kingbird north of Mexico. Field Notes 52:6-13.

Mlodinow, S. G. 1999. Common and King Eiders: Vagrancy patterns in western North America. Birders Journal 8:234-42.

Mlodinow, S. G. 2001. The Sharp-tailed Sandpiper (*Calidris acuminata*) in North America. Birding 33:330-41.

Mlodinow, S. G. 2002. A Little Curlew (*Numenius minutus*) at Leadbetter Point: a first Washington record. Wash. Birds 8:58-61.

Mlodinow, S. G. 2004. A Bean Goose (*Anser fabalis*) at Hoquiam, Washington: A first state record. N. Amer. Birds 58:298-300.

Mlodinow, S. G., and D. Duffy. 2000. Whiskered Auklet (*Aethia pygmaea*) on Whidbey Island: A first record for the contiguous United States. Wash. Birds, in press.

Mlodinow, S. G., S. Feldstein, and B. Tweit. 1999. The Bristle-thighed Curlew landfall of 1998: Climactic factors and notes on identification. West. Birds 30: 133-55.

Mlodinow, S. G., and M. O'Brien. 1996. America's 100 most wanted birds. Falcon Press, Helena, MT.

Mlodinow, S. G., and S. Pink. 2000. Red-faced Cormorant (*Phalacrocorax urile*) in Clallam County: A first record for the contiguous United States. Wash. Birds 7: 46-50.

Moldenhauer, R. R., and E. D. Bawdon. 1962. Mockingbird collected in Washington. Murrelet 43:15.

Monda, M. J. 2000. Washington waterfowl breeding populations and production. Agency report for Wash. Dept. of Fish and Wildl. Olympia,.

Monda, M. J., and J. T. Ratti. 1988. Niche overlap and habitat use by sympatric duck broods in eastern Washington. J. Wildl. Manage. 52: 95-103.

Monroe, B. L., Jr. and C. G. Sibley. 1993. A world checklist of birds. Yale Univ. Press, New Haven, CT. 393 pp.

Moore, K. R., and C. J. Henney. 1983. Nest site characteristics of three coexisting Accipiter hawks in northeastern Oregon. Raptor Research 17(3):65-76.

Moore, K. R., and C. J. Henney. 1984. Age-specific productivity and nest site charactersitics of Cooper's Hawks (*Accipiter cooperii*). Northwest Science 58(4):290-99.

Moore, W. S. 1995. Northern Flicker (*Colaptes auratus*). *In* A. Poole and F. Gill, eds. The birds of North America, No. 166. Acad. of Nat. Sci., Philadelphia, and Amer. Ornithol. Union, Washington, D.C.

Morgan, K. H., K. Vermeer, and R. W. McKelvey. 1991. Atlas of pelagic birds of western Canada. Occas. paper No. 72. Can. Wildl. Serv.

Moring, J. 2002. Early American naturalists: exploring the American West, 1804-1900. Cooper Square Press, New York.

Morlan, J. 1981. Status and identification of forms of White Wagtail in western North America. Continental Birdlife 2:37-50.

Morrison, M., B. Marcot, and R. Mannan. 1992. Wildlife-habitat relationships – concepts and applications. Univ. of Wiscosin Press, Madison.

Morrison, M. L. 1982. The structure of western warbler assemblages: Ecomorphological analysis of Black-throated Gray and Hermit warblers. Auk 99: 503-13.

Morrison, M. L. 1990. Morphological and vocal variation in the Black-throated Gray Warbler in the Pacific Northwest. Northwestern Naturalist 71: 53-58.

Morrison, R. I. G., C. Downes, and B. Collins. 1994. Population trends of shorebirds on fall migration in eastern Canada. Wilson Bull. 106:431-47.

Morrison, R. I. G., R. E. Gill, Jr., B. A. Harrington, S. Skagen, G. W. Page, C. L. Gratto-Trevor, and S. M. Haig. 2000. Population estimates of nearctic shorebirds. Waterbirds 23:337-52.

Morse, B. 1996. A birder's guide to Ocean Shores, Washington. R. W. Morse Company, Olympia, WA. 51 pp.

Morse, R. W. 2001. A birder's guide to coastal Washington. R. W. Morse Co., Olympia, WA.

Moskoff, W. 1995. Veery (*Catharus fuscescens*). *In* A. Poole and F. Gill, eds. The birds of North America. Number 142. Acad. of Nat. Sci., Philadelphia, and Amer. Ornithol. Union, Washington, D.C.

Moskoff, W. 2003. Monk Parakeets in urban areas outside Chicago. Birding 35: 274-77.

Moskoff, W., and L. R. Bevier. 2002. Mew Gull (*Larus canus*). *In* A. Poole and F. Gill, eds. The birds of North America, No. 687. Acad. of Nat. Sci., Philadelphia, and Amer. Ornithol. Union, Washington, D.C.

Mosconi, S. L, and R. L. Hutto. 1982. The effect of grazing on the landbirds of western Montana riparian habitat. Pp. 221-33 in L. Nelson and J. M. Peek, eds. Proceedings of the wildlife-livestock relationships symposium. For. Wildl. and Range Exper. Station, Moscow, ID.

Motschenbacker, M. D. 1984. The feasibility of restoring a breeding white pelican population in the state of Washington. Thesis, Washington State Univ., Pullman, Washington.

Mowbray, T. B., F. Cooke, and B. Ganter. 2000. Snow Goose (*Chen caerulescens*). *In* A. Poole and F. Gill, eds. The birds of North America, No. 514. Acad. of Nat. Sci., Philadelphia, and Amer. Ornithol. Union, Washington, D.C.

Mueller, H. 1999. Common Snipe (*Gallinago gallinago*). *In* A. Poole and F. Gill, eds. The birds of North America. No. 417. Acad. of Nat. Sci., Philadelphia, and Amer. Ornithol. Union, Washington, D.C.

Muller, M. J. 1995. Pied-billed Grebes nesting on Green Lake, Seattle, Washington. Wash. Birds 4:35-39.

Munro, J. A. 1935. Recent records from British Columbia. Condor 37:178-79.

Munro, J. A., and I. McT.-Cowan. 1947. A review of the bird fauna of British Columbia. British Columbia Prov. Mus. Spec. Publ. no. 2, Victoria.

Myers, A. M., D. D. Roby, K. Collis, D. E. Lyons, and J. Y. Adkins. 2002. Diet composition of Double-crested Cormorants nesting at East Sand Island in the Columbia River Estuary. Abstr. Pacific Seabirds 29:56.

Myers, J. P., C. T. Schick, and C. J. Hohenberger. 1984. Notes on the 1983 distribution of Sanderlings along the United States' Pacific coast. Wader Study Group Bull. 40:22-26.

National Audubon Society. 2002. The Christmas Bird Count historical results [Online]. http://www.audubon.org/bird/cbc.

National Geographic Society. 1999. Field guide to the birds of North America. Third ed. Nat. Geog. Soc., Washington, D.C. 480 pp.

Nehls, H. 1989. A review of the status and distribution of dowitchers in Oregon. Oreg. Birds 15:97-102.

Nehls, H. B. 1994. Oregon shorebirds: their status and movements. Tech. rep. 94-1-02, Oreg. Dept. Fish and Wildl.

Nehls, H. B. 1995. The records of the Oregon Bird Records Committee, 194-195. Oreg. Birds 21:103-5.

Neitro, W. A., V. W. Binkley, S. P. Cline, R. W. Mannan, B. G. Marcot, D. Taylor, and F. F. Wagner. 1985. Snags (wildlife trees). Pp. 129-69 *in* E. R. Brown, tech. ed. Management of wildlife and fish habitats in forests of western Oregon and Washington. USDA For. Serv. Publ. R6-F&WL-192-1985. Portland, OR. 32 p.

Nelson, S. K. 1997. Marbled Murrelet (*Brachyramphus marmoratus*). *In* A. Poole and F. Gill, eds. The birds of North America, No. 276. Acad. of Nat. Sci., Philadelphia, and Amer. Ornithol. Union, Washington, D.C.

Nicholls, T. H., and D. W. Warner. 1972. Barred Owl habitat use as determined by radio telemetry. J. Wildl. Manage. 36:213-24.

Nilsson, L. 1969. Food consumption of diving ducks wintering at the coast of south Sweden in relation to food resources. Oikos 20: 128-35.

Nol, E., and M. S. Blanken. 1999. Semipalmated Plover (*Charadrius semipalmatus*). *In* A. Poole and F. Gill, eds. The birds of North America, No. 444. Acad. of Nat. Sci., Philadelphia, and Amer. Ornithol. Union, Washington, D.C.

Nolan, V., Jr., E. D. Ketterson, D. A. Cristol, C. M. Rogers, E. D. Clotfelter, R. C. Titus, S. J. Schoech, and E. Snajor. 2002. Dark-eyed Junco (*Junco hyemalis*). *In* A. Poole and F. Gill, eds. The birds of North America, No. 716. Acad. of Nat. Sci., Philadelphia, and Amer. Ornithol. Union, Washington, D.C.

Norment, C. J., and S. A. Shackleton. 1993. Harris' Sparrow (*Zonotrichia querula*). *In* A. Poole and F. Gill, eds. The birds of North America, No. 429. Acad. of Nat. Sci., Philadelphia, and Amer. Ornithol. Union, Washington, D.C.

Nysewander, D. R. 1977. Reproductive success of the Black Oystercatcher in Washington state. M.S. Thesis, Univ. Wash., Seattle.

Nysewander, D., and J. R. Evenson. 1998. Status and trends for selected diving duck species examined by the marine bird component, Puget Sound Ambient Monitoring Program (PSAMP), Washington Department of Fish and Wildlife. Proc. 1998 Puget Sound Res. Conf. publ. by Puget Sound Water Qual. Action Team, Olympia. Vol. 1: pp 847-867.

Nysewander, D. R., J. R. Evenson, B. L. Murphie, and T. A. Cyra. 2001. Report of marine bird and marine mammal component, Puget Sound Ambient Monitoring Program, for July 1992 to January 2000 Period. Agency rep., Wash. Dept. Fish and Wildl., Olympia.

O'Connell, M. A., J. G. Hallet, and D. K. Beutler. 1997. Bird populations in managed forests of northeastern Washington. *In*: Wildlife use of managed forests: A landscape perspective. Vol. 3. Wash. Dept. Nat. Res., Olympia.

Oberholser, H. C. 1932. Description of new birds from Oregon, chiefly from the Warner Valley region. Cleveland Mus. Nat. Hist. Sci. Publ. 4:8.

Ogilvie, M., and S. Young. 1998. Photographic handbook of the wildfowl of the world. New Holland Publishers, London, U.K.

Olendorff, R. R. 1973. Raptorial birds of the U.S.A.E.C. Hanford Reservation, south-central Washington. Unpubl. rep., Battelle Northwest Lab., Richland, Washington.

Olendorff, R. R., A. D. Miller, and R. N. Lehman. 1981. Suggested practices for raptor protection on power lines. The state of the art in 1981. Raptor Res. Rep. No. 4

Olsen, K. M., and H. Larsson. 1997. Skuas and jaegers: a guide to the skuas and jaegers of the world. Yale Univ. Press, New Haven, CT. 190 pp.

O'Neil, T. A., K. A. Bettinger, M. Vander Heyden, B. G. Marcot, C. Barrett, T. K. Mellen, W. M. Vanderhaegen, D. H. Johnson, P. J. Doran, L. Wunder, and K. M. Boula. 2001. Structural conditions and habitat elements in Oregon and Washington. Pp. 115-39 *in* D. H. Johnson and T. A. O'Neil, dirs. Wildlife-habitat relationships in Oregon and Washington. Oreg. State Univ. Press, Corvallis.

Opperman, H. 1992. Bushtits in Kittitas County, Washington. Wash. Birds 2:25-31.

Opperman, H. 2003. A birder's guide to Washington. Amer. Birding Assoc. 636 pp.

Oring, L. W., L. Neel, and K. E. Oring. 2000. United States shorebird conservation plan: Intermountain west regional shorebird plan. Version 1.0. Ornithol. Newsl. 2002.146:3.

Orr, R. T. 1962. The Tufted Duck in California. Auk 79:482-83.

Orthmeyer, D. L., J. Y. Takekawa, C. R. Ely, M. L. Wege and W. E. Newton. 1995. Morphological differences in Pacific coast populations of greater white-fronted geese (*Anser albifrons*). Condor 97:123-32.

Pacific Flyway Council (PFC). 1987. Pacific flyway management plan for the Greater White-fronted Goose. USFWS. Portland, OR.

Pacific Flyway Council (PFC). 1998. Pacific flyway management plan for Northwest Oregon - Southwest Washington Canada Goose agricultural depredation control. Unpubl. report. Canada Goose agricultural depredation working group, Pacific Flyway Study Committee (c/o USFWS), Portland, OR.

Pacific Flyway Council (PFC). 2000. Pacific flyway midwinter waterfowl survey, USFWS. Portland, OR.

Pacific Flyway Study Committee (PFSC). 1992. Pacific flyway management plan for the Wrangel Island population of Lesser Snow Geese. Unpubl. rep. c/o USFWS, MBMO, Portland, OR.

Pacific Flyways Study Committee (PFSC). 2003. Pacific flyway managment plan for the Pacific flyway population of greater white-fronted geese. Unpubl. rep. Pacific Flyway Study Committee (c/o USFWS), Portland, OR.

Page, G. W., and R. E. Gill, Jr. 1994. Shorebirds in western North America: Late 1800s to late 1900s. Studies in Avian Biol. 15:147-60.

Palmer, R. S, ed. 1962. Handbook of North American birds: Vol. 1. Loons through Flamingos. Yale Univ. Press, New Haven, CT. 567 pp.

Palmer, R. S. 1967. Species account *in* G. D. Stout, ed. The shorebirds of North America. The Viking Press, New York.

Palmer, R. S. 1988. Northern Goshawk *Accipiter gentilis*. Handbook of North American birds. Vol. 4: Diurnal raptors. Yale Univ. Press, New Haven, CT.

Parkhurst, J. A., R. P. Brooks, and D. E. Arnold. 1987. A survey of wildlife depredation and control techniques at fish-rearing facilities. Wildl. Soc. Bull. 15:386-94.

Parrish, J. K. 1995. Oral presentation: Altering ecological interactions by habitat modification: a restoration technique for Common Murres. Pacific Seabirds 22:40.

Parsons, K. C., and T. L. Master. 2000. Snowy Egret (*Egretta thula*). *In* A. Poole and F. Gill, eds. The birds of North America, No. 489. Acad. of Nat. Sci., Philadelphia, and Amer. Ornithol. Union, Washington, D.C.

Pashley, D. N., C. J. Beardmore, J. A. Fitzgerald, R. P. Ford, W. C. Hunter, M. S. Morrison, and K. V. Rosenberg. 2000. Partners in flight: Conservation of the land birds of the United States. Amer. Bird Conserv., The Plains, VA.

Patten, M. A., and J. C. Burger. 1998. Spruce budworm outbreaks and the incidence of vagrancy in eastern North American wood-warblers. Can. J. Zool. 76:433-39.

Patten, M. A., and C. A. Marantz. 1996. Implications of vagrant southeastern vireos and warblers in California. Auk 113:911-23.

Patten, M. A., and K. Radamaker. 1991. A fall record of Sharp-tailed Sparrow for interior California. West. Birds 22:37-38.

Patten, S., and A. R. Weisbrod. 1974. Sympatry and interbreeding of Herring and Glaucous-winged gulls in southeastern Alaska. Condor 76:343-44.

Patterson, M. 1998. Birds and other wildlife on the Columbia River estuary. Oreg. Field Ornithologists' Spec. Publ. No. 11. 86 pp.

Paulson, D. 1979. Dotterel at Ocean Shores, Washington. Continental Birdlife 1:109-11.

Paulson, D. R. 1992a. The distribution of the northern Fork-tailed Storm-Petrel. Wash. Birds 2:6-7.

Paulson, D. R. 1992b. Northwest bird diversity: from extravagant past to changing present to precarious future. Northwest Env. Jour. 8:71-118.

Paulson, D. R. 1993. Shorebirds of the Pacific Northwest. Univ. Wash. Press, Seattle. 406 pp.

Paulson, D. R., and D. S. Lee. 1992. Wintering of Lesser Golden-Plovers in eastern North America. J. Field Ornithol. 63:121-28.

Paulson, D. R., and P. W. Mattocks, Jr. 1992. Eastern Phoebes in Washington. Wash. Birds 2: 20-22.

Payne, R. B. 1979. Ardeidae *in* E. Mayr, G.W. Cottrell, eds. Checklist of birds of the world. Mus. Comparative Zool. 193-244.

Pearce, J. M. 2002. First record of a Greater Shearwater in Alaska. West. Birds 33:121-22.

Pearson, R. R., and K. B. Livezey. 2003. Distribution, numbers and site characteristics of Spotted Owls and Barred Owls in the Cascade Mountains of Washington. J. Raptor Res. 37:265-76.

Pearson, S. F. 1997. Hermit Warbler (*Dendroica occidentalis*). *In* A. Poole and F. Gill, eds. The birds of North America, No. 303. Acad. of Nat. Sci., Philadelphia, and Amer. Ornithol. Union, Washington, D.C.

Pearson, S. F. 2000. Behavioral asymmetries in a moving hybrid zone. Behavioral Ecol. 11: 84-92.

Pearson, S.F . 2003. Breeding phenology, nesting success, habitat selection, and census methods for the Streaked Horned Lark in the Puget lowlands of Washington. Natural Areas Report 2003-2. Wash. Dept. Nat.l Res,, Olympia.

Pearson, S. F., and D. A. Manuwal. 2000. Influence of niche overlap and territoriality on hybridization between Hermit Warblers and Townsend's Warblers. Auk 117: 175-83.

Penland, S. 1981. Natural history of the Caspian Tern in Grays Harbor, Washington. Murrelet 62:66-72.

Penland, S. 1982. Distribution and status of the Caspian Tern in Washington State. Murrelet 63:73-79.

Penland, S. T., and S. J. Jeffries. 1977. New breeding records for the Ring-billed Gull in Washington. Murrelet 58:86-87.

Peterjohn, B. 1991. Summaries of Breeding Bird Survey data for Washington, 1966-1991. Unpubl. ms.

Peterjohn, B. G., and J. R. Sauer. 1993. North American Breeding Bird Survey annual summary 1990-1991. Bird Populations 1:1-15.

Peterson, R. T. 1941. Field guide to western birds. Houghton Mifflin Co., Boston, MA. 240 pp.

Peterson, R. T. 1961. Field guide to western birds. Second ed. Houghton Mifflin Co., Boston, MA. 366 pp.

Peterson, R. T. 1980. Field guide to the birds. Fourth ed. Houghton Mifflin Co., Boston, MA. 384 pp.

Peterson, R. T. 1990. Field guide to western birds. Third ed. Houghton Mifflin Co., Boston, MA. 432 pp.

Pfister, C., B. A. Harrington, and M. Levine. 1992. The impact of human disturbance on shorebirds at a migration staging area. Biol. Conserv. 60(2):115-26.

Phillips, A. R. 1991. The known birds of North and middle America. Part 2. Allan R. Phillips, Denver, CO.

Piatt, J. F., and A. S. Kitaysky. 2002. Tufted Puffin (*Fratercula cirrhata*). *In* A. Poole and F. Gill, eds . The birds of North America. No. 708. Acad. of Nat. Sci., Philadelphia, and Amer. Ornithol. Union, Washington, D.C.

Piatt, J. F., C. J. Lensink, W. Butler, M. Kendziorek, and D. K. Nysewander. 1990. Immediate impact of the 'Exxon Valdez' oil spill on marine birds. Auk 107:387-97.

Pierotti, R. J., and C. A. Annett. 1995. Western Gull (*Larus occidentalis*). *In* A. Poole and F. Gill, eds. The birds of North America. No. 174. Acad. of Nat. Sci., Philadelphia, and Amer. Ornithol. Union, Washington, D.C.

Piersma, T., and N. Davidson. 1992. The migrations and annual cycles of five subspecies of Knots in perspective. Wader Study Group Bull. 64, Suppl.: 187-97.

Pitman, R. L. 1985. The marine birds of Alijos Rocks, Mexico. West. Birds 16:81-92.

Pitman, R. L. 1986. Atlas of seabird distribution and relative abundance in the eastern tropical Pacific. Admin. Rep. LJ-86-02C. NMFS Southwest Fisheries Center, LaJolla, Calif.

Pitman, R. L., and L. T. Ballance. 2002. The changing status of marine birds breeding at San Benedicto Island, Mexico. Wilson Bull. 114:11-19.

Pitman, R. L., and M. R. Graybill. 1985. Horned Puffin sightings in the eastern Pacific. West. Birds 16:99-102.

Pitman, R. L., W. A. Walker, W. T. Everett and J. P. Gallo-Reynoso. 2002. Population status, foods and foraging of Laysan Albatrosses nesting on Guadalupe Island, Mexico. Abstr. Pacific Seabirds 29:60.

Pitt, W. C., and M. R. Conover. 1996. Predation at intermountain west fish hatcheries. J. Wildl. Manage. 60:616-24.

Pope, P. 1948. European Starling at Walla Walla, Washington. Murrelet 29:29.

Post, P. W., and R. H. Lewis. 1995. The Lesser Black-backed Gull in the Americas, part 1. Birding 27:282-90.

Poulin, R. G., S. D. Grindal, and R. M. Brigham. 1996. Common Nighthawk (*Chordeiles minor*). *In* A. Poole and F. Gill, eds. The birds of North America, No. 213. Acad. of Nat. Sci., Philadelphia, and Amer. Ornithol. Union, Washington, D.C.

Powell, H. 2000. The influence of prey density on post-fire habitat use of the Black-backed Woodpecker. MS Thesis. Univ. Montana, Missoula. 99 pp.

Pratt, H. D., P. L. Bruner, and D. G. Berrett. 1987. The Birds of Hawaii and the tropicalPacific. Princeton Univ. Press, Princeton, NJ.

Pravosudov, V. V., and T. C. Grubb, Jr. 1993. White-breasted Nuthatch (*Sitta carolinensis*). *In* A. Poole and F. Gill, eds. The birds of North America, No. 54. Acad. of Nat. Sci., Philadelphia, and Amer. Ornithol. Union, Washington, D.C.

Preston, C. R., and R. D. Beane. 1993. Red-tailed Hawk (*Buteo jamaicensis*). *In* A. Poole and F. Gill, eds. The birds of North America, No. 52. Acad. of Nat. Sci., Philadelphia, and Amer. Ornithol. Union, Washington, D.C.

Price, F. E., and C. E. Bock. Population ecology of the Dipper (*Cinclus mexicanus*) in the Front Range of Colorado. Stud. Avian Biol., No. 7. 84 pp.

Price, J. 1995. Ranges of North American breeding birds visualizing long-term population changes in North America breeding birds. Northern Prarie Wildlife Research Center, Jamestown, ND. Home Page. Http:// www.npwrc.usgs.gov/resource/ distr/birds/breedrng/breedrng.htm (Version 16 JUL97).

Price, J., S. Droege, and A. Price. 1995. Summer atlas of North American birds. Academic Press, London, U.K.

Pyle, P. 1997. Identification guide to central interior North American birds, Part 1. Slate Creek Press, Bolinas, CA.

Pyle, P., and S. N. G. Howell. 1996. Spizella Sparrows: Intraspecific variation and identification. Birding 28:374-87.

Ralph, C. J., G. L. Hunt, Jr., M. G. Martin, and J. F. Piatt, eds. 1995. Ecology and conservation of the Marbled Murrelet. Gen. Tech. Rep. PSW-GTR-152. Pacific Southwest Res. Sta., For. Serv., U.S. Dept. Agriculture; Albany CA. 420 p.

Ralph, C. J., L. L. Long, B. P. O'Donnell, M. Raphael, S. Miller, and S. Courteney. 1996. Population and productivity of Marbled Murrelets during 1995 in the San Juan islands, Washington. Draft rep. to National Council of the Paper Industry for Air and Stream Improvement. 106 p.

Ransom, W. H. 1948. European Starling taken in Cowlitz County, western Washington. Murrelet 29:28-29.

Raphael, M. G., D. Evans Mack, J. M. Marzluff, and J. M. Luginbuhl. 2002. The potential effects of forest fragmentation on populations of the Marbled Murrelet. Studies in Avian Biol. 25:221-35.

Rathbun, S. F. 1902. A list of the land birds of Seattle, Washington, and vicinity. Auk 19: 131-41.

Rathbun, S. F. 1911. Notes on the birds of Seattle, Washington, and vicinity. Auk 28:492-94.

Rathbun, S. F. 1915. List of water and shore birds in the Puget Sound region in the vicinity of Seattle. Auk 32:459-65.

Rathbun, S. F. 1916. The Lake Crescent region, Olympic Mountains, Washington, with notes regarding its avifauna. Auk 33: 357-70.

Ratoosh, E. S. 1995. Birds of the Montlake Fill, Seattle, Washington (1978-1983). Wash. Birds 4:1-34.

Ratti, J. T. 1981. Identification and distribution of Clark's Grebe. West. Birds 12:41-46.

Reed, A., M. A.Davison, and D. K. Kraege. 1989. Segregation of Brent Geese *Brant bernicla* wintering and staging in Puget Sound and the Strait of Georgia. Wildfowl 40:22-31.

Reed, A., D. H. Ward, D. V. Derksen, and J. S. Sedinger. 1998. Brant (*Branta bernicla*). *In* A. Poole and F. Gill, eds. The birds of North America, No. 337). Acad. of Nat. Sci., Philadelphia, and Amer. Ornithol. Union, Washington, D.C.

Reilly, E. M., and O. S. Pettingill. 1968. The Audubon illustrated handbook of American birds. McGraw-Hill Book Co., N.Y.

Remsen, J. V., Jr., and L. C. Binford. 1975. Status of the Yellow-billed Loon (*Gavia adamsii*) in the western United States and Mexico. West. Birds 6:7-20.

Reynolds, R. T. 1989. Accipiters. Pp. 92-101 *in* B. G. Pendleton, ed. Proc. of the Western Raptor Management Symposium and Workshop. Natl. Wildl. Fed. Tech. Series, No. 12.

Reynolds, R. T., E. C. Meslow, and H. M. Wright. 1982. Nesting habitat of coexisting accipiters in Oregon. J. Wildl. Manage. 46(1):124-38.

Reynolds, R. T., and H. M. Wight. 1978. Distribution, density, and productivity of accipiter hawks breeding in Oregon. Wilson Bull. 90:182-96.

Rhoads, S. N. 1891. The wild pigeon (*Ectopistes migratorius*) on the Pacific coast. Auk 8:310-12.

Ribic, C. A., S. W. Johnson, and C. A. Cole. 1997. Distribution, type, accumulation, and source of marine debris in the United States, 1989-1993. Pp. 35-47 *in* J. M. Coe and D. B. Rogers, eds. Marine debris: Sources, impacts, and solutions. Springer Verlag, New York.

Rich, T. D., C. J. Beardmore, H. Berlanga, P. J. Blancher, M. F. W. Bradstreet, G. S. Butcher, D. W. Demarest, E. H. Dunn, W. C. Hunter, E. E. Inigo-Elias, J. A. Kennedy, A. M. Martell, A. O. Panjabi, D. N. Pashley, K. V. Rosenberg, C. M. Rustay, J. S. Wendt, and T. C. Will. 2004. Partners in Flight North American landbird conservation plan. Cornell Lab. Ornithology, Ithaca, NY.

Richardson, F. 1960. Breeding of the Fork-tailed Petrel off the Washington coast. Condor 62:140.

Richardson, F. 1961. Breeding biology of the Rhinoceros Auklet on Protection Island, Washington. Condor 63:456-73.

Richardson, S. 1997. East Bay bird guide. Black Hills Aud. Soc., Olympia, WA. 85 pp.

Richardson, S. A. 1996. Washington state recovery plan for the Ferruginous Hawk. Wash. Dept. Fish and Wildl., Olympia.

Ridgway, R. 1902. Birds of middle and North America, part II. Bull. U.S. Natl. Mus. 50:1-834.

Ridgway, R. 1916. The birds of North and middle America. VII. Cuculidae, Psittacidae, and Columbidae. U.S. Natl. Mus. Bull. 50:1-543.

Rimmer, C. C., and K. P. McFarland. 1998. Tennessee Warbler *(Vermivora peregrina)*. *In* A. Poole and F. Gill, eds. The birds of North America, No. 350. Acad. of Nat. Sci., Philadelphia, and Amer. Ornithol. Union, Washington, D.C.

Rising, J. D. 1996. The Sparrows of the United States and Canada. Academic Press, San Diego, CA.

Rising, J. D., and P. L. Williams. 1999. Bullock's Oriole (*Icterus bullockii*). *In* A. Poole and F. Gill, eds. The birds of North America, No. 416. Acad. of Nat. Sci., Philadelphia, and Amer. Ornithol. Union, Washington, D.C.

Robbins, C. S., D. Bystrack, and P. H. Geissler.1986. The Breeding Bird Survey, its first fifteen years,1965-1979. USFWS Res. Publ. 157. Washington, D.C.

Roberson, D. 1980. Rare birds of the west coast. Woodcock Publ. 496 pp.

Roberson, D. 1996. Identifying Manx Shearwaters in the northeastern Pacific. Birding 28:18-33.

Roberson, D., and L. F. Baptista. 1988. White-shielded coots in North America: a critical evaluation. Amer. Birds: 42(5):1241-46.

Robertson, R. J, B. J. Stuchbury, and R. R. Cohen. 1992. Tree Swallow (*Tachycineta bicolor*). *In* A. Poole and F. Gill, eds. The birds of North America, No. 11. Acad. of Nat. Sci., Philadelphia, and Amer. Ornithol. Union, Washington, D.C.

Roby, D. D., K. Collis, D. E. Lyons, D. P. Craig, J. Y. Adkins, A. M. Myers, and R. M. Suryan. 2002. Effects of colony relocation on diet and productivity of Caspian Terns in the Columbia River estuary. Abstr. Pacific Seabirds 29:62.

Roby, D. D., D. P. Craig, K. Collis, and S. Adamany. 1998. Avian predation on juvenile salmonids in the lower Columbia River. 1997 annual report. Oreg. State Univ. Corvallis. 25 pp.

Rodrick, E., and R. Miller. 1996. Priority habitats and species—Bald Eagle. Hab. Progr., Olympia, WA.

Rogers, R. E., Jr. 2000. The status and microhabitat selection of Streaked Horned Lark, Western Bluebird, Oregon Vesper Sparrow and Western Meadowlark in western Washington. M.S. thesis, The Evergreen State College, Olympia, WA.

Rohwer, S., D. F. Martin, and G. G. Benson. 1979. Breeding of the Black-necked Stilt in Washington. Murrelet 60:67-71.

Rohwer, S., and C. Wood. 1998. Three hybrid zones Between Hermit and Townsend's warblers in Washington and Oregon. Auk 115: 284-310.

Rohwer, S., C. Wood, and E. Bermingham. 2000. A new hybrid warbler (*Dendroica nigrescens X D. occidentalis*) and diagnosis of similar *D. townsendi X D. occidentalis* recombinants. Condor 102:713-18.

Romagosa, C. M., and T. McEneaney. 1999. Eurasian Collared-Dove in North America and the Caribbean. N. Amer. Birds 53:348-53.

Root, T. L., and J. D. Weckstein. 1994. Changes in distribution patterns of select wintering North American birds from 1901 to 1989. Pp. 191-201 *in* J. R. Jehl, Jr., and N. K. Johnson, eds. A century of avifaunal change in western North America. Studies in Avian Biol. 15.

Root, T. L. 1988. Atlas of wintering North American birds: an analysis of Christmas Bird Count data. Univ. Chicago Press, Chicago.

Roselaar, C. S. 1983. Subspecies recognition in Knot *Calidris canutus* and occurrence of races in western Europe. Beaufortia 33 (7): 97-109.

Rosenberg, K. V., and M. G. Raphael. 1986. Effects of forest fragmentation on vertebrates in Douglas-fir forests. Pp. 263-72 *in* J. Verner, M. L. Morrison, and C. J. Ralph, eds. Wildlife 2000: Habitat relationships of terrestrial vertebrates. Univ. of Wisconsin Press, Madison.

Rotenberry, J. T., M. A. Patten, and K. L. Preston. 1999. Brewer's Sparrow (*Spizella breweri*). *In* A. Poole and F. Gill, eds. The birds of North America, No. 390. Acad. of Nat. Sci., Philadelphia, and Amer. Ornithol. Union, Washington, D.C.

Rothstein, S. I. 1994. The cowbird's invasion of the far west: History, causes, and consequences experienced by host species. Studies in Avian Biol. 15:301-15.

Rubega, M. A., and J. A. Robinson. 1997. Water salinization and shorebirds: Emerging issues. International Wader Studies 9:45-54.

Runde, D. E., L. A. Dickson, S. M. Desimone, and J. B. Buchanan. 1999. Notes on habitat relationships for selected forest wildlife species in southwestern Washington. (3 volumes and compact disc). Weyerhaeuser Co., Tacoma, WA. 1180 pages.

Rusch, D. H., S. DeStefano, M. C. Reynolds, and D. Lauten. 2000. Ruffed Grouse (*Bonasa umbellus*). *In* A. Poole and F. Gill, eds. The birds of North America, No. 515. Acad. of Nat. Sci., Philadelphia, and Amer. Ornithol. Union, Washington, D.C.

Russell, R. W. 2002. Pacific Loon (*Gavia pacific*), Arctic Loon (*Gavia arctica*). *In* A. Poole and F. Gill, eds. The birds of North America, No. 657. Acad. of Nat. Sci., Philadelphia, and Amer. Ornithol. Union, Washington, D.C.

Russell, S. M. 1996. Anna's Hummingbird (*Calypte anna*). *In* A. Poole and F. Gill, eds. The birds of North America, No. 226. Acad. of Nat. Sci., Philadelphia, and Amer. Ornithol. Union, Washington, D.C.

Ryan, T. P., and C. T. Collins. 2000. White-throated Swift (*Aeronautes saxatalis*). *In* A. Poole and F. Gill, eds. The birds of North America, No. 526. Acad. of Nat. Sci., Philadelphia, and Amer. Ornithol. Union, Washington, D.C.

Ryder, J. P. 1993. Ring-billed Gull (*Larus delawarensis*). *In* A. Poole and F. Gill, eds. The birds of North America, No. 33. Acad. of Nat. Sci., Philadelphia, and Amer. Ornithol. Union, Washington, D.C.

Ryder, J. P., and R. T. Alisauskas. 1995. Ross' Goose (*Chen rossii*). *In* A. Poole and F. Gill, eds. The birds of North America, No. 130. Acad. of Nat. Sci., Philadelphia, and Amer. Ornithol. Union, Washington, D.C.

Ryder, R. R., and D. E. Manry. 1994. White-faced Ibis (*Plegadis chihi*). *In* A. Poole and F. Gill, eds. The birds of North America, No. 130. Acad. of Nat. Sci., Philadelphia, and Amer. Ornithol. Union, Washington, D.C.

Sakai, H. F. 1987. Response of Hammond's and Western flycatchers to different-aged Douglas-fir stands in northwestern California. M.S. thesis, Humboldt St. Univ., Arcata, CA.

Sallabanks, R., and F. C. James. 1999. American Robin (*Turdus migratorius*). *In* A. Poole and F. Gill, eds. The birds of North America, No. 462. Acad. of Nat. Sci., Philadelphia, and Amer. Ornithol. Union, Washington, D.C.

Sallabanks, R., B. G. Marcot, R. A. Riggs, C. A. Mehl, and E. B. Arnett. 2001. Wildlife of eastside (interior) forests and woodlands. Pp. 213-38 *in* D. H. Johnson and T. A. O'Neil, eds. Wildlife-habitat relationships in Oregon and Washington. Oreg. State Univ. Press, Corvallis.

Sanders, T. A., and R. L. Jarvis. 2000. Do Band-tailed Pigeons seek a calcium supplement at mineral sites? Condor 102:855-63.

Sanger, G. A. 1965. Observations of wildlife off the coast of Washington and Oregon in 1963, with notes on the Laysan Albatross *(Diomedea immutabilis*) in the area. Murrelet 46:1-6.

Sanger, G. A. 1970. The seasonal distribution of some seabirds off Washington and Oregon; with notes on their ecology and behavior. Condor 72:339-57.

Sanger, G. A. 1972. Checklist of bird observations from the eastern North Pacific ocean. Murrelet 53:16-21.

Sanger, G. A. 1973. Pelagic records of Glaucous-winged and Herring gulls in the North Pacific Ocean. Auk 90:384-93.

Sanger, G. A. 1974. Laysan Albatross *(Diomedea immutabilis)*. Pp. 129-153 *in* W. B. King, ed. Pelagic studies of seabirds in the central and eastern Pacific Ocean. Smithsonian Contrib. to Zool. Washington D.C. 158 pp.

Sater, D. M. 1999. Distribution and habitat associations of the Northern Pygmy-Owl in Oregon. Master's thesis. Oreg. State Univ., Corvallis.

Sauer, J. R., J. E. Hines, and J. Fallon. 2001. The North American Breeding Bird Survey, results and analysis 1966 - 2000. Version 2001.2, USGS Patuxent Wildl. Res. Ctr, Laurel, MD.

Sauer, J. R., J. E. Hines, and J. Fallon. 2003a. The North American Breeding Bird Survey, results and analysis 1966-2002. Version 2003.1, USGS Patuxent Wildl. Res. Ctr., Laurel, MD. http://www.mbr-pwrc.usgs.gov/bbs/bbs.html.

Sauer, J. R., J. E. Hines, I. Thomas, J. Fallon, and G. Gough. 1999a. The North American Breeding Bird Survey, results and analysis 1966-1998. Version 98.1. USGS Patuxent Wildl. Res. Ctr, Laurel, MD.

Sauer, J. R., J. E. Hines, I. Thomas, J. Fallon, and G. Gough. 1999b. USGS Breeding Bird Survey. Trend data: Estimating equation results. Http://www.mbr-pwrc.usgs.gov/bbs/htm96.

Sauer, J. R., J. E. Hines, I. Thomas, J. Fallon, and G. Gough. 2000. The North American Breeding Bird Survey, results and analysis 1966-1999. Version 98.1, USGS Patuxent Wildlife Research Center, Laurel, MD. Available at: http://www.mbr-pwrc.usgs.gov/bbs/bbs.html.

Sauer, J. R., J. E. Hines, G. Gough, I. Thomas, and B. G. Peterjohn. 1997. The North American Breeding Bird Survey: results and analysis. Versions 96.3 and 96.4 Patuxent Wildl. Res. Ctr, Laurel, MD.

Sauer, J. R., and W. H. Link. 2002. Using Christmas Bird Count data in analysis of population change. Amer. Birds, June.

Sauer, J. R., W. A. Link, and J. A.Royle. 2003b. Estimating population trends with a linear model: technical comments. Condor 106:435-40.

Sauer, J. R., S. Schwartz, and B. Hoover. 1996. The Christmas Bird Count homepage. Version 95.1 Patuxent Wildl. Res. Ctr, Laurel, MD.

Scheuering, E. J., and G. L. Ivey. 1995. First nesting of Great-tailed Grackle in Oregon. Wilson Bull. 107:562-63.

Schirato, G., and J. Hardin. 1998. Washington Harlequin Duck demographics. Proceedings of the fourth Harlequin Duck Symposium. Harlequin Duck Working Group, Delta, B.C.

Schreiber, R. W., and R. L. DeLong. 1969. Brown Pelican status in California. Aud. Field Notes 23:57-59.

Schroeder, M. A., C. L. Aldridge, A. D. Apa, J. R. Bohne, C. E. Braun, S. D. Bunnell, J. W. Connelly, P. A. Deibert, S. C. Gardner, M. A. Hilliard, G. D. Kobriger, S. M. McAdam, C. W. McCarthy, J. J. McCarthy, D. L. Mitchell, E. V. Rickerson and S. J. Stiver. 2004. Distribution of Sage-Grouse in North America. Condor 106:363-76.

Schroeder, M. A., J. R. Young, and C. E. Braun. 1999. Sage Grouse (*Centrocercus urophasianus*). *In* A. Poole and F. Gill, eds. The birds of North America, No. 425. Acad. of Nat. Sci., Philadelphia, and Amer. Ornithol. Union, Washington, D.C.

Schukman, J. M., and B. O. Wolf. 1998. Say's Phoebe (*Sayornis saya*). *In* A. Poole and F. Gill, eds. The birds of North America, No. 374. Acad. of Nat. Sci., Philadelphia, and Amer. Ornithol. Union, Washington, D.C.

Schultz, Z. M. 1970. The Occurrence of the Yellow-billed Loon in Washington. Murrelet 51:23.

Schultz, Z. M. 1971. Sight records of the Tufted Duck at Seattle, Washington. Murrelet 51:25.

Scott, J. M. 1971. Interbreeding of the Glaucous-winged Gull and Western Gull in the Pacific Northwest. Calif. Birds 2:129-33.

Sedgwick, J. A. 1993. Dusky Flycatcher (*Empidonax oberholseri*). *In* A. Poole and F. Gill, eds. The birds of North America, No. 78. Acad. of Nat. Sci., Philadelphia, and Amer. Ornithol. Union, Washington, D.C.

Sedgwick, J. A. 2000. Willow Flycatcher (*Empidonax traillii*). *In* A. Poole and F. Gill, eds. The birds of North America, No. 288. Acad. of Nat. Sci., Philadelphia, and Amer. Ornithol. Union, Washington, D.C.

Sekercioğlu, C. H. 2002. Impacts of birdwatching on human and avian communities. Env. Conserv. 29:282-89.

Selander, R. K. 1954. A systematic review of the booming nighthawks of western North America. Condor 56:57-82.

Semenchuk, G. P., ed. 1992. The atlas of breeding birds of Alberta. Fed. of Alberta Nat., Edmonton.

Senner, S. E., and B. J. McCaffery. 1997. Surfbird (*Aphriza virgata*). *In* A. Poole and F. Gill, eds. The birds of North America, No. 266. Acad. of Nat. Sci., Philadelphia, and Amer. Ornithol. Union, Washington, D.C.

Sharp, B. E. 1992. Neotropical migrants on national forests of the Pacific Northwest. Prepared by Ecological Perspectives, Portland, OR, for USDA Forest Service, Washington, D.C. (No. PB93-128825; available from Natl. Tech. Info. Serv., Springfield, VA.)

Sharp, D. U. 1989. Range extension of the Barred Owl in western Washington and first breeding record on the Olympic Peninsula. J. Raptor Res. 23:179-80.

Sharpe, F. A. 1993. Olympic Peninsula birds: The songbirds. Unpubl. Ms.

Sherony, D. F., and K. J. Brock. 1997. Jaeger migration on the Great Lakes. Birding 29:372-85.

Short, L. L. 1982. Woodpeckers of the world. Delaware Mus. Nat. Hist., Greenville, DE. 676 pp.

Shuford, W. D., G. W. Page, J. G. Evens, and L. E. Stenzel. 1989. Seasonal abundance of waterbirds at Point Reyes: A coastal California perspective. West. Birds 20:137-265.

Shugart, G. W., and M. Tirhi 2001. Nesting by Caspian terns in lower Puget Sound, Washington, during 1999. Northwestern Naturalist. 82:32-35.

Sibley, D. A. 2000. The Sibley guide to birds. Alfred A. Knopf, New York.

Sibley, D. A. 2003. The Sibley field guide to the birds of eastern North America. Alfred A. Knopf, New York.

Sibley, D. A., and S. N. G. Howell. 1998. Identification of White and Black-backed wagtails in basic plumage. West. Birds 29:180-98.

Siegel-Causey, D., and N. M. Litvinenko. 1993. Status, ecology, and conservation of shags and cormorants of the temperate North Pacific. Pp. 122-130 *in* K. Vermeer, K. T. Briggs, K. H. Morgan, K.H., and D. Siegel-Causey, eds. The status, ecology and conservation of marine birds of the North Pacific. Can. Wildl. Serv. Spec. Publ., Ottawa.

Skaar, P. D. 1992. Montana bird distribution, 4th ed. Self-publ.

Skriletz, J. 1996. First Washington record of "Long-billed" Marbled Murrelet. Wash. Birds 5: 53-54.

Slipp, J. W. 1942a. The tube-nosed swimmers of Puget Sound. Murrelet 23:54-59.

Slipp, J. W. 1942b. Franklin's Gull in the state of Washington. Murrelet 23:18.

Slipp, J. W. 1942c. Vagrant occurrences of *Tyrannus melancholicus* in North America. Auk 59:310-12.

Slipp, J. W. 1943. Further notes on the Franklin's Gull in the Pacific Northwest. Condor 45:38-39.

Slipp, J. 1952. A record of the Tasmanian White-capped Albatross, *Diomedea cauta cauta,* in American North Pacific waters. Auk 69:458-59.

Sloanaker, J. L. 1925. Notes from Spokane. Condor 27:73-74.

Small, A. 1994. California birds: Their status and distribution. Ibis Publ. Co., Vista, CA.

Smith, J. P. 2002. Fall raptor migration studies at Chelan Ridge, Washington: Fall 2002 season summary. Hawk Watch International, Inc., Salt Lake City, UT.

Smith, J. P., and P. Grindrod. 1998. Fall 1998 raptor migration study at Diamond Head, Washington. Hawk Watch International, Inc., Salt Lake City, UT.

Smith, K. G. 1979. Migrational movements of Blue Jays west of the 100th meridian. N. Amer. Bird Bander 4: 49-52.

Smith, M. R., P. W. Mattocks, Jr., and K. M. Cassidy. 1997. Breeding birds of Washington state. Vol. 4 *in* K. M. Cassidy, C. E. Grue, M. R. Smith and M. K. Dvornich, eds. Washington State Gap Analysis - Final Report. Seattle Aud. Soc. Publ. in Zool. No. 1. 538 pp.

Smith, P. W. 1987. The Eurasian Collared-Dove arrives in the Americas. Amer. Birds 41:1370-79.

Snell, R. R. 2002. Iceland Gull (*Larus glaucoides*), Thayer's Gull (*Larus thayeri*). *In* A. Poole and F. Gill, eds. The birds of North America, No. 699. Acad. of Nat. Sci., Philadelphia, and Amer. Ornithol. Union, Washington, D.C.

Snow, C. 1973. Golden Eagle. USDI Bureau of Land Management, Denver, CO. Habitat Manage. Series for Unique or Endangered Species Rep. No. 7.

Snyder, D. E. 1961. First record of the Least Frigatebird (*Fregata ariel*) in North America. Auk 78:265.

Sodhi, N. S., L. W. Oliphant, P. C. James, and I. G. Warkentin. 1993. Merlin (*Falco columbarius*). *In* A. Poole and F. Gill, eds. The birds of North America, No. 44. Acad. of Nat. Sci., Philadelphia, and Amer. Ornithol. Union, Washington, D.C.

Spear, L. B., M. J. Lewis, M. T. Myers, and R. L. Pyle. 1988. The recent occurrence of Garganey in North America and the Hawaiian Islands. Amer. Birds 42:385-92.

Speich, S. M., and S. P. Thompson. 1987. Impacts on waterbirds from the 1984 Columbia River and Whidby Island, Washington, oil spills. West. Birds18:109-16.

Speich, S. M., and T. R. Wahl. 1989. Catalog of Washington seabird colonies. U.S. Dept. Interior, Fish and Wildl. Serv., Biol. Rept. 88(6). MMS 89-0054.

Speich, S. M., and T. R. Wahl. 1995. Marbled Murrelet populations of Washington - marine habitat preferences and variability of occurrence. Pp. 313-26 *in* Ralph et al., eds. 1994. Marbled Murrelet conservation assessment. U.S. Dep. Agriculture, Forest Serv., Pacific Southwest For. Experiment Sta., Gen. Tech. Rep.

Speich, S. M., T. R. Wahl, and D. A. Manuwal. 1992. The numbers of Marbled Murrelets in Washington marine waters. *In* H. R. Carter and M. L. Morrison, eds. Status and conservation of the Marbled Murrelet in North America. Proc. Western Found. Vertebrate Zool. 5(1):48-60.

Spies, T. A., and S. P. Cline. 1988. Coarse woody debris in forests and plantations of coastal Oregon. Pp. 4-24 in C. Maser, R. F. Tarrant, J. M. Trappe, and J. F. Franklin, eds. From the forest to the sea: A story of fallen trees. U.S. Dept. Agriculture For. Serv. Genl. Tech. Rep. PNW-GTR-229.

Spitmann, P. 1996. Blackburnian Warbler sighting—Fort Nelson area, British Columbia. Birders J. 5:24-26.

Stein, R. C., and M. C. Michener. 1961. Least Flycatchers in northwestern Washington and central British Columbia. Condor 63:181-82.

Stepniewski, A. 1996. Boreal Owls found nesting in Washington in 1992. Wash. Birds 5:55-60.

Stepniewski, A. 1999. The birds of Yakima County. A. Stepniewski. 278 pp.

Stepniewski, A., and K. Woodruff. 1997. Great Gray Owls breeding in Washington. Wash. Birds 6:83-87.

Sterling, J. C. 1999. Gray Flycatcher (*Empidonax wrightii*). *In* A. Poole and F. Gill, eds. The birds of North America, No. 458. Acad. of Nat. Sci., Philadelphia, and Amer. Ornithol. Union, Washington, D.C.

Stewart, A. C., R. W. Campbell, and S. Dickin. 1996. Use of dawn vocalizations for detecting breeding Cooper's Hawks in an urban environment. Wildl. Soc. Bull. 24(2):291-93.

Storer, R. W., and G. L. Nuechterlein. 1992. Western Grebe (*Aechmophorus occidentalis*), Clark's Grebe (*Aechmophorus clarkii*). *In* A. Poole and F. Gill, eds. The birds of North America. No. 26. Acad. of Nat. Sci., Philadelphia, and Amer. Ornithol. Union, Washington, D.C.

Stouffer, P. C., and R. T. Chesser. 1998. Tropical Kingbird (*Tyrannus melancholicus*). *In* A. Poole and F. Gill, eds. The birds of North America, No. 358. Acad. of Nat. Sci., Philadelphia, and Amer. Ornithol. Union, Washington, D.C.

Suckley, M. D., and J. G. Cooper. 1860. The natural history of Washington Territory and Oregon. Balliere Bros. New York.

Summers, K., and M. Gebauer. 1995. Status of the Vaux's Swift in British Columbia. B. C. Environ. Wildl. Working Rep. No. WR-67.

Sustainable Ecosystems Institute. 1997. Seabird surveys in Puget Sound 1996. Rep. to Northwest Indian Fish. Comm. 36 p.

Suydam, R. S., D. L. Dickson, J. B. Fadely, and L. T. Quackenbush. 2000. Population declines of King and Common eiders of the Beaufort Sea. Condor 102: 219-22.

Swarth, H., and A. Brooks. 1925. The Timberline Sparrow. A new species from northwestern Canada. Condor 27:27-69.

Swisher, O. D. 1978. Poor-wills nesting in southwestern Oregon. N. Amer. Bird Bander 3:152-55.

Sydeman, W. J., E. R. McLaren, and P. Pyle. 1994. Oral presentation: ENSO 1992 and ENSO 1983: biological consequences and seabird population regulation in the California Current. Pacific Seabirds 21:50.

Tacha, T. C., S. A. Nesbitt, and P. A. Vohs. 1992. Sandhill Crane (*Grus canadensis*). *In* A. Poole and F. Gill, eds. The birds of North America. No. 31. Acad. of Nat. Sci., Philadelphia, and Amer. Ornithol. Union, Washington, D.C.

Tacha, T. C., S. A. Nesbitt, and P. A. Vohs. 1994. Sandhill Crane. In T.C. Tacha and eds. Migratory shore and upland game bird management in North America. International Assoc. of Fish and Wildl. Agencies, Washington, D.C.

Takekawa, J. 1992. Regional reports, Washington/ British Columbia. PSG Bull. 19:18.

Tate, J., Jr. 1986. The blue list for 1986. Amer. Birds 40:227-35.

Tate, J., and D. J. Tate. 1982. The blue list for 1982. Amer. Birds 36:126-35.

Taverner, P. A. 1936 Taxonomic comments on Red-tailed Hawks. Condor 38:66-71.

Taylor, B. 1998. Rails: A guide to the rails, crakes,gallinulcs, and coots of the world. Yale Univ. Press, New Haven, CT.

Telfair, R. C. II. 1994. Cattle Egret (*Bubulcus ibis*). *In* A. Poole and F. Gill, eds. The birds of North America, No. 113. Acad. of Nat. Sci., Philadelphia, and Amer. Ornithol. Union, Washington, D.C.

Terres, J. K. 1980. The Audubon Society encyclopedia of North American birds. Alfred Knopf, N.Y.

Terres. K. L. 1991. The Audubon Society Encyclopedia of North American Birds. Wings Books, New York.

Thompson, B. C. 1977c. Behavior of Vaux's Swifts nesting and roosting in a chimney. Murrelet 58:73-77.

Thompson, C. W. 1997a. Distribution and abundance of Marbled Murrelets and Common Murres on the outer coast of Washington - 1997 completion report to the Tenyo Maru Trustee's Council. Wash. Dept. Fish and Wildl. 91 pp.

Thompson, C. W. 1997b. Distribution and abundance of Marbled Murrelets on the outer coast of Washington, Winter 1996-1997. Wash. Dept. Fish and Wildl. 16 pp.

Thompson, C. W., E. R. Donelan, M. M. Lance, and A. E. Edwards. 2002. Diet of Caspian Terns in Commencement Bay, Washington. Waterbirds 25:78-85.

Thompson, S. P., and D. K. McDermond. 1983. Summary of recent Northern Harrier nesting in western Washington. Murrelet 66: 82-84.

Thompson, W. L. 2002. Towards reliable bird surveys: accounting for individuals present but not detected. Auk 119:18-25.

Thompson-Hanson, P. A. 1984. Nesting ecology of Northern Harriers on the Hanford Site, south-central Washington. M.S. Thesis. Wash. State Univ., Pullman. 102 pp.

Thoresen, A. C. 1981. Midsummer occurrence of the Horned Puffin in Rosario Strait, Washington. West. Birds 12:56.

Titman, R. D. 1999. Red-breasted Merganser (*Mergus serrator*). *In* A. Poole and F. Gill, eds. The birds of North America, No. 443. Acad. of Nat. Sci., Philadelphia, and Amer. Ornithol. Union, Washington, D.C.

Tobalske, B. W. 1997. Lewis' Woodpecker (*Melanerpes lewis*). *In* A. Poole and F. Gill, eds. The birds of North America, No. 284. Acad. of Nat. Sci., Philadelphia, and Amer. Ornithol. Union, Washington, D.C.

Tomback, D. F. 1998. Clark's Nutcracker (*Nucifraga columbiana*). *In* A. Poole and F. Gill, eds. The birds of North America, No. 331. Acad. of Nat. Sci., Philadelphia, and Amer. Ornithol. Union, Washington, D.C.

Tomkovich, P. S. 1992. An analysis of the geographic variability in Knots *Calidris canutus* based on museum skins. Wader Study Group Bull. 64, Suppl.: 17-23.

Toochin, R. 1998. Seasonal status of the Birds of Vancouver, B.C. checklist area. Vancouver Nat. Hist. Soc., Vancouver, B.C.

Trost, C. 1991. A Fork-tailed Flycatcher in Idaho! Idaho Wildl. 45:24.

Trost, C. H. 2000. Black-billed Magpie (*Pica pica*). *In* A. Poole and F. Gill, eds. The birds of North America, No. 389. Acad. of Nat. Sci., Philadelphia, and Amer. Ornithol. Union, Washington, D.C.

Trost, R. E. 1998. 1998 Pacific Flyway Data Book. Unpublished report. USFWS-OMBM, Portland, OR. 126 pp.

Tweit, B., and B. Flores. White-faced Ibis in Washington in 2001: A significant incursion and attempted breeding. Unpubl.

Tweit, B., and D. R. Paulson 1994. First report of the Washington Bird Records Committee. Wash. Birds 3:11-41.

Tweit, B., and J. Skriletz.1996. Second report of the Washington Bird Records Committee. Wash. Birds 5:6-28.

U.S. Department of Agriculture. 2000. Environmental assessment of bird damage management in the state of Washington.

U.S. Department of Agriculture. 2001. Final environmental assessment and finding of no significant impact and decision for alternative strategies for the management of damage caused by migratory birds in the State of Washington. USDA Animal and Plant Health Inspection Serv., Wildl. Serv. 36 pp.

U.S. Department of Agriculture. 2002. Management of damage caused by migratory birds. Monitoring rep., USDA Animal and Plant Health Inspection Serv., Wildl. Serv. 11 pp.

U.S. Department of Agriculture. 2003. Piscivorous bird damage management for the protection of juvenile salmonids on the mid-Columbia River. USDA Animal and Plant Health Inspection Serv., Wildl. Serv. 153 pp.

U.S. Department of Agriculture and U.S. Department of Interior. 1994. Record of decision for amendments to Forest Service and Bureau of Land Management planning documents within the range of the northern spotted owl. USDA For. Serv. and USDI Bureau of Land Manage., Portland, OR.

U.S. Department of Interior. 1992. Final draft recovery plan for the Northern Spotted Owl, 2 vols. USDI, Washington, D.C.

U.S. Fish and Wildlife Service. 1985. Guidelines for the management of the Purple Martin, Pacific coast population. Portland, OR.

U.S. Fish and Wildlife Service. 1995. 12-month finding for a petition to list the Queen Charlotte Goshawk as endangered. Fed. Register 60:33784-33785.

U.S. Fish and Wildlife Service. 1997a. 90-day finding for a petition to list the Northern Goshawk in the contiguous United States west of the 100th meridian. Fed. Register 62:50892-50896.

U.S. Fish and Wildlife Service (USFWS). 1997b. Recovery plan for the Marbled Murrelet (*Brachyramphus marmoratus*) in Washington, Oregon and California. U.S. Fish and Wildl. Serv., Portland, OR.

U.S. Fish and Wildlife Service. 1999. An American success story. http://www.fws.gov/r3pao/eagle/success/index.html.

U.S. Fish and Wildlife Service (USFWS). 2000a. Pacific Flyway data book. U.S. Fish and Wildl. Serv. Portland, OR.

U.S. Fish and Wildlife Service (USFWS). 2000b. Pacific Flyway midwinter waterfowl survey. U.S. Fish and Wildl. Serv. Portland, OR.

U.S. Fish and Wildlife Service (USFWS). 2000c. Waterfowl population status, 2000. U.S. Dept. Interior, Washington, DC.

U.S. Fish and Wildlife Service (USFWS)/Canadian Wildlife Service. 1996. Waterfowl population status 1996. U.S. Fish and Wildl. Serv., Washington DC.

Van Der Geld, A. 1997. Pacific Northwest. HMANA [Hawk Migration Assoc. of N. Amer.] Hawk Migration Studies 23 (1):68-70.

Van Der Geld, A. 1998a. Pacific Northwest. HMANA [Hawk Migration Assoc. of N. Amer.] Hawk Migration Studies 23 (2):5-6.

Van Der Geld, A. 1998b. Pacific Northwest. HMANA [Hawk Migration Assoc. of N. Amer.] Hawk Migration Studies 24:18-20.

Van Pelt, T. I., and J. F. Piatt. 1995. Oral presentation: A wreck of Common Murres in the Gulf of Alaska during early 1993, and methods used to estimate total mortality. Pacific Seabirds 22:46.

Van Velzen, A. C. 1973. Seasonal fluctuations of sandpipers in western Washington. Murrelet 54:1-3.

Vander Haegen, W. M., F. C. Dobler, and D. J. Pierce. 2000. Shrubsteppe bird response to habitat and landscape variables in eastern Washington, USA. Conserv. Biol. 14:1145-60.

Vander Haegen, W. M., S. M. McCorquodale, C. R. Peterson, G. A. Green, and E. Yenson. 2001. Wildlife of eastside shrubland and grassland habitats. Pp. 292-316 *in* D. H. Johnson and T. A. O'Neil, eds. Wildlife-habitat relationships in Oregon and Washington. Oreg. State Univ. Press, Corvallis.

Vander Haegen, W. M., M. A. Schroeder, and R. M. DeGraaf. 2002. Predation on real and artificial nests in shrubsteppe landscapes fragmented by agriculture. Condor 104:496-506.

Veit, R. R., J. A. McGowan, D. G. Ainley, T. R. Wahl, and P. Pyle. 1997. Apex marine predator declines ninety percent in association with changing oceanic climate. Global Change Biol. 3:23-28.

Verbeek, N. A. M., and R. W. Butler. 1999. Northwestern Crow (*Corvus caurinus*). *In* A. Poole and F. Gill, eds. The birds of North America, No. 407. Acad. of Nat. Sci., Philadelphia, and Amer. Ornithol. Union, Washington, D.C.

Verbeek, N. A. M., and C. Caffrey. 2002. American Crow (*Corus brachyhynchus*). *In* A. Poole and F. Gill, eds. The birds of North America, No. 647. Acad. of Nat. Sci., Philadelphia, and Amer. Ornithol. Union, Washington, D.C.

Verbeek, N. A. M., and P. Hendricks. 1994. American Pipit (*Anthus rubescens*). *In* A. Poole and F. Gill, eds. The birds of North America, No. 95. Acad. of Nat. Sci., Philadelphia, and Amer. Ornithol. Union, Washington, D.C.

Vermeer, K., R. Hay, and L. Rankin. 1987. Pelagic seabird populations off southwestern Vancouver Island. Can. Tech. Rep. Hydrogr. and Ocean Sci. No. 87. Inst. Ocean Sci. Dept. Fish. and Oceans. Sidney, B.C. 26 pp.

Vermeer, K., D. B. Irons, E. Velarde, and Y. Watanuki. 1993. Status, conservation and management of nesting *Larus* gulls in the North Pacific. Pp. 131-39 *in* K. Vermeer, K. T. Briggs, K. H. Morgan, and D. Siegel-Causey, eds. The status, ecology and conservation of marine birds of the North Pacific. Can. Wildl. Serv. Spec. Publ., Ottawa.

Vermeer, K., and C. D. Levings. 1977. Populations, biomass and food habits of ducks on the Fraser Delta intertidal area, British Columbia. Wildfowl 28: 39-60.

Vermeer, K., and K. H. Morgan, eds. 1997. The ecology, status and conservation of marine and shoreline birds of the Queen Charlotte Islands. Can. Wildl. Serv. Occas. Pap. No. 93, Ottawa.

Vermeer, K., K. H. Morgan, G. E. J. Smith, and R. Hay. 1989. Fall distribution of pelagic birds over the shelf off southwestern Vancouver Island. Colonial Waterbirds 12:207-14.

Vermeer, K., and L. Rankin. 1984. Population trends in nesting Double-crested and Pelagic cormorants in Canada. Murrelet 65:1-9.

Verner, J. 1974. Notes on the current status of some birds of central Washington. Murrelet 55:19-22.

Vogel, C. A., and K. P. Reese. 1995. Habitat conservation assessment for mountain quail (*Oreoryx pictus*). Unpubl Rep. Idaho Dept. Fish and Game, Boise.

Voous, K. H. 1988. Owls of the Northern Hemisphere. MIT Press, Cambridge, MA.

Vrooman, A. G. 1901. Discovery of the egg of the black swift (Cypseloides niger borealis). Auk 18:394-95.

Wagenknecht, J. F., S. P. Finn, and L. Stream. 1998. Northern Goshawk breeding ecology in the Upper Yakima River basin, 1994 to 1996. Unpubl. Rep., Wash. Dept. Fish and Wildl., Region 3, Yakima. 33 p.

Wahl, T. R. 1970. A Short-tailed Albatross record for Washington state. Calif. Birds 1:113-15.

Wahl, T. R. 1973. A Bar-tailed Godwit record for Washington. West. Birds 4:89-90.

Wahl, T. R. 1975. Seabirds in Washington's offshore zone. West. Birds 6:117-34.

Wahl, T. R. 1977. Notes on behavior of California Gulls and South Polar Skuas off the Washington coast. Murrelet 58:47-49.

Wahl, T. R. 1982. Identification of Sooty and Short-tailed shearwaters in the north Pacific Ocean. Sea Swallow 31:42-44.

Wahl, T. R. 1984. Distribution and abundance of seabirds over the continental shelf off Washington (including information on selected marine mammals). Wash. State Dept. Ecology.

Wahl, T. R. 1985. The distribution of Buller's Shearwater (*Puffinus bulleri*) in the North Pacific Ocean. Notornis 32:109-17.

Wahl, T. R. 1986. Notes of the feeding behavior of Buller's Shearwater. West. Birds. 17:45-47.

Wahl, T. R. 1995. Birds of Whatcom County. T. R. Wahl. Bellingham. 184 pp.

Wahl, T. R. 1996. Waterbirds in Washington's inland marine waters: Some high counts from systematic censusing. Wash. Birds.5: 29-50.

Wahl, T. R. 2002. Trends in numbers of marine birds wintering on Bellingham Bay. Wash. Birds 8:29-40.

Wahl, T. R., D. G. Ainley, A. H. Benedict, and A. R. DeGange. 1989. Associations between seabirds and water masses in the northern Pacific Ocean in summer. Marine Biol. 103:1-17.

Wahl, T. R., and D. Heinemann. 1979. Seabirds and fishing vessels: co-occurrence and attraction. Condor 81:390-96.

Wahl, T. R., K. H. Morgan, and K. Vermeer. 1993. Seabird distribution off British Columbia and Washington. Pp. 39-47 *in* K. Vermeer, K. T. Briggs, K. H. Morgan, and D. Siegel-Causey, eds. The status, ecology and conservation of marine birds of the North Pacific. Can. Wildl. Serv. Spec. Publ., Ottawa.

Wahl, T. R., and D. R. Paulson. 1991. A guide to bird finding in Washington. T. R. Wahl, Bellingham, WA. 177 pp.

Wahl, T. R., and S. M. Speich. 1983. First winter survey of marine birds in Puget Sound and Hood Canal, December 1982 and February 1983. Report for the Washington Department of Game Nongame Wildl. Progr. 34 pp.

Wahl, T. R., and S. M. Speich. 1984. Survey of marine birds in Puget Sound, Hood Canal and waters east of Whidbey Island, Washington, in summer 1982. West. Birds 15:1-14.

Wahl, T. R., and S. M. Speich. 1994. Distribution of foraging Rhinoceros Auklets in the Strait of Juan de Fuca, Washington. Northwest. Naturalist 75:63-69.

Wahl, T. R., S. M. Speich, D. A. Mauwal, K. V. Hirsch, and C. Miller. 1981. Marine bird populations of the Strait of Juan de Fuca, Strait of Georgia and adjacent waters in 1978 and 1979. U.S. Environmental Protection Agency, DOC/EPA interagency Energy/Environment R&D Progr. Rep. EPA/600/f-81/156. 789 pp.

Wahl, T. R., and B. Tweit. 2000a. Seabird abundances off Washington, 1972-1998. West. Birds 31:69-88.

Wahl, T. R., and B. Tweit. 2000b. Where do Pigeon Guillemots from California go for the winter? West. Birds 31:203-6.

Wahl, T. R., and H. E. Wilson. 1971. Nesting record of European Skylark in Washington State. Condor 73:254.

Walsh, R. G. 1986. Recreation economic decisions: comparing benefits and costs. Venture Publishing, State College, PA.

Ward, L. Z., D. K. Ward, and T. J. Tibbitts. 1992. Canopy density analysis at goshawk nesting territories on the North Kaibab Ranger District, Kaibab National Forest. P.O. #43-8156-0-0487. Arizona Game and Fish Dept., Phoenix, AZ.

Warheit, K. 1997. Draft report, Common Murre. Wash. Dept. Fish and Wildl. 12 pp.

Warheit, K. I., and C. Thompson. 2002. Common Murre (*Uria aalge*). *In* E.M. Larson, J. M. Azerrad, and N. Nordstrom, eds. Managment recommendations for Washington's Priority Species, Vol. IV: Birds. Wash. Dept. of Fish and Wildl. Olympia, WA.

Watson, J. W., D. W. Hays, and D. J. Pierce. 1999. Efficacy of Northern Goshawk broadcast surveys in Washington state. J. Wildl. Manage. 63:98-106.

Watson, J. W., and D. J. Pierce. 2000. Migration and winter ranges of Ferruginous Hawk from Washington. Annual Rep.. Wash. Dept. Fish and Wildl., Olympia.

Washington Department of Fish and Wildlife. 1979. Game birds of Washington. Washington Department of Game. Olympia. 32 pp. Library of Congress # wln93668214 W 061.

Washington Department of Fish and Wildlife. 1995a. Washington state management plan for Sage Grouse. Game Div., Wash. Dept. Fish and Wildl., Olympia. 101 pp.

Washington Department of Fish and Wildlife. 1995b. Washington State management plan for Columbian Sharp-tailed Grouse. Game Div., Wash. Dept. Fish and Wildl., Olympia. 99 pp.

Washington Department of Fish and Wildlife. 1995c. Washington State recovery plan for the Snowy Plover. Wash. Dept. Fish and Wildl., Olympia.

Washington Department of Fish and Wildlife. 1996. Washington State recovery plan for the Ferruginous Hawk. Washington Dept. Fish and Wildl., Olympia.

Washington Department of Fish and Wildlife. 1997a. Status report, White Pelican *Pelecanus erythrorhynchos* in Washington State. Wash. Dept. Fish and Wildl. Olympia.

Washington Department of Fish and Wildlife. 1997b. Status report, Osprey *Pandion haliaetus* in Washington State. Wash. Dept. Fish and Wildl. Olympia.

Washington Department of Fish and Wildlife. 1997c. Management recommendations: Band-tailed Pigeon *Columba fasciata*. Wash. Dept. Fish and Wildl. Olympia.

Washington Department of Fish and Wildlife. 1997d. Status report, Loggerhead Shrike *Lanius ludovicianus*. Wash. Dept. Fish and Wildl. Olympia.

Washington Department of Fish and Wildlife. 1998. Snow Goose management plan. Unpubl. rep. Wash. Dept. Fish and Wildl., Waterfowl Section, Olympia.

Washington Department of Fish and Wildlife. 1999. Management recommendations: American White Pelican *Pelecanus erythrorhynchos*. Wash. Dept. Fish and Wildl. Olympia.

Washington Department of Fish and Wildlife. 2000a. 1961-2000 Breeding waterfowl survey. Wash. Dept. Fish and Wildl. Olympia.

Washington Department of Fish and Wildlife. 2000b. 2000 Game status and trend. Wildl. Prog. Wash. Dept Fish and Wildl., Olympia. 120 pp.

Washington Department of Fish and Wildlife. 2001. 2000 Game harvest report. Wildl. Prog. Wash. Dept Fish and Wildl., Olympia. 120 pp.

Washington Department of Fish and Wildlife. 2003. Status report, Northern Goshawk *Accipiter gentilis*. Wash. Dept. Fish and Wildl. Olympia.

Washington Department of Natural Resources. 1997. Final habitat conservation plan. Wash. Dept. Nat. Res. Olympia.

Washington Department of Natural Resources. 1998. Our changing nature: Natural resource trends in Washington State. 75 pp. Wash. Dept. Nat. Res. Olympia.

Washington Department of Ecology. 2000. Neah Bay rescue tug: Report to the Washington state legislature. Publication 00-08-023. Wash. Dept. Ecol., Olympia.

Weber, J. W. 1977. Blue Jay influx into Washington during 1976-77 winter. Murrelet 58: 84-86.

Weber, J. W. 1981. Status of the Semipalmated Sandpiper in Washington and Northern Idaho. Continental Birdlife 2:150-53.

Weber, J. W. 1985. First specimen record of the Short-billed Dowitcher from eastern Washington; subspecific identification of Idaho specimens. Murrelet 66:31-34.

Weber, J. W., and R. F. Fitzner. 1986. Nesting of the Glaucous-winged Gull east of the Washington Cascades. Amer. Birds 40:567-69.

Weber, J. W., and E. J. Larrison. 1977. Birds of southeastern Washington. Univ. Press of Idaho, Moscow.

Weber, W. C., and E. S. Hunn. 1978. First record of the Little Blue Heron for British Columbia and Washington. West. Birds 9: 33-34.

Weber, W. C., and J. Ireland. 1992. Tidewater breeding records of the Western Grebe near Vancouver, British Columbia. West. Birds 23:33-34.

Wehtje, W. 2001. Range expansion of the Great-tailed Grackle in western North America. West. Birds 32:141-43.

Weisberg, S. 1983. Winter behavior and ecology of the surbird (*Aphriza virgata*) in northern Puget Sound. M.S. thesis, Western Wash. Univ.

Weisbrod, A. R., and W. F. Stevens. 1974. The Skylark in Washington. Auk 91:832-35.

Whelton, B. D. 1989. Distribution of the Boreal Owl in eastern Washington and Oregon. Condor 91:712-16.

Whittow, G. C. 1993. Laysan Albatross (*Diomedea immutabilis*). *In* A. Poole and F. Gill, eds. The birds of North America, No. 66. Acad. of Nat. Sci., Philadelphia, and Amer. Ornithol. Union, Washington, D.C.

Wick, W. Q. 1958. A nine year bird list from Eliza and Protection islands, Washington. Murrelet 39:1-9.

Widén, P. 1997. How, and why, is the goshawk (*Accipiter gentilis*) affected by modern forest management in Fennoscandia? J. Raptor Res. 31(2):107-13.

Widrig, R. S. 1979. The shorebirds of Leadbetter Point. R. S. Widrig, Ocean Park, WA. 63 pp.

Widrig, R. S. 1983. A Bristle-thighed Curlew at Leadbetter Point, Washington. West. Birds 14: 203-4.

Widrlechner, M. P., and S. K. Dragula. 1984. Relationship of cone-crop size to irruptions of four seed-eating birds in California. Amer. Birds 38:840-46.

Wiens, J. A. 1996. Oil, seabirds, and science: the effects of the *Exxon Valdez* oil spill. Bioscience 46:587-97.

Wilbur, S. R. 1973. The California Condor in the Pacific Northwest. Auk 90: 196-98.

Wilcove, D. S., and J. Lee. 2004. Using economic and regulatory incentives to restore endangered species: lessons learned from three new programs. Conserv. Biol. 18:639-45.

Wiley, H. R., and D. S. Lee. 2000. Pomarine Jaeger (*Stercorarius pomarinus*). *In* A. Poole and F. Gill, eds. The birds of North America, No. 483. Acad. of Nat. Sci., Philadelphia, and Amer. Ornithol. Union, Washington, D.C.

Williams, B. D. C. 1998. Distribution, habitat associations, and conservation of Purple Martins breeding in California. California State Univ., Sacramento.

Williams, J. M. 1996. Nashville Warbler (*Vermivora ruficapilla*). *In* A. Poole and F. Gill, eds. The birds of North America, No. 205. Acad. of Nat. Sci., Philadelphia, and Amer. Ornithol. Union, Washington, D.C.

Wilson, U. W. 1991. Response of three seabird species to El Niño events and other warm episodes on the Washington coast, 1979-1990. Condor 93:853-58.

Wilson, U. W. 2003. 2002 common murre colony surveys with management recommendations for Point Grenville. Final report submitted to the *Tenyo Maru* Trustee Committee, U. S. Fish and Wildl. Service, Sequim, WA.

Wilson, U., A. McMillan, and F. Dobler. 2000. Nesting, population trend and breeding success of Peregrine Falcons on the Washington Outer Coast, 1980-98. J. Rapt. Res. 34(2):67-74.

Wing, L. 1943. The Starling in eastern Washington. Condor 45:159.

Wing, L. 1949. Breeding birds of virgin Palouse Prairie. Auk 66:38-41.

Wing, L. 1957. Summer bird studies in the Okanogan Cascades. Murrelet 25:3-8.

Winkler, H. D., D. A. Christie, and D. Nurney. 1995. Woodpeckers: A guide to the woodpeckers of the world. Houghton Mifflin Co., Boston, MA.

Winter, J. 1986. Status, distribution, and ecology of the Great Gray Owl (*Strix nebulosa)* in California. M.S. Thesis. San Francisco State Univ., San Francisco.

Witmer, G. W., and J. C. Lewis. 2001. Introduced wildlife of Oregon and Washington. Pp. 423-43 in D. H. Johnson and T. A. O'Neil, eds. Wildlife-habitat relationships in Oregon and Washington. Oreg. State Univ. Press, Corvallis.

Witmer, M. C., D. J. Mountjoy, and L. Elliot. 1997. Cedar Waxwing (*Bombycilla cedrorum*). *In* A. Poole and F. Gill, eds. The birds of North America, No. 309. Acad. of Nat. Sci., Philadelphia, and Amer. Ornithol. Union, Washington, D.C.

Woodbridge, B. 1997. Tracking the migration of Swainson's Hawks: Conservation lessons in a global classroom. HMANA Conference VIII, Snowbird, Utah.

Woodbridge, B., and P. J. Detrich. 1994. Territory occupancy and habitat patch size of Northern Goshawks in the southern Cascades of California. Studies in Avian Biol. 16:83-87.

Woodby, D. A. 1976. Winter owl records for eastern Washington. Murrelet 57:16-17.

Woodruff, R. D. 1995. Purple Martins in the Inland Northwest. Purple Martin Update 6(1): 24-26. Purple Martin Conserv. Assoc., Edinboro, PA.

Wootton, J. T. 1996. Purple Finch *(Carpodacus purpureus*). *In* A. Poole and F. Gill, eds. The birds of North America, No. 208. Acad. of Nat. Sci., Philadelphia, and Amer. Ornithol. Union, Washington, D.C.

Wildlife Resources Data Systems. 2001. Unpubl. Database. Wash. Dept. Fish and Wildl., Olympia.

Wright, A. L., and G. D. Hayward. 1998. Barred Owl range expansion into the central Idaho wilderness. J. Raptor Res. 32:77-81.

Wright, S. K., D. D. Roby, and R. G. Anthony. 2002. California Brown Pelicans nesting in the Pacific Northwest?: Potential for a major northward expansion in breeding range. Abstr. Pacific Seabirds 29:71-72.

Yaich, J. A., and E. J. Larrison. 1973. Nesting record and behavioral observations on the Gray Flycatcher in Washington. Murrelet 54:14-16.

Yasukawa, K, and W. A. Searcy. 1995. Red-winged Blackbird (*Agelaius phoeniceus*). *In* A. Poole and F. Gill, eds. The birds of North America, No. 184. Acad. of Nat. Sci., Philadelphia, and Amer. Ornithol. Union, Washington, D.C.

Yocom, C. F. 1949. Sight records of Loons in Southeast Washington. Murrelet 30:20.

Yocom, C. F. 1950. Pacific Loon found near Ellensburg, Washington. Murrelet 31:47.

Yocum, C. F. 1956. Re-establishment of breeding populations of Long-billed Curlews in Washington. Wilson Bull. 68:228-231.

Yocom, C. F. 1966. Western White-throated Swift nesting under Spanish-type tile roof edge. Murrelet 47:20-21.

Zarnowitz, J. E., and D. A. Manuwal. 1985. The effects of forest management on cavity-nesting birds in northwestern Washington. J. Wildl. Manag. 49:255-63.

Zimmerman, D. A. 1973. Range expansion of Anna's Hummingbird. Amer. Birds 27:827-35.

Zink, R. M. 1986. Patterns and evolutionary significance of geographic variation in the *Schistacea* group of the Fox Sparrow (*Passeralla iliaca*). Ornithol. Monogr., v.40. Amer. Ornithol. Union, Washington, D.C.

Zink, R. M. 1994. The geography of mitochondrial DNA variation, population structure, hybridization, and species limits in Fox Sparrow (*Passerella iliaca*). Evolution 48:96-111.

Zink, R. M., and A. E. Kessen. 1999. Species limits in the Fox Sparrow. Birding 31:508-17.

Appendix

Regularly occurring species' habitat associations, derived from Johnson and O'Neil (2001). Established introduced species and uncommon migrants included; species occurring predominantly only in Oceanic and Outer Shelf habitats not included.

Habitats

1 Westside Lowland Conifer-Hardwood Forest
2 Westside Oak and Dry Douglas-fir Forest & Woodlands
4 Montane Mixed Conifer Forest
5 Eastside (Interior) Mixed Conifer Forest
6 Lodgepole Pine Forest & Woodlands
7 Ponderosa Pine and Eastside White Oak Forest & Woodlands
8 Upland Aspen Forest
9 Subalpine Parkland
10 Alpine Grasslands and Shrublands
11 Westside Grasslands
13 Westside Juniper and Mountain Mahogany Woodlands
14 Eastside (Interior) Canyon Shrublands
15 Eastside (Interior) Grasslands
16 Shrub-steppe
17 Dwarf Shrub-steppe
18 Desert Playa and Salt Scrub
190 Agriculture, Pasture, and Mixed Environs (westside)
191 Agriculture, Pasture, and Mixed Environs (eastside)
200 Urban and Mixed Environs (westside)
201 Urban and Mixed Environs (eastside)
21 Lakes, Rivers, Ponds, and Reservoirs
22 Herbaceous Wetlands
23 Westside Riparian - Wetlands
24 Montane Coniferous Wetlands
25 Eastside (Interior) Riparian Wetlands
26 Coastal Dunes and Beaches
27 Coastal Headlands and Islets
28 Bays and Estuaries
29 Inland Marine Deeper Waters
30 Marine Nearshore

Habitat use

B - feeds and reproduces
R - reproduces
F - feeds
O - other

Habitat	*1*	*2*	*4*	*5*	*6*	*7*	*8*	*9*	*10*	*11*	*13*	*14*	*15*	*16*	*17*	*18*	*190*	*191*	*200*	*201*	*21*	*22*	*23*	*24*	*25*	*26*	*27*	*28*	*29*	*30*
Greater White-fronted Goose																					F	F						F		
Snow Goose																	F				F	F						F		
Ross's Goose																					F	F						F		
Brant																												F	O	F
Canada Goose													F	B	B						F	B						F		
Trumpeter Swan																					F	B						F		
Tundra Swan																	F	F			F	F						F		
Wood Duck	R																F	F			F		B		B					
Gadwall										R			R				B	B			F	B						F		
Eurasian Wigeon																					F	F						F		
American Wigeon													R					B			F	B	F					F		F
Mallard													R	R			B	B	B	B	F	B	B		B			F		F
Blue-winged Teal																	B	B			F	B								
Cinnamon Teal																	F	F			F	B								
Northern Shoveler										R			R				R	R			F	B								
Northern Pintail													R					F			F	B						F		F
Green-winged Teal										R			R				B	B			F	B	F		F			F		
Canvasback																					F	B						F	F	F
Redhead																					F	B						F	F	F
Ring-necked Duck																					F	B	B		B					
Greater Scaup																					F							F		F

Habitat	*1*	*2*	*4*	*5*	*6*	*7*	*8*	*9*	*10*	*11*	*13*	*14*	*15*	*16*	*17*	*18*	*190*	*191*	*200*	*201*	*21*	*22*	*23*	*24*	*25*	*26*	*27*	*28*	*29*	*30*
Lesser Scaup																					F	B						F		
Harlequin Duck																					F		B		B			F		
Surf Scoter																												F	F	F
White-winged Scoter																												F	F	F
Black Scoter																												F	F	F
Long-tailed Duck																													F	F
Bufflehead			R																		F	F	B	B	B			F		F
Common Goldeneye				R																	F							F		F
Barrow's Goldeneye			R						F												F	F		B				F		F
Hooded Merganser	R																				F	F	B		B			F		
Common Merganser	R																				F		B		B			F		
Red-breasted Merganser																												F		F
Ruddy Duck																					F	B						F		
Chukar												B	F	F																
Gray Partridge						F						F	B	B				B												
Ring-necked Pheasant										B		F	B	B				B	B	B	B		B		B	B				
Ruffed Grouse	B	B		F		B	B	F									B	B					B	B	B					
Greater Sage Grouse													B	B	B	B		B												
Spruce Grouse				B	B			B	B															B						
White-tailed Ptarmigan								B	B																					
Blue Grouse	B	B	B	B	B	B	B	B						B										B	B					

Habitat	1	2	4	5	6	7	8	9	10	11	13	14	15	16	17	18	190	191	200	201	21	22	23	24	25	26	27	28	29	30
Sharp-tailed Grouse													B	B	B			B							F					
Wild Turkey	B	B		B		B				F							F	F					O		O					
Mountain Quail	B	B	B	B		B						B											B		B					
California Quail	F	B				B				B		B	B	B			B	B	B	B										
Northern Bobwhite		B								B							B													
Red-throated Loon																												F	F	F
Pacific Loon																												F	F	F
Common Loon																					B	B						F	=	F
Yellow-billed Loon																					F							F	=	F
Pied-billed Grebe																					B	B	B		B			F		
Horned Grebe																					B	B						F		F
Red-necked Grebe																					B	B						F	F	F
Eared Grebe																					B	B								
Western Grebe	B	B												F		F		F												
Clark's Grebe																					B	B						F	F	F
Sooty Shearwater																											F	F	F	
Fork-tailed Storm-petrel																											R			
Leach's Storm-petrel																											R			
American White Pelican																					B	F								
Brown Pelican																										O	O	F		F
Brandt's Cormorant																											R	F	F	F

Habitat	*1*	*2*	*4*	*5*	*6*	*7*	*8*	*9*	*10*	*11*	*13*	*14*	*15*	*16*	*17*	*18*	*190*	*191*	*200*	*201*	*21*	*22*	*23*	*24*	*25*	*26*	*27*	*28*		*30*
Sharp-shinned Hawk	B	B	B	B	B	B			F		B											F								
Cooper's Hawk	B	B	B	B		B	B												B	B		F	B	B	B					
Northern Goshawk	B		B	B	B	B	B															F	F	B	B					
Swainson's Hawk									F		B		B	B	B	F		B				F			B					
Red-tailed Hawk	B	B	B	B		B	B			B	B	B	B	B	B	F	B	B				F	B		B					
Ferruginous Hawk									F			B	B	B	B	B		B												
Rough-legged Hawk		F								F			F	F	F		F	F				F				F				
Golden Eagle						B					B	B	B	B	B		F	F							B					
American Kestrel	B	B	B	B	B	B				B	B	B	B	B	B	F	B	B	B	B		F	B		B					
Merlin	B																		F		F		B			F		F		F
Gyrfalcon																	F	F			F	F				F		F		
Peregrine Falcon	B	B	B	B	B	B	F			F	B	B	B	B					B		F	F	F	F	B	F	B	F		F
Prairie Falcon											B		B	B	B	F		B												
Double-crested Cormorant																					B	R			B	O	R	F		F
Pelagic Cormorant																											R	F		F
American Bittern																	B	B				B						B		
Great Blue Heron	R	R								F							F	F	B	B	F	F	B		B	F		F		
Great Egret										F						F		F			F	F	B		B			F		
Cattle Egret																						F			B					
Green Heron																					F	F	B					B		
Black-crowned Night-heron																					F	F	F		B					

Habitat	*1*	*2*	*4*	*5*	*6*	*7*	*8*	*9*	*10*	*11*	*13*	*14*	*15*	*16*	*17*	*18*	*190*	*191*	*200*	*201*	*21*	*22*	*23*	*24*	*25*	*26*	*27*	*28*	*29*	*30*
Turkey Vulture	B	B	B	B	B	B				F	B	B	B	B	B	F	B	B				F	B		B		F	F		
Osprey	R	R	R	R	R	R								R					R	R	F		B		B		R	B		F
White-tailed Kite										B							B									F				
Bald Eagle	R	R	R	R	R	R											F		B		F	F	B		B		B	B	F	F
Northern Harrier									F	B			B	B	B	B	B	B				B				B		F		
Virginia Rail																	B	B				B						B		
Sora																	B	B				B						B		
American Coot																	B	B	B	B	F	B						F		
Sandhill Crane													F					B				B			B					
Black-bellied Plover																F	F				F					F		F		
American Golden-Plover																					F					F		F		
Pacific Golden-Plover																	F				F					F		F		
Snowy Plover																B					F					B		F		
Semipalmated Plover																F					F					F		F		
Killdeer										B			B	B	B	B	B	B	B	B	B	B	B		B	B	B	B		
Black Oystercatcher																										F	B			
Black-necked Stilt													B	B		B		B			F	B								
American Avocet													B	B		B		B			F	B								
Greater Yellowlegs																F	F	F			F	F	F		F	F		F		
Lesser Yellowlegs																F	F	F			F	F	F		F	F		F		
Solitary Sandpiper																	F	F			F	F	F		F					

Habitat	*1*	*2*	*4*	*5*	*6*	*7*	*8*	*9*	*10*	*11*	*13*	*14*	*15*	*16*	*17*	*18*	*190*	*191*	*200*	*201*	*21*	*22*	*23*	*24*	*25*	*26*	*27*	*28*	*29*	*30*
Willet														B	B	B					F	B						F		
Wandering Tattler																					F					F	F			
Spotted Sandpiper													F	F	F	F	F	F			B	B	B		B	F		B		
Whimbrel																	F									F	F	F		
Long-billed Curlew													B	B	B	B	B	B				F						F		
Marbled Godwit																F					F					F		F		
Ruddy Turnstone																										F	F	F		
Black Turnstone																					F					F	F	F		
Surfbird																										F	F			
Red Knot																	F									F		F		
Sanderling																F										F		F		
Semipalmated Sandpiper																					F							F		
Western Sandpiper																F	F	F			F	F				F		F		
Least Sandpiper									F							F	F	F			F	F						F		
Baird's Sandpiper																F	F	F			F	F				F		F		
Pectoral Sandpiper																	F	F			F	F						F		
Sharp-tailed Sandpiper																					F							F		
Rock Sandpiper																										F	F			
Dunlin																F	F				F	F				F		F		
Stilt Sandpiper																					F									
Short-billed Dowitcher																	F				F					F		F		

Habitat	*1*	*2*	*4*	*5*	*6*	*7*	*8*	*9*	*10*	*11*	*13*	*14*	*15*	*16*	*17*	*18*	*190*	*191*	*200*	*201*	*21*	*22*	*23*	*24*	*25*	*26*	*27*	*28*	*29*	*30*
Long-billed Dowitcher																F	F	F			F	F						F		
Wilson's Snipe										B							B	B				B						F		
Wilson's Phalarope																F					F	B								
Red-necked Phalarope																					F							F	F	F
Parasitic Jaeger																														F
Franklin's Gull																					F	B				F		F		F
Little Gull																					F							F	F	F
Bonaparte's Gull																	F		F		F					F	F	F	F	F
Heermann's Gull																										F	F	F	F	F
Mew Gull																					F					F		F	F	F
Ring-billed Gull																	B	B	F	F	B	B				F	F	B	F	F
California Gull																		F		F	B	B				F	F	F	F	F
Herring Gull																	F	F	F	F	F	F				F	F	F	F	F
Thayer's Gull																	F	F	F	F	F	F				F	F	F	F	F
Western Gull																										F	B	F		F
Glaucous-winged Gull																										F	B	F	F	F
Glaucous Gull																					F	F				F	F	F	F	F
Sabine's Gull																										F				F
Black-legged Kittiwake																														
Caspian Tern																					B	F				O	O	F		F
Elegant Tern																										O	F	F		F

Habitat	*1*	*2*	*4*	*5*	*6*	*7*	*8*	*9*	*10*	*11*	*13*	*14*	*15*	*16*	*17*	*18*	*190*	*191*	*200*	*201*	*21*	*22*	*23*	*24*	*25*	*26*	*27*	*28*	*29*	*30*
Common Tern																					F					O	O	F	F	F
Arctic Tern																										O	O		F	F
Forster's Tern																					B	B				O		F		F
Black Tern																		F			F	B								
Common Murre																											R	F	F	F
Pigeon Guillemot																											R	F	F	F
Marbled Murrelet	R																											F	F	F
Ancient Murrelet																													F	F
Cassin's Auklet																											R		F	F
Rhinoceros Auklet																											R	F	F	F
Tufted Puffin																											R		F	F
Rock Dove										F		B	F				B	B	B	B										
Band-tailed Pigeon	B	B	B														F	F	B	B			B			F	F	F		
Mourning Dove	B	B				B				B	B	B	B	B	B	F	B	B	B	B			B		B					
Barn Owl										F			B	B		F	B	B	B	B		F	B		B					
Flammulated Owl				B	B	B	B																		B					
Western Screech-owl	B	B		B	B	B	B				B						B	B	B	B			B		B					
Great Horned Owl	B	B	B	B	B	B	B			B	B	B	B	B		F	B	B	B	B		F	B		B					
Snowy Owl																	F	F								F				
Northern Pygmy-owl	B	B		B	B	B		B			B												B	B	B					
Burrowing Owl													B	B	B	B		B				F								

Habitat	*1*	*2*	*4*	*5*	*6*	*7*	*8*	*9*	*10*	*11*	*13*	*14*	*15*	*16*	*17*	*18*	*190*	*191*	*200*	*201*	*21*	*22*	*23*	*24*	*25*	*26*	*27*	*28*	*29*	*30*
Spotted Owl	B		B	B		B																								
Barred Owl	B			B																			B							
Great Gray Owl				B	B	B																F		B						
Long-eared Owl				B	B	B	B				B	B	B	B	B			B				F			B					
Short-eared Owl										B			B	B				B				B								
Boreal Owl			B	B			B																							
Northern Saw-whet Owl	B	B	B	B	B	B	B				B						F	F						B						
Common Nighthawk	B	B	B	B	B	B	B			B	B	B	B	B	B	B	B	B	B	B	F	F	F	F	F	F	F	F		
Common Poorwill		B		B	B	B					B		B	B	B	B		B												
Black Swift	F		F	F				F													F		B	F	B	F	B			F
Vaux's Swift	B	B	B	B		B				B									B	B	F	F	B	B						
White-throated Swift				B		B					B	B	B	B	B	F		F			F	F			F					
Black-chinned Hummingbird						B						B		B	B			B		B		F			B					
Anna's Hummingbird	B	B																								F	F			
Calliope Hummingbird						B	B		B			B												B	B					
Rufous Hummingbird	B	B	B	B		B	B	B		F		B					B	B	B	B			B	B	B	B	B			
Belted Kingfisher																					B		B		B	B	B	F		F
Lewis's Woodpecker		F		B		B				B								B							B					
Acorn Woodpecker		B				B																								
Williamson's Sapsucker			B	B	B	B	B	B																B	B					
Red-naped Sapsucker			B	B		B	B	B																	B					

Habitat	*1*	*2*	*4*	*5*	*6*	*7*	*8*	*9*	*10*	*11*	*13*	*14*	*15*	*16*	*17*	*18*	*190*	*191*	*200*	*201*	*21*	*22*	*23*	*24*	*25*	*26*	*27*	*28*	*29*	*30*
Red-breasted Sapsucker	B	B	B																				B	B	B					
Downy Woodpecker	B	B					B										B	B	B	B			B		B					
Hairy Woodpecker	B	B	B	B	B	B	B										B	B	B	B			B	B	B					
White-headed Woodpecker				B		B																			B					
Three-toed Woodpecker			B	B	B	B		B																B						
Black-backed Woodpecker			B	B	B	B		B																B						
Northern Flicker	B	B	B	B	B	B	B	B		B	B	B					B	B	B	B			B	B	B	F				
Pileated Woodpecker	B		B	B		B													B	B			B	B	B					
Olive-sided Flycatcher	B		B	B	B	B		B															B	B	B					
Western Wood-pewee	B	B		B		B	B			B							B	B	B	B			B		B					
Willow Flycatcher	B	B	B																				B	B	B					
Least Flycatcher																							B		B					
Hammond's Flycatcher	B	B	B	B		B																								
Gray Flycatcher					B	B					B			B																
Dusky Flycatcher			B	B	B	B	B				B													B	B					
Pacific-slope Flycatcher	B	B	B																				B	B						
Cordilleran Flycatcher				B																					B					
Say's Phoebe						B					B		B	B	B	F	B	B	B	B					B					
Ash-throated Flycatcher		B								B	B						B	B	B	B										
Western Kingbird						B				B	B	B	B	B		B	B	B	B	B										
Eastern Kingbird						B					F	B	F	F	F	F		B		B		F			B					

Habitat	*1*	*2*	*4*	*5*	*6*	*7*	*8*	*9*	*10*	*11*	*13*	*14*	*15*	*16*	*17*	*18*	*190*	*191*	*200*	*201*	*21*	*22*	*23*	*24*	*25*	*26*	*27*	*28*	*29*	*30*
Loggerhead Shrike											B		F	B	F	B		B												
Northern Shrike										F	F	F	F	F	F	F	F	F								F				
Cassin's Vireo	B	B		B		B	B																		B					
Hutton's Vireo	B	B																					B							
Warbling Vireo	B	B	B	B		B	B																B	B	B					
Red-eyed Vireo																							B		B					
Gray Jay	B		B	B	B	F		B															B	B	B					
Steller's Jay	B	B	B	B		B					F								B	B			B	B	B	F				
Blue Jay						B												B		B										
Western Scrub-Jay		B				B				B							B	B	B	B										
Clark's Nutcracker			B	B		B		B																						
Black-billed Magpie						B					B	B	F	F	F		B	B	B	B					B					
American Crow	B	B		B		B											B	B	B	B	F		B		B	B	B			
Common Raven	B	B	B	B	B	B	B	B	B		B	B	B	F	F	F	B	B	B	B		F	B	B	B	F	B	F		
Horned Lark								B		B			B	B	B	B	B	B	B	B						B				
Purple Martin	B	B																	B		F	F	B			F	F	B		
Tree Swallow		B					B			B							B	B	B	B	F	F	B	B	B					
Violet-green Swallow	B	B	B	B		B	B			B	B	B					B	B	B	B	F	F	B	B	B					
N. Rough-winged Swallow	F	F	F	F	F	F		F		F	F	F	F	F	F	F					B	F	B		B	F	B	B		F
Bank Swallow				F		F			B		F	F	F	F	F	F					B	F			B		B			
Cliff Swallow	B	B	B	B	B	B		B		F	B	B	B	B	B	F	B	B	B	B	F	F	B		B	F	B	F		

Habitat	*1*	*2*	*4*	*5*	*6*	*7*	*8*	*9*	*10*	*11*	*13*	*14*	*15*	*16*	*17*	*18*	*190*	*191*	*200*	*201*	*21*	*22*	*23*	*24*	*25*	*26*	*27*	*28*	*29*	*30*
Barn Swallow	B	B				B				F	B	B	B	B	B	F	B	B	B	B	F	F	B		B	F	B	F	F	F
Black-capped Chickadee	B	B				B													B	B			B		B					
Mountain Chickadee			B	B	B	B	B	B			B	B												B	B					
Chestnut-backed Chickadee	B		B	B															B	B			B	B						
Boreal Chickadee			B					B																B						
Bushtit	B	B				B					B						B	B	B	B			B		B	B				
Red-breasted Nuthatch	B	B	B	B	B	B		B											B	B			B	B	B					
White-breasted Nuthatch		B				B	B			B									B	B			B		B					
Pygmy Nuthatch						B																			B					
Brown Creeper	B	B	B	B	B	B													B	B			B	B	B					
Rock Wren		B		B		B					B	B	B	B	B	B														
Canyon Wren			B	B	B	B					B	B	B	B	B										B					
Bewick's Wren	B	B															B		B				B							
House Wren	B	B		B		B	B												B	B			B		B					
Winter Wren	B		B	B				B															B	B	B	F				
Marsh Wren																						B						B		
American Dipper																					B		B		B					
Golden-crowned Kinglet	B	B	B	B		B		B															B	B	B					
Ruby-crowned Kinglet	F	F	B	B	B	B	F	B			B	F					F	F	F	F			F	B		F	F			
Western Bluebird	B	B	B	B		B				B	B		B				B	B	B	B					B					
Mountain Bluebird			B	B	B	B	B	B					B	F	F		B	B							B					

Habitat	*1*	*2*	*4*	*5*	*6*	*7*	*8*	*9*	*10*	*11*	*13*	*14*	*15*	*16*	*17*	*18*	*190*	*191*	*200*	*201*	*21*	*22*	*23*	*24*	*25*	*26*	*27*	*28*	*29*	*30*
Townsend's Solitaire	B		B	B	B	B	B	B			F														F					
Veery																									B					
Swainson's Thrush	B		B	B			B				B												B	B	B					
Hermit Thrush	B	F							B		B	F					F	F	F	F			F	B	B	F	F			
American Robin	B	B	B	B	B	B	B	B		F		B	F				B	B	B	B		F	B	B	B	B				
Varied Thrush	B	F	B	B		F		B			F						F	F	F	F				B						
Gray Catbird											B														B					
Northern Mockingbird																	F	F	F	F										
Sage Thrasher														B	B	B														
European Starling	B	B		B		B				B		B	B	F			B	B	B	B			F		F			F		
American Pipit								B			F						F	F								F		F		
Bohemian Waxwing			F	F								F						F		F				F	F					
Cedar Waxwing	B										B								B	B			B	B	B					
Orange-crowned Warbler	B	B		B	B	B	B	F			B	B											B		B	B	B			
Nashville Warbler	B	B	B	B		B	B																B	B	B					
Yellow Warbler							B																B		B					
Yellow-rumped Warbler	B	B	B	B	B	B	B	B			B						F	F	F	F			F	B	B	F	F			
Black-throated Gray Warbler	B	B		B		B																	B		B					
Townsend's Warbler	B	B	B	B			F	B											F	F			B	B						
Hermit Warbler	B	B	B																				B	B						

Habitat	1	2	4	5	6	7	8	9	10	11	13	14	15	16	17	18	190	191	200	201	21	22	23	24	25	26	27	28	29	30
Palm Warbler																						F	F			F	F			
American Redstart																									B					
Northern Waterthrush																							B		B					
Macgillivray's Warbler	B	B	B	B		B	B	B															B	B	B					
Common Yellowthroat	B	B															B	B				B	B		B			B		
Wilson's Warbler	B	B	B	B																			B	B	B					
Yellow-breasted Chat	B	B																B					B		B					
Western Tanager	B	B	B	B	B	B	B	B			B												B	B	B					
Green-tailed Towhee			B	B	B	B	B				B	B	B	B																
Spotted Towhee	B	B		B		B						B					B	B	B	B			B		B	B				
American Tree Sparrow											B							F							F					
Chipping Sparrow		B	B	B	B	B	B	B		B							B	B							B					
Clay-colored Sparrow											B		B																	
Brewer's Sparrow						B					B		B	B	B	B														
Vesper Sparrow								B		B	B		B	B	B		B	B												
Lark Sparrow						B								B	B															
Black-throated Sparrow											B			B	B	B														
Sage Sparrow									B		B			B	B															
Savannah Sparrow								B		B			B	B	B		B	B	B	B		B	B		B	B	B	B		
Grasshopper Sparrow										B			B	B	B			B												
Fox Sparrow	F	F	B	B	B	B		B															F		B	B	B			

Habitat	1	2	4	5	6	7	8	9	10	11	13	14	15	16	17	18	190	191	200	201	21	22	23	24	25	26	27	28	29	30
Song Sparrow	B	B	B	B			B			B							B	B	B	B		B	B	B	B	B	B	B		
Lincoln's Sparrow								B														B	B		B					
Swamp Sparrow																						F	F					F		
White-throated Sparrow	F	F															F	F	F	F			F		F					
Harris's Sparrow		F				F				F			F	F			F	F												
White-crowned Sparrow	B	B	B	B		B	B	B	B	F	F					F	B	B	B	B		F	B		B	B	B	B		
Golden-crowned Sparrow	F	F	F			F		B		F	B											F	F		F	F	F			
Dark-eyed Junco	B	B	B	B	B	B	B	B				B					B	B	B	B			B	B	B	B				
Lapland Longspur													F				F	F								F		F		
Snow Bunting													F					F								F				
Black-headed Grosbeak	B	B				B	B				B												B		B					
Lazuli Bunting	B	B				B	B			B		B					B	B					B		B					
Bobolink													B					B				B			B					
Red-winged Blackbird										B	B						B	B				B	B		B			B		
Western Meadowlark										B			B	B	B		B	B								F	F	F		
Yellow-headed Blackbird											B											B								
Rusty Blackbird																	F	F					F		F					
Brewer's Blackbird		B				B	B			B		B	B	B	B	B	B	B	B	B		B	B		B	B	B	B		
Brown-headed Cowbird	R	R		R		B	B			B		B	B	B	B	B	B	B	B	B		B	B		B	B				
Bullock's Oriole		B							B								B	B	B	B			B		B					
Gray-crowned Rosy-Finch								F																						

Habitat	1	2	4	5	6	7	8	9	10	11	13	14	15	16	17	18	190	191	200	201	21	22	23	24	25	26	27	28	29	30
Pine Grosbeak			B		B			B																B						
Purple Finch	B	B	B	B		B		F			B								B	B			B							
Cassin's Finch			B	B	B	B	B	B																B	B					
House Finch										F							B	B	B	B										
Red Crossbill	B		B	B	B	B		B																B						
White-winged Crossbill			B	B				B	F																					
Common Redpoll			F	F				F																F	F					
Pine Siskin	B	B	B	B	B	B	B	B											B	B			B	B	B					
Lesser Goldfinch		B				B				B								B					B							
American Goldfinch	B	B		B		B				B							B	B	B	B		F	B		B					
Evening Grosbeak	B		B	B	B	B		B															B	B	B					
House Sparrow										F		F	F	F	F	F	B	B	B	B			F		F			F		

Author Biographies

Clifford M. (Bud) Anderson. Born in Seattle in 1945, Bud observed his first wild raptor at age eight. He attended the University of Washington and later studied under Dr. Steve Herman at The Evergreen State College. In 1985 he founded the nonprofit Falcon Research Group and has worked with a wide variety of raptors for over 40 years. His primary interest is the Peregrine Falcon, which he has studied worldwide. He currently resides in Bow, Washington.

Keith Aubry is a research wildlife biologist with the USDA Forest Service, Pacific Northwest Research Station in Olympia, Washington. He has been studying the ecology, management, and conservation of terrestrial wildlife in the Pacific Northwest for over 25 years. His current research includes field studies of the Pileated Woodpecker, fisher, and Canada lynx; historical biogeography of the wolverine in North America; and the application of genetic information to wildlife research and conservation.

Devorah Bennu has birded the west coast, bred and raised parrots, volunteered as a public speaker as well as a bird guide, and written many articles about birds, nature, and science for most of her life. Devorah earned her Ph.D. in Zoology (Ornithology) at the University of Washington in 2002 and currently works as a Chapman Postdoctoral Fellow at the American Museum of Natural History in New York City, researching the evolution, speciation, and biogeography of parrots.

Thomas Bosakowski received his Ph.D. degree in Zoology from Rutgers University. Since then, he worked for five years as a wildlife biologist for Beak Consultants in the Seattle area, conducting surveys on threatened and endangered wildlife species and producing GIS-based habitat models. He has published over 50 scientific papers and two books on raptors, breeding birds, small mammals, amphibians, and fish. He is senior author of the book *Raptors of the Pacific Northwest* and author of *The Northern Goshawk*.

Joseph B. Buchanan is a wildlife biologist with the Washington Department of Fish and Wildlife. Joe graduated from The Evergreen State College in 1981 and received an M.S. degree from the University of Washington in 1991. He has been involved in Spotted Owl research and management for over 15 years. Joe has studied aspects of the ecology and behavior of several other bird species, and is especially interested in shorebird ecology and migration.

Kelly Cassidy has a B.S. in math and physics from the University of Texas in Austin, an M.S. in biology from the University of North Texas in Denton, and a Ph.D. in botany from Washington State University in Pullman. She has worked on several projects involved with conservation in Washington State, and has written articles and contributed to a number of publications about birds and habitats. She is currently curator of the Conner Vertebrate Museum at Washington State University in Pullman.

Christopher B. Chappell has been birding in Washington for the past 32 years. He studied ornithology and earned a B.S. at The Evergreen State College. He worked on several ornithological research projects before returning to school for an M.S. in Forest Ecology at the University of Washington. Since 1992, he has been an ecologist for the Washington Natural Heritage Program, Department of Natural Resources, where he is lead ecologist for plant communities/ecological systems in western Washington.

Paul DeBruyn is a raptor biologist who studied under Dr. Steven Herman at The Evergreen State College. He has studied Red-tailed Hawks in Whatcom County, Peregrine Falcons throughout Washington and British Columbia, and Golden Eagles in Okanogan County. He is co-compiler of the Bellingham Christmas Bird Count. He divides his time between northern Whatcom County, Okanogan County, and northeastern California. He is a falconer who raises and flies Gyrfalcons, Peregrines, and Merlins.

Steven M. Desimone has worked extensively in Washington and Oregon studying forest birds, primarily raptors. His M.S. thesis research was on the Northern Goshawk. He is currently a wildlife biologist with Washington Department of Fish and Wildlife specializing in forest wildlife and forest management issues. He earned a B.S. in Zoology at the University of Washington and an M.S. in Wildlife Science at Oregon State University.

Ann M. Eissinger is a professional wildlife biologist and owner of Nahkeeta Northwest Wildlife Services in Bow, Washington. She holds a B.S. from The Evergreen State College and has worked as an investigative researcher, consultant, educator, and conservationist. Over the past 20 years, Ann has facilitated key habitat protection for Great Blue Heron, Marbled Murrelet, and Vaux's Swift. She is currently researching the upland habitat relationships of the Great Blue Heron and monitors heron populations throughout the Salish Sea.

Tracy L. Fleming is a research wildlife biologist with the National Council for Air and Stream Improvement. He has worked extensively with Spotted Owls and most other raptor species. He also has worked with Marbled Murrelets and participated in many other wildlife research projects, including studies that examined environmental contaminants in birds. Tracy is a master bird bander, and in the past 15 years he and subpermittees have banded numerous birds in Washington, including 1200 Spotted Owls. His interests include avian predator-prey dynamics, radio telemetry research techniques, development of literature databases for numerous bird species, and shrub-steppe ecology.

Rich Fredrickson is a doctoral candidate at Arizona State University in Tempe studying ecology, evolution, and genetics. His doctoral research is focused on individual and population-level consequences of inbreeding and outbreeding in Mexican and red wolves. He expects to complete his degree in 2005. Prior to this he worked for 13 years with wildlife associated with late-successional forests, including Spotted Owls, Marbled Murrelets, and American marten in research and management venues.

Tom Gleason has a life-long interest in raptors. He has spent innumerable hours searching for raptor nests over much of the state. Tom is particularly interested in Merlins. He lives and works in Ellensburg.

Gregory A. Green is a consultant wildlife ecologist based out of Bothell, Washington. He received his M.S. degree at Oregon State University, where he studied the ecology of Burrowing Owls in the Columbia Basin. Most of his career has focused on shrub-steppe wildlife, forest ecology, and marine mammal surveys. He is currently developing conservation assessments for various wildlife species, including Willow Flycatchers, in the Sierra Nevada, and is an associate editor for the *Journal of Wildlife Management*.

John Grettenberger is a biologist with the U.S. Fish and Wildlife Service in Lacey, Washington, where he supervises the Consultation Branch of the Endangered Species Division. He has worked with Snowy Plovers, Bald Eagles, and Marbled Murrelets, and participated in the development of the Washington Partners in Flight plans. He also worked with the Fish and Wildlife Service in Iowa and Michigan, the Corps of Engineers in Oregon, and as a Peace Corps biologist in Niger.

Steven G. Herman received his Ph.D. from the University of California at Davis and has been teaching natural history at The Evergreen State College since 1971. His research interests have centered on raptors, shorebirds, and pesticide-wildlife relationships. In California in 1970, he conducted the first post-decline Peregrine survey in North America. He and his students have studied Peregrines in Washington for some 30 years, and initiated the campaign to establish Grays Harbor NWR.

Michael Husak is a biologist at Cameron University. He has a master's degree from Angelo State University and Ph.D. from Mississippi State University. His field research and numerous publications focus on avian ecology and forest bird communities and, especially pertinent to this book, woodpeckers.

Stan Kostka, naturalist and amateur ornithologist, has been studying Washington's recovering Purple Martin population since 1999, installing nestboxes, collecting data on reproduction and morphology; he began banding martins in 2000, assisted by Washington Department of Fish and Wildlife and various private individuals. He co-authored *Productivity and Inter-Colony Movements of the Western Purple Martin* and is currently preparing *The Purple Martin in Washington State*, a comprehensive history and status report.

Don Kraege is Washington Department of Fish and Wildlife statewide section manager of waterfowl and other migratory gamebirds. During a 24-year career at WDFW, he has worked as a field biologist and conducted a number of wildlife surveys. He has served on a number of multi-agency, international, and citizens' committees related to waterfowl. He is author and contributor to a number of publications dealing with biology and management of wildlife, particularly Brant, Canada Geese, and swans.

Robert (Bob) Kuntz II, born in Tulsa, Oklahoma, received a B.S. degree in Biology at Eastern Washington University in 1979. Since 1980, he has worked for the Bureau of Land Management, U.S. Forest Service, U.S. Fish and Wildlife Service, and for the past 17 years for the National Park Service at North Cascades NP as a bird biologist. He currently serves as head of the park's wildlife section. Bob's interest in birds spans the past 30 years. He is a member of the Cooper Ornithological Society and has several publications to his credit.

Rose Leach started birding in high school, and has studied or monitored birds in western Montana and on the Yakama Indian Reservation, Toppenish, Washington. Her interests are in old-growth associates, forest management, and species of concern—including birds, skippers, and gray squirrels—in central Washington. She is currently the NEPA Compliance Program Manager for the Confederated Salish and Kootenai Tribes of the Flathead Indian Reservation in Pablo, Montana.

Jeff Lewis is a wildlife biologist with the Washington Department of Fish and Wildlife, where he works on ecoregional conservation planning and endangered species recovery projects. He received his B.S. degree in Forest Biology from the SUNY College of Environmental Science and Forestry, and his M.S. in Wildlife Biology from Humboldt State University. His research and management experience includes work with Spotted Owls, passerines, California Least Terns, Caspian Terns, waterfowl, eagles, and urban avifauna.

Diann MacRae is the director of the Olympic Vulture Study and North American correspondent for the Vulture Study Group. She is a past chairman of the Hawk Migration Association of North America and has served on the board and as an editor for the Washington Ornithological Society. Her undergraduate work was at Oregon State University and graduate work at the University of Washington. She works as a consulting biologist and writer/editor.

Kelly McAllister is a district wildlife biologist with the Washington Department of Fish and Wildlife. He was born in Olympia, a descendant of Northwest Indians, a Hudson's Bay Company employee, and American pioneers who arrived by covered wagon. In college, he majored in Fisheries and received a B.S. degree from the University of Washington in 1979. He has worked for the state since January 1980, mostly within the endangered species section of the agency.

Steven Mlodinow has been birding since 1973. He moved from Chicago to Washington, via Maine and California, in 1992. Steve is currently on the Washington Bird Records Committee and is a technical editor for North American Birds. He has authored two previous books: *Chicago Area Birds* and *America's 100 Most Wanted Birds.* He also has written a number of articles dealing with identification, distribution, and occurrence of birds.

Matt Monda received his bachelor's degree in Wildlife Biology from Washington State University in 1982. He studied duck brood ecology for his M.S. degree from Eastern Washington University between 1983 and 1984. Waterfowl continued to be his focus when he began his Ph.D. research on Tundra Swans on the Arctic NWR in 1987. He received his Ph.D. from the University of Idaho in 1991. Washington Department of Fish and Wildlife hired him as state waterfowl biologist in 1994 and he continued in that position until he was promoted to wildlife program manager for north central Washington in 2001.

Martin Muller was born in 1953 in Holland where he was trained as a teacher of Biology and Health Education. In 1983 he moved to Seattle, where he practiced his bird watching mainly at Green Lake. His detailed observations of the local Pied-billed Grebes resulted in publication of an article in *Washington Birds.* Consequently he was invited to co-author the Pied-billed Grebe account of the *Birds of North America* publication.

Noelle Nordstrom is a forest habitats biologist with the Washington Department of Fish and Wildlife. She holds a long-time fascination with Vaux's Swifts, not only in forested landscapes, but also in urban/suburban areas, such as her Olympia neighborhood, where Vaux'sSwifts sometimes nest and roost in chimneys. Noelle earned her Bachelor of Environmental Science degree from The Evergreen State College in 1991.

Dave Nysewander has been project leader with Washington Department of Fish and Wildlife since 1992 for the marine bird component of the Puget Sound Ambient Monitoring Program. He received his M.S. in Wildlife Science from University of Washington in 1977. His work experience 1975-92 in Alaska with U. S. Fish and Wildlife Service included monitoring seabird colonies, at-sea surveys of marine birds and mammals, reintroduction of endangered species, and resource damage assessments related to oil spills.

Roger Orness, born in Puyallup, Washington, August 10, 1941, lived in Edgewood and Milton, and graduated from Fife High School in 1959. A year of vocational school in Tacoma and a year at Yakima Valley College led to a job at the Boeing Company. Two years of active duty in the Navy reserves was sandwiched in during Vietnam, then back to Boeing until he retired in 2003. He was introduced to hawks by Bud Anderson, and captured some of his passion for raptors 11 years ago, filling many journals of observations..

Ian Paulsen is a free-lance wildlife biologist, a co-founder of the Washington Ornithological Society, and an avid book collector, and has been studying the birds of Washington State, particularly the *Rallidae*, since 1976.

Catherine Raley is a wildlife biologist with the USDA Forest Service, Pacific Northwest Research Station in Olympia, Washington. She received a B.S. in Wildlife Management and Zoology from Humboldt State University, and a M.S. in Zoology from the University of Wyoming. She has been studying the ecology and management of forest bird communities, Pileated Woodpeckers, and fishers in the Pacific Northwest since 1988.

Scott Richardson is a lifelong birdwatcher who has studied or monitored birds in the Pacific Northwest and New England, on the Aleutian Islands, and at Midway Atoll. Scott edited the Washington Ornithological Society newsletter for three years and was an endangered species biologist for Washington State in the 1990s. He is now communications director for Laudholm Trust, a nonprofit organization supporting the Wells National Estuarine Research Reserve in southern Maine.

Russell Rogers is a South Carolina native who holds a Bachelor of Fine Arts in painting from the Tyler School of Art, Temple University, Philadelphia, and a Master of Environmental Studies from The Evergreen State College. He has spent much of his career studying habitat selection and population ecology in birds and invasive species biology. He is currently the statewide coordinator for shellfish disease pest and predator control for the Washington Department of Fish and Wildlife and lives in Sequim.

Greg Schirato has worked for the Washington Department of Fish and Wildlife for the past 20 years. Most of this time he has worked as the district wildlife biologist for the eastern Olympics. Greg has conducted research on Harlequin Ducks with several graduate students for about 10 years. More recently his emphasis has been on protection and restoration of estuary habitats.

Michael A. Schroeder has been the upland bird research biologist for the Washington Department of Fish and Wildlife since 1992. He received his Ph.D. from Colorado State University, his M.S. degree from the University of Alberta, and his B.S. degree from Texas A&M University. Mike's research is primarily focused on the behavior, population dynamics, and management of grouse.

Jennifer Seavey is a Ph.D. student in a joint program at Smith College and University of Massachusetts in Massachusetts. She obtained her B.S. at Lewis and Clark College in Portland, Oregon and an M.S. degree at the University of Washington in Seattle, Washington. During the past 14 years, she has studied various aspects of ornithology worldwide. She is currently studying the ecology of the Piping Plover on New York's barrier islands.

Derek Stinson has a B.S. in Biology from Framingham State College in Massachusetts and an M.S. in Zoology from Washington State University. He currently is employed as an endangered species biologist with the Washington Dept of Fish and Wildlife. Prior work experience includes four years as a biologist in the Northern Mariana Islands and he has published several papers on the forest, wetland, and migrant birds of the Marianas.

Patricia Thompson received a B.S. degree in Wildlife Science from Utah State University and an M.S. degree from Washington State University in Zoology Wildlife Biology. Her research concentrated on the natural history, breeding, and wintering biology, including home range assessment, of the Northern Harrier on the Hanford Site, Richland. Much of her field experience has been with raptors, including work on the Snake River Birds of Prey National Conservation Area as a field biologist. She has been a biologist for the Washington Department of Fish and Wildlife for 17 years.

Bill Tweit is a native of Illinois who came to Washington in the early 1970s and graduated from The Evergreen State College. He lives in Olympia and is a Washington Department of Fish and Wildlife fisheries biologist. He is a long-time member of the Washington Bird Records Committee, author of a number of papers on Washington birds, and a regional editor of *North American Birds* seasonal reports. He has for years conducted Breeding Bird Surveys and acquired data during offshore bird trips from Grays Harbor and research cruises off Washington.

Matthew Vander Haegen is a senior research scientist with Washington Department of Fish and Wildlife. He has 20 years of experience studying wildlife populations and habitat associations in forested and rangeland ecosystems. Much of his recent research has focused on determining landscape and local effects on abundance and productivity of shrub-steppe obligate passerines in Washington. He holds B.S. and M.S. degrees in Wildlife Ecology from the University of Massachusetts and a Ph.D. in Wildlife Science from the University of Maine.

Terence Wahl organized seabird trips off Washington for 35 years, conducted seabird surveys over the North Pacific, was co-principal investigator on the Marine Ecosystems Analysis program in 1978-79, ran the Newhalem Breeding Bird Survey route for 30 years, contributed to the state Breeding Bird Atlas, was a bird bander, and has been a Christmas Bird Count compiler for 34 years. He has published a number of papers on marine bird distribution. He wrote *Birds of Whatcom County* and co-authored *A Guide to Bird Finding in Washington.*

Dave Ware graduated from Washington State University with a B.S. in Range/Wildlife Management in 1978. He has worked for the Department of Fish and Wildlife for 22 years. One summer while attending college, he worked on a game farm raising pheasants, Bobwhite quail, and tinamoe. He worked on and ultimately managed the Snake River Wildlife Area developing upland bird and waterfowl habitat. He managed the habitat development program in central Washington, working with farmers to enhance habitat primarily for pheasant and Sharp-tailed and Sage grouse. In 1991, he became the small game section manager; and is currently the game division manager.

Dawn Wilkins received her B.S. in Natural Resources Management with an emphasis on Wildlife Biology from the University of Tennessee at Martin, her M.S. in Biology with an emphasis on Applied Ecology from Eastern Kentucky University, and her Ph.D. in Biological Sciences from Mississippi State University. She is currently an assistant professor of Biology at the University of Tennessee at Martin.

Doug Wood is a Ph.D. and assistant professor of Zoology at Southeastern Oklahoma State University, where he teaches Ornithology, Zoology, and Conservation Biology. His research interests include the ecology of cavity-nesting bird species such as the Red-cockaded Woodpecker and Prothonotary Warbler; nocturnal bird monitoring; wintering ecology of migrant songbirds; and box turtle spatial ecology. He frequently visits the Puget Sound area to visit family.

Kent Woodruff has worked as a biologist for state, federal and private land-management organizations in Colorado, Oregon, Arizona, Idaho, Missouri, and Washington since 1976. He is currently a district wildlife biologist for the Okanogan and Wenatchee National Forests, where he has been for the last 15 years. For much of his career his specialty has been bats and birds. Owls especially have intrigued him. His work includes publications on barn owls and boreal owls, and an ongoing attempt to understand more about Washington's Great Gray Owls.

Index of Birds